A STUDENT'S COMPANION

J. RICHARD CHRISTMAN

U.S. Coast Guard Academy
New London, CT

to accompany

FUNDAMENTALS OF
PHYSICS

FOURTH EDITION
INCLUDING EXTENDED CHAPTERS

DAVID HALLIDAY
University of Pittsburgh

ROBERT RESNICK
Rensselaer Polytechnic Institute

JEARL WALKER
Cleveland State University

JOHN WILEY & SONS, INC.
New York   Chichester   Brisbane   Toronto   Singapore

TO THE STUDENT:
HOW TO USE A STUDENT'S COMPANION
TO FUNDAMENTALS OF PHYSICS

A Student's Companion to Fundamentals of Physics is designed to be used closely with the text *FUNDAMENTALS OF PHYSICS* by Halliday, Resnick, and Walker. There are 4 overview chapters, corresponding to major sections of the text: mechanics, thermodynamics, electricity and magnetism (including optics), and modern physics. Modern physics chapters are included in the extended version of the text only. Read the appropriate overview when you start to study a section and refer back to it as your study proceeds. Read it again when you finish the section. The overviews are designed to help you understand how the important topics fit together and how the text is organized.

Other chapters in the Companion correspond to chapters in the text and should be read along with the text. Most of the Companion chapters are divided into 3 sections: Basic Concepts, Problem Solving, and Notes. Some chapters also contain a Mathematical Skills section.

BASIC CONCEPTS. This section deals with two important types of information you should obtain from your reading of the text. The first consists of definitions of physical concepts used in the chapter; the second consists of the laws of physics (relationships between the concepts).

A firm understanding of the basic concepts is important for understanding how nature behaves, for working problems, for doing well on exams, and for understanding following chapters. Rather than just list the definitions and laws, this section asks you to do the important part of the work—write the key phrases by filling in blanks. To derive the greatest benefit, don't copy information from the text. Rather, read the text first, then try to fill in the blanks with your own words, without reference to the text. Thinking about what to write and writing it well should help you retain the important information. Comparing what you have written with what the text says will help you pinpoint any misconceptions you might have. If you have trouble expressing an idea you probably don't understand it very well. Go back and study the appropriate section of the text.

Try to write your responses carefully. The section, as completed by you, will serve later as a review. Before working a problem assignment and while studying for an exam, read over the completed section. If there are parts you don't understand you might want to write them more carefully. The better the job you do, the better this section will serve you when you review.

The concepts of physics are best learned in small doses. Try to obtain a firm understanding of each concept before moving on to the next. The Basic Concepts section will help you.

PROBLEM SOLVING. You cannot claim to understand a definition or law of physics unless you can apply it in a variety of different situations. The purpose of the problems is to provide you with various situations so you can test your understanding. There are three main parts to solving a physics problem. The first and probably the most difficult for many students is identifying the physical concepts or physical laws involved in the problem. Once the concepts and laws are identified, you are ready for the second part, writing the equations associated with them. The third part of problem solving involves carrying out mathematical manipulations to obtain an answer.

This section of the Companion concentrates on helping you identify the concepts and laws involved in a problem and on helping you write the appropriate equations. The various types of problems that can be associated with the concepts of the chapter are discussed and classified. As you read a Problem Solving section be sure you also study the Sample Problems (in the text) that are referenced. As you work problems, you will learn more details about the concepts and laws involved. If necessary, go back to the Basic Concepts section and revise your responses there.

MATHEMATICAL SKILLS. Here you will find a list of the mathematical skills required to solve the problems of the chapter. You will have learned most of these in algebra, trigonometry, geometry, and calculus classes. Review the list and study the brief explanations to see if you need to brush up on any of the skills. If you do, consult your math texts for more complete discussions and for practice problems. Your goal should be to become facile enough with the required mathematical techniques that the math does not hinder your ability to solve problems.

NOTES. Part or all of a page is left blank for you to record any additional notes you think will be beneficial when you review. You might write some detail of a definition or law that is not covered in the Basic Concepts section and that you have trouble remembering. You might also record any details of problem solving that give you special trouble. Try to write your notes so you can understand them later when you review.

EXAM REVIEWS. The last few pages of the Companion can be used to keep a list of the important concepts and laws you need to know for each exam. Here you reduce the important ideas to brief sentences so each page becomes an outline that can be read to remind you of the details you learned while reading the text, filling in the BASIC CONCEPTS sections, and solving problems. You might mark the ideas you had trouble with. Start your review for the final exam by reading the EXAM pages.

The author owes a debt of gratitude to the good people of Wiley who helped with this Student's Companion, especially Cliff Mills, Joan Kalkut, and Catherine Donovan. He is extremely thankful for the support and encouragement of his wife, Mary Ellen.

TABLE OF CONTENTS

OVERVIEW I
MECHANICS

Wherever we look, from the submicroscopic world of fundamental particles to the grand scale of galaxies, we see objects in motion, influencing the motions of other objects.

Think of a gardener pushing a wheelbarrow. Each of the objects involved (the surface of the earth, the gardener, the wheelbarrow, and the air) influences the motion of the others. If the gardener stops pushing the wheelbarrow might coast for a while but it eventually stops, chiefly because of friction and air resistance. To keep himself and the wheelbarrow going the gardener must push against the ground with his feet. Both the ground and wheelbarrow push on him. For the wheels to turn, the ground must also push on them.

From a microscopic viewpoint, each of these objects consists of a myriad of particles, mainly electrons, protons, and neutrons, in continual motion and continually exerting an influence over the other particles.

On a grander scale the earth makes its yearly trip around the sun because the sun influences its motion. The sun itself travels around the center of our galaxy, the Milky Way, under the influence of other stars. The presence of other galaxies makes the motion of our galaxy different from what it otherwise would be.

Mechanics is the study of motion. The goal is to understand exactly what aspects of the motion of one object are changed by the presence of other objects and exactly what properties objects must have in order to influence the motion of each other. The fundamental problem of mechanics is: given the relevant properties of a group of interacting objects, what are their motions?

Mechanics divides neatly into two parts: kinematics, a study of the description of motion, and dynamics, a study of the causes of motion.

First you will learn to describe the motion of an object and to use the ideas of displacement, velocity, and acceleration to predict changes in the position of an object. You will find that if you know the position and velocity of an object at some initial time and the acceleration of the object at all times you can predict its position and velocity at all times. In Chapter 2 you concentrate on motion in one dimension so you can master the important concepts without the geometric complications of more complex motions. In Chapter 4 the concepts are extended to describe motion in two dimensions.

Displacement, velocity, and acceleration in more than one dimension have direction as well as magnitude. They behave like mathematical quantities called vectors. Vectors are so important for understanding physics that the whole of Chapter 3 is devoted to explaining their properties and describing how they are manipulated mathematically.

Newton's laws of motion, introduced in Chapter 5, are at the heart of classical mechanics. Here you will learn that the interaction between two objects can be described in terms of a force and that the net force acting on an object accelerates it. That is, the net force changes the velocity of the object. The fundamental problem splits into two parts: (1) given the properties of the objects, what forces do they exert on each other and (2) given the net force on an object, what is its motion?

You will be introduced to a few simply described forces: the gravitational force of the earth on an object near its surface and the force exerted by a spring on an object in contact with one end, for example. Details about the gravitational force are described in Chapter 15 and later on you will learn about other forces: electrical force in Chapter 23 and magnetic force in Chapter 30.

You will also learn to calculate the acceleration of objects in a wide variety of situations. The study of Newton's laws of motion is continued in Chapter 6, where you will concentrate on frictional forces and the force required to make an object move on a circular path.

One important quantity that characterizes a system of interacting objects is its energy. Energy may take several forms: kinetic energy is introduced in Chapter 7, potential energy and internal energy in Chapter 8, and kinetic energy of rotation in Chapter 11. As objects move and exert forces on each other the values of their individual energies may change and the form of the total energy may change, from kinetic to potential, for example. You will learn that the mechanism for these changes is the work done by the forces of interaction. You will learn how to compute the work done by a force and also the changes in the energies of the objects.

Energy is important because, under certain conditions, the total energy of a collection (or system) of objects does not change. As the objects move the energy may change form but if certain conditions hold the total remains the same. This is one of the great conservation principles of mechanics and in

Chapter 8 you will learn when it applies and when it does not. You will also learn to use it to answer questions about the motions of objects.

When the total energy of a system does change the change can be accounted for by the work done on objects in the system by objects outside the system. We may think of energy flowing into or out of the system as objects outside interact with objects inside.

Another quantity that behaves in much the same way is the momentum of a system. In Chapter 9 momentum is defined and the conditions for which it is conserved are discussed. In Chapter 10 the principle of momentum conservation is applied to collisions between objects.

Chapters 11 and 12 are devoted to rotational motion. Here Newton's laws are applied to wheels spinning on fixed axes and to gyroscopes, for example. The plan is like the one followed for linear motion: first kinematics, then dynamics. You will also learn about the kinetic energy of rotation and how it changes when work is done on the rotating object.

The most important concept introduced here is angular momentum, another quantity that is conserved in certain situations. You should learn to identify those situations and to use the principle of angular momentum conservation to help with your understanding of rotational motion.

In Chapter 13 Newton's laws for linear and rotational motion are used to discuss the special case of an object at rest. Here you will learn to calculate the forces that must act on an object to hold it at rest. You will also learn about the deformation of an object by the forces acting on it. These topics are enormously important for engineering applications. Bridges, buildings, and automobiles, for example, must be designed to withstand the loads to which they are subjected during use.

Motions that repeat, called oscillations, are discussed in Chapter 14 using Newton's laws and the conservation of energy principle. Oscillations are prevalent in nature and among man-made artifacts. The swaying of trees, buildings, and bridges, the motion of a clock pendulum, and the bouncing of a car as it rides over a pothole are all examples. In addition, what you learn here will be of great use when you study waves.

In Chapter 15 you will study the force of gravity, one of the fundamental forces. This discussion provides an excellent example of the dependence of a force on properties of the interacting bodies. The principles of dynamics are then applied to the motions of objects moving under the influence of gravity: planets, satellites, and spacecraft. You will bring to bear many of the concepts you learned earlier, most notably conservation of energy and angular momentum. Electrical and gravitational forces are mathematically quite similar. Much of what you study here will be put to use when you study Chapter 23 and later chapters.

Chapter 16 deals with fluids, both at rest and in motion. Density and pressure are defined, then the principles of mechanics are used to understand the variation of pressure with depth in a body of water and with altitude in the earth's atmosphere, as well as the variation of pressure along a pipe in which a fluid is flowing. You will also be able understand, for example, why some objects float while others sink when they are placed in a fluid and why the fluid speed increases when the nozzle opening of a hose is decreased.

Wave motion, in which a disturbance created at one place is propagated to another, is the basis for transmitting information (the form of the disturbance) and energy. Sound, light, radio signals, x rays, and microwaves are all examples of waves. Fundamental concepts developed in Chapter 17 are applied to sound waves in Chapter 18 and later on, in Chapter 38, the ideas are used to discuss electromagnetic radiation.

Chapter 1
MEASUREMENT

I. BASIC CONCEPTS

Physics is an experimental science and relies strongly on accurate measurements of physical quantities. All measurements are comparisons, either direct or indirect, with standards. This means that for every quantity you must not only have a qualitative understanding of what the quantity represents but also an understanding of how it is measured. A length measurement is a familiar example. You should know that the length of an object represents its extent in space and also that length might be measured by comparison with a meter stick, say, whose length is accurately known in terms of the SI standard for the meter. Make a point of understanding both aspects of each new quantity as it is introduced.

Systems of units and standards. A system of units consists of a unit for each physical quantity, organized so that all can be derived from a small number of independent base units. Standards are associated with base units but not with derived units. Ideally a standard should have the following properties: _____

_____

The three International System <u>base</u> <u>units</u> used in mechanics are:

length: _____ (abbreviation: ___)
mass: _____ (abbreviation: ___)
time: _____ (abbreviation: ___)

The <u>standards</u> for these units are:

length: _____

mass: _____

time: _____

Notice that the SI unit for length is defined in terms of the speed of light and the time standard. The speed of light is by definition exactly $c =$ _____ m/s. Assume you have an instrument that accurately measures any time interval. Briefly explain how you can, in principle, calibrate a meter stick in terms of SI standards: _____

_____

At the atomic level the second, non-SI, standard for mass is _____

_____ .

The atomic mass unit is related to the kilogram by 1 u = _____ kg.

In the table below list several examples of quantities with <u>derived</u> <u>units</u> in the International System. For each, give the units as a combination of SI base units.

QUANTITY	UNITS
_____	_____
_____	_____
_____	_____
_____	_____

To appreciate the magnitudes of quantities discussed in this course you should have an intuitive feeling for the size of a meter, kilogram, and second. Search tables 3, 4, and 5 for familiar objects and try to visualize them as you remember their sizes. Use the tables or seek elsewhere for quantities that are about 1 m long, 1 kg in mass, or 1 s in duration. List them here and remember them as examples:

objects about 1 m long: _____

objects with about 1 kg of mass: _____

time intervals of about 1 s: _____

SI prefixes. SI prefixes are used extensively throughout this course. The following are used the most. For each of them write the associated power of ten and the symbol used as a prefix.

PREFIX	POWER OF TEN	SYMBOL
kilo:	10	_____
mega:	10	_____
centi:	10	_____
milli:	10	_____
micro:	10	_____
nano:	10	_____
pico:	10	_____

Memorize them. When evaluating an algebraic expression, substitute the value using the appropriate power of ten. That is, for example, if a length is given as 25 μm, substitute 25×10^{-6} m. One catch: the SI unit for mass is the <u>kilogram</u>. Thus a mass of 25 kg is substituted directly, while a mass of 25 g is substituted as 25×10^{-3} kg.

Unit conversion. Carefully study Section 1–3 to see how a quantity given in one system of units is converted to another. Turn to Appendix F and become familiar with the conversion tables there. A good habit to cultivate is to say the words associated with a conversion. Suppose you want to convert 50 ft to meters. The length table in the appendix tells you that 1 ft is equivalent to 0.3048 m. Say "If 1 ft is equivalent to 0.3048 m, then 50 ft must be equivalent to (50 ft) $\times$ (0.3048 m/ft) = 15 m".

For practice verify the following:

2.90 in = 73.7 mm	4.50 ft = 1.37 m	2.10 mi = 3.38 km
36.0 mi/h = 57.9 km/hr	36.0 mi/h = 16.1 m/s	45.0 ft/s = 13.7 m/s
32.0 ft/s^2 = 9.75 m/s^2	100 lb = 445 N	5.10 slugs = 74.4 kg

II. PROBLEM SOLVING

Most of the problems at the end of this chapter are exercises in unit conversion, powers of 10 arithmetic, and SI prefixes. In addition, some deal with calculations of areas and volumes. See the Mathematical Skills section below for a discussion of powers of ten arithmetic and some useful equations from geometry. Here are some examples from the collection at the end of Chapter 1 of the text. Their identifiers in the text are given below in parentheses next to problem numbers.

PROBLEM 1 (7E).   Calculate the number of kilometers in 20.0 mi using only the following conversion factors: 1 mi = 5280 ft, 1 ft = 12 in., 1 in. = 2.54 cm, 1 m = 100 cm, and 1 km = 1000 m.

STRATEGY: Multiply or divide 20.0 mi by appropriate conversion factors until all units cancel but kilometers.

SOLUTION: Here's the chain of conversions:

$$20.0\,\text{mi} = 20.0\,\text{mi} \times \frac{5280\,\text{ft}}{\text{mi}} \times \frac{12\,\text{in.}}{\text{ft}} \times \frac{2.54\,\text{cm}}{\text{in.}} \times \frac{1\,\text{m}}{100\,\text{cm}} \times \frac{1\,\text{km}}{1000\,\text{m}}$$

$$[\text{ans: } 32.2\,\text{km}]$$

Notice that units in the chain cancel, leaving only km. Also notice that we may rightfully retain only 3 significant figures. The conversion factor 1 ft = 12 in. is exact so its use does not reduce the answer to 2 figure accuracy.

PROBLEM 2 (11P).   A room is 20 ft, 2 in. long and 12 ft, 5 in. wide. What is the floor area in (*a*) square feet and (*b*) square meters?

STRATEGY: The area is the product of the width and length. Each of the dimensions is given in feet and inches and must be converted to feet alone. Once the area is found in square feet, convert to square meters.

SOLUTION: (*a*) 2 in. is equivalent to 2/12 = 0.17 ft so the length is 20.17 ft. 5 in. is equivalent to 5/12 = 0.42 ft so the width is 12.42 ft. Since the dimensions are given to the nearest inch the dimensions in feet should have only one significant figure after the decimal point, but this is an intermediate calculation and we carry one more figure. Now multiply the length and width to get the area in square feet. Retain 3 significant figures.

(*b*) Use 1 ft = 0.3046 m to convert ft^2 to m^2.

$$[\text{ans: } (a)\ 250\,\text{ft}^2;\ (b)\ 23.3\,\text{m}^2]$$

If the ceiling is 12 ft, $2\frac{1}{2}$ in. above the floor, what is the volume of the room in (*c*) cubic feet and (*d*) cubic meters?

STRATEGY: The volume is the product of the floor area and ceiling height. First convert the ceiling height, given in feet and inches, to feet alone.

SOLUTION:

$$[\text{ans: } (c)\ 3060\,\text{ft}^3;\ (d)\ 86.8\,\text{m}^3]$$

PROBLEM 3 (15P). A certain brand of house paint claims a coverage of 460 ft²/gal. (*a*) Express this quantity in square meters per liter. (*b*) Express this quantity in SI base units (see Appendices A and F). (*c*) What is the inverse of the original quantity, and what is its physical significance?

STRATEGY: (*a*) Multiply or divide 460 ft²/gal by appropriate conversion factors to obtain m²/L. (*b*) Convert liters to cubic meters.

SOLUTION: (*a*) Use 1 ft = 0.3048 m, 1 gal = 231 in.³, and 1 in.³ = 1.639 × 10⁻² L. The conversion chain is

$$460\,\text{ft}^2/\text{gal} = 460\,\frac{\text{ft}^2}{\text{gal}}\left(\frac{0.3048\,\text{m}}{\text{ft}}\right)^2\left(\frac{1\,\text{gal}}{231\,\text{in.}^3}\right)\left(\frac{1\,\text{in.}^3}{1.639 \times 10^{-2}\,\text{L}}\right).$$

(*b*) To convert to SI base units use 1 L = 1.00 × 10⁻³ m³.

$$\left[\text{ans: } (a)\ 11.3\,\text{m}^2/\text{L}; (b)\ 1.13 \times 10^4\,\text{m}^{-1}; (c)\ 2.17 \times 10^{-3}\,\text{gal/ft}^2\right]$$

PROBLEM 4 (16P). Astronomical distances are so large compared to terrestrial ones that much larger units of length are used for easy comprehension of the relative distances of astronomical objects. An *astronomical unit* (AU) is equal to the average distance from the Earth to the sun, about 92.9 × 10⁶ mi. A *parsec* (pc) is the distance at which 1 AU would subtend an angle of exactly 1 second of arc (Fig. 1–8). A *light-year* (ly) is the distance that light, traveling through a vacuum with a speed of 186, 000 mi/s, would cover in 1.0 year. (*a*) Express the distance from the Earth to the sun in parsecs and in light-years. (*b*) Express 1 ly and 1 pc in miles. Although "light-year" appears frequently in popular writing, the parsec is preferred by astronomers.

STRATEGY: First find the number of miles in a parsec and in a light-year, then use these results as conversion factors to convert 92.9 mi to parsecs and to light-years.

SOLUTION: First find the number of miles in a parsec. There are 60 seconds of arc in a minute of arc and 60 minutes in a degree, so there are 360 × 60 × 60 = 1.296 × 10⁶ seconds in a circle. If the radius of the circle is *r* the circumference is 2π*r* and the length of the arc subtended by 1 s is 2π*r*/1.296 × 10⁶. Set this equal to 92.9 × 10⁶ mi and solve for *r*. You should get 1.92 × 10¹³ mi. The distance from the Earth to the sun in parsecs is (92.9 × 10⁶ mi) × (1 pc/1.92 × 10¹³ mi). Now find the number of miles in a light-year. There are 365 days in a year, 24 hours in a day, 60 minutes in an hour, and 60 seconds in a minute, so there are 365 × 24 × 60 × 60 = 3.15 × 10⁷ seconds in a year. In a year light travels 186, 000 × 3.15 × 10⁷ = 5.87 × 10¹² mi. The distance from the Earth to the sun is given in light-years by (92.9 × 10⁶ mi) × (1 ly/5.87 × 10¹² mi).

$$\left[\text{ans: } (a)\ 4.85 \times 10^{-6}\,\text{pc},\ 1.58 \times 10^{-5}\,\text{ly}; (b)\ 1\,\text{ly} = 5.87 \times 10^{12}\,\text{mi},\ 1\,\text{pc} = 1.91 \times 10^{13}\,\text{mi}\right]$$

PROBLEM 5 (18P*). The standard kilogram (see Fig. 1–7) is in the shape of a circular cylinder with its height equal to its diameter. Show that, for a circular cylinder of fixed volume, this equality gives the smallest surface area, thus minimizing the effects of surface contamination and wear.

STRATEGY: Let *r* be the radius and ℓ the length of the cylinder. Its total surface area (including the two ends) is given by $A = 2\pi r\ell + 2\pi r^2$. The radius must be chosen so this is the smallest possible value for a given volume. Since the volume is fixed, *r* and ℓ are related: a decrease in ℓ for example, necessitates an increase in *r*. The volume is given by $V = \pi r^2\ell$, so $\ell = V/\pi r^2$. Thus $A = 2V/r + 2\pi r^2$. Set $dA/dr = 0$ and solve for r^3. You should get $r^3 = V/2\pi$. Now substitute $\pi r^2\ell$ for *V* and solve for *r*.

SOLUTION:

$$\left[\text{ans: } r = \ell/2 \text{ or } 2r = \ell\right]$$

PROBLEM 6 (21E).   Enrico Fermi once pointed out that a standard lecture period (50 min) is close to 1 microcentury. How long is a microcentury in minutes and what is the percent difference from Fermi's approximation?

STRATEGY: There are 100 years in a century, 365 days in a year, 24 hours in a day, and 60 minutes in an hour, so there are $100 \times 365 \times 24 \times 60 = 5.256 \times 10^7$ minutes in a century. A microcentury is this divided by 10^6. If T is a microcentury in minutes then the percent difference from Fermi's approximation is $100(T - 50)/50$.

SOLUTION:

[ans: 52.6 min; 5.2%]

III. MATHEMATICAL SKILLS

Powers of 10 arithmetic.   Although the topic is not covered in the text, you should be quite facile with powers of 10 arithmetic. If you are not, practice until you can carry out operations quickly. Later on, while trying to work a problem, you will be able to concentrate more on the problem itself and less on this detail.

When you multiply two numbers expressed as powers of ten, multiply the numbers in front of the tens, then multiply tens themselves. This last operation is carried out by adding the powers. Thus $(1.6 \times 10^3) \times (2.2 \times 10^2) = (1.6 \times 2.2) \times (10^3 \times 10^2) = 3.5 \times 10^5$ and $(1.6 \times 10^3) \times (2.2 \times 10^{-2}) = (1.6 \times 2.2) \times (10^3 \times 10^{-2}) = 3.5 \times 10 = 35$.

When you divide two numbers, divide the numbers in front of the tens, then divide the tens. The last operation is carried out by subtracting the power in the denominator from the power in the numerator. Thus $(1.6 \times 10^3)/(2.2 \times 10^2) = (1.6/2.2) \times (10^3/10^2) = 0.73 \times 10 = 7.3$ and $(1.6 \times 10^3)/(2.2 \times 10^{-2}) = (1.6/2.2) \times (10^3/10^{-2}) = 0.73 \times 10^5 = 7.3 \times 10^4$.

When you add or subtract two numbers first convert them so the powers of ten are the same, then add or subtract the numbers in front of the tens and multiply the result by 10 to the common power. Thus $1.6 \times 10^3 + 2.2 \times 10^2 = 1.6 \times 10^3 + 0.22 \times 10^3 = 1.8 \times 10^3$.

This means you must know how to write the same number with different powers of ten. Remember that multiplication by 10 is equivalent to moving the decimal point one digit to the right and division by 10 is equivalent to moving the decimal point one digit to the left. Thus $1.6 \times 10^3 = 16 \times 10^2 = 0.16 \times 10^4$. In the first case we multiplied 1.6 by 10 and divided 10^3 by 10. In the second we divided 1.6 by 10 and multiplied 10^3 by 10.

You should be able to verify the following without difficulty:

$$512 \times 10^2 = 5.12 \times 10^4$$
$$0.00512 = 5.12 \times 10^{-3}$$
$$(3.4 \times 10^2) \times (2.0 \times 10^4) = 6.8 \times 10^6$$
$$(3.4 \times 10^2)/(2.0 \times 10^4) = 1.7 \times 10^{-2}$$
$$(3.4 \times 10^4) + (2.0 \times 10^3) = (3.4 \times 10^4) + (0.20 \times 10^4) = 3.6 \times 10^4$$

Geometry.   You should be familiar with some geometric concepts.

   a.   The circumference of a rectangle is given by $2(a + b)$, where a and b are the lengths of its sides.

b. The area of a rectangle is given by ab.

c. The area of a triangle is given by $\frac{1}{2}\ell h$, where ℓ is the length of one side and h is the length of the line from the vertex opposite that side to the side (the altitude).

d. The volume of a rectangular solid is given by abc, where a, b, and c are the lengths of its sides. The volume of a cube is given by a^3, where a is the length of one of its sides.

e. The circumference of a circle is given by $2\pi r$, where r is its radius. The value of π is about 3.14159.

f. The area of a circle is πr^2, where r is its radius.

g. The surface area of a sphere is given by $4\pi r^2$, where r is its radius.

h. The volume of a sphere is given by $\frac{4}{3}\pi r^3$.

i. The area of the curved surface of a right circular cylinder is the product of the circumference of an end and the cylinder length: $2\pi r\ell$. The ends are circles and each has an area of πr^2.

j. The volume of a cylinder is the product of the area of an end and the length: $\pi r^2\ell$.

Carefully note that all circumferences have units of length, all areas have units of length squared, and all volumes have units of length cubed.

IV. NOTES

Chapter 2
MOTION ALONG A STRAIGHT LINE

I. BASIC CONCEPTS

This chapter introduces you to some of the concepts used to describe motion; most important are those of position, velocity, and acceleration. Pay particular attention to their definitions and to the relationships between them.

Definitions. This chapter of the text deals with <u>kinematics</u>. Tell what this term means:

_____

_____

In this section of the text objects are treated as particles. Tell in your own words what a particle is. Pay particular attention to the properties of the motion. Can a particle rotate? Can a particle have parts that move relative to each other? _____

_____

_____

An object can be treated as a particle if _____

_____ .

If an extended object can be treated as a particle we may pick one point on the object and follow its motion. The position of a crate, for example, means the position of the point on the crate we have chosen to follow, perhaps one of its corners.

The motion of a particle in one dimension can be described by giving its coordinate x as a function of time t. You must carefully distinguish between an <u>instant</u> of time and an <u>interval</u> of time. The symbol t represents an instant and has no extension. Thus t might be *exactly* 12 min, 2.43 s after noon on a certain day. At any other time, no matter how close, t has a different value. On the other hand, an interval extends from some initial time to some final time: *two* instants of time are required to describe it. Note that a value of the time may be positive or negative, depending on whether the instant is after or before the instant designated as $t = 0$.

Similarly, a value of the coordinate x specifies a <u>point</u> on the x axis. It has no extension in space. On the other hand, two values are required to specify a <u>displacement</u>. A coordinate may be positive or negative, depending on where the point lies relative to the origin.

Suppose a particle has coordinate x_1 at time t_1 and coordinate x_2 at a later time t_2. Then its displacement Δx over the interval from t_1 to t_2 is given by

$$\Delta x =$$

The magnitude of the displacement during a time interval may be quite different from the distance traveled during the interval. Suppose a particle starts at time $t_1 = 0$ with coordinate $x_1 = 5.0$ m, arrives at $x_2 = 15.0$ m at time $t_2 = 2.0$ s, then turns around and arrives at $x_3 = 10.0$ m at time $t_3 = 3.0$ s. Note that at first it travels in the positive x direction but later it travels in the negative x direction. From t_1 to t_2 its displacement is $(15.0 - 5.0) = 10.0\,\mathrm{m}$; from t_2 to t_3 its displacement is $(10.0 - 15.0) = -5.0\,\mathrm{m}$; and from t_1 to t_3 its displacement is $(10.0 - 5.0) = 5.0\,\mathrm{m}$. On the other hand, the total distance traveled from t_1 to t_3 is $15\,\mathrm{m}$.

If a particle goes from x_1 at time t_1 to x_2 at time t_2, its <u>average</u> velocity $\bar{v}$ in the interval from t_1 to t_2 is given by

$$\bar{v} =$$

Write down in words the steps you would take to find the average velocity in the interval from t_1 to t_2 if you are given the function $x(t)$: _____

_____

On a graph of x vs. t the average velocity over the interval from t_1 to t_2 is related to a certain line you might draw. Describe the line and tell what property gives the average velocity:

_____

You must distinguish average velocity and <u>average speed</u>. The average speed over a time interval Δt is defined by

$$\bar{s} = \frac{d}{\Delta t},$$

where d is _____ .

Describe the limiting process used to obtain the <u>instantaneous velocity</u> at time t by applying it to a series of average velocities: _____

_____

If the function $x(t)$ is known, the instantaneous velocity at any time t_1 is found by _____
_____ and evaluating the result for $t = $ _____. On a graph of x vs. t, the instantaneous velocity at any time t_1 is related to a line you might draw. Describe the line and tell what property gives the instantaneous velocity: _____

_____

The term "velocity" means instantaneous velocity. The modifier "instantaneous" is implied.

For each of the functions $x(t)$ shown graphically below, tell if the average velocity is positive or negative in the interval from t_1 to t_2. Also give the sign of the instantaneous velocity at t_1 and t_2 or state that it is zero if it is.

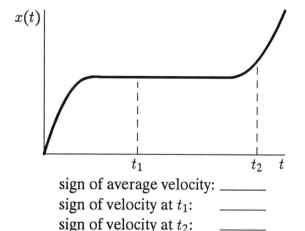

sign of average velocity: _____
sign of velocity at t_1: _____
sign of velocity at t_2: _____

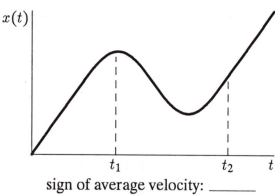

sign of average velocity: _____
sign of velocity at t_1: _____
sign of velocity at t_2: _____

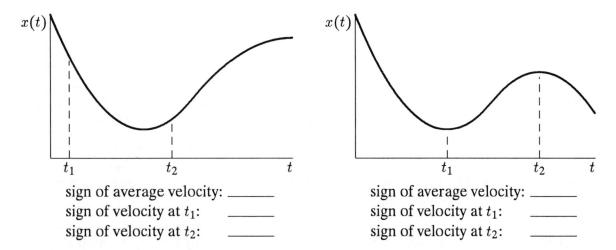

sign of average velocity: _____
sign of velocity at t_1: _____
sign of velocity at t_2: _____

sign of average velocity: _____
sign of velocity at t_1: _____
sign of velocity at t_2: _____

On each of the first three diagrams indicate a time t_3 such that the average velocity from t_1 to t_3 is zero.

On the axes below sketch graphs of the velocity $v(t)$ for the motions represented by the first two of the x vs. t graphs above. Be sure you get the sign of v correct. Also be sure your graphs indicate roughly where the magnitude of v is large and where it is small or zero.

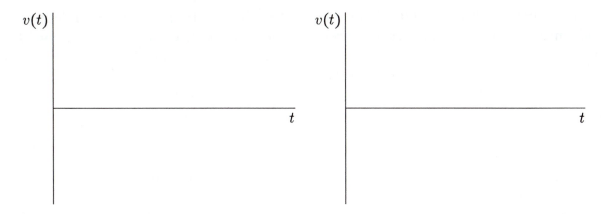

The sign of the velocity indicates the direction of travel. Describe the direction of particle motion if the velocity is positive; if the velocity is negative. Warning! Do not assume the x axis is positive to the right.

positive velocity: _____
_____
negative velocity: _____
_____
Define speed: _____

Carefully note that speed and average speed may be quite different since the displacement over a time interval may be quite different from the distance traveled over that interval.

If the velocity of a particle changes from v_1 at time t_1 to v_2 at a later time t_2 then its average acceleration $\bar{a}$ over the interval from t_1 to t_2 is given by

$$\bar{a} =$$

Write down in words the steps you would take to find the average acceleration in the interval from t_1 to t_2, given the function $x(t)$: _____

_____

On a graph of v vs t, the average acceleration over the interval from t_1 to t_2 is related to a line you might draw. Describe the line and tell what property gives the average acceleration:

_____

Describe the limiting process used to obtain the <u>instantaneous acceleration</u> at time t by applying it to a series of average accelerations: _____

_____

If the function $x(t)$ is known, the instantaneous acceleration at any time t_1 is found by _____
_____ and evaluating the result for $t =$ _____. On a graph of v vs. t, the instantaneous acceleration at any time t_1 is related to a line you might draw. Describe the line and tell what property gives the instantaneous acceleration: _____

_____

The term "acceleration" means instantaneous acceleration. The modifier "instantaneous" is implied.

Note that a positive acceleration does not necessarily mean the particle speed is increasing and a negative acceleration does not necessarily mean the particle speed is decreasing. The speed increases if the velocity and acceleration have the same sign and decreases if they have opposite signs, no matter what the signs are.

The graph to the left below shows the velocity as a function of time for an object moving along a straight line. On the t axis mark two times (t_1 and t_2, say) such that the acceleration is negative in the interval between them and label that portion of the curve "$a < 0$". Mark two times such that the acceleration is positive between them and label that portion of the curve "$a > 0$". Mark two times such that the acceleration is zero between them and label that portion of the curve "$a = 0$". There is one time for which the acceleration is zero only for that instant and is not zero for neighboring times. Label it. On the axes to the right below sketch the acceleration as a function of time.

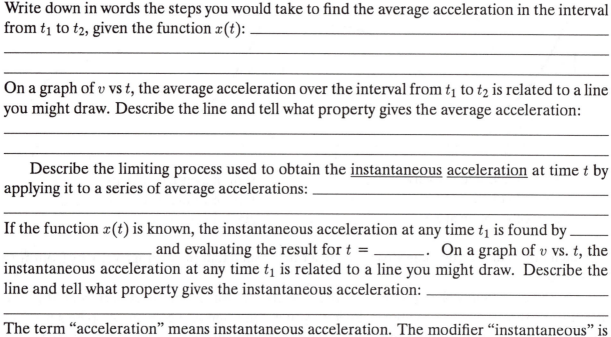

On the coordinate axes below draw possible graphs of $x(t)$ for a particle moving along the x axis with constant acceleration. Take the initial position to be $x = 0$ in all cases. For the

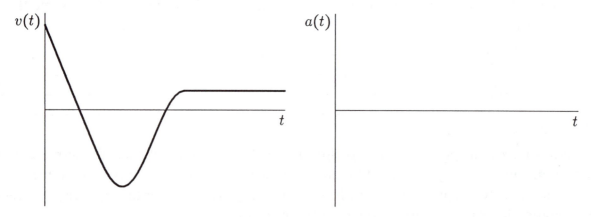

first curve suppose the particle starts with a positive velocity and has a positive acceleration; for the second suppose it starts with a negative velocity and has a positive acceleration; for the third suppose it starts with a positive velocity and has a negative acceleration; and for the fourth suppose it starts with a negative velocity and has a negative acceleration. Label each graph with the signs of the initial velocity and the acceleration. Also label points where the particle momentarily stops to start moving in the opposite direction.

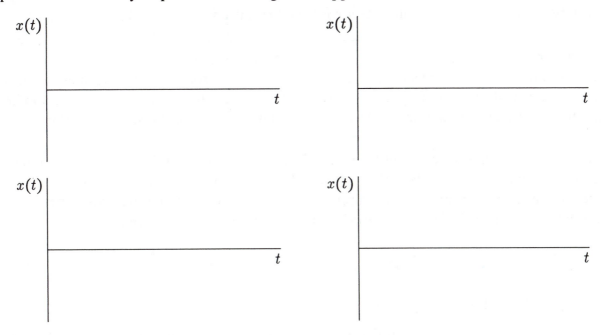

At the instant a particle is momentarily at rest its acceleration is not necessarily 0. To remind yourself that $v = 0$ does not imply $a = 0$ write down a function $x(t)$ for which $v = 0$ but $a \neq 0$ at $t = 0$:

$$x(t) =$$

A particle is initially moving in the positive x direction. Describe in words what its motion might be if at some instant it has zero velocity but non-zero acceleration: _____

_____

For comparison, describe in words what its motion might be if at some instant it has zero velocity and thereafter it has zero acceleration: _____

_____

Motion with constant acceleration. If a particle is moving along the x axis with constant acceleration a, its coordinate is given as a function of time by

$$x(t) =$$

and its velocity is given as a function of time by

$$v(t) =$$

The coordinate and velocity of the particle at time $t = 0$ should appear in your equations. Give the symbol used for each: coordinate at $t = 0$: _____ ; velocity at $t = 0$: _____ .

Eq. 2–14 of the text is also extremely useful. Write it here:

$$v^2 =$$

It can be obtained from the equations for $x(t)$ and $v(t)$ you wrote above by _____
_____ .

You should memorize these three equations. Since the second is the derivative with respect to time of the first and so can be derived quickly, you probably need to memorize only the first and third. Before using these equations always ask if the problem specifies or implies that the acceleration is constant. They are not valid if the acceleration varies with time.

Free fall. Specialize the constant acceleration kinematic equations for the case of an object in free fall, subject only to the pull of gravity. Take the y axis to be positive in the upward direction, away from the earth, and suppose the particle moves along that axis. Then the y coordinate and the velocity of the particle, as functions of time, are given by

$$y(t) =$$

$$v(t) =$$

Here $g =$ _____ m/s^2 = _____ ft/s^2 is the magnitude of the acceleration due to gravity near the surface of the earth.

Remember that, in the absence of air resistance, *all* objects in free fall have the same acceleration, regardless of their masses. Also remember their acceleration is the same throughout their motion from the time they are released or thrown to the time they hit something. Their acceleration is g downward while they are going up, when they are at their highest points, and while they are going down.

These equations are for constant acceleration only. To emphasize this point describe a situation in which an object is thrown upward or downward near the surface of the earth and these equations are *not* valid: _____
_____
_____

II. PROBLEM SOLVING

Many of the problems at the end of this chapter can be solved using the definitions of average and instantaneous velocity and acceleration. Some also deal with the interpretation of $x(t)$ versus t and $v(t)$ versus t graphs. Here are a few examples.

PROBLEM 1 (3E).   On average, a blink lasts about 100 ms. How far does a MIG-25 "Foxbat" fighter travel during a pilot's blink if the plane's average velocity is 2110 mi/h?

STRATEGY: Solve $\bar{v} = \Delta x / \Delta t$ for Δx. Substitute $\bar{v} = 2110/(60 \times 60) = 0.5861$ mi/s and $\Delta t = 100 \times 10^{-3}$ s.

SOLUTION:

$$\big[\text{ans: } 0.0586 \text{ mi } (310 \text{ ft})\big]$$

PROBLEM 2 (9P).   Compute your average velocity in the following two cases. (*a*) You walk 240 ft at a speed of 4.0 ft/s and then run 240 ft at a speed of 10 ft/s along a straight track. (*b*) You walk for 1.0 min at a speed of 4.0 ft/s and then run for 1.0 min at 10 ft/s along a straight track.

STRATEGY: (*a*) You know the displacement for the whole trip is $\Delta x = 480$ ft. You need to find the total time for the trip. The time for the first part of the trip is $\Delta t_1 = \Delta x_1/v_1 = 240/4 =$ _____s. The time for the second part is $\Delta t_2 = \Delta x_2/v_2 = 240/10 =$ _____s. The total time is $\Delta t = \Delta t_1 + \Delta t_2 =$ _____s. Use $\bar{v} = \Delta x / \Delta t$ to calculate the average velocity.

(*b*) Now you know the total time for the trip is 120 s, but you must find the displacement. For the first part of the trip the displacement is $\Delta x_1 = v_1 \Delta t_1 = 4.0 \times 60 =$ _____ft and for the second part the displacement is $\Delta x_2 = v_2 \Delta t_2 = 10 \times 60 =$ _____ft. The total displacement is $\Delta x = \Delta x_1 + \Delta x_2 =$ _____ft. Again use $\bar{v} = \Delta x / \Delta t$ to compute the average velocity.

SOLUTION:

$$\big[\text{ans: } (a)\ 5.7 \text{ ft/s}; (b)\ 7.0 \text{ ft/s}\big]$$

(*c*) Graph x versus t for both cases and indicate how the average velocity is found on the graph.

SOLUTION: Here are some axes you might use. Each graph will consist of 2 straight line segments. Draw them and then draw the line with a slope that represents the average velocity.

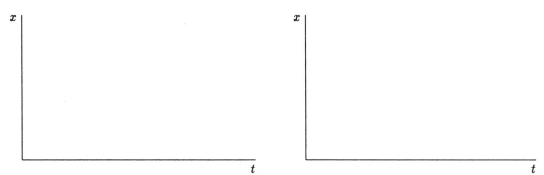

PROBLEM 3 (11P).   You drive on Interstate 10 from San Antonio to Houston, half the *time* at 35 mi/h (= 56 km/h) and the other half at 55 mi/h (= 89 km/h). On the way back you travel half the *distance* at 35 mi/h and the other half at 55 mi/h. What is your average speed (*a*) from San Antonio to Houston, (*b*) from Houston back to San Antonio, and (*c*) for the entire trip?

STRATEGY: Let d be the total displacement and Δt be the time for either of the one-way trips. The trip is composed of two parts, with different speeds: $d = d_1 + d_2$ and $\Delta t = \Delta t_1 + \Delta t_2$. The partial displacements and times are related by $d_1 = v_1 \Delta t_1$ and $d_2 = v_2 \Delta t_2$.

  For part *a*, $\Delta t_1 = \Delta t_2$ so $d = (v_1 + v_2)\Delta t_1$. Divide by $\Delta t = \Delta t_1 + \Delta t_2 = 2\Delta t_1$. For part *b*, $d_1 = d_2$, so $d = 2d_1$ and $\Delta t = [(1/v_1) + (1/v_2)]d_1$. To compute the average speed divide the expression for d by the expression for Δt. (*c*) To find the average speed over the round trip divide the total distance $2d$ by the total time. Let $\bar{s}_t$ be the average speed to Houston and $\bar{s}_f$ be the average speed from Houston. Then the time for the first one-way trip is $\Delta t_t = d/\bar{s}_t$, the time for the second one-way trip is $t_f = d/\bar{s}_f$, and the total time is $[(1/\bar{s}_t) + (1/\bar{s}_f)]d$.

SOLUTION:

$$\left[\text{ans: } (a) \text{ 45 mi/h (73 km/h); } (b) \text{ 43 mi/h (69 km/h); } (c) \text{ 44 mi/h (71 km/h)}\right]$$

(d) What is your average velocity for the entire trip?

SOLUTION: The displacement for the entire trip is _____ , so the average velocity for the entire trip is _____ .

$$\left[\text{ans: } 0\right]$$

(e) Graph x versus t for part (a), assuming the motion is all in the positive x direction. Indicate how the average velocity can be found on the graph.

SOLUTION: Use the axes below. Your graph will consist of two straight line segments with different slopes. When you have drawn them, draw the line with a slope that represents the average velocity.

PROBLEM 4 (13P). The position of a particle moving along the x axis is given in centimeters by $x = 9.75 + 1.50t^3$, where t is in seconds. Consider the time interval $t = 2.00$ s to $t = 3.00$ s and calculate (a) the average velocity; (b) the instantaneous velocity at $t = 2.00$ s; (c) the instantaneous velocity at $t = 3.00$ s; (d) the instantaneous velocity at $t = 2.50$ s; and (e) the instantaneous velocity when the particle is midway between its positions at $t = 2.00$ s and $t = 3.00$ s.

STRATEGY: (a) The average velocity is given by $\bar{v} = \Delta x/\Delta t = (x_f - x_i)/\Delta t$, where x_i is the initial coordinate ($x(t)$ evaluated for $t = 2.00$ s) and x_f is the final coordinate ($x(t)$ evaluated for $t = 3.00$ s).

(b), (c), and (d) Differentiate $x(t)$ with respect to t to find an expression for the velocity as a function of time. Evaluate the expression for the 3 given times.

(e) First you must find the coordinate of the midpoint. It is given by $x_m = (x_f + x_i)/2$. Now find the time the particle is at the midpoint: solve $x_m = 9.75 + 1.50t^3$ for t. This is a cubic equation. Look up a technique for solving it or else use a trial and error method. Systematically substitute various value of t into $9.75 + 1.50t^3$ until you get an answer that is close to x_m. Keep a record of the results to help you decide on your next trial value. Once you have found a value for t substitute it into your expression for the velocity.

18   *Chapter 2: Motion Along a Straight Line*

SOLUTION:

[ans: (a) 28.5 cm/s; (b) 18.0 cm/s; (c) 40.5 cm/s; (d) 28.1 cm/s; (e) 30.3 cm/s]

Notice that all these velocities are different. The average velocity is not the same as the velocity at the midpoint in position nor is it the same as the velocity at the midpoint in time. It also is not the average of the velocities at the end points. If the acceleration were constant the average velocity over the interval, the average of the initial and final velocities, and the instantaneous velocity at the midpoint in time, would all be the same and they would be different from the instantaneous velocity at the midpoint in position.

(f) Graph x versus t and indicate your answers graphically.

SOLUTION: Use the axes below to graph $x(t)$ from $t = 0$ to $t = 3.00$ s. Draw the appropriate lines, with slopes equal to the various average and instantaneous velocities requested.

PROBLEM 5 (16E). (a) If a particle's position is given by $x = 4 - 12t + 3t^2$ (where t is in seconds and x is in meters), what is its velocity at $t = 1$ s? (b) Is it moving toward increasing or decreasing x just then? (c) What is its speed just then? (d) Is the speed larger or smaller at later times?

STRATEGY: (a) Differentiate $x(t)$ with respect to t to obtain an expression for the velocity at any time. Substitute $t = 1$ s into this expression.

(b) If v is positive the particle is moving in the positive x direction, if v is negative it is moving in the negative x direction.

(c) The speed is the magnitude of the velocity.

(d) Look at the expression for the velocity as a function of time and ask what happens as t increases from 1 s. You will find several answers: over the short term the speed is doing one thing, at longer times it is doing another.

SOLUTION:

[ans: (a) −6 m/s; (b) negative x direction; (c) 6 m/s; (d) speed decreases until $t = 2$ s, then increases]

(Try answering the next two questions without further calculation.) (*e*) Is there ever an instant when the velocity is zero? (*f*) Is there a time after $t = 3$ s when the particle is moving leftward on the x axis?

STRATEGY: (*e*) Look at the expression for the velocity as a function of time and notice that the initial velocity is negative but that the velocity is increasing algebraically (the acceleration is positive). (*f*) Notice that the velocity is zero at $t = 2$ s and then continues to increase without change in sign.

SOLUTION:

[ans: (*e*) Yes; (*f*) No]

PROBLEM 6 (23E). The graph of x versus t in Fig. 2–20*a* is for a particle in straight-line motion. (*a*) State, for each of the intervals AB, BC, CD, and DE, whether the velocity v is positive, negative, or 0 and whether the acceleration a is positive, negative, or 0. (Ignore the end points of the intervals.)

STRATEGY: Determine the sign of the slope of the curve in each of the intervals. If the tangent line goes upward with increasing t the slope and the velocity are positive, if the tangent line goes down the slope and velocity are negative, and if the tangent line is horizontal the slope and the velocity are zero. To determine the sign of the acceleration in an interval determine how the slope changes with increasing time. If the slope is positive and becomes more positive or if it is negative and become less negative, the acceleration is positive; if it is positive and becomes less positive or if it is negative and becomes more negative, the acceleration is negative.

SOLUTION:

[ans: AB: v is positive, a is 0; BC: v is positive, a is negative; CD: v is 0, a is 0; DE: v is negative, a is positive]

(*b*) From the curve, is there any interval over which the acceleration is obviously not constant?

STRATEGY: Here you must look for an interval in which the slope is changing at a different rate in different parts of the interval.

SOLUTION:

[ans: no]

(*c*) If the axes are shifted upward together such that the time axis ends up running along the dashed line, do any of your answers change?

STRATEGY: Note that all your answers depend on the slope of the curve at various points. These certainly do not change if the axes are shifted.

SOLUTION:

[ans: no]

At any given time the position of the particle is different for the two cases but the velocity and acceleration are both the same.

PROBLEM 7 (32P).   In an arcade video game, a spot is programmed to move across the screen according to $x = 9.00t - 0.750t^3$, where x is distance in centimeters measured from the left edge of the screen and t is time in seconds. When the spot reaches a screen edge, at either $x = 0$ or $x = 15$ cm, t is reset to 0 and the spot starts moving again according to $x(t)$. (*a*) At what time after starting is the spot instantaneously at rest?

STRATEGY: Differentiate $x(t)$ with respect to t to find an expression for the velocity as a function of time. Set the expression equal to 0 and solve for t.

SOLUTION:

[ans: 2.00 s]

(*b*) Where does this occur? (*c*) What is its acceleration when this occurs? (*d*) In what direction is it moving just prior to coming to rest? (*e*) Just after?

STRATEGY: (*b*) Evaluate $x(t) = 9.00t - 0.750t^3$ for the value of t you found in part *a*.

(*c*) Differentiate $v(t)$ to find an expression for the acceleration as a function of time and evaluate the expression for the value of t you found in part *a*. Notice that at that instant the velocity is zero but the acceleration is not.

(*d*) Imagine that you substitute into the expression for $v(t)$ a value for t that is slightly less than the value you found in part *a* and decide if the result is positive or negative. Note that t^2 is slightly less so the expression tells you to subtract a smaller number from a constant.

(*e*) Do the same for a value of t slightly larger than the value you found in part *a*.

SOLUTION:

[ans: (*b*) 12.0 cm; (*c*) −9.00 m/s²; (*d*) to the right; (*e*) to the left]

(*f*) When does it first reach an edge of the screen after $t = 0$?

STRATEGY: The spot changes its direction of motion only once for $t > 0$. This is at $t = 2.00$ s, $x = 12.0$ cm, without first reaching the right edge of the screen. Notice that $v = 0$ has only one solution for $t > 0$. This means that the screen edge first reached by the spot is the left edge, at $x = 0$. Solve $9.00t - 0.750t^2 = 0$ for t. Since you are not interested in the solution $t = 0$ you may cancel a t from each term.

SOLUTION:

[ans: 3.46 s]

All constant acceleration problems can be solved using only the equations $x(t) = x_0 + v_0 t + \frac{1}{2}at^2$ and $v(t) = v_0 + at$. Six quantities appear in these equations: $x(t)$, $v(t)$, x_0, v_0, a, and t. In most cases all but two are given and you are asked to solve for one or both of the others. Mathematically a typical constant acceleration problem involves identifying the known and unknown quantities, then simultaneously solving the two kinematic equations for the unknowns.

Other kinematics equations are derived in the text by eliminating one or another of the kinematic quantities from the expressions for $x(t)$ and $v(t)$. See Table 2–2 of the text. You may use them or solve the two fundamental equations simultaneously. The problems below show you how to use the simultaneous equation approach.

Kinematics problems that deal with a single object describe two events, each of which has a time, a coordinate, and a velocity associated with it. The acceleration is the additional quantity that enters the problem. It is often helpful in solving a kinematics problem to fill in a table such as:

	FIRST EVENT	SECOND EVENT	
time:	_____	_____	acceleration: _____
coordinate:	_____	_____	
velocity:	_____	_____	

For any problem write numerical values next to given or assigned quantities and question marks next to unknown quantities. When you are finished you should have no more than 2 question marks. If you do you missed some information.

You should be aware that you may assign the value 0 to the time when the particle is at *any* point along its trajectory: when the particle starts out, when it reaches a particular point, when it has a particular velocity, or any other point. A negative value for t simply means a time before the instant you chose as $t = 0$. In the kinematic equations x_0 is the coordinate of the object and v_0 is its velocity at time $t = 0$, not necessarily when the particle starts out.

You usually have the option of selecting x_0 to be zero; this selection simply places the origin of the coordinate system at the position of the object when $t = 0$. If you do this you need deal with only the other 5 kinematic quantities.

Here are some examples from the end-of-chapter problems.

PROBLEM 8 (38E). A jumbo jet must reach a speed of 360 km/h (= 225 mi/h) on the runway for takeoff. What is the least constant acceleration needed for takeoff from a 1.80-km runway?

STRATEGY: At the first event the jet is at rest on the runway at the instant it starts its run. For this event take the time to be 0 and place the coordinate system with its origin at the jet. The second event is the takeoff at the end of the runway. Then the jet is 1.80 km further along the runway and its velocity is 360 km/h. Convert these quantities to standard SI units and fill in the following table:

	FIRST EVENT	SECOND EVENT	
time:	_____	_____	acceleration: _____
coordinate:	_____	_____	
velocity:	_____	_____	

You should have question marks after the time of the second event and the acceleration. All other quantities should have numbers associated with them. In the kinematics equations x_0 is the coordinate and v_0 is the velocity at the first event because you selected the time to be zero for that event. The time t, coordinate x, and velocity v are associated with the second event.

SOLUTION: Now solve the simultaneous equations $x = \frac{1}{2}at^2$ and $v = at$, obtained by setting $x_0 = 0$ and $v_0 = 0$ in the usual kinematic equations. Use the second equation to eliminate the unknown t from the first. That is, find an expression for t in terms of a and substitute it for t in the first equation. You should get $x = v^2/2a$. Now solve for a.

$$[\text{ans: } 2.8\,\text{m/s}^2]$$

You might also have used the kinematic equation $v^2 = v_0^2 + 2a(x - x_0)$. Here t has already been eliminated for you.

PROBLEM 9 (43E).   On a dry road a car with good tires may be able to brake with a deceleration of $4.92\,\text{m/s}^2$ (assume that it is constant). (*a*) How long does such a car, initially traveling at $24.6\,\text{m/s}$, take to come to rest? (*b*) How far does it travel in this time?

STRATEGY: The first event occurs at the instant the brakes are first applied, the second at the instant the car is first at rest. Take the time to be 0 at the first event and place the origin of the coordinate system at the position of the car then. Fill in the following table:

	FIRST EVENT	SECOND EVENT		
time:	_____	_____	acceleration:	_____
coordinate:	_____	_____		
velocity:	_____	_____		

The coordinate and time of second event should have question marks after them, all other quantities should have numbers after them. If you took the velocity at the first event to be positive (assumed the car is traveling in the positive x direction), then your value for the acceleration should be negative. Since the car is slowing down the initial velocity and the acceleration must be in opposite directions. In the kinematic equations x_0 and v_0 correspond to the first event; t, x, and v correspond to the second.

SOLUTION: (*a*) Solve $x = v_0 t + \frac{1}{2}at^2$ and $v = v_0 + at$ for t. You need only the second equation. (*b*) Evaluate the first equation.

$$[\text{ans: } (a)\ 5.00\,\text{s};\ (b)\ 61.5\,\text{m}]$$

(*c*) Graph x versus t and v versus t for the deceleration.

SOLUTION: Use the axes below.

PROBLEM 10 (46P). A certain drag racer can accelerate from 0 to 60 km/h in 5.4 s. (*a*) What is its average acceleration, in m/s^2, during this time?

STRATEGY: The average acceleration over the interval is given by $\bar{a} = (v_f - v_i)/\Delta t$. Convert $v_f = 60$ km/h to m/s before substituting.

SOLUTION:

[ans: 3.1 m/s^2]

(*b*) How far will it travel during the 5.4 s, assuming its acceleration to be constant?

STRATEGY: Evaluate $x = v_0 t + \frac{1}{2} a t^2$, with $a = 3.1$ m/s^2.

SOLUTION:

[ans: 45 m]

(*c*) How much time would be required to go a distance of 0.25 km if the acceleration could be maintained at the same value?

STRATEGY: Solve $x = v_0 t + \frac{1}{2} a t^2$ for t. Convert $x = 0.25$ km to m before making the substitution. There are two solutions, one positive and one negative. Only the positive solution has physical significance for this problem.

SOLUTION:

[ans: 13 s]

PROBLEM 11 (53P). (*a*) If the maximum acceleration that is tolerable for passengers in a subway train is 1.34 m/s^2, and subway stations are located 806 m apart, what is the maximum speed a subway train can attain between stations?

STRATEGY: The train speeds up as it leaves a station and reaches its maximum speed at the point halfway to the next station. It then slows down. The first event occurs at the instant a train leaves a station. Take the time to be zero then and place the origin of the coordinate system at the position of the train then. The second event is the arrival of the train at the halfway point, 403 m away. Fill in the following table:

	FIRST EVENT	SECOND EVENT		
time:	_____	_____	acceleration:	_____
coordinate:	_____	_____		
velocity:	_____	_____		

In the kinematic equations x_0 and v_0 correspond to the first event; t, x, and v correspond to the second.

SOLUTION: The unknowns are v and t. Solve $x = \frac{1}{2} a t^2$ for t, then $v = at$ for v.

[ans: 32.9 m/s]

(*b*) What is the travel time between stations?

STRATEGY: It takes as much time to slow to a stop from the midpoint as it does to reach maximum velocity. Double the time found as an intermediate step in part *a*.

SOLUTION:

$$\left[\text{ans: } 49.1\,\text{s}\right]$$

(*c*) If the subway train stops for a 20-s interval at each station, what is the maximum average speed of a subway train?

STRATEGY: The time elapsed from the leaving of one station to the leaving of the next is $49.1 + 20 = 69.1\,\text{s}$. The distance traveled is $806\,\text{m}$.

SOLUTION:

$$\left[\text{ans: } 11.7\,\text{m/s}\right]$$

(*d*) Graph x, v, and a versus t.

STRATEGY: $x(t)$ should increase with upward curvature until the midpoint is reached, then it should increase with downward curvature until the next station is reached. It should be constant for the 20 s the train is in the station. $v(t)$ should be a straight line with positive values and positive slope until the midpoint is reached. Then it is a straight line with positive values and negative slope until the next station is reached, when $v = 0$. While in the station the velocity of the train is 0. $a(t)$ is a positive constant until the midpoint is reached, a negative constant with the same magnitude from the midpoint to the next station, and 0 while the train is in the station.

SOLUTION: Draw the curves on the axes below.

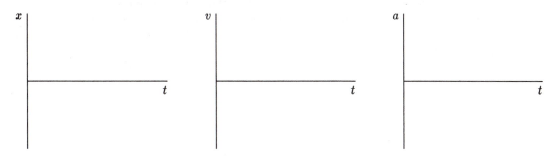

Some problems deal with two objects. You must now write down two sets of kinematic equations, one for each object: $x_1(t) = x_{01} + v_{01}t + \frac{1}{2}a_1t^2$ and $v_1(t) = v_{01} + a_1t$ for object 1 and $x_2(t) = x_{02} + v_{02}t + \frac{1}{2}a_2t^2$ and $v_2(t) = v_{02} + a_2t$ for object 2. At $t = 0$ object 1 is at x_{01} and has velocity v_{01} while object 2 is at x_{02} and has velocity v_{02}. These equations can be solved for 4 unknown quantities.

PROBLEM 12 (54P). At the instant the traffic light turns green, an automobile starts with a constant acceleration a of $2.2\,\text{m/s}^2$. At the same instant a truck, traveling with a constant speed of $9.5\,\text{m/s}$, overtakes and passes the automobile. (*a*) How far beyond the traffic light will the automobile overtake the truck?

STRATEGY: Place the origin of a coordinate system at the traffic light with the positive x axis in the direction of travel of the vehicles. Take $t = 0$ to be the time the light changes to green. Then the coordinate of the automobile as function of time is given by $x_a(t) = \frac{1}{2}at^2$ and the coordinate of the truck is given by $x_t(t) = v_t t$. The automobile overtakes the truck when $x_a = x_t$. Solve $\frac{1}{2}at^2 = v_t t$ for t, the time when this occurs, then evaluate $x_t = v_t t$ to find where it occurs.

SOLUTION:

[ans: 82 m from the light]

(*b*) How fast will the car be traveling at that instant?

STRATEGY: Evaluate $v_a = at$.

SOLUTION:

[ans: 19 m/s]

PROBLEM 13 (56P). As a high-speed passenger train traveling at 100 mi/h rounds a bend, the engineer is shocked to see that a locomotive has improperly entered onto the track from a siding 0.42 mi ahead; see Fig. 2–23. The locomotive is moving at 18 mi/h. The engineer of the passenger train immediately applies the brakes. (*a*) What must be the magnitude of the resulting constant acceleration if a collision is to be just avoided?

STRATEGY: A collision is avoided if the passenger train slows to the speed of the locomotive before its coordinate is the same as that of the locomotive. Place the coordinate system with its origin at the front of the train when the engineer sees the locomotive and let $t = 0$ at this instant. Let v_{0T} be the initial speed of the train and let a be its acceleration. Then the coordinate of the train is given by $x_T = v_{0T}t + \frac{1}{2}at^2$ and its velocity is given by $v_T = v_{0T} + at$. Let x_{0L} be the coordinate of the back of the locomotive when it is spotted by the engineer and let v_L be the velocity of the locomotive. The coordinate of the locomotive is then given by $x_L = x_{0L} + v_Lt$. We want $v_T = v_L$ while $x_T < x_L$.

SOLUTION: Solve $v_{0T} + at = v_L$ for t. This gives the time when the velocities of the train and locomotive are the same, in terms of the acceleration of the train. Substitute this expression into $x_T = v_{0T}t + \frac{1}{2}at^2$ and into $x_L = x_{0L} + v_Lt$. Substitute the resulting expressions into $x_T < x_L$ and solve for the acceleration. You should get $a < -(v_L - v_{0T})^2/2x_{0L}$ or, since a is negative $|a| > (v_L - v_{0T})^2/2x_{0L}$. The minimum magnitude of the acceleration that will do the job is given by $(v_L - v_{0T})^2/2x_{0L}$.

Convert the given velocities to ft/s and the given value for x_{0L} to ft, then substitute.

[ans: 3.26 ft/s²]

(*b*) Assume that the engineer is at $x = 0$ when, at $t = 0$, he first spots the locomotive. Sketch the $x(t)$ curves representing the locomotive and train for the situation in which a collision is just avoided. Add a curve to represent what happens if the braking rate is insufficient to avoid a crash.

STRATEGY: The graph for the locomotive should be a straight line with a positive slope of 18 mi/h, starting from $x = 0.42$ mi at $t = 0$. The curve for the train should start at $x = 0$ and have an initial slope of 100 mi/h (much greater than that for the locomotive) but its slope should uniformly decrease, indicating a negative acceleration. The slope of the curve representing the train should just graze the line representing the locomotive and at that point it should have the same slope as the line. The line is tangent to the curve.

SOLUTION: Draw the curves on the axes to the right.

26 *Chapter 2: Motion Along a Straight Line*

Free-fall problems are exactly the same as other constant-acceleration kinematics problems in which the acceleration is given. The notation is different because we choose the y axis to be vertical, so the object moves along that axis rather than the x axis. You should realize that this is an extremely superficial difference. The acceleration is always known (g, downward) and is usually not given explicitly in the problem statement.

PROBLEM 14 (60E).   (*a*) With what speed must a ball be thrown vertically up in order to rise to a maximum height of 50 m?

STRATEGY: The first event is the throwing of the ball. The second occurs when the ball reaches its highest point. You should recognize that its velocity is 0 at the instant it is at its highest point.

SOLUTION: Take the y axis to be positive in the upward direction and place the origin at the point where the ball is thrown. Fill in the following table with numerical values for known quantities and question marks for unknowns.

	FIRST EVENT	SECOND EVENT	
time:	_____	_____	acceleration: _____
coordinate:	_____	_____	
velocity:	_____	_____	

You should have question marks after the velocity at the first event (the initial velocity) and the time of the second event, numerical values after all other quantities. In the kinematic equations y_0 and v_0 are the coordinate and velocity at the first event: t, y, and v are the time, coordinate, and velocity at the second.

Now solve $y = v_0 t - \frac{1}{2}gt^2$ and $v = v_0 - gt$ for v_0. From the second equation $t = v_0/g$. Substitute this expression into the first equation to obtain $y = \frac{1}{2}v_0^2/g$. Solve for v_0.

$$\left[\text{ans: } 31\,\text{m/s}\right]$$

(*b*) How long will it be in the air?

STRATEGY: The ball will land when $y = 0$. Solve $0 = v_0 t - \frac{1}{2}gt^2$ for t. You want the non-zero solution.

SOLUTION:

$$\left[\text{ans: } 6.4\,\text{s}\right]$$

(*c*) Sketch graphs of y, v. and a versus t. On the first two graphs, indicate the time at which 50 m is reached.

SOLUTION: Use the axes below.

PROBLEM 15 (68P).   A model rocket is fired vertically and ascends with a constant vertical acceleration of $4.00\,\text{m/s}^2$ for $6.00\,\text{s}$. Its fuel is then exhausted and it continues as a free-fall particle. (*a*) What is the maximum altitude reached?

STRATEGY: When the fuel runs out the acceleration changes from $4.00\,\text{m/s}^2$ up to $9.8\,\text{m/s}^2$ down. Work the problem in two parts: first find the altitude and velocity of the rocket when its fuel is gone, then use these values as initial conditions for the free fall part of the motion.

SOLUTION: Take the y axis to be positive in the upward direction and place the origin to be at the launch site. Set $t = 0$ at launch. Use $y = \frac{1}{2}at^2$ and $v = at$ to find the coordinate and velocity at the end of $6.,00\,\text{s}$. Here $a = 4.00\,\text{m/s}^2$. You should get $72.0\,\text{m}$ and $24.0\,\text{m/s}$.

   For the second part of the motion reset the time to 0 and set $y_0 = 72.0\,\text{m}$ and $v_0 = 24.0\,\text{m/s}$. Solve $y = y_0 + v_0 t - \frac{1}{2}gt^2$ and $v = v_0 - gt$ for the coordinate when $v = 0$.

$$[\,\text{ans: }101\,\text{m}\,]$$

(*b*) What is the total time elapsed from takeoff until the rocket strikes the Earth?

STRATEGY: Again take the initial conditions to be $y_0 = 72.0\,\text{m}$ and $v_0 = 24.0\,\text{m/s}$. Solve $y = y_0 + v_0 t - \frac{1}{2}gt^2$ for the time when $y = 0$. This is a quadratic equation and has two solutions, one positive and one negative. You want the positive solution. To it you must add the $6.00\,\text{s}$ of the powered part of the motion.

SOLUTION:

$$[\,\text{ans: }13.0\,\text{s}\,]$$

PROBLEM 16 (76P).   A stone is thrown vertically upward. On the way up it passes point A with speed v and point B, $3.00\,\text{m}$ higher than A, with speed $\frac{1}{2}v$. Calculate (*a*) the speed v and (*b*) the maximum height reached by the stone above point B.

STRATEGY: (*a*) Let the y axis be positive in the upward direction and place the origin at point A. Take the time to be 0 when the stone passes point A. Let ℓ (= $3.00\,\text{m}$) be height of B above A. The first event occurs when the ball passes A and the second when the ball passes point B. At point A the time is 0, the coordinate is 0, and the velocity is v. At B the time is t, the coordinate is ℓ, and the velocity is $\frac{1}{2}v$. Fill in the following table with appropriate algebraic symbols.

	FIRST EVENT	SECOND EVENT		
time:	_____	_____	acceleration:	_____
coordinate:	_____	_____		
velocity:	_____	_____		

Solve the kinematic equations for v. The time t is the other unknown.

SOLUTION: The kinematic equations become $\ell = vt - \frac{1}{2}gt^2$ and $\frac{1}{2}v = v - gt$. Solve the second for t in terms of v. You should get $t = v/2g$. Substitute this expression into the first equation and solve for v. You should get $v = \sqrt{8g\ell/3}$.

$$[\text{ans: } 8.85 \text{ m/s}]$$

STRATEGY: *(b)* Use the same coordinate system and initial conditions. Solve for the coordinate of the highest point. The second event occurs when the velocity vanishes. Fill in the following table.

	FIRST EVENT	SECOND EVENT		
time:	_____	_____	acceleration:	_____
coordinate:	_____	_____		
velocity:	_____	_____		

The unknowns are the coordinate of the highest point and the time the ball reaches the highest point.

SOLUTION: Use $v - gt = 0$ to find an expression for t and substitute the expression into $y = vt - \frac{1}{2}gt^2$. Substitute numerical values. The solution gives the coordinate of the highest point measured from A. To find the height from B subtract 3.00 m.

$$[\text{ans: } 1.00 \text{ m}]$$

Nearly all the problems at the end of this chapter can be solved using one or more of the following:

    a. definition of average velocity
    b. definition of instantaneous velocity
    c. definition of average acceleration
    d. definition of instantaneous acceleration
    e. equations for the coordinate and velocity of a particle with constant acceleration (including the free fall equations)

Cultivate the good habit of classifying a problem according to the principle or definition that it illustrates. Look at what is given and what is asked, then see if all the ingredients are present for any given classification. Read through all the homework problems you have been assigned for this chapter and classify each part by writing the problem number and part (e.g. 2a) in the appropriate space below.

CLASSIFICATION	PROBLEM NUMBER AND PART
definition of average velocity	_____
definition of instantaneous velocity	_____
definition of average acceleration	_____
definition of instantaneous acceleration	_____
constant acceleration kinematics equations	_____
other	_____

III. MATHEMATICAL SKILLS

The following is a listing of mathematical knowledge you will need to understand and solve problems in this chapter. Refer to a mathematics text if you are rusty on any of it.

Significant figures.   Always express your answers to problems using the proper number of significant figures. Some students unthinkingly copy all 8 or 10 figures displayed by their calculator, thus demonstrating a lack of understanding. A calculated value cannot be more precise than the data that went into the calculation. Here is what you must remember about significant figures:

 a. Leading zeros are not counted as significant. Thus 0.00034 has _____ significant figures.
 b. Following zeros after the decimal point count. Thus 0.000340 has _____ significant figures.
 c. Following zeros before the decimal point may or may not be significant. Thus 500 might contain 1, 2, or 3 significant figures. Use powers of ten notation to avoid ambiguities: 5.0×10^2, for example, unambiguously contains 2 significant figures.
 d. When two numbers are added or subtracted, the number of significant figures in the result is obtained by _____
    _____.
 e. When two numbers are multiplied or divided, the number of significant figures in the result is the same as _____
    _____.

Functions.   You should understand the notation used for a function. The symbol $f(t)$, for example, indicates that the value of f depends on the value of t. The coordinate of a moving object, for example, is different at different times so it is a function of time. Sometimes the functional dependence is given by means of an equation, such as $x(t) = v_0 t + \frac{1}{2}at^2$. Sometimes it is given by means of a graph or a table of values.

If you know the dependence of x on t you can select a value for t and find the corresponding value for x. Since you can choose the value for t, t is called the <u>independent</u> <u>variable</u>. Since the value for x is determined through the functional relationship by the value of t, x is called the <u>dependent</u> <u>variable</u>.

You must carefully distinguish between a function and its value for a particular value of the independent variable. If $x(t) = 5 - 7t$, for example, then $x(2) = 5 - 7 \times 2 = -9$. If you know the coordinate of an object as a function time and wish to find its velocity at $t = 2$ s, say, you must first find the velocity as a function of time by differentiating $x(t)$ with respect to time. Only when you have done this do you substitute numerical values into the expression. An evaluation of $x(t)$ for $t = 2$ s tells you nothing about the velocity. Similarly, an evaluation of the expression $v(t)$ for a particular instant of time tells you nothing about the acceleration.

Derivatives. You should understand the meaning of a derivative as a limit of a ratio. This is useful for a solid understanding of instantaneous velocity and acceleration. The formula $v(t) = dx/dt$ gives the velocity at any instant of time. To find the velocity at a given instant you must substitute a value for t. If the object is accelerating its velocity just before or just after the given instant is different from the velocity at the instant.

You should be able to write down derivatives of polynomials. For example, $d(A + Bt + Ct^2 + Dt^3)/dt = B + 2Ct + 3Dt^2$ if A, B, C, and D are constants. Your instructor may also require you to know how to evaluate the derivative with respect to t of other functions such as e^{At}, $\sin(At)$, and $\cos(At)$. The derivatives are Ae^{At}, $A\cos(At)$, and $-A\sin(At)$ respectively.

You should know the product and quotient rules for differentiation:

$$\frac{d}{dt}[f(t)g(t)] = \frac{df(t)}{dt}g(t) + f(t)\frac{dg(t)}{dt}$$

and

$$\frac{d}{dt}\left[\frac{f(t)}{g(t)}\right] = \frac{1}{g(t)}\frac{df(t)}{dt} - \frac{f(t)}{g^2(t)}\frac{dg(t)}{dt}$$

You should also know the chain rule. Suppose $u(t)$ is a function of t and $f(u)$ is a function of u. Then f depends on t through u and

$$\frac{df}{dt} = \frac{df}{du}\frac{du}{dt}$$

The chain rule was used above to find the derivatives of e^{At}, $\sin(At)$, and $\cos(At)$. In each case u was taken to be At.

You should be able to interpret the slope of a line tangent to a curve as a derivative. The instantaneous velocity is the slope of the coordinate as a function of time, the instantaneous acceleration is the slope of the velocity as a function of time.

Simultaneous equations. You should be able to solve two simple simultaneous equations for 2 unknowns. For example, given any four of the algebraic symbols in $x = x_0 + v_0t + \frac{1}{2}at^2$ and $v = v_0 + at$, you should be able to solve for the other two.

One way is to solve one of the equations algebraically for one of the unknowns, thereby obtaining an expression for the chosen unknown in terms of the second unknown. Substitute the expression into the second equation, replacing the first unknown wherever it occurs. You now have a single equation with only one unknown. Solve in the usual way, then go back to the expression you obtained from the first equation and evaluate it for the first unknown. Several examples have been given in the Problem Solving section above.

With a little practice you will learn some of the shortcuts. If the problem asks for only one of the two unknowns, eliminate the one that is *not* requested. If one equation contains only one of the unknowns, solve it immediately and use the result in the second equation to obtain the value of the second unknown. If one equation is linear in the unknown you wish to eliminate but the other equation is quadratic, use the linear equation to eliminate the unknown from the quadratic equation, rather than vice versa.

Quadratic equations. You should be able to solve algebraic equations that are quadratic in the unknown. If $At^2 + Bt + C = 0$ then

$$t = \frac{-B \pm \sqrt{B^2 - 4AC}}{2A}$$

When the quantity under the radical sign does not vanish there are two solutions. Always examine both to see what physical significance they have, then decide which is required to answer the particular problem you are working. If the quantity under the radical sign is negative the solutions are complex numbers and probably have no physical significance for problems in this course. Check to be sure you have not made a mistake.

IV. NOTES

Chapter 3
VECTORS

I. BASIC CONCEPTS

You will deal with vector quantities throughout the course. In this chapter you will learn about their properties and how they can be manipulated mathematically. A solid understanding of this material will pay handsome dividends later.

Definitions. What properties distinguish a vector from a scalar? _____
_____

List below some examples of physical quantities that are scalars and some that are vectors:

SCALARS	VECTORS
_____	_____
_____	_____
_____	_____
_____	_____

A vector is represented graphically by an arrow in the direction of the vector, with length proportional to the magnitude of the vector (according to some scale). As an algebraic symbol, a vector is always written in boldface (**a**, for example) or with an arrow over the symbol ($\vec{a}$). The magnitude of **a** is written a, in italics (or not bold and without an arrow) or as $|\mathbf{a}|$ (or $|\vec{a}|$). Be sure you follow this convention. It helps you distinguish vectors from scalars and components of vectors. It helps you communicate properly with your instructors and exam graders. Do *not* write $a + b$ when you mean **a** + **b**, for example. They have entirely different meanings!

Displacement vectors are used as examples of vectors in this chapter. Tell in words what a displacement vector is: _____
_____

Note that a displacement vector tells us nothing about the path of the object, only the relationship between the initial and final positions. When you need an example to illustrate addition or subtraction of vectors, think of displacement vectors.

Graphical vector addition and subtraction. You will need to know the physical significance of the sum and difference of two vectors as well as how to carry out vector addition and subtraction using both graphical and analytical techniques.

When two displacement vectors, one from point A to point B and the other from point B to point C, are added, the result is the displacement vector from point _____ to point _____. Except in special circumstances, the magnitude of the resultant vector is *not* the sum of the magnitudes of the vectors entering the sum and the direction of the resultant vector is *not* in the direction of any of the vectors entering the sum.

Suppose the incomplete diagram on the right is meant to demonstrate the addition of two vectors **a** and **b**. Place arrows on two sides of the triangle and label them $\vec{a}$ and $\vec{b}$. Place an arrow on the third side and label it $\vec{a} + \vec{b}$. Be sure the arrows are placed correctly so the triangle represents vector addition.

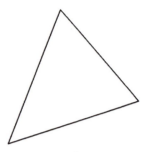

Now describe in words the steps you must take to add two vectors graphically. Be sure you mention how the vectors must be placed relative to each other and how the resultant vector is drawn: _____

_____

_____

When three or more vectors are added, the order of performing the sum is unimportant: **a** + (**b** + **c**) = (**a** + **b**) + **c**. Also remember you can reposition a vector as long as you do not change its magnitude and direction. Thus if two vectors you wish to add graphically do not happen to be placed with the tail of one at the head of the other, simply move one into the proper position.

The idea of the negative of a vector is used to define vector subtraction. How are the magnitude and direction of the negative of a vector related to the magnitude and direction of the original vector?

magnitude: _____

direction: _____

If **c** = **a** − **b**, then **c** is found by adding the negative of **b** to **a**: **c** = **a** + (−**b**). This defines vector subtraction. Write down the steps to subtract two vectors graphically (be sure to include a description of how they are positioned relative to each other): _____

_____

_____

Notice that vector subtraction is defined so that if **a** + **b** = 0 then **a** = −**b** and if **c** = **a** + **b**, then **a** = **c** − **b**. Just subtract **b** from both sides of each equation. Vector subtraction is clearly useful for solving vector equations.

In the space on the right draw two vectors such that the magnitude of their sum equals the sum of their magnitudes.

In the space on the right draw two vectors such that the magnitude of their sum equals the difference of their magnitudes.

In the space to the right draw two vectors such that the magnitude of their difference equals the sum of their magnitudes.

In the space to the right draw two vectors such that the magnitude of their difference equals the difference of their magnitudes.

Analytic vector addition and subtraction. To carry out vector addition and subtraction analytically you will need to know how to find the components of a vector. For each of the vectors shown below, illustrate the x and y components by marking their lengths along the axes. Write expressions that give the components in terms of the magnitude and angle shown. Evaluate the expressions. Notice that the components of a vector can be positive or negative.

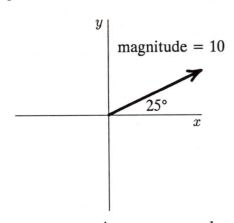

magnitude = 10

25°

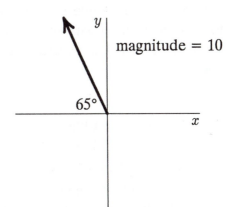

magnitude = 10

65°

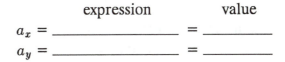

	expression	value
$a_x =$	_____	= _____
$a_y =$	_____	= _____

	expression	value
$a_x =$	_____	= _____
$a_y =$	_____	= _____

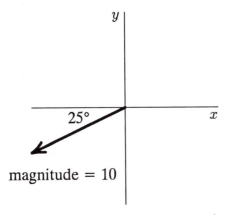

25°

magnitude = 10

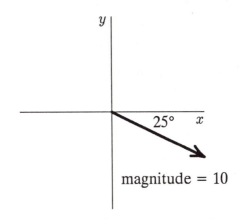

25°

magnitude = 10

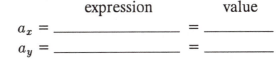

	expression	value
$a_x =$	_____	= _____
$a_y =$	_____	= _____

	expression	value
$a_x =$	_____	= _____
$a_y =$	_____	= _____

Describe in words the steps you can take to find the components of a vector, given its magnitude and direction and given a coordinate system: _____
_____
_____

You must also be able to find the magnitude and orientation of a vector when you are given its components. Suppose a vector **a** lies in the xy plane and its components a_x and a_y are given. In terms of the components the magnitude of **a** is given by

$a =$

and the angle **a** makes with the x axis is given by

$$\theta =$$

The unit vectors **i**, **j**, and **k** are used when a vector is written in terms of its components. These vectors have magnitude 1 and are in the positive x, y, and z directions respectively. $a_x\,\mathbf{i}$ is a vector parallel to the x axis with x component a_x, $a_y\,\mathbf{j}$ is a vector parallel to the y axis with y component a_y, and $a_z\,\mathbf{k}$ is a vector parallel to the z axis with z component a_z. The vector **a** is given by

$$\mathbf{a} = a_x\,\mathbf{i} + a_y\,\mathbf{j} + a_z\,\mathbf{k},$$

where the rules of vector addition apply. Are units associated with the unit vectors **i**, **j**, and **k**? _____ Are units associated with the components of a vector? _____ $a_x\,\mathbf{i}$, $a_y\,\mathbf{j}$, and $a_z\,\mathbf{k}$ are sometimes called the <u>vector</u> <u>components</u> of **a**. Do not confuse them with the components a_x, a_y, and a_z. When unit vectors are handwritten the symbols $\hat{\imath}$, $\hat{\jmath}$, and $\hat{k}$ are used.

Suppose **c** is the sum of two vectors **a** and **b** (i.e. **c** = **a** + **b**). In terms of the components of **a** and **b**: $c_x =$ _____, $c_y =$ _____, and $c_z =$ _____. Suppose **c** is the negative of **a** (i.e. **c** = −**a**). In terms of the components of **a**: $c_x =$ _____, $c_y =$ _____, and $c_z =$ _____. Suppose **c** is the difference of two vectors **a** and **b** (i.e. **c** = **a** − **b**). In terms of the components of **a** and **b**: $c_x =$ _____, $c_y =$ _____, and $c_z =$ _____.

Multiplication involving vectors. Vectors can be multiplied by scalars. Let **a** be a vector and s a scalar. Then $s\mathbf{a}$ is a vector. If s is positive its direction is _____ and its magnitude is _____. If s is negative the direction of $s\mathbf{a}$ is _____ and its magnitude is _____. If **c** = $s\mathbf{a}$ then in terms of components $c_x =$ _____, $c_y =$ _____, and $c_z =$ _____. Division of a vector by a scalar is, of course, just multiplication by the reciprocal of the scalar.

The <u>scalar</u> <u>product</u> (or dot product) of two vectors is defined in terms of the magnitudes of the two vectors and the angle between them when they are drawn with their tails at the same point: $\mathbf{a} \cdot \mathbf{b} = ab\cos\phi$. For each of two cases shown below write an expression for the scalar product of **a** and **b** in terms of the given quantities. To evaluate the expressions take $a = 10$ and $b = 5$.

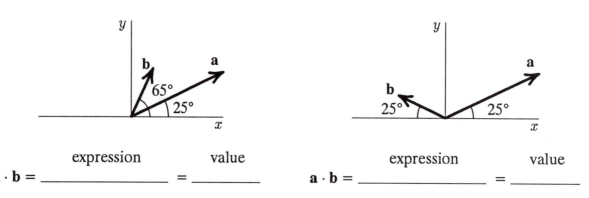

	expression	value		expression	value
$\mathbf{a} \cdot \mathbf{b} =$	_____	= _____	$\mathbf{a} \cdot \mathbf{b} =$	_____	= _____

The scalar product can be interpreted in terms of the component of one vector along the direction of the other. For the scalar product $\mathbf{a} \cdot \mathbf{b}$, write that interpretation in words:

_____

_____

In terms of components the scalar product is given by $\mathbf{a} \cdot \mathbf{b} = $ _____ .
Notice that this expression gives $\mathbf{a} \cdot \mathbf{a} = a_x^2 + a_y^2 = a^2$ for a vector in the xy plane.

Let ϕ be the angle between $\mathbf{a}$ and $\mathbf{b}$ when they are drawn with their tails at the same point. The sign of the scalar product $\mathbf{a} \cdot \mathbf{b}$ is positive if ϕ is in the range from _____ to _____ and is negative if ϕ is in the range from _____ to _____. Also $\mathbf{a} \cdot \mathbf{b} = 0$ if ϕ is _____. Note that ϕ is always 180° or less.

Write the equation that gives the magnitude of the <u>vector</u> <u>product</u> $\mathbf{a} \times \mathbf{b}$ in terms of the magnitudes of $\mathbf{a}$ and $\mathbf{b}$: $|\mathbf{a} \times \mathbf{b}| = $ _____ . Carefully describe how to determine the angle that occurs in this expression: _____

_____

_____

For two vectors with given magnitudes the magnitude of $\mathbf{a} \times \mathbf{b}$ is the greatest when the angle between them is _____ and is zero when the angle between them is _____ .

Describe the right hand rule used to find the direction of $\mathbf{a} \times \mathbf{b}$: _____

_____

_____

In terms of the cartesian components of $\mathbf{a}$ and $\mathbf{b}$, $\mathbf{a} \times \mathbf{b} = $ _____

_____ .

Remember that the direction of the vector product depends on the order in which the vectors appear: $\mathbf{a} \times \mathbf{b}$ and $\mathbf{b} \times \mathbf{a}$ are in opposite directions.

The magnitude of the vector product can also be interpreted in terms of the component of one vector along a certain direction perpendicular to the other vector. The direction is in the plane defined by the two vectors in the product. For the vector product $\mathbf{a} \times \mathbf{b}$ write the interpretation in words: _____

_____

Remember that the scalar product of two vectors is a scalar and has no direction associated with it while the vector product is a vector and does have a direction associated with it.

II. PROBLEM SOLVING

You should know how to add and subtract vectors graphically. Review the rules for drawing the representative arrows to scale and in the proper positions.

PROBLEM 1 (1E).   Consider two displacements, one of magnitude 3 m and another of magnitude 4 m. Show how the displacement vectors may be combined to get a resultant displacement of magnitude (a) 7 m, (b) 1 m, and (c) 5 m.

STRATEGY: Draw the 4 m displacement vector. Now imagine the 3 m displacement vector drawn with its tail at the head of the longer vector. Rotate it about its tail and for each orientation check the magnitude of the resultant, a vector from the tail of the 4 m vector to the head of the 3 m vector. Observe the orientation when the resultant has each of the desired magnitudes.

SOLUTION: The magnitude of the resultant is 7 m when the vectors are in the same direction. It has a magnitude of 1 m when the vectors are in opposite directions. It has a magnitude of 5 m when the vectors make a right angle with each other. The orientations are shown in the diagrams below.

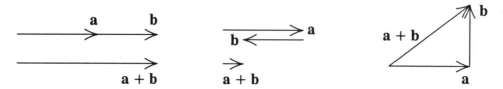

PROBLEM 2 (6P). Vector **a** has a magnitude of 5.0 units and is directed east. Vector **b** is directed 35° west of north and has a magnitude of 4.0 units. Construct vector diagrams for calculating **a** + **b** and **b** − **a**. Estimate the magnitudes and directions of **a** + **b** and **b** − **a** from your diagrams.

STRATEGY: To find **a** + **b** draw the vectors to scale and in the proper directions, with the tail of **a** at the head of **b**. Draw the resultant vector and measure its length. Use a protractor to measure the angle it makes with a compass direction. To find **b** − **a** draw the vector −**a** with its tail at the head of **b**, then draw the resultant and measure its length and the angle it makes with a compass direction.

SOLUTION: The vectors **a**, **b**, and **a** + **b** are drawn to scale on the diagram below. Each 0.25 in. represents 1 unit. Measure the resultant and the angle it makes with the northerly direction. Then carry out similar steps to find **b** − **a**.

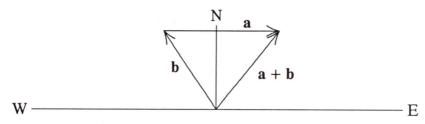

[ans: 4.2 units, 40° east of north; 8.0 units, 24° north of west]

All of the problems at the end of the chapter deal with vector manipulations: addition, subtraction, finding components, finding magnitude and direction, and the various kinds of multiplication. The vectors are given either in terms of magnitude and direction or in terms of components and answers may be requested in either of these forms. This means you may need to convert from the given form to a form suitable for the vector operation, then convert again to obtain the form required for the answer.

If a is the magnitude of a vector **a** in the xy plane and θ is the angle that the vector makes with the positive x axis, then the components of **a** are $a_x = a\cos\theta$ and $a_y = a\sin\theta$. If you know the components you can find the magnitude and the angle with the positive x axis. The magnitude is given by

$$a = \sqrt{a_x^2 + a_y^2}$$

and the angle is given by

$$\theta = \arctan(a_y/a_x)$$

When you have found values for a and θ check to be sure $a\cos\theta$ has the proper sign for the x component and $a\sin\theta$ has the proper sign for the y component. If they have the wrong signs add 180° to the value you used for the angle.

The basic prescription for vector addition is: if $\mathbf{c} = \mathbf{a} + \mathbf{b}$ then $c_x = a_x + b_x$, $c_y = a_y + b_y$, and $c_z = a_z + b_z$. The basic prescription for vector subtraction is: if $\mathbf{c} = \mathbf{a} - \mathbf{b}$ then $c_x = a_x - b_x$, $c_y = a_y - b_y$, and $c_z = a_z - b_z$.

PROBLEM 3 (13E).   The minute hand of a wall clock measures 10 cm from axis to tip. What is the displacement vector of its tip (a) from a quarter after the hour to half past, (b) in the next half hour, and (c) in the next hour?

STRATEGY: You want to find the vector from (a) the 3 to the 6, (b) from the 6 to the 12, and (c) from the 12 to the 12 on the face of the clock.

SOLUTION: (a) Choose the coordinate system shown. The x component of the displacement vector is -10 cm and the y component is -10 cm. The length of the vector is the square root of the sum of the squares of the components and the vector makes an angle of 45° with the negative y axis. (b) and (c) Now draw the other displacements and find their lengths and directions.

SOLUTION:

$$\left[\text{ans: } (a) \text{ 14 cm, 45° to the left of straight down; } (b) \text{ 20 cm, vertically up; } (c)\ 0\right]$$

PROBLEM 4 (23E).   Calculate the x and y components, magnitudes, and directions of (a) $\mathbf{a} + \mathbf{b}$ and (b) $\mathbf{b} - \mathbf{a}$ if $\mathbf{a} = 3.0\mathbf{i} + 4.0\mathbf{j}$ and $\mathbf{b} = 5.0\mathbf{i} - 2.0\mathbf{j}$.

STRATEGY: To find $\mathbf{c} = \mathbf{a} + \mathbf{b}$, use $c_x = a_x + b_x$ and $c_y = a_y + b_y$. To find $\mathbf{c} = \mathbf{b} - \mathbf{a}$, use $c_x = b_x - a_x$ and $c_y = b_y - a_y$. In each case the magnitude is given by $c = \sqrt{c_x^2 + c_y^2}$ and the angle θ that $\mathbf{c}$ makes with the x axis is given by $\tan\theta = c_y/c_x$. When you evaluate the inverse tangent on your calculator be sure to verify that $\cos\theta$ has the same sign as c_x and $\sin\theta$ has the same sign as c_y. If they do not, add 180° to the result.

SOLUTION:

$$\left[\text{ans: } (a)\ \mathbf{c} = 8.0\mathbf{i} + 2.0\mathbf{j}, c = 8.2, \theta = 14°; (b)\ \mathbf{c} = 2.0\mathbf{i} - 6.0\mathbf{j}, c = 6.3, \theta = -72°\right]$$

PROBLEM 5 (26P).   If $\mathbf{a} - \mathbf{b} = 2\mathbf{c}$, $\mathbf{a} + \mathbf{b} = 4\mathbf{c}$, and $\mathbf{c} = 3\mathbf{i} + 4\mathbf{j}$, then what are $\mathbf{a}$ and $\mathbf{b}$?

STRATEGY: Add the first two equations to obtain $2\mathbf{a} = 6\mathbf{c}$ or $\mathbf{a} = 3\mathbf{c}$. Subtract the first from the second to obtain $2\mathbf{b} = 2\mathbf{c}$ or $\mathbf{b} = \mathbf{c}$. Now substitute $3\mathbf{i} + 4\mathbf{j}$ for $\mathbf{c}$.

SOLUTION:

$$\left[\text{ans: } \mathbf{a} = 9\mathbf{i} + 12\mathbf{j};\ \mathbf{b} = 3\mathbf{i} + 4\mathbf{j}\right]$$

PROBLEM 6 (29P). A radar station detects an airplane approaching directly from the east. At first observation, the range to the plane is 1200 ft at 40° above the horizon. The plane is tracked for another 123° in the vertical east-west plane, the range at final contact being 2580 ft. See Fig. 3–27. Find the displacement of the plane during the period of observation.

STRATEGY: Place the origin of a coordinate system at the radar, with the positive x axis to the east and the positive y axis upward. Let $\mathbf{r}_1$ be the position vector from the radar to the plane at the first observation and let $\mathbf{r}_2$ be the position vector from the radar to the plane at the last observation. Note that the vectors $\mathbf{r}_1$, $\mathbf{r}_1$, and $\Delta\mathbf{r}$ form a vector addition triangle: $\mathbf{r}_1 + \Delta\mathbf{r} = \mathbf{r}_2$. Thus the displacement vector is given by $\Delta\mathbf{r} = \mathbf{r}_2 - \mathbf{r}_1$. Find the components of $\mathbf{r}_1$ and $\mathbf{r}_2$ and use them to calculate the components of $\Delta\mathbf{r}$.

SOLUTION: The x component of $\mathbf{r}_1$ is $x_1 = 1200\cos 40°$ and the y component is $y_1 = 1200\sin 40°$. The x component of $\mathbf{r}_2$ is $x_2 = 2580\cos 163°$ and the y component is $y_2 = 2580\sin 163°$. The x component of the displacement is given by $\Delta x = x_2 - x_1$ and the y component is given by $\Delta y = y_2 - y_1$.

$$\left[\,\text{ans: } \Delta x = 3390\,\text{ft}, \Delta y = -20\,\text{ft (essentially 0)}\,\right]$$

PROBLEM 7 (31P). A particle undergoes three successive displacements in a plane, as follows: 4.00 m southwest, 5.00 m east, and 6.00 m in a direction 60.0° north of east. Choose the y axis pointing north and the x axis pointing east and find (*a*) the components of each displacement, (*b*) the components of the resultant displacement, (*c*) the magnitude and direction of the resultant displacement, and (*d*) the displacement that would be required to bring the particle back to the starting point.

STRATEGY: Draw the displacements on the axes to the right, with the tail of the first displacement at the origin and with each of the last two positioned so their tails are at the head of the previous displacement. Label the displacements and the angles they make with the positive x direction. Also draw and label the resultant displacement. Use $\Delta x = |\Delta\mathbf{r}|\cos\theta$ and $\Delta y = |\Delta\mathbf{r}|\sin\theta$ to calculate the components of each displacement. Here $|\Delta\mathbf{r}|$ is the magnitude of the displacement and θ is the angle it makes with the positive x axis. Add the x components to find the x component of the resultant, add the y components to find the y components of the resultant. To find the angle Θ between the resultant and the positive x axis use $\tan\Theta = R_y/R_x$, where $\mathbf{R}$ is the resultant. The displacement that brings the particle back to its starting point is $-\mathbf{R}$.

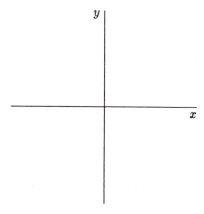

SOLUTION: Fill in the following table by writing the expressions for the various components of the displacements and their numerical values. Then sum to find the components of the resultant.

	x component	value	y component	value
displacement #1				
displacement #2				
displacement #3				
sum				

Now find the magnitude of the resultant by evaluating the square root of the sum of the squares of the components. Find the angle it makes with the positive x axis by calculating the ratio of the y component to the x component and finding the inverse tangent of the result. To answer part d find a vector with the same magnitude as the resultant but in the opposite direction.

$\big[$ans: (a) $(-2.83\,\mathbf{i}-2.83\,\mathbf{j})$ m, $(5.00\,\mathbf{j})$ m, $(3.00\,\mathbf{i}+5.20\,\mathbf{j})$ m; (b) $(5.17\,\mathbf{i}+2.37\,\mathbf{j})$ m; (c) 5.69 m, $24.6°$ N of E; (d) 5.69 m, $24.6°$ S of W $\big]$

Here are some problems dealing with multiplication. You must know how to carry out the multiplication of a vector by a scalar, the scalar product of two vectors, and the vector product of two vectors.

PROBLEM 8 (37E).   A vector $\mathbf{d}$ has a magnitude of 2.5 m and points north. What are the magnitudes and directions of the vectors (a) $4.0\mathbf{d}$ and (b) $3.0\mathbf{d}$?

STRATEGY: The product of a vector and a scalar is another vector, with a magnitude that is the product of the magnitude of the scalar and the magnitude of the original vector. If the scalar is positive the product vector is in the same direction as the original vector; if the scalar is negative it is in the opposite direction.

SOLUTION:

$\big[$ans: (a) 10 m, north; (b) 7.5 m, north$\big]$

PROBLEM 9 (44E).   Two vectors, $\mathbf{r}$ and $\mathbf{s}$, lie in the xy plane. Their magnitudes are 4.50 and 7.30 units, respectively, and their directions are 320° and 85.0°, respectively, as measured counterclockwise from the positive x axis. What are the values of (a) $\mathbf{r}\cdot\mathbf{s}$ and (b) $\mathbf{r}\times\mathbf{s}$?

STRATEGY: Use $\mathbf{r}\cdot\mathbf{s}=rs\cos\phi$ and $|\mathbf{r}\times\mathbf{s}|=rs\sin\phi$, where ϕ is the angle between the two vectors when they are drawn with their tails at the same point. Use the right-hand rule to find the direction of $\mathbf{r}\times\mathbf{s}$.

SOLUTION: Draw the vectors on the axes to the right. Label the angles the vectors make with the positive x axis. Calculate the angle between the vectors. You should get 125°. Complete the calculation. For purposes of giving the direction of the vector product note that the positive z axis comes out of the page.

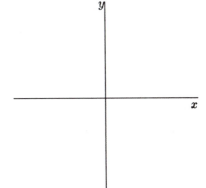

$\big[$ans: (a) -18.8; (b) 26.9, positive z direction$\big]$

Note that the scalar product is negative because the angle between the vectors is greater than $90°$.

PROBLEM 10 (49P). (*a*) Determine the components and magnitude of $\mathbf{r} = \mathbf{a} - \mathbf{b} + \mathbf{c}$ if $\mathbf{a} = 5.0\mathbf{i} + 4.0\mathbf{j} - 6.0\mathbf{k}$, $\mathbf{b} = -2.0\mathbf{i} + 2.0\mathbf{j} + 3.0\mathbf{k}$, and $\mathbf{c} = 4.0\mathbf{i} + 3.0\mathbf{j} + 2.0\mathbf{k}$.

STRATEGY: This is a standard vector addition and subtraction problem. Use $r_x = a_x - b_x + c_x$, $r_y = a_y - b_y + c_y$, and $r_z = a_z - b_z + c_z$. The magnitude is the square root of the sum of the squares of the components.

SOLUTION:

$$\left[\text{ans: } \mathbf{r} = 11\mathbf{i} + 5.0\mathbf{j} - 7.0\mathbf{k}, r = 14\right]$$

(*b*) Calculate the angle between $\mathbf{r}$ and the positive z axis.

STRATEGY: Let ϕ be the angle between $\mathbf{r}$ and the positive z axis. According to the definition of the scalar product $\mathbf{r} \cdot \mathbf{k} = r \cos \phi$. Evaluate $\mathbf{r} \cdot \mathbf{k}$ and solve $\cos \phi = \mathbf{r} \cdot \mathbf{k}/r$ for ϕ.

SOLUTION: Use $\mathbf{r} \cdot \mathbf{k} = (11\mathbf{i} + 5\mathbf{j} - 7\mathbf{k}) \cdot \mathbf{k} = 11(\mathbf{i} \cdot \mathbf{k}) + 5(\mathbf{j} \cdot \mathbf{k}) - 7(\mathbf{k} \cdot \mathbf{k})$. Now $\mathbf{i} \cdot \mathbf{k} = 0$ because the angle between these vectors is $90°$. Similarly, $\mathbf{j} \cdot \mathbf{k} = 0$. The last scalar product $\mathbf{k} \cdot \mathbf{k}$ is 1 because each vector has magnitude 1 and the angle between them is 0. Thus $\mathbf{r} \cdot \mathbf{k} = -7$. Complete the calculation.

$$\left[\text{ans: } 120°\right]$$

PROBLEM 11 (51P). Two vectors are given by $\mathbf{a} = 3.0\mathbf{i} + 5.0\mathbf{j}$ and $\mathbf{b} = 2.0\mathbf{i} + 4.0\mathbf{j}$. Find (*a*) $\mathbf{a} \times \mathbf{b}$, (*b*) $\mathbf{a} \cdot \mathbf{b}$, and (*c*) $(\mathbf{a} + \mathbf{b}) \cdot \mathbf{b}$.

STRATEGY: (*a*) Write out the vector product: $\mathbf{a} \times \mathbf{b} = (3.0\mathbf{i} + 5.0\mathbf{j}) \times (2.0\mathbf{i} + 4.0\mathbf{j}) = 3.0 \times 2.0 (\mathbf{i} \times \mathbf{i}) + 3.0 \times 4.0 (\mathbf{i} \times \mathbf{j}) + 5.0 \times 2.0 (\mathbf{j} \times \mathbf{i}) + 5.0 \times 4.0 (\mathbf{j} \times \mathbf{j})$. Use $\mathbf{i} \times \mathbf{i} = 0, \mathbf{j} \times \mathbf{j} = 0, \mathbf{i} \times \mathbf{j} = \mathbf{k}$, and $\mathbf{j} \times \mathbf{i} = -\mathbf{k}$.

SOLUTION:

$$\left[\text{ans: } 2.0\mathbf{k}\right]$$

STRATEGY: (b) Write out the scalar product: $\mathbf{a} \cdot \mathbf{b} = (3.0\mathbf{i} + 5.0\mathbf{j}) \cdot (2.0\mathbf{i} + 4.0\mathbf{j}) = 3.0 \times 2.0\,(\mathbf{i} \cdot \mathbf{i}) + 3.0 \times 4.0\,(\mathbf{i} \cdot \mathbf{j}) + 5.0 \times 2.0\,(\mathbf{j} \cdot \mathbf{1}) + 5.0 \times 4.0\,(\mathbf{j} \cdot \mathbf{j})$. Use $\mathbf{i} \cdot \mathbf{i} = 1, \mathbf{j} \cdot \mathbf{j} = 1, \mathbf{i} \cdot \mathbf{j} = 0,$ and $\mathbf{j} \cdot \mathbf{i} = 0$.

SOLUTION:

[ans: 26]

STRATEGY: (c) Use $(\mathbf{a} + \mathbf{b}) \cdot \mathbf{b} = \mathbf{a} \cdot \mathbf{b} + b^2$.

SOLUTION:

[ans: 46]

PROBLEM 12 (56P). Show that the area of the triangle contained between the vectors $\mathbf{a}$ and $\mathbf{b}$ in Fig. 3–30 is $\frac{1}{2}|\mathbf{a} \times \mathbf{b}|$, where the vertical bars signify magnitude.

STRATEGY: The area of a triangle is one half the product of a side and the altitude of the opposite vertex from that side. Note that $b \sin \phi$ is the altitude of the vertex opposite side a in the figure. The magnitude of a vector product is given by $|\mathbf{a} \times \mathbf{b}| = ab \sin \phi$, where ϕ is the angle between the vectors when they are drawn with their tails at the same point.

SOLUTION:

PROBLEM 13 (58P). The three vectors shown in Fig. 3–31 have magnitudes $a = 3.00, b = 4.00,$ and $c = 10.0$. (a) Calculate the x and y components of these vectors.

SOLUTION:

[ans: $\mathbf{a} = 3.00\,\mathbf{i}, \mathbf{b} = 3.46\,\mathbf{i} + 2.00\,\mathbf{j}, \mathbf{c} = -5.00\,\mathbf{i} + 8.66\,\mathbf{j}$]

(*b*) Find the numbers p and q such that $\mathbf{c} = p\mathbf{a} + q\mathbf{b}$.

STRATEGY: Notice that $\mathbf{b}$ and $\mathbf{c}$ are perpendicular to each other. Take the scalar product of $p\mathbf{a} + q\mathbf{b}$ with $\mathbf{c}$ to obtain $c^2 = p\mathbf{a} \cdot \mathbf{c}$. Thus $p = c^2/\mathbf{a} \cdot \mathbf{c}$. Evaluate c^2 and $\mathbf{a} \cdot \mathbf{c}$, then calculate p. Take the scalar product of $p\mathbf{a} + q\mathbf{b}$ with $\mathbf{j}$ to obtain $c_y = pa_y + qb_y$. Solve for q.

SOLUTION:

$$[\text{ans: } p = -6.67, q = 4.33]$$

To practice setting up problems, read the problems you have been assigned for homework, then fill in the table by naming the form in which the vectors are given (mag & dir or comp, say), the operation to be performed (addition, subtraction, multiplication by a scalar, scalar multiplication, or vector multiplication), the form you wish to write the vectors to perform the operation, and the form required for the answer. Vectors may be given in different forms in the same problem. If this is the case give both forms in the second column or just write "mixed". If the answer is a scalar just write "scalar" in the last column.

PROBLEM	GIVEN FORM	OPERATION	FORM FOR OPERATION	FORM FOR ANSWER
___	___	___	___	___
___	___	___	___	___
___	___	___	___	___
___	___	___	___	___

III. MATHEMATICAL SKILLS

Trigonometry is a large portion of the mathematics used in this chapter. Here is a listing of the important elements you should know very well.

The Pythagorean theorem The square of the hypotenuse of a right triangle equals the sum of squares of the other two sides. In the diagram $C^2 = A^2 + B^2$. The theorem is true only if the triangle contains a right angle (90°). The theorem is used, for example, to calculate the magnitude of a vector in the xy plane given its x and y components.

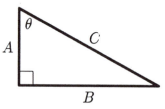

Trigonometric functions. For the triangle shown $A = C\cos\theta$ and $B = C\sin\theta$. Remember these relations in the following form: the length of the side adjacent to an interior angle is the product of the hypotenuse and the cosine of the angle and the length of the opposite side is the product of the hypotenuse and the sine of the angle. The relations follow directly from the definition of the sine and cosine and are used to find the components of a vector, given the magnitude and the angle it makes with an axis.

Also know that for the triangle above $\tan\theta = B/A$. In a right triangle the tangent of an interior angle is the length of the opposite side divided by the adjacent side. This relationship, in the form $\theta = \arctan(a_y/a_x)$, is used to find the angle a vector makes with a coordinate axis.

WARNING! For any values of a_x and a_y the equation $\theta = \arctan(a_y/a_x)$ has two solutions for θ. If you use a calculator to evaluate θ, it will give the solution closest to 0. The other solution is the one given by the calculator plus or minus 180°. When solving for θ, first make a sketch of the vector pointing in roughly the right direction, so its components have the correct signs, as given, then check the answer displayed by your calculator. If necessary, add 180° to the calculator answer or subtract 180° from it to make the answer agree with the sketch. Alternatively, once θ has been found, calculate $a\cos\theta$ and $a\sin\theta$ to be sure they give the original components and not their negatives.

Memorize these special values of the trigonometric functions:

$\cos(0) = 1$	$\cos(90°) = 0$	$\cos(180°) = -1$	$\cos(270°) = 0$
$\sin(0) = 0$	$\sin(90°) = 1$	$\sin(180°) = 0$	$\sin(270°) = -1$
$\tan(0) = 0$	$\tan(90°) = \pm\infty$	$\tan(180°) = 0$	$\tan(270°) = \pm\infty$

Take special care that you don't get them confused.

Trigonometric identities.   You should also know the following trigonometric identities:

$$\sin(-A) = -\sin A$$
$$\cos(-A) = \cos A$$
$$\sin(A + B) = \sin(A)\cos(B) + \cos(A)\cos(B)$$
$$\cos(A + B) = \cos(A)\cos(B) - \sin(A)\sin(B)$$

Other trigonometric relationships are listed in an appendix of the text.

The identities given above are useful for evaluating $\sin(180° - \phi)$, $\sin(90° - \phi)$, $\sin(90° + \phi)$, $\sin(180° + \phi)$, and the corresponding cosines, for example. Using these relations you should be able to show that the components of the vector in the diagram are $a_x = a\cos(90° + \phi) = -a\sin\phi$ and $a_y = a\sin(90° + \phi) = a\cos\phi$.

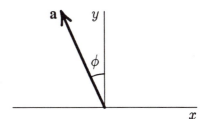

Radian measure.   You are probably familiar with the degree as a measure of angle: an arc that is the circumference of a circle divided by 360 subtends an angle of 1 degree. A radian is another measure of angle. The angle subtended by an arc, in radians, is the arc length divided by the radius of the circle. Since the circumference of a circle with radius r is given by $2\pi r$, 360° is equivalent to $2\pi r/r = 2\pi$ rad. You should also know that 180° is equivalent to π rad, 90° is equivalent to $\pi/2$ rad, and 45° is equivalent to $\pi/4$ rad. 1 rad is equivalent to $180/\pi \approx 57.30°$.

Since an angle in radians is one length (the arc length) divided by another (the radius) the radian is actually unitless. However, instead of writing simply a number, without a unit, for an angle, the characters "rad" are often appended to remind you that the angle is given in radians.

IV. NOTES

Chapter 4
MOTION IN TWO
AND THREE DIMENSIONS

I. BASIC CONCEPTS

The ideas of position, velocity, and acceleration that were introduced earlier in connection with one-dimensional motion are now extended. You should pay close attention to the definitions and relationships discussed in this chapter. There are three main topics: projectile motion, circular motion, and relative motion. First, however, some general kinematic concepts are discussed.

Definitions. Consider a particle moving in 2 or 3 dimensions. The fundamental concept used to describe its motion is its <u>position</u> <u>vector</u>. The tail of this vector is always at _____ _____ and at any instant the head is at _____ . The cartesian components of the position vector are the coordinates of the particle. As the particle moves its position vector changes and so is a function of time.

A <u>displacement</u> <u>vector</u> is used to describe a change in a position vector. If the particle has position vector $\mathbf{r}_1$ at time t_1 and position vector $\mathbf{r}_2$ at a later time t_2 then the displacement vector for this interval is

$$\Delta\mathbf{r} =$$

If the particle has coordinates x_1, y_1, z_1 at time t_1 and coordinates x_2, y_2, z_2 at time t_2 then the components of the displacement vector are given by

$$(\Delta\mathbf{r})_x = \Delta x = \qquad\qquad (\Delta\mathbf{r})_y = \Delta y = \qquad\qquad (\Delta\mathbf{r})_z = \Delta z =$$

In writing these equations be sure to get the order of the subscripts right. A displacement vector is a position vector at a *later* time minus a position vector at an *earlier* time.

In terms of the displacement vector $\Delta\mathbf{r}$ the <u>average</u> <u>velocity</u> of the particle in the interval from t_1 to t_2 is

$$\bar{\mathbf{v}} =$$

The average velocity has components given by

$$\bar{v}_x = \qquad\qquad\qquad \bar{v}_y = \qquad\qquad\qquad \bar{v}_z =$$

To use the definition to calculate the average velocity over the time interval from t_1 to t_2 you must know _____ for the beginning and end of the interval. Just as for one–dimensional motion, it is important to realize that the components of the position vector are coordinates and represent points on a coordinate axis. They are not necessarily related in any way to the distance traveled by the particle.

The <u>instantaneous velocity</u> **v** at any time t is the limit of the average velocity over a time interval that includes t, as the duration of the interval becomes vanishingly small. In terms of the position vector, it is given by the derivative

$$\mathbf{v} =$$

In terms of the particle coordinates its components are

$$v_x = \qquad\qquad\qquad v_y = \qquad\qquad\qquad v_z =$$

You should be aware that the instantaneous velocity, unlike the average velocity, is associated with a single instant of time. At any other instant, no matter how close, the instantaneous velocity might be different. The term "instantaneous" is usually implied: "velocity" means "instantaneous velocity".

To use the definition to calculate the instantaneous velocity you must know the position vector as a function of time. This is identical to knowing the coordinates as functions of time. The information may be given in algebraic form or as a graph. You should remember that the instantaneous velocity vector at any time is tangent to the path at the position of the particle at that time. If you are asked for the direction the particle is traveling at a certain time, you automatically calculate the components of its _____ for that time. <u>Speed</u> is the magnitude of the instantaneous velocity and, if the velocity components are given, can be calculated using $v = $ _____ .

In terms of the velocity $\mathbf{v}_1$ at time t_1 and the velocity $\mathbf{v}_2$ at a later time t_2 the <u>average acceleration</u> over the interval from t_1 to t_2 is given by

$$\bar{\mathbf{a}} =$$

In terms of velocity components the components of the average acceleration are

$$\bar{a}_x = \qquad\qquad\qquad \bar{a}_y = \qquad\qquad\qquad \bar{a}_z =$$

To use the definition to calculate the average acceleration over the interval from t_1 to t_2 you must know _____ .

The <u>instantaneous acceleration</u> **a** at any time t is the limit of the average acceleration over an interval that includes t, as the duration of the interval becomes vanishingly small. In terms of the velocity vector, it is given by

$$\mathbf{a} =$$

In terms of the velocity components its components are

$$a_x = \qquad\qquad\qquad a_y = \qquad\qquad\qquad a_z =$$

The terms "instantaneous acceleration" and "acceleration" mean the same thing. To use the definition to calculate the acceleration you must know the velocity vector as a function of time.

A non-zero velocity indicates that the _____ vector of the particle is changing with time. A non-zero acceleration indicates that the _____ vector of the particle is changing with time. Remember that these changes may be changes in magnitude, in direction, or both. Describe a possible motion in which the magnitude of the position vector does not change but the velocity does not vanish: _____

_____

Describe a possible motion in which the speed does not change but the acceleration does not vanish: _____

_____

Motion with constant acceleration. Suppose a particle starts, at time $t = 0$, with position vector $\mathbf{r}_0$ and velocity vector $\mathbf{v}_0$ and it has constant acceleration $\mathbf{a}$. Then at any time t its position and velocity vectors are given by:

$$\mathbf{r}(t) =$$

$$\mathbf{v}(t) =$$

In component form:

$$x(t) = \qquad\qquad\qquad\qquad v_x(t) =$$

$$y(t) = \qquad\qquad\qquad\qquad v_y(t) =$$

$$z(t) = \qquad\qquad\qquad\qquad v_z(t) =$$

Notice that the x component of the acceleration influences the x component of the velocity and the x coordinate but it does not influence the other velocity components or coordinates. Similar statements about the y and z components of the acceleration are also true.

Projectile motion. Projectile motion with negligible air resistance is an important example of constant acceleration kinematics in 2 dimensions. The acceleration (magnitude and direction) of a particle in projectile motion with negligible air resistance is _____

_____ .

Consider a projectile near the surface of the earth, moving with negligible air resistance. Assume it moves in the xy plane, that the y axis is vertical with the positive direction upward, and that the x axis is horizontal in the plane of the motion. Take the initial coordinates to be x_0 and y_0 and components of the initial velocity to be v_{0x} and v_{0y}. Then the coordinates and velocity components at time t are given by

$$x(t) = \qquad\qquad\qquad\qquad v_x(t) =$$

$$y(t) = \hspace{8cm} v_y(t) =$$

Check to be sure these equations are consistent with the magnitude and direction of the acceleration. Note that the horizontal component of the velocity does not change as the projectile moves. Also note carefully that once the projectile is fired its acceleration does not change until it hits the ground or a target.

Another important equation results when $v_y = v_{0y} - gt$ is used to eliminate t from $y = y_0 + v_{0y}t - \frac{1}{2}gt^2$. It is

$$v_y^2 - v_{0y}^2 = -2g(y - y_0)$$

This relates the y coordinate and the y component of velocity.

According to the projectile motion equations the motion of a projectile may be considered to be a combination of two independent motions that take place simultaneously: as far as the y components of its position and velocity vectors are concerned, it is in free fall; as far as the x components are concerned, it is moving with constant velocity.

Sometimes the initial speed v_0 and the launch angle θ_0 are known, rather than the x and y components of the initial velocity. Describe the launch angle: _____

_____

If the initial speed and launch angle are given, components of the initial velocity can be found using $v_{0x} =$ _____ and $v_{0y} =$ _____ .

The trajectory of a projectile is shown below. Suppose $t = 0$ when the projectile is launched. On the graph label the initial coordinates x_0 and y_0. Draw an arrow to show the initial velocity and label it $\vec{v}_0$. Label the launch angle. Draw an arrow to show the velocity of the projectile just before it hits the target and label it $\vec{v}_f$. Draw both the velocity and acceleration vectors at the point marked with a dot and at the highest point on the trajectory; label them $\vec{v}$ and $\vec{a}$, as appropriate.

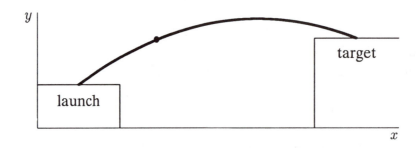

There are two special conditions you should remember. At the highest point of its trajectory the velocity of a projectile is horizontal and $v_y =$ _____ . To find the time when the projectile is at its highest point you solve _____ for t. When a projectile returns to the original launch height, $y =$ _____ . To find the time when the projectile returns to the launch height you solve _____ for t. The magnitude of the displacement from the launch point to the point at launch height is called the _____ .

Uniform circular motion. Uniform circular motion, in which a particle moves around a circle with constant speed, is the second important example of motion in a plane. Remember that the velocity vector is always tangent to the path and therefore continually changes direction. This means the acceleration is *not* zero.

If the radius of the circle is r and the speed is v, the acceleration of the particle has magnitude

$$a =$$

and always points toward _____ . This means the direction of the acceleration continually changes as the particle moves around the circle. The term "centripetal acceleration" is applied to this acceleration to indicate its direction. You should not forget it is the rate of change of velocity, as are all accelerations.

Suppose a particle travels counterclockwise with constant speed around the circular path shown to the right. At each of the points A, B, and C draw a vector that gives the direction of its velocity and another that gives the direction of its acceleration. Label the vectors **v** and **a**, as appropriate.

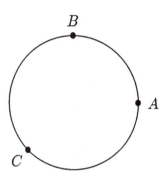

You should be aware that the magnitude of the acceleration is proportional to the *square* of the speed. If, for example, the speed is doubled without changing the orbit radius, the acceleration increases by a factor of 4.

Non-uniform circular motion. If the acceleration and velocity are always perpendicular to each other, as they are for uniform circular motion, the speed does *not* change. If the acceleration has a vector component in the direction of the velocity ($\mathbf{a} \cdot \mathbf{v} > 0$) then the speed increases and if the acceleration has a vector component in the direction opposite to the velocity ($\mathbf{a} \cdot \mathbf{v} < 0$) then the speed decreases. Illustrate these three cases below by drawing appropriately oriented acceleration vectors with tails at the particle. Label them **a**.

Speed does not change   Speed is increasing   Speed is decreasing

Be sure to remember when the speed changes and when it does not. The ideas will be useful when you study work and energy.

Relative motion. This topic deals with a comparison of the values obtained when the position, velocity, or acceleration of a particle is measured using two coordinate systems (or reference frames) that are moving relative to each other. Given the position, velocity, and acceleration of the particle in one frame and the relative motion of the frames, you should be able to calculate the position, velocity, and acceleration in the other frame.

According to the diagram on the right, at any instant of time the position of the particle relative to the origin of coordinate frame A is given by $\mathbf{r}_{PA}$ = _____ in terms of its position $\mathbf{r}_{PB}$ relative to the origin of frame B and the position vector $\mathbf{r}_{BA}$ of the origin of frame B relative to the origin of frame A.

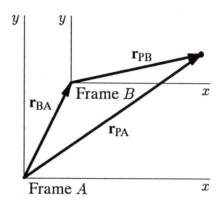

The above expression can be differentiated with respect to time to obtain $\mathbf{v}_{PA}$ = _____ for the velocity of the particle in frame A in terms of the particle velocity $\mathbf{v}_{PB}$ in frame B and the velocity $\mathbf{v}_{BA}$ of frame B as measured in frame A. This expression is valid even if the two frames are accelerating relative to each other.

Take special care with the subscripts. The first names an object and the second names the coordinate frame used to measure the position or velocity of the object. You should say all the words as you read the symbols. That is, when you see $\mathbf{r}_{BA}$ you should say "the position vector of the origin of frame B relative to the origin of frame A." You will then get acquainted with the notation fast and won't get it mixed up later.

The expression for the velocity can be differentiated with respect to time to obtain $\mathbf{a}_{PA}$ = _____ for the acceleration of the particle in frame A in terms of the particle acceleration $\mathbf{a}_{PB}$ in frame B and the acceleration $\mathbf{a}_{BA}$ of frame B as measured in frame A. Specialize this equation to the case when the two frames are *not* accelerating with respect to each other: $\mathbf{a}_{PA}$ = _____ . Now the acceleration of the particle is the same in both frames.

If a particle has a different velocity in two reference frames, then the frames must be moving relative to each other. Similarly, if a particle has a different acceleration in two frames, then the frames must be accelerating relative to each other.

To remember what each of the symbols mean, invent a specific example that you can visualize easily. Here's one you might try. Draw an airplane on a small piece of paper and a straight line across a larger piece of paper. Lay the large paper on a table and move the airplane along the line with constant speed. The large paper represents the air and what you see is the motion of the airplane relative to the air. The speed of the airplane relative to the air can be found simply by dividing the length of the line by the time the airplane takes to fly it. Now move the large paper across the table top with constant velocity as you move the airplane along the line on the paper. You are now viewing the airplane from the ground as the wind blows. Its ground speed can be found by measuring the distance it moves on the table and dividing by the time. You might mark the starting and ending positions of the airplane with small pieces of tape on the table, then measure the distance between the pieces of tape. Convince yourself that the velocity of the airplane relative to the ground is the vector sum of its velocity relative to the air and the velocity of the air relative to the ground. Try various directions for the airplane and wind velocities.

Airplanes flying in moving air or ships sailing in moving water are often used as examples of relative motion. The airplane or ship is the particle, one coordinate frame moves with the air or water, and the other coordinate frame is fixed to the earth. The heading of the airplane or ship is in the direction of its velocity as measured relative to the air or water, *not* relative to the ground. Use your paper airplane to convince yourself of this. Its long axis is parallel to the line on the moving paper and is not necessarily parallel to the line of motion on the table.

The diagram on the right shows the velocity $\mathbf{v}_{PA}$ of an airplane relative to the air and the velocity $\mathbf{v}_{AG}$ of the air relative to the ground. Draw the vector that represents the velocity $\mathbf{v}_{PG}$ of the airplane relative to the ground. Near the midpoint of this vector draw a small airplane oriented correctly; that is, with its long axis parallel to $\mathbf{v}_{PA}$.

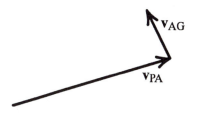

Relative motion at high speeds. If an object is moving at nearly the speed of light or if we compare its velocity as measured in two reference frames that are moving at nearly the speed of light relative to each other, then the results given above fail and we must use a relativistically correct equation. Consider an object P that is moving along the x axis with velocity $\mathbf{v}_{PB}$, as measured in reference frame B. If B is moving along the x axis with velocity $\mathbf{v}_{BA}$, as measured in another frame A, the velocity of P, as measured in A, is given by

$$\mathbf{v}_{PA} =$$

Here c is _____ and has the value _____ m/s.

You should be able to show that if $\mathbf{v}_{PB} = c$, then $\mathbf{v}_{PA} = c$. If something moves with the speed of light in one frame then it moves with the speed of light in all frames. You should also be able to show that if $\mathbf{v}_{PB}$ and $\mathbf{v}_{BA}$ are both much less than the speed of light then the correct non-relativistic result is obtained.

II. PROBLEM SOLVING

Some problems deal with the definitions of average and instantaneous velocity and average and instantaneous acceleration. Here are a few.

PROBLEM 1 (6E). A train moving at a constant speed of 60.0 km/h moves east for 40.0 min, then in a direction 50.0° east of north for 20.0 min, and finally west for 50.0 min. What is the average velocity of the train during this run?

STRATEGY: Find the total displacement and divide by the total time. The total displacement is the sum of three displacements. The length of each of them can be found by multiplying the speed by the time.

SOLUTION: The length of the first displacement is (60.0 km/h) × (40.0/60.0)h = 40.0 km. Similarly show that the lengths of the second and third displacements are 20.0 km and 50.0 km, respectively. Take the x axis to be from west to east and the y axis to be from south to north and fill in the following table with expressions for the displacement components and their values:

	x component	value	y component	value
displacement #1	_____	_____	_____	_____
displacement #2	_____	_____	_____	_____
displacement #3	_____	_____	_____	_____
sum	_____	_____	_____	_____

The total time is 110 min. Divide by this to find the average velocity. Finally find the magnitude and direction of the average velocity.

[ans: 7.59 km/h, 22.5° east of north]

PROBLEM 2 (12E).   The position **r** of a particle moving in the xy plane is given by $\mathbf{r} = (2.00t^3 - 5.00t)\mathbf{i} + (6.00 - 7.00t^4)\mathbf{j}$. Here **r** is in meters and t is in seconds. Calculate (a) **r**, (b) **v**, and (c) **a** when $t = 2.00$ s.

STRATEGY: To find the position vector substitute $t = 2.00$ s into the expression for **r**. To find the velocity, differentiate the expression for **r** with respect to time, then substitute $t = 2.00$ s. To find the acceleration, differentiate the expression for the velocity with respect to time, then substitute $t = 2.00$ s.

SOLUTION:

[ans: (a) $(6.00\mathbf{i} - 106\mathbf{j})$ m; (b) $(19.0\mathbf{i} - 224\mathbf{j})$ m/s; (c) $(24.0\mathbf{i} - 336\mathbf{j})$ m/s^2]

(d) What is the orientation of a line that is tangent to the particle's path at $t = 2.00$ s?

STRATEGY: The velocity vector is tangent to the path so the slope of the line is given by v_y/v_x. This is the tangent of the angle between the line and the positive x axis.

SOLUTION:

[ans: 85.2° below the x axis]

Here's a problem in which an object moves with constant acceleration in a plane and you must find its direction of motion. You should realize you can apply the usual equations of constant acceleration kinematics, but you must allow the acceleration to have both x and y components.

PROBLEM 3 (14P).   A particle A moves along the line $y = 30$ m with a constant velocity **v** ($v = 3.0$ m/s) directed parallel to the positive x axis (Fig. 4–28). A second particle B starts at the origin with zero speed and constant acceleration **a** ($a = 0.40$ m/s^2) at the same instant that particle A passes the y axis. What angle θ between **a** and the positive y axis would result in a collision between these two particles?

STRATEGY: You want to solve for the value of θ such that the position vectors of the two particles are the same at some instant of time.

SOLUTION: The position vector of particle A is given by $\mathbf{r}_A = vt\,\mathbf{i} + d\,\mathbf{j}$, where $d = 30$ m. The position vector of particle B is given by $\mathbf{r}_B = \frac{1}{2}a_xt^2\mathbf{i} + \frac{1}{2}a_yt^2\mathbf{j} = \frac{1}{2}at^2\sin\theta\,\mathbf{i} + \frac{1}{2}at^2\cos\theta\,\mathbf{j}$, where $a\sin\theta$ was substituted for a_x and $a\cos\theta$ was substituted for a_y. A collision occurs if $\frac{1}{2}at^2\sin\theta = vt$ and $\frac{1}{2}at^2\cos\theta = d$. Solve the first equation for t and substitute the resulting expression into the second. You should obtain $2v^2\cos\theta = ad\sin^2\theta$, an equation to be solved for θ. Use the trigonometric identity $\sin^2\theta = 1 - \cos^2\theta$ to obtain an equation that contains only $\cos\theta$. It is $\cos^2\theta + (2v^2/ad)\cos\theta - 1 = 0$. Solve this quadratic equation for $\cos\theta$ and the result for θ.

$$[\text{ans: } 60°]$$

Sometimes you must solve the kinematic equations for the time for which the particle has a certain position, velocity, or acceleration.

PROBLEM 4 (16P). The velocity **v** of a particle moving in the xy plane is given by $\mathbf{v} = (6.0t - 4.0t^2)\mathbf{i} + 8.0\mathbf{j}$. Here **v** is in meters per second and $t\ (> 0)$ is in seconds. (a) What is the acceleration when $t = 3.0$ s?

STRATEGY: Differentiate the expression for **v** to find an expression for the acceleration. Evaluate it for $t = 3.0$ s.

SOLUTION:

$$[\text{ans: } -18\,\mathbf{i}\,\text{m/s}^2]$$

(b) When (if ever) is the acceleration zero?

STRATEGY: Set the expression for **a** equal to 0 and solve for t.

SOLUTION:

$$[\text{ans: } 0.75\,\text{s}]$$

(c) When (if ever) is the velocity zero?

STRATEGY: Set the expression for **v** equal to 0 and solve for t. Both components must vanish for the same value of t.

SOLUTION:

$$[\text{ans: never}]$$

(d) When (if ever) does the speed equal 10 m/s?

STRATEGY: Develop an expression for the speed as a function of time. It is the square root of the sum of the squares of the components. Set the expression equal to 10 m/s and solve for t.

SOLUTION: You should obtain $\sqrt{(6.0t - 4.0t^2)^2 + 8.0^2} = 10$. Square both sides and obtain $(6.0t - 4.0t^2)^2 = 36$. Take the square root of both sides to obtain $-4.0t^2 + 6.0t = \pm 6$. The choice of one sign will lead to an imaginary solution. Use the other and solve the quadratic equation. One solution is negative and is not allowed by the conditions of the problem. The other solution is the correct solution.

$$[\,\text{ans: } 2.2\,\text{s}\,]$$

Here are some projectile motion problems. Just as you did for one-dimensional kinematics problems you should be able to identify 2 events. Take the time to be 0 for one of them, the launching of the projectile, for example. Take the y axis to be vertically upward and the x axis to be horizontal in the plane of the motion. The coordinates and velocity components are x_0, y_0, v_{0x}, and v_{0y} for the event at time 0. Let t be the time of the other event. The coordinates and velocity components are x, y, v_x, and v_y for that event. Identify the known and unknown quantities, then solve $x = x_0 + v_{0x}t$, $y = y_0 + v_{0y}t - \frac{1}{2}gt^2$, and $v_y = v_{0y} - gt$ simultaneously for the unknowns. As an alternative you might use $v_y^2 - v_{0y}^2 = -2g(y - y_0)$ instead of the equation for y or the equation for v_y. Remember $v_x = v_{0x}$.

PROBLEM 5 (17E). A dart is thrown horizontally toward the bull's eye, point P on the dart board of Fig. 4–29, with an initial speed of 10 m/s. It hits at point Q on the rim, vertically below P, 0.19 s later. (a) What is the distance PQ?

STRATEGY: Place the origin of the coordinate system at the throwing point and take the time to be 0 when the dart is thrown. The y axis is vertically upward and the x axis is horizontal. Let x and y be the coordinates of point Q and let t be the time the dart hits there. The vertical component of the initial velocity is 0 so $y = -\frac{1}{2}gt^2$. Evaluate this expression.

SOLUTION:

$$[\,\text{ans: } 18\,\text{cm}\,]$$

(b) How far away from the dart board did the dart thrower stand?

STRATEGY: Evaluate $x = v_{0x}t$.

SOLUTION:

$$[\,\text{ans: } 1.9\,\text{m}\,]$$

PROBLEM 6 (27E). You throw a ball with a speed of 25.0 m/s at an angle of 40.0° above the horizontal directly toward a wall as shown in Fig. 4–30. The wall is 22.0 m from the release point of the ball. (a) How long is the ball in the air before it hits the wall?

STRATEGY: Place the coordinate system with its origin at the release point of the ball. Take the y axis to be positive in the upward direction and the x axis to be horizontal in the plane of the ball's motion. Take the time to be 0 at the instant the ball is released. Assume the ball hits the wall before it hits the ground and take x and y to be the coordinates of the point where it hits and t to be the time when it hits. You already know $x = 22.0$ m, $v_{0x} = 25.0 \cos 40°$, and $v_{0y} = 25.0 \sin 40°$. You must solve for t.

SOLUTION: Use $x = v_{0x} t$.

[ans: 1.15 s]

(b) How far above the release point does the ball hit the wall?

STRATEGY: Evaluate $y = v_{0y} t - \frac{1}{2} g t^2$.

SOLUTION:

[ans: 12.0 m]

(c) What are the horizontal and vertical components of its velocity as it hits the wall?

STRATEGY: Evaluate $v_x = v_{0x}$ and $v_y = v_{0y} - gt$.

SOLUTION:

[ans: $v_x = 19.2$ m/s, $v_y = 4.80$ m/s]

(d) Has it passed the highest point on its trajectory when it hits?

STRATEGY: Notice the sign of the vertical component of the velocity when the ball hits the wall. Is the ball rising or falling then?

SOLUTION: Evaluate $v_y = v_{0y} - gt$. If the result is positive the ball has not yet reached its highest point. If the result is negative it has.

[ans: ball has not yet reached the highest point]

PROBLEM 7 (39P). A third baseman wishes to throw to first base, 127 ft distant. His best throwing speed is 85 mi/h. (a) If he throws the ball horizontally 3.0 ft above the ground, how far from first base will it hit the ground?

STRATEGY: Place the coordinate system with its origin on the ground directly under the release point, with the y axis vertically upward and the x axis horizontal in the plane of the ball's motion. Take the time to be 0 when the ball is thrown. You then know $x_0 = 0$, $y_0 = 3.0$ ft, $v_{0x} = 85$ mi/h $= 125$ ft/s, $v_{0y} = 0$, and $y = 0$. The unknowns are t, x, and v_y. Here you must solve for x and subtract the result from 127 ft.

SOLUTION: Solve $y = y_0 - \frac{1}{2} g t^2 = 0$ for t and substitute the resulting expression into $x = v_{0x} t$. You should get $x = v_{0x} \sqrt{2 y_0 / g}$. Since y_0 is given in feet and v_{0y} is given in ft/s, use $g = 32$ ft/s^2.

[ans: 73 ft]

(*b*) At what upward angle must the third baseman throw the ball if the first baseman is to catch it 3.0 ft above the ground?

STRATEGY: Write $v_{0x} = v_0 \cos\theta$ and $v_{0y} = v_0 \sin\theta$, where $v_0 = 125$ ft/s and θ is the angle above the horizontal at which the ball is thrown. $x \, (= 127\,\text{ft})$ and $y \, (= 3.0\,\text{ft})$ are also known. You must solve for θ.

SOLUTION: Solve $x = v_0 t \cos\theta$ for t and substitute the resulting expression into $y = y_0 + v_0 t \sin\theta - \frac{1}{2}gt^2$. You should get $gx = 2v_0^2 \sin\theta \cos\theta$. Trial and error can be used to solve this for θ. Another method is to use the trigonometric identity $\sin\theta \cos\theta = \frac{1}{2}\sin(2\theta)$ to obtain $gx = v_0^2 \sin(2\theta)$, which can be solved easily for 2θ.

$$[\,\text{ans: } 7.6°\,]$$

(*c*) What will be the time of flight in that case?

STRATEGY: Evaluate $t = x/v_0 \cos\theta$.

SOLUTION:

$$[\,\text{ans: } 1.0\,\text{s}\,]$$

PROBLEM 8 (40P). During volcanic eruptions, chunks of solid rock can be blasted out of the volcano; these projectiles are called *volcanic bombs*. Fig. 4–35 shows a cross section of Mt. Fuji, in Japan. (*a*) At what initial speed would a bomb have to be ejected, at 35° to the horizontal, from the vent at A in order to fall at the foot of the volcano at B? Ignore, for the moment, the effects of air on the bomb's travel.

STRATEGY: Take the dotted lines in the figure to be the coordinate axes, with the x axis horizontal and the y axis vertical. Take the time to be 0 when the block is launched. Suppose the block lands at the point with coordinates x and y, at time t. You then know $x_0 \, (= 0)$, $y_0 \, (= 3.30\,\text{km})$, $\theta_0 \, (= 35°)$, $x \, (= 9.4\,\text{km})$, and $y \, (= 0)$. The unknowns are v_0, t, v_x, and v_y. You are asked to solve for v_0.

SOLUTION: First solve $x = v_0 t \cos\theta_0$ for t, then substitute the resulting expression into $y = y_0 + v_0 t \sin\theta_0 - \frac{1}{2}gt^2$. Solve for v_0^2. You should get $v_0^2 = gx^2/2(y_0 + x\tan\theta_0)\cos^2\theta_0$. Evaluate this expression.

$$[\,\text{ans: } 260\,\text{m/s}\,]$$

(*b*) What would be the time of flight?

STRATEGY: Solve $x = v_0 t \cos\theta_0$ for t.

SOLUTION:

$$[\,\text{ans: } 45\,\text{s}\,]$$

(c) Would the effect of the air increase or decrease your answer in (a)?

SOLUTION:

[ans: increase; the block must have a greater initial speed to go the same horizontal distance]

PROBLEM 9 (52P). (a) During a tennis match, a player serves at 23.6 m/s (as recorded by a radar gun), the ball leaving the racquet horizontally 2.37 m above the court surface. By how much does the ball clear the net, which is 12 m away and 0.90 m high?

STRATEGY: You want to calculate the height of the ball when it has moved 12 m horizontally. Place the origin of a coordinate system at the point on the court surface under the place where the ball leaves the racquet. Take the y axis to be vertical and the x axis to be horizontal. Take the time to be 0 when the ball leaves the racquet. Then $x_0 = 0$, $y_0 = 2.37$ m, $v_{0x} = 23.6$ m/s, and $v_{0y} = 0$. Let t be the time when the ball is directly over the net. Then $x = 12$ m; t, y, and v_y are unknown. You want to solve for y.

SOLUTION: Solve $x = v_{0x}t$ for t and substitute the resulting expression into $y = y_0 - \frac{1}{2}gt^2$. Evaluate the result and subtract 0.90 m from it.

[ans: 0.20 m]

(b) Suppose the player serves the ball as before except that the ball leaves the racquet at 5.00° below the horizontal. Does the ball clear the net now?

STRATEGY: The procedure is the same as before except that the components of the initial velocity are $v_{0x} = 23.6\cos 5.0°$ and $v_{0y} = -23.6\sin 5.0°$. Substitute for t from $x = v_{0x}t$ into $y = y_0 + v_{0y}t - \frac{1}{2}gt^2$.

SOLUTION:

[ans: the ball hits the net 0.044 m above the court surface]

All centripetal acceleration problems are solved using $a = v^2/r$. This equation contains 3 quantities. Two must be given, either directly or indirectly. Remember that the acceleration vector points toward the center of the circle if the speed is constant. Here are some examples.

PROBLEM 10 (59E). (a) What is the acceleration of a sprinter running at 10 m/s when rounding a bend with a turn radius of 25 m?

STRATEGY: The sprinter has an acceleration because the direction of his velocity is changing with time. The magnitude is given by $a = v^2/r$. Evaluate this expression.

SOLUTION:

[ans: 4.0 m/s²]

(*b*) To what does the acceleration **a** point?

SOLUTION:

[ans: toward the center of the circular path]

PROBLEM 11 (62E).   An Earth satellite moves in a circular orbit 640 km above the Earth's surface. The time for one revolution is 98.0 min. (*a*) What is the speed of the satellite?

STRATEGY: The satellite travels a distance equal to the circumference of its orbit in 98 min. Thus $v = 2\pi r/T$, where r is the radius and $T = 98.0$ min. Calculate r by adding the radius of the Earth to the altitude of the satellite: $r = 6.37 \times 10^6 + 640 \times 10^3 = 7.01 \times 10^6$ m. Convert the time to seconds.

SOLUTION:

[ans: 7.49×10^3 m/s]

(*b*) What is the free-fall acceleration at the orbit?

STRATEGY: This is the same as the acceleration of the satellite and is given by $a = v^2/r$.

SOLUTION:

[ans: 8.00 m/s^2]

PROBLEM 12 (65E).   The fast train known as the TGV (Train à Grande Vitesse) that runs south from Paris in France has a scheduled average speed of 216 km/h. (*a*) If the train goes around a curve at that speed and the acceleration experienced by the passengers is to be limited to 0.050*g*, what is the smallest radius of curvature for the track that can be tolerated?

STRATEGY: Solve $a = v^2/r$ for r. The acceleration is $a = 0.050 \times 9.8 = 0.49$ m/s^2. Convert the speed to meters per second.

SOLUTION:

[ans: 7.3 km]

(*b*) If there is a curve with a 1.00-km radius, to what speed must the train be slowed to keep the acceleration below the limit?

STRATEGY: Solve $a = v^2/r$ for v.

SOLUTION:

[ans: 22 m/s (80 km/h) or less]

All relative motion problems are essentially vector addition problems. Identify the two reference frames of interest. Identify the velocities in the equation $\mathbf{v}_{PA} = \mathbf{v}_{PB} + \mathbf{v}_{BA}$, write the equation in component form and solve for the unknown quantities. Sometimes the equation must be rewritten in terms of magnitudes and angles rather than cartesian components. We start with a one-dimensional problem, then move on to more complicated situations.

PROBLEM 13 (77E).   The airport terminal in Geneva, Switzerland has a "moving sidewalk" to speed passengers through a long corridor. Peter, who walks through the corridor but does not use the moving sidewalk, take 150 s to do so. Paul, who simply stands on the moving sidewalk, covers the same distance in 70 s. Mary not only uses the sidewalk but walks along it. How long does Mary take? Assume that Peter and Mary walk at the same speed.

STRATEGY: Let v_{Peter}, v_{Paul}, and v_{Mary} be the speeds of the three people. Since Mary's speed relative to the moving sidewalk is v_{Peter} and the speed of the sidewalk is v_{Paul}, Mary's speed is given by $v_{\text{Mary}} = v_{\text{Peter}} + v_{\text{Paul}}$. Let ℓ be the length of the sidewalk and t_{Peter}, t_{Paul}, and t_{Mary} be the times taken by the three people. Then $v_{\text{Peter}} = \ell/t_{\text{Peter}}$, $v_{\text{Paul}} = \ell/t_{\text{Paul}}$, and $v_{\text{Mary}} = \ell/t_{\text{Mary}}$. Thus $\ell/t_{\text{Mary}} = (\ell/t_{\text{Peter}}) + (\ell/t_{\text{Paul}})$. Notice that ℓ cancels from the equation and solve for t_{Mary}.

SOLUTION:

$$[\text{ans: } 48\,\text{s}]$$

PROBLEM 14 (84P).   Two ships, A and B, leave port at the same time. Ship A travels northwest at 24 knots and ship B travels at 28 knots in a direction 40° west of south. (1 knot = 1 nautical mile per hour; see Appendix F.) (a) What are the magnitude and direction of the velocity of ship A relative to B?

STRATEGY: The velocity $\mathbf{v}_{AE}$ of ship A relative to the Earth is the velocity $\mathbf{v}_{BE}$ of ship B relative to the Earth plus the velocity $\mathbf{v}_{AB}$ of ship A relative to ship B. This means $\mathbf{v}_{AB} = \mathbf{v}_{AE} - \mathbf{v}_{BE}$. Carry out the vector subtraction component by component.

SOLUTION: On the axes to the right draw the vector velocities of the two ships and label the given angles with their values. Take the positive y axis to be to the north and the positive x axis to be to the east and fill in the following table with the values of the components.

	x component	y component
$\mathbf{v}_{AE}$		
$\mathbf{v}_{BE}$		
$\mathbf{v}_{AB}$		

Now compute the magnitude and angle with the y axis of $\mathbf{v}_{AB}$.

$$[\text{ans: } 38\,\text{knots}; \ 1.5°\text{ east of north}]$$

(b) After what time will they be 160 nautical miles apart?

STRATEGY: The rate at which they are moving apart is given by their *relative* speed. Solve $d = v_{AB}t$ for t.

SOLUTION:

[ans: 4.2 h]

(c) What will be the bearing of B from A at that time?

STRATEGY: The displacement vector of B relative to A is $\mathbf{r}_{BA} = -\mathbf{r}_{AB} = -\mathbf{v}_{AB}t$. Find the components of this vector, then the angle it makes with one of the coordinate directions.

SOLUTION: You should get $\mathbf{r}_{BA} = -4.29\,\mathbf{i} - 160\,\mathbf{j}$ nautical miles.

[ans: 1.5° west of south]

PROBLEM 15 (85P). A light plane attains an airspeed of 500 km/h. The pilot sets out for a destination 800 km to the north but discovers that the plane must be headed 20.0° east of north to fly there directly. The plane arrives in 2.00 h. What was the wind velocity vector?

STRATEGY: Let $\mathbf{v}_{PG}$ be the velocity of the plane relative to the ground, $\mathbf{v}_{PA}$ be the velocity of the plane relative to the air, and $\mathbf{v}_{AG}$ be the velocity of the air relative to the ground. Then $\mathbf{v}_{PG} = \mathbf{v}_{PA} + \mathbf{v}_{AG}$. Solve this for $\mathbf{v}_{AG}$.

SOLUTION: Draw the vectors $\mathbf{v}_{PG}$, $\mathbf{v}_{PA}$, and $\mathbf{v}_{AG}$ on the axes to the right. Take the y axis to be northward and the x axis to be eastward. $\mathbf{v}_{PG}$ is straight north and has a magnitude of 800 km/2.00 h $= 400$ km/h. $\mathbf{v}_{PA}$ is 20.0° east of north and has a magnitude of 500 km/h. $\mathbf{v}_{AG}$ completes the vector addition triangle. Find the components of $\mathbf{v}_{PG}$ and $\mathbf{v}_{PA}$, then compute the components of $\mathbf{v}_{AG}$. The following table might help.

	x component	y component
$\mathbf{v}_{PG}$	_____	_____
$\mathbf{v}_{PA}$	_____	_____
$\mathbf{v}_{AG}$	_____	_____

[ans: $(-171\,\mathbf{i} - 70.0\,\mathbf{j})$ km/h or 185 km/h, 22.2° south of west]

PROBLEM 16 (88P). A woman can row a boat at 4.0 mi/h in still water. (a) If she is crossing a river where the current is 2.0 mi/h, in what direction must her boat be headed if she wants to reach a point directly opposite her starting point?

STRATEGY: Let $\mathbf{v}_{BS}$ be the velocity of the boat relative to the shore, $\mathbf{v}_{BW}$ be the velocity of the boat relative to the water, and $\mathbf{v}_{WS}$ be the velocity of the water relative to the shore. Then $\mathbf{v}_{BS} = \mathbf{v}_{BW} + \mathbf{v}_{WS}$. You know $\mathbf{v}_{BS}$ is perpendicular to the shore but you do not know its magnitude. You know both the magnitude and direction of $\mathbf{v}_{WS}$: it parallel to the shore and has a magnitude of 2.0 mi/h. You know the magnitude of $\mathbf{v}_{BW}$ is 4.0 mi/h and you wish to solve for its direction.

SOLUTION: Draw the three velocity vectors on the axes to the right. Take the current to be left to right and suppose the boat is to travel along the y axis. Suppose $\mathbf{v}_{BW}$ makes the angle θ with the positive y axis. Then the x component of $\mathbf{v}_{BS} = \mathbf{v}_{BW} + \mathbf{v}_{WS}$ is $0 = -v_{BW}\sin\theta + v_{WS}$ and the y component is $v_{BS} = v_{BW}\cos\theta$. The second of these can be solved for θ.

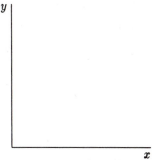

[ans: 30° upstream]

(b) If the river is 4.0 mi wide, how long will it take her to cross the river?

STRATEGY: Now you need to know the speed of the boat relative to the shore. Solve the first of the equations above for v_{BS}. The time is given by $t = d/v_{BS}$, where d is the width of the river.

SOLUTION:

[ans: 1.2 h (69 min)]

(c) Suppose that instead of crossing the river she rows 2.0 mi *down* the river and then back to her starting point. How long will she take?

STRATEGY: On the way downstream her speed relative to the shore is given by $v_{BS} = v_{BW} + v_{WS} = 4.0 + 2.0 = 6.0$ mi/h. Calculate the time for this portion of the trip. On the way upstream her speed relative to the shore is given by $v_{BS} = v_{BW} - v_{WS} = 4.0 - 2.0 = 2.0$ mi/h. Calculate the time for this portion of the trip and add the two times.

[ans: 1.3 h (80 min)]

(d) How long will she take to row 2.0 mi *up* the river and then back to her starting point?

SOLUTION:

[ans: 1.3 h (80 min)]

(*e*) In what direction should she head the boat if she wants to cross in the shortest possible time, and what is that time?

STRATEGY: She wants the y component of $\mathbf{v}_{BS}$ to be as great as possible. Since $\mathbf{v}_{BS} = \mathbf{v}_{BW} + \mathbf{v}_{WS}$ and $\mathbf{v}_{WS}$ is in the x direction, this means she wants the y component of $\mathbf{v}_{BW}$ to be as great as possible. Once you have found the maximum value of the y component for a magnitude of 4.0 mi/h, use it to compute the crossing time.

SOLUTION:

[ans: head directly across; 1.0 h (60 min)]

III. NOTES

Chapter 5
FORCE AND MOTION — I

I. BASIC CONCEPTS

The emphasis now changes from kinematics to dynamics as you start to study how objects influence the motion of each other. This is the central chapter of the mechanics section of the text. Be sure you understand the concepts of force and mass and pay particular attention to the relationship between the net force on an object and its acceleration.

Dynamics.   The fundamental problem of dynamics is to find the acceleration of an object, given the object and its environment. The problem is split into two parts, connected by the idea of a force: the environment of an object exerts forces on the object and the net force on it causes it to accelerate. The first part of the problem is to find the net force exerted on the object, given the relevant properties of the object and its environment. An expression for a force in terms of the properties of interacting objects is called a force law. The second part of the problem is to find the acceleration of the object, given the net force. In this chapter we concentrate on the second part although, of necessity, some force laws are discussed. Other force laws are described in detail at appropriate points in the text.

Newton's first law.   The text gives two statements of Newton's first law. Write both of them here and learn them:

Statement #1: _____

_____

_____

Statement #2: _____

_____

_____

The first statement in the text is closer to Newton's words, the second is closer to the modern interpretation of the law.

The first law helps us define inertial reference frames. Suppose we have found a particle on which zero net force acts and we attach reference frame S to it. Clearly the acceleration of the particle, as measured in S, is zero. Describe a reference frame S' in which the acceleration of the particle is not zero: _____

_____

Which of these frames is an inertial frame? _____
Why? _____
How is any other inertial frame related to the one attached to the particle? _____

_____

Newton's second law. This is the central law of classical mechanics. It gives the relationship between the net force $\sum \mathbf{F}$ acting on an object and the acceleration $\mathbf{a}$ of the object:

$$\sum \mathbf{F} = m\mathbf{a},$$

where m is the mass of the object. To understand the law you must understand the definitions of force and mass.

A <u>force</u> is measured, in principle, by applying it to the standard (1 kg) mass and measuring the _____ of the standard mass. If SI units are used, the magnitudes of these quantities are numerically equal. Both are vectors in the same direction. That forces obey the laws of vector addition can be checked by simultaneously applying two forces in different directions and verifying that the result is the same as when the resultant of the forces is applied as a single force. All measurements must be made using an inertial reference frame.

The SI unit of force is _____ and is abbreviated _____. In terms of the SI base units (kg, m, s) it is _____.

The <u>mass</u> of an object is measured, in principle, by comparing the accelerations of the object and the standard mass when the same force is applied to them. In particular, the mass of the object is given by $m = $ _____, where a is the magnitude of the acceleration of the object, a_0 is the magnitude of the acceleration of the mass standard, and m_0 is the mass of the mass standard. The accelerations must be measured using an inertial frame.

Note that small masses acquire a _____ acceleration than large masses when the same force is applied. Mass is said to measure <u>inertia</u> or resistance to changes in motion.

Mass is a scalar and is always positive. The mass of two objects in combination is the sum of the individual masses.

Newton's second law $\sum \mathbf{F} = m\mathbf{a}$ is a vector equation. It is equivalent to the three component equations

$$\sum F_x = $$

$$\sum F_y = $$

$$\sum F_z = $$

You must be aware that in these equations $\sum \mathbf{F}$ is the *total* (or *net*) force acting on the object, the vector sum of all the individual forces. This means that in any given situation you must identify all the forces acting on the object and then sum them *vectorially*.

Note that $\sum \mathbf{F} = 0$ implies $\mathbf{a} = 0$. If the resultant force vanishes then the object does not accelerate; its velocity as observed in an inertial reference frame is constant in both magnitude and direction. The resultant force may vanish because no forces act on the object or because the forces that act sum to 0. For some situations you may know that 3 forces act but are given only 2 of them. If you also know that the acceleration vanishes you can solve $\mathbf{F}_1 + \mathbf{F}_2 + \mathbf{F}_3 = 0$ for the third force.

Newton's third law. Newton's third law tells us something about forces. If object A exerts a force $\mathbf{F}_{BA}$ on object B, then according to the third law, the force exerted by object B on object A is given by $\mathbf{F}_{AB} = $ _____. Compared to the force of A on B, the force of B on A is _____ in magnitude and _____ in direction. You should also be aware that these two forces are of the same type. That is, if object A exerts a *gravitational* force on B, then B exerts a *gravitational* force on A.

The third law is useful in solving problems involving more than one object. If two objects exert forces on each other, we immediately use the same symbol to represent their magnitudes and, in writing the second-law equations, we remember the forces are in opposite directions. In addition, we remember that the forces act on different objects. When we want to write Newton's second law for object A, one of the forces we include is the force of B on A, but emphatically *NOT* the force of A on B. The force of A on B, in addition to the other forces on B, determines the acceleration of B, not A.

Gravitational force. One force law you will use a great deal gives the force of gravity on an object. This force is called the weight and its magnitude is given by $W = $ _____, where m is the mass of the object and g is the magnitude of the acceleration due to gravity at the position of the object. Near the surface of the earth the direction of the weight is _____. Be sure you understand that mass and weight are quite different concepts. Mass is a property of an object and does not change as the object is moved from place to place or even into outer space. It is a scalar. Weight, on the other hand, is a force. It varies as the object moves from place to place and vanishes when the object is far from all other objects, as in outer space. This is because $\mathbf{g}$, not the mass, varies from place to place.

Remember that the weight of an object is $m\mathbf{g}$ regardless of its acceleration. Weight is a force and, if appropriate, is included in the sum of all forces acting on the object. This sum equals $m\mathbf{a}$ and if other forces act then $\mathbf{a}$ is different from $\mathbf{g}$.

The SI unit of weight is _____ .

Forces of strings. If a string with negligible mass connects two objects it pulls on each with a force of the same magnitude, called the _____ in the string. You may think of the string as simply transmitting a force from one object to the other; the situation is exactly the same if the objects are in contact and exert forces on each other. Strings pull, not push, along their lengths, so a string serves to define the direction of the force. Pay careful attention to the way forces of strings are handled in the sample problems of the text and in the examples given below.

Normal forces. When an object is in contact with a surface the surface may exert a force on it. If the surface is frictionless that force must be perpendicular to the surface. Thus it is called a normal force. Unless some adhesive is between the object and surface a normal force can only push on the object. It must be directed away from the surface toward the interior of the object.

If the surface is at rest or moving with constant velocity the normal force adjusts until the component of the object's acceleration perpendicular to the surface vanishes. We often use this condition to solve for the normal force. Set the sum of the perpendicular components of the forces acting on the object equal to zero. Since the normal force is one of these the resulting equation can be solved for it in terms of the perpendicular components of the other forces.

II. PROBLEM SOLVING

Some problems deal with the definitions of force and mass. If a force is applied to the standard kilogram and, as a result, the standard kilogram has an acceleration a_s then the magnitude of the force in newtons is numerically equal to the magnitude of the acceleration. Force is a vector and is in the direction of the acceleration. The net force acting on an object is the *vector* sum of the individual forces acting on it. If identical forces are applied to the standard kilogram and another object and their accelerations are a_s and a_o, respectively, then the mass in kg of the object is given by $m_0 = a_s/a_o$.

PROBLEM 1 (3P). Suppose that the 1-kg standard body accelerates at $4.00\,\text{m/s}^2$ at $160°$ from the positive direction of the x axis, owing to two forces, one of which is $\mathbf{F}_1 = (2.50\,\text{N})\,\mathbf{i} + (4.60\,\text{N})\,\mathbf{j}$. What is the other force (*a*) in unit-vector notation and (*b*) as a magnitude and direction?

STRATEGY: The vector sum of the two forces is given by $\mathbf{F}_1 + \mathbf{F}_2 = m\mathbf{a}$ so $\mathbf{F}_2 = m\mathbf{a} - \mathbf{F}_1$. Find the x and y components of $m\mathbf{a}$, then carry out the vector subtraction. Use $F_2 = \sqrt{F_{2x}^2 + F_{2y}^2}$ and $\tan\theta = F_{2y}/F_{2x}$ to find the magnitude and the angle the force makes with the positive x axis.

SOLUTION:

$$[\text{ans: } (a)\ (-6.26\,\mathbf{i} - 3.23\,\mathbf{j})\,\text{N}; \ (b)\ 7.0\,\text{N}, 27° \text{ below the negative } x \text{ axis}]$$

You should be very careful to distinguish between weight and mass. Weight is a force, with magnitude given by mg, where g is the acceleration due to gravity at the site of the object. Mass is a property of an object and determines the acceleration it is given by any applied force. Here's a problem to help you with the difference.

PROBLEM 2 (13E). A space traveler whose mass is 75 kg leaves the Earth. Compute his weight (*a*) on Earth, (*b*) on Mars, where $g = 3.8\,\text{m/s}^2$, and (*c*) in interplanetary space, where $g = 0$.

STRATEGY: Use $W = mg$, where m is the same at each location and g is the local value of the acceleration due to gravity.

SOLUTION:

$$[\text{ans: } (a)\ 740\,\text{N}; \ (b)\ 290\,\text{N}; \ (c)\ \text{zero}]$$

(*d*) What is his mass at each of these locations?

SOLUTION:

[ans: 75 kg at all of them]

You should know how to handle objects connected by strings. Recognize that a massless string pulls on the objects attached at each end with the same force, equal to the tension in the string.

PROBLEM 3 (17E).   A three-piece mobile, strung by massless cord, is shown in Fig. 5–43. The masses of the highest and lowest pieces are given. The tension in the top cord is 199 N. What is the tension in (*a*) the lowest cord and (*b*) the middle cord?

STRATEGY: The net force on each piece of the mobile is zero. You know this since the pieces are not accelerating. The force of gravity and the force of the lowest cord on the lowest piece must sum to 0. This condition will enable you to find the tension in the lowest cord. The force of the highest cord, the middle cord, and gravity on the highest piece must also sum to 0. This condition will enable you to find the tension in the middle cord.

SOLUTION: Consider first the lowest piece of the mobile. Take the tension in the lowest cord to be T_1. The force of gravity on the piece is $m_1 g$. If the positive direction is upward, then $T_1 - m_1 g = 0$. Solve for T_1.

Now consider the highest piece of the mobile. The force of gravity is $m_3 g$, down. Take the tension in the middle cord to be T_2 and the tension in the upper cord to be T_3. The upper cord pulls up and the lower cord pulls down, so $T_3 - T_2 - m_3 g = 0$. Solve for T_2.

[ans: (*a*) 54 N; (*b*) 152 N]

The mass of the middle piece can be found from the given data. Do you see how?

Sometimes the acceleration is given indirectly by giving information that can be used in the kinematic equations to calculate it.

PROBLEM 4 (22E).   A car traveling at 53 km/h hits a bridge abutment. A passenger in the car moves forward a distance of 65 cm (with respect to the road) while being brought to rest by an inflated air bag. What magnitude of force (assumed constant) acts on the passenger's upper torso, which has a mass of 41 kg?

STRATEGY: You know the initial speed and the stopping distance so you can calculate the acceleration of the person's upper torso. Either solve $x = v_0 t + \frac{1}{2} a t^2$ and $0 = v_0 + at$ simultaneously or else use $2ax = -v_0^2$. Once a value has been found for a use $F = ma$ to calculate the force.

SOLUTION:

$[\,$ans: $6.8 \times 10^3\,\mathrm{N}\,]$

A definite procedure has been devised to solve dynamics problems. It ensures that you consider only one object at a time, reminds you to include all forces acting on the object you are considering, and guides you in writing Newton's second law in an appropriate form. Follow it closely. Use the list below as a check list until the procedure becomes automatic.

1. Identify the object to be considered. It is usually the object on which the given forces act or about which a question is posed.

2. Represent the object by a dot on a diagram. Do not include the environment of the object since this is replaced by the forces it exerts on the object.

3. On the diagram draw arrows to represent the forces exerted by the environment on the object. Try to draw them in roughly the correct directions. The tail of each arrow should be at the dot. Label each arrow with an algebraic symbol to represent the magnitude of the force, regardless of whether or not a numerical value is given in the problem statement.

   The hard part is getting all the forces. If appropriate, don't forget to include the weight of the object, the normal force of a surface on the object, and the forces of any strings or rods attached to the object. These special forces are explained in the text. Carefully go over the sample problems in the text and the end-of-chapter problems worked out below to see how to handle these forces.

   Some students erroneously include forces that are not acting on the object. For each force you include you should be able to point to something in the environment that is exerting the force. This simple procedure should prevent you from erroneously including a normal force, for example, when the object you are considering is not in contact with a surface.

4. Draw a coordinate system on the diagram. In principle, the placement and orientation of the coordinate system do not matter as far as obtaining the correct answer is concerned but some choices reduce the work involved. If you can guess the direction of the acceleration place one of the axes along that direction. The acceleration of an object sliding on a surface such as a table top or inclined plane, for example, is parallel to the surface. Once the coordinate system is drawn, label the angle each force makes with a coordinate axis. This will be helpful in writing down the components of the forces later.

   The diagram, with all forces shown but without the coordinate system, is called a free-body diagram (or a force diagram). We add the coordinate system to help us carry out the next step in the solution of the problem.

5. Write Newton's second law in component form: $\sum F_x = ma_x$, $\sum F_y = ma_y$, and, if necessary, $\sum F_z = ma_z$. The left sides of these equations should contain the appropriate components of the forces you drew on your diagram. You should be able to write the

equations by inspection of your diagram. Use algebraic symbols to write them, not numbers; most problems give or ask for force magnitudes so you should usually write each force component as the product of a magnitude and the sine or cosine of an appropriate angle.

6. If more than one object is important, as when two objects are connected by a string, you must carry out the steps above separately for each object. There is then usually additional conditions you must consider. In many cases when two objects are connected by a string, for example, the magnitudes of their accelerations are the same. Be aware of these situations as you study the sample problems of the text and the examples below.

6. Identify the known quantities and solve for the unknowns.

Here are some examples for you to try.

PROBLEM 5 (27E).   Refer to Fig. 5–20 and suppose the two masses are $m = 2.0\,\text{kg}$ and $M = 4.0\,\text{kg}$. (a) Decide without any calculations which of them should be hanging if the magnitude of the acceleration is to be the largest.

STRATEGY: The inertia of the system $(m + M)$ is the same no matter which mass is hanging. The force of gravity, however, is greater if the larger mass is hanging and the smaller is on the horizontal surface.

SOLUTION:

[ans: the larger mass should be hanging]

What are (b) the magnitude and (c) the associated tension in the cord?

STRATEGY: For each mass draw a free-body diagram and write Newton's second law in terms of the forces acting on that mass. Use the condition that the acceleration of each mass has the same magnitude and solve the second law equations.

SOLUTION: In the space to the right draw a free-body diagram for mass m, on the horizontal surface. The forces acting are: the force of gravity mg, down; the normal force of the surface N, up; and the tension T in the string, to the right. Take the coordinate axis to be positive to the right. Then the horizontal component of Newton's second law is: $T = ma$, where a is the acceleration of m.

Now draw a free-body diagram for the hanging mass. The forces acting on it are: the force of gravity Mg, down and the tension T in the string, up. Take the positive axis to be downward. Then Newton's second law is:

Solve the two equations simultaneously for a and T.

[ans: (b) $6.5\,\text{m/s}^2$; (c) $13\,\text{N}$]

Notice that because the force of the string on m is the same in magnitude as its force on M the same symbol was used. The difference in direction was taken into account in the second law equations: T entered the horizontal component of the equation for m and the vertical component of the equation for M.

Also note the choices of positive axes for the two objects. You are free to choose any axes whatsoever for each object but what you choose determines how you write the relationship between the accelerations. If the acceleration of m is to the right then the acceleration of M is downward. Because you chose the axis for m to be positive to the right and the axis for M to be positive downward you represented both accelerations by the same symbol a. Had you chosen the axis for M to be positive upward then you would take a for the acceleration of m and $-a$ for the acceleration of M.

PROBLEM 6 (45P). An object is hung from a spring balance attached to the ceiling of an elevator. The balance reads 65 N when the elevator is standing still. (*a*) What is the reading when the elevator is moving upward with a constant speed of 7.6 m/s?

STRATEGY: The reading on the balance gives the magnitude of the upward force on the on the object. In addition, there is the downward force of gravity. The combination produces the acceleration of the object, in this case 0.

SOLUTION: Draw a free body diagram of the object. Use T to label the upward force of the balance on the object and mg to label the downward force of gravity. Newton's second law becomes $T - mg = ma$, where up was chosen to be the positive direction. Solve for T. The mass of the object can be computed using $T_r = mg$, where T_r is the reading on the balance when the elevator is at rest.

[ans: 65 N]

(*b*) What is the reading of the balance when the elevator is moving upward with a speed of 7.6 m/s while decelerating at a rate of 2.4 m/s²?

SOLUTION: The analysis is just as before but now $a = -2.4\,\mathrm{m/s^2}$.

[ans: 49 N]

PROBLEM 7 (48P). A 80-kg man jumps down to a concrete patio from a window ledge only 0.50 m above the ground. He neglects to bend his knees on landing, so that his motion is arrested in a distance of 2.0 cm. (*a*) What is the average acceleration of the man from the time his feet first touch the patio to the time he is brought fully to rest?

STRATEGY: This is strictly a kinematics problem. Assume the acceleration is constant and use $2ax = -v_0^2$, where v_0 is his velocity as his feet first touch the patio and x is the distance traveled in coming to rest. Down was taken to be positive. The acceleration is upward.

You will need to find v_0. This is the final velocity of the falling portion of his jump. Use $2gd = v_0^2$, where x (= 0.50 m) is his displacement during the fall.

SOLUTION:

[ans: 250 m/s²]

(*b*) With what force does this jump jar his bone structure?

STRATEGY: Use $F = ma$.

SOLUTION:

[ans: $2.0 \times 10^4 \text{N}$]

PROBLEM 8 (49P).   Three blocks are connected, as shown in Fig. 5-48, on a horizontal frictionless table and pulled to the right with a force $T_3 = 65.0 \text{N}$. If $m_1 = 12.0 \text{kg}$, $m_2 = 24.0 \text{kg}$, and $m_3 = 31.0 \text{kg}$, calculate (*a*) the acceleration of the system and (*b*) the tensions T_1 and T_2.

STRATEGY: Think of the blocks as a single object with mass $m_1 + m_2 + m_3$, being pulled by the force T_3. Use Newton's second law to compute the acceleration. To find the tensions use the second law equations for the individual blocks.

SOLUTION: (*a*) Take the axis to be positive to the right and write the horizontal component of the second law: $T_3 = (m_1 + m_2 + m_3)a$. Solve for a. (*b*) The horizontal component of Newton's second law for m_3 is $T_3 - T_2 = m_3 a$. Solve for T_2. The horizontal component of Newton's second law for m_2 is $T_2 - T_1 = m_2 a$. Solve for T_1. You can check your solution by solving the second law equation for m_1 ($T_1 = m_1 a$).

[ans: (*a*) 0.970m/s^2; (*b*) $T_1 = 11.6 \text{N}$, $T_2 = 34.9 \text{N}$]

PROBLEM 9 (50P).   Figure 5–49 shows four penguins that are being playfully pulled along very slippery (frictionless) ice by a curator. The masses of three penguins and the tension in two of the cords are given. Find the mass that is not given.

STRATEGY: Because one of masses is not given you cannot find the acceleration of the four penguins considered as a single object, as you found the acceleration of the blocks in the previous problem. You must resort to solving the four Newton's second law equations, one for each penguin.

SOLUTION: Take the axis to be positive to the right and label the tensions in the cords T_1, T_2, T_3, and T_4, from left to right. You have been given T_2 and T_4. Similarly, label the masses of the penguins m_1, m_2, m_3, and m_4, again from left to right. Write down the second law equations: $T_1 = m_1 a$, $T_2 - T_1 = m_2 a$, $T_3 - T_2 = m_3 a$, and $T_4 - T_3 = m_4 a$. Systematically eliminate T_1, T_3, and a until you are left with a single equation in a single unknown, namely m_2. Solve for this quantity. You should get $m_2 = [(m_1 + m_3 + m_4)T_2 - m_1 T_4]/[T_4 - T_2]$.

[ans: 23kg]

PROBLEM 10 (56P). A 52-kg circus performer is to slide down a rope that will snap if the tension exceeds 425 N. (*a*) What happens if the performer hangs stationary on the rope?

STRATEGY: Use Newton's second law to calculate what the tension in the rope would be if it could hold the performer stationary.

SOLUTION: Take down to be positive and write $mg - T = ma$. Since $a = 0$, $T = mg$, where m is the mass of the performer.

[ans: the tension would be 510 N, greater than 425 N; the rope snaps]

(*b*) What is the magnitude of the least acceleration the performer can have so as to avoid breaking the rope?

STRATEGY: Examine the Newton's second law equation and note that as the acceleration of the performer increases from 0 the tension in the rope decreases from 510 N. If the acceleration is great enough the tension becomes 425 N or less and the rope does not break. Put $T = 425$ N and solve for a.

SOLUTION:

[ans: $1.6 \, \text{m/s}^2$]

PROBLEM 11 (62P). A 100-kg crate is pushed at constant speed up the frictionless 30.0° ramp shown in Fig. 5–53. (*a*) What horizontal force **F** is required?

STRATEGY: The acceleration of the crate is 0 so the net force acting on it is 0. Solve this condition for the applied force.

SOLUTION: Use the space to the right to draw a free-body diagram for the crate. Label the normal force N. It is perpendicular to the ramp. Label the force of gravity mg, where m is the mass of the crate. It is down. Label the applied force F. It is horizontal. Take the x axis to be parallel to and up the ramp. Take the y axis to be in the direction of the normal force. Write the x component of Newton's second law: $F \cos\theta - mg\sin\theta = 0$, where θ is the ramp angle. This equation does not contain N, the second unknown. Solve it for F.

[ans: 566 N]

(*b*) What force is exerted by the ramp on the crate?

SOLUTION: Write the y component of Newton's second law: $N - F\sin\theta - mg\cos\theta = 0$. Solve for N.

[ans: $1130\,\mathrm{N}$]

PROBLEM 12 (63P). A 10-kg monkey climbs up a massless rope that runs over a frictionless tree limb and back down to a 15-kg package on the ground (Fig. 5–54). (*a*) What is the magnitude of the least acceleration the monkey must have if it is lift the package off the ground?

STRATEGY: The greater the upward acceleration of the monkey the greater the tension in the rope. To lift the package off the ground the tension in the rope must be at least the weight of the package. Note that the third law is at work here. To obtain an upward acceleration the monkey pushes down on the rope. The rope then pushes up on the monkey. If the package moves upward with constant speed the rope is pulling up on the monkey with a force that equals the weight of the package.

SOLUTION: First draw a free-body diagram for the package. Use T to label the force of the rope on it and N to label the normal force of the ground on it. Both these forces are upward. Use $m_P g$ to label the force of gravity on it. Write the vertical component of Newton's second law: $T + N - m_P g = m_p a$. While the package is on the ground $a = 0$. After the package has lifted off $N = 0$. You want to consider the situation for which T is just large enough to lift the package off the ground and move it upward without acceleration. Set $a = 0$ and $N = 0$ and find that $T = m_P g$.

Now draw a free-body diagram for the monkey. Use T to label the force of the rope on the monkey. It acts upward. Use mg to label the downward force of gravity on the monkey. Write the vertical component of Newton's second law: $T - mg = ma$. Place $T = m_P g$ and solve for a.

[ans: $4.9\,\mathrm{m/s^2}$]

If, after the package is lifted, the monkey stops its climb and holds onto the rope, what are (*b*) its acceleration and (*c*) the tension in the rope?

STRATEGY: Solve the two second law equation simultaneously for the acceleration of the monkey and the tension in the rope. The monkey and package have the same magnitude acceleration.

SOLUTION: Notice that we took up to be the positive direction for both the monkey and package. If the monkey has a positive acceleration (up) the package must have a negative acceleration (down). If a is the acceleration of the monkey, then $-a$ is the acceleration of the package. The second law equations become: $T - mg = ma$ and $T - m_P g = -m_P a$. Solve these for a and T.

[ans: (b) 2.0 m/s²; (c) 120 N]

PROBLEM 13 (68P).   In earlier days, horses pulled barges down canals in the manner shown in Fig. 5–58. Suppose that the horse pulls on the rope with a force of 7900 N at an angle of 18° to the direction of motion of the barge, which is headed straight along the canal. The mass of the barge is 9500 kg, and its acceleration is 0.12 m/s². Calculate the force exerted by the water on the barge.

STRATEGY: The acceleration is parallel to the canal so the force of the water must be such that the perpendicular component of the total force vanishes and the parallel component leads to the given acceleration. Use Newton's second law to find the components of the force of the water.

SOLUTION: Draw a free-body diagram for the barge. Label the force of the rope F_R and the force of the water F_W. You do not know the direction of this force but you suspect it is somewhere in the third quadrant, roughly opposite the force of the rope. Use θ_R to label the angle between the rope and the forward direction. Take the forward direction to be the positive x axis and take the y axis to be across the canal. Write Newton's second law in component form. The x component is $F_R \cos\theta_R + F_{Wx} = ma$; the y component is $F_R \sin\theta_R + F_{Wy} = 0$. Solve for F_{Wx} and F_{Wy}. You might also wish to calculate the magnitude of the force and the angle it makes with the canal.

[ans: $F_{Wx} = -6300$ N, $F_{Wy} = -2400$ N ($F_W = 6800$ N, 21° below the negative x direction)]

PROBLEM 14 (73P).   Figure 5–61 shows a man sitting in a bosun's chair that dangles from a massless rope, which runs over a massless, frictionless pulley and back to the man's hand. The combined mass of the man and chair is 95.0 kg. (a) With what force must the man pull on the rope for him to rise at constant speed?

STRATEGY: Suppose the man pulls downward on the rope with a force F. The rope then pulls upward on his hand with a force of the same magnitude. The tension in the rope is F so, in addition, the rope pulls upward on the chair with force F. You want to calculate the value of F so the acceleration of the man and chair is 0.

SOLUTION: Draw a free-body diagram for the man and chair, considered as a single object. The total force of the rope is $2F$, upward; the force of gravity is mg, downward. Take the positive direction to be upward and write the vertical component of Newton's second law: $2F - mg = ma$. Set $a = 0$ and solve for F.

[ans: 466 N; the force of the man on the rope is downward]

(*b*) What force is needed for an upward acceleration of 1.30 m/s^2?

SOLUTION:

[ans: 527 N; the force of the man on the rope is downward]

(*c*) Suppose, instead, that the rope on the right is held by a person on the ground. Repeat parts (*a*) and (*b*) for this new situation.

STRATEGY: If the person on the ground pulls down with a force of magnitude F, the force of the rope on the chair and man is F, upward. $2F$ is replaced by F in the second law equations.

SOLUTION:

[ans: 931 N, 1050 N]

(*d*) In each of the four cases, what is the force exerted on the ceiling by the pulley system?

STRATEGY: Now consider the pulley. The rope is pulling down on it and the ceiling is pulling up. Since the pulley does not accelerate the vector sum of these forces must vanish. Solve for the force of the ceiling on the pulley. The force of the pulley on the ceiling, of course, has the same magnitude but is in the opposite direction.

SOLUTION: Draw a free-body diagram for the pulley. If F is the tension in the rope then the force of the rope on the pulley is $2F$. It is downward. Use F_c to label the force of the ceiling. Since the pulley is massless there is no force of gravity. Write Newton's second law: $F_c - 2F = 0$. Calculate F_c for each of the four cases.

[ans: 931 N; 931 N; 1860 N; 2100 N; in all cases the force of the pulley on the ceiling is downward]

III. NOTES

Chapter 6
FORCE AND MOTION — II

I. BASIC CONCEPTS

This chapter contains a great many applications of Newton's laws, with special emphasis on frictional and centripetal forces. Here's where your understanding of the fundamentals begins to pay off!

Friction. Two macroscopic objects in contact may exert frictional forces on each other. Friction is unavoidable when the objects are sliding on each other, although lubricants and air films may make it small. Even when the objects are stationary with respect to each other, they exert frictional forces if other forces present would otherwise cause them to slide. Explain in words the mechanism that gives rise to a frictional force: _____

_____

_____

_____

Although all frictional forces arise from the same fundamental phenomenon two types are discussed in the text: static and kinetic. Tell how you can identify the situation in which each is acting.

static: _____

kinetic: _____

When the two objects are not moving relative to each other the force of static friction is determined by the condition that their accelerations be equal. Usually, but not always, an object rests on a surface (a table top or an inclined plane, for example) that is as rest. Then the force of static friction on the object is just sufficient to hold it at rest. Mathematically the frictional force is determined, via Newton's second law, by the condition that the component of acceleration parallel to the surface is zero. This condition is analogous to the condition used to determine the normal force. The difference is that the normal force of one object on another is perpendicular to the surface of contact while the frictional force is parallel to it.

The magnitude f_s of the force of static friction exerted by one surface on another must be less than a certain value, determined by the nature of the surfaces and by the magnitude of the normal force one surface exerts on the other. In particular, $f_s \leq$ _____, where μ_s is the coefficient of static friction and N is the magnitude of the normal force. If the force of friction required to hold the surfaces at rest with respect to each other is greater than the maximum allowed, then the surfaces slide over each other. Once this happens the magnitude of the force of friction is given by $f = \mu_k N$, where μ_k is called _____.

List some properties of the objects in contact that determine μ_s and μ_k:

_____

_____

List some properties that you might think determine μ_s and μ_k but in fact do not:

_____

_____

_____

The normal force that appears in the expressions for the force of kinetic friction and the maximum force of static friction must be computed for each situation using Newton's second law. As you know by now the magnitude of the normal force depends on the directions and magnitudes of other forces acting.

Air resistance. When an object moves in air the air exerts a force on it. When the relative speed of the object and air is so great that the air flow around the object is turbulent the force of the air on the object is proportional to the square of the relative speed. In fact, the magnitude of the force is given by:

$$D =$$

where v is _____, A is _____, and C is _____. The direction of the drag force is _____ .

When an object falls in air it approaches a constant speed, called the _____ speed. You can use Newton's second law to find a value for this speed. The force of gravity is mg, down and the drag force is $\frac{1}{2}C\rho Av^2$, up. At terminal speed these sum to zero. Thus

$$v_t =$$

The larger the combination $C\rho A$ the _____ the terminal speed and the shorter the time taken to reach that speed from rest.

You should realize that when the object is dropped from rest its acceleration is g at first. As it picks up speed the drag force increases, thereby reducing the acceleration. The object continues to gain speed but at a lesser rate. At terminal speed its acceleration vanishes. From then on the acceleration remains zero, the speed does not change, and the drag force remains constant.

Look at Table 6–1 to see the terminal speed of some objects in air.

Uniform circular motion. An object in uniform circular motion has a non-zero acceleration because the direction of its velocity changes with time. A force must be applied to the object in order to produce its acceleration. If m is the mass of the object, then the applied force must have magnitude $F =$ _____, where r is the radius of the orbit and v is the speed of the object. The force is directed _____ and, because of its direction, is called a _____ force. Acquire the habit of pointing out to yourself the object in the environment that exerts the force. It might be a string, for example.

If a force **F** is applied to an object of mass m traveling at speed v and the force is maintained perpendicular to the velocity then the object will travel with constant speed in a circle of radius $r =$ _____ . If the magnitude of the force is decreased, the radius of the path will _____ .

If you are sitting in a car without a restraining belt and the car rounds a curve, the force that pulls you around the curve with the car is provided by _____ between you and the car seat. If this force is not great enough you slide toward the outside of the curve; you are going to a larger-radius path. If the force is zero (a very slippery seat) you will travel in a straight line as the seat moves out from under you around the curve.

Forces of nature. There seem to be only a small number of fundamental forces in nature, from which all other forces can be derived. Give their names and for each list one or more examples of phenomena in which the force plays an important role.

Force: _____
Examples: _____
_____

Force: _____
Examples: _____
_____

Force: _____
Examples: _____
_____

Force: _____
Examples: _____
_____

Physicists believe that these forces are actually all different manifestations of the same more fundamental force. The weak and electromagnetic forces have been shown to be different aspects of the same force, called the electroweak force. Earlier, electric and magnetic forces were shown to be different aspects of the same force, now called the electromagnetic force. The number of forces you listed above depends on whether or not you took these connections into account.

II. PROBLEM SOLVING

Many problems of this chapter deal with frictional forces. Proceed as before: draw the free-body diagram and write down Newton's second law in component form, just as for any other problem. Use an algebraic symbol, f say, for the frictional force. You must now decide if the frictional force is static or kinetic. If static friction is involved, f is probably an unknown but the acceleration is known or is related to other known quantities in the problem. If the object is at rest on a stationary surface its acceleration is zero. If it is at rest relative to an accelerating surface its acceleration is the same as that of the surface. Kinetic friction is involved if one surface is sliding on the other. Then the magnitude of the frictional force is given by $\mu_k N$.

If you do not know that the object is at rest relative to the surface, assume it is and use Newton's second law, with the acceleration of the object equal to the acceleration of the surface, to calculate both the force of static friction f_{rest} that will hold it at rest and the normal force N. Compare f_{rest} with $\mu_s N$. If $f_{rest} < \mu_s N$ the object remains at rest relative to the surface and the force of friction has the value you computed. That is, $f = f_{rest}$. If $f_{rest} > \mu_s N$ then the object moves relative to the surface. Go back to the second law equations and set $f = \mu_k N$, then solve for the acceleration.

PROBLEM 1 (5E).   A 100-N force is applied at an angle θ above the horizontal to a 25.0-kg chair sitting on the floor. (*a*) For each of the following angles θ, calculate the magnitude of the normal force of the floor on the chair and the horizontal component of the applied force: (*i*) 0°, (*ii*) 30°, (*iii*) 60°.

STRATEGY: Use the vertical component of Newton's second law, along with the condition that the vertical component of the acceleration vanishes, to find values for the normal force. The horizontal component of the applied force, of course, is given by $F \cos \theta$, where F is its magnitude.

SOLUTION: Draw a free-body diagram for the chair. The force of gravity is mg, down; the normal force is N, up; the applied force is F, at the angle θ above the horizontal; the force of friction is f, horizontal and opposite the horizontal component of the applied force. The vertical component of Newton's second law is $N + F \sin \theta - mg = 0$. Solve for N. Calculate the horizontal component of the applied force.

$$\left[\text{ans: } 0°: 245\,\text{N}, 100\,\text{N}; 30°: 195\,\text{N}, 86.6\,\text{N}; 60°: 158\,\text{N}, 50.0\,\text{N}\right]$$

(*b*) Take the coefficient of static friction between the chair and the floor to be 0.420 and, for each of the values of θ, decide if the chair remains at rest or slides.

STRATEGY: If the chair remains at rest the horizontal component of the net force must vanish. This means $f = F \cos \theta$. For each angle compare $F \cos \theta$ with $\mu_s N$. If $F \cos \theta \leq \mu_s N$ the chair remains at rest. Otherwise the chair slides.

SOLUTION: The following table might help.

θ	$F \cos \theta$	$\mu_s N$	conclusion
_____	_____	_____	_____
_____	_____	_____	_____
_____	_____	_____	_____

$$\left[\text{ans: remains at rest; slides; remains at rest}\right]$$

Do you understand what is happening? When the applied force is horizontal the normal force is large. The horizontal component of the applied force has its greatest value for this angle but it is not enough to start the chair moving. When the angle is increased the normal force decreases, so the chair will start moving with a smaller horizontal component of the applied force. This component has been reduced, of course, but not as much as the normal force. At larger angles the horizontal component is reduced more for a given increase in angle than the normal force. Eventually the horizontal component of the applied force is too small to move the chair.

PROBLEM 2 (12E).   A 49-kg rock climber is climbing a "chimney" between two rock slaps as shown in Fig. 6–19. The static coefficient of friction between her shoes and the rock is 1.2; between her back and the rock it is 0.80. She has reduced her push against the rock until her back and her shoes are on the verge of slipping. (*a*) What is her push against the rock?

STRATEGY: The climber pushes against the rock with her back and the rock pushes on her back with a force of equal magnitude. The rock at her back provides a frictional force that helps hold her against the downward force of gravity. Since she is one the verge of slipping the magnitude of the force of friction on her back is given by $f_{back} = \mu_{back}N_{back}$, where N_{back} is the normal force of the rock on her back. A similar statement can be made about the forces at her feet: $f_{shoes} = \mu_{shoes}N_{shoes}$. Solve Newton's second law, with the acceleration equal to zero, for N_{back} and N_{shoes}.

SOLUTION: Draw a free-body diagram for the climber. Indicate and label the two normal forces N_{back} and N_{shoes}, the two forces of friction f_{back} and f_{shoes}, and the force of gravity mg. Take the x axis to be horizontal and the y axis to be vertical. Write Newton's second law in component form: $N_{shoes} - N_{back} = 0$ and $f_{shoes} + f_{back} - mg = 0$. Substitute $f_{shoes} = \mu_{shoes}N_{shoes}$ and $f_{back} = \mu_{back}N_{back}$, then solve for N_{shoes} and N_{back}.

[ans: each is 240 N]

(*b*) What fraction of her weight is supported by the frictional force on her shoes?

SOLUTION: Calculate f_{shoes}/mg.

[ans: 0.60]

PROBLEM 3 (17P).   A worker wishes to pile a cone of sand onto a circular area in his yard. The radius of the circle is R and no sand is to spill onto the surrounding area (Fig. 6–22). If μ_s is the static coefficient of friction between each layer of sand along the slope and the sand beneath it (along which it might slip), show that the greatest volume of sand that can be stored in this manner is $\pi\mu_s R^3/3$. (The volume of a cone is $Ah/3$, where A is the base area and h is the cone height.)

STRATEGY: As sand is piled on, the height of the cone increases and the sides become steeper. Eventually they are so steep the sand slides down them and the radius of the base increases beyond R. For a given base radius find the greatest height for which the sand does not slide.

SOLUTION: Draw a free-body diagram for a single grain of sand on the surface of the cone. Let θ be the angle made with the horizontal by the surface and label the force of gravity, the normal force, and the force of friction. Take the x axis to parallel to the surface and the y axis to be normal to the surface, at the grain. Take the acceleration of the grain to be zero and write Newton's second law in component form. If the surface is as steep as possible without the grain sliding then $f = \mu_s N$, where f is the magnitude of the frictional force and N is the magnitude of the normal force. Show that the second law and this condition lead to $\tan\theta = \mu_s$.

If h is the height of the cone then $\tan\theta = h/R$. Thus $h = R\tan\theta = R\mu_s$. Substitute $R\mu_s$ for h in the expression $V = \pi R^2 h/3$ for the cone volume. You should obtain the desired result.

PROBLEM 4 (20P).   A railroad flatcar is loaded with crates having a coefficient of static friction of 0.25 with the floor. If the train is moving at 48 km/h, in how short a distance can the train be stopped at a constant deceleration without causing the crates to slide?

STRATEGY: As the train decelerates, the force of friction on the crates causes them to decelerate. The maximum deceleration occurs when the force of friction has its maximum value: $f = \mu_s N$, where N is the normal force of the floor. Use this condition and Newton's second law to find the acceleration, then use the kinematic equations to find the stopping distance.

SOLUTION: Draw a free-body diagram for a crate. Show the normal force, the force of gravity, and the force of friction. Take the x axis to opposite the direction of motion and take the y axis to be vertical. Write Newton's second law in component form: $f = ma$ and $N - mg = 0$. Substitute $f = \mu_s N$ and solve for a.
Once a has been found use $2ax = v^2 - v_0^2$ to find the stopping distance $|x|$.

[ans: 36 m]

PROBLEM 5 (21P).   A block slides down an inclined plane of slope angle θ with constant velocity. It is then projected up the same plane with an initial speed v_0. (*a*) How far up the incline will it move before coming to rest?

STRATEGY: The problem centers on finding the force of friction. Its magnitude is, of course, the same for sliding up as for sliding down the incline so use the data given for sliding down to find it. Next find the acceleration when the crate is sliding up the incline and use it to find the stopping distance.

SOLUTION: Draw a free-body diagram for the crate when it is sliding down the incline. Label the force of friction; it is up the incline. Also label the normal force and the force of gravity. Take the x axis to be down the incline and the y axis to be normal to the incline. Write Newton's second law in component form: $mg\sin\theta - f = 0$ and $N - mg\cos\theta = 0$. Solve the first equation for f.

Now draw a free-body diagram for the crate sliding up the incline. It is the same as before except the force of friction is down the incline. Write Newton's second law: $f + mg \sin \theta = ma$ and $N - mg \cos \theta = 0$. Solve the first of these equations for a and substitute $mg \sin \theta$ for f. You should get $a = 2g \sin \theta$. Use $2ax = v^2 - v_0^2$ to find an expression for the stopping distance $|x|$.

[ans: $v_0^2 / 4g \sin \theta$]

(b) Will it slide down again? Give an argument to back your answer.

STRATEGY: The coefficient of kinetic friction can be found from the data given for sliding down the incline. It is $\mu_k = f/N = mg \sin \theta / mg \cos \theta = \tan \theta$. On the other hand, the angle for which sliding starts from rest is given by $\theta_s = \tan^{-1} \mu_s$, a somewhat greater angle because the coefficient of static friction is somewhat greater than the coefficient of kinetic friction.

[ans: no, it remains at rest]

To decide on the direction of a force of static friction, first decide which way the object would move if the force were absent. The frictional force is in the opposite direction. Consider an object on an inclined plane that is tilted so the object will slide down if you do not exert a force on it. Suppose, however, you pull on it with a force F that is parallel to the plane and directed up the plane. You will find that you can apply a fairly wide range of forces without having the object move. If F is small the force of friction is up the plane; if F is large the force of friction is down the plane. The static frictional force can have any value from $\mu_s N$ down the plane to $\mu_s N$ up the plane (including 0), depending on the value of F.

PROBLEM 6 (29P). A block weighing 80 N rests on a plane inclined at 20° to the horizontal (Fig. 6–29). The coefficient of static friction is 0.25 and the coefficient of kinetic friction is 0.15. (a) What is the minimum magnitude of the force **F**, parallel to the plane, that will prevent the block from slipping down the plane?

STRATEGY: When the applied force is a minimum the force of friction is up the plane (helping the applied force) and has its maximum value $\mu_s N$. Use Newton's second law, with the acceleration equal to zero, to solve for F.

SOLUTION: Draw a free-body diagram for the block. Indicate and label the force of friction f, the applied force F, the normal force N, and the force of gravity mg. Take the x axis to be up the plane and the y axis to be normal to the plane. Newton's second law yields $N - mg \cos \theta = 0$ and $f + F - mg \sin \theta = 0$. Use $f = \mu_s N$ and solve for F.

[ans: 8.6 N]

(*b*) What is the minimum magnitude F that will start the block moving up the plane?

STRATEGY: As F is increased the force of friction diminishes to zero, then it increases in the other direction (down the plane) until it reaches its greatest possible magnitude, $\mu_s N$. Then the block starts to slide. Repeat the calculation above, but with the force of friction down the plane.

[ans: 46 N]

(*c*) What value of F is required to move the block up the plane at constant velocity?

STRATEGY: The Newton's second law equations are exactly like those for part (*b*) (the frictional force is down the plane) but now $f = \mu_k N$.

SOLUTION:

[ans: 39 N]

PROBLEM 7 (32P). Two blocks are connected over a pulley as shown in Fig. 6–32. The mass of block A is 10 kg and the coefficient of kinetic friction is 0.20. Block A slides down the incline at constant speed. What is the mass of block B?

STRATEGY: Block A is pulled down the incline by the force of gravity and is restrained by the force of friction and the tension in the rope. The components of these forces parallel to the incline must sum to zero since the velocity of the block is constant. Similarly block B is pulled upward by the tension in the rope and is restrained by the force of gravity. These forces must also sum to zero. The tension in the rope depends on the mass of block B: the greater the mass the greater the tension. Thus the mass of B must be just right to bring about the proper tension to just balance all other forces on A.

SOLUTION: Draw a free-body diagram for block B. Show the force of gravity ($m_B g$, down) and the tension in the rope (T, up). Take the axis to be upward and write the Newton's second law equation: $T - m_B g = 0$.

Draw a free-body diagram for block A. Show the force of gravity ($m_A g$, down), the normal force (N, perpendicular to the plane), the force of friction (f, parallel to the plane, up the plane), and the tension in the rope (T, parallel to the plane, up the plane). Take the x axis to be parallel to the plane and down the plane; take the y axis to be perpendicular to the plane. Write the second law equations: $m_A g \sin\theta - T - f = 0$ and $N - m_A g \cos\theta = 0$, where θ is the angle of the plane above the horizontal.

Replace f with $\mu_k N$ and solve the 3 equations for m_B. From the first equation $T = m_B g$. Substitute this expression into the other two equations. You should now have $N - m_A g \cos\theta = 0$ and $m_A g \sin\theta - m_B g -$

$\mu_k m_A g \cos \theta = 0$. Substitute $N = m_A g \cos \theta$, from the first, into the second and solve for m_B. You should get $m_B = (\sin \theta - \mu_k \cos \theta)m_A$. Evaluate this expression.

[ans: 3.3 kg]

In some situations the surface of contact is moving. Consider, for example, a crate on the bed of a moving pick-up truck. If the crate and truck move together, the force of friction acting on the crate is whatever is necessary to give the crate the same acceleration as the truck. This must be less than $\mu_s N$, where N is the normal force of the truck on the crate. The next problem is a similar example. In the one after that the surface is vertical but is again moving. Remember that frictional forces are parallel to the surface where they are exerted. If the surface is vertical, the force of friction is vertical.

PROBLEM 8 (36P). A 4.0-kg block is put on top of a 5.0-kg block. To cause the top block to slip on the bottom one, which is held fixed, a horizontal force of at least 12 N must be applied to the top block. The assembly of blocks is now placed on a horizontal, frictionless table (Fig. 6–35). Find (*a*) the magnitude of the maximum horizontal force F that can be applied to the lower block so that the blocks will move together and (*b*) the resulting acceleration of the blocks.

STRATEGY: When the force is applied to the lower block the force of friction acting on the upper block causes it to accelerate. The two blocks have the same acceleration as long as the force of friction required to keep them together is less than $\mu_s N$, where N is the normal force of the lower block on the upper block. You need to find the coefficient of static friction μ_s. The information given in the first part of the problem statement enables you to do this.

SOLUTION: Draw a free-body diagram for the upper block. Let F_0 be the applied force. It is horizontal. Also include the normal force N of the lower block on the upper block, the force of gravity $m_u g$, and the force of friction f. Take the x axis to be horizontal and the y axis to be vertical. Write Newton's second law. Substitute $f = \mu_s N$ and solve for μ_s. You should get 0.306.

Now draw the free-body diagram for the upper box when the force F is applied to the lower box. The forces acting are the force of friction f, the normal force N, and the force of gravity $m_u g$. Newton's second law equations are $f = m_u a$ and $N - m_u g = 0$. The greatest acceleration occurs if f has the greatest possible value. Replace f with $\mu_s N = \mu_s m_u g$ and solve for a. You should get $a = \mu_s g$.

Now calculate the force applied to the lower box. Consider the two blocks as a single object, so $F = (m_u + m_l)a$, where m_l is the mass of the lower block. Calculate values for F and a.

[ans: 27 N; 3.0 m/s^2]

PROBLEM 9 (38P).   Two blocks (with $m = 16\,\mathrm{kg}$ and $M = 88\,\mathrm{kg}$) shown in Fig. 6–37 are not attached. The coefficient of static friction between the blocks is $\mu_s = 0.38$, but the surface beneath M is frictionless. What is the minimum magnitude of the horizontal force **F** required to hold m against M?

STRATEGY: A large force F clearly holds m against M. The force of static friction between the blocks balances the force of gravity on the left-hand block. If F is reduced the force of the left block on the right block is reduced, the reaction force of the right block on the left is reduced, and the maximum allowable force of static friction is reduced. When it becomes less than the weight of the left block, that block falls. If the minimum force is applied the force of friction is given by $f = \mu_s N$, where N is the (horizontal) normal force of the right block on the left.

SOLUTION: Draw a free-body diagram for the left block. The forces acting on this block are the force of gravity mg (down), the force of friction f (up), the applied force F (right), and the normal force N exerted by the right block (left). Take the x axis to be horizontal and the y axis to be vertical. Write the Newton's second law equations: $F - N = ma$ and $f - mg = 0$. Since $f = mg$ and $f = \mu_s N$, $N = mg/\mu_s$ and $F - mg/\mu_s = ma$.

Now consider the two blocks as a single object and write $F = (m + M)a$. Solve this equation and $F - mg/\mu_s = ma$ simultaneously for F.

[ans: 490 N]

Uniform circular motion problems are solved in much the same way as any other second law problem. Carry out the set of instructions given in the previous chapter of this Student's Companion. Draw a free-body diagram. Place the coordinate system so one of the axes is in the direction of the acceleration, pointing from the object toward the center of its orbit. For most problems you will want to substitute v^2/r for the magnitude of the acceleration. Here v is the speed of the object and r is the radius of its orbit.

To see how it is done you should carefully study the examples discussed in the text. Always identify the source of the centripetal force that pulls the object around the circle. For a conical pendulum it is the horizontal component of the tension in the string; for a car rounding an unbanked road it is the frictional force of the road on the tires; and for a car rounding a banked curve it is, perhaps in addition, the horizontal component of the normal force of the road on the car.

PROBLEM 10 (50E).   A car weighing $10.7\,\mathrm{kN}$ and traveling at $13.4\,\mathrm{m/s}$ attempts to round an unbanked curve with a radius of $61.0\,\mathrm{m}$. (*a*) What force of friction is required to keep the car on its circular path?

STRATEGY: The force of friction is the only force with a horizontal component acting on the car. It must produce the appropriate acceleration v^2/r to hold the car on the curve.

SOLUTION: Evaluate $f = mv^2/r$.

[ans: 3210 N]

(*b*) If the coefficient of static friction between the tires and road is 0.35, is the attempt at taking the curve successful?

STRATEGY: Since the sum of the forces in the vertical direction must vanish, the normal force is given by $N = mg$. Compare f with $\mu_s N$.

SOLUTION:

$$\left[\text{ans: } f < \mu_s N \text{ so the attempt is successful}\right]$$

PROBLEM 11 (52E).   A banked circular highway curve is designed for traffic moving at 60 km/h. The radius of the curve is 200 m. Traffic is moving along the highway at 40 km/h on a stormy day. What is the minimum coefficient of friction between the tires and road that will allow cars to negotiate the turn without sliding off the road?

STRATEGY: If the traffic were going at 60 km/h it could negotiate the curve without the aid of friction. It is going slower so we expect, in the absence of friction, a car would slide toward the center of the circle. That is, the horizontal component of the normal force is too large for uniform circular motion. Thus the force of friction must be up the incline. If the coefficient of friction is the smallest possible the cars must be on the verge of slipping at this speed, so the magnitude of the frictional force is given by $f = \mu_s N$, where N is the normal force. Solve Newton's second law for μ_s. You will need to know the banking angle. It can be found knowing that friction is not needed at 60 km/h.

SOLUTION: First find the banking angle. Draw a free-body diagram of a car going at 60 km/h on a road banked at the angle θ. Only two forces act: the normal force N and the force of gravity mg. Take the x axis to be horizontal and positive toward the center of the circle. Take the y axis to vertical. Write the Newton's second law equations: $N \cos\theta - mg = 0$ and $N \sin\theta = mv^2/r$. Solve for θ. You should get 8.07°.

Now draw a free-body for a car going at 40 km/h on the same road. In addition to the other forces a force of friction f is exerted up the incline. Use the same coordinate system and write the second law equations: $N \cos\theta + f \sin\theta - mg = 0$ and $N \sin\theta - f \cos\theta = mv^2/r$. Substitute $\mu_s N$ for f and solve for μ_s.

To do this use $N(\cos\theta + \mu_s \sin\theta) - mg = 0$ to find an expression for N. Substitute the expression into $N(\sin\theta - \mu_s \cos\theta) = mv^2/r$ and solve for μ_s. You should get $\mu_s = [g \sin\theta - (v^2/r) \cos\theta]/[g \cos\theta + (v^2/r) \sin\theta]$. Don's forget to convert 40 km/h to m/s when evaluating this expression.

$$\left[\text{ans: } 0.078\right]$$

PROBLEM 12 (57E).   A stuntman drives a car over the top of a hill, the cross section of which can be approximated by a circle of radius 250 m, as in Fig. 6–40. What is the greatest speed at which he can drive without the car leaving the road at the top of the hill?

STRATEGY: At the top of the hill the forces acting on the car are the force of gravity (down) and the normal force of the road (up). These must combine to give the product of the mass and acceleration (mv^2/r, down). If the car goes faster, a greater centripetal force is required to get it over the hill. The normal force becomes less and the net force becomes greater. A limit is reached, however, when the normal force is zero. Then a downward force greater than mg is required but the road can only push up on the car, not pull down. At any speed greater than this the car leaves the road.

SOLUTION: Draw a free-body diagram for the car at the top of the hill. Show the normal force (N, up) and the force of gravity (mg, down). Take the axis to positive in the downward direction and write Newton's second law: $mg - N = mv^2/r$. Set $N = 0$ and solve for v.

$$\left[\text{ans: } 49.5\,\text{m/s } (178\,\text{km/h})\right]$$

PROBLEM 13 (58P). A small coin is placed on a flat, horizontal turntable. The turntable is observed to make three revolutions in 3.14 s. (*a*) What is the speed of the coin when it rides without slipping at a distance 5.0 cm from the center of the turntable?

STRATEGY: Find the distance traveled in 3.14 s and divide by the time. Since the circumference of a circle is given by $2\pi r$, where r is the radius, the distance is $6\pi r$ in 3.14 s. Here r is the distance from the center of the turntable to the coin.

SOLUTION:

$$\left[\text{ans: } 0.30\,\text{m/s}\right]$$

(*b*) What is the acceleration (magnitude and direction) of the coin?

SOLUTION: Use $a = v^2/r$.

$$\left[\text{ans: } 1.8\,\text{m/s}^2, \text{ horizontally toward the center of the turntable}\right]$$

(*c*) What is the magnitude of the frictional force acting on the coin if the coin has a mass of 2.0 g?

SOLUTION: The force of friction is the only force with a horizontal component. It alone is responsible for the acceleration. Use $f = ma$.

$$\left[\text{ans: } 3.6 \times 10^{-3}\,\text{N}\right]$$

(*d*) What is the coefficient of static friction between the coin and the turntable is the coin is observed to slide off the turntable when it is more than 10 cm from the center of the turntable?

STRATEGY: Calculate the force of friction required to hold the coin on the turntable in a circular path with a radius of 10 cm. The calculation is exactly like that above, only the radius is changed. The normal force of the turntable on the coin is given by $N = mg$. Since the coin is on the verge of slipping $\mu_s = f/N$.

SOLUTION:

$$\left[\text{ans: } 0.37\right]$$

PROBLEM 14 (63P).   A stone at the end of a string is whirled around in a vertical circle of radius R. Find the critical speed below which the string would become slack at the highest point.

STRATEGY: Two forces act on the stone: the force of gravity (down) and the tension in the string. At the top of the swing both point downward, toward the center of the circular path. When the stone is moving fast both forces are required to get it around the path; the force of gravity is not enough. As it slows down less force is required and the tension in the string is reduced. For some speed the force of gravity, acting alone, is just right. The stone cannot go any slower and get around the circle since the force of gravity is then too large and the string would need to push out on the stone at the top of the swing. The string can only pull, not push, so the stone falls and the string becomes slack.

SOLUTION: Draw a free-body diagram for the stone at the top of the swing. Show the force of gravity and the tension in the string. Take the axis to be positive toward the center of the circular path and write the second law equation: $mg - T = mv^2/R$, where T is the tension in the string. Set $T = 0$ and solve for v.

$$\left[\,\text{ans: } v = \sqrt{gR}\,\right]$$

PROBLEM 15 (67P).   A model airplane of mass 0.75 kg is flying at constant speed in a horizontal circle at one end of a 30-m cord and at a height of 18 m. The other end of the cord is tethered to the ground. The airplane circles 4.4 times per minute, and has its wings horizontal so that the air is pushing vertically upward. (*a*) What is the acceleration of the plane?

STRATEGY: Use $a = v^2/R$, where v is its speed and R is the radius of its circular path.

SOLUTION: The path radius R, the path height h, and the cord length L form a right triangle with the cord as the hypotenuse, so $R = \sqrt{L^2 - h^2}$. You should get $R = 24$ m. Since the circumference of a circle is given by $2\pi R$ and the plane goes around 4.4 times in 60 s, its speed is $4.4 \times 2\pi R/60 = 11.1$ m/s. Now you can calculate a.

$$\left[\,\text{ans: } 5.1\text{ m/s}^2\text{, radially inward}\,\right]$$

(*b*) What is the tension in the cord?

STRATEGY: The force of the cord on the plane is the only force with a horizontal component, so this component must be responsible for the acceleration.

SOLUTION: Draw a free-body diagram for the plane. Show the force of gravity (mg, down), the lifting force of the air on the wings (F_a, up), and the tension in the cord (T, at an angle θ below the horizontal). Take the x axis to be horizontal and positive toward the center of the circular path. Take the y axis to vertical. Write the Newton's second law equations: $T\cos\theta = ma$ and $F_a - T\sin\theta - mg = 0$. Solve the first equation for T. The right triangle formed by the cord length, the path radius, and the path height gives $\cos\theta = R/L$.

$$\left[\,\text{ans: } 4.8\text{ N}\,\right]$$

(*c*) What is the total upward force (*lift*) on the plane's wings?

SOLUTION: Solve the second of the Newton's second law equations for F_a. Use $\sin \theta = h/L$.

$$\left[\, \text{ans: } 10\,\text{N} \,\right]$$

III. NOTES

Chapter 7
WORK AND KINETIC ENERGY

I. BASIC CONCEPTS

The central concept of this chapter is the idea of <u>work</u>. You should learn and understand the definition of work, you should learn how to calculate the work done by forces in various situations, and you should learn how work is related to the change in the kinetic energy of a particle (the work-energy theorem).

Definition of work. Several definitions of work are given, for situations of increasing complexity. You should remember them and the situations to which they apply.

1. The particle moves through a displacement **d**. The force **F** being considered is constant and makes the angle ϕ with the displacement when **F** and **d** are drawn with their tails at the same point. Then the work done by **F** is given by

   $$W =$$

   Alternatively, if the particle has displacement Δx, along the x axis, and the force has x component F_x then the expression for the work can be written

   $$W =$$

2. The particle moves along a straight line (the x axis). The force **F** is parallel to the x axis and is not constant. Then the work it does as the particle moves from x_1 to x_2 is given by the integral

   $$W =$$

   Be sure your definition allows for a force in the same direction as the displacement and for one in the opposite direction. To evaluate the integral, the x component of the force must be known as a function of _____. The work done by the force is the area under the graph of _____ vs. _____.

3. A particle moves in a plane, subjected to a variable force **F**. Write the integral definition of the work here:

   $$W =$$

   Explain how the integral can be evaluated, in principle, by dividing the path into a large number of segments. Don't forget to give the quantity to be evaluated for each segment.

   _____

   _____

This expression, generalized slightly to three dimensions, is the general definition of work. The expressions you wrote above in 1 and 2 for special situations can be derived from it.

If several forces act on the particle you may calculate the work done by each separately. The total work done by all forces is the sum of the individual works and is the same as the work done by the resultant force.

A person carrying a heavy box horizontally with constant velocity does no work on the box because _____.
The normal force of a stationary surface on a sliding crate does no work, no matter what the orientation of the surface, because _____

_____.

A string used to whirl an object around a circle with constant speed does no work because

_____.

Work can be positive or negative. Each of the four diagrams below show a block moving on a table top. On each of the upper two show how you would apply a force so it does positive work. On each of the lower two show how you would apply a force so it does negative work. In each case direct the force so it is not parallel to the velocity and assume the block continues to move in the same direction, at least for a short while after the force is applied.

Force **F** does
positive work

Force **F** does
negative work

Work is a scalar. It does not have a direction associated with it. When several forces act on an object and you want to find the total work done, you simply add the works done by the individual forces. You must, of course, include the appropriate sign for each work. The direction of each force and the direction of the displacement are important for calculating the work done by a force but neither these directions or any others must be taken into account when the individual works are summed.

The SI unit of work is _____. In terms of SI base units it is _____. To discuss atomic phenomena physicists often use a unit of work called an electron volt. In terms of the SI unit for work 1 eV is _____ J.

Work done by gravity. You should know how to calculate the work done by the force of gravity and the work done by the force of an ideal spring. In addition to being excellent examples of a constant and a variable force, respectively, they enter many problems.

When an object of mass m falls from height y_i to height y_f near the earth's surface gravity does work $W =$ _____. When the mass is raised from height y_i to height y_f gravity does work $W =$ _____. The two expressions you wrote should be identical. On the way down the sign of the work done by gravity is _____ while on the way up it is _____. If a ball is thrown into the air, falls, and is caught at the height from which it was thrown, the total work done by gravity on the round trip is _____.

You should know that the work done by gravity is the same no matter what path is taken between the initial and final points. In addition, all that counts is the initial and final altitudes. The two positions need not have the same horizontal coordinate.

Remember that the expression you wrote for the work done by gravity is valid even if the object experiences air resistance. If air resistance is present this expression does not, of course, give the total work done by all forces.

Work done by an ideal spring. The force exerted by an ideal spring on an object is a variable force. Its direction depends on whether the spring is extended or compressed and its magnitude depends on the amount of extension or compression.

Assume the spring is along the x axis with one end fixed and the other attached to the object, as shown. When the object is at $x = 0$ the spring is neither extended or compressed. This is the equilibrium configuration. When the object is at any coordinate x, the force exerted by the spring is given by $F_s =$ _____. Here k is the <u>spring constant</u> of the spring. The _____ the spring, the larger the spring constant.

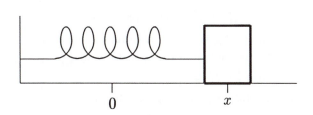

When the spring shown above is extended the sign of the force it exerts is _____ and the force tends to pull the object toward _____. When the spring is compressed the sign of the force is _____ and the force tends to push the object toward _____. This behavior is summarized by calling the force a <u>restoring force</u>.

Sample Problem 7–7 of the text shows how the spring constant of an ideal spring can be measured by applying a force F to the mass and measuring the elongation x of the spring with the mass at rest. In terms of F, x, and k the net force on the mass is given by _____ and since the mass is in equilibrium this must be zero. Thus $k =$ _____.

The SI units of a spring constant are _____.

As the object moves from some initial coordinate x_i to some final coordinate x_f the work done by the spring is

$$W_s = \int_{x_i}^{x_f} -kx \, dx =$$

Give an example in which the spring does positive work: _____

_____

Give an example in which the spring does negative work: _____

_____

In general, the spring does positive work whenever the object attached to it is moving toward the equilibrium point and does negative work whenever the object is moving away from the equilibrium point.

Suppose a mass is attached to a horizontal spring and is free to move on a frictionless table top, as in the diagram above. Now you pull on the mass with a force given by $F_{\text{ext}} = +kx$.

As the mass goes from $x = x_i$ to $x = x_f$ the work you do is given by $W_{ext} =$ _____.
Of course, you may pull on the mass with any force you like, not necessarily $+kx$. The external force $+kx$, however, has special significance because the net force on the mass is then _____ and the mass does not accelerate during the pulling. When this external force is applied, the total work done by the spring and your force, taken together, is _____.

Work-energy theorem. The significance of work is found in the work-energy theorem, which shows us that work tends to change the kinetic energy of a particle. First, the kinetic energy of a particle with mass m and speed v is defined by

$$K =$$

Kinetic energy is a scalar and so does not have a direction associated with it. Sometimes you will be given the velocity components v_x and v_y for an object moving in the xy plane and asked to compute the kinetic energy. In terms of the velocity components the kinetic energy is given by $K =$ _____. You cannot interpret the two terms in this expression as components of a vector. They are not.

The work-energy theorem is: during any portion of a particle's motion the *total* work done by *all* forces acting on the particle equals the change in the particle's kinetic energy. Let W_{total} be the total work done on a particle during some portion of its motion. If K_i is the initial kinetic energy and K_f is the final kinetic energy for that portion, then $W_{total} =$ _____. Be sure you perform the subtraction in the correct order.

If the total work is negative, then the kinetic energy and speed of the particle both _____ _____; if the total work is positive, then the kinetic energy and speed _____; and if the total work is zero then the kinetic energy and speed _____. Go back and review the exercise in Chapter 5 of this Student's Companion showing that a particle speeds up when the acceleration has a vector component in the direction of the velocity and slows down when the acceleration has a vector component opposite the velocity. In the first case the net force is doing positive work and in the second it is doing negative work.

You should be able to carry out the proof of the work-energy theorem for a particle moving in one dimension. Start with $F = ma$ and consider the velocity v to be a function of coordinate x, which in turn is a function of time t. Use the chain rule of the calculus to show that $F = mv\,dv/dx$:

Then substitute into $W = \int_i^f F\,dx$ and show that $W = \int_i^f mv\,dv$:

Finally evaluate the integral to show that $W = \frac{1}{2}mv_f^2 - \frac{1}{2}mv_i^2$:

This derivation shows that the work-energy theorem is a direct consequence of Newton's second law. It also explains why the *total* work (done by *all* forces acting on the particle) enters the theorem.

If two observers use different inertial frames (moving relative to each other) to measure the change in kinetic energy of an object over some interval, they will obtain different values. They will also obtain different values for the total work done on the object, but the work-energy theorem will still be valid for both frames. Each will conclude that the change in kinetic energy equals the total work done.

You should understand that the work-energy theorem is valid only for a single particle or an object that can be treated as a particle. If the object becomes distorted or the positions of individual particles within the object change relative to each other then the theorem cannot be applied directly to the object as a whole.

Power. In words, <u>power</u> is _____ .
Suppose the function $W(t)$ represents the work done by a force from time 0 to time t. Then the instantaneous power delivered by the force is given by $P = $ _____. The SI unit of power is _____. In terms of SI base units the unit for power is _____ .

Consider a particle that at some instant of time is moving with velocity **v** and is acted on by a force **F**. The power delivered to the particle by the force is given by $P = $ _____. Be sure your equation is valid even if the force is not parallel to the velocity.

Relativistic kinetic energy. When an object is moving with a speed near the speed of light c another definition of kinetic energy must be used. If m is the mass and v is the speed then the kinetic energy is defined by

$$K = $$

At speeds that are much less than the speed of light this expression reduces to $K = $ _____ .

II. PROBLEM SOLVING

You should know how to calculate the work done by a force if the force is constant or if its component along the path is given as a function of the object's position. In some cases you might need to solve a Newton's second law equation to find the force. In other cases the force as a function of position might be given as a graph and you should be able to obtain the work from the area under the curve. The force in question might be the only force acting on an object or one of several.

PROBLEM 1 (5E).   A 45-kg block of ice slides down a frictionless incline 1.5 m long and 0.91 m high. A worker pushes up on the ice parallel to the incline so that it slides down at constant speed. (*a*) Find the force exerted by the worker.

STRATEGY: Newton's second law can be used to find the force. Since the block moves with constant velocity the net force is 0.

SOLUTION: Draw a free-body diagram for the block. Show the force of the worker (F, up the plane), the force of gravity (mg, down), and the normal force (N, perpendicular to the surface). Choose the x axis to be parallel to the plane and the y axis to be perpendicular to it. Write the second law equations: $F - mg \sin\theta = 0$ and $N - mg \cos\theta = 0$. Solve for F. A little trigonometry shows that $\sin\theta = h/L$, where h is the height of the incline and L is its length.

$$\left[\text{ans: } 270\,\text{N}\right]$$

Find the work done on the block by (*b*) the worker, (*c*) the weight of the block, (*d*) the normal force exerted by the surface of the incline on the block, and (*e*) the resultant (or net) force on the block.

STRATEGY: Each of the forces is constant so you can calculate the work by multiplying the magnitude of the displacement (1.5 m) by the component of the force along the direction of the displacement. For the force of the worker this component is $-270\,\text{N}$, for the force of gravity it is $mg\sin\theta$ (or mgh/L), for the normal force it is 0, and for the resultant force it is 0.

SOLUTION:

$$\left[\text{ans: } -400\,\text{J}; \ +400\,\text{J}; \ 0; \ 0\right]$$

PROBLEM 2 (7E).   A particle moves along a straight path through a displacement $\mathbf{d} = (8\,\text{m})\mathbf{i} + c\mathbf{j}$ while a force of $\mathbf{F} = (2\,\text{N})\mathbf{i} - (4\,\text{N})\mathbf{j}$ acts on it. (Other forces also act on the particle.) What is the value of c if the work done by $\mathbf{F}$ on the particle is (*a*) zero, (*b*) positive, and (*c*) negative?

STRATEGY: Evaluate the expression $W = \mathbf{F} \cdot \mathbf{d} = F_x d_x + F_y d_y$ in terms of c then solve for c or the range of values it might have for each of the desired conditions.

SOLUTION: The result of the scalar product is $W = 16 - 4c$.

$$\left[\text{ans: } (a)\ c = 4\,\text{m}; \ (b)\ c < 4\,\text{m}; \ (c)\ c > 4\,\text{m}\right]$$

PROBLEM 3 (8E).   In Fig. 7–26, a cord runs around two massless, frictionless pulleys; a canister with mass $m = 20\,\text{kg}$ hangs from one pulley; and you exert a force $\mathbf{F}$ on the free end of the cord. (*a*) What must be the magnitude of $\mathbf{F}$ if you are to lift the canister at a constant speed?

STRATEGY: The tension in the cord is uniform so the cord pulls on the movable pulley with a force of magnitude $2F$, up. This force is transmitted to the canister. The force of gravity pulls down on the canister with a force of magnitude mg. Since the canister moves with constant velocity, these forces must have equal magnitudes.

SOLUTION:

$$\left[\text{ans: } 98\,\text{N}\right]$$

(b) To lift the canister by 2.0 cm, how far must you pull the free end of the cord?

STRATEGY: If the canister rises a distance d the portion of the cord between the movable pulley and the ceiling is shorter by d; the portion of the cord between the two pulleys is also shorter by the same amount.

SOLUTION:

[ans: 4.0 cm]

During that lift, what is the work done on the canister by (c) you and (d) the weight mg of the canister?

STRATEGY: You actually do work on the rope, pulling on it with a force of 98 N as it moves 4.0 cm. The work done by the force of gravity is given by $-mgd$, where d is the distance the canister moves.

SOLUTION:

[ans: (c) 3.9 J; (d) −3.9 J]

PROBLEM 4 (13E). A 5.0-kg block moves in a straight line on a horizontal frictionless surface under the influence of a force that varies with position as shown in Fig. 7–28. How much work is done by the force as the block moves from the origin to $x = 8.0$ m?

STRATEGY: Calculate the area under the curve, taking care about the sign of the work.

SOLUTION: From $x = 0$ to $x = 2$ m the force is constant and the work done is the product of the force and the magnitude of the displacement: $10 \times 2.0 = 20$ J. The area under the curve from $x = 2.0$ m to $x = 4.0$ m can be computed if you know that the area of a triangle is half the product of the base and altitude: $\frac{1}{2} \times 10 \times 2.0 = 10$ J. From $x = 4.0$ m to $x = 6.0$ m the force is 0 and does no work. From $x = 6.0$ m to $x = 8.0$ the force is opposite to the displacement and the work is negative. Use the formula for the area of a triangle. Add the results for the various segments.

[ans: 25 J]

Now consider the work done by a force when the motion is in a plane.

PROBLEM 5 (17E).   What work is done by a force $\mathbf{F} = (2x\,\mathbf{i} + 3\,\mathbf{j})$ N, where x is in meters, that is exerted on a particle while it moves from a position of $\mathbf{r}_i = (2\,\text{m})\,\mathbf{i} + (3\,\text{m})\,\mathbf{j}$ to a position of $\mathbf{r}_f = -(4\,\text{m})\,\mathbf{i} - (3\,\text{m})\,\mathbf{j}$?

STRATEGY: The path is not given so we will pick one. Suppose the particle moves along the line $y = 3$ m from $x = 2$ m to $x = -4$ m, then along the line $x = -4$ m from $y = 3$ m to $y = -3$ m. Along the first portion the work done is $W_1 = \int_2^{-4} F_x\,dx$, while along the second the work done is $W_2 = \int_3^{-3} F_y\,dy$. The total work, of course, is the sum of W_1 and W_2.

SOLUTION:

[ans: −6 J]

For this force you will get the same result for any path between the same two points. You might want to try some other paths.

A spring provides a good example of a variable force. If one end is fixed and the other end is moved so the spring is either extended or compressed from its equilibrium length, then the force exerted by the spring is given by $F = -kx$, where k is the spring constant. The coordinate x of the spring end is measured with the origin at the position of the movable end when the spring has its equilibrium length. If the end of the spring is moved from x_i to x_f the work done by the spring is $-\frac{1}{2}(x_f^2 - x_i^2)$.

PROBLEM 6 (18E). A spring with a spring constant of 15 N/cm has a cage attached to one end, as in Fig. 7–31. (*a*) How much work does the spring do on the cage if the spring is stretched from its relaxed length by 7.6 mm? (*b*) How much additional work is done by the spring if it is stretched by an additional 7.6 mm?

STRATEGY: Use $W = -\frac{1}{2}k(x_f^2 - x_i^2)$, where for (*a*) $x_i = 0$ and $x_f = 7.6$ mm and for (*b*) $x_i = 7.6$ mm and $x_f = 15.2$ mm. Convert $k = 15$ N/cm to N/m.

SOLUTION:

$$\left[\text{ans: } (a) -4.3 \times 10^{-2}\,\text{J}; (b) -0.13\,\text{J}\right]$$

Notice that because the force is not uniform the work done is not the same during the two displacements although the displacements are the same.

The work-energy theorem tells us that the total work W done on a particle is equal to the change in the kinetic energy of the particle. That is, $W = \Delta K$, or since the kinetic energy is given by $\frac{1}{2}mv^2$, $W = \frac{1}{2}m(v_f^2 - v_i^2)$. Here m is the mass of the particle, v_i is its speed at the beginning of the interval, and v_f is its speed at the end of the interval.

PROBLEM 7 (27E). A firehose (Fig. 7–34) is uncoiled by pulling the end of the hose horizontally along a frictionless surface at the steady speed of 2.3 m/s. The mass of 1.0 m of the hose is 0.25 kg. How much kinetic energy is imparted in uncoiling 12 m of hose?

STRATEGY: At the beginning none of the hose is moving; its kinetic energy is 0. At the end 12 m of hose is moving with the same speed and the rest of the hose, if there is any, is at rest. The kinetic energy is $\frac{1}{2}mv^2$, where m is the mass of 12 m of hose: $m = 12 \times 0.25 = 3.0$ kg.

SOLUTION:

$$\left[\text{ans: } 7.9\,\text{J}\right]$$

You should recognize that this is also the work that must be done to bring 12 m of hose from rest to 2.3 m/s.

PROBLEM 8 (28E). From what height would an automobile weighing 12,000 N have to fall to gain the kinetic energy equivalent to what it would have when going 89 km/h? Why does the answer not depend on the weight of the car?

STRATEGY: The kinetic energy is given by $K = \frac{1}{2}mv^2$. To gain this energy during a fall from rest the force of gravity must do work equal to K. Since the work done by gravity is given by mgh, where h is the distance fallen, $K = mgh$. Solve for h. Don't forget to convert $v = 89$ km/h to m/s.

SOLUTION:

$$\left[\text{ans: } 31\,\text{m}\right]$$

Notice that both the work done by gravity and the final kinetic energy are proportional to the mass of the car. Thus m cancels from the equation and the height fallen to gain a given kinetic energy does not depend on the mass.

PROBLEM 9 (33P).   A force acts on a 3.0-kg particle in such a way that the position of the particle as a function of time is given by $x = 3.0t - 4.0t^2 + 1.0t^3$, where x is in meters and t is in seconds. Find the work done by the force from $t = 0$ to $t = 4.0$ s.

STRATEGY: Differentiate the coordinate with respect to time to find an expression for the velocity. Evaluate the expression for $t = 0$ to find the initial velocity v_i, evaluate it for $t = 4.0$ s to find the final velocity v_f, then use the work-energy theorem: $W = K_f - K_i$.

SOLUTION:

$$[\text{ans: } 530 \, \text{J}]$$

PROBLEM 10 (37P).   A crate with a mass of 230 kg hangs from the end of a 12.0-m rope. You push horizontally on the crate with a varying force **F** to move it 4.00 m to the side (Fig. 7–36). (*a*) What is the magnitude of **F** when the crate is in this final position?

STRATEGY: The crate is at rest so the vector sum of the forces acting on it is 0. Use this condition to find **F**.

SOLUTION: Draw a free-body diagram of the crate in its final position. Suppose the rope makes the angle θ with the horizontal. Show the applied force (F, horizontal), the force of gravity (mg, down), and the tension in the rope (T, at θ above the horizontal). Write the Newton's second law equations: $F - T\cos\theta = 0$ and $T\sin\theta - mg = 0$. Eliminate T and solve for F. Since the rope is 12 m long and has been pushed aside 4.00 m, $\cos\theta = 4.00/12 = 0.333$ and $\sin\theta = \sqrt{1 - \cos^2\theta} = 0.943$.

$$[\text{ans: } 797 \, \text{N}]$$

During the crate's displacement, what are (*b*) the total work done on it, (*c*) the work done by the weight of the crate, and (*d*) the work done by the pull on the crate from the rope?

STRATEGY: The crate starts and ends at rest. Use the work-energy theorem to find the total work done on it. The work done by the force of gravity is $-mgh$, where h is the vertical distance through which the crate is lifted. A little geometry should show you that $h = L(1 - \sin\theta)$, where L is the length of the rope. The pull of the rope is always perpendicular to the circular path followed by the crate.

SOLUTION:

$$[\text{ans: } (b) \, 0; \, (c) \, -1550 \, \text{J}; \, (d) \, 0]$$

(e) From the answers to (b), (c), and (d) and the fact that the crate is motionless before and after its displacement, find the work you do on the crate.

STRATEGY: The net work is the sum of the work done by you, by gravity, and by the pull of the rope. It must be 0.

SOLUTION:

[ans: $+1550\,\mathrm{J}$]

(f) Why is your work not equal to the product of the horizontal displacement and the answer to (a)?

SOLUTION: Recall that $W = \mathbf{F} \cdot \mathbf{d}$ gives the work done by $\mathbf{F}$ only if the force is *constant*. A little thought should convince you that a constant force cannot be used to swing the crate aside *and* have it stop.

[ans: the force is variable]

PROBLEM 11 (38P). A 250-g block is dropped onto a vertical spring with spring constant $k = 2.5\,\mathrm{N/cm}$ (Fig. 7–37). The block becomes attached to the spring, and the spring compresses 12 cm before momentarily stopping. While the spring is being compressed, what work is done (a) by the block's weight and (b) by the spring?

STRATEGY: Let x be the distance the spring is compressed. Then the work done by gravity is given by $W_g = mgx$ and the work done by the spring is given by $W_s = -\frac{1}{2}kx^2$.

SOLUTION:

[ans: (a) $0.29\,\mathrm{J}$; (b) $-1.80\,\mathrm{J}$]

(c) What was the speed of the block just before it hits the spring?

STRATEGY: Use the work-energy theorem. The initial kinetic energy is $K_i = \frac{1}{2}mv_i^2$, where v_i is the speed of the block just before it hits the spring. The final kinetic energy is 0. The net work is the sum of the work done by the spring and the work done by gravity: $W_{\mathrm{net}} = W_s + W_g = -1.80 + 0.29 = -1.51\,\mathrm{J}$.

SOLUTION:

[ans: $3.5\,\mathrm{m/s}$]

(d) If the speed at impact is doubled, what is the maximum compression of the spring? Assume the friction is negligible.

STRATEGY: The work-energy theorem yields $mgx - \frac{1}{2}kx^2 = -\frac{1}{2}mv_i^2$, where $v_i = 7.0\,\mathrm{m/s}$. Solve this quadratic equation for x.

SOLUTION:

[ans: $23\,\mathrm{cm}$]

Some problems involve calculations of the power delivered by forces. In some cases you evaluate $P = dW/dt$ while in others you evaluate $P = \mathbf{F} \cdot \mathbf{v}$.

PROBLEM 12 (42E).   (*a*) At a certain instant, a particle experiences a force of $\mathbf{F} = (4.0\,\text{N})\,\mathbf{i} - (2.0\,\text{N})\,\mathbf{j} + (9.0\,\text{N})\,\mathbf{k}$ while having a velocity of $\mathbf{v} = -(2.0\,\text{m/s})\,\mathbf{i} + (4.0\,\text{m/s})\,\mathbf{k}$. What is the instantaneous rate at which the force does work on the particle?

STRATEGY: Evaluate the scalar product $\mathbf{F} \cdot \mathbf{v} = F_x v_x + F_y v_y + F_z v_z$.

SOLUTION:

$$[\,\text{ans: } 28\,\text{W}\,]$$

(*b*) At some other time, the velocity consists of only a $\mathbf{j}$ component. If the force is unchanged, and the instantaneous power is $-12\,\text{W}$, what is the velocity of the particle just then?

STRATEGY: Let $\mathbf{v} = v_y\,\mathbf{j}$ and set $\mathbf{F} \cdot \mathbf{v} = -12\,\text{W}$. Solve for v_y.

SOLUTION:

$$[\,\text{ans: } (6.0\mathbf{j})\,\text{m/s}\,]$$

PROBLEM 13 (46P).   A 2.0-kg object accelerates uniformly from rest to a speed of $10\,\text{m/s}$ in $3.0\,\text{s}$. (*a*) How much work is done on the object during the 3.0-s interval?

STRATEGY: Use the work-energy theorem. You can calculate the initial and final kinetic energy.

SOLUTION:

$$[\,\text{ans: } 100\,\text{J}\,]$$

What is the rate at which work is done on the object (*b*) at the end of the interval and (*c*) at the end of the first half of the interval?

STRATEGY: Use $P = Fv$. Use $F = ma$ to compute the force, with $a = v/t = 10/3.0 = 3.33\,\text{m/s}^2$. At the end of the interval $v = 10\,\text{m/s}$ while at the midpoint of the interval $v = at = 3.33 \times 1.5 = 5.0\,\text{m/s}$.

SOLUTION:

$$[\,\text{ans: } (b)\ 67\,\text{W};\ (c)\ 33\,\text{W}\,]$$

PROBLEM 14 (49P).   The force (but not the power) required to tow a boat at constant velocity is proportional to the speed. If a speed of $2.5\,\text{mi/h}$ requires $10\,\text{hp}$ how much horsepower does a speed of $7.5\,\text{mi/h}$ require?

STRATEGY: Take the force to be given by $F = bv$, where b is a constant of proportionality. The power is then given by $P = Fv = bv^2$. The ratio of the power for two different speeds is given by $P_1/P_2 = v_1^2/v_2^2$. Solve for P_2, given P_1, v_1, and v_2.

SOLUTION:

$$[\,\text{ans: } 90\,\text{hp}\,]$$

III. MATHEMATICAL SKILLS

Scalar products.   These are used extensively in this chapter to calculate the work done and the power delivered by a force: $W = \mathbf{F} \cdot \mathbf{d}$ for a constant force and $P = \mathbf{F} \cdot \mathbf{v}$ for any force. Be sure you know how to evaluate them. In particular know that $\mathbf{F} \cdot \mathbf{d} = Fd \cos \phi$, where ϕ is the angle between $\mathbf{F}$ and $\mathbf{d}$ when they are drawn with their tails at the same point. Also know how to evaluate a scalar product in terms of components: $\mathbf{F} \cdot \mathbf{d} = F_x d_x + F_y d_y + F_z d_z$. Review the appropriate sections of Chapter 3 of the text and this Student Companion.

Binomial theorem.   You will also need to know the binomial theorem to derive the non-relativistic expression for the kinetic energy from the relativistic expression. The first few terms of $(A + B)^n$ are

$$(A + B)^n = A^n + nA^{n-1}B + \frac{n(n-1)}{2}A^{n-2}B^2 + \frac{n(n-1)(n-2)}{2 \cdot 3}A^{n-3}B^3 + \ldots$$

and, in general, term r in the sum is

$$\frac{n!}{r!(n-r)!}A^{n-r}B^r$$

where $r!$ is the factorial of r: $r! = 2 \cdot 3 \cdot 4 \ldots r$.

The exponent n may be any number, positive or negative. To estimate the square root of $A + B$ take $n = 1/2$ and to estimate the reciprocal of the square root take $n = -1/2$.

The series has a finite number of terms if n is a positive integer, the last term being B^n. Otherwise the series never ends. Notice that in successive terms of the series the exponent of A decreases while the exponent of B increases. Since you will want a sum in which successive terms are smaller than previous terms, take A to be the larger of the two quantities and B the smaller.

IV. NOTES

Chapter 8
CONSERVATION OF ENERGY

I. BASIC CONCEPTS

The closely related concepts of a conservative force and a potential energy are central to this chapter. Pay close attention to their definitions. If *all* forces exerted by objects in a system on each other are conservative and no net work is done on them by an outside agent then the mechanical energy of the system (the sum of the kinetic and potential energies) does not change. When non-conservative forces act another energy, called the internal energy of the system, must be included in the sum for the principle of energy conservation to hold.

Definitions Section 4 of the text gives two ways to test a force to see if it is <u>conservative</u>. One of them is: _____

_____

_____

The other is: _____

_____

_____

The two tests are equivalent to each other. If a force meets either of them then it automatically meets the other. If it fails either of them then it automatically fails the other.

Give some examples of conservative forces: _____

_____

Give at least one example of a non-conservative force: _____

_____

Consider a block attached to a horizontal spring and on a rough horizontal surface. Suppose the system starts with the spring neither extended nor compressed. An external force is applied so the block moves from its initial position to a position for which the spring extension is d, then back to its initial position. You should be able to show that the spring does zero work during this motion and that the force of friction does non-zero work.

work done by spring on outward trip: $W_s = $ _____
work done by spring on inward trip: $W_s = $ _____
sign of the work done by friction on outward trip: _____
sign of the work done by friction on inward trip: _____

Because a force of friction is actually the sum of a large number of forces, acting at the welds that form between two surfaces, and the welds move through displacements that are different from the displacement of the object, the work done by friction is *not* given by $\int \mathbf{f} \cdot \mathbf{dr}$, where $\mathbf{dr}$ is an infinitesimal displacement of the object. You cannot calculate the work done by friction without a detailed model of the surfaces but you should be able to argue that the work done by friction cannot be zero over a round trip. You should also be able to argue that the work done by friction on an object sliding on a stationary surface is negative.

A <u>potential</u> <u>energy</u> <u>function</u> can be associated with a conservative force. Here we consider forces that are given as functions of the positions of the interacting objects and not, for example, as functions of their velocities or the time. For a given force we consider the system consisting of the two objects that are interacting via the force (the Earth and a projectile, a spring and a mass, for example). Some configuration of the system, called the reference configuration, is selected and the potential energy is arbitrarily assigned the value 0 for this configuration. The potential energy for any other configuration is taken to be the negative of the work done by the force as the system moves from the reference configuration to the configuration being considered. For the force of gravity a configuration is specified by giving the positions of the objects relative to each other. For the force of a spring a configuration is specified by giving the extension or compression of the spring or, alternatively, the position of the object attached to the spring. Later on, when you study rotational motion, a specification of configuration may also include the orientations of the objects.

The work done by a conservative force as the system goes from any initial to any final configuration is the negative of the change in the potential energy. That is,

$$W =$$

where U_i is the potential energy associated with the initial configuration and U_f is the potential energy associated with the final configuration. The path taken is immaterial since the work done by a conservative force is independent of the path. Every time the system reaches the same configuration it has the same potential energy. This property is at the very heart of the potential energy concept. Carefully note that a potential energy is associated with at least two objects, not with a single object.

To test your understanding of potential energy, explain in words why a potential energy cannot be associated with a frictional force. After all, the force does work. Why not just define the potential energy to be the negative of the work? _____

_____

_____

Potential energy is a scalar. If two or more conservative forces act, the total potential energy is simply the sum of the individual potential energies.

The SI unit for potential energy is _____ and is abbreviated _____. In terms of SI base units this unit is _____ .

Only *changes* in potential energy have physical meaning. If the same arbitrary value is added to the potential energy for every configuration of the system, the motions of the objects in the system do not change. This is why almost any configuration can be selected as the reference configuration. Selection of a new reference configuration simply adds the same value to the potential energy for every configuration.

Because potential energy is a function of configuration a system may be considered to store energy as potential energy. The potential energy of the system is increased when positive work is done the system by an external agent without changing its kinetic energy. This stored energy can be converted to kinetic energy. As you study this chapter and its sample problems or work end-of-chapter problems notice when work being done by an external agent is changing the potential energy of a system and when potential energy is being converted to kinetic energy or vice versa.

Gravitational potential energy.   The gravitational potential energy of a system consisting of the Earth and a mass m, near its surface, is given by

$$U =$$

where y is the vertical coordinate of the object, measured relative to the Earth, and the reference configuration is

_____ .

Gravitational potential energy is often ascribed to m alone, but it is actually a property of the Earth-mass system.

Remember that the gravitational potential energy depends only on the altitude of the object above the surface of the Earth, even if the object moves horizontally as well as vertically. When the object is raised a distance Δy and simultaneously moved horizontally by Δx, the potential energy increases by $mg\Delta y$. You should be able to show this, starting with $\Delta U = -\mathbf{F} \cdot \mathbf{d}$, where $\mathbf{d}$ is the displacement $\Delta x\,\mathbf{i} + \Delta y\,\mathbf{j}$. Here the positive y direction was taken to be upward.

If the altitude of an object above the Earth is increased, the sign of the work done by gravity is _____ and the sign of the potential energy change is _____ . If the altitude is decreased, the sign of the work done by gravity is _____ and the sign of the potential energy change is _____ . See Sample Problem 8–1 of the text for a calculation of gravitational potential energy and carefully note the sign of the potential energy change. The sample problem also shows that the choice of reference configuration does not influence *changes* in the potential energy.

When an object is lifted with constant speed in the Earth's gravitational field, the _____ energy of the Earth-object system increases. When the object is released and falls _____ energy is converted to _____ energy by the action of the gravitational force.

Spring potential energy.   The potential energy of a system consisting of a mass attached to an ideal spring with spring constant k is given by

$$U =$$

where the coordinate x of the mass is measured relative to an origin at _____ .
The reference configuration is _____ .
When a spring is elongated by pulling the mass outward from its equilibrium position the sign of the work done by the spring is _____ and the sign of the change in the spring potential energy is _____ . When the spring is compressed by pushing the mass inward from its equilibrium position the sign of the work done by the spring is _____ and the sign of the change in the spring potential energy is _____ . An external agent can increase the spring potential energy by compressing or extending the spring. This stored energy is converted to kinetic energy when the spring is released.

Calculation of force from potential energy. You have learned how to compute the potential energy associated with a given conservative force. The reverse calculation can also be carried out. If the potential energy function is known, the force can be computed by evaluating its derivatives with respect to coordinates. Consider an object moving along the x axis and acted on by a conservative force F. If $U(x)$ is the potential energy as a function of the object's coordinate, then

$$F =$$

If you are given a graph of the potential energy as a function of x and asked for the force on the particle when its coordinate has a given value, you measure the _____ of the line that is _____ to the curve at that point. Be careful about the signs here: a positive slope indicates a force in the _____ x direction and a negative slope indicates a force in the _____ x direction.

Conservation of mechanical energy. We consider a system of objects that interact with each other via conservative forces and suppose no net work is done on objects of the system by outside agents. The change in the total potential energy as the system changes configuration is the negative of the total work done by all forces. According to the work-energy theorem the total work is also equal to the change in _____. Thus if all forces are conservative the sum of the _____ and _____ energies does not change as the system changes configuration. In symbols, $\Delta K + \Delta U = 0$.

The <u>mechanical</u> energy of a system is defined by $E =$ _____ + _____ and if all forces are conservative then $\Delta E = 0$. You should recognize that, in general, K is the sum of the kinetic energies of all objects in the system. When, for example, an object moves in the Earth's gravitational field, K is strictly the sum of the kinetic energies of the object and the Earth. If, however, the object is much less massive than the Earth then the kinetic energy of the Earth does not change significantly and can be omitted from the energy equation. Similarly, an ideal spring is considered to be massless and so does not contribute to the kinetic energy of any system of which it is a part.

Conservation of energy can be used to solve some problems. If you know the potential and kinetic energies for one configuration of the system, then you can calculate the total mechanical energy for that configuration. If mechanical energy is conserved then it has the same value for all configurations. Now suppose you are given or can calculate the potential energy for a second configuration. Then the conservation of energy principle allows you to compute the kinetic energy for the second configuration. Likewise, if you know the kinetic energy for the second configuration then the conservation of energy principle allows you to compute the potential energy.

For example, suppose a mass m is attached to a spring with spring constant k. Initially it is pulled out so the spring is elongated by x_i and it is given a speed v_i to start its motion. In terms of these quantities the mechanical energy of the spring-mass system is $E =$ _____. Suppose the speed of the mass is v_f when it is at x_f. In terms of these quantities the mechanical energy is $E =$ _____. Equate the two expressions for the mechanical energy to obtain $\frac{1}{2}mv_i^2 + \frac{1}{2}kx_i^2 = \frac{1}{2}mv_f^2 + \frac{1}{2}kx_f^2$. This equation can be solved for any one of the quantities that appear in it.

Suppose a ball of mass m is thrown into the air with an initial speed v_0. Take the zero of gravitational potential energy to be at the release point and neglect the motion of the Earth. The mechanical energy of the Earth-ball system is then $E =$ _____. What is the speed of the ball when it a distance y above the release point? The potential energy is then $U(y) =$ _____ and if we let v be the speed of the ball its kinetic energy is $K =$ _____. Equate the two expressions for the mechanical energy to obtain $\frac{1}{2}mv_0^2 = \frac{1}{2}mv^2 + mgy$. This can be solved for v. The speed is the same whether it is going up or coming down. You can also find how high the ball goes before falling back again. Put $v = 0$ and solve for y.

Potential energy curves. You should know how to obtain information about the motion of an object from a potential energy curve, the potential energy plotted as a function of its co-ordinate. In thinking about these curves you may assume that the agent exerting the force is stationary. Only the object whose motion is being considered has kinetic energy.

Consider the potential energy function shown below for an object that moves along the x axis. Suppose the object has mechanical energy E, represented by a dotted line, and suppose further that mechanical energy is conserved. Indicate on the graph the potential and kinetic energies of the object when it is at x_1. Use x_{min} to label the minimum coordinate of the object in its motion and use x_{max} to label the maximum coordinate of the object in its motion. Look at the slope of the potential energy function and mark with an arrow the directions of the force when the object is at x_{min} and at x_{max}. x_{min} and x_{max} are called the _____ of the motion.

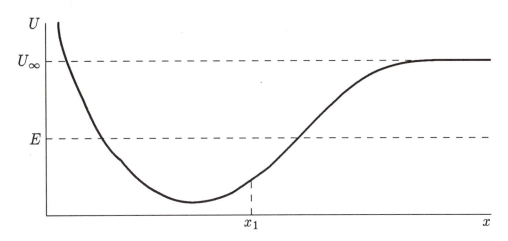

Suppose the object starts at x_1 and travels in the negative x direction. Qualitatively describe its subsequent motion, telling where it is speeding up, where it is slowing down, where it momentarily stops, and what it does after it stops: _____

_____

_____

If the energy of the object is increased slightly, what happens to the minimum and max-imum coordinates? The minimum coordinate _____ and the maximum coordinate

_____.

Suppose now that the object is far to the right and is traveling in the negative x direction with a mechanical energy that is slightly greater than U_∞ on the graph. Describe its subsequent motion: _____

_____

Equilibrium points are coordinates for which the _____ vanishes. The slope of the potential energy curve is _____ at these points. When an object is released near a point of stable equilibrium its subsequent motion is _____. When an object is released near a point of unstable equilibrium its subsequent motion is _____.
When an object is released near a point of neutral equilibrium its subsequent motion is

_____.

Given a potential energy function $U(x)$ for one dimensional motion we can find the equilibrium points by solving _____ for x. We can check to see if the equilibrium is stable or unstable by evaluating _____ at the equilibrium point. The value is _____ for a stable equilibrium point, _____ for an unstable equilibrium point, and _____ for a neutral equilibrium point.

The graph above shows an equilibrium point just to the left of x_1. Is it a point of stable, unstable, or neutral equilibrium? _____

External work. Consider a system of objects that may interact with each other and with other objects outside the system. Its mechanical energy is not conserved if an external force does work on any object in the system. Suppose, for example, a crate is carried up some stairs. The force of the person doing the carrying is an external force if the system is taken to consist of the Earth and the crate. The work done by this force increases the mechanical energy of the Earth-crate system. If the speed of the crate does not change the increase in energy appears as potential energy; the crate is further away from the Earth than previously. If the crate's speed does change the increase in energy is both potential and kinetic.

When external forces do work on objects in the system the energy equation is written

$$\Delta K + \Delta U = W,$$

where W is the total work done *on* particles of the system by outside agents. The potential energy U arises from interactions of objects in the system with each other and K is the total kinetic energy of objects in the system.

Internal energy. When we take into account the individual particles that make up an object we may need to add another term to the energy, a term called the internal energy and denoted by E_{int}. This is the sum of the kinetic and potential energies associated with motions of the particles and with their mutual interactions, as distinct from the motion of the object as a whole or its interaction with other objects.

You must include E_{int} whenever work done on an object results in different changes in the motions of different particles in the object; that is, when the object as a whole cannot be treated as a particle. When a gas is compressed, for example, its internal energy changes. Non-conservative forces, such as friction, always change the internal energy of the object on which they act because the welds between the surfaces deform and the microscopic surface structure changes as time goes on. Thus both the macroscopic kinetic energies and the internal energy of the system change. Similarly, air resistance results from collisions of many air molecules with particles in an object. These change both the macroscopic kinetic energy of the object and its internal energy.

When internal energy is included the energy equation becomes

$$\Delta K + \Delta U + \Delta E_{\text{int}} = W.$$

Here K is the sum of the kinetic energies of objects in the system, U is the sum of the potential energies associated with interactions between objects in the system, E_{int} is the total internal energy of the system, and W is the total work done on particles of the system by outside agents.

Carefully note that if no net work is done on objects of the system by outside agents, then $\Delta K + \Delta U + \Delta E_{\text{int}} = 0$. Energy, in the form $K + U + E_{\text{int}}$, is conserved. Forces between particles of the system may change the kinetic energies of the objects, the potential energies of their interactions, or the internal energy, but the sum retains the same value. Energy may change form but the total does not change.

Consider a block sliding on a table. Because friction is present it slows and stops. No external work is done on the system composed of the block and table so its total energy is conserved. All the kinetic energy originally possessed by the block ends up as _____ energy.

If **f** is the force of friction acting on the block and **d** is the displacement of the block then **f** · **d** does not give the work done by friction. Nevertheless, **f** · **d** is used, along with the work done by any other forces, to determine the change in the kinetic energy of the block. This conclusion follows directly from Newton's second law: **f** is included with all other forces to determine the net force acting on the block and the integral $\int \mathbf{F}_{\text{net}} \cdot d\mathbf{s}$ gives ΔK. The work done by friction is algebraically greater than **f** · **d** and the additional work appears as an increase in the internal energy of the block and the table on which it slides.

An increase in the internal energy of an object may result in an increase in the temperature of that object. When the temperature of an object is different from the temperature of its surroundings, energy is transferred as heat and still another term must be included in the energy equation given above. You will learn more about this when you study Chapter 20.

Whenever the total mechanical energy is not conserved physicists have been able to restore the principle of energy conservation by defining another form of energy and including it in the energy balance, just as internal energy restores the energy conservation principle when an object is deformed or frictional forces act. The energy associated with electromagnetic radiation (visible light, for example) and the energy associated with the rest mass of a particle are other examples of energies that sometimes must be taken into account. The change in energy associated with a mass change Δm is given by $\Delta E = $ _____.

Energy quantization. A particle is said to be bound if $E \geq U$ only in a finite region of space. Quantum mechanically the energy of a bound particle is quantized: it may have only certain discrete values; other values are not allowed. For macroscopic spring-mass systems the separation between allowed energy values is too small to be detected, but for atomic masses it has profound effects. Fig. 8–17 shows some of the allowed energy values for a sodium atom.

The energy of an atomic system can change from one allowed value to another with the emission or absorption of a quantum of electromagnetic radiation, called a photon. The relation between the energy of the photon and the frequency f of the wave associated with it is given by $E =$ _____. Here h is the Planck constant and has the value _____ J·s. It is a constant of nature.

When an atom changes energy from some initial value E_i to a lower final value E_f with the emission of a photon, conservation of energy relates the change in the energy of the atom to the frequency of the light emitted:

$$E_i - E_f =$$

II. PROBLEM SOLVING

Problems involving the conservation of mechanical energy are relatively straightforward. You must first select a system of objects. Normally you will pick the system so that no outside agents do work on objects in the system. You must then decide if all the forces of interaction between objects of the system are conservative. The force of gravity and the force of an ideal spring are; the force of friction and drag forces are not. In one dimension forces that are uniform over all possible paths are conservative. For other forces you may have to apply one of the tests for a conservative force.

If all forces that do work are conservative, then conservation of energy yields $K_1 + U_1 = K_2 + U_2$, where K_1 and U_1 are the kinetic and potential energies for one configuration and K_2 and U_2 are the kinetic and potential energies for another. You must identify the two configurations from the problem statement. You will be given enough information to calculate three of the four energies that appear in the energy equation and can use the conservation of energy to compute the fourth.

In many problems you are asked to find some parameter that occurs in the expression for the kinetic or potential energy at the second configuration. For example, you may be asked for the compression of a spring, a spring constant, the speed of an object, or its mass. You first find the relevant energy (kinetic or potential), then solve for the parameter.

PROBLEM 1 (6E). An ice flake is released from the edge of a hemispherical frictionless bowl whose radius is 22 cm (Fig. 8–22). How fast is the ice moving at the bottom of the bowl?

STRATEGY: Take the system to consist of the ice flake, the bowl, and the Earth. No external forces act and the only internal force that does work (the force of gravity) is conservative. Other internal forces acts, the normal forces of the bowl on the flake and the flake on the bowl, but neither of these do work. In the initial configuration the ice flake is at rest at the top of the bowl. If the gravitational potential energy is taken to be 0 when the flake is at the bottom of the bowl the potential energy for the initial configuration is given by $U = mgr$, where r is the radius of the bowl. The kinetic energy is 0. When the flake is at the bottom of the bowl the potential energy is 0 and the kinetic energy is $\frac{1}{2}mv^2$, where v is its speed. Conservation of energy leads to $mgr = \frac{1}{2}mv^2$. Solve for v.

SOLUTION:

[ans: 2.1 m/s]

PROBLEM 2 (7E).   A frictionless roller-coaster car tops the first hill in Fig. 8–23 with speed v_0. What is its speed at (*a*) point A, (*b*) point B, and (*c*) point C?

STRATEGY: Take the system to consist of the car, the track, and the Earth. No external forces do work and the internal force that does work (the force of gravity) is conservative. The normal forces of the track on the car and the car on the track do no work. Mechanical energy is conserved. Take the zero of gravitation potential energy to be at ground level. The mechanical energy of the car at the top of the first hill is given by $E = mgh + \frac{1}{2}mv_0^2$. At each of the other points it is given by $mgy + \frac{1}{2}mv^2$, where y is the height of the car above the ground and v is its speed. The values are $y = h$ at A, $y = h/2$ at B, and $y = 0$ at C. In each case solve $mgh + \frac{1}{2}mv_0^2 = mgy + \frac{1}{2}mv^2$ for v.

SOLUTION:

$$\left[\text{ans: } (a)\ v_0;\ (b)\ \sqrt{v_0^2 + gh};\ (c)\ \sqrt{v_0^2 + 2gh}\ \right]$$

(*d*) How high will it go on the last hill, which is too high for it to cross?

STRATEGY: It will go until it momentarily stops and the kinetic energy is zero. If this point is y above the ground then the potential energy is mgy. Solve $mgh + \frac{1}{2}mv_0^2 = mgy$ for y.

SOLUTION:

$$\left[\text{ans: } (v_0^2/2g) + h\ \right]$$

PROBLEM 3 (14E).   A thin rod whose length is $L = 2.00$ m and whose mass is negligible is pivoted at one end so that it can rotate in a vertical circle. A heavy ball with mass m is attached to the lower end. The rod is pulled aside through an angle $\theta = 30.0°$, as shown in Fig. 8–27, and then released. How fast is the ball moving at its lowest point?

STRATEGY: Take the system to be the ball, the rod, and the Earth. No external forces do work and the internal force (gravity) that does work is conservative, so the mechanical energy of the system is conserved. The force of the rod on the ball is always perpendicular to the velocity of the ball so does 0 work. Take the gravitational potential energy to be 0 when the ball is at the bottom of its swing. Then at the initial configuration $U = mgh$ and $K = 0$. A little geometry shows that $h = L(1 - \cos\theta)$, where $\theta = 30°$. At the bottom $U = 0$ and $K = \frac{1}{2}mv^2$, where v is the speed of the ball. Solve $mgL(1 - \cos\theta) = \frac{1}{2}mv^2$ for v.

SOLUTION:

$$\left[\text{ans: } 2.29\ \text{m/s}\ \right]$$

PROBLEM 4 (16P).   A 2.0-kg block is placed against a spring on a frictionless 30° incline (Fig. 8–29). The spring, whose spring constant is 19.6 N/cm, is compressed 20 cm and then released. How far along the incline does it send the block?

STRATEGY: Take the system to consist of the block, the spring, the incline, and the Earth. No external forces do work and the only internal force that does work (the force of gravity) is conservative, so the mechanical energy of the system is conserved. Take the gravitational potential energy to be 0 when the block is in its initial configuration, with the spring compressed. The spring potential energy is given by $\frac{1}{2}kx^2$, where k is the spring constant and x is the distance the spring is compressed. The kinetic energy is zero. After release the block leaves the spring when it has its equilibrium length. That is when the spring would start pulling on the block if they were attached; since they are not the block flies off. The final configuration is when the block comes to rest a distance d along the incline. The spring potential energy is 0 and the gravitational potential energy is mgh, where h is the height of the block above its starting point. The kinetic energy is again 0. A little trigonometry will show you that $h = d \sin \theta$, where θ is the angle the incline makes with the horizontal. Solve $\frac{1}{2}kx^2 = mgd \sin \theta$ for d.

SOLUTION:

$$[\text{ans: } 4.00\,\text{m}]$$

PROBLEM 5 (24P).   One end of a vertical spring is fastened to the ceiling. An object is attached to the other end and slowly lowered to its equilibrium position, where the upward force exerted by the spring on the object balances the weight of the object. Show that the loss of gravitational potential energy of the object equals twice the gain in the spring's potential energy.

STRATEGY: The gain in spring potential energy is given by $\frac{1}{2}kx^2$, where k is the spring constant and x is the elongation of the spring. The loss in gravitational potential energy is mgx. Since the forces of the spring and gravity sum to 0, $kx = mg$ and $x = mg/k$. Substitute mg/k for x in the expression for the potential energies. You should get $(mg)^2/k$ for the magnitude of the change in the gravitational potential energy and $(mg)^2/2k$ for the magnitude of the change in the spring potential energy.

SOLUTION:

Why are these two quantities not equal?

SOLUTION: The mechanical energy of the system consisting of the block, the spring, and the Earth is not conserved because an external agent did work in lowering the object. In fact, the calculation above shows that $(mg)^2/2k$ energy is lost. The work done by the agent was negative and accounts for the loss. Take the external force to be $F = mg - kx$, where up is positive, and show that the work done by the agent was $-(mg)^2/2k$. This external force will lower the block without changing its kinetic energy.

PROBLEM 6 (27P).   Two children are playing a game in which they try to hit a small box on the floor with a marble fired from a spring-loaded gun that is mounted on a table. The target box is 2.20 m horizontally from the edge of the table; see Fig. 8–34. Bobby compresses the spring 1.10 cm, but the center of the marble falls 27.0 cm short of the center of the box. How far should Rhoda compress the spring to score a direct hit?

STRATEGY: The distance from the table to the place where the marble hits is related the speed of the marble as it leaves the gun and this, of course, is related to the compression of the spring. The first relationship can be found using kinematics, the second using conservation of energy.

SOLUTION: Suppose the gun is a distance h above the floor, the marble has a speed v_0 as it leaves the gun, and it hits a distance ℓ from the table. Use $0 = h - \frac{1}{2}gt^2$ and $\ell = v_0 t$ to show that $\ell = v_0\sqrt{2h/g}$. The spring originally had potential energy $\frac{1}{2}kx^2$, where k is the spring constant and x is the compression of the spring; the marble had kinetic energy $\frac{1}{2}mv_0^2$ as it left the gun. Since energy was conserved, $\frac{1}{2}kx^2 = \frac{1}{2}mv_0^2$. Use this to show that $v_0 = x\sqrt{k/m}$. Combine the two results to show that $\ell = x\sqrt{2hk/mg}$.

Suppose Bobby compressed the spring by x_1 and the marble landed a distance $\ell_1 = x_1\sqrt{2hk/mg}$ from the table. Rhoda compressed the spring by x_2 and the marble landed $\ell_2 = x_2\sqrt{2hk/mg}$ from the table. Divide one equation by the other to obtain $\ell_1/\ell_2 = x_1/x_2$ or $x_2 = (\ell_2/\ell_1)x_1$. Use $x_1 = 1.10$ cm, $\ell_1 = 2.20 - .27 = 1.93$ m, and $\ell_2 = 2.20$ m. Solve for x_2.

[ans: 1.25 cm]

Some problems of this chapter involve the calculation of a potential energy change, given the force and the end points of the path. Since the change in potential energy is just the negative of the work done by the force, the calculation proceeds like the calculations of work in the last chapter. If you are asked for the potential energy itself, the reference configuration (for which the potential energy is zero) must be identified, either in the problem statement or by you. You then calculate the negative of the work done on the system as the system changes from the reference configuration to the configuration of interest.

PROBLEM 7 (29P).   A 20-kg object is acted on by a force given by $F = -3.0x - 5.0x^2$, where F is in newtons if x is in meters. As the force causes the object to move, assume that the object's mechanical energy is conserved. Also assume that the potential energy U of the object is zero at $x = 0$. (*a*) What is the potential energy of the object at $x = 2.0$ m?

STRATEGY: The potential energy function associated with the force is the negative of the work done by the force and so is given by $U(x) = -\int F\,dx$. Evaluate this integral and use the condition that $U(0) = 0$ to find the value of the constant of integration. You should get $U(x) = \frac{3}{2}x^2 + \frac{5}{3}x^3$. Finally, evaluate this expression for $x = 2.0$ m.

SOLUTION:

[ans: 19 J]

(b) If the object has a velocity of 4.0 m/s in the negative x direction when it is at $x = 5.0$ m, what is its speed as it passes through the origin?

STRATEGY: Let $K_1 = \frac{1}{2}mv_1^2$ be the kinetic energy and U_1 the potential energy when the object is at $x = 5.0$ m and let $K_2 = \frac{1}{2}mv_2^2$ be the kinetic energy and U_2 the potential energy when it is at the origin. Since mechanical energy is conserved, $K_1 + U_1 = K_2 + U_2$. Find values for the potential energy by evaluating the expression you developed in (a). Calculate K_1. Finally, solve the conservation of energy equation for v_2.

SOLUTION:

$$\big[\text{ans: } 6.4 \text{ m/s}\big]$$

(c) Next assume that the potential energy of the object is -8.0 J at $x = 0$ and reanswer parts (a) and (b).

STRATEGY: Simply add -8.0 J to all of the potential energies computed in parts (a) and (b). The expression for the potential energy is now $U(x) = \frac{3}{2}x^2 + \frac{5}{3}x^3 - 8.0$, in joules for x in meters. Since the same quantity is added to both U_1 and U_2 in (b) the final kinetic energy is the same as before.

SOLUTION:

$$\big[\text{ans: } 11 \text{ J}; 6.4 \text{ m/s}\big]$$

PROBLEM 8 (30P).   A small block of mass m can slide along the frictionless loop-the-loop track shown in Fig. 8–35. (a) The block is released from rest at point P. What is the net force acting on it at point Q?

STRATEGY: The are two forces: the normal force of the track mv^2/R, to the left, and the force of gravity mg, down. Use conservation of energy to find mv^2.

SOLUTION: Take the zero of gravitational potential energy to be at the lowest point on the track. Then the initial potential energy is $5mgR$. The initial kinetic energy is zero. The potential energy at Q is mgR. Write $\frac{1}{2}mv^2$ for the kinetic energy at Q. Then $5mgR = mgR + \frac{1}{2}mv^2$. Solve for mv^2 and substitute the result into $N = mv^2/R$. You should get $8mg$ for the normal force of the track.

Since the two forces are perpendicular the net force has a magnitude that is given by the square root of the sum of their squares. Also find the angle made by the net force with the horizontal. It is given by $\tan\theta = mg/8mg = 0.125$.

$$\big[\text{ans: } 8.1mg, 7.1° \text{ down from horizontal}\big]$$

(b) At what height above the bottom of the loop should the block be released so that it is on the verge of losing contact with the track at the top of the loop?

STRATEGY: You must use Newton's second law to find the speed at the top for which the block is on the verge of losing contact. This is the speed for which the normal force of the track vanishes. Then use conservation of energy to find the height of the release point for which the block has the desired speed when it gets to the top.

SOLUTION: At the top the second law yields $N + mg = mv^2/R$, where N is the normal force of the track. The block is on the verge of losing contact if $N = 0$ since N cannot be negative. Thus $mv^2 = mgR$ and the block should have a kinetic energy of $K = mgR/2$ at the top. The potential energy is $2mgR$. Suppose the block starts from rest at a height h above the bottom of the loop. The potential energy is then mgh and the kinetic energy is 0. Conservation of energy leads to $mgh = mgR/2 + 2mgR$. Solve for h.

[ans: $5R/2$]

Sometimes energy concepts can be used to carry out calculations involving power. Here is an example.

PROBLEM 9 (34P).   An escalator joins one floor with another one 8.0 m above. The escalator is 12 m long and moves along its length at 60 cm/s. (*a*) What power must its motor deliver if the escalator is required to carry a maximum of 100 persons per minute, of average mass 75 kg?

STRATEGY: Take the system to be the people on the escalator, the escalator, and the Earth. The motor supplies an external force and, as a result, increases the potential energy of the system by $Nmgh$ when N people, each of mass m, are carried from one floor to the next. Since the kinetic energy of the people does not change during the trip the work done by the motor must be $W = Nmgh$. The power delivered by the motor is $P = dW/dt = (dN/dt)mgh$. $dN/dt = 100\,\text{min}^{-1} = 1.67\,\text{s}^{-1}$.

SOLUTION:

[ans: $9.8\,\text{kW}$]

(*b*) An 80-kg man walks up the escalator in 10 s. How much work does the escalator do on him?

STRATEGY: The force of the escalator on the man has magnitude mg and the power exerted by the escalator on the man is $P = mgv_v$, where v_v is the vertical component of the escalator's velocity. The work done by the escalator is $W = Pt = mgv_v t$, where t is the time of the trip. Use $v_v = (h/L)v$, where h is the distance between floors and L is the length of the escalator.

SOLUTION:

[ans: $3.1\,\text{kJ}$]

(*c*) If this man turned around and walked down the escalator so as to stay at the same level in space, would the escalator do work on him? If so, what power is required of it?

SOLUTION: The work is given by the scalar product of the force of the escalator on the man and the displacement of the man.

[ans: no, his displacement is 0]

Even though the escalator does no work on the man the motor must deliver power to the escalator. The escalator exerts a force on the man: it just balances the force of gravity so its component along the escalator is $mg\sin\theta$, where θ is the angle between the escalator and the horizontal. The man exerts a force of equal magnitude but opposite direction on the escalator. Since the escalator is moving this force does work on the escalator. It is negative and tends to slow the escalator. To keep the escalator running at constant speed the motor must do an equal amount of positive work. The power delivered by the motor is $P = Fv$, where v is the escalator speed. Since $F = mg\sin\theta$, this is $P = mgv\sin\theta$.

You should be able to interpret a $U(x)$ graph for a particle. You can read the potential energy for any particle coordinate directly from the graph. If you know the total energy you can easily find the kinetic energy for any coordinate by using $E = K + U$. Sometimes the mass and speed are given for one coordinate and requested for another. Use the value of K for the first coordinate and the value of U read from the graph to calculate E. Read the graph for the value of U for the second coordinate, then use $K = E - U$ to calculate the kinetic energy. You should also know how to obtain the force from the graph. At any coordinate it is the negative of the slope of the curve at that coordinate. Here's an example.

PROBLEM 10 (40P). A particle of mass 2.0 kg moves along the x axis through a region in which its potential energy $U(x)$ varies as shown in Fig. 8–41. When the particle is at $x = 2.0$ m its velocity is -2.0 m/s. (a) What force acts on it at this position?

STRATEGY: Since the force and potential energy are related by $F = -dU/dx$ you calculate the slope of the curve at $x = 2.0$ m. The force is the negative of the value you find.

SOLUTION:

[ans: $+4.7$ N]

(b) Between what limits of x does the particle move?

STRATEGY: Fist calculate the total mechanical energy. Add the potential and kinetic energies when the particle is at $x = 2.0$ m. Read the potential energy from the graph, calculate the kinetic energy using $K = \frac{1}{2}mv^2$. You should get about -3.5 J. Now draw a horizontal line across the graph at -3.5 J and note the values of x where it crosses the curve. At these points the kinetic energy is 0 and the particle reverses its direction of motion.

[ans: 1.2 m, 14 m]

(c) What is its speed at $x = 7.0$ m?

STRATEGY: The total energy is $E = \frac{1}{2}mv^2 + U$. Read the value of U at $x = 7.0$ m from the graph. Use $E = -3.5$ J and calculate v.

SOLUTION:

[ans: 3.6 m/s]

Some problems deal with the dissipation of mechanical energy by friction or air resistance. Here are some examples.

PROBLEM 11 (47E). A 75-g frisbee is thrown from a height of 1.1 m above the ground with a speed of 12 m/s. When it has reached a height of 2.1 m, its speed is 10.5 m/s. (a) How much work was done on the frisbee by its weight?

STRATEGY: The initial and final heights are given so you use $W = -mg(h_f - h_i)$.

SOLUTION:

[ans: -0.74 J]

(*b*) How much of the frisbee's mechanical energy was dissipated by air drag?

STRATEGY: Calculate the initial and final mechanical energies and find their difference. Take the system to consist of the frisbee and the Earth and include both kinetic and mechanical energies.

SOLUTION:

[ans: 0.53 J dissipated]

PROBLEM 12 (62E). A projectile whose mass is 9.4 kg is fired vertically upward. On its upward flight, an energy of 68 kJ is dissipated because of air drag. How much higher would it have gone if the air drag had been made negligible (for example, by streamlining the projectile)?

STRATEGY: Take the system to consist of the projectile and the Earth. Take the initial potential energy to be $U_i = 0$ and let K_1 be the initial kinetic energy. If it reaches a height h then its final potential energy is $U_f = mgh$ and its final kinetic energy is 0. The change in mechanical energy is given by $\Delta E = U_f + K_f - U_i - K_i = mgh - K_i$. When air resistance is present $\Delta E = -6.8 \times 10^4$ J. When air resistance is negligible $\Delta E = 0$. Take the maximum height to be h' then and write $0 = mgh' - K_i$ for the conservation of energy equation. Solve the two expressions for $h' - h$. You should get $h' - h = -\Delta E/mg$.

SOLUTION:

[ans: 740 m]

PROBLEM 13 (70P). A factory worker accidentally releases a 400-lb crate that was being held at rest at the top of a 12-ft-long ramp inclined at 39° to the horizontal. The coefficient of kinetic friction between the crate and the ramp, and between the crate and the horizontal factory floor is 0.28. (*a*) How fast is the crate moving as it reaches the bottom of the ramp?

STRATEGY: Take the system to consist of the crate, the ramp, and the earth and let d be the distance the crate slides down the ramp. The potential energy changes by $\Delta U = -mgh$ and the kinetic energy changes by $\Delta K = \frac{1}{2}mv^2$, where h is the original height of the crate above the factory floor and v is its speed at the bottom of the ramp. The force of friction changes the mechanical energy by $-fd$ so $-mgh + \frac{1}{2}mv^2 = -fd$. A little geometry shows that $h = d\sin\theta$, where θ is the angle between the ramp and the horizontal. Furthermore, $f = \mu_k N = \mu_k mg\cos\theta$, where N $(= mg\cos\theta)$ is the normal force of the ramp on the crate. Solve $-mgd\sin\theta + \frac{1}{2}mv^2 = -\mu_k mgd\cos\theta$ for v. Notice that the mass m cancels from the equation. Don't forget to use 32 ft/s² for g.

SOLUTION:

[ans: 18 ft/s]

(*b*) How far will it subsequently slide across the factory floor? (Assume that the crate's kinetic energy does not change as it moves from the ramp onto the floor.)

STRATEGY: As it starts across the floor its kinetic energy is $\frac{1}{2}mv^2$, where v is its speed at the bottom of the ramp (the answer to *a*). When it stops sliding its kinetic energy is 0. The potential energy does not change because the floor is horizontal. If the crate slides a distance d across the floor the mechanical energy dissipated by the force of friction is given by $-fd = -\mu_k Nd = -\mu_k mgd$. Thus $-\frac{1}{2}mv^2 = -\mu_k mgd$. Solve for d. Again m can be canceled from the equation.

SOLUTION:

$$[\text{ans: } 18\,\text{ft}]$$

PROBLEM 14 (88P). The resistance to motion of an automobile depends on road friction, which is almost independent of speed, and on air drag, which is proportional to speed-squared. For a car with a weight of 12,000 N, the total resistant force F is given by $F = 300 + 1.8v^2$, where F is in newtons and v is in meters per second. Calculate the power required to accelerate the car at 0.92 m/s^2 when the speed is 80 km/h.

STRATEGY: The power supplied increases the kinetic energy of the car and is dissipated by resistive forces. The rate at which kinetic energy is increasing is given by $dK/dt = d(\frac{1}{2}mv^2)/dt = mva$. The rate of dissipation is given by $Fv = v(300 + 1.8v^2)$. So the power required is $P = mva + v(300 + 1.8v^2)$. The mass of the car is $W/g = 12000/9.8 = 1.22 \times 103$ kg and its speed is 80 km/h = 22.2 m/s.

SOLUTION:

$$[\text{ans: } 5.1 \times 10^4\,\text{W } (69\,\text{hp})]$$

III. NOTES

Chapter 9
SYSTEMS OF PARTICLES

I. BASIC CONCEPTS

In this chapter you consider systems of more than one particle. The center of mass is important for describing the motion of the system as a whole and its velocity is closely related to the total momentum of the system. External forces acting on the system accelerate the center of mass and when the net external force vanishes the velocity of the center of mass is constant. The total momentum of the system is then conserved.

Coordinates of the center of mass. You should know how to calculate the coordinates of the center of mass of a collection of particles. If particle i has mass m_i and coordinates x_i, y_i, and z_i then the coordinates of the center of mass are given by

$$x_{cm} =$$

$$y_{cm} =$$

$$z_{cm} =$$

where M is the total mass of the system. In vector notation the position vector of the center of mass is given by

$$\mathbf{r}_{cm} =$$

where $\mathbf{r}_i$ is the position vector of particle i. The center of mass is not necessarily at the position of any particle in the system. Sample Problem 9–1 of the text shows you how to calculate the coordinates of the center of mass of a small collection of particles.

For a *continuous* distribution of mass the coordinates of the center of mass are given by the integrals

$$x_{cm} =$$

$$y_{cm} =$$

$$z_{cm} =$$

If the mass is distributed symmetrically about some point or line then the center of mass is at that point or on that line. For example, the center of mass of a uniform spherical shell is at its _____, the center of mass of a uniform sphere is at its _____, the center of mass of a uniform cylinder is at the midpoint of its _____, the center of mass of a uniform rectangular plate is at its _____, and the center of mass of a uniform triangular plate is at the intersection of _____ and halfway through its thickness.

Sometimes an object with a complicated shape can be thought of as a group of parts such that the center of mass of each part can be found easily. Then each part is replaced by a particle with mass equal to the mass of the part, positioned at the center of mass of the part. The center of mass of the whole object is at the position of the center of mass of these particles. This idea can also be used to find the center of mass of an object with a hole. See Sample Problem 9–3 of the text.

Velocity and acceleration of the center of mass. As the particles that comprise the system move, the coordinates of the center of mass might change. Consider a system of discrete particles and differentiate the expression for the center of mass position vector to find the velocity of the center of mass in terms of the individual particle velocities:

$$\mathbf{v}_{cm} =$$

where $\mathbf{v}_i$ is the velocity of particle i.

Consider two objects moving along the x axis, one with mass m_A and coordinate given by $x_A(t) = x_{A0} + v_A t$ and the other with mass m_B and coordinate given by $x_B(t) = x_{B0} + v_B t$, where v_A and v_B are constant velocities. Use the definition of the center of mass to find an expression for the coordinate of the center of mass as a function of time: $x_{cm}(t) =$ _____ _____. Now differentiate this expression to find an expression for the velocity of the center of mass: $v_{cm} =$ _____. How should the velocities of the objects be related for the center of mass to be at rest? _____

As the particles accelerate the center of mass might accelerate. Differentiate the expression for the velocity of the center of mass to find the acceleration of the center of mass in terms of the accelerations of the individual particles:

$$\mathbf{a}_{cm} =$$

Newton's second law for the center of mass. Since each individual particle obeys Newton's second law, $\sum \mathbf{F}_i = M\mathbf{a}_{cm}$, where $\mathbf{F}_i$ is the vector sum of all forces on particle i, some of which might be due to other particles in the system and some of which might be due to the environment of the system. The sum $\sum \mathbf{F}_i$ is the vector sum of all forces acting on all particles of the system but it reduces to $\sum \mathbf{F}_{ext}$, the sum over all *external* forces acting on particles of the system, i.e. those due to the environment of the system. Explain in words why these two sums are equal: _____

_____

_____

Thus the center of mass obeys a Newton's second law: $\sum \mathbf{F}_{ext} = M\mathbf{a}_{cm}$. The mass that appears in the law is the total mass of the system (the sum of the masses of the individual particles) and the force that appears is the vector sum of all *external* forces acting on all particles of the system.

If the total external force acting on a system is zero, then the acceleration of the center of mass is _____ and the velocity of the center of mass is _____. If the center of mass of the system is initially at rest and the total external force acting on the system is zero, then the velocity of the center of mass is always _____ and the center of mass remains at its initial position.

These statements are true no matter what forces the particles of the system exert on each other. The particles might, for example, be fragments that are blown apart in an explosion or they might be objects that collide violently. On the other hand, they might interact with each other from afar via gravitational or electrical forces and never touch.

Momentum. The momentum of a particle of mass m, moving with a velocity $\mathbf{v}$ (assumed to be much less than the speed of light) is given by $\mathbf{p} = $ _____. Momentum is a vector. Its SI units are _____.

In terms of the magnitude of its momentum the kinetic energy of a particle is given by $K = $ _____. This is identical to $K = \frac{1}{2}mv^2$. In terms of momentum, Newton's second law for a particle is $\sum \mathbf{F} = $ _____, where $\sum \mathbf{F}$ is _____. You should be able to prove this formulation is exactly the same as $\sum \mathbf{F} = m\mathbf{a}$.

If the particle speed is close to the speed of light, relativistically correct expressions must be used. The momentum of a particle of mass m moving with velocity $\mathbf{v}$ is then given by $\mathbf{p} = $ _____ and the kinetic energy is then given by $K = $ _____.

The total momentum $\mathbf{P}$ of a system of particles is the vector sum of the individual momenta. Since, for non-relativistic particles, $\mathbf{P} = m_1\mathbf{v}_1 + m_2\mathbf{v}_2 + \ldots$, the total momentum is given by $\mathbf{P} = M\mathbf{v}_{cm}$, where M is the total mass of the system. That is, the total momentum of the system is identical to the momentum of a single particle with mass equal to the total mass of the system, moving with a velocity equal to the velocity of the center of mass. Furthermore the net external force changes the total momentum of the system according to

$$\sum \mathbf{F}_{ext} = $$

Conservation of momentum. State in words the principle of conservation of total momentum for a system of particles. Your statement should have the form "If ... , then ... ." _____

_____

_____

Since momentum is a vector this principle is equivalent to the three equations: $P_x = $ constant if $\sum F_{ext\,x} = 0$, $P_y = $ constant if $\sum F_{ext\,y} = 0$, and $P_z = $ constant if $\sum F_{ext\,z} = 0$. One component of momentum may be conserved even if the others are not. See, for example, Sample Problem 9–8 of the text.

If the momentum of a system is conserved then the acceleration of the center of mass of the system is _____ and the velocity of the center of mass is _____.

Variable mass systems. $\sum \mathbf{F}_{ext} = M\mathbf{a}_{cm}$ and $\sum \mathbf{F}_{ext} = d\mathbf{P}/dt$ hold only for systems of constant mass. Recall that when they were derived, the number of particles in the system was held constant. In Section 9–7 the derivation of an expression for $d\mathbf{P}/dt$ is carefully carried out for a system that loses mass. To understand it you should carefully carry out the steps.

Suppose that at time t a rocket with mass M is moving with velocity $\mathbf{v}$. It will eject fuel of mass $-\Delta M$ (a positive quantity) in time Δt. Originally the fuel is traveling with the rocket at velocity $\mathbf{v}$ but after it is ejected it has velocity $\mathbf{U}$ and the rocket has velocity $\mathbf{v} + \Delta\mathbf{v}$. Carefully note that $\mathbf{v}$, $\mathbf{v} + \Delta\mathbf{v}$, and $\mathbf{U}$ are all measured relative to an inertial frame and that ΔM is taken to be negative, so the mass of the rocket after ejection is $M + \Delta M$. Also note that the rocket exerts a force on the fuel to eject it and the fuel exerts a force on the rocket of the same magnitude but in the opposite direction.

The rocket and the ejected fuel form a constant mass system, so $\sum \mathbf{F}_{\text{ext}} = d\mathbf{P}/dt$ is valid for it. The total momentum of the system at the initial time t, in terms of M and $\mathbf{v}$, is given by $\mathbf{P}_i = $ _____ . The total momentum of the system at time $t + \Delta t$, in terms of M, ΔM, $\mathbf{v}$, $\Delta\mathbf{v}$, and $\mathbf{U}$, is given by $\mathbf{P}_f = $ _____ . .

Use the expressions you gave above to write $(\mathbf{P}_f - \mathbf{P}_i)/\Delta t$ in terms of M, $\mathbf{v}$, $\mathbf{U}$, $\Delta\mathbf{v}$, and ΔM, then evaluate the expression in the limit as $\Delta t \to 0$. The term $\Delta M \Delta\mathbf{v}/\Delta t$ vanishes in this limit since it is proportional to an infinitesimal. Because Newton's second law is valid, the result equals the net external force on the object. In the space below carry out the steps to show that $\sum \mathbf{F}_{\text{ext}} = M\,d\mathbf{v}/dt - (\mathbf{U} - \mathbf{v})dM/dt$:

Now $\mathbf{U} - \mathbf{v}$ is the velocity of the fuel relative to the rocket. This is usually constant during firing so we write $\mathbf{U} - \mathbf{v} = \mathbf{u}$. The text also makes the substitution $R = -dM/dt$ for the rate at which mass is ejected. This is a positive quantity. Then the acceleration of the rocket is given by $M\mathbf{a} = \mathbf{F}_{\text{ext}} - R\mathbf{u}$. The second term on the right is called the thrust of the rocket. Notice that if there are no external forces acting the acceleration is opposite in direction to the relative velocity of the fuel. To accelerate forward a rocket ejects fuel toward the rear.

Translational and internal energy. The total kinetic energy of a system of particles can be divided into two parts. One is associated with motion of the center of mass and is called translational kinetic energy. The other is associated with the motions of the various particles relative to the center of mass. It, along with the potential energy associated with mutual interactions of particles in the system, make up the internal energy.

For a system of particles with total mass M and center of mass velocity $\mathbf{v}_{\text{cm}}$, the translational kinetic energy of the system is given by $K_{\text{cm}} = $ _____ . Changes in this kinetic energy are related to the net external force acting on the system and the displacement of the center of mass. For purposes of calculating ΔK_{cm} the system is replaced by a single particle of mass M, located at the position of the center of mass and acted on by the net external force

$\sum \mathbf{F}_{\text{ext}}$. The change in the translational kinetic energy then equals the work W_{cm} done by this force on this particle.

Suppose the center of mass of a system moves from x_i to x_f along the x axis, during which motion the net external force is F_{ext}, a constant. The work done by this force on the "particle" that replaces the system is given by $W_{\text{cm}} = \underline{\hspace{2in}}$ and the change in the translational kinetic energy of the system is given by $\Delta K_{\text{cm}} = \underline{\hspace{1.2in}}$.

Be sure you realize that W_{cm} is not necessarily the actual work done by the resultant external force. In some cases the actual work done by $\sum \mathbf{F}_{\text{ext}}$ may change the internal energy of the system as well as change the translational kinetic energy. In other cases $\sum \mathbf{F}_{\text{ext}}$ may not actually do any work but W_{cm} is not zero and the kinetic energy changes. This occurs, for example when the point of application of an external force does not move but the center of mass does. Then the actual work that changes the kinetic energy is done by internal forces. If you are asked for a change in energy use the energy equation $\Delta K_{\text{cm}} + \Delta U + \Delta E_{\text{int}} = W$, where W is the actual work done by external forces. If you are asked about a change in the translational kinetic energy use $W_{\text{cm}} = \Delta K_{\text{cm}}$.

Be sure you understand the motion of a skater pushing off from a railing, as discussed in the text. Assume the skater's hands do not leave the railing. When filling in the blanks below consider what happens from the point of view of an observer at rest with respect to the railing.

The force that accelerates the skater is: $\underline{\hspace{3.5in}}$

Does this force do any work? $\underline{\hspace{1in}}$

The forces that do work are: $\underline{\hspace{3in}}$

Does the translational kinetic energy of the skater change? $\underline{\hspace{1in}}$

II. PROBLEM SOLVING

When asked to calculate the coordinates of the center of mass, first decide if the system consists of discrete particles, with the masses and coordinates of each given. If it does, the calculation proceeds by straightforward evaluation of $\mathbf{r}_{\text{cm}} = (1/M) \sum m_i \mathbf{r}_i$.

If, on the other hand, the system can be considered to be a continuous distribution of mass, next decide if the mass is uniformly distributed. Look for symmetry if it is. The object may have a plane of symmetry, such as the plane that bisects a plate halfway through its thickness or the plane that is the perpendicular bisector of a cylinder axis. The center of mass must be on such a plane. If there is more than one, the center of mass is somewhere on their intersection.

Also look for axes of symmetry. Every diameter of a uniform sphere or uniform spherical shell is an axis of symmetry as is the axis of a uniform cylinder or cylindrical shell. The center of mass must lie on these. If two axes of symmetry intersect or if one intersects a plane of symmetry the center of mass is at the intersection.

Perhaps you can partition the object into several smaller objects, each with an easily identifiable center of mass. Replace each of these with a single particle, with mass equal to the mass of the object it replaces and located at the center of mass of the object. Then find the center of mass of the particles.

PROBLEM 1 (8P). In the ammonia (NH_3) molecule (see Fig. 9–26), the three hydrogen (H) atoms form an equilateral triangle; the center of the triangle is 9.40×10^{-11} m from each hydrogen atom. The nitrogen (N) atom is at the apex of a pyramid, the three hydrogens forming the base. The nitrogen-to-hydrogen distance is 10.14×10^{-11} m, and the nitrogen-to-hydrogen atomic mass ratio is 13.9. Locate the center of mass relative to the nitrogen atom.

STRATEGY: First consider the hydrogen atoms alone. Their center of mass is at the center of the triangle. To see this replace two of the atoms with an atom of twice the mass at the midpoint of a triangle side (at A). Now draw the line from the opposite vertex to A. Since there is twice as much mass at A as at the vertex, the center of mass of the 3 atoms lies on the dotted line, one third the distance to the vertex (that is, at B). This is directly beneath the nitrogen atom.

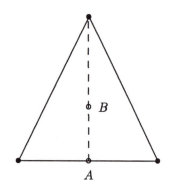

Replace the three atoms with one of three times the mass, at B. Find the coordinates of the center of mass of this atom and the nitrogen atom. Place the origin of a coordinate system at the nitrogen atom and take the z axis to be positive downward. If d is the distance from the nitrogen to the plane of hydrogens below, the z coordinate of the center of mass is given by $3m_H d/(3m_H + m_N)$. The distance d can be found by applying the Pythagorean theorem to the triangle formed by the nitrogen, the center of the base triangle and a hydrogen atom.

SOLUTION:

$$\left[\,\text{ans: } 6.75 \times 10^{-12}\,\text{m directly below the nitrogen atom}\,\right]$$

PROBLEM 2 (10P). Two pieces of sheet-metal, each in the shape of a right triangle with height $H = 2.0$ cm and length $L = 3.5$ cm, are shown in Fig. 9–28. (*a*) What are the coordinates of the center of mass of the composite structure?

STRATEGY: Replace each triangle with a particle located at the center of mass of the triangle. Then use $\mathbf{r}_{cm} = (M\mathbf{r}_1 + M\mathbf{r}_2)/(M + M)$ to find the center of mass of the composite. Here M is the mass of either triangle, $\mathbf{r}_1$ is the position of the center of mass of one triangle, and $\mathbf{r}_2$ is the position of the center of mass of the other.

SOLUTION: The center of mass of a triangle is at the intersection point of the lines from the vertices to the midpoints of opposite sides. In the space to the right draw a right triangle with the right-angle vertex at the origin, the side of length L along the x axis, and the side of length H along the y axis. Draw the line from the right-angle vertex to the midpoint of the opposite side. Its equation is $y = (H/L)x$. Now draw the line from the vertex at $x = 0$, $y = H$ to the midpoint of the opposite side. Its equation is $y = H - (2H/L)x$. Find the coordinates of the intersection by solving these equations simultaneously. You should get $x = L/3$ and $y = H/3$.

Now consider the composite as drawn in the figure. The coordinates of the center of mass of the triangle on the left are $x = L/3$, $y = H/3$ while the coordinates of the center of mass of the triangle on the right are

$x = 2L - L/3 = 5L/3, y = H/3.$

$$\left[\text{ans: } x_{\text{cm}} = 3.5 \,\text{cm}, y_{\text{cm}} = 0.67 \,\text{cm} \right]$$

(b) If each piece is reversed left-for-right so that the 2.0 cm sides are against each other, what are the coordinates of the center of mass of the composite structure?

SOLUTION:

$$\left[\text{ans: } x_{\text{cm}} = 3.5 \,\text{cm}, y_{\text{cm}} = 0.67 \,\text{cm} \right]$$

PROBLEM 3 (12P). A right cylindrical can with mass M, height H, and uniform density is initially filled with pop of mass m (Fig. 9–30). We punch small holes in the top and bottom to drain the pop; we then consider the height h of the center of mass of the can and any pop within it. What is h (a) initially, and (b) when all the pop has drained?

STRATEGY: Since the can is uniform its center of mass is at its geometric center. When the can is filled the center of mass of the pop is also at the geometric center. This is at height $H/2$.

SOLUTION:

$$\left[\text{ans: } (a) \, H/2; (b) \, H/2 \right]$$

(c) What happens to h during the draining of the pop?

SOLUTION:

$$\left[\text{ans: at first it drops from } H/2 \text{ then it rises again to } H/2 \right]$$

(d) If x is the height of the remaining pop at any given instant, find x (in terms of M, H, and m) when the center of mass reaches its lowest point.

STRATEGY: The center of mass of the pop is at height $x/2$. The mass m_ℓ of the pop left in the can is proportional to x and is m when $x = H$, so $m_\ell = (m/H)x$. The height of the center of mass of the can and pop is given by $h = [(m/2H)x^2 + MH/2]/[(m/H)x + M]$. Find the value of x for which the center of mass is at the lowest point by differentiating h with respect to x, setting the result equal to 0, and solving for x.

SOLUTION:

$$\left[\text{ans: } (HM/m)(\sqrt{1 + m/M} - 1) \right]$$

Some problems deal with two-body systems for which the center of mass is initially at rest and the net external force vanishes. One of the bodies changes position and the problem asks for the change in position of the other. Since the net external force is zero the velocity of the center of mass is a constant and since it is initially zero it remains zero. The center of mass does not move. Write an expression for the center of mass in terms of the initial coordinates of the bodies, write a second expression in terms of the final coordinates, and equate the two expressions. Then solve for the final coordinates of the second body. In a variant of this type problem the two objects end up at nearly the same place. This place must be the center of mass.

PROBLEM 4 (13E). Two skaters, one with mass 65 kg and the other with mass 40 kg, stand on an ice rink holding a pole with a length of 10 m and a negligible mass. Starting from the ends of the pole, the skaters pull themselves along the pole until they meet. How far will the 40-kg skater move?

STRATEGY: The net external force acting on the system consisting of the two skaters and the pole is zero, so the velocity of the center of mass does not change. The forces of the skaters on the pole and of the pole on the skaters are all internal forces and sum to zero. Since the center of mass was at rest to begin with it remains at the same position. The skaters meet at the center of mass. To find how far the 40-kg skater moves calculate the original distance from that skater to the center of mass.

SOLUTION: Place the origin of the coordinate system at the 40-kg skater. The 65-kg skater is then at $x = 10$ m. Calculate the coordinate of the center of mass.

[ans: 6.2 m]

PROBLEM 5 (17E). A cannon and a supply of cannonballs are inside a sealed railroad car of length L, as in Fig. 9–32. The cannon fires to the right; the car recoils to the left. Fired cannon balls remain in the car after hitting the far wall and landing on the floor there. (*a*) After all the cannonballs have been fired, what is the greatest distance the car can have moved from its original position?

STRATEGY: If the total mass of all the cannonballs is much greater than the mass of the car the center of mass of the system consisting of the balls and car is at the position of the balls, at the left end of the car before firing and at the right end when all have been fired. The net external force has no horizontal component and the cannon and balls were initially at rest. This means the center of mass remains at the same place and the car moves a distance L. Calculate the distance moved no matter what the mass of the balls.

SOLUTION: Place the origin of a coordinate system at the center of mass of the car before firing and take the x axis to be horizontal. Let M_b be the total mass of the balls. They are at $x = -L/2$. The center of mass of the system is at $x = -(L/2)M_b/(M_b + M_c)$, where M_c is the mass of the car. Let x_c be the coordinate of the center of mass of the car after firing. The balls are now at $x_c + L/2$ and the coordinate of the center of mass is given by $[x_cM_c + (x_c + L/2)M_b]/[M_b + M_c]$. Equate the two expression for the center of mass and solve for x_c. You should get $x_c = -LM_b/(M_b + M_c)$. If $M_b \gg M_c$ then $x_c = -L$.

[ans: L]

(b) What is the speed of the car just after all the cannonballs have been fired?

SOLUTION: The velocity of the center of mass is initially 0 and, since the net external force has no horizontal component, it remains 0. When the balls have landed and are at rest with respect to the car, the balls and car must all have velocity 0.

[ans: 0]

PROBLEM 6 (18E).   A stone is dropped at $t = 0$. A second stone, with twice the mass of the first, is dropped from the same point at $t = 100$ ms. (a) Where is the center of mass of the two stones at $t = 300$ ms? Neither stone has yet reached the ground.

STRATEGY: Use kinematics to find the coordinates of the stones, then calculate the coordinate of the center of mass.

SOLUTION: Take the y axis to be downward with the origin at the release point. The coordinate of the first stone is then given by $y_1 = \frac{1}{2}gt^2$ and the coordinate of the second stone is given by $y_2 = \frac{1}{2}g(t - t_0)^2$, where $t_0 = 200$ ms. Evaluate these expressions for $t = 300$ ms, then use $y_{cm} = (m_1y_1 + m_2y_2)/(m_1 + m_2)$, with $m_2 = 2m_1$.

[ans: 28 cm]

(b) How fast is the center of mass of the two-stone system moving at that time?

STRATEGY: The velocity of the first stone is given by $v_1 = gt$ and that of the second is given by $v_2 = g(t - t_0)$. Evaluate these expressions for $t = 300$ ms, then use $v_{cm} = (m_1v_1 + m_2v_2)/(m_1 + m_2)$.

SOLUTION:

[ans: 2.3 m/s]

Some problems deal with the definition of momentum and the equation of motion $\sum \mathbf{F} = d\mathbf{p}/dt$ for a particle. Here are two examples.

PROBLEM 7 (28P).   An object is tracked by a radar station and found to have a position vector given by $\mathbf{r} = (3500 - 160t)\mathbf{i} + 2700\mathbf{j} + 300\mathbf{k}$, with $\mathbf{r}$ in meters and t in seconds. The radar station's x axis points east, its y axis north, and its z axis vertically up. If the object is a 250-kg missile warhead, what are (a) its linear momentum, (b) its direction of motion, and (c) the net force on it?

STRATEGY: For parts a and b use $\mathbf{p} = m\mathbf{v} = m\,d\mathbf{r}/dt$ and for part c use $\mathbf{F} = d\mathbf{p}/dt$. The direction of the momentum is the direction of motion.

SOLUTION:

[ans: (a) $(-4.0 \times 10^4\,\mathbf{i})$ kg · m/s; (b) west; (c) 0]

PROBLEM 8 (29P). A 0.165-kg cue ball with an initial speed of 2.00 m/s bounces off the rail in a game of pool, as shown from overhead in Fig. 9–35. For x and y axes located as shown, the bounce reverses the y component of the ball's velocity but does not alter the x component. (a) What is θ in Fig. 9–35?

STRATEGY: Let $\mathbf{v}_b = v_x \mathbf{i} + v_y \mathbf{j}$ be the velocity of the ball before it bounces. Then $\mathbf{v}_a = v_x \mathbf{i} - v_y \mathbf{j}$ is its velocity afterwards. From the diagram both 30° and θ have the same tangent, namely v_x / v_y. Furthermore they are in the same quadrant.

SOLUTION:

[ans: 30°]

(b) What is the change in the ball's linear momentum in unit-vector notation?

STRATEGY: Calculate the initial momentum: $\mathbf{p}_b = mv_x \mathbf{i} + mv_y \mathbf{j} = (mv \sin \theta) \mathbf{i} + (mv \cos \theta) \mathbf{j}$. Calculate the final momentum: $\mathbf{p}_a = (mv \sin \theta) \mathbf{i} - (mv \cos \theta) \mathbf{j}$. Finally, calculate the difference $\mathbf{p}_a - \mathbf{p}_b$.

SOLUTION:

[ans: $(-0.572\,\mathbf{j})\,\text{kg} \cdot \text{m/s}$]

Many problems of this chapter can be worked using the principle of momentum conservation. If you suspect the principle can be used, first check for external forces: if there are none or if they add to zero, total momentum is conserved. For some problems only one component of momentum is conserved. If the problem asks about motion in that direction, then the principle can be used.

PROBLEM 9 (34E). Two blocks of masses 1.0 kg and 3.0 kg connected by a spring rest on a frictionless surface. They are given velocities toward each other such that the 1.0-kg block travels initially at 1.7 m/s toward the center of mass, which remains at rest. What is the initial velocity of the other block?

STRATEGY: Since the velocity of the center of mass does not change, the momentum of the system consisting of the two blocks is conserved. Since the center of mass is at rest, the momentum of the system is 0. Solve $m_1 v_1 + m_2 v_2 = 0$ for v_2.

SOLUTION:

[ans: 0.57 m/s toward the other block]

PROBLEM 10 (41P). A 2140-kg railroad flatcar, which can move with negligible friction, is motionless next to a platform. A 242-kg sumo wrestler runs at 5.3 m/s along the platform (parallel to the track) and then jumps onto the flatcar. What is the speed of the flatcar if he then (a) stands on it, (b) runs at 5.3 m/s relative to it in his original direction, and (c) turns and runs at 5.3 m/s relative to the flatcar opposite his original direction?

STRATEGY: Take the system to be the flatcar and the wrestler. If we can neglect any friction between the flatcar and the tracks, the net external force vanishes and the total momentum of the system is conserved. Let m_w be the mass of the wrestler, m_c be the mass of the car, v_{wi} be the initial velocity of the wrestler, v_{wf} be the final velocity of the wrestler, and v_{cf} be the final velocity of the car. Then $m_w v_{wi} = m_w v_{wf} + m_c v_{cf}$. You are given the final velocity of the wrestler relative to the flatcar. This is $v_{rel} = v_{wf} - v_{cf}$. Substitute $v_{wf} = v_{rel} + v_{cf}$

into the conservation of momentum equation and solve for v_{cf}. Take the initial velocity of the wrestler to be positive. Then $v_{rel} = 0$ for part a, $= +5.3$ m/s for part b, and $= -5.3$ m/s for part c.

SOLUTION:

[ans: (a) 0.54 m/s; (b) 0; (c) 1.1 m/s]

PROBLEM 11 (42P).   A rocket sled with a mass of 2900 kg moves at 250 m/s on a set of rails. At a certain point, a scoop on the sled dips into a trough of water located between the tracks and scoops water into an empty tank on the sled. By applying the conservation of linear momentum, determine the speed of the sled after 920 kg of water have been scooped up. Ignore any retarding force on the scoop.

STRATEGY: Take the system to consist of the sled and the water scooped up. The net external force is zero so its total momentum is conserved. Let m_s be the mass of the sled, m_w be the mass of the water, v_{si} be the initial velocity of the sled, and v_f be the final velocity of the sled and water. The initial velocity of the water, of course, is 0. The conservation of momentum equation is $m_s v_{si} = m_f(m_s + m_w)$. Solve for v_f.

SOLUTION:

[ans: 190 m/s]

PROBLEM 12 (44P).   A body of mass 8 kg is traveling at 2 m/s with no external force acting. At a certain instant an internal explosion occurs, splitting the body into two chunks of 4-kg mass each. The explosion gives the chunks an additional 16 J of kinetic energy. Neither chunk leaves the line of original motion. Determine the speed and direction of motion of each of the chunks after the explosion.

STRATEGY: Since no external forces act total momentum is conserved. Let m be the original mass and let v be its velocity. Take the masses of the chunks to be m_1 and m_2 with velocities v_1 and v_2 after the explosion. Then $mv = m_1 v_1 + m_2 v_2$. If Q is the energy added in the explosion then $\frac{1}{2}m_1 v_1^2 + \frac{1}{2}m_2 v_2^2 = \frac{1}{2}mv^2 + Q$. Solve these equations simultaneously for v_1 and v_2.

SOLUTION: The algebra is simplified considerably if you first make the substitutions $m_1 = m/2$ and $m_2 = m/2$. Use the momentum equation to substitute for one of the unknown velocities in the energy equation. You will obtain a quadratic equation for the other velocity. Either solution is valid.

[ans: 0; 4.0 m/s, in the original direction of motion]

PROBLEM 13 (46P). A 1400-kg cannon, which fires a 70.0-kg shell with a speed of 556 m/s relative to the muzzle, is set at an elevation angle of 39.0° above the horizontal. The cannon is mounted on frictionless rails, so that it recoils freely. (*a*) What is the speed of the shell with respect to the Earth?

STRATEGY: Take the shell and cannon to be the system. Since the net external force has no horizontal component, the horizontal component of the total momentum is conserved. Since the shell and cannon were initially at rest it is 0. Take the x axis to be horizontal and let v_c be the final velocity of the cannon, v_{sx} the x component of the final velocity of the shell, and v_{sy} the y component of the final velocity of the shell. Conservation of momentum gives $0 = m_c v_c + m_s v_{sx}$. This must be written in terms of the muzzle velocity. If v_m is the muzzle speed then $v_{sx} = v_c + v_m \cos \theta$, where θ is the elevation angle of the cannon. Use this expression to eliminate v_c from the momentum equation, then solve for v_{sx}. The shell velocity also has a y component. Since the cannon does not recoil perpendicularly to the surface, it is $v_{sy} = v_m \sin \theta$. Compute the shell speed from the components of its velocity.

SOLUTION:

[ans: 540 m/s]

(*b*) At what angle with the ground is the shell projected?

STRATEGY: If the angle is ϕ then $\tan \phi = v_{sy}/v_{sx}$. Solve for ϕ.

SOLUTION:

[ans: 40.4°]

Here are some examples of variable-mass problems.

PROBLEM 14 (47E). A rocket is moving away from the solar system at a speed of 6.0×10^3 m/s. It fires its engine, which ejects exhaust with a velocity of 3.0×10^3 m/s relative to the rocket. The mass of the rocket at this time is 4.0×10^4 kg, and it experiences an acceleration of 2.0 m/s^2. (*a*) What is the thrust of the engine?

STRATEGY: No external forces act on the rocket-fuel system so $T = Ma$, where T is the thrust, M is the mass, and a is the acceleration. Solve for T.

SOLUTION:

[ans: 8.0×10^4 N]

(*b*) At what rate, in kilograms per second, was exhaust ejected during the firing?

STRATEGY: The thrust is given by $T = Ru$, where R is the rate of fuel exhaust and u is its relative speed. Solve for R.

SOLUTION:

[ans: 27 kg/s]

PROBLEM 15 (52E).   A railroad car moves at a constant speed of 3.20 m/s under a grain elevator. Grain drops into it at the rate of 540 kg/min. What force must be applied to the railroad car, in the absence of friction, to keep it moving at constant speed?

STRATEGY: Suppose that at some instant the mass of the car and its load is M. In the short time interval Δt suppose grain of mass ΔM is added to the car. The horizontal component of the grain's velocity is initially 0 but during loading becomes v, the velocity of the car. The momentum of the car-grain system initially is Mv and, after time Δt, is $(M + \Delta M)v$. The rate of change of momentum is $\Delta P/\Delta t = v\Delta M/\Delta t$ and in the limit as Δt becomes small $dP/dt = v\,dM/dt$. The external force that must be applied is $F = dP/dt = v\,dM/dt$.

SOLUTION:

[ans: 28.8 N]

PROBLEM 16 (53P).   A single-stage rocket, at rest in a certain inertial reference frame, has mass M when the rocket engine is ignited. Show that, when the mass has decreased to $0.368M$, the gases streaming out of the rocket engine at that time will be at rest in the original reference frame.

STRATEGY: Use the second rocket equation, $v = -u\ln(m/M)$, where v is the rocket speed, m is the mass of rocket and unspent fuel, M is the original mass of rocket and fuel when the rocket was at rest, and u is the relative exhaust speed. Substitute $m/M = 0.368$ and find v in terms of u. The exhaust speed in the original frame is given by $U = v - u$.

SOLUTION:

[ans: $u = 0$]

Some problems deal with the energy of a system of particles. Most of these involve the relationship between the net external force and the change in the translational kinetic energy: $\int \sum F_{\text{ext}}\,dx = K_f - K_i$, in one dimension.

PROBLEM 17 (58E).   An automobile with passengers has weight 16, 400 N and is moving at 113 km/h when the driver brakes to a stop. The road exerts a frictional force of 8230 N on the wheels. Using energy considerations, find the stopping distance.

STRATEGY: If f is the frictional force and d is the stopping distance then the change in the translational kinetic energy is given by $K_f - K_i = -fd$. The initial translational kinetic energy is $K_i = \frac{1}{2}mv_i^2$ and the final translational kinetic energy is $K_f = 0$. Solve for d.

SOLUTION:

[ans: 100 m]

PROBLEM 18 (59E).   You crouch from a standing position, lowering your center of mass 18 cm in the process. Then you jump vertically. The average force exerted on you by the floor while you jump is three times your weight. What is your upward speed as you pass through your standing position in leaving the floor?

STRATEGY: Let F be the force of the floor. If up is taken to be positive the net force on you while jumping is $F - mg$ and the change in your translational kinetic energy is given by $K_f - K_i = Fd - mgd$. Here d is the magnitude of the displacement of your center of mass. Substitute $F = 3mg$, $K_i = 0$, and $K_f = \frac{1}{2}mv^2$. Solve for v.

SOLUTION:

[ans: 2.7 m/s]

III. NOTES

Chapter 10
COLLISIONS

I. BASIC CONCEPTS

Here you will apply the principle of momentum conservation to collisions between objects. You should pay special attention to the role played by the impulses the objects exert on each other. You should also take notice of when energy conservation can be invoked to solve a problem and when it cannot.

Definitions. As succinctly as you can, tell in words what a collision is. Be sure to include the important characteristics. _____
_____
_____

During a collision two objects exert forces on each other. What is important is not the force alone or its duration alone but a combination called the <u>impulse</u> of the force. If one body acts on the other with a force $\mathbf{F}(t)$ for a time interval from t_i to t_f, the impulse of the force is given by the integral

$$\mathbf{J} =$$

Don't confuse impulse and work. Impulse is an integral of force with respect to _____ and work is an integral of force with respect to _____ . Impulse is a vector, work is a scalar. The SI unit of impulse is _____ ; the SI unit of work is _____ .

In terms of the average force $\overline{\mathbf{F}}$ that acts during the collision, the impulse is given by $\mathbf{J} = $ _____ , where Δt is _____ . This expression is often useful for calculating the average force, which in turn is useful as an estimate of the strength of the interaction.

Because Newton's second law is valid, the total impulse acting on an object gives the change in the _____ of the object. Because Newton's third law is valid the impulse of one object on the other is the negative of the impulse of the second object on the first. If external impulses can be neglected the total _____ of the two objects is conserved during a collision.

For most collisions the impulse of one colliding object on the other is usually much greater than any external impulse and we may neglect impulses exerted by the environment of the colliding bodies. We may, for example, neglect the effects of gravity and air resistance during the time a baseball is in contact with a bat.

Collisions. Total momentum is conserved during the collisions we consider. Since momentum is conserved during a collision, the velocity of the center of mass of the colliding bodies is _____ .

Kinetic energy may or may not be conserved in a collision. If it is, the collision is said to be _____. If it is not, the collision is said to be _____. The distinguishing characteristic of a two-body <u>completely</u> inelastic collision is _____

_____.

During a completely inelastic collision the loss in total kinetic energy is as large as conservation of momentum allows. That is, it is impossible for the bodies to lose a larger fraction of their original kinetic energy and together still retain all the original total momentum.

Explosions, in which an object splits into two or more parts as a result of internal forces, can be handled in exactly the same manner as a collision. During an explosion the total momentum is conserved but the total kinetic energy increases.

Series of collisions.   Here you consider a stream of objects that collide, one after the other, with a another object. Examples are a rapid succession of bullets hitting a target or the molecules of a gas hitting the walls of a container. Suppose each of the "bullets" has mass m and each suffers the same change in velocity $\Delta \mathbf{v}$. If n "bullets" hit in time Δt then during that interval the change in momentum of the "target" is given by

$$\Delta \mathbf{P} =$$

and the average force exerted by the "bullets" on the "target" is given by

$$\overline{\mathbf{F}} =$$

Elastic one-dimensional collisions.   Consider an elastic two-body collision in one dimension and suppose both objects move along the x axis. Object 1 has mass m_1, moves with velocity v_{1i} before the collision, and moves with velocity v_{1f} after the collision. Object 2 has mass m_2, moves with velocity v_{2i} before the collision, and moves with velocity v_{2f} after the collision. The equation that expresses conservation of momentum during the collision is

The equation that expresses conservation of kinetic energy during the collision is

Using the equations of momentum and energy conservation you should be able to show that the relative velocity with which the objects approach each other before the collision is the same as the relative velocity with which they separate after the collision. In symbols, _____. The proof follows immediately when you divide the kinetic energy conservation equation, in the form $m_1(v_{1f}^2 - v_{1i}^2) = -m_2(v_{2f}^2 - v_{2i}^2)$, by the momentum conservation equation, in the form $m_1(v_{1f} - v_{1i}) = -m_2(v_{2f} - v_{2i})$. You must recognize that $A^2 - B^2 = (A - B)(A + B)$ for any quantities A and B. Carry out the steps in the space below:

In the space below solve $m_1 v_{1i} + m_2 v_{2i} = m_1 v_{1f} + m_2 v_{2f}$ and $(v_{1f} - v_{2f}) = -(v_{1i} - v_{2i})$ for v_{1f} and v_{2f} in terms of the other quantities. Draw a box around your answers and check them with the text.

Using these general expressions, you should be able to obtain the results for some special elastic collisions. Write the results below without looking them up and explain what has happened during the collision. Unless specifically stated otherwise do not assume either object is initially at rest.

a. If the two masses are the same, then $v_{1f} = $ _____ and $v_{2f} = $ _____ .
   In words, what has happened is _____
   _____

b. If the two masses are the same and object 2 is initially at rest, then $v_{1f} = $ _____ and
   $v_{2f} = $ _____ .
   In words, what has happened is _____
   _____

c. If the target object, object 2 say, is very massive compared to the incident object, object 1, then $v_{1f} = $ _____ and $v_{2f} = $ _____ .
   In words, what has happened is _____
   _____

d. If the target object, object 2, is very massive compared to the incident object, object 1, and is initially at rest, then $v_{1f} = $ _____ and $v_{2f} = $ _____ .
   In words, what has happened is _____
   _____

e. If object 1 is very massive compared to object 2 and object 2 is initially at rest, then $v_{1f} = $ _____ and $v_{2f} = $ _____ .
   In words, what has happened is _____
   _____

For all these collisions the total kinetic energy (the sum of the individual kinetic energies of the colliding objects) is the same after the collision as before. However, kinetic energy is nearly always transferred from one object to the other. Sometimes you will want to know how much. The fractional loss of kinetic energy by object 1 is given by $(K_{1i} - K_{1f})/K_{1i}$ and, since $K = \frac{1}{2}mv^2$, this is $1 - v_{1f}^2/v_{1i}^2$.

Completely inelastic one-dimensional collisions. Consider a completely inelastic two-body collision in one dimension. Both objects move along the x axis. Object 1 has mass m_1 and moves with velocity v_{1i} before the collision. Object 2 has mass m_2 and moves with velocity v_{2i} before the collision. Both objects move with velocity v_f after the collision. The equation than that expresses conservation of momentum during the collision is

If the masses and initial velocities are known, conservation of momentum is sufficient to determine the common final velocity. This velocity is given by

$$v_f =$$

An example of this type collision is _____
_____.

If object 2 is initially at rest then

$$v_f =$$

Notice that the final speed of the combination *must* be less than the initial speed of object 1. Explain why in words: _____
_____

Kinetic energy is lost during a completely inelastic collision, transformed to internal energy, radiation, etc. Consider a completely inelastic collision with object 2 initially at rest. In terms of the masses and velocities the initial total kinetic energy is $K_i =$ _____ and the final total kinetic energy is $K_f =$ _____. In the space below make the substitutions to express the kinetic energies in terms of the velocities, carry out the algebra, and obtain the result given below for the fractional energy loss of the two body system:

$$\left[\text{ans: } 1 - [(m_1 + m_2)/m_1](v_f^2/v_{1i}^2)\right]$$

Use conservation of momentum to substitute for v_f^2 in terms of v_{1i}^2 and find an expression for the fractional energy loss in terms of the masses alone:

$$\left[\text{ans: } m_2/(m_1 + m_2)\right]$$

To make the fractional energy loss large the mass of object _____ should be made much larger than the mass of the other object. To make the loss small the mass of _____ should be made much larger than the mass of the other object.

Two-dimensional collisions.   Consider the two-dimensional elastic collision diagramed below.

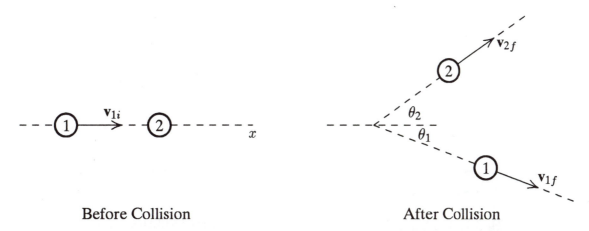

Before Collision   After Collision

Object 2 is initially at rest and object 1 impinges on it with speed v_{1i} along the x axis. Object 1 leaves the collision with speed v_{1f} along a line that is below the x axis and makes the angle θ_1 with that axis. Object 2 leaves the collision with speed v_{2f} along a line that is above the x axis and makes the angle θ_2 with that axis. Take the y axis to be upward in the diagram. In terms of the masses, speeds, and angles between the velocities and the x axis, conservation of total momentum **P** leads to the equations:

conservation of P_x:

conservation of P_y:

Suppose the collision is elastic and write the equation that expresses conservation of kinetic energy in terms of the masses and speeds:

conservation of K:

These equations can be solved for three unknowns. If the masses (or their ratio) are given then, of the five other quantities (v_{1i}, v_{1f}, v_{2f}, θ_1, and θ_2) that enter the equation, two must be given.

Note that the masses and the initial velocity of object 1 alone do not determine the outcome of a two-dimensional elastic collision. The magnitude and direction of the impulse acting on each object during the collision also play important roles. Often these are not known, so the outcome cannot be predicted. Experimenters usually measure the speed or the orientation of the line of motion of one of the objects after the collision then use the equations to calculate other quantities. You should know how to solve for any three of the quantities, given the others. Some hints are given in the Problem Solving section below.

Here's an important result you should remember: if a particle collides elastically with another particle of the same mass, initially at rest, then the two velocity vectors after the collision are perpendicular to each other. In this case, $\mathbf{v}_{1i} = \mathbf{v}_{1f} + \mathbf{v}_{2f}$ and $v_{1i}^2 = v_{1f}^2 + v_{2f}^2$. The three vectors form a triangle and the square of one side equals the sum of the squares of the other two sides. The triangle must be a right triangle with $\mathbf{v}_{1i}$ as the hypotenuse.

If a two-dimensional collision is completely inelastic, the objects stick together after the collision and kinetic energy is not conserved. We now consider collisions for which both objects are initially moving. The geometry is shown below. Before the collision mass m_1 has velocity $\mathbf{v}_{1i}$ and mass m_2 has velocity $\mathbf{v}_{2i}$. After the collision both masses have velocity $\mathbf{v}_f$.

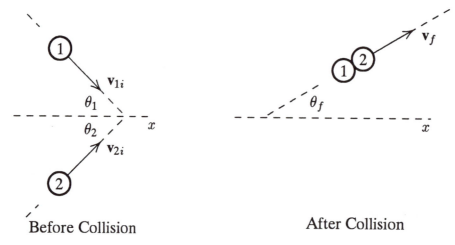

Before Collision   After Collision

Since $v_{1f} = v_{2f}$, we write v_f for both of speeds. Take the y axis to be upward in the diagram. The conservation of momentum equations become

conservation of P_x:

conservation of P_y:

These two equations contain eight quantities (m_1, m_2, v_{1i}, v_{2i}, θ_1, θ_2, v_f, and θ_f) and can be solved for 2 unknowns. If the masses (or their ratio) are given, then 4 additional quantities must be given. Unlike an elastic collision, the masses and initial conditions completely determine the outcome. That is, knowledge of m_1, m_2, v_{1i}, v_{2i}, θ_1, and θ_2 allow us to solve for v_f and θ_f. Notice that the six given quantities allow us to find the velocity of the center of mass. Since momentum is conserved and the objects stick together this is the same as the final velocity of both objects.

Change in kinetic energy during an inelastic collision.   Sometimes the gain or loss of kinetic energy during an inelastic collision is known. The energy equation is then useful. Suppose energy Q is gained by the system during a collision. Then, for either one- or two-dimensional collisions

$$\frac{1}{2}m_1 v_{1f}^2 + \frac{1}{2}m_2 v_{2f}^2 = \frac{1}{2}m_1 v_{1i}^2 + \frac{1}{2}m_2 v_{2i}^2 + Q.$$

This equation is also valid if kinetic energy is lost; Q is then negative. Q has a positive value, for example, when internal energy is converted to kinetic energy during the collision, perhaps through an explosion. On the other hand, Q is negative if kinetic energy is converted to internal energy or to another form. In a ballistic pendulum experiment, for example, Q is negative; most of the "lost" kinetic energy becomes internal energy or energy of deformation. In any event, the energy equation is solved simultaneously with the conservation of momentum equations.

The ideas discussed here are valid for atomic and subatomic collisions and for spontaneous decay of subatomic particles as well as for collisions of macroscopic objects. In many nuclear and subatomic collision processes the internal energy and mass energy may change, so they must be included in the conservation of energy equation. The number and type of particles may also change. Changes in particle identities are taken into account in the energy equation by including the energy associated with mass. In the energy equation a mass decrease is represented by a positive value for Q while a mass increase is represented by a negative value. To work problems involving changes in mass you must recall that the energy associated with mass m is _____, where c is the speed of light.

When the problem involves subatomic particles relativistically valid expressions for momentum and energy must usually be used rather than the non-relativistic expressions we have been working with. If a particle has mass m and velocity $\mathbf{v}$ its momentum is given by

$$\mathbf{p} =$$

and its kinetic energy is given by

$$K =$$

II. PROBLEM SOLVING

Most calculations of impulse are rather straightforward. You might be given the force acting on an object as a function of time and asked to find the impulse over a given time interval. You simply evaluate the integral that defines impulse. Remember that impulse is a vector quantity and you must, for example, use the x component of the force to find the x component of the impulse. For one-dimensional motion the force might be given graphically as a function of time. You must then recognize that the impulse is the area under the curve. You might also be given sufficient information to calculate the initial and final momentum of a particle. Then use $\mathbf{J} = \mathbf{p}_f - \mathbf{p}_i$ to calculate the impulse.

PROBLEM 1 (11E).   A golfer hits a golf ball, giving it an initial velocity of magnitude 50 m/s directed 30° above the horizontal. Assuming that the mass of the ball is 46 g and the club and ball are in contact for 1.7 ms, find (a) the impulse on the ball, (b) the impulse on the club, (c) the average force exerted on the ball by the club, and (d) the work done on the ball.

STRATEGY: The impulse on the ball is given by the change in the momentum of the ball: $\mathbf{J} = \mathbf{p}_f - \mathbf{p}_i$. Since the ball is initially at rest the magnitude of the impulse is $J = mv$, where v is the final speed of the ball. The impulse is in the direction of the final velocity. The impulse on the club is equal in magnitude and opposite in direction. The average force is found from $\mathbf{J} = \overline{\mathbf{F}}\Delta t$, where Δt is the time of contact. Finally the work done on the ball can be found from the change in its kinetic energy.

SOLUTION:

[ans: (a) 2.3 N·s, 30° above the horizontal; (b) 2.3 N·s, opposite the final velocity of the ball; (c) 1.4 kN, 30° above the horizontal; (d) 58 J]

Notice that impulse and work are quite different concepts.

PROBLEM 2 (21P).   Figure 10–30 shows an approximate plot of force versus time during the collision of a 58-g tennis ball with a wall. The initial velocity of the ball is 34 m/s perpendicular to the wall; it rebounds with the same speed, also perpendicular to the wall. What is F_{max}, the maximum value of the contact force during the collision?

STRATEGY: Calculate the impulse by finding the area under the curve. Your result will be in terms of F_{max}. Equate this to the magnitude of the change in momentum, which is $2mv$, where v is the speed of the ball. Solve for F_{max}.

SOLUTION: The area of a triangle is half the base times the altitude so the impulse from $t = 0$ to $t = 2$ ms is $2 \times 10^{-3} \times F_{max}/2$. The area of a rectangle is the product of its base and altitude so the impulse from 2 ms to 4 ms is $2 \times 10^{-3} F_{max}$. Finally the impulse from 4 ms to 6 ms is $2 \times 10^{-3} F_{max}/2$. The total impulse is the sum of these, or $4 \times 10^{-3} F_{max}$. In each case the impulse is in N·s for F_{max} in N.

[ans: 990 N]

PROBLEM 3 (22P).   A ball having a mass of 150 g strikes a wall with a speed of 5.2 m/s and rebounds with only 50% of its initial kinetic energy. (*a*) What is the speed of the ball immediately after rebounding?

STRATEGY: The initial and final speeds are related by $\frac{1}{2}mv_f^2 = \frac{1}{2}\frac{1}{2}mv_i^2$ so $v_f = v_i/\sqrt{2}$.

SOLUTION:

[ans: 3.7 m/s]

(*b*) What was the magnitude of the impulse of the ball on the wall?

STRATEGY: It is equal in magnitude and opposite in direction to the impulse of the wall on the ball. If we assume the ball hits the wall perpendicularly and bounces straight back that impulse is given by $J = p_f - p_i = m(v_f - v_i)$. Don't forget that these velocities have opposite signs. If the positive direction is toward the wall, v_i is positive and v_f is negative.

SOLUTION:

[ans: 1.3 N·s, away from the wall]

(*c*) If the ball was in contact with the wall for 7.6 ms, what was the magnitude of the average force exerted by the wall on the ball during this time interval?

STRATEGY: Use $J = \overline{F}\Delta t$.

SOLUTION:

[ans: 180 N, away from the wall]

Collision problems are more complicated. First decide if the collision is one- or two-dimensional. A head-on collision (one for which the impact parameter vanishes) is always one-dimensional. So is a completely inelastic two-body collision with one object initially at rest. If the objects move along different lines, whether initially or finally, the collision is two dimensional.

We consider one-dimensional collisions first. Since total momentum is always conserved in collisions, nearly every problem solution begins by writing the equation for momentum conservation. There is only one such equation for a one dimensional collision. To write it use the component, not the magnitude, of the momentum of each object and write the component as the product of the appropriate mass and velocity component. Always use symbols, not numbers, even for given quantities.

Make a list of the quantities given in the problem statement and a list of the unknowns. If there is only one unknown the momentum conservation equation can be solved immediately for it. Once the equation has been solved, the result can be used in other calculations. You might be asked to test for the conservation of kinetic energy, for example, and classify the collision as elastic or inelastic. You might also be asked to calculate the impulse exerted by one object on the other.

If more than one quantity is unknown, look for more information in the problem statement. If it specifies that the collision is elastic, write the equation for the conservation of kinetic energy in terms of the masses and speeds of the objects. If it specifies that the collision is completely inelastic, equate the final velocities of the objects to each other and use a single symbol to denote them. If it specifies that kinetic energy was gained or lost during the collision, write the energy equation in the form $K_f = K_i + Q$ (using masses and speeds to write K_i and K_f, of course).

Look for special conditions. If, for example, the problem statement tells you that the masses of the objects are the same, use the same symbol to represent both masses. If $Q = 0$ the masses then cancel from the equations and the algebra simplifies greatly.

PROBLEM 4 (29P).   The blocks in Fig. 10–34 slide without friction. (*a*) What is the velocity v of the 1.6-kg block after the collision?

STRATEGY: Use conservation of momentum. Let v_{1i} be the velocity of the light block before the collision and v_{1f} be its velocity after. Let v_{2i} be the velocity of the heavy block before the collision and v_{2f} be its velocity after. Then $m_1 v_{1i} + m_2 v_{2i} = m_1 v_{1f} + m_2 v_{2f}$. There is only one unknown; you can immediately solve for v_{1f}.

SOLUTION:

[ans: 1.9 m/s, to the right]

(*b*) Is the collision elastic?

STRATEGY: Calculate the total kinetic energy before the collision: $K_i = \frac{1}{2} m_1 v_{1i}^2 + \frac{1}{2} m_2 v_{2i}^2$. Calculate the total kinetic energy after the collision: $K_f = m_1 v_{1f}^2 + \frac{1}{2} m_2 v_{2f}^2$. Compare the two values. If they are the same the collision was elastic, otherwise it was not.

SOLUTION:

[ans: yes; the total kinetic energy was 31.7 J, before and after]

(c) Suppose the initial velocity of the 2.4-kg block is the reverse of what is shown. Can the velocity **v** of the 1.6-kg block after the collision be in the direction shown?

STRATEGY: The collision is elastic so

$$v_{1f} = \frac{m_1 - m_2}{m_1 + m_2} v_{1i} + \frac{2m_2}{M_1 + m_2} v_{2i} .$$

Here $v_{1i} = 5.5$ m/s and $v_{2i} = -2.5$ m/s. Calculate v_{1f}.

SOLUTION: Notice that as long as $m_1 < m_2$ and v_{2i} is to the left, v_{1f} must be to the left.

[ans: no]

PROBLEM 5 (31E). An electron collides elastically with a hydrogen atom initially at rest. (All motions are along the same straight line.) What percentage of the electron's initial kinetic energy is transferred to the hydrogen atom? The mass of the hydrogen atom is 1840 times the mass of the electron.

STRATEGY: To obtain the fraction f of the original kinetic energy that is transferred, divide the kinetic energy lost by the electron in the collision by the original kinetic energy: $f = (\frac{1}{2}m_{ei}^2 - \frac{1}{2}m_e v_{ef}^2)/\frac{1}{2}m_e v_{ei}^2 = 1 - v_{ef}^2/v_{ei}^2$. Since the collision is elastic and the hydrogen atom is initially at rest the final velocity of the electron is given by $v_{ef} = v_{ei}(m_e - m_H)/(m_e + m_H)$. Substitute this expression and $m_H = 1840m_e$ into the equation for f.

SOLUTION:

[ans: 0.22%]

PROBLEM 6 (35P). A steel ball of mass 0.500 kg is fastened to a cord 70.0 cm long and fixed at the far end, and is released when the cord is horizontal (Fig. 10–35). At the bottom of its path, the ball strikes a 2.50-kg steel block initially at rest on a frictionless surface. The collision is elastic. Find (a) the speed of the ball and (b) the speed of the block, both just after the collision.

STRATEGY: You need to know the speed of the ball just before it strikes the block. Use conservation of energy. If it has a potential energy of 0 just before striking the block its initial potential energy is $m_{ball}gL$, where L is the length of the cord. Its kinetic energy is $\frac{1}{2}m_{ball}v^2$. Solve $m_{ball}gL = \frac{1}{2}m_{ball}v^2$ for v. You should get 3.70 m/s. This is the initial velocity for the collision. Since the collision is elastic and the block is initially at rest, the final velocity of the ball is given by $v_{ball} = v(m_{ball} - m_{block})/(m_{ball} + m_{block})$ and the final velocity of the block is given by $v_{block} = 2m_{ball}v/(m_{ball} + m_{block})$.

SOLUTION:

[ans: ball: 2.47 m/s; block: 1.23 m/s]

PROBLEM 7 (38P).   A ball of mass m is aligned above a ball of mass M (with a slight separation, as in Fig. 10–25a), and the two are dropped simultaneously from height h. (Assume the radius of each ball is negligible compared to h.) (a) If M rebounds elastically from the floor and then m rebounds elastically from M, what ratio m/M results in M stopping upon its collision with m? (The answer is approximately the mass ratio of a baseball to a basketball, as in Question 13.)

STRATEGY: First use conservation of energy to find the speed of M just before it hits the floor. Take the potential energy to be 0 at the floor. Then its initial potential energy is Mgh. Its initial kinetic energy is 0, so $Mgh = \frac{1}{2}Mv^2$ and $v = \sqrt{2gh}$. After rebounding M is going upward with speed $\sqrt{2gh}$ while m is going downward with the same speed. The collision of the two balls is elastic, so $v_{Mf} = [-\sqrt{2gh}(M - m)/(M + m)] + [2m\sqrt{2gh}/(M + m)]$, where the positive direction was taken to downward. Set this equal to 0 and solve for m.

SOLUTION:

$$\left[\text{ans: } m/M = 1/3\right]$$

(b) What height does m then reach?

STRATEGY: The velocity of m after the collision is given by $v_{mf} = [-2M\sqrt{2gh}/(M + m)] + [\sqrt{2gh}(m - M)/(M + m)]$. Substitute $m = M/3$. You should obtain $V_{mf} = -2\sqrt{2gh}$. The negative sign indicates it is going upward. Use conservation of energy to find its maximum height.

SOLUTION:

$$\left[\text{ans: } 4h\right]$$

PROBLEM 8 (44E).   Two 2.0-kg masses, A and B, collide. The velocities before the collision are $\mathbf{v}_A = 15\,\mathbf{i} + 30\,\mathbf{j}$ and $\mathbf{v}_B = -10\,\mathbf{i} + 5.0\,\mathbf{j}$. After the collision, $\mathbf{v}'_A = -5.0\,\mathbf{i} + 20\,\mathbf{j}$. All speeds are given in meters per second (a) What is the final velocity of B?

STRATEGY: Use conservation of momentum: $m_A\mathbf{v}_A + m_B\mathbf{v}_B = m_A\mathbf{v}'_A + m_B\mathbf{v}'_B$. Solve for $\mathbf{v}'_B$.

SOLUTION:

$$\left[\text{ans: } (10\,\mathbf{i} + 15\,\mathbf{j})\,\text{m/s}\right]$$

(b) How much kinetic energy was gained or lost in the collision?

STRATEGY: The initial kinetic energy is given by $K_i = \frac{1}{2}m_A v_A^2 + \frac{1}{2}m_B v_B^2$ and the final kinetic energy is given by $K_f = \frac{1}{2}m_A(v'_A)^2 + \frac{1}{2}m_B(v'_B)^2$. Calculate the difference. Don't forget that the square of the speed is the sum of the squares of the components of the velocity.

SOLUTION:

[ans: 500 J lost]

PROBLEM 9 (48P).   Two cars A and B slide on an icy road as they attempt to stop at a traffic light. The mass of A is 1100 kg and the mass of B is 1400 kg. The coefficient of kinetic friction between the locked wheels of both cars and the road is 0.13. Car A succeeds in coming to rest at the light, but car B cannot stop and rear-ends car A. After the collision, A comes to rest 8.2 m ahead of the impact point and B 6.1 m ahead: see Fig. 10–37. Both drivers had their brakes locked throughout the incident. (*a*) From the distances each car moved after the collision, find the speed of each car immediately after impact.

STRATEGY: For each car the change in kinetic energy while sliding is given by $-fd$ where f is the force of friction of the road on the car and d is the distance the car slid. Write $K = \frac{1}{2}mv^2$ for the kinetic energy just after impact and $f = \mu_k N = \mu_k mg$ for the force of friction. Then $-\frac{1}{2}mv^2 = -\mu_k mgd$. Solve for v.

SOLUTION:

$$[\text{ans: } v_{Af} = 4.6 \text{ m/s}; \; v_{Bf} = 3.9 \text{ m/s}]$$

(*b*) Use conservation of momentum to find the speed at which car B struck car A. On what grounds can the use of linear momentum conservation be criticized here?

STRATEGY: Here we assume the impulse exerted by friction during impact was much less than the impulse of each car on the other, so momentum was conserved. Write $m_B v_{Bi} = m_A v_{Af} + m_B v_{Bf}$ and solve for V_{Bi}.

SOLUTION:

$$[\text{ans: } 7.5 \text{ m/s}]$$

If the cars are in contact an appreciable time the impulse of friction on them might be significant. Then momentum would not be conserved.

PROBLEM 10 (49P).   A 3.0-ton weight falling through a distance of 6.0 ft drives a 0.50-ton pile 1.0 in. into the ground. Assuming that the weight-pile collision is completely inelastic, find the average force of resistance exerted by the ground.

STRATEGY: Use conservation of energy to find the speed of the weight just before hitting the pile: $mgh = \frac{1}{2}mv_{wi}^2$. Use conservation of momentum to find the speed of the weight and pile after the collision: $m_w v_{wi} = (m_w + m_p)v_f$. If $\overline{F}$ is the average force of the ground on the pile then $-\overline{F}d$ gives the change in the kinetic energy of the weight-pile system as the pile sinks into the ground. Thus $-\overline{F}d = -\frac{1}{2}(m_w + m_p)v_f^2$. The mass in slugs is given by the weight in tons times 2000 lb/ton divided by 32 lb/slug.

146   *Chapter 10: Collisions*

SOLUTION:

PROBLEM 11 (55P).   A block of mass $m_1 = 2.0$ kg slides along a frictionless table with a speed of 10 m/s. Directly in front of it, and moving in the same direction, is a block of mass $m_2 = 5.0$ kg moving at 3.0 m/s. A massless spring with spring constant $k = 1120$ N/m is attached to the near side of m_2, as shown in Fig. 10–39. When the blocks collide, what is the maximum compression of the spring?

STRATEGY: When the compression of the spring is maximum the two blocks have the same velocity and the conservation of momentum equation becomes $m_1 v_{1i} + m_2 v_{2i} = (m_1 + m_2)v_f$. Solve for v_f. Whatever kinetic energy was lost by the blocks is stored as potential energy in the spring. If x is the compression of the spring then conservation of energy yields $\frac{1}{2}m_1 v_{1i}^2 + \frac{1}{2}m_2 v_{2i}^2 = \frac{1}{2}(m_1 + m_2)v_f^2 + \frac{1}{2}kx^2$. Solve for x.

SOLUTION:

[ans: 25 cm]

For two-dimensional collisions there are two momentum conservation equations, one for each component. The amount of algebraic manipulation can usually be reduced significantly if one of the coordinate axes is placed along the direction of the total momentum. For some problems, however, this direction cannot be found from the given data. In any event, select a coordinate system and write the momentum conservation equations in terms of the masses, speeds, and angles between the velocities and a coordinate axis.

If the line of motion of one of the objects, either before or after the collision, makes the angle θ with a coordinate axis, then you will find $\sin\theta$ in one of the components of the momentum conservation equation and $\cos\theta$ in the other. To eliminate θ, use the trigonometric identity $\cos^2\theta + \sin^2\theta = 1$. That is, solve one equation for $\sin\theta$ and the other for $\cos\theta$, square both results, add, and set the sum equal to 1.

Sometimes the angle of the line of motion is sought. You then solve the momentum conservation equations for the sine or cosine of the angle and evaluate the inverse trigonometric function. Sometimes the equation for $\sin\theta$ or $\cos\theta$ contains an unknown and you must eliminate it before solving for the angle. An unknown can often be eliminated by dividing the expression for $\sin\theta$ by the expression for $\cos\theta$. The resulting equation for $\tan\theta$ is then solved.

Suppose the collision is elastic or Q is given for an inelastic collision and you wish to eliminate one of the speeds. Solve one of the momentum conservation equations for the speed to be eliminated and substitute the resulting expression in the energy equation, rather than vice versa. If you solve the energy equation first, the resulting expression will contain a square root and probably will be difficult to manipulate.

PROBLEM 12 (61E).   A proton (atomic mass 1 u) with a speed of 500 m/s collides elastically with another proton at rest. The original proton is scattered 60° from its initial direction. (*a*) What is the direction of the velocity of the target proton after the collision?

STRATEGY: Since the collision is elastic and the particles have the same mass their scattering angles must sum to 90°. If the first proton is scattered by the angle θ_1 above the line of its original motion then the second is scattered by $\theta_2 = 90° - \theta_1$ below.

SOLUTION:

$[$ ans: 30° $]$

(*b*) What are the speeds of the two protons after the collision?

STRATEGY: The two momentum conservation equations are $v_{1i} = v_{1f} \cos\theta_1 + v_{2f} \cos\theta_2$ and $0 = v_{1f} \sin\theta_1 - v_{2f} \sin\theta_2$. Solve these simultaneously for v_{1f} and v_{2f}.

SOLUTION:

$[$ ans: $v_{1f} = 250$ m/s, $v_{2f} = .430$ m/s $]$

PROBLEM 13 (67P).   Two balls A and B, having different but unknown masses, collide. A is initially at rest and B has speed v. After collision, B has speed $v/2$ and moves perpendicularly to its original motion. (*a*) Find the direction in which ball A moves after collision.

STRATEGY: The conservation of momentum equations are $m_B v_{Bi} = m_A v_{AF} \cos\theta_A$ and $0 = m_A v_{Af} \sin\theta_A - m_B v/2$, where the scattering angles θ_A and θ_B are on opposite sides of the original direction of motion of B. Solve for θ_A.

SOLUTION: One way to do this is to use one of the equations to eliminate v_{AF} from the other. You should get $\tan\theta_A = 1/2$.

$[$ ans: 26.6° $]$

(*b*) Can you determine the speed of A from the information given? Explain.

SOLUTION: Solve the momentum conservation equations for v_{Af}. You should get $v_{Af} = m_B v/2m_A \sin\theta_A$. What must you be given to obtain a value for v_{Af}?

$[$ ans: no; the ratio of the masses must be given $]$

148   *Chapter 10: Collisions*

PROBLEM 14 (71P).   A billiard ball moving at a speed of 2.2 m/s strikes an identical stationary ball a glancing blow. After the collision, one ball is found to be moving at a speed of 1.1 m/s in a direction making a 60° angle with the original line of motion. (*a*) Find the velocity of the other ball.

STRATEGY: Label the incident ball 1 and the target ball 2. Then the conservation of momentum equations are $v_{1i} = v_{1f} \cos \theta_1 + v_{2f} \cos \theta_2$ and $0 = v_{1f} \sin \theta_1 - v_{2f} \sin \theta_2$, where θ_2 is on the opposite side of the original line of motion from θ_1. Solve these simultaneously for v_{2f} and θ_2.

SOLUTION: First eliminate θ_2. Solve the first equation for $\cos \theta_2$ and the second for $\sin \theta_2$. Square the results and add them. Use $\cos^2 \theta_2 + \sin^2 \theta_2 = 1$ and obtain $v_{2f} = \sqrt{v_{1i}^2 - 2v_{1i}v_{1f} \cos \theta_1 + v_{1f}^2}$. After evaluating this expression use $\sin \theta_2 = (v_{1f}/v_{2f}) \sin \theta_1$ to find θ_2.

$$\left[\text{ans: } 1.9 \text{ m/s}, 30° \text{ from the original line of motion}\right]$$

(*b*) Can the collision be inelastic, given these data?

STRATEGY: Compare the kinetic energy before the collision with the kinetic energy after.

SOLUTION: Since the masses are the same all you need to do is compare v_{1i}^2 with $v_{1f}^2 + v_{2f}^2$.

$$\left[\text{ans: no, the collision is elastic or close to it}\right]$$

PROBLEM 15 (73P).   A barge with mass 1.50×10^5 kg is proceeding down river at 6.2 m/s in heavy fog when it collides broadside with a barge heading directly across the river; see Fig. 10–44. The second barge has mass 2.78×10^5 kg and was moving at 4.3 m/s. Immediately after impact, the second barge finds its course deflected by 18° in the down river direction and it speed increased to 5.1 m/s. The river current was practically zero at the time of the accident. (*a*) What are the speed and direction of motion of the first barge immediately after the collision?

STRATEGY: Take the x axis to be downstream and the y axis to be across the river, in the original direction of motion of the second barge. Let A represent the barge originally going downstream, and let B represent the barge going across the river. Then the conservation of momentum equations become $m_A v_{Ai} = m_B v_{Bf} \sin \theta_B + m_A v_{Ax}$ and $m_B v_{Bi} = m_B v_{Bf} \cos \theta_B + m_A v_{Ay}$, where v_{Ax} and v_{Ay} are the components of the velocity of barge A after the collision and $\theta_B = 18°$. Solve these for v_{Ax} and v_{Ay}. Find the speed of A using $v_{Af}^2 = v_{Ax}^2 + v_{Ay}^2$ and the angle the velocity makes with the downstream direction using $\tan \theta_A = v_{Ay}/v_{Ax}$.

SOLUTION:

$$\left[\text{ans: } 3.4 \text{ m/s}, -17.3° \text{ with the downstream direction}\right]$$

(b) How much kinetic energy is lost in the collision?

STRATEGY: Calculate the kinetic energy before and after the collision and find the difference.

SOLUTION: Use $K_i = \frac{1}{2}m_A v_{Ai}^2 + \frac{1}{2}m_B v_{Bi}^2$ and $K_f = \frac{1}{2}m_A v_{Af}^2 + \frac{1}{2}m_B v_{Bf}^2$.

$$[\text{ans: } 9.6 \times 10^5 \, \text{J}]$$

III. NOTES

Chapter 11
ROTATION

I. BASIC CONCEPTS

You now begin the study of rotational motion. The pattern is similar to that used for the study of linear motion: you first study kinematics, the description of the motion, and then the dynamics of the motion. Important new concepts include angular displacement, angular velocity, angular acceleration, torque, and rotational inertia. Take special care to understand their definitions and the roles they play in rotational motion. Also concentrate on rotational kinetic energy and the work-energy theorem for rotation.

Definitions. If a rigid body undergoes pure rotation about a fixed axis then the path followed by each point on the body is a _____, centered on the _____.

The <u>angular</u> <u>position</u> of the body is described by giving the angle between a <u>reference</u> <u>line</u>, fixed to the body, and a non-rotating coordinate axis. Suppose the body rotates around the z axis and describe how a reference line might be chosen. In particular, identify the plane it is parallel to and the angle it makes with the axis of rotation. _____

Illustrate the idea in the space to the right by drawing the outline of a rotating body, with the axis of rotation, the z axis, out of the page. Mark the axis with a circled dot ($\odot$). Show x and y axes and a reference line, all in the plane of the page. Usually the angle θ that gives the angular position is measured counterclockwise from a coordinate axis. Show θ on your diagram.

The angular position θ is often measured in radians. Define a radian: _____

You should know the relationship between radian, degree, and revolution measure: 1 radian = _____ degrees = _____ revolution. Some often used angles are:

$360° = $ _____ radians $ = $ _____ revolution

$270° = $ _____ radians $ = $ _____ revolution

$180° = $ _____ radians $ = $ _____ revolution

$90° = $ _____ radians $ = $ _____ revolution

$45° = $ _____ radians $ = $ _____ revolution

Since an angle in radians is the ratio of two lengths it is dimensionless.

If the body rotates from θ_1 to θ_2, then its <u>angular</u> <u>displacement</u> is $\Delta\theta = $ _____. If θ_2 is greater than θ_1 then the angular displacement is positive. According to the convention stated above for measuring θ, the body rotated in the _____ direction. If θ_2 is less than θ_1 then the angular displacement is negative and the body rotated in the _____ direction.

If the body rotates through more than 1 revolution, θ continues to increase beyond 2π rad. Be sure you understand that an angle of 2π rad is NOT equivalent to 0. A rotation from $\theta = 0$ to $\theta = 2\pi$ rad represents an angular displacement of 2π rad, not 0. An angular displacement of 2 revolutions represents an angular displacement of _____ rad.

If the body undergoes an angular displacement $\Delta\theta$ in time Δt, its <u>average angular velocity</u> during the interval is $\overline{\omega} =$ _____ . Its <u>instantaneous angular velocity</u> at any time is given by the derivative $\omega =$ _____ . According to the convention used above to measure the angular position, a positive value for ω indicates rotation in the _____ direction and a negative value for ω indicates rotation in the _____ direction. Instantaneous angular velocity is usually called simply angular velocity.

If the angular velocity of the body changes with time then the body has a non-vanishing angular acceleration. If the angular velocity changes from ω_1 to ω_2 in the time interval Δt, then the <u>average angular acceleration</u> in the interval is $\overline{\alpha} =$ _____ . The <u>instantaneous angular acceleration</u> at any time t is given by the derivative $\alpha =$ _____ . Instantaneous angular acceleration is usually called simply angular acceleration.

Common units of angular velocity are deg/s, rad/s, rev/s, and rev/min. Corresponding units of angular acceleration are _____ , _____ , _____ , and _____ .

You should understand that a positive angular acceleration does NOT necessarily mean the rotational speed is increasing and a negative angular acceleration does NOT necessarily mean the rotational speed is decreasing. Remember that the rotational speed decreases if ω and α have opposite signs and increases if they have the same sign, no matter what the sign of α. If ω is negative and α is positive, for example, then ω becomes less negative as time goes on and its magnitude becomes smaller. If the positive acceleration continues beyond $\omega = 0$, of course, the magnitude of ω starts increasing with time.

The values of $\Delta\theta$, ω, and α are the same for every point in a rigid body. The angular positions of different points may be different, of course, but when one point rotates through any angle, all points rotate through the same angle. All points rotate through the same angle in the same time and their angular velocities change at the same rate.

Angular velocity and angular acceleration, but NOT angular position or displacement, can be considered to be vectors along the axis of rotation. To determine the direction of ω, use the right hand rule: _____
_____

The diagram on the right shows a rigid rod rotating around the z axis, with points to the right of the axis moving into the page and points to the left of the axis moving out of the page. Show the angular velocity vector ω on the diagram. If the body were rotating in the opposite direction the vector angular velocity would be in the _____ direction.

z

moving out of page | moving into page

If the body rotates faster as time goes on, then α and ω are in the same direction. If it rotates more slowly, then α and ω are in opposite directions. Consider the situation illustrated above (right side moving into page, left side moving out of page) and suppose the body rotates faster as time goes on. Then the angular acceleration vector α is in the _____ direction. If the body rotates more slowly as time goes on, α would be in the _____ direction.

Although directions can be assigned arbitrarily to angular displacements, they do not add as vectors. The result of two successive angular displacements around different axes that are not parallel, for example, depends on the order in which the angular displacements are carried out. This means angular displacements are not vectors.

Rotation with constant angular acceleration. If a body is rotating around a fixed axis with constant angular acceleration α, then as functions of time its angular velocity is given by

$$\omega(t) =$$

and its angular position is given by

$$\theta(t) =$$

Here ω_0 is the _____ and θ_0 is the _____ .

These two equations are solved simultaneously to find answers to constant angular acceleration kinematics problems. You should memorize them or, better yet, learn to derive them quickly using techniques of the integral calculus. Another useful equation can be obtained by using $\omega = \omega_0 + \alpha t$ to eliminate t from $\theta = \omega_0 t + \frac{1}{2}\alpha t^2$. It is

$$\omega^2 - \omega_0^2 = 2\alpha\theta \,,$$

where θ_0 was taken to be 0.

Table 11–1 of the text lists all possible results when one of the kinematic equations given above is used to eliminate a variable from the other. You may use these equations to solve rotational kinematics problems or else simultaneously solve the two you wrote above.

Any consistent set of units can be used. Thus θ in degrees, ω in degrees/s, and α in degrees/s^2 as well as θ in radians, ω in radians/s, and α in radians/s^2 or θ in revolutions, ω in revolutions/s, and α in revolutions/s^2 can be used. Be careful not to mix units, however.

Relationships between linear and angular quantities. Each point in a rotating body has a velocity and acceleration and these are related to the angular variables. Suppose the body is rotating around a fixed axis and consider a point in the body a perpendicular distance r from the axis. If the body turns through the angle θ (in radians) the point moves a distance $s =$ _____ along its circular path. The speed of the point is $v = ds/dt$ and the angular velocity of the body is $\omega = d\theta/dt$, so v is related to ω by $v =$ _____ , where the units of ω MUST be _____ . The tangential component of the acceleration of the point is $a_t = dv/dt$ and the angular acceleration is $\alpha = d\omega/dt$ so a_t is related to α by $a_t =$ _____ , where the units of α MUST be _____ .

Because the point is moving in a circular path its acceleration also has a radial component. You learned in Chapter 4 that $a_r = v^2/r$. Now write the relationship in terms of the angular velocity instead of the speed: $a_r = $ _____ .   In terms of the components a_r and a_t, the magnitude of the total acceleration is given by $a = $ _____ and, in terms of ω, α, and r, is given by $a = $ _____ .   Recall that an angle in radians is really unitless and check the units on the two sides of each equation you just wrote to be sure they are consistent.

For a point in a body rotating with constant angular acceleration, the tangential component of the acceleration is also constant. If s represents the distance traveled by the point along its circular path from time 0 to time t, then $s(t) = v_0 t + \frac{1}{2} a_t t^2$. The speed v of the point is given by $v(t) = v_0 + a_t t$. Divide these equations by the radius r of the orbit to obtain the kinematic equations for $\theta(t)$ and $\omega(t)$.

Be sure you understand the relationships between the angular and linear quantities. Every point on a rotating body turns through the same angle in the same time, has the same angular velocity, and has the same angular acceleration. This means that compared to a point on the rim of a rotating wheel, for example, a point halfway out travels _____ the distance, has _____ the speed, and has _____ the tangential acceleration.

Rotational inertia.   Rotational inertia is a property of a rotating body that depends on the distribution of mass in the body and on the axis of rotation. It determines the angular acceleration and change in kinetic energy produced by the net torque acting. That is, the same torque acting on different bodies produces different angular accelerations and different changes in kinetic energy because the bodies have different rotational inertias.

The rotational inertia I of a rigid body consisting of N particles, rotating about a given fixed axis, is defined by the sum over all particles:

$$I = $$

where m_i is the mass of particle i and r_i is the distance of particle i from the axis of rotation. This distance is measured along a line that is perpendicular to the axis of rotation, from the axis to the particle.

The sum is difficult to carry out for a body with more than a few particles. Some bodies however, can be approximated by a continuous distribution of mass and techniques of the integral calculus can be used to find the rotational inertia. For a body with a continuous distribution of mass the rotational inertia is given by the integral

$$I = $$

In Sample Problem 11–8 of the text the integral is evaluated for a thin rod rotating about an axis through it center. Table 11–2 of the text gives the rotational inertias of several bodies, some for two different axes. You will need to refer to this figure as you work through problems.

The equation that defines the rotational inertia as a sum over the particles of a body is very useful when we need to think about the rotational inertia of an object. For example, the sum shows us that the rotational inertia depends on the square of the distance of each particle from the axis as well as on the mass of each particle. A particle that is far from the axis contributes more to the rotational inertia than a particle of equal mass close to the axis.

Look at the expressions given in Table 11–2 for the rotational inertias of a hoop about the cylinder axis and for a solid cylinder, also about the cylinder axis. The rotational inertia of a solid cylinder is less than the rotational inertia of a hoop with the same mass and radius because _____ .

The defining equation also shows that the rotational inertia depends on the orientation and position of the axis of rotation. Look at Table 11–2 and compare the expression for the rotational inertia of a thin rod rotating about an axis through its center to the equation for the rotational inertia of the same rod but rotating about an axis through one end. The rotational inertia is smaller when the axis is through the center because _____

_____

The defining equation for the rotational inertia is a sum over all particles in the body. If we like, we can consider the body to be composed of two or more parts and calculate the rotational inertia of each part about the same axis, then add the results to obtain the rotational inertia of the complete body. As an example, suppose two identical solid uniform balls, each of mass m and radius r, are attached to each other at a point on their surfaces and the composite body rotates around the line that joins their centers. Both balls rotate about a diameter so each contributes $2mr^2/5$ to the rotational inertia of the complete body. Thus for the composite $I = 4mr^2/5 = 2Mr^2/5$, where $M = 2m$ is the total mass.

The defining equation for the rotational inertia is used to prove the parallel-axis theorem. Here we consider two identical bodies. One is rotating about an axis through the center of mass and the other is rotating about an axis that is parallel to the first axis but is displaced from the center of mass by a distance h (measured along a line that is perpendicular to the axis). The rotational inertia I for the second body is related to the rotational inertia I_{cm} for the first by $I =$ _____, where M is the total mass of the body.

An example is given in Table 11–2 of the text. Compare the equations for the two cases of a thin rod. The distance between the end and center of the rod is $L/2$, so for a rod rotating about one end $I = I_{cm} + ML^2/4 = ML^2/12 + ML^2/4 = ML^2/3$.

Now you try one. Consider a uniform sphere of radius r and mass M, rotating about an axis that is tangent to its surface. The distance between this axis and a parallel axis through the center is _____. Look up the expression for the rotational inertia of a uniform solid sphere about an axis through its center and apply the parallel-axis theorem. The rotational inertia about the axis tangent to the surface is given by $I =$ _____ + _____ = _____. You should get $I = 7Mr^2/5$.

The parallel-axis theorem tells us of all the places we can position the axis of rotation the one that leads to the smallest rotational inertia is the one through the center of mass and that the rotational inertia increases as the axis moves away from the center of mass (remaining parallel to its original orientation, of course).

Rotational kinetic energy. Suppose a rigid body, rotating about a fixed axis, is made up of N particles. The kinetic energy of the particles is given by $K = \sum \frac{1}{2} m_i v_i^2$, where m_i is the mass of particle i and v_i is its speed. Substitute $v_i = r_i \omega$, where r_i is the distance of particle i from the axis, and obtain $K =$ _____. Notice that the sum over particles for the rotational inertia appears in the expression for K. For a rigid body with rotational inertia I, rotating with angular velocity ω about a fixed axis, the total kinetic energy of all the particles in the body is given by $K =$ _____. For this expression to be valid ω must be in rad/s.

A particle far from the axis of rotation contributes more to the rotational kinetic energy than a particle of the same mass closer to the axis. Use $v = r\omega$ to explain: _____

Torque. The net torque acting on a rotating body gives the body an angular acceleration. It does work and changes the rotational kinetic energy of the body.

If a force is applied to a body free to rotate about a fixed axis the torque (about the axis) associated with the force is intimately related to the force component tangent to the circle through the point of application and centered at the axis. In fact, $\tau =$ _____, where r is the radius of the circle (the distance from the origin to the point of application) and F_t is the tangential component of the force.

The diagram on the right shows a particle traveling counterclockwise around a circle of radius r. A force **F** acting on it at some instant of time is also shown. On the diagram show the tangential and radial components of the force. The radial component holds it in the circle. The tangential component produces an angular acceleration. In terms of the magnitude F of the force and the angle ϕ shown, the radial component is $F_r =$ _____ and the tangential component is $F_t =$ _____. The magnitude of the torque is _____.

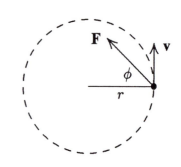

The diagram on the right shows a body subjected to three forces. The axis of rotation is through O. The forces and their points of application are all in the plane of the page. For each force draw the line from the axis to the point of application and label it r_1, r_2, or r_3, as appropriate. For each, label the angle ϕ that appears in $\tau = Fr\sin\phi$.

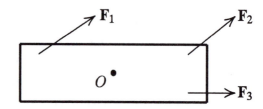

Notice that the magnitude of a torque depends on the distance from the axis to the point of application. A tangential force produces a greater angular acceleration if it is applied far from the axis than if it applied closer.

For rotation about a fixed axis we may take a torque to be positive if it tends to turn the object counterclockwise and negative if it tends to turn the body clockwise. This convention is consistent with the one introduced in the last chapter for the signs of the angular velocity and acceleration. For the situation illustrated in the diagram above, the sign of τ_1 is _____, the sign of τ_2 is _____, and the sign of τ_3 is _____.

Newton's second law for rotation about a fixed axis. This law relates the net torque $\sum \tau$ acting on a rigid body to the angular acceleration α of the body. Quantitatively the law is

$$\sum \tau =$$

where I is the rotational inertia of the body about the axis of rotation. It is as important to the study of rotational motion about a fixed axis as $\sum F = ma$ is to the study of one-dimensional translational motion. Concentrate on learning to use it.

The law is easy to derive for a particle moving in a circle and subjected to a force **F**. The tangential component of the force produces a tangential acceleration a_t according to $F_t =$ _____ , where m is the mass of the particle. Since the tangential and angular accelerations are related by $a_t = r\alpha$, you can write $F_t =$_____. Now multiply both sides by r to obtain $\tau =$ _____ and notice that the rotational inertia mr^2 appears as a factor. For a body composed of many particles, acted on by more than one torque, the derivation is slightly more complicated but the result is the same: $\sum \tau = I\alpha$.

Be careful that you use the total torque in this equation. You must identify and sum all the torques that are acting and you must be careful about signs. You must also be careful to use radian measure for the angular acceleration α.

Work done by a torque. Consider a particle traveling counterclockwise in a circular orbit, subjected to a force with tangential component F_t. As it moves through an infinitesimal arc length ds the magnitude of the work done by the force is d$W =$ _____. Since d$s = r\,$dθ, where dθ is the infinitesimal angular displacement, and $\tau = F_t r$, the expression for the work becomes d$W =$ _____. As the particle travels from θ_1 to θ_2 the work done by the torque is given by the integral

$$W =$$

In terms of the torque τ and angular velocity ω the power supplied by the torque is given by $P = dW/dt =$ _____.

The work done by a torque may be positive or negative. If the torque and angular displacement are in the same direction, both clockwise or both counterclockwise, then the work is _____ ; if they are in opposite directions, one clockwise and one counterclockwise, then the work is _____.

On the left-hand diagram below, show a force that does positive work on the particle and give the signs of the torque and the angular velocity, using the convention described above (positive for counterclockwise, negative for clockwise). On the right hand diagram, do the same for a force that does negative work.

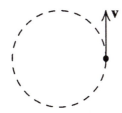

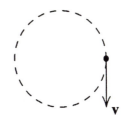

Force does positive work
Sign of torque: _____
Sign of angular velocity: _____

Force does negative work
Sign of torque: _____
Sign of angular velocity: _____

The same results hold for an extended body. If more than one torque acts, sum the works done by the individual torques to find the total work.

A work-energy theorem is valid for rotational motion: the total work done by all torques acting on a body equals the change in its rotational kinetic energy. Suppose the net torque τ is constant. Further suppose that in a certain time interval the body turns through the angular displacement $\Delta\theta$ and its angular velocity changes from ω_0 to ω. In terms of these quantities the work-energy theorem becomes: _____ = _____ , where I is the rotational inertia of the body.

II. PROBLEM SOLVING

Many problems of this chapter are mathematically identical to those of one-dimensional linear kinematics, discussed in Chapter 2: ϕ replaces x, ω replaces v, and α replaces a.

PROBLEM 1 (4E). The angular position of a point on the rim of a rotating wheel is given by $\theta = 4.0t - 3.0t^2 + t^3$, where θ is in radians if t is given in seconds. (a) What are the angular velocities at $t = 2.0$ s and $t = 4.0$ s?

STRATEGY: Find an expression for the angular velocity ω as a function of time by differentiating $\theta(t)$. Evaluate the expression for the two values of the time.

SOLUTION: The angular velocity is given by $\omega(t) = 4.0 - 6.0t + 3.0t^2$.

[ans: 4.0 rad/s, 28 rad/s]

(b) What is the average angular acceleration for the time interval that begins at $t = 2.0$ s and ends at $t = 4.0$ s?

STRATEGY: Use the definition of the average angular acceleration: $\overline{\alpha} = (\omega_f - \omega_i)/\Delta t$.

SOLUTION:

[ans: 12 rad/s^2]

(c) What are the instantaneous angular accelerations at the beginning and end of this time interval?

STRATEGY: Differentiate $\omega(t)$ to obtain an expression for the angular acceleration as a function of time. Evaluate the expression for the two values of the time.

SOLUTION: You should get $\alpha(t) = -6.0 + 6.0t$.

[ans: 6.0 rad/s^2, 18 rad/s^2]

PROBLEM 2 (8P). A good baseball pitcher can throw a baseball toward home plate at 85 mi/h with a spin of 1800 rev/min. How many revolutions does the baseball make on its way to home plate? For simplicity, assume that the 60-ft trajectory is a straight line.

STRATEGY: The time taken by the ball to reach home plate is given by $t = d/v$, where d is the distance and v is the speed of the ball. The angle through which the ball rotates in this time is given by $\theta = \omega t$, where ω is its angular velocity.

SOLUTION: Be careful about units here. Since d is given in ft, convert 85 mi/h to ft/s. There are 5280 feet in a mile and 3600 seconds in an hour. Also convert the angular velocity 1800 rev/min to revolutions per second.

[ans: 14 rev]

PROBLEM 3 (10P). A wheel has eight spokes and a radius of 30 cm. It is mounted on a fixed axle and is spinning at 2.5 rev/s. You want to shoot a 20-cm long arrow parallel to this axle and through the wheel without hitting any of the spokes. Assume that the arrow and the spokes are very thin; see Fig. 11–26. (*a*) What minimum speed must the arrow have?

STRATEGY: The minimum speed will just allow the arrow to travel its own length in the time the wheel turns through one-eighth of a revolution. This time is given by $t = \theta/\omega$, where $\theta = 1/8$ rev. The minimum speed of the arrow is given by $v = \ell/t = \omega\ell/\theta$, where ℓ is the length of the arrow.

SOLUTION:

[ans: 4.0 m/s]

(*b*) Does it matter where between the axle and rim of the wheel you aim? If so, where is the best location?

STRATEGY: Near the rim there is more space but the spokes are moving faster than near the axle. In fact, the both the spoke speed and the arc length between spokes are proportional to the distance from the axle.

SOLUTION:

[ans: no]

If a problem involves rotation with constant angular acceleration, first write the kinematic equations $\theta(t) = \theta_0 + \omega_0 t + \frac{1}{2}\alpha t^2$ and $\omega(t) = \omega_0 + \alpha t$, then solve them simultaneously for the unknowns. As an alternative, search Table 11–1 for the equation containing the quantities that are given in the problem statement.

Recall that when you solve one-dimensional kinematic problems you can always place the origin at any point you choose. In particular you might select the origin to be at the position of the particle when time $t = 0$ and thus take x_0 to be zero. Similarly, to describe rotational motion you can always orient the fixed reference frame and select the reference line in the body so the initial angular position θ_0 is zero. This is usually helpful because it reduces the number of algebraic symbols you must carry in the calculation.

PROBLEM 4 (13E). The angular speed of an automobile engine is increased from 1200 rev/min to 3000 rev/min in 12 s. (*a*) What is its angular acceleration in rev/min², assuming it to be uniform?

STRATEGY: Use $\omega = \omega_0 + \alpha t$.

SOLUTION: Keep the angular velocities in revolutions per minute and convert 12 s to minutes.

[ans: 9000 rev/min²]

(*b*) How many revolutions does the engine make during this time?

STRATEGY: Use $\theta = \omega_0 t + \frac{1}{2}\alpha t^2$.

SOLUTION:

[ans: 420 rev]

PROBLEM 5 (17E).   A pulley wheel 8.0 cm in diameter has a 5.6-m long cord wrapped around its periphery. Starting from rest, the wheel is given a constant angular acceleration of 1.5 rad/s². (*a*) Through what angle must the wheel turn for the cord to unwind?

STRATEGY: The cord unwinds a distance $2\pi r$ for every revolution of the wheel. If ℓ is the length of the cord then when it is completely unwound the wheel has turned through the angle $\theta = \ell/2\pi r$ in revolutions or $\theta = \ell/r$, in radians.

SOLUTION:

[ans: 22 rev (140 rad)]

(*b*) How long does it take?

STRATEGY: Solve $\theta = \frac{1}{2}\alpha t^2$ for t.

SOLUTION:

[ans: 14 s]

PROBLEM 6 (22P).   At $t = 0$, a flywheel has an angular velocity of 4.7 rad/s, an angular acceleration of -0.25 rad/s², and a reference line at $\theta_0 = 0.0$. (*a*) Through what maximum angle θ_{max} will the reference line turn in the positive direction?

STRATEGY: The line will turn with the wheel in the positive direction until its angular velocity is zero. Then θ is a maximum. Solve $0 = \omega_0 + \alpha t$ for t, then evaluate $\theta = \omega_0 t + \frac{1}{2}\alpha t^2$. As an alternative, put $\omega = 0$ in $\omega^2 - \omega_0^2 = 2\alpha\theta$ and solve for θ.

SOLUTION:

[ans: 44 rad]

At what times t will the line be at (*b*) $\theta = \frac{1}{2}\theta_{max}$, and (*c*) $\theta = -10.5$ rad (consider both positive and negative values of t)?

STRATEGY: Solve $\theta = \omega_0 t + \frac{1}{2}\alpha t^2$ for t. This is a quadratic equation and you should get two solutions for each value of θ.

SOLUTION: The general solution to the quadratic equation is $t = -(\omega_0/\alpha) \pm \sqrt{(\omega_0/\alpha)^2 + (2\theta/\alpha)}$. Evaluate this expression for $\theta = 22$ rad and for $\theta = -10.5$ rad.

[ans: (*b*) 5.5 s, 32 s; (*c*) −2.1 s, 40 s]

(*d*) Graph θ versus t, and indicate the answers to (*a*), (*b*), and (*c*) on the graph.

SOLUTION: Use the axes drawn below.

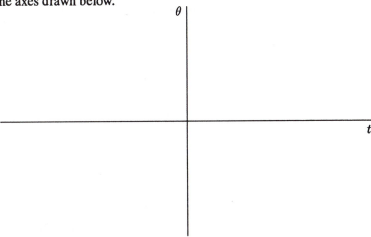

PROBLEM 7 (23P). A disk rotates about a fixed axis starting from rest and accelerates with constant angular acceleration. At one time it is rotating at 10 rev/s. After undergoing 60 more complete revolutions its angular speed is 15 rev/s. Calculate (*a*) the angular acceleration, (*b*) the time required to complete the 60 revolutions mentioned, (*c*) the time required to attain the 10-rev/s angular speed, and (*d*) the number of revolutions from rest until the time the disk attained the 10-rev/s angular speed.

STRATEGY: (*a*) Solve $\omega^2 - \omega_0^2 = 2\alpha\theta$ for α. Substitute $\omega = 15$ rev/s, $\omega_0 = 10$ rev/s, and $\theta = 60$ rev. (*b*) Solve $\omega = \omega_0 + \alpha t$ for t. Use the value for α found in the last part. (*c*) Solve $\omega = \omega_0 + \alpha t$ for t. Now $\omega = 10$ rev/s and $\omega_0 = 0$. The angular acceleration has the same value as before. (*d*) Use $\theta = \omega_0 t + \frac{1}{2}\alpha t^2$, with $\omega_0 = 0$. The time has the value found in part (*c*).

SOLUTION:

[ans: (*a*) 1.0 rev/s²; (*b*) 4.8 s; (*c*) 9.6 s; (*d*) 48 rev]

Problems in another category deal with relationships between linear and angular variables. Given the radius of the orbit of a point in a rotating body you should be able to obtain values for the linear variables from values for the angular variables and vice versa. Use $\Delta s = r\Delta\theta$, $v = r\omega$, $a_t = r\alpha$, and $a_r = r\omega^2$. Don't forget that the angular variables must be expressed in radian measure. This may require a change in units.

If you are asked to calculate the acceleration of a point in a spinning body, remember it has both tangential and radial components. You need to know ω to compute the radial component and α to compute the tangential component. These might be given or might be solutions to a kinematic problem. If you are asked for the magnitude of the acceleration use $a = \sqrt{a_r^2 + a_t^2}$.

PROBLEM 8 (32E). An astronaut is being tested in a centrifuge. The centrifuge has a radius of 10 m and, in starting, rotates according to $\theta = 0.30t^2$, where t in seconds gives θ in radians. When $t = 5.0$ s, what are the astronaut's (*a*) angular velocity, (*b*) linear speed, (*c*) tangential acceleration (magnitude only), and (*d*) radial acceleration (magnitude only)?

STRATEGY: (*a*) Differentiate $\theta = 0.30t^2$ to obtain an expression for the angular velocity as a function of time. Evaluate the expression for $t = 5.0$ s. (*b*) Use $v = r\omega$, where r is the radius of the centrifuge and ω is the answer to the last part. (*c*) Differentiate the angular velocity with respect to time to find the value of the angular acceleration, then use $a_t = r\alpha$ to find the tangential component of the acceleration. (*d*) Use $a_r = r\omega^2$ to find the radial component of the acceleration.

SOLUTION:

[ans: (*a*) 3.0 rad/s; (*b*) 30 m/s; (*c*) 6.0 m/s^2; (*d*) 90 m/s^2]

PROBLEM 9 (40P). A car starts from rest and moves around a circular track of radius 30.0 m. It speed increases at the constant rate of 0.500 m/s^2. (*a*) What is the magnitude of its *net* linear acceleration 15.0 s later?

STRATEGY: The tangential component of its acceleration is 0.500 m/s^2. Find the radial component. First use $\omega = \alpha t$, with $\alpha = a_t/r$, to find its angular speed, then use $a_r = r\omega^2$ to find its radial acceleration. Here r is the radius of the track. Finally, use $a^2 = a_t^2 + a_r^2$ to find the magnitude of the acceleration.

SOLUTION:

[ans: 1.94 m/s^2]

(*b*) What angle does this net acceleration vector make with the car's velocity at this time?

STRATEGY: Since the velocity is tangent to the track, the angle is given by $\tan\phi = a_r/a_t$. Since the car is going around a circle, the radial component is inward and, since it is speeding up, the tangential component is in the direction of the velocity.

SOLUTION:

[ans: 75.1°, from the direction of the velocity, toward the center of the circle]

PROBLEM 10 (42P). Four pulleys are connected by two belts as shown in Fig. 11–30. Pulley A (radius 15 cm) is the drive pulley, and it rotates at 10 rad/s. Pulley B (radius 10 cm) is connected by belt 1 to pulley A. Pulley B' (radius 5 cm) is concentric with pulley B and is rigidly attached to it. Pulley C (radius 25 cm) is connected by belt 2 to pulley B'. Calculate (*a*) the linear speed of a point on belt 1, (*b*) the angular speed of pulley B, (*c*) the angular speed of pulley B', (*d*) the linear speed of a point on belt 2, and (*e*) the angular speed of pulley C.

STRATEGY: (*a*) Belt 1 has the same speed as a point on the rim of pulley A. Use $v_1 = r_A\omega_A$ to calculate that speed. (*b*) The rim of pulley B has the same speed as belt 1 so $v_1 = r_B\omega_B$. Calculate ω_B. (*c*) Since B' is rigidly attached to B they rotate with the same angular speed. (*d*) Belt 2 has the same speed as a point on the rim of pulley B'. Use $v_2 = r_{B'}\omega_{B'}$. (*e*) A point on the rim of pulley C has the same speed as belt 2. Use $v_2 = r_C\omega_C$ to compute ω_C.

SOLUTION:

You should know how to compute the rotational inertia of a system composed of a small number of discrete particles. Given the masses and positions of the particles you must evaluate the sum $I = \sum m_i r_i^2$, where r_i is the perpendicular distance of particle i from the axis of rotation. You may need to use some geometry to obtain r_i. If, for example, the z coordinate axis is the axis of rotation then $r_i^2 = x_i^2 + y_i^2$. For many problems you will want to make use of Table 11–2. For some you will need to use the parallel-axis theorem.

PROBLEM 11 (49E).   The masses and coordinates of four particles are as follows: $50\,\text{g}$, $x = 2.0\,\text{cm}$, $y = 2.0\,\text{cm}$; $25\,\text{g}$, $x = 0$, $y = 4.0\,\text{cm}$; $25\,\text{g}$, $x = -3.0\,\text{cm}$, $y = -3.0\,\text{cm}$; $30\,\text{g}$, $x = -2.0\,\text{cm}$, $y = 4.0\,\text{cm}$. What is the rotational inertia of this collection with respect to the (a) x, (b) y, and (c) z axes?

STRATEGY: The rotational inertia of a collection of particles is given by the sum $\sum m_i r_i^2$, where m_i is the mass of particle i and r_i is its distance from the axis of rotation. If the axis of rotation is the x axis then $r_i^2 = y_i^2 + z_i^2 = y_i^2$. The last equality holds for the four particles of this system because all are in the xy plane. If the axis of rotation is the y axis then $r_i^2 = x_i^2 + z_i^2 = x_i^2$ and if the axis of rotation is the z axis then $r_i^2 = x_i^2 + y_i^2$.

SOLUTION:

(d) If the answers to (a) and (b) are A and B respectively, then what is the answer to (c) in terms of A and B?

STRATEGY: Notice that $A = \sum m_i y_i^2$, $B = \sum m_i x_i^2$, and the answer to (c) is $\sum m_i (x_i^2 + y_i^2)$.

SOLUTION:

PROBLEM 12 (57P).   Figure 11–34 shows a uniform solid block of mass M and edge lengths a, b, and c. Calculate its rotational inertia about an axis through one corner and perpendicular to the large faces of the block.

STRATEGY: Table 11–2 gives the rotational inertia about an axis through the center of the face. It is $I_{cm} = M(a^2 + b^2)/12$. The axis through the corner is displaced from the center by $h = \sqrt{(a/2)^2 + (b/2)^2}$. Use the parallel-axis theorem: $I = I_{cm} + Mh^2$.

SOLUTION:

PROBLEM 13 (59P).   Delivery trucks that operate by making use of energy stored in a rotating flywheel have been used in Europe. The trucks are charged by using an electric motor to get the flywheel up to its top speed of 200π rad/s. One such flywheel is a solid, homogeneous cylinder with a mass of 500 kg and a radius of 1.0 m. (*a*) What is the kinetic energy of the flywheel after charging?

STRATEGY: Use $K = \frac{1}{2}I\omega^2$, with $I = \frac{1}{2}MR^2$ (see Table 11–2).

SOLUTION:

$$\left[\text{ans: } 4.9 \times 10^7\,\text{J}\right]$$

(*b*) If the truck operates with an average power requirement of 8.0 kW, for how many minutes can it operate between chargings?

STRATEGY: Use $\overline{P} = \Delta K/\Delta t$. Since the wheel starts fully charged and ends at rest, ΔK is the kinetic energy found in part (*a*). Solve for Δt.

SOLUTION:

$$\left[\text{ans: } 6.2 \times 10^3\,\text{s } (100\,\text{min})\right]$$

Some problems ask you to compute the torque associated with a given force. Simply find the force component tangent to the circular orbit at the application point, then multiply by the perpendicular distance from the axis to the application point. For fixed axis rotation you may treat the torque as a scalar. Assign a sign to it according to whether it tends to cause rotation in the counterclockwise ($+$) or clockwise ($-$) direction.

PROBLEM 14 (62E).   The length of a bicycle pedal arm is 0.152 m and a downward force of 111 N is applied by the foot. What is the magnitude of the torque about the pivot point when the arm makes an angle of (*a*) 30°, (*b*) 90°, and (*c*) 180° with the vertical?

STRATEGY: Use $\tau = F\ell\sin\phi$, where ℓ is the length of the pedal and ϕ is the angle between the pedal and the force. Since the force is downward this is the same as the angle made by the pedal with the vertical.

SOLUTION:

$$\left[\text{ans: } (a)\ 8.4\,\text{N·m};\ (b)\ 17\,\text{N·m};\ (c)\ 0\right]$$

PROBLEM 15 (64P).   The body in Fig. 11–36 is pivoted at O. Three forces act on it in the directions shown on the figure: $F_A = 10$ N at point A, 8.0 m from O; $F_B = 16$ N at point B, 4.0 m from O; and $F_C = 19$ N at point C, 3.0 m from O. What is the net torque about O?

STRATEGY: Use $\tau = F_t\ell$, where F_t is the tangential component of the force and ℓ is the distance from O to the point of application of the force. Attach appropriate signs to the torques and sum them.

SOLUTION: For force A the tangential component is $F_A\cos(135° - 90°)$, for force B the tangential component is F_B, and for force C the tangential component is $F_C\cos(160° - 90°)$. Forces A and C tend to turn the body counterclockwise, so τ_A and τ_C are positive; force B tends to turn the body clockwise, so τ_B is negative. Sum the torques, with their signs, to obtain the net torque.

[ans: 12 N·m]

Many problems deal with the rotational motion of a rigid body about a fixed axis and can be solved using $\sum \tau_{\text{ext}} = I\alpha$. Don't forget that $\sum \tau_{\text{ext}}$ is the *net* torque on the body and that the signs of the individual torques are important.

PROBLEM 16 (67E). A cylinder having a mass of 2.0 kg can rotate about an axis through its center O. Forces are applied as in Fig. 11–37: $F_1 = 6.0$ N, $F_2 = 4.0$ N, $F_3 = 2.0$ N, and $F_4 = 5.0$ N. Also, $R_1 = 5.0$ cm and $R_2 = 12$ cm. Find the magnitude and direction of the angular acceleration of the cylinder, assuming that during the rotation, the forces maintain their same angles relative to the cylinder.

STRATEGY: Use $\sum \tau = I\alpha$. Calculate the individual torques, being careful about the signs, then sum them. Use $I = \frac{1}{2} M R_2^2$ to find the rotational inertia of the cylinder (see Table 11–2).

SOLUTION: You should get $\tau_1 = 0.72$ N·m (counterclockwise), $\tau_2 = 0.48$ N·m (clockwise), $\tau_3 = 0.10$ N·m (clockwise), and $\tau_4 = 0$. You should also get $I = 1.44 \times 10^{-2}$ kg·m^2.

[ans: 9.7 rad/s^2, counterclockwise]

PROBLEM 17 (68E). A small object with mass 1.30 kg is mounted on one end of a rod 0.780 m long and of negligible mass. The system rotates in a horizontal circle about the other end of the rod at 5010 rev/min. (*a*) Calculate the rotational inertia of the system about the axis of rotation.

STRATEGY: Since the mass of the rod can be neglected, only the object at the end contributes to the rotational inertia. Use $I = m\ell^2$, where ℓ is the length of the rod and m is the mass of the object on its end.

SOLUTION:

[ans: 0.791 kg·m^2]

(*b*) Air resistance exerts a force of 2.30×10^{-2} N on the object, directed opposite its direction of motion. What torque must be applied to the system to keep it rotating at constant speed?

STRATEGY: Since the rod and object have a constant angular speed the net torque acting on it must be 0. Let τ_{app} be the applied torque and τ_{air} be the torque of air resistance. Then $\tau_{\text{app}} + \tau_{\text{air}} = 0$. The force exerted by the air is always tangent to the circular path of the object so $\tau_{\text{air}} = -F_{\text{air}}\ell$, where the direction of rotation was taken to be positive.

SOLUTION:

[ans: 1.79×10^{-2} N·m, in direction of motion]

Some problems involve several bodies, some in translation and some in rotation. An example is a mass tied to a string that is wrapped around a pulley and allowed to fall.

To solve this type problem draw a free-body diagram for each body. For a rotating body, choose the direction of positive rotation (the counterclockwise direction, say) and write the sum of the torques, taking their directions into account. Set the sum equal to $I\alpha$. For a translating body, choose a coordinate system with one axis in the direction of the acceleration, if possible. Write the sum of the force components for each coordinate direction and set each sum equal to the product of the mass of the body and the appropriate acceleration component. Write all equations in terms of the magnitudes of the forces and appropriate angles. Use algebraic symbols, not numbers.

Don't forget tensions in strings, forces of gravity, normal forces, and frictional forces, if they act. If a string is wrapped around a disk that is free to rotate and if the string does not slip on the disk, then the force exerted by the string on the disk is the tension T in the string. Since the string must be tangent to the disk, the torque it exerts has magnitude TR, where R is the radius of the disk. If a string passes over a pulley then the tension in the string will be different on different sides. If T_1 is the tension on one side and T_2 is the tension on the other side, then $\pm(T_1 - T_2)R$ is the net torque exerted by the string on the pulley. The sign, of course, depends on which direction of rotation is taken to be positive.

The accelerations of the translating bodies and the angular accelerations of rotating bodies are usually related. Perhaps a string runs from a translating body and around a rotating body. Then, if the string does not slip or stretch, the magnitude of the tangential acceleration of a point on the rim of the rotating body must be the same as the magnitude of the acceleration of the translating body. You write $a = \pm R\alpha$. Which sign you use depends on which directions you chose for positive acceleration and positive angular acceleration. If you use the $+$ sign you are saying that a positive acceleration of the translating body is consistent with a positive angular acceleration of the rotating body. If you use the $-$ sign you are saying a positive acceleration of the translating body is consistent with a negative angular acceleration of the rotating body.

The second-law equations (for translation and rotation) and the equation or equations that link the accelerations of translating bodies and the angular accelerations of rotating bodies are solved simultaneously for the unknowns.

PROBLEM 18 (75P).   Two identical blocks, each of mass M, are connected by a massless string over a pulley of radius R and rotational inertia I (Fig. 11–40). The string does not slip on the pulley; it is not known whether or not there is friction between the table and the sliding block; the pulley's axis is frictionless. When the system is released, it is found that the pulley turns through an angle θ in time t and the acceleration of the blocks is constant. (a) What is the angular acceleration of the pulley?

STRATEGY: This can be found using kinematics: $\theta = \frac{1}{2}\alpha t^2$.

SOLUTION:

[ans: $2\theta/t^2$]

(b) What is the acceleration of the two blocks?

STRATEGY: Since they are connected by a string that does not slip on the pulley the magnitudes of their accelerations are the same and are the same as the magnitude of the tangential acceleration of a point on the rim of the pulley. Use $a = R\alpha$ and substitute the expression you found above for α.

SOLUTION:

[ans: $2R\theta/t^2$]

(c) What are the tensions in the upper and lower sections of the string?

STRATEGY: Use the equation of motion for the falling block to find the tension in the lower section. To find the tension in the upper section use the equation of motion for the pulley. For each of these bodies the acceleration or angular acceleration is known. In addition all forces are known except the tension that is sought.

SOLUTION: Draw a free body diagram for the falling block. Show the tension T_1 in the string (up) and the force of gravity (down). Take the downward direction to be positive and write Newton's second law: $Mg - T_1 = Ma$. Substitute the expression you found above for a and solve for a.

Draw a force diagram for the pulley. Show the force of the lower section of rope on it (T_1, down) and the force of the upper section (T_2, left). Both forces are tangent to the pulley rim. Write Newton's second law for rotation: $(T_1 - T_2)R = I\alpha$. Substitute the expression you found above for α and solve for T_2.

[ans: $Mg - 2MR\theta/t^2$, $Mg - (2\theta/t^2)(MR + I/R)$]

Some problems can be solved using the work-energy theorem for rotation or the principle of energy conservation. Here is an example.

PROBLEM 19 (78E). (a) If $R = 12$ cm, $M = 400$ g, and $m = 50$ g in Fig. 11–19, find the speed of m after it has descended 50 cm starting from rest. Solve the problem using energy-conservation principles.

STRATEGY: Take the zero of potential energy to be at the bottom of the 50-cm fall. Then the initial potential energy is $U_i = mgh$, where h is the distance fallen. The initial kinetic energy is $K_i = 0$, since the mass and disk are initially at rest. Suppose that after m has fallen a distance h it has speed v and the disk has angular speed ω. Then the final kinetic energy is $\frac{1}{2}mv^2 + \frac{1}{2}I\omega^2$. Conservation of energy yields $mgh = \frac{1}{2}mv^2 + \frac{1}{2}I\omega^2$. Since the string does not slip on the disk, $v = R\omega$. Substitute $\omega = v/R$ into the energy equation. Also substitute $I = \frac{1}{2}MR^2$. Solve for v. You should get $\sqrt{4mgh/(M + 2m)}$.

SOLUTION:

[ans: 1.4 m/s]

(b) Repeat (a) with $R = 5.0$ cm.

STRATEGY: Look at the algebraic expression you derived for v. It does not contain R.

SOLUTION:

[ans: 1.4 m/s]

III. NOTES

Chapter 12
ROLLING, TORQUE, AND ANGULAR MOMENTUM

I. BASIC CONCEPTS

This chapter starts with a discussion of objects that roll; that is, simultaneously translate and rotate. Pay careful attention to the role played by friction between the object and the surface on which it rolls. Learn the relationship between the velocity of the center of mass and the angular velocity when an object rolls without slipping. Next you study rotational motion using a new concept, that of angular momentum. Learn the definition for a single particle and learn how to calculate the total angular momentum of a collection of particles. Pay particular attention to the result for a rigid body rotating about a fixed axis. You will also study the change in the angular momentum of a system brought about by the total external torque applied to the system. Angular momentum is at the heart of one of the great conservation laws of mechanics. Pay attention to the conditions for which this law is valid.

Combined rotational and translational motion. In this section you study an object, such as a rolling wheel, that rotates as its center of mass moves along a line. The center of mass obeys Newton's second law: _____, where $\sum \mathbf{F}$ is the total force acting on the object, M is the total mass of the object, and $\mathbf{a}_{cm}$ is the acceleration of the center of mass. Rotation about an axis through the center of mass is governed by Newton's second law for rotation: _____, where $\sum \tau$ is the total torque acting on the object, I is the rotational inertia of the object, and α is the angular acceleration of the object.

You should understand that these two motions are related by the forces that act on the wheel. The total force accelerates the center of mass and the total torque (derived from the forces) produces an angular acceleration. If, in addition to the mass and rotational inertia of the body, you know all the forces acting and their points of application you can calculate the acceleration of the center of mass and the angular acceleration about the center of mass.

Rolling without slipping is a special case. We usually do not know the force of friction exerted by the surface on the body at the point of contact but we do know what that force does. If the object rolls without slipping, it causes the point on the wheel in contact with the surface to have a velocity of _____. This means that the speed v_{cm} of the center of mass and the angular speed ω around the center of mass are related by $v_{cm} =$ _____, where R is the radius of the wheel. Furthermore, since the wheel accelerates without slipping, the magnitude of the acceleration of the center of mass a_{cm} is related to the magnitude of the angular acceleration around the center of mass by $a_{cm} =$ _____.

You should understand the relationship between v_{cm} and $R\omega$ for a rolling wheel. The velocity of the point in contact with the ground is given by the vector sum of the velocity due to the motion of the center of mass and the velocity due to rotation: $v = v_{cm} - R\omega$, where the forward direction was taken to be positive. If the wheel slips then $v \neq 0$ and v_{cm} is different from $R\omega$. If it does not slip then $v = 0$ and $v_{cm} = R\omega$.

For a wheel rolling without slipping along a horizontal surface there are 3 basic equations containing the forces, torques, and accelerations. They are:

_____ for translation of the center of mass

_____ for rotation around the center of mass

_____ for the condition of no slipping.

These can be solved, for example, for the linear and angular accelerations and for the force of friction if the other quantities are known.

To test if the wheel slips or not, you will also need the vertical component of Newton's second law for the center of mass. This is used to solve for the normal force of the surface on the wheel. Imagine that the wheel does not slip and calculate the force of friction required to prevent slipping. Compare the magnitude of this force with $\mu_s N$, where μ_s is the coefficient of static friction. If it less, the wheel does not slip and the force of friction is as computed. If it is greater, the wheel does slip and the force of friction is $\mu_k N$, where μ_k is the coefficient of kinetic friction.

A rolling wheel, whether slipping or not, has both translational and rotational kinetic energy. If the center of mass has velocity v_{cm} and the wheel is rotating with angular velocity ω about an axis through the center of mass, then the total kinetic energy is given by $K =$ _____ . If the wheel is not slipping then both terms of the kinetic energy can be written in terms of v_{cm}. Use $v_{cm} = R\omega$ to eliminate ω in favor of v_{cm}. The result is $K =$ _____ + _____ . The kinetic energy can also be written in terms of ω: _____ + _____ .

If an object is rolling without slipping the point in contact with the surface is instantaneously at rest. As a result, the frictional force acting there does zero work.

An object that is rolling without slipping on a surface may be considered to be in pure rotation about an axis through the point of contact with the surface. Consider a wheel of radius R, rolling on a horizontal surface. If its center of mass is at its center and its rotational inertia about an axis through that point is I_{cm}, then, according to the parallel axis theorem, its rotational inertia about the point of contact is given by $I =$ _____ . The total kinetic energy of the wheel is now just its rotational kinetic energy $\frac{1}{2}I\omega^2$. Substitute for I in terms of I_{cm} and obtain the result you wrote in the last paragraph, $K =$ _____ .

Torque.   When a force **F** is applied to a an object at a point with position vector **r**, relative to some origin, then the torque τ associated with the force is given by the vector product

$$\tau =$$

Review the parts of Chapter 3 that deal with vector product. The magnitude of the torque is given by $\tau =$ _____ , where ϕ is the angle between _____ and _____ when they are drawn with their tails at the same point. The direction of the torque is given by the right-hand rule: _____

_____ .

You should be able to make a connection with the definition used in the last chapter. The diagram shows a particle going around a circle of radius R in a plane perpendicular to the plane of the page. Only the edge of the circle is seen. Suppose a force, tangent to the circle and into the page, is applied to the particle. Take the origin to be at O on the axis and draw a vector to indicate the torque. Since $\mathbf{r}$ and $\mathbf{F}$ are perpendicular to each other, the magnitude of the torque is given by $\tau = $ _____ .

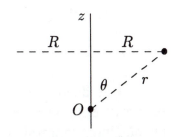

In terms of the angle θ shown the z component of τ is $\tau_z = $ _____ . Notice that $r \sin \phi = R$, the radius of the circle. Thus, in terms of R, $\tau_z = $ _____ . In the last chapter we were interested in motion about a fixed axis and we needed only the component of the torque along that axis. We used $\tau = F_t R$. Now you see this is only one component of the vector torque. If the origin is at the center of the circle it is the only non-vanishing component.

Angular momentum. If a particle has momentum $\mathbf{p}$ and position vector $\mathbf{r}$ relative to some origin, then its <u>angular momentum</u> is defined by the vector product

$$\boldsymbol{\ell} = $$

Notice that the angular momentum depends on the choice of origin. In particular, if the origin is picked so $\mathbf{r}$ and $\mathbf{p}$ are parallel at some instant of time then $\boldsymbol{\ell} = $ _____ at that instant.

You should be familiar with some frequently occurring special cases. If a particle of mass m is traveling around a circle of radius R with speed v and the origin is placed at the center of the circle, then the magnitude of the angular momentum is $\ell = $ _____ . If the orbit is in the plane of the page and the particle is traveling in the counterclockwise direction, the direction of its angular momentum is _____ . If it is traveling in the clockwise direction, the direction is _____ .

For rotation about a fixed axis the component of the angular momentum along the axis of rotation is, in most cases, the component of greatest interest. Nevertheless you should be able to compute the angular momentum vector and its component along any axis.

Suppose a particle of mass m is traveling around a circle of radius R parallel to the xy plane and centered on the z axis above the origin. The diagram on the right shows the xz plane as the particle crosses the plane going into the page. Draw the position vector $\mathbf{r}$ and the angular momentum vector $\boldsymbol{\ell}$. In terms of m, v, and r (the distance from the origin), the magnitude of $\boldsymbol{\ell}$ is $\ell = $ _____ .

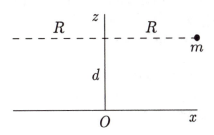

Let θ be the angle between $\mathbf{r}$ and the positive x axis. In terms of θ the z component of $\boldsymbol{\ell}$ is $\ell_z = $ _____ and the x component is $\ell_x = $ _____ . In terms of m, R, and v the z component of $\boldsymbol{\ell}$ is given by $\ell_z = $ _____ . In terms of m, d, and v the x component of $\boldsymbol{\ell}$ is given by $\ell_x = $ _____ . The component of $\boldsymbol{\ell}$ in the plane of the orbit is radial and follows the particle around the circle, always pointing toward or away from the center of the circle, depending on the direction of travel. Notice that ℓ and ℓ_x depend on the position of the origin along the z axis but ℓ_z does not.

Even a particle moving in a straight line may have a non-zero angular momentum. Consider a particle of mass m traveling with velocity **v** in the negative y direction along the line $x = d$.

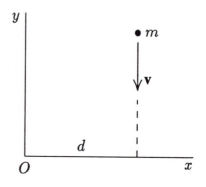

On the diagram draw the vector **r** from the origin to the particle. In terms of m, v, and d, the magnitude of its angular momentum about the origin is $\ell =$ _____ and the direction of its angular momentum is _____. Show the angular momentum vector on the diagram by drawing an × in a small circle (⊗) if it is into the page and a dot in a small circle (⊙) if it is out of the page. Label it $\boldsymbol{\ell}$. If the particle is traveling along the same line but in the positive y direction, the direction of its angular momentum is _____. If the particle is traveling along a line through the origin its angular momentum is _____.

The total angular momentum **L** of a system of particles is the vector sum of the individual angular momenta of all the particles in the system. For an object rotating with angular velocity ω about the z axis the z component of the angular momentum, written in terms of ω, the mass m_i, and distance r_i of each particle from the axis, is the sum

$$L_z =$$

In terms of the rotational inertia I of the object it is

$$L_z =$$

Notice that this equation is also valid for a single particle traveling in a circular orbit. Then the rotational inertia is mR^2.

If the mass is distributed symmetrically about the axis of rotation the total angular momentum is along the axis of rotation and the magnitude of L_z is also the magnitude of **L**. The direction of **L** is given by the right hand rule: curl the fingers in the direction of _____, then the thumb will point in the direction of _____.

A "dumbbell" consists of two masses connected by a light rod of length 2d, spinning with angular velocity ω about the z axis, through the center of the rod and perpendicular to it. Take the plane of rotation to be a distance z from the origin. The diagram shows the situation when mass m_1 is moving into the page and mass m_2 is moving out of the page. At first suppose the masses are not equal. For mass 1, the magnitude of the angular momentum is _____, the z component is _____, and the x component is _____.

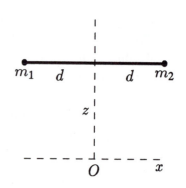

For mass 2 the magnitude of the angular momentum is _____, the z component is _____, and the x component is _____. The z component of the total angular momentum is _____ and the x component is _____.

Now specialize the results to the symmetric case: the masses are equal. Then the z component of the total angular momentum is _____, while the x component vanishes. If the masses are unequal the rotational inertia of this object about the z axis is $I =$ _____. Notice that $L_z = I\omega$, whether or not the masses are equal.

Newton's second law for rotation. If a net torque acts on a particle, then its angular momentum changes with time. Recall that, in terms of the momentum of a particle, Newton's second law can be written $\mathbf{F} = d\mathbf{p}/dt$, where $\mathbf{F}$ is the net force acting on the particle. Take the vector product of this equation with $\mathbf{r}$ to obtain $\mathbf{r} \times \mathbf{F} = \mathbf{r} \times d\mathbf{p}/dt$. Note that $\mathbf{r} \times \mathbf{F} = \boldsymbol{\tau}$, the net torque acting on the particle. With a small amount of mathematical manipulation you can show that $\mathbf{r} \times (d\mathbf{p}/dt) = d(\mathbf{r} \times \mathbf{p})/dt - (d\mathbf{r}/dt) \times \mathbf{p} = d\boldsymbol{\ell}/dt$. The second term, $(d\mathbf{r}/dt) \times \mathbf{p}$, vanishes because _____ is in the same direction as _____. Thus, in terms of torque and angular momentum, Newton's second law becomes $\boldsymbol{\tau} =$ _____.

For a system of particles the time rate of change of the angular momentum of each particle equals the net torque acting on the particle. This means that the time rate of change of the total angular momentum for the system is the vector sum of all torques acting on all particles. The sum includes the torque exerted by any particle in the system on any other particle in the system as well as external torques, exerted on particles in the system by objects in the environment of the system.

According to Newton's third law, the sum of the torques exerted by particles of the system on each other is _____. (This result follows if the force of each object on each of the others is along the line joining the particles.) Only torques exerted by objects outside the system can change the total angular momentum of the system. Be sure you understand this. One particle in the system can change the angular momentum of another particle in the system but the second particle simultaneously changes the angular momentum of the first and the two changes are equal in magnitude and opposite in direction. Internal interactions may change the angular momenta of individual particles but they do not change the total angular momentum of the system as a whole.

Newton's second law for the angular momentum of the system is

$$\sum \tau_{\text{ext}} =$$

where $\sum \tau_{\text{ext}}$ is the total *external* torque exerted on all particles of the system and $\mathbf{L}$ is the total angular momentum of the system.

This is the fundamental equation for a system of particles. You should learn to apply it to as many situations as you can. Remember it is a three-dimensional equation. In component form $\sum \tau_{\text{ext x}} =$ _____, $\sum \tau_{\text{ext y}} =$ _____, and $\sum \tau_{\text{ext z}} =$ _____. You must use the same origin to compute the angular momentum and torques.

For a body rotating about the z coordinate axis, we are usually interested in only the z components of the torque and angular momentum. If an external torque with other components is applied, the bearings must apply whatever additional torque is required to hold the axis fixed. If the spinning body is symmetric about the axis of rotation then **L** is along that axis and the bearings do not need to exert a torque to hold the axis fixed.

Since $L_z = I\omega$ for a body rotating about the z axis, Newton's second law for fixed axis rotation becomes $\tau_{\text{ext } z} = $ _____. In some cases the body is rigid and I does not change with time. Then Newton's second law for fixed axis rotation becomes

$$\tau_{\text{ext } z} = $$

where α is the angular acceleration of the body. This relationship was derived in the last chapter. In this chapter you will encounter some situations for which I does change.

Conservation of angular momentum. If the total external torque acting on a body is zero then the change over any time interval of the angular momentum of the body is _____. Angular momentum is then said to be conserved. This is a vector law. One component of the angular momentum may be conserved while other components are not. To see if angular momentum is conserved or not, calculate the total external torque acting on the body.

If the z component of the total external torque acting on a body rotating about the z axis vanishes then $I\omega$ is a constant. Consider an ice skater, initially rotating on the points of her skates with her arms extended. When she drops her arms to her side, her angular velocity increases. By dropping her arms she decreases her _____ and her angular velocity must increase for _____ to remain the same.

Suppose the skater is spinning in the counterclockwise direction when viewed from above, so her angular velocity vector points upward. Consider her torso and arms as two individual objects. As she drops her arms the angular momentum of her torso increases, so you know the torque exerted by her arms on her torso is in the _____ direction. Her torso exerts a torque of equal magnitude but opposite direction on her arms. This means the magnitude of the angular momentum of her arms _____ as they drop.

Precession. If the axis of rotation is not fixed a torque that has components perpendicular to the axis might change the orientation of the axis. One example is discussed in the text: a wheel spinning on a long horizontal axle. For an origin placed on the axle the angular momentum **L** of the wheel is along the axle. Now one end of the axle is placed on a support and the other end is released with the axle horizontal. The force of gravity, downward on the wheel, exerts a torque. Describe the direction of the torque relative to the vertical and to the axis of rotation:

_____

_____

Describe the change in the direction of **L** with time: _____

_____

If the wheel is spinning fast, the axle moves so as to remain in the direction of **L**.

Carefully note that τ remains perpendicular to **L**, so the magnitude of **L** is constant as **L** rotates. To convince yourself that L is constant, take the scalar product of **L** with both sides of $\tau = d\mathbf{L}/dt$ to obtain $\mathbf{L} \cdot \tau = \mathbf{L} \cdot d\mathbf{L}/dt$. Since τ remains perpendicular to **L**, $\mathbf{L} \cdot \tau = 0$. Furthermore $\mathbf{L} \cdot d\mathbf{L}/dt = \frac{1}{2}dL^2/dt$, so $dL^2/dt = 0$ and L^2 is constant. This means L is constant.

Suppose τ remains perpendicular to **L** and, over some time interval Δt, **L** rotates through the angle $\Delta\phi$, as shown on the right. The magnitude of the change in **L** is given by $|\Delta \mathbf{L}| = 2L\sin(\Delta\phi/2)$, as some simple trigonometry shows. In the limit as the angle becomes small the sine of any angle equals the angle itself in radians. Thus when $\Delta\phi$ is small $|\Delta \mathbf{L}| = $ _____ if $\Delta\phi$ is measured in radians. Divide by Δt, take the limit as Δt becomes small, and equate the result to the magnitude of τ.

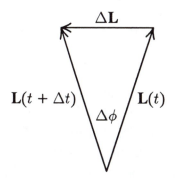

In terms of $d\phi/dt$, $\tau = $ _____ . Ω $(= d\phi/dt)$ is called the <u>precessional angular velocity</u>. It is the angular velocity with which the angular momentum vector precesses. If the angular momentum of the wheel is large it is also the angular velocity with which the rotation axis precesses. For the equation that relates Ω to L and τ to be valid, the units of Ω must be _____ .

The angular momentum changes magnitude if the applied torque has a non-vanishing component along a line that is _____ ; it changes direction if the applied torque has a non-vanishing component along a line that is _____ .

II. PROBLEM SOLVING

We start with a problem that examines the no-slipping condition: the point on the rolling object in contact with the surface is instantaneously at rest with respect to the surface

PROBLEM 1 (9P).   Consider a 66-cm diameter tire on a car traveling at 80 km/h on a level road in the direction of increasing x. As seen by a passenger in the car, what are the linear velocity and the magnitude of the linear acceleration of (a) the center of the wheel, (b) a point at the top of the tire, and (c) a point at the bottom of the tire?

STRATEGY: As viewed by a passenger the tire is simply rotating. Every point on the rim has speed $v = v_c$, where v_c is the speed of the car. The point on top is moving forward, the point on the bottom is moving backward. Their accelerations have magnitude v^2/R, toward the center. The center of the wheel has velocity 0.

SOLUTION:

[ans: (a) 0, 0; (b) 80 km/h forward; 1500 m/s^2 down; (c) 80 km/h backward, 1500 m/s^2 up]

(d) Repeat (a) through (c), in the same order, for a stationary observer alongside the road.

STRATEGY: Simply add vectorially v_c in the forward direction to each of the velocities found above. Since the observer is not accelerating with respect to the car, the accelerations are the same.

SOLUTION:

[ans: center: 80 km/h, 0; top: 160 km/h forward, 1500 m/s^2 down; bottom: 0, 1500 m/s^2 up]

Problems involving rolling objects can be solved dynamically by considering the forces acting on the object. Many can be also solved using the energy method.

A dynamical solution starts with a force diagram. See Sample Problem 12–4 and Fig. 12–8 of the text for an illustrative example. Show all forces and label them with algebraic symbols. Since the application points of forces are important for rotation you cannot represent the body as a point. Instead, sketch the body and place the tail of each force vector at the point where the force is applied. Choose a coordinate system to describe the motion of the center of mass, with one axis in the direction of the acceleration, if possible. Sum the force components for each coordinate direction, again taking care with the signs. Equate the sum to the product of the mass and the corresponding component of the center of mass acceleration. Choose a direction for positive rotation. For each force write an expression that gives the torque about the axis of rotation. Be careful about the signs. Sum the torques and equate the sum to $I\alpha$.

When a wheel accelerates without slipping a frictional force is present at the point of contact between the body and the surface on which it rolls. Don't forget to include it, even if it is not mentioned in the problem statement. You then write an additional equation relating the acceleration of the center of mass to the angular acceleration about the center of mass: $a_{cm} = \pm R\alpha$. The sign depends on the choices made for the directions of positive translation and rotation.

If the body is slipping the magnitude of the force of friction is given by $f = \mu_k N$, where μ_k is the coefficient of kinetic friction for the body and surface and N is the magnitude of the normal force exerted by the surface on the body. In some cases you can determine the direction of the force before solving the problem and can take it into account in drawing the force diagram and writing Newton's second law. In other cases you must arbitrarily select a direction, then examine the solution to see if the direction of rotation is consistent with your selection. If it is not, rework the problem using the opposite direction.

The energy method makes use of either the work-energy theorem or conservation of mechanical energy. If external forces do work on the object, use $W = \Delta K$, where W is the total work done by all forces and ΔK is the change in the kinetic energy. If all forces that do work are conservative use $U_i + K_i = U_f + K_f$, where U_i and U_f are the initial and final potential energies and K_i and K_f are the initial and final kinetic energies. In either case don't forget to include both the kinetic energy of translation and the kinetic energy of rotation. If the object is not slipping, include the auxiliary equation $v_{cm} = \pm R\omega$. Remember that gravitational potential energy is given by MgH, where h is the height of the center of mass above the reference height.

Notice that the energy method requires fewer equations than the dynamical method. Less information is required and fewer unknowns can be evaluated. Directions of forces and torques enter only insofar as they determine the work or potential energy. The energy method cannot be used to solve for directions except in special cases. If the work done by forces or torques is given, rather than the forces or torques themselves, the energy method *must* be used.

PROBLEM 2 (6E). A uniform sphere rolls down an incline. (*a*) What must be the incline angle if the linear acceleration of the center of the sphere is to be $0.10g$?

STRATEGY: Here an acceleration is given, not a speed, so we use the dynamical method. Write Newton's second law equations for the motion of the center of mass and for rotation about the center of mass. Apply the condition $a = R\alpha$ and solve for the incline angle.

SOLUTION: Draw a force diagram for the sphere. Take the x axis to be parallel to the incline and down the incline. Take the y axis to be perpendicular to the incline. Draw the sphere and show the forces: the force of friction f, up the incline and acting at the point of contact of the sphere with the plane; the normal force N, in the y direction and acting at the point of contact, and the force of gravity mg, down and acting at the center of the sphere. Let θ be the angle of incline and write the x component of the second law: $mg \sin\theta - f = ma$, where m is the mass of the sphere and a is its acceleration.

Choose the direction for positive rotation, consistent with rolling down the plane, and write the second law for rotation: $fR = I\alpha$, where I is the rotational inertia of the sphere. The normal force and the force of gravity do not exert torques about the center of the sphere; only the force of friction does. Finally write the condition for no slipping: $a = R\alpha$, where R is the radius of the sphere. Systematically eliminate α and f, then solve for $\sin\theta$. You should get $\sin\theta = (1 + I/mR^2)a/g$. Substitute $I = \frac{2}{5}mR^2$ (see Table 11–2) and calculate θ.

$$[\text{ans: } 8.1°]$$

(b) For this angle, what would be the acceleration of a frictionless block sliding down the incline?

STRATEGY: Use Newton's second law and solve for a or put $I = 0$ in the result for the sphere. You should get $a = g \sin\theta$.

SOLUTION:

$$[\text{ans: } 1.4\,\text{m/s}^2]$$

PROBLEM 3 (7E). A solid sphere of weight 8.00 lb rolls up an incline with an inclination angle of 30.0°. At the bottom of the incline the center of mass of the sphere has a translational speed of 16.0 ft/s. (a) What is the kinetic energy of the sphere at the bottom of the incline?

STRATEGY: The sphere has both translational and rotational kinetic energy. Use $K_i = \frac{1}{2}mv^2 + \frac{1}{2}I\omega^2$, where v is the speed of the center of mass and ω is the angular speed of rotation about the center of mass. Since the sphere is not slipping, substitute $\omega = v/R$, where R is the radius of the sphere. Also substitute $I = \frac{2}{5}mR^2$. Use $m = (8.00\,\text{lb})/(32\,\text{ft/s}^2) = 0.250\,\text{slugs}$ for the mass.

SOLUTION:

$$[\text{ans: } 44.8\,\text{ft·lb}]$$

(*b*) How far does the sphere travel up the incline? Does the answer to(*b*) depend on the weight of the sphere?

STRATEGY: Because friction does not do work on the sphere, mechanical energy is conserved. If h is the vertical component of the displacement of the center of mass, then $mgh = K_i$. Solve for h, then use $h = d\sin\theta$, where θ is the inclination angle, to find the distance d traveled.

SOLUTION:

$$[\text{ans: } 11.2\,\text{ft}]$$

PROBLEM 4 (13P).   A small solid marble of mass m and radius r rolls without slipping along the loop-the-loop track shown in Fig. 12–30, having been released from rest somewhere on the straight section of track. (*a*) From what minimum height h above the bottom of the track must the marble be released in order that it not leave the track at the top of the loop? (The radius of the loop-the-loop is R; assume $R \gg r$.)

STRATEGY: First find the speed the marble must have at the top in order to stay on the track. At the top the marble experiences the smallest possible inward force when the normal force of the track is 0 and the only force acting is the force of gravity mg. The inward force is greater than this if the track exerts a normal force but it cannot be less than this. Thus the minimum speed v of the center of mass at the top is given by $mg = mv^2/(R-r)$. Note that the radius of the circle traversed by the center of mass, $R - r$, is used. Now use conservation of energy to find the initial height h of the center of mass that will result in speed v at the top. If the zero of gravitational potential energy is at the bottom of the loop, then the initial potential energy is mgh. The initial kinetic energy is 0 since the marble starts from rest. The potential energy at the top of the loop is $mg(2R - r)$ since the center of mass is then $2R - r$ above the bottom. The kinetic energy is given by $\frac{1}{2}mv^2 + \frac{1}{2}I\omega^2$, where I is the rotational inertia of the marble. When the condition $R \gg r$ is used and when $I = \frac{2}{5}mr^2$ (see Table 11–2) and $\omega = v/r$ (no slipping) are substituted, the energy equation becomes $mgh = (7/10)mv^2 + 2mgR$. Use $mv^2 = mgR$ to eliminate v from $mgh = (7/10)mv^2 + 2mgR$, then solve for h.

SOLUTION:

$$[\text{ans: } 2.7R]$$

(*b*) If the marble is released from height $6R$ above the bottom of the track, what is the horizontal component of the force acting on it at point Q?

STRATEGY: The horizontal component (the normal force of the track) is given by $F = mv^2/R$, where v is the speed of the marble when it is at Q. $R \gg r$ was assumed. Use energy conservation, with $h = 6R$, to find v.

SOLUTION:

$$[\text{ans: } 7.1mg]$$

PROBLEM 5 (15P). An apparatus for testing the skid resistance of automobile tires is constructed as shown in Fig. 12–32. The tire is initially motionless and is held in a framework with negligible mass that is pivoted at point B. The rotational inertia of the wheel about its axis is $0.750\,\text{kg·m}^2$, its mass is $15.0\,\text{kg}$, and its radius is $0.300\,\text{m}$. The tire is free to rotate around point A. The tire is placed on the surface of a conveyor belt that is moving with a surface velocity of $12.0\,\text{m/s}$. (*a*) If the coefficient of kinetic friction between the tire and the conveyor belt is 0.60 and if the tire does not bounce, what time will be required for the wheel to reach its final angular velocity?

STRATEGY: The force of friction gives the tire an angular acceleration until the speed of a point on the rim matches the speed of the belt. Thereafter the tire does not slip on the belt. The frictional torque is fR, where f is the force of friction and R is the radius of the tire. The normal force acting on the tire is mg so $f = \mu_k mg$ and the torque is $\mu_k mgR$. The angular acceleration of the tire is $\alpha = \tau/I = \mu_k mgR/I$, where I is the rotational inertia of the tire. The angular speed of the tire is given by $\omega = \alpha t$ and the speed of a point on the rim is given by $v = R\omega = R\alpha t = \mu_k mgR^2 t/I$. Set v equal to the speed of the belt and solve for t.

SOLUTION:

$$[\text{ans: } 1.13\,\text{s}]$$

(*b*) What will be the length of the skid mark on the conveyor belt?

STRATEGY: This is same as the distance traveled by a point on the belt during the time the tire is skidding.

SOLUTION:

$$[\text{ans: } 13.6\,\text{m}]$$

Some problems simply ask for a calculation of a torque or angular momentum using the general definitions. Identify the point to be used as the origin, then evaluate the appropriate vector product: to find a torque take **r** to be the vector from the origin to the point of application of the force and evaluate **r** × **F**; to find the angular momentum of a particle take **r** to be the vector from the origin to the particle, then evaluate **r** × **p** or m**r** × **v**. You must be given the momentum of the particle or else its mass and velocity.

PROBLEM 6 (22P).   Force **F** $= (2.0\,\text{N})$**i** $- (3.0\,\text{N})$**k** acts on a pebble with position vector **r** $= (0.50\,\text{m})$**j** $- (2.0\,\text{m})$**k**, relative to the origin. What is the resulting torque acting on the pebble about (a) the origin and (b) a point with coordinates $(2.0\,\text{m}, 0, -3.0\,\text{m})$?

STRATEGY: Evaluate (a) $\tau = $ **r** × **F**. (b) The position vector of the pebble relative to the point is given by **r**$' = $ **r** $-$ **R**, where **R** is the position vector of the point relative to the origin. That is **r**$' = (0.50$**j** $- 2.0$**k**$)\,\text{m} - (2.0$**i** $- 3.0$**k**$)\,\text{m} = (-2.0$**i** $+ 0.50$**j** $+ $**k**$)\,\text{m}$. Evaluate $\tau' = $ **r**$'$ × **F**.

SOLUTION: Use **r** × **F** $= (x$**i** $+ y$**j** $+ z$**k**$) × (F_x$**i** $+ F_z$**k**$) = yF_z$**i** $+ (zF_x - xF_z)$**j** $- yF_x$**k** (see the Mathematical Skills section.).

$$\left[\,\text{ans:}\ (a)\ (-1.5\,\textbf{i} - 4.0\,\textbf{j} - \textbf{k})\,\text{N} \cdot \text{m};\ (b)\ (-1.5\,\textbf{i} - 4.0\,\textbf{j} - \textbf{k})\,\text{N} \cdot \text{m}\,\right]$$

Usually the torque around different points has different values. In this case the values are the same because the displacement of the second point from the origin is parallel to the force. That is, **R** × **F** $= 0$.

PROBLEM 7 (26E).   A 1200-kg airplane is flying in a straight line at 80 m/s, 1.3 km above the ground. What is the magnitude of its angular momentum with respect to a point on the ground directly under the path of the plane?

STRATEGY: If the line from the point on the ground to the plane has length r and makes the angle θ with the horizontal then the magnitude of the angular momentum of the plane is given by $\ell = mrv \sin\theta$, where v is the speed and m is the mass of the plane. Notice that $r \sin\theta = h$, the plane's altitude, so $\ell = mvh$.

SOLUTION:

$$\left[\,\text{ans:}\ 1.3 \times 10^8\,\text{kg·m}^2\text{/s}\,\right]$$

PROBLEM 8 (31P).   Calculate the angular momentum, about the Earth's center, of an 84-kg person on the equator of the rotating Earth.

STRATEGY: Use $L = I\omega$. The rotational inertia of the person is $I = mR^2$, where m is the person's mass and R is the radius of the Earth. The Earth rotates 2π rad in 24 h or $24 \times 60 \times 60 = 8.64 \times 10^4$ s, so $\omega = 2\pi/8.64 \times 10^4 = 7.27 \times 10^{-5}$ rad/s. The radius of the Earth is $R = 6.37 \times 10^6$ m.

SOLUTION:

$$\left[\,\text{ans:}\ 2.5 \times 10^{11}\,\text{kg·m}^2\text{/s}\,\right]$$

Dynamics problems for a rotating body are based on the vector equation $\tau = dL/dt$.

PROBLEM 9 (40P).   At $t = 0$, a 2.0-kg particle has position vector $\mathbf{r} = (4.0\,\text{m})\,\mathbf{i} - (2.0\,\text{m})\,\mathbf{j}$ m relative to the origin. Its velocity is given by $\mathbf{v} = (-6.0t^4\,\text{m/s})\,\mathbf{i} + (3.0\,\text{m/s})\,\mathbf{j}$. About the origin and for $t > 0$, what are (a) the particle's angular momentum, and (b) the torque acting on the particle?

STRATEGY: (a) Evaluate $\boldsymbol{\ell} = m\mathbf{r} \times \mathbf{v}$. See the Mathematical Skills section for details. (b) Use $\tau = d\boldsymbol{\ell}/dt$ to find an expression for the torque.

SOLUTION:

$$\left[\,\text{ans: } (a)\ (24 - 24t^4)\,\mathbf{k}\ \text{kg·m}^2/\text{s};\ (b)\ (-96t^3\,\mathbf{k})\ \text{N·m}\,\right]$$

(c) Repeat (a) and (b) for a point with coordinates $(-2.0\,\text{m}, -3.0\,\text{m}, 0)$ instead of the origin.

STRATEGY: The position vector of the particle relative to the new point is $\mathbf{r}' = (4.0\mathbf{i} - 2.0\mathbf{j})\,\text{m} - (-2.0\mathbf{i} - 3.0\mathbf{j})\,\text{m} = (6.0\mathbf{i} + \mathbf{j})\,\text{m}$. Evaluate $\boldsymbol{\ell}' = m\mathbf{r}' \times \mathbf{v}$ and its derivative.

SOLUTION:

$$\left[\,\text{ans: } (36 + 12t^4)\,\mathbf{k}\ \text{kg·m}^2/\text{s};\ (48t^3\,\mathbf{k})\ \text{N·m}\,\right]$$

Notice that the angular momentum and torque have different values in (c) than in (a) and (b). Nevertheless, $\tau = dL/dt$ no matter what origin is used to calculate these quantities.

PROBLEM 10 (41P).   A projectile of mass m is fired from the ground with an initial speed v_0 and an initial angle θ_0 above the horizontal. (a) Find an expression for its angular momentum about the firing point as a function of time.

STRATEGY: Take the x axis to be horizontal and the y axis to be upward. Let x and y be the coordinates of the projectile, relative to the firing point, and v_x and v_y be the components of its velocity. Then its angular momentum is given by $\boldsymbol{\ell} = m\mathbf{r} \times \mathbf{v} = m(x\,\mathbf{i} + y\,\mathbf{j}) \times (v_x\,\mathbf{i} + v_y\,\mathbf{j}) = m(xv_y - yv_x)\,\mathbf{k}$. Recall from your study of projectile motion that $x = v_0 t \cos\theta_0$, $y = v_0 t \sin\theta_0 - \frac{1}{2}gt^2$, $v_x = v_0 \cos\theta_0$, and $v_y = v_0 \sin\theta_0 - gt$. Make the substitutions.

SOLUTION:

$$\left[\,\text{ans: } -\tfrac{1}{2}mv_0 gt^2 \cos\theta\,\mathbf{k}\,\right]$$

(*b*) Find the rate at which the angular momentum changes with time.

STRATEGY: Differentiate the result of (*a*) with respect to time.

SOLUTION:

$$\left[\text{ans: } -mv_0gt\cos\theta\,\mathbf{k}\right]$$

(*c*) Evaluate $\mathbf{r}\times\mathbf{F}$ directly and compare the result with (*b*). Why should the results be identical?

STRATEGY: The force is $-mg\,\mathbf{i}$ so the torque is $\boldsymbol{\tau} = (x\,\mathbf{i} + y\,\mathbf{j})\times(-mg\,\mathbf{i}) = -mgx\,\mathbf{k}$.

SOLUTION:

$$\left[\text{ans: } -mv_0gt\cos\theta\right]$$

Some problems can be solved using the principle of angular momentum conservation. To see if the principle can be used you must select a system and examine the net external torque acting on it to see if it vanishes. If it does then angular momentum is conserved. If one component of the total torque vanishes then that component of the angular momentum is conserved, regardless of whether or not the other components change.

Some problems deal with two objects that exert torques on each other, thereby changing each other's angular momentum. These changes have the same magnitude and opposite directions. Their sum vanishes. If the net external torque is zero then the total angular momentum remains constant. Write the total angular momentum in terms of the quantities that describe the motions before the objects interact (their initial velocities or angular velocities); write the angular momentum in terms of the quantities that describe the motions after the interaction (their final velocities or angular velocities). Equate the two expressions and solve for the unknown. In some problems one of the objects is moving along a straight line, either before or after the interaction. Don't forget to include the angular momentum of this object.

Some problems deal with an object whose rotational inertia changes at some time while it is spinning about a fixed axis. The problem statement gives information that can be used to calculate the initial values I_0 and ω_0, before the rotational inertia changes, and asks for either I or ω at a later time, after they change. If no external torques act, angular momentum is conserved and $I_0\omega_0 = I\omega$. This can be solved for one of the quantities in terms of the others.

PROBLEM 11 (52E). Two disks are mounted on low-friction bearings on the same axle and can be brought together so that they couple and rotate as a unit. (*a*) The first disk, with rotational inertia $3.3\,\text{kg}\cdot\text{m}^2$, is set spinning at 450 rev/min. The second disk, with rotational inertia $6.6\,\text{kg}\cdot\text{m}^2$, is set spinning at 900 rev/min in the same direction as the first. They then couple together. What is the angular speed after coupling?

STRATEGY: Take the system to consist of the two disks. During the coupling process each exerts a torque on the other and their individual angular momenta change until they come to the same rotational speed. If friction in the bearings can be neglected no external torques act on the system so the total angular momentum of the system is conserved during the coupling. Let I_1 be the rotational inertia and ω_1 be the original angular velocity of disk 1; let I_2 be the rotational inertia and ω_2 be the original angular velocity of disk 2. Then $I_1\omega_1 + I_2\omega_2 = (I_1 + I_2)\omega_f$, where ω_f is the rotational velocity of the coupled disks. Solve for ω_f. ω_1 and ω_2 have the same sign.

SOLUTION:

[ans: 750 rev/min]

(*b*) If instead the second disk is set spinning at 900 rev/min in the direction opposite the first disk's rotation, what is the angular speed after coupling?

STRATEGY: The conservation of angular momentum equation is the same but now ω_1 and ω_2 have opposite signs.

SOLUTION:

[ans: 450 rev/min, in original direction of rotation of disk 2]

PROBLEM 12 (57P).   With center and spokes of negligible mass, a certain bicycle wheel has a thin rim of radius 1.14 ft and weight 8.36 lb; it can turn on its axle with negligible friction. A man holds the wheel above his head with the axis vertical while he stands on a turntable free to rotate without friction; the wheel rotates clockwise, as seen from above, with an angular speed of 57.7 rad/s, and the turntable is initially at rest. The rotational inertia of wheel + man + turntable about the common axis of rotation is 1.54 slug·ft². The man's hand suddenly stops the rotation of the wheel (relative to the turntable). Determine the resulting angular velocity (magnitude and direction) of the system.

STRATEGY: Consider a system consisting of the man, the wheel, and the turntable. The torque of the man on the wheel is an internal torque and, although it changes the angular momentum of the wheel, it is balanced by the torque of the wheel on the man and the total angular momentum of the system does not change. Let I_w be the rotational inertia of the wheel and ω_w be its initial angular velocity. Since all the mass of the wheel is located at its rim $I = mR^2$, where R is the radius of the wheel. The angular momentum is given by $I_w\omega_w$. Let I be the rotational inertia of the system as a whole and ω_f be its final angular velocity, after all parts are spinning together. Then $I\omega_f$ also gives the angular momentum. Conservation of angular momentum yields $I_w\omega_w = I\omega_f$. Solve for ω_f. The mass of the wheel is $m = W/g = (8.36\,\text{lb})/(32\,\text{ft/s}^3) = 0.261\,\text{slugs}$.

SOLUTION:

[ans: 12.7 rad/s]

PROBLEM 13 (59P).   Two children, each with mass M, sit on opposite ends of a narrow board with length L and mass M (the same as the mass of each child). The board is pivoted at its center and is free to rotate in a horizontal circle without friction. (Treat it as a thin rod.) (*a*) What is the rotational inertia of the board plus the children about a vertical axis through the center of the board?

STRATEGY: The rotational inertia is the sum of three parts: the board has rotational inertia $ML^2/12$ (see Table 11–2) and each of the two children have rotational inertia $M(L/2)^2$ since each is a distance $L/2$ from the axis of rotation.

SOLUTION:

$$[\text{ans: } (7/12)ML^2]$$

(b) What is the angular momentum of the system if it is rotating with angular speed ω_0 in a clockwise direction as seen from above? What is the direction of the angular momentum?

STRATEGY: The total angular momentum of the system consisting of the board and the two children is given by $I\omega_0$. Use the right-hand rule to find the direction.

SOLUTION:

$$[\text{ans: } (7/12)ML^2\omega_0, \text{ down}]$$

(c) While the system is rotating, the children pull themselves toward the center of the board until they are half as far from the center as before. What is the resulting angular speed in terms of ω_0?

STRATEGY: As the children pull themselves no net external torque acts and the angular momentum of the system is conserved. If I' is the new rotational inertia of the system and ω is the rotational velocity then $I\omega_0 = I'\omega$. Since each child is now a distance $L/4$ from the axis of rotation $I' = ML^2/12 + M(L/4)^2 + M(L/4)^2 = (5/24)ML^2$. Solve for ω.

SOLUTION:

$$[\text{ans: } (14/5)\omega_0]$$

(d) What is the change in kinetic energy of the system as a result of the children changing their positions? (From where does the added kinetic energy come?)

STRATEGY: The initial kinetic energy is $K_i = \frac{1}{2}I\omega_0^2$ and the final is $K_f = \frac{1}{2}I'\omega^2$. Calculate $\Delta K = K_f - K_i$.

SOLUTION:

$$[\text{ans: it decreases by } 0.25ML^2\omega_0^2]$$

The internal forces of the children on the board and the board on the children do negative net work.

PROBLEM 14 (61P). A cockroach of mass m runs counterclockwise around the rim of a lazy Susan (a circular dish mounted on a vertical axle) of radius R and rotational inertia I with frictionless bearings. The cockroach's speed (relative to the Earth) is v, whereas the lazy Susan turns clockwise with angular speed ω_0. The cockroach finds a bread crumb on the rim and, of course, stops. (a) What is the angular speed of the lazy Susan after the cockroach stops?

STRATEGY: Since the net external torque is 0 the total angular momentum of the system consisting of the lazy Susan and the cockroach is conserved. Before the cockroach stops, the angular momentum of the lazy Susan is $I\omega_0$ and the angular momentum of the cockroach is mRv. They are in opposite directions so the total angular momentum of the system is given by $L = -I\omega_0 + mRv$. Let ω be the angular speed of the lazy Susan and cockroach after the bug stops (relative to the lazy Susan). Then the angular momentum is given by $(I + mR^2)\omega$. Conservation of angular momentum yields $-I\omega_0 + mRv = (I + mR^2)\omega$. Solve for ω.

SOLUTION:

$$\left[\text{ans: } (mRv - I\omega_0)/(I + mR^2)\right]$$

(b) Is mechanical energy conserved?

STRATEGY: The kinetic energy before the cockroach stops is given by $K_i = \frac{1}{2}I\omega_0^2 + \frac{1}{2}mv^2$ and the kinetic energy after it stops is given by $K_f = \frac{1}{2}(I + mR^2)\omega^2$. Substitute for ω in terms of ω_0 and find an expression for $\Delta K = K_f - K_i$. You should obtain $-\frac{1}{2}(2mRIv\omega_0 + mR^2I\omega_0^2 + mIv^2)/(I + mR^2)$. This is inherently negative.

SOLUTION:

$$\left[\text{ans: no}\right]$$

If $\boldsymbol{\tau}$ is always perpendicular to $\mathbf{L}$, then $\mathbf{L}$ rotates (or precesses) without change in magnitude. The magnitude τ of the net torque, the magnitude L of the angular momentum, and the precessional angular velocity Ω are related by $\tau = L\Omega$. Given any two of these quantities you can solve for the third.

PROBLEM 15 (67P).   A gyroscope consists of a rotating uniform disk with a 50-cm radius suitably mounted at the center of a 12-cm long axle (of negligible mass) so that the gyroscope can spin and precess freely. Its spin rate is 1000 rev/min. Find the rate of precession (in rev/min) if the axle is supported at one end and is horizontal.

STRATEGY: The torque is due to gravity and has magnitude $\tau = Mgr$, where r is the distance from the support to the center of the disk and M is the mass of the disk. If the disk is spinning fast the magnitude of the angular momentum is given by $L = I\omega = \frac{1}{2}MR^2\omega$, where I is the rotational inertia of the disk and ω is its spin angular speed. $I = \frac{1}{2}MR^2$ was found in Table 11–2. Substitute into $\Omega = \tau/L$ and evaluate the resulting expression. Observe that r is *half* the length of the axle. You must convert $\omega = 1000$ rev/min to rad/s. Multiply by $2\pi/60$. Your answer, automatically in rad/s, must be converted to rev/min.

SOLUTION:

$$\left[\text{ans: } 0.43\,\text{rev/min}\right]$$

III. MATHEMATICAL SKILLS

You should know how to find the magnitude and direction of vector products, such as $\mathbf{a} \times \mathbf{b}$. Recall that the magnitude is $ab \sin \phi$, where ϕ is the smallest angle between $\mathbf{a}$ and $\mathbf{b}$ when they are drawn with their tails at the same point. You should also remember the right hand rule for finding the direction. The product is perpendicular to the plane of $\mathbf{a}$ and $\mathbf{b}$. Curl the fingers of your right hand so they rotate $\mathbf{a}$ toward $\mathbf{b}$ through the angle ϕ. Then the thumb will point in the direction of the vector product.

You also need to know how to compute the vector product in terms of the components of the factors:

$$(\mathbf{a} \times \mathbf{b})_x = a_y b_z - a_z b_y$$

$$(\mathbf{a} \times \mathbf{b})_y = a_z b_x - a_x b_z$$

$$(\mathbf{a} \times \mathbf{b})_z = a_x b_y - a_y b_x$$

IV. NOTES

Chapter 13
EQUILIBRIUM AND ELASTICITY

I. BASIC CONCEPTS

This chapter consists largely of applications of Newton's second laws for center of mass motion and for rotation, specialized to situations in which the total force and total torque both vanish. However, several new ideas are introduced to deal with the torque exerted by gravity and with elastic forces: the center of gravity and the elastic moduli. Pay careful attention to what they are and how they are used.

Equilibrium conditions. A rigid body is said to be in <u>static</u> <u>equilibrium</u> if its center of mass is not moving and if it is not rotating. Its total momentum and its total angular momentum both vanish. In terms of the external forces and torques acting on it, equilibrium means

$$\sum \mathbf{F} =$$

and

$$\sum \boldsymbol{\tau} =$$

where $\sum \mathbf{F}$ is the vector sum of all external forces and $\sum \boldsymbol{\tau}$ is the vector sum of all external torques.

You will use these equations in applications to determine the external forces and torques on static objects. In some cases you will be interested in the force or torque exerted by one part of an object on another part. In that event, simply consider the system to be the part of the object on which the force or torque acts and treat the force or torque as an external force or torque.

In nearly all examples and problems in this chapter the forces considered are in the same plane. If you pick an origin in this plane every torque is perpendicular to the plane and the number of equations to be solved is reduced to three. If the plane of the forces is the xy plane then the three equations are $\sum F_x = $ _____, $\sum F_y = $ _____, $\sum \tau_z = $ _____. They can be solved for three unknowns. Usually the last equation is written using τ to stand for τ_z.

Recall that the value of a torque depends on the origin used to specify the point of application of the force. When an object is in equilibrium, however, the torques sum to zero no matter which point is used as the origin. The choice is a matter of convenience, not necessity, but the same origin must be used to compute all torques. If an origin is not specified by the problem statement, place it in the plane of the forces. Carefully study the sample problems of Section 13–4 to see how equilibrium problems are solved.

Center of gravity. The force of gravity is often one of the forces acting on an object. Although each particle of the object experiences a gravitational force, for purposes of determining equilibrium these forces can be replaced by a single force acting at a point called _____. If the acceleration **g** due to gravity is uniform throughout the object, the magnitude of the replacement force is _____, where M is the total mass of the object, and the direction of the replacement force is _____. In addition, the point where the replacement force is applied then coincides with _____. In this case the position of the center of gravity does not depend on the orientation of the object.

As you read the sample problems of Section 13–4 in the text, carefully note how gravitational forces are taken into account. In each case their sum is replaced by a single force of magnitude Mg, directed downward and applied at the center of mass.

Even if the acceleration due to gravity is not uniform over the object, the sum of the gravitational forces can still be replaced by a single force applied at the center of gravity. Then, however, the center of gravity does not coincide with the center of mass. Furthermore, the position of the center of gravity depends on the orientation of the object. Practically speaking, it is not as useful a concept then.

Elasticity. In many cases the conditions that the net force and torque both vanish are not sufficient to determine the forces acting on an object. From a mathematical viewpoint there are too many unknowns or not enough equations. As a result, the equations have an infinite number of solutions and additional information is needed to decide which one is physically correct.

Contact forces, like the normal force of a surface on an object, are <u>elastic</u> in nature. That is, they arise from slight deformations of the objects in contact. In many circumstances the additional information you need to understand equilibrium is how elastic forces depend on deformation.

You should know the special terminology that is used in the study of elasticity. A force per unit area is called a _____ and a fractional deformation is called a _____ .

There are two regimes of deformation. If the stress is less than the <u>yield</u> strength then, when the stress is removed, _____
_____ .

If the stress is greater than the yield strength but less than the <u>ultimate</u> strength then, when the stress is removed, _____
_____ .

If the stress is greater than the ultimate strength, then _____
_____ .

For stresses well below the yield strength, the strain is proportional to the stress and the material is said to be elastic. Elastic materials behave like springs.

Suppose forces that are equal in magnitude and opposite in direction are applied to the opposite ends of a rod. If the forces are perpendicular to the rod faces, the rod will be elongated or compressed, depending on the directions of the forces. Suppose the forces are pulling outward and let ΔL be the elongation of the rod. Also suppose the magnitude of each force is F, the area of each face is A, and the undeformed rod length is L. Then the stress is given by _____ and the <u>strain</u> is given by _____ . If the object remains elastic (small stress), then the stress is proportional to the strain. The proportionality is written

$$\frac{F}{A} = \frac{E\,\Delta L}{L},$$

where the modulus of elasticity E is called _____ . E is a property of the material and does not depend on F, A, L, or ΔL.

Values of E for some materials are listed in Table 13–1 of the text. A large value of E means a large force is required to compress or elongate a given length of sample by a given amount. If E is small and the length and area are the same, only a small force is required.

If the forces are opposite in direction but parallel to the rod faces, then shear occurs. Suppose one face moves a distance Δx relative to the other, along a line that is parallel to the faces. Also suppose the magnitude of each force is F, the area of each face is A, and the rod length is L. The stress is still given by _____ and the strain is still given by _____ . Stress is proportional to strain and the proportionality is written

$$\frac{F}{A} = \frac{G\,\Delta x}{L}.$$

The modulus of elasticity G is called _____ . G is a property of the material. Materials with a small modulus of elasticity shear more easily than materials with a large modulus of elasticity provided, of course, the lengths and areas of the samples are the same.

If the object is placed in a fluid and pressure is applied uniformly on its surface, then hydraulic compression occurs. In this case the stress is the _____ in the fluid. This is the same as the force per unit _____ acting on the object. The strain is _____ where V is the original volume and ΔV is the change in volume. The modulus of elasticity is called the _____ modulus and is denoted by B. In symbols $B =$ _____ where p is the pressure in the fluid.

II. PROBLEM SOLVING

Except for the ideas of center of gravity and of elasticity, this chapter chiefly covers applications of concepts discussed in earlier chapters and the problem solving techniques you will need are not new to you.

We consider situations in which all forces acting on the object under consideration are in the xy plane. Furthermore, in each case we select the origin for the computation of torques to be in that plane. All torques are then along the z axis, perpendicular to the plane. The method can easily be extended to situations in which forces with z components and torques with x and y components act.

Here are the steps you should use to solve problems. First select the object to be considered. Draw a force diagram showing as arrows all external forces acting on the object. Don't forget normal and frictional forces exerted by any objects in contact with the selected object. Use an algebraic symbol to designate the magnitude of each force, known or unknown. Use either Mg or W to label the force of gravity, depending on whether the weight or mass of the object is given.

Draw each force arrow in the correct direction, if you know it. Gravity acts downward, a string pulls along its length, a frictionless surface pushes along the perpendicular to the surface. Sometimes you will know the line of a force but not its direction. You might know, for example, that a rigid rod acts along its length but you may not know if it is pushing or pulling. In that event, draw the arrow in either direction. If, after you solve the problem, the value turns out to be positive then you picked the correct direction; if it turns out to be negative the force is actually opposite the direction you picked. In other cases you may know only that the force acts in the xy plane. Pick an arbitrary direction, not along one of the coordinate axes. You will then solve for both components.

Because you will need to compute torques, be sure you indicate the point of application of each force by placing the tail or head of the arrow at the appropriate place on the diagram. Gravity acts at the center of gravity. For the situations considered here this is the same as the center of mass. A string or rod acts at its point of attachment. In many cases an object makes contact only at a point and that is the point of application of the force it exerts.

Choose a coordinate system for calculating force components. If you know the direction of one of the unknown forces, you can usually simplify the algebra by selecting one of the coordinate axes to be in that direction. Write $\sum F_x = 0$ and $\sum F_y = 0$ in terms of the individual forces.

Choose an origin for calculating torques. The choice is completely arbitrary but you can usually reduce the algebra considerably by choosing an origin so that one or more torques vanish. Try to place the origin on the line of action of a torque associated with an unknown force. Write $\sum \tau = 0$ in terms of the forces and their lever arms. Pay special attention to the signs of the torques.

PROBLEM 1 (2E). A rigid square object of negligible weight is acted on by three forces that pull on its corners as shown, to scale, in Fig. 13–20. (a) Is the first requirement of equilibrium satisfied?

STRATEGY: Take the x axis to be horizontal and the y axis to be vertical. Sum the x components of the forces and see if the result is 0. Do the same for the y components.

SOLUTION:

[ans: yes]

(b) Is the second requirement of equilibrium satisfied?

STRATEGY: Sum the torques. To make the job easy place the origin at the lower right corner. Then two of the torques vanish.

SOLUTION:

[ans: no]

(c) If the answer to either (a) or (b) is no, could a fourth force restore the equilibrium of the object? If so, specify the magnitude, direction, and point of application of the needed force.

STRATEGY: You need to find a single force with magnitude 0 and a net torque associated with it.

SOLUTION:

[ans: no]

PROBLEM 2 (9E).   The system in Fig. 13–24 is in equilibrium, but it begins to slip if any additional mass is added to the 5.0-kg object. What is the coefficient of static friction between the 10-kg block and the plane on which it rests?

STRATEGY: Choose the object to be the knot where the three ropes are joined and write the equilibrium conditions in terms of the two masses and the force of friction on the upper mass. Since the system is on the verge of slipping the force of friction is given by $f = \mu_s N$, where N is the normal force of the surface on the block.

SOLUTION: Draw a free-body diagram for the knot. Let T_1 be the tension in the cord pulling to the left and m_1 be the 10-kg mass at its end. Let T_2 be the tension in the cord pulling down and m_2 be the 5.0-kg mass at its end. Let T_3 be the tension in the last cord. Choose the x axis to be toward the right and the y axis to be upward. The equilibrium conditions are $T_3 \sin 30° - T_1 = 0$ and $T_3 \cos 30° - T_2 = 0$. Take the origin to be at the knot. Then no torques are associated with the forces.

Since the 10-kg block is in equilibrium $T_1 = f = \mu_s N$ and $N = m_1 g$. Since the 5.0-kg block is in equilibrium $T_2 = m_2 g$. Thus the equilibrium conditions for the knot become $T_3 \sin 30° - \mu_s m_1 g = 0$ and $T_3 \cos 30° - m_2 g = 0$. Eliminate T_3 and solve for μ_s.

[ans: 0.29]

PROBLEM 3 (12E).   A 160-lb man is walking across a level bridge and stops one-fourth of the way from one end. The bridge is uniform and weighs 600 lb. What are the vertical forces exerted on the bridge by its supports at (a) the far end and (b) the near end?

STRATEGY: Use the equilibrium conditions for bridge to solve for the forces.

SOLUTION: Draw a force diagram for the bridge. The force of gravity W_b on the bridge acts at its center, the force of the man on the bridge is given by his weight W_m and acts at the point $L/4$ from the near end. Here L is the length of the bridge. Also show the upward force F_1 of the support at the near end and the upward force F_2 of the support at the far end.

Take the x axis to be horizontal and the y axis to be vertical. Place the origin at the near end of the bridge. Then the equilibrium conditions are $F_1 + F_2 - W_b - W_m = 0$ and $F_2 L - W_b L/2 - W_m L/4 = 0$. Solve for F_1

and F_2.

[ans: (a) 340 lb; (b) 420 lb]

PROBLEM 4 (18E).   A uniform cubical crate is 0.750 m on each side and weighs 500 N. It rests on the floor with one edge against a very small, fixed obstruction. At what height above the floor must a horizontal force of 350 N be applied to the crate to just tip it?

STRATEGY: When the crate is about to tip both the normal force of the floor and the horizontal force of the obstruction are applied at the edge in contact with the obstruction. In addition, a force of gravity acts at the center of the crate. Calculate the position at which the 350-N force must be applied to hold the crate in equilibrium.

SOLUTION: Draw a force diagram for the crate. Show it in cross section, as a square, with the obstruction at the lower right corner and the applied force F acting horizontally a distance d from the floor at the left face. Show the normal force N of the floor acting upward at the lower right corner, the force F_o of the obstruction acting to the left at the lower right corner, and the force of gravity W acting downward at the center.

    If you pick the lower right corner as the origin all you need to solve the problem is the condition for rotational equilibrium: $WL/2 - Fd = 0$, where L is the length of one side of the cube. Solve for d.

[ans: 0.536 m]

PROBLEM 5 (19P).   Two identical, uniform, frictionless spheres, each of weight W, rest in a rigid rectangular container as shown in Fig. 13–28. Find, in terms of W, the forces acting on the spheres due to (a) the container surfaces and (b) one another, if the line of centers of the spheres makes an angle of 45° with the horizontal.

STRATEGY: Since the spheres are frictionless the forces of the container walls on them must be normal to the walls; the force of the container bottom must be vertically upward; and the force of each sphere on the other must be along the line joining their centers. Write the equations for the equilibrium of each sphere and solve them for the unknowns.

SOLUTION: Draw a force digram for the upper sphere. Show the force of gravity W down, the force of the wall F_1 horizontally to the left, and the force of the lower sphere F from the point of contact toward the center. Take the x axis to be horizontal and the y axis to be vertical. Write the equilibrium conditions: $F\sin 45° - W = 0$ and $F\cos 45° - F_1 = 0$.

Draw a force diagram for the lower sphere. Show the force of gravity down, the force of the wall F_2 to the right, the force of the container bottom F_3 upward, the force of gravity W downward, and the force of the upper sphere F from the point of contact toward the center. Write the equilibrium condition equations: $F_2 - F\cos 45° = 0$ and $F_3 - W - F\sin 45° = 0$.

In each case the center of the sphere is chosen as the origin. Since the line of action of each force acting on a sphere goes through its center each torque is 0.

Solve for F_1, F_2, F_3, and F.

$$\left[\,\text{ans: } (a)\text{ right side: } W,\text{ left side: } W,\text{ bottom: } 2W;\ (b)\ \sqrt{2}W\,\right]$$

PROBLEM 6 (21P). The system in Fig. 13–30 is in equilibrium with the string in the center exactly horizontal. Find (a) tension T_1, (b) tension T_2, (c) tension T_3, and (d) angle θ.

STRATEGY: Consider the equilibrium equations for each of the two knots.

SOLUTION: Draw a force diagram for the knot on the left. Show the forces of the three strings that are tied there. Two of the tensions are unknown. The third equals the weight W_1 of the mass hung from the knot. If the x axis is horizontal and the y axis is vertical, the equilibrium equations are $T_1\cos 35° - W_1 = 0$ and $T_1\sin 35° - T_3 = 0$. All torques about the knot vanish.

Draw a force diagram for the knot on the right. Show the forces of the three strings tied there. One tension is the weight W_2 of the mass hung there, the others are unknown. The angle θ is also unknown. The equilibrium equations are $T_2\sin\theta - T_3 = 0$ and $T_2\cos\theta - W_2 = 0$. Again all torques about the knot vanish.

The equations for the left knot can be solved for T_1 and T_3, then the equations for the right knot can be solved for T_2 and θ. Use $\cos^2\theta + \sin^2\theta = 1$ to eliminate θ when you solve for T_2. Divide one equation by the other to eliminate T_2 and obtain an equation for $\tan\theta$.

$$\left[\,\text{ans: } (a)\ 49\,\text{N};\ (b)\ 28\,\text{N};\ (c)\ 57\,\text{N};\ (d)\ 29°\,\right]$$

PROBLEM 7 (23P).   A balance is made up of a rigid rod with mass M, supported at and free to rotate about a point not at the center of the rod. It is balanced by unequal weights placed in the pans at each end of the rod. When an unknown mass m is placed in the left-hand pan, it is balanced by a mass m_1 in the right-hand pan; and when the mass m is placed in the right-hand pan, it is balanced by a mass m_2 in the left-hand pan. Show that $m = \sqrt{m_1 m_2}$.

STRATEGY: Suppose the pivot point is a distance ℓ from the left-hand pan and a distance $L - \ell$ from the right-hand pan. Here L is the length of the rod. Write the equilibrium conditions for the situation when m is in the left-hand pan. Then write the conditions for the situation when it is in the right-hand pan. Solve for m.

SOLUTION: Draw the force diagram for the rod with m in the left-hand pan and m_1 in the right-hand pan. Show the force mg downward at the left end of the rod, the force $m_1 g$ downward at the right end of the rod, and the force F of the pivot upward at a distance ℓ from the left end of the rod. Write the condition for rotational equilibrium. Take the origin to be at the pivot. Then $mg\ell - m_1 g(L - \ell) = 0$. You do not need the equations for equilibrium of the center of mass since you are not asked for F.

Draw the force diagram for the rod with m_2 in the left-hand pan and m in the right-hand pan. Show the forces. Again take the origin to be at the pivot point and write the condition for rotational equilibrium: $m_2 g\ell - mg(L-\ell) = 0$. Eliminate ℓ and solve for m. You should get the result $m = \sqrt{m_1 m_2}$.

PROBLEM 8 (25P).   A 50.0-kg uniform square sign, 2.00 m on a side, is hung from a 3.00-m rod of negligible mass. A cable is attached to the end of the rod and to a point on the wall 4.00 m above the point where the rod is fixed to the wall, as shown in Fig. 13–33. (*a*) What is the tension in the cable?

STRATEGY: Solve the equilibrium condition equations for the rod.

SOLUTION: Draw a force diagram for the rod. The wall exerts a vertical force F_v and a horizontal force F_h. Take the vertical force to be upward and the horizontal force to be to the right. The sign exerts a force equal to its weight $m_s g$, at the midpoint of its length, a distance $L - \ell/2$ from the left end of the rod. Here L is the length of the rod and ℓ is the length of the sign. Include the tension in the cable, acting at the end of the rod and at an angle θ above the horizontal. The rod is essentially massless, so the force of gravity on it is ignored. A little trigonometry shows that if d is the distance between the left end of the rod and the point where the cable is attached then $\sin\theta = d/\sqrt{d^2 + L^2}$ and $\cos\theta = L/\sqrt{d^2 + L^2}$.

Take the x axis to be horizontal and the y axis to be vertical. The conditions for equilibrium are then $T\sin\theta + F_v - m_s g = 0$, $F_h - T\cos\theta = 0$, and $TL\sin\theta - m_s g(L - \ell/2) = 0$. The last equation can easily be solved for T.

$\big[$ans: 408 N$\big]$

What are the (b) horizontal and (c) vertical components of the force exerted by the wall on the rod?

SOLUTION: The first of the equations you wrote can be solved for F_v and the second can be solved for F_h.

$\big[$ans: (b) $F_h = 245$ N to the right; (c) $F_v = 164$ N up$\big]$

PROBLEM 9 (26P). In Fig. 13–34, a 55-kg rock climber is in a lie-back climb along a fissure, with hands pulling on one side of the fissure and feet pressed against the opposite side. The fissure has width $w = 0.20$ m, and the center of mass of the climber is a horizontal distance of $d = 0.40$ m from the fissure. The coefficient of static friction between hands and rock is $\mu_1 = 0.40$, and between boots and rock it is $\mu_2 = 1.2$. (a) What is the least horizontal pull by the hands and push by the feet that will keep him stable?

STRATEGY: Use the equilibrium conditions for the rock climber. Since he is on the verge of slipping the frictional force of the rock on his hands has magnitude given by $f_1 = \mu_1 N_1$ and the frictional force of the rock on his feet has magnitude given by $f_2 = \mu_2 N_2$, where N_1 is the (horizontal) normal force of the rock on his hands and N_2 is the (horizontal) normal force of the rock on his feet.

SOLUTION: Draw a force diagram for the rock climber. Show the two normal forces, the one on his hands to the right and the one on his boots to the left, and the two frictional forces, both upward. Also show the force of gravity mg downward. Take the x axis to be horizontal and the y axis to be vertical. Since the net force on the climber vanishes $\mu_1 N_1 + \mu_2 N_2 - mg = 0$ and $N_1 - N_2 = 0$. Take the origin to be at his boots. Then the condition for rotational equilibrium becomes $N_1 h + \mu_1 N_1 w - mg(d + w) = 0$, where h is the vertical distance from his boots to his hands.

The first two equations can be solved for N_1 and N_2. The second tells you that $N_1 = N_2$. The first then yields $N_1 = mg/(\mu_1 + \mu_2)$.

$\big[$ans: 340 N$\big]$

(b) For the horizontal pull of (a), what must be the vertical distance h between hands and feet?

SOLUTION: Solve the third equilibrium condition equation for h. You should get $h = [mg(d + w) - \mu_1 N_1 w]/N_1$.

$\big[$ans: 0.88 m$\big]$

(c) If the climber encounters wet rock, so that μ_1 and μ_2 are reduced, what happens to the answer to (a) and (b), respectively?

SOLUTION: You derived an algebraic expression for N_1. What happens to its value if the friction coefficients are reduced? To find out what happens to h, substitute the expression for N_1 into the expression you obtained for h. You should get $h = \mu_1(d + w) + \mu_2(d + w)$. What happens when the coefficients are reduced?

[ans: N_1 and N_2 increase, h decreases]

PROBLEM 10 (28P). Forces F_1, F_2, and F_3 act on the structure of Fig. 13–36 as shown in an overhead view. We wish to put the structure in equilibrium by applying a force, at a point such as P, whose vector components are F_h and F_v. We are given that $a = 2.0\,\mathrm{m}$, $b = 3.0\,\mathrm{m}$, $c = 1.0\,\mathrm{m}$, $F_1 = 20\,\mathrm{N}$, $F_2 = 10\,\mathrm{N}$, and $F_3 = 5.0\,\mathrm{N}$. Find (a) F_h, (b) F_v, and (c) d.

STRATEGY: Use the conditions for equilibrium of the structure.

SOLUTION: Draw a force diagram for the structure. Show all the forces in the diagram of the text. Since the net force on the structure vanishes, $F_v - F_1 - F_2 = 0$ and $F_h - F_3 = 0$. Since the net torque on the structure vanishes, $F_v(d + c) - F_1 c - F_2(b + c) - F_3 a = 0$, where the origin was taken to be at the left end of the structure. Solve the first equation for F_v, the second for F_h, and the third for d.

[ans: (a) 5.0 N; (b) 30 N; (c) 1.3 m]

PROBLEM 11 (36P). In Fig. 13–41, suppose the length L of the uniform bar is 3.0 m and its weight is 200 N. Also, let $W = 300\,\mathrm{N}$ and $\theta = 30°$. The wire can withstand a maximum tension of 500 N. (a) What is the maximum possible distance x before the wire breaks?

STRATEGY: Consider the conditions for equilibrium of the bar with W placed an arbitrary distance x from the left end. Take the tension in the wire to be the maximum, 500 N, and calculate x.

SOLUTION: Draw a force diagram for the bar. Show W acting downward a distance x from the left end; the force of gravity W_b acting downward at the center of the bar, a distance $L/2$ from the left end; the wall pushing toward the right with a force F_h and upward with a force F_v; and the tension in the wire, at an angle θ above the horizontal.

Take the x axis to be horizontal and the y axis to be vertical. The horizontal components of the forces sum to zero: $F_h - T\cos\theta = 0$. The vertical components sum to zero: $F_v + T\sin\theta - W - W_b = 0$. The torques sum to zero: $TL\sin\theta - Wx - W_bL/2 = 0$, where the left end of the rod was taken to be the origin. Solve the third equation for x.

[ans: $1.50\,\text{m}$]

With W placed at this maximum x, what are the (b) horizontal and (c) vertical components of the force exerted on the bar by the pin at A?

SOLUTION: Solve the two force equations for F_h and F_v.

[ans: (b) $433\,\text{N}$; (c) $250\,\text{N}$]

PROBLEM 12 (44P). A crate, in the form of a cube with edge lengths of 4.0 ft, contains a piece of machinery whose design is such that the center of gravity of the crate and its contents is located 1.0 ft above its geometrical center. The crate rests on a ramp which makes an angle θ with the horizontal. As θ is increased from zero, an angle will be reached at which the crate will either start to slide down the ramp or tip over. Which event will occur (a) when the coefficient of static friction is 0.60 and when it is (b) 0.70? In each case, give the angle at which the event occurs.

STRATEGY: For each value of the coefficient μ_s find the angle at which sliding occurs and the angle at which tipping occurs. The smaller of the two determines which event occurs.

SOLUTION: Consider sliding first. Draw a force diagram for the crate. The force of gravity mg acts downward, the normal force N acts perpendicular to the plane, and the force of friction f acts parallel to the plane, up the plane. Take the x axis to be down the plane and the y axis to be perpendicular to the plane. Newton's second law yields $mg\sin\theta - f = 0$ and $mg\cos\theta - N = 0$. Since the crate is on the verge of sliding $f = \mu_s N$. Eliminate N and solve for θ.

You should get $\tan\theta = \mu_s$. For each of the two values of μ_s find the value of θ at which sliding starts:

You should have gotten $31°$ for $\mu_s = 0.60$ and $35°$ for $\mu_s = 0.70$.

Chapter 13: Equilibrium and Elasticity **197**

Now suppose the crate is on the verge of tipping over. Draw a force diagram. The normal force N and the friction force f now act at the front edge of the crate, the force of gravity acts at the center of gravity, a distance d ($= 3.0\,\text{ft}$) from the bottom face and a distance $L/2$ from the front face. Here L is the length of a cube edge. Place the origin at the front edge and write the equation for rotational equilibrium: $mgd\sin\theta - mg(L/2)\cos\theta = 0$. Solve for θ. You should get $34°$.

Now decide which event occurs for each value of μ_s.

[ans: (a) sliding occurs, at $\theta = 31°$; (b) tipping occurs, at $\theta = 34°$]

Some elasticity problems are simply direct applications of Eqs. 13–29, 30, or 31 in the text. You may be asked to calculate the stress or the strain, given the other. In variants you may be asked for the change in length under compression or the relative displacement of the faces under shear. The appropriate modulus must be given or else listed in Table 13–1. In some cases you may be asked to calculate the modulus, given the other quantities in the equations. These problems are all straightforward. In some situations the elastic properties of objects determine the forces at equilibrium. To solve these, more complicated problems, you must simultaneously apply one of the equations of elasticity and the equations of equilibrium.

PROBLEM 13 (48E).   A mine elevator is supported by a single steel cable 2.5 cm in diameter. The total mass of the elevator cage plus occupants is 670 kg. By how much does the cable stretch when the elevator is (a) at the surface, 12 m below the elevator motor, and (b) at the bottom of the 350-m deep shaft? (Neglect the mass of the cable.)

STRATEGY: Use the defining equation for Young's modulus: $F/A = E\Delta L/L$, where F is the force exerted on one end of the cable, A is the cable cross-sectional area πR^2, L is the cable length, ΔL is the amount the cable stretches, and E is Young's modulus for steel ($200 \times 10^9\,\text{N/m}^2$, from Table 13–1). Here R is the radius of the cable (1.25 cm).

SOLUTION: Solve for ΔL. In (a) $L = 12\,\text{m}$, in (b) $L = 350 + 12 = 363\,\text{m}$.

[ans: (a) 0.80 mm; (b) 2.4 cm]

PROBLEM 14 (51E).   A solid copper cube has an edge length of 85.5 cm. How much pressure must be applied to the cube to reduce the edge length to 85.0 cm? The bulk modulus of copper is $1.4 \times 10^{11}\,\text{N/m}^2$.

STRATEGY: Use $p = B\Delta V/V$, where p is the pressure, V is the original volume, ΔV is the change in volume, and B is the bulk modulus.

SOLUTION: The original volume is given by $V = a^3$, where a is the original cube edge (85.5 cm). The final volume is given by $V' = (a')^3$, where a' is the final cube edge (85.0 cm). The change in volume is $\Delta V = V - V'$. Calculate p.

$$[\text{ans: } 2.4 \times 10^9 \, \text{N/m}^2]$$

PROBLEM 15 (54P). In Fig. 13–51, a lead brick rests horizontally on cylinders A and B. The areas of the top faces of the cylinders are related by $A_A = 2A_B$; the Young's moduli of the cylinders are related by $E_A = 2E_B$. The cylinders had identical lengths before the brick was placed on them. (a) What fraction of the brick's weight is supported by cylinder A, and (b) what fraction by cylinder B?

STRATEGY: The force exerted by cylinder A is given by $F_A = E_A A_A \Delta L_A/L_A$, where E_A is the Young's modulus, A_A is the cylinder's cross-sectional area, L_A is the length of the cylinder, and ΔL_A is the amount by which the cylinder is compressed when the brick is in place. Similarly the force exerted by cylinder B is given by $F_B = E_B A_B \Delta L_B/L_B$. In addition, the brick is in equilibrium, so $F_A + F_B = W$, where W is the weight of the brick. The cylinders start with the same length and the brick is horizontal, so $L_A = L_B$ and $\Delta L_A = \Delta L_B$. Use the conditions $A_A = 2A_B$ and $E_A = 2E_B$, then solve for F_A and F_B.

SOLUTION: First calculate the ratio $F_A/F_B = E_A A_A/E_B A_B$. You should get 4. Next substitute $F_A = 4F_B$ into $F_A + F_B = W$ and solve for F_B. Multiply by 4 to obtain F_A.

$$[\text{ans: } F_A = 4W/5, \; F_B = W/5]$$

The horizontal distances between the center of mass of the brick and the center lines of the cylinders are d_A for cylinder A and d_B for cylinder B. (c) What is the ratio d_A/d_B?

STRATEGY: Take the forces of the cylinders to act at the center lines. The brick is in rotational equilibrium, so $F_A d_A = F_B d_B$, where the origin was taken to be at the center of mass of the brick. Solve for d_A/d_B.

SOLUTION:

$$[\text{ans: } 1/4]$$

PROBLEM 16 (55P). In Fig. 13–52, a 103-kg uniform log hangs by two steel wires, A and B, both of radius 1.20 mm. Initially, wire A was 2.50 m long and 2.00 mm shorter than wire B. The log is now horizontal. What forces are exerted on it by (a) wire A and (b) wire B?

STRATEGY: The force exerted by wire A is given by $F_A = EA\Delta L_A/L_A$, where E is Young's modulus for steel ($200 \times 10^9 \, \text{N/m}^2$, from Table 13–1), A is the cross-sectional area (πR^2), L_A is the original length, and ΔL_A is the amount the wire stretches. Similarly the force exerted by wire B is given by $F_B = EA\Delta L_B/L_B$. In addition, the log is in equilibrium, so $F_A + F_B = W$, where W is the weight of the log. Let δ be the amount by which wire A was originally shorter than wire B. Then $L_B = L_A + \delta$. After the log is hung the wires have the same length so $\Delta L_B = \Delta L_A - \delta$. Thus $F_A = EA\Delta L_A/L_A$ and $F_B = EA(\Delta L_A - \delta)/(L_A + \delta)$. Solve these equations and $F_A + F_B = W$ simultaneously for F_A and F_B. You must eliminate the other unknown, ΔL_A.

SOLUTION:

$\left[\,\text{ans:}\ (a)\ F_A = 867\,\text{N};\ (b)\ F_B = 143\,\text{N}\,\right]$

(c)What is the ratio d_A/d_B?

STRATEGY: The log is also in rotational equilibrium, so $F_A d_A = F_B d_B$, where the origin was taken to be at the center of mass. Solve for d_A/d_B.

SOLUTION:

$\left[\,\text{ans:}\ 0.165\,\right]$

III. NOTES

Chapter 14
OSCILLATIONS

I. BASIC CONCEPTS

This chapter is about periodic motion: the motion of an object that moves back and forth between two points, its motion being the same during every cycle. Pay attention to the meanings of the terms used to describe simple harmonic motion: amplitude, period, frequency, angular frequency, and phase constant; pay attention to the transfer of energy from kinetic to potential and back again as the object moves; and also pay attention to the form of the force law that leads to this type of motion.

Simple harmonic motion. Concentrate on simple harmonic motion (SHM), for which the position variable (angle or coordinate) is a sinusoidal function of time. For a model we take an object of mass m attached to a spring with spring constant k, moving along the x axis with its equilibrium position at $x = 0$. Take the potential energy to be zero when the spring is neither stretched nor compressed. Then, as functions of the coordinate, the force on the mass is given by

$$F(x) =$$

and the potential energy of the system is given by

$$U(x) =$$

You should recognize that these equations are valid only if x = 0 is the equilibrium point. If the origin is placed elsewhere other terms must be added to the right sides.

The force and potential energy for the spring-mass system have certain distinguishing characteristics in common with all systems executing simple harmonic motion. The object, of course, does not move along a line for all SHM systems. For example, some objects have rotational motions and for them position is described by an angle rather than by a linear coordinate.

If we place the origin at the point for which the force or torque is zero then for SHM to occur the force or torque must always be proportional to _____ and if, in addition, we choose the potential energy to be zero when the position variable is zero then this energy is proportional to _____ .

The sign in the force law is important. When the coordinate is positive the force is _____ and when the coordinate is negative the force is _____ . A force that always pushes an object toward its equilibrium position is called a _____ force.

The force law is substituted into Newton's second law to obtain the equation of motion. For the spring-mass system it is

$$\frac{\mathrm{d}^2 x}{\mathrm{d}t^2} =$$

The most general solution to this equation is

$$x(t) = x_m \cos(\omega t + \phi),$$

where x_m, ω, and ϕ are constants. If the function given here is to obey the differential equation then the constant ω, in terms of m and k, must be $\omega = $ _____ .

The maximum value of the coordinate x is denoted by _____ . The object moves back and forth between $x = $ _____ and $x = $ _____ . x_m is called the _____ of the oscillation.

The constant ω is called the _____ of the oscillation. Since ωt is measured in radians, the units of ω are _____ .

The angular frequency is related to the frequency f of the oscillation: $\omega = $ _____ . The physical significance of the frequency is: _____
_____

It has the dimension _____ and its SI unit is: _____ .

The angular frequency and frequency are both related to the period T of the motion: $\omega = $ _____ , and $f = $ _____ . The physical significance of the period is _____
_____

It has the dimension _____ and its SI unit is _____ .

As a fraction of the period T, the time taken by the mass to go from $x = 0$ to $x = x_m$ for the first time is _____ and the time taken to go from $x = -x_m$ to $x = +x_m$ for the first time is _____ .

Suppose an oscillation has a period of 5.0 s. Then its frequency is _____ Hz and its angular frequency is _____ rad/s. The motion repeats every _____ s or, in other words, it repeats _____ times per s.

You should remember that the angular frequency is determined by the ratio of k to m. Consider two springs, one with spring constant k and the other with spring constant $2k$. An object of mass m is attached to the first spring and set into oscillation. We can obtain an oscillation of exactly the same angular frequency by attaching an object of mass _____ to the second spring. These two systems take the same time to complete a cycle of their motions.

An expression for the velocity of the object as a function of time can be found by differentiating the expression for $x(t)$ with respect to time. The result is

$$v(t) = $$

In terms of x_m and ω, the maximum speed of the object is _____ . The object has maximum speed when its coordinate is $x = $ _____ . The velocity of the object is zero when its coordinate is $x = $ _____ and also when its coordinate is $x = $ _____ .

An expression for the acceleration of the object as a function of time can be found by differentiating $v(t)$ with respect to time. It is

$$a(t) = $$

In terms of x_m and ω, the magnitude of the maximum acceleration is _____ . The mass has maximum acceleration when its coordinate is $x = $ _____ and also when its coordinate is $x = $ _____ . This is when the force has maximum magnitude and the velocity vanishes. The acceleration is zero when the coordinate is _____. This is when the force vanishes and the speed is a maximum.

The argument of the trigonometric function, $\omega t + \phi$, is called the _____ of the oscillation and ϕ is called the _____.

For a given system, x_m and ϕ are determined by the initial conditions (at $t = 0$). Since $x(t) = x_m \cos(\omega t + \phi)$ the initial coordinate of the object is given by $x_0 =$_____ and the initial velocity of the object is given by $v_0 = $ _____.

x_0 is positive for ϕ between _____ and _____.

x_0 is negative for ϕ between _____ and _____.

v_0 is positive for ϕ between _____ and _____.

v_0 is negative for ϕ between _____ and _____.

The equations $x_0 = x_m \cos \phi$ and $v_0 = -\omega x_m \sin \phi$ can be solved for x_m and ϕ. To obtain an expression for x_m solve the first equation for $\cos \phi$ and the second for $\sin \phi$, then use $\cos^2 \phi + \sin^2 \phi = 1$. To obtain an expression for ϕ, divide the second equation by the first and solve for $\tan \phi$. The results are

$$x_m =$$

$$\phi =$$

Be careful when you evaluate the expression for ϕ. There are always two angles that are the inverse tangent of any quantity but your calculator only gives the one closest to 0. The other is 180° or π radians away. Always check to be sure the values you obtain for x_m and ϕ give the correct initial coordinate and velocity. That is, $x_m \cos \phi$ must give the correct initial coordinate and $-\omega x_m \sin \phi$ must give the correct initial velocity. If they do not, add π to your answer for ϕ sand check again.

On the axes below draw a graph of $x(t) = x_m \cos(\omega t + \phi)$. Take ϕ to be 0 and x_m to have the value marked on the vertical axis. Draw the function so the period is T, as marked on the time axis. On your graph label the times for which magnitude of the velocity is a maximum, the times for which it is a minimum, the times for which the magnitude of the acceleration is a maximum, and the times for which it is a minimum.

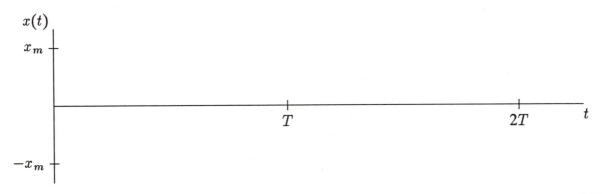

Energy considerations. An expression for the potential energy as a function of time can be found by substituting $x(t) = x_m \cos(\omega t + \phi)$ into $U = \frac{1}{2}kx^2$. The result is

$$U(t) =$$

An expression for the kinetic energy as a function of time can be found by substituting $v = -\omega x_m \sin(\omega t + \phi)$ into $K = \frac{1}{2}mv^2$. The result is

$$K(t) =$$

or, if $\omega^2 = k/m$ is used,

$$K(t) =$$

Both the potential and kinetic energies vary with time. As the object moves away from the equilibrium point it slows down; kinetic energy is converted to potential energy and stored in the extended spring. As the object moves toward the equilibrium point the stored potential energy is converted to kinetic energy and the object speeds up.

The potential energy is a maximum when the coordinate is $x =$ _____ and also when it is $x =$ _____. Then the speed of the object is $v =$ _____ and the kinetic energy vanishes. The kinetic energy is a maximum when the coordinate is $x =$ _____. Then the potential energy vanishes. Notice that the maximum kinetic energy has exactly the same value as the maximum potential energy.

Although both the potential and kinetic energies vary with time, the total mechanical energy $E = K + U$ is constant, as you can see by adding the two expressions you wrote above. If you know that the object has speed v when its coordinate is x, then you can use $E =$ _____ to compute the total mechanical energy. If you know the amplitude x_m of the oscillation then you can compute the total mechanical energy using $E =$ _____. If you know the maximum speed v_m of the object then you can compute the total mechanical energy using $E =$ _____. Since total mechanical energy is conserved these three expressions must have the same value.

Applications. Several other systems that oscillate in simple harmonic motion are discussed in the text. For each of them use one of the tables below to describe the displacement variable (linear or angular coordinate) that is oscillating and give its symbol. Then give an expression for the force or torque, as appropriate. For the torsional pendulum the torque is proportional to the angular displacement. For the other pendulums the force or torque is not proportional to the displacement unless the amplitude of the oscillation is small. Give both the exact expression and the small amplitude approximation. In all cases tell the physical meaning of the symbols that appear in the constant of proportionality that relates the force or torque and the displacement. Give the equation of motion obeyed by the displacement, in a form similar to that given above for a spring-mass system. Write the general form of the solution to the equation of motion and give the expressions for the angular frequency ω and the period T in terms of properties of the system.

a. Torsional pendulum
displacement variable: _____
_____

torque law: _____
meaning of symbols: _____
_____

equation of motion: _____
general solution: _____
$\omega = $ _____ $T = $ _____

b. Simple pendulum
displacement variable: _____
_____

force law (exact): _____
force law (small amplitude): _____
meaning of symbols: _____
_____

equation of motion: _____
general solution: _____
$\omega = $ _____ $T = $ _____

c. Physical pendulum
displacement variable: _____
_____

torque law (exact): _____
torque law (small amplitude): _____
meaning of symbols: _____
_____

equation of motion: _____
general solution: _____
$\omega = $ _____ $T = $ _____

Remember that the small angle approximation $\sin\theta \approx \theta$ is valid only if θ is measured in radians. See the Mathematical Skills section for details.

SHM and uniform circular motion.   When a particle moves around a circle with constant speed, the projection of its position vector on the x axis performs simple harmonic motion. Suppose the angular speed of the particle is ω and the radius of the circle is R. Put the origin of the coordinate system at the center of the circle and measure angles counterclockwise from the x axis. If the initial angular position of the particle is ϕ then the x component of its position vector is given by $x(t) = $ _____. You can identify the angular speed ω of the particle with the _____ of the oscillation and the radius of the circle with the _____ of the oscillation.

The projection of the particle's position vector on the y axis also performs simple harmonic motion. Since $y(t) = R\sin(\omega t + \phi) = R\cos(\omega t + \phi - \pi/2)$, the two coordinates are out of phase by $\pi/2$. Looked at another way, we can say that if a particle simultaneously performs simple harmonic motions with the same amplitude and frequency along two perpendicular axes and the motions are out of phase by $\pi/2$, then its trajectory is a _____ .

Damped and forced harmonic motions. Many oscillating systems in nature are damped by a force that is proportional to the velocity. Take the damping force to be $-b(dx/dt)$ and write the equation of motion obeyed by the displacement x of a mass m on a spring with spring constant k:

$$\frac{d^2x}{dt^2} =$$

If the damping coefficient b is small you may think of the motion as simple harmonic with an exponentially decreasing amplitude. The angular frequency is not quite the same as in the absence of damping and is now denoted by ω_d. Give the solution in terms of the angular frequency ω_d and the damping constant b:

$$x(t) =$$

Give an expression for the angular frequency in terms of k, m, and b:

$$\omega_d =$$

Note that if b is small the angular frequency is nearly the same as the angular frequency for undamped motion, namely $\sqrt{k/m}$.

If $(b/2m)^2 < k/m$ then the mass oscillates with decreasing amplitude. If $(b/2m)^2 \geq k/m$ then the mass does not oscillate, but instead simply returns to its equilibrium point. The solution is then usually written in another form.

Sketch $x(t)$ on the graph below for $(b/2m)^2 < k/m$. Take the phase constant to be zero. Be sure to show both the oscillations and the decay of the amplitude.

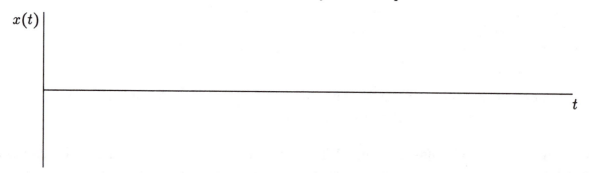

The total mechanical energy of a damped oscillator is not constant. As time goes on, energy is dissipated by _____

_____ .

Oscillations can be driven by applying an external sinusoidal force. Suppose the external force is given by $F_m \cos \omega_f t$ and write the equation of motion for the displacement variable x, including a damping term proportional to dx/dt:

$$\frac{d^2x}{dt^2} =$$

The angular frequency of a forced oscillation is the same as the angular frequency of the driving force and is NOT necessarily the natural angular frequency of the oscillator. Also know that the amplitude of the oscillation depends on the driving frequency, as well as on the driving amplitude F_m. The driving frequency for which the velocity amplitude is a maximum is called the _____ frequency. This is the same as the natural frequency of the oscillator. It is nearly the same as the frequency for which the amplitude is a maximum.

The amplitude of the motion does not decay with time even though a resistive force is present and energy is being dissipated. The mechanism that drives the oscillator and provides the external force does work on the system and supplies the energy required to keep it going at constant amplitude.

Study Fig. 14–21 of the text to see how the amplitude depends on the driving frequency. For a given value of the damping coefficient b, the amplitude decreases as the driving frequency moves away from the resonance frequency in either direction. As b increases, the amplitude at resonance _____. Also the width of the curve, measured at an amplitude that is half the resonance amplitude, increases. Resonance becomes less sharp. This has important implications for the design of bridges and other structures that might resonate under the influence of cars, people, wind, or other external forces. It also has important implications for musical instruments, for tuning radios and television sets, for the emission of light from atoms, and in fact for almost every branch of science and engineering.

II. PROBLEM SOLVING

Some problems make use of the relationships between angular frequency, frequency, and period for simple harmonic motion: $\omega = 2\pi f$, $f = 1/T$, and $\omega = 2\pi/T$. Occasionally the period is given indirectly by describing a time interval. You must then know, for example, that the time the oscillator takes to go from maximum displacement in one direction to maximum displacement in the other direction is $T/2$ or the time it takes to go from maximum displacement to zero displacement is $T/4$. If these time intervals or others are given you should be able to calculate the period, frequency, and angular frequency. You should also know how to find the maximum speed and maximum acceleration in terms of the angular frequency and amplitude: $v_m = \omega x_m$ and $a_m = \omega^2 x_m$. Some problems require you to know the relationship between the angular frequency and the appropriate physical properties of the oscillating system: $\omega = \sqrt{k/m}$ for an undamped spring-mass system.

PROBLEM 1 (1E). An object undergoing simple harmonic motion takes 0.25 s to travel from one point of zero velocity to the next such point. The distance between those points is 36 cm. Calculate (a) the period, (b) the frequency, and (c) the amplitude of the motion.

STRATEGY: The object has zero velocity at the extreme points of its motion, so the time given is half a period and the distance given is twice the amplitude. Use $f = 1/T$ to calculate the frequency.

SOLUTION:

[ans: (a) 0.50 s; (b) 2.0 Hz; (c) 18 cm]

PROBLEM 2 (8E). The scale of a spring balance which reads from 0 to 32.0 lb is 4.00 in. long. A package suspended from the balance is found to oscillate vertically with a frequency of 2.00 Hz. (a) What is the spring constant?

STRATEGY: When a 32-lb weight is hung from the balance it extends 4.00 in. from its equilibrium position. Then the force of gravity on the weight balances the force of the spring, so $kx = W$ or $k = W/x$. Convert 4.00 in. to feet.

SOLUTION:

$$[\text{ans: } 96.0 \,\text{lb/ft}]$$

(b) How much does the package weigh?

STRATEGY: The frequency of oscillation, in terms of the mass m of the package and the spring constant k, is given by $f = (1/2\pi)\sqrt{k/m}$. Solve for m. The weight of the package is mg.

$$[\text{ans: } 19.5 \,\text{lb}]$$

A problem might give you an expression for the displacement as a function of time and ask for the amplitude, angular frequency, and phase constant (or related quantities). Simply identify the various constants in the given expression. It might also ask for the coordinate, velocity, and acceleration at some specific time. Simply evaluate the expression and its first and second derivatives for that value of the time.

PROBLEM 3 (16E). A body oscillates with simple harmonic motion according to the equation

$$x = (6.0 \,\text{m}) \cos[(3\pi \,\text{rad/s})t + \pi/3 \,\text{rad}].$$

At $t = 2.0$ s, what are (a) the displacement, (b) the velocity, (c) the acceleration, and (d) the phase of the motion?

STRATEGY: The displacement is given by the expression above, the velocity is given by

$$v = \frac{dx}{dt} = -(18\pi \,\text{m/s}) \sin[(3\pi \,\text{rad/s})t + \pi/3 \,\text{rad}],$$

and the acceleration is given by

$$a = \frac{dv}{dt} = -(54\pi^2 \,\text{m/s}^2) \cos[(3\pi \,\text{rad/s})t + \pi/3 \,\text{rad}].$$

The phase is given by $(3\pi \,\text{rad/s})t + \pi/3 \,\text{rad}$. Evaluate these expression for $t = 2.0$ s. Don't forget to put your calculator in radian mode.

$$[\text{ans: } (a) \, 3.0 \,\text{m}; \ (b) \, -49 \,\text{m/s}; \ (c) \, -270 \,\text{m/s}^2; \ (d) \, 20 \,\text{rad}]$$

Also what are (e) the frequency and (f) the period of the motion?

STRATEGY: The angular frequency is $\omega = 3\pi$ rad/s. Use $\omega = 2\pi f$ to calculate the frequency and $T = 1/f$ to calculate the period.

$$[\text{ans: } (e) \ 1.5\,\text{Hz}; \ (f) \ 0.67\,\text{s}]$$

PROBLEM 4 (23P). A 0.10-kg block oscillates back and forth along a straight line on a frictionless horizontal surface. Its displacement from the origin is given by

$$x = (10\,\text{cm})\cos[(10\,\text{rad/s})t + \pi/2\,\text{rad}] \,.$$

(a) What is the oscillation frequency?

STRATEGY: The angular frequency is $\omega = 10$ rad/s. Use $\omega = 2\pi f$ to calculate the frequency f.

SOLUTION:

$$[\text{ans: } 1.6\,\text{Hz}]$$

(b) What is the maximum speed acquired by the block? At what value of x does this occur?

STRATEGY: Use $v_m = \omega x_m$, where x_m is the amplitude (10 cm). The block has its maximum speed as it passes the equilibrium point.

SOLUTION:

$$[\text{ans: } 1.0\,\text{m/s, at } x = 0]$$

(c) What is the maximum acceleration of the block? At what value of x does this occur?

STRATEGY: Use $a_m = \omega^2 x_m$. The block has its maximum acceleration when it is instantaneously stopped, at the ends of its motion.

SOLUTION:

$$[\text{ans: } 10\,\text{m/s}^2, \text{ at } x = \pm 10\,\text{cm}]$$

(d) What force, applied to the block, results in the given oscillation?

STRATEGY: Use $F = ma = -m\omega^2 x$.

SOLUTION:

$$[\text{ans: } -(10\,\text{N/m})x]$$

The examples above covered most of the fundamentals of simple harmonic motion. Other problems make use of these.

PROBLEM 5 (27P). A block is on a piston that is moving vertically with simple harmonic motion. (*a*) If the SHM has period 1.0 s, at what amplitude of motion will the block and piston separate?

STRATEGY: There are two forces acting on the block: the force of gravity, downward, and the force of the piston, upward. Since the block is not attached the piston cannot pull down on the block. The block is in simple harmonic motion as long as the piston is not required to pull downward to produce the motion.

SOLUTION: Let N be the normal force of the piston on the block and m be the mass of the block. If up is taken to be positive then Newton's second law for the block is $N - mg = ma$. If the block is in simple harmonic motion then $a = -\omega^2 x_m \cos(\omega t + \phi)$ and $N = mg - m\omega^2 x_m \cos(\omega t + \phi)$. Since N cannot be negative $\omega^2 x_m$ must be less than g or x_m must be less than g/ω^2. Use $\omega = 2\pi/T$ to calculate the angular frequency.

[ans: 25 cm]

(*b*) If the piston has an amplitude of 5.0 cm, what is the maximum frequency for which the block and piston will be in contact continuously?

SOLUTION: The same analysis shows that ω must be less than $\sqrt{g/x_m}$ or the frequency must be less than $(1/2\pi)\sqrt{g/x_m}$.

[ans: 2.2 Hz]

PROBLEM 6 (28P). An oscillator consists of a block attached to a spring ($k = 400$ N/m). At some time t, the position (measured from the system's equilibrium location), velocity, and acceleration of the block are $x = 0.100$ m, $v = -13.6$ m/s, and $a = -123$ m/s^2. Calculate (*a*) the frequency, (*b*) the mass of the block, and (*c*) the amplitude of oscillation for the motion.

STRATEGY: Write $x = x_m \cos(\omega t + \phi)$. Then $v = -\omega x_m \sin(\omega t + \phi)$ and $a = -\omega^2 x_m \cos(\omega t + \phi)$. Notice that $a/x = -\omega^2$. Use $f = \omega/2\pi = (1/2\pi)\sqrt{-a/x}$ to calculate the frequency. Use $\omega^2 = k/m$ to calculate the mass of the block. Now square the first two equations to obtain $x^2 + (v/\omega)^2 = x_m^2[\cos^2(\omega t + \phi) + \sin^2(\omega t + \phi)] = x_m^2$. Thus $x_m = \sqrt{x^2 + (v/\omega)^2}$.

SOLUTION:

[ans: (*a*) 5.58 Hz; (*b*) 0.325 kg; (*c*) 0.400 m]

PROBLEM 7 (34P). Suppose that the two springs in Fig. 14–27 have different spring constants k_1 and k_2. Show that the frequency f of oscillation of the block is then given by

$$f = \sqrt{f_1^2 + f_2^2},$$

where f_1 and f_2 are the frequencies at which the block would oscillate if connected only to spring 1 or only to spring 2.

STRATEGY: Suppose the block is displaced by x from its equilibrium position and calculate the force exerted on it by the two springs. If the net force is proportional to x and the constant of proportionality is negative then the block will oscillate in simple harmonic motion and the square of its angular frequency will be the constant of proportionality divided by the mass of the block.

SOLUTION: Suppose the block is at x_1 when the force of spring 1 on it vanishes and suppose it is at x_2 when the force of spring 2 on it vanishes. Put the origin at the position of the block when the net force on it (due to the two springs together) is zero. Then $k_1 x_1 + k_2 x_2 = 0$. Now displace the block to x. The force of spring 1 on it is $-k_1(x - x_1)$ and the force of spring 2 on it is $-k_2(x - x_2)$. The net force is $F = -k_1(x - x_1) - k_2(x - x_2) = -(k_1 + k_2)x$, where the relation between x_1 and x_2, derived above, was used. The angular frequency is therefore $\omega = \sqrt{(k_1 + k_2)/m}$ or, since $\omega_1 = \sqrt{k_1/m}$ and $\omega_2 = \sqrt{k_2/m}$, $\omega = \sqrt{\omega_1 + \omega_2}$. Divide by 2π to obtain the desired result.

Some problems can be solved using the principle of energy conservation. For a mass-spring system the total mechanical energy E, the speed v, and the coordinate x are related by $E = \frac{1}{2}mv^2 + \frac{1}{2}kx^2$. If the mass has speed v_1 when it is at x_1 and speed v_2 when it is at x_2 then conservation of energy yields $\frac{1}{2}mv_2^2 + \frac{1}{2}kx_2^2 = \frac{1}{2}mv_1^2 + \frac{1}{2}kx_1^2$. Either of these equations can be solved for one of the quantities that appear in them.

PROBLEM 8 (42E). A 5.00-kg object on a horizontal frictionless surface is attached to a spring with spring constant 1000 N/m. The object is displaced 50.0 cm horizontally and given an initial velocity of 10.0 m/s back toward the equilibrium position. (*a*) What is the frequency of the motion?

STRATEGY: Use $f = (1/2\pi)\sqrt{k/m}$.

SOLUTION:

[ans: 2.25 Hz]

What are (*b*) the initial potential energy of the block-spring system, (*c*) the initial kinetic energy, and (*d*) the amplitude of the oscillation?

STRATEGY: Use $U = \frac{1}{2}kx^2$ to calculate the potential energy and $K = \frac{1}{2}mv^2$ to calculate the kinetic energy. The total energy is the sum of the two. When the block reaches x_m the potential energy is $\frac{1}{2}kx_m^2$ and the kinetic energy is 0. Set $E = \frac{1}{2}kx_m^2$ and solve for x_m.

SOLUTION:

[ans: (*b*) 125 J; (*c*) 250 J; (*d*) 86.6 cm]

PROBLEM 9 (46P).   A 3.0-kg particle is in simple harmonic motion in one dimension and moves according to the equation

$$x = (5.0\,\text{m}) \cos[(\pi/3\,\text{rad/s})t - \pi/4\,\text{rad}].$$

(*a*) At what value of x is the potential energy of the particle equal to half the total energy?

STRATEGY: The total energy is given by $E = \frac{1}{2}kx_m^2$ and the potential energy when the particle is at x is given by $U = \frac{1}{2}kx^2$. Set $U = E/2$ and solve for x. x_m, of course, is 5.0 m.

SOLUTION:

$$\left[\text{ans: } \pm 5.0/\sqrt{2} = \pm 3.5\,\text{m}\right]$$

(*b*) How long does it take the particle to move to this position x from the equilibrium position?

STRATEGY: Solve $5.0\cos(\pi t/3 - \pi/4) = 0$ for the times t when the particle is at the equilibrium position. Solve $5.0\cos(\pi t/3 - \pi/4) = 5.0/\sqrt{2}$ for the times when the potential energy is half the total energy.

SOLUTION: The equation $\cos\phi = 0$ has the solutions $\phi = \pi/2 + n\pi$, where n is a positive or negative integer or zero. Solve $\pi t/3 - \pi/4 = \pi/2 + n\pi$ for t. Some results are $t = 2.25, 5.25, 8.25, \ldots$ s. The equation $\cos\phi = 1/\sqrt{2}$ has the solutions $\phi = \pi/4 + n\pi/2$, where n is a positive or negative integer or zero. Solve $\pi t/3 - \pi/4 = \pi/4 + n\pi/2$ for t. Some results are $t = 0, 1.5, 3, 4.5, 6, \ldots$ s. Pick any one of the times in the first group and calculate the interval to the next time in the second group.

$$\left[\text{ans: } 0.75\,\text{s}\right]$$

PROBLEM 10 (48P).   A massless spring with spring constant 19 N/m hangs vertically. A body of mass 0.20 kg is attached to its free end and then released. Assume that the spring was unstretched before the body was released. Find (*a*) how far below the initial position the body descends, and (*b*) the frequency and (*c*) the amplitude of the resulting motion, assumed to be simple harmonic.

STRATEGY: (*a*) Place the zero of both the gravitational and spring potential energies at the original position of the body. If it descends a distance h its final potential energy is given by $U = \frac{1}{2}kh^2 - mgh$. Since the kinetic energy was zero both initially and finally, $\frac{1}{2}kh^2 - mgh = 0$. Solve for h. (*b*) Use $f = (1/2\pi)\sqrt{k/m}$. (*c*) The amplitude is half the distance found in part (*a*).

SOLUTION:

$$\left[\text{ans: } (a)\ 0.21\,\text{m};\ (b)\ 1.6\,\text{Hz};\ (c)\ 0.10\,\text{m}\right]$$

Some problems deal with the other oscillating systems discussed in the text: the torsional pendulum, the simple pendulum, and the physical pendulum. In each case, you should know how the angular frequency depends on properties of the oscillating body: $\sqrt{\kappa/I}$ for a torsional pendulum, $\sqrt{g/\ell}$ for a simple pendulum, and $\sqrt{mgd/I}$ for a physical pendulum.

PROBLEM 11 (53P).   An engineer wants to find the rotational inertia of an odd-shaped object of mass 10 kg about an axis through its center of mass. The object is supported with a wire along the desired axis. The wire has a torsion constant $\kappa = 0.50\,\mathrm{N\cdot m}$. If this torsion pendulum oscillates through 20 complete cycles in 50 s, what is the rotational inertia of the odd-shaped object?

STRATEGY: The period of a torsion pendulum is given by $T = 2\pi\sqrt{I/\kappa}$. Solve for I. The period is $T = (50\,\mathrm{s})/20 = 2.5\,\mathrm{s}$.

SOLUTION:

[ans: $0.079\,\mathrm{kg\cdot m^2}$]

PROBLEM 12 (59E).   Two oscillating systems that you have studied are the block-spring and the simple pendulum. There is an interesting relation between them. Suppose that you hang a weight on the end of a spring, and when the weight is at rest, the spring is stretched a distance h. Show that the frequency of this block-spring system is the same as that of a simple pendulum whose length is h. See Fig. 14–34.

STRATEGY: When the weight is at rest the force of the spring pulling up on it has the same magnitude as the force of gravity pulling down. Thus $mg = kh$. Find an expression for the frequency: $f = (1/2\pi)\sqrt{k/m}$. Substitute g/h for k/m and show this frequency is $f = (1/2\pi)\sqrt{g/h}$, the frequency of a simple pendulum with length h.

SOLUTION:

PROBLEM 13 (64E).   A physical pendulum consists of a uniform solid disk (of mass M and radius R) supported in a vertical plane by a pivot located a distance d from the center of the disk (Fig. 14–35). The disk is displaced by a small angle and released. Find an expression for the period of the resulting simple harmonic motion.

STRATEGY: The period is given by $T = 2\pi\sqrt{I/mgd}$, where I is the rotational inertia of the disk about the pivot point. Use the parallel axis theorem to find an expression for I. Since the rotational inertia of a disk about an axis through its center of mass is $I_{\mathrm{cm}} = \frac{1}{2}MR^2$ (see Table 11–2) and the center of mass is a distance d from the pivot, $I = \frac{1}{2}MR^2 + Md^2$.

SOLUTION:

[ans: $2\pi\sqrt{(R^2 + 2d^2)/2gd}$]

PROBLEM 14 (69P).   A stick with length L oscillates as a physical pendulum, pivoted about point O in Fig. 14–37. (a) Derive an expression for the period of the pendulum in terms of L and x, the distance from the point of support to the center of mass of the pendulum.

STRATEGY: Use $T = 2\pi\sqrt{I/mgx}$. Use the parallel axis theorem to find an expression for the rotational inertia I. The rotational inertia of the stick about an axis through its center of mass is $I_{\mathrm{cm}} = ML^2/12$ (see Table 11–2). The pivot is a distance x from the center of mass, so $I = ML^2/12 + Mx^2$.

SOLUTION:

[ans: $2\pi\sqrt{(L^2 + 12x^2)/12gx}$]

(*b*) For what value of x/L is the period a minimum?

STRATEGY: Set the derivative of T with respect to x equal to 0 and solve for x/L. Easier yet, solve $dT^2/dx = 0$ for x/L.

SOLUTION:

$$\left[\text{ans: } x/L = \sqrt{1/12} = 0.289\right]$$

(*c*) Show that if $L = 1.00$ m and $g = 9.80$ m/s^2, this minimum is 1.53 s.

STRATEGY: Substitute $x/L = \sqrt{1/12}$ into the expression you found for T.

SOLUTION:

Some problems deal with damped oscillations. The position of the oscillating object is then given by $x = x_m e^{-bt/2m} \cos(\omega_d t + \phi)$, where $\omega_d^2 = (k/m) - (b^2/4m^2)$. Here the damping force is given by $F_d = -bv$. Other problems deal with forced oscillations. You should remember that the amplitude is a maximum when the frequency of the applied force is roughly the natural frequency of oscillation.

PROBLEM 15 (83P). A damped harmonic oscillator consists of a block ($m = 2.00$ kg), a spring ($k = 10.0$ N/m), and a damping force $F = -bv$. Initially, it oscillates with an amplitude of 25.0 cm; because of the damping, the amplitude falls to three-fourths of this initial value at the completion of four oscillations. (*a*) What is the value of b?

STRATEGY: The time to complete four cycles is $4T$, where T is the period. Replace t with $4T$ and x with $3x_m/4$ in the expression for x. You should also substitute $T = 2\pi/\omega_d = 2\pi/\sqrt{(k/m) - (b^2/4m^2)}$, then solve for b. To save considerable algebra, however, try $T = 2\pi\sqrt{m/k}$ as a first approximation. Once you have computed b, compare $b^2/4m^2$ with k/m. If it is much smaller your approximation is valid. If it is not much smaller go back and substitute the exact expression for T and solve for b again.

SOLUTION:

$$\left[\text{ans: } 0.102 \text{ kg/s}\right]$$

(b) How much energy has been "lost" during these four oscillations?

STRATEGY: Use $E = \frac{1}{2}kA^2$, where A is the amplitude. Initially, $A = x_m$. At the end of the four cycles, $A = 3x_m/4$.

SOLUTION:

[ans: 0.137 J]

PROBLEM 16 (87P). A 2200-lb car carrying four 180-lb people is traveling over a rough "washboard" dirt road. The corrugations are 13 ft apart. The car is observed to bounce with maximum amplitude when its speed is 10 mi/h. The car now stops and the four people get out. By how much does the car body rise on its suspension owing to this decrease in weight?

STRATEGY: When the car is stationary the force of the suspension balances the force of gravity: $W = kx$, where W is the weight of the car and any passengers, k is the spring constant of the suspension, and x is the compression of the suspension. When the weight changes by ΔW as the passengers leave the car, the car rises by $\Delta x = \Delta W/k$. You need to find the spring constant k. The time interval between bumps is given by $T = d/v$, where d is the distance between the corrugations and v is the speed of the car. Since the oscillation of the car has maximum amplitude we take this to be the resonance period of the suspension. The resonance angular frequency is $\omega = 2\pi/T = 2\pi v/d$. Use $k = m\omega^2$, where m is the mass of the car and passengers, to find k.

SOLUTION:

[ans: 0.16 ft (1.9 in.)]

III. MATHEMATICAL SKILLS

Derivatives of sinusoidal functions. You need to know how to differentiate $\sin(\omega t + \phi)$ and $\cos(\omega t + \phi)$ to verify the solutions to several of the equations of motion in the chapter and to calculate the velocity and acceleration of an oscillating body. Remember how to use the chain rule. Let $\omega t + \phi = u$. Then $d\sin(\omega t + \phi)/dt = (d\sin u/du)(du/dt) = (\cos u)(\omega) = \omega\cos(\omega t + \phi)$. Similarly $d\cos(\omega t + \phi) = -\omega\sin(\omega t + \phi)$.

Small angle approximation. Several of the oscillators discussed in this chapter are harmonic only if the amplitude is small. Simple and physical pendulums are examples. For the motion to be considered harmonic the angle θ of swing must be sufficiently small that $\sin\theta$ may be replaced by θ in radians without generating unacceptable error.

The Maclaurin series for $\sin\theta$ is

$$\sin\theta = \sum_{n=0}^{\infty}(-1)^n\frac{\theta^{2n+1}}{(2n + 1)!}$$

and its first three terms are

$$\sin\theta = \theta - \frac{\theta^3}{6} + \frac{\theta^5}{120} - \cdots .$$

The series is valid only if θ is in radians. Notice that if θ is small each term in the series is less in magnitude than the previous term. The small angle approximation amounts to using only the first term of the series.

If θ is small the error generated by the small angle approximation is nearly the second term $\theta^3/6$ and the fractional error is roughly $\theta^2/6$. For example, the error is less than 1 per cent if $\theta^2/6 < 0.01$ or $\theta < 0.2$ radians. In fact the error for $\theta = 0.2$ radians is 0.7 per cent.

To check the approximation use your calculator to fill in the following table. Use $100\,(\theta - \sin\theta)/\sin\theta$ to compute the percent error.

θ (rad)	$\sin\theta$	percent error
1	_____	_____
0.1	_____	_____
0.01	_____	_____
0.001	_____	_____

IV. NOTES

Chapter 15
GRAVITATION

I. BASIC CONCEPTS

You will learn about the force law that describes the gravitational attraction of two particles for each other. By considering the attraction of every particle in one object for every particle in a second object the law is extended to deal with objects containing many particles. Then the law is applied to the motion of planets and satellites. Here you will bring to bear some of the concepts you studied in earlier chapters, chiefly Newton's second law and the laws of energy and momentum conservation.

Newton's law of gravity. According to Newton's law of gravity, two particles with masses m_1 and m_2, separated by a distance r, each attract the other with a force of magnitude

$$F =$$

where the universal gravitational constant G has the value $G =$ _____ $N \cdot m^2/kg^2$.

Gravitational forces, like all other forces, are vectors. The force that particle 1 exerts on particle 2 is toward _____, along the line that joins the particles. The force that particle 2 exerts on particle 1 is toward _____, along the same line. Note that the force obeys Newton's third law: the force of particle 1 on particle 2 has the same magnitude as the force of particle 2 on particle 1, but is in the opposite direction. Also note that the magnitude of the force is inversely proportional to the square of the particle separation. If that distance is doubled the gravitational force is _____ as great.

Experimental evidence indicates that the masses in the law of gravity are identical to those in Newton's second law of motion. If the gravitational force exerted by particle 1 is the only force acting on particle 2 then the acceleration of particle 2 is given by $a =$ _____, toward _____. Notice that the acceleration of particle 2 depends on m_1 but is independent of m_2.

When more than two particles are present, the force on any one of them is the <u>vector</u> sum of the individual forces the other particles exert on it. If an object can be considered a continuous distribution of mass, the gravitational force it exerts on a particle is computed as the integral $\mathbf{F} =$ _____, where d$\mathbf{F}$ is the force exerted by an infinitesimal volume element of the distribution. The integral represents the vector sum of _____ _____.

Two very important consequences of Newton's law of gravity are stated in Sections 15–2 and 15–6. Both are concerned with the force of gravity exerted by a uniform spherical shell of mass M on a particle of mass m. In the first case the particle is *outside* the shell, a distance r from the shell center. Then the magnitude of the force of the shell on the particle is given by

$$F =$$

The direction of the force is _____. In the second case the particle is *inside* the shell. Then the magnitude of the force of the shell on the particle is given by $F =$ _____, no matter where the particle is located inside.

The first of these theorems can be used to show that the force of attraction of two objects with spherically symmetric mass distributions is given by an equation that is identical to Newton's law of gravity for particles, provided we interpret r as _____

_____.

If the force of gravity exerted by the Earth is the only force acting on a particle of mass m at its surface, we may write $F = ma_g$, where a_g is the acceleration due to gravity at the Earth's surface. Assume the Earth is a sphere with a spherically symmetric mass distribution and equate F to the force given by Newton's gravitational law to obtain an expression for a_g in terms of the mass M_e and radius R_e of the Earth:

$$a_g =$$

This expression can be used to calculate the mass of the Earth, once a_g, R_e, and G are known. A similar expression (with R_e replaced by the distance between the center of the Earth and a particle) can be used to estimate the acceleration due to gravity at any given distance above the Earth's surface. This calculation is an estimate because we have made the assumption that the Earth has a spherically symmetric mass distribution, an assumption that is not precisely valid.

Even if deviations from a spherical mass distribution can be ignored the acceleration due to gravity is not the same as the acceleration g of an object in free-fall because the Earth is spinning. At the Earth's surface the difference between the two is $a_g - g = \omega^2 R$, where ω is the angular velocity of _____ and R is the radius of the circle traversed by the object as it spins around with the Earth. At the equator R is the radius of _____ and numerically $a_g - g =$ _____ m/s^2. The weight of the object, as read by a spring balance, is _____. Carefully read Sample Problem 15–3, in which the same ideas are applied to a spinning pulsar.

The shell theorems can also be used to find an expression for the gravitational force on a particle located *within* a spherically symmetric mass distribution. Consider a particle somewhere inside the Earth, a distance r from the center. If the Earth's mass distribution were spherically symmetric, we could calculate the force of the Earth on the particle by considering only that part of the Earth's mass that is less than r from the center. All parts of the Earth attract the particle but the forces due to parts further from the center than the point mass sum to zero.

To test your understanding, assume the Earth's mass is uniformly distributed and use the space below to show that the magnitude of the force on the particle is given by $(GM_e m/R_e^3)r$, where M_e is the mass of the Earth, m is the mass of particle, and r is the distance from the center of the Earth to the particle. First show that the mass inside r is $M_e r^3/R_e^3$, then substitute this expression for one of the masses in Newton's law of gravity.

Notice that the force on a particle at the center of the Earth is zero.

In many cases the dimensions of an extended body are small enough that the acceleration due to gravity is very nearly uniform over the body. Then, for purposes of studying the motion of the body, we may treat it as a particle located at the center of mass. Thus objects, including satellites in the Earth's gravitational field and planets in the sun's gravitational field, are discussed in the text as if they were particles.

Gravitational potential energy.   The force of gravity is a conservative force. Go back to Chapter 8 and review the tests for conservative forces. In general, when a system changes configuration how does the work done by a conservative force depend on the paths taken by parts of the system? _____

Since the gravitational force is conservative a potential energy function is associated with it. If two point masses m and M are separated by a distance r, their gravitational potential energy is given by

$$U(r) =$$

where the potential energy was taken to be zero for infinite separation. This expression also gives the potential energy of two non-overlapping bodies with spherically symmetric mass distributions. Then r is interpreted as the separation of their _____. It is important to recognize that this energy is associated with the *pair* of masses, NOT with either mass alone.

As you know, the potential energy is a scalar. To calculate the gravitational potential energy of a system of particles, sum the potential energies of each *pair* of particles in the system. The total potential energy of a collection of particles is the work done by an external agent to assemble the particles from the reference configuration. The particles start from rest and are placed at rest in their final positions. If the particles are brought from infinite separation the external agent must do negative work since the mass already in place attracts any new mass being brought in and the agent must pull back on it. Also remember that the potential energy is the *negative* of the work done by the gravitational forces of the particles on each other as the system is assembled.

If the particles of a system interact only via gravitational forces and no external forces act, then the total mechanical energy of the system is conserved. The sum of the kinetic energies of all the particles and the total gravitational potential energy remains the same as the particles move.

For a two-particle system the total mechanical energy consists of three terms, corresponding to the kinetic energy of each particle and the potential energy of their interaction. Suppose that at some instant particle 1 (with mass m_1) has speed v_1, particle 2 (with mass m_2) has speed v_2, and the particles are a distance r apart. Then the total mechanical energy of the system is given by

$$E =$$

At a later time v_1, v_2, and r may have different values but the sum of the three terms will have the same value. You can equate the expressions for the energy in terms of the speeds and separation at two different times and use the resulting equation to solve for one of the quantities that appear in the equation.

If we consider a satellite in orbit around the Earth or a planet in orbit around the sun, then one body is much more massive than the other. We may usually place the origin at the center of the more massive body and take its velocity to be zero. Then the total mechanical energy is the sum of two terms, the kinetic energy of the less massive body and the gravitational potential energy.

Suppose an Earth satellite with mass m (much less than that of the Earth) has an initial speed v_0 when it is a distance r_0 from the Earth's center and has speed v when it is a distance r from the Earth's center. Write the conservation of energy equation in terms of m, M_e, r_0, v_0, r, and v:

This equation can be used to solve for any one of the quantities in it, given the others.

The escape speed is the initial speed that an object must be given at the surface of the Earth (or other large mass) in order to _____

_____ .

To use the conservation of energy equation to calculate the escape speed for an object on the Earth's surface, set $r_0 =$ _____ , $r =$ _____ , and $v =$ _____ , then solve for v_0. The result is: $v_0 =$ _____ .

What other conservation principles are valid for the gravitational interaction of two spherically symmetric bodies? _____

_____

Planetary motion. The motions of planets are controlled by gravity, due chiefly to the sun. Here we consider only the gravitational force of the sun and neglect the influence of other planets. The motion of a planet is then at least partially described by Kepler's three laws. State the laws in words:

1. Law of orbits: _____

    _____

2. Law of areas: _____

    _____

    This law is a direct consequence of the principle of _____ conservation.

3. Law of periods: _____

    _____

    A planetary orbit can be described by its semimajor axis, which is _____

_____

and its eccentricity, which is _____

_____ .

The point of closest approach to the sun is called the _____ and, in terms of the semimajor axis a and eccentricity e, the distance of this point from the sun is given by $R_p =$ _____. The point of maximum distance from the sun is called the _____ and this distance is given by $R_a =$ _____. For circular orbits $e =$ _____ and $R_a = R_p = a$. An eccentricity of nearly 1 corresponds to an ellipse that is much longer than it is wide. The same expressions are valid for a satellite in Earth orbit but the point of closest approach is called the _____ and the point of maximum distance is called the _____.

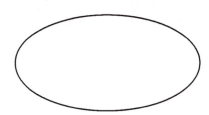

An elliptical orbit for a planet is shown to the right. Label the position of the sun, the semimajor axis, aphelion, and perihelion.

The law of areas provides us with a means of relating a planet's speed at one point to its speed at another point. The simplest relationship holds for points that are the greatest and least distances from the sun because at these points the velocity is perpendicular to the position vector from the sun.

In terms of the mass m of the planet, distance r from the sun, and speed v of the planet, the magnitude of the angular momentum at one of these points is given by $\ell =$ _____, where the origin was placed at the sun. Let R_p and v_p be the distance from the sun and speed at perihelion and let R_a and v_a be the distance from the sun and speed at aphelion. Then conservation of angular momentum leads to the equality: _____.

For a circular orbit the period T is related to the radius r by

$$T^2 =$$

where M is the mass of the central body (the sun). This expression can be used, for example, to calculate the mass of the sun. The equation is also valid for elliptical orbits if r is replaced by the semimajor axis a.

You should recognize that asteroids and recurring comets in orbit around the sun and satellites (including the moon) in orbit around the Earth or another planet also obey Kepler's laws. When the two bodies have comparable mass, as for example the two stars in a binary star system, each travels in an elliptical orbit around the _____.

Consider a body of mass m in a circular orbit about a much more massive sun (mass M), essentially at rest. The speed of the body and the energy of the system are closely related to the radius of the orbit. Gravity produces a centripetal force of magnitude $F = GmM/r^2$ and this must equal mv^2/r, so $v =$ _____. You can use this equation to find the speed of a body in a given circular orbit.

As a function of the orbit radius r the kinetic energy is $K = \frac{1}{2}mv^2 =$ _____. If we take the potential energy to be zero for infinite separation then the potential energy for an orbit of radius r is

$$U =$$

and the total mechanical energy, as a function of r, is

$$E =$$

The negative value of the total energy indicates that the system is bound. The orbiting body does not have enough kinetic energy to escape. This expression for the total energy is valid for elliptical orbits if we replace r with a, the semimajor axis. You should be aware that the kinetic energy is *not* given by $GmM/2a$ for non-circular orbits but the total energy *is* given by $-GmM/2a$.

To emphasize these ideas, consider a planet of mass m in an elliptical orbit with semimajor axis a, around a sun of mass M, essentially at rest. No matter where the planet is in its orbit, the total energy is given by $E = $ _____ . When it is a distance r from the sun, the potential energy is given by $U = $ _____ , and the kinetic energy is given by

$$K = E - U = GmM \left(\frac{1}{r} - \frac{1}{2a} \right).$$

This expression can be used to find the speed of the planet if M, r, and a are given.

As this discussion indicates, the speed of the orbiting body cannot be changed without changing the semimajor axis. Carefully review Sample Problem 15–13 of the text to see how a change in speed alters the orbit.

II. PROBLEM SOLVING

Some problems are straightforward applications of Newton's law of gravity. The law relates the force between two particles to their masses and separation. In some situations you are asked for the force of two or more masses on another. Then you must evaluate the vector sum of the individual forces. In some variants you are given the force and asked for the position of one of the masses. If one of the objects is not a particle but is an extended body and can be approximated by a continuous mass distribution, then you may need to evaluate an integral. Look for any simplification brought about by symmetry.

PROBLEM 1 (7E). A spaceship is on a straight-line path between the Earth and the moon. At what distance from the Earth is the net gravitational force on the spaceship zero?

STRATEGY: Both the Earth and the moon pull on the spaceship, but in opposite directions. Write an equation for the net force in terms of the distance from the Earth, equate the net force to zero, and solve for the distance. Look in an appendix for the values of the mass of the Earth, the mass of the moon, and the Earth-moon distance.

SOLUTION: Let M_m be the mass of the moon, M_e be the mass of the Earth, and M_s be the mass of spaceship. If R is the center-to-center distance from the Earth to the moon and r is the distance from the center of the Earth to the spaceship, then the magnitude of the net force on the spaceship is given by $F = GM_s[M_e/r^2 - M_m/(R-r)^2]$. $F = 0$ yields a quadratic equation for r: $(M_e - M_m)r^2 - 2M_eRr + M_eR^2 = 0$. Solve for r. You will get two solutions. Reject the one that corresponds to the far side of the moon.

[ans: 3.4×10^8 m]

PROBLEM 2 (10P).   In Fig. 15–28, four spheres form the corners of a square whose side is 2.0 cm long. What are the magnitude and direction of the net gravitational force from them on a central sphere with mass $m_5 = 250$ kg?

STRATEGY: Vectorially sum the four forces acting on the central mass, one due to each of the masses at the corners. Notice that two of the corners are occupied by identical masses, the same distance from the center and along the same line but on opposite sides. The forces due to these two masses cancel. The other masses pull on the central mass in opposite directions. Use $F = Gmm_5/r$, where $r = a/\sqrt{2}$ and a is the edge length, to compute each individual force.

SOLUTION:

$$[\text{ans: } 1.7 \times 10^{-2}\,\text{N, toward the 300 kg mass}]$$

PROBLEM 3 (14P).   A thin rod with mass M is bent in a semicircle of radius R, as in Fig. 15–30. (a) What is its gravitational force (both magnitude and direction) on a particle with mass m at P, the center of curvature?

STRATEGY: Treat the rod as a continuous distribution of mass, divided into infinitesimal segments, and evaluate the integral that sums the forces exerted by the segments. Since the forces are vectors, this must be done component by component.

SOLUTION: If ds is the length of a segment and dm is its mass then, since the mass per unit length of the rod is $M/\pi R$, d$m = (M/\pi R)\,$ds. The segment exerts of force of magnitude d$F = (GmM/\pi R^3)\,$ds, along the line joining the segment to the center of curvature. If this line makes an angle θ with the horizontal the vertical component of the force is given by d$F_y = (GmM/\pi R^3)\sin\theta\,ds$. It is easy to carry out the integration if the angle is taken to be the variable of integration. Use d$s = R\,$dθ to show that $F_y = (GmM/\pi R^2)\int_0^\pi \cos\theta\,d\theta$. Symmetry indicates that the horizontal component vanishes: for every segment on one side of the center you can find a segment on the other side such that the horizontal components of the forces exerted by the segments have the same magnitudes and opposite signs. You may also want to carry out the integration: $F_x = (GmM/\pi R^2)\int_0^\pi \sin\theta\,d\theta$.

$$[\text{ans: } 2GmM/\pi R^2]$$

(b) What would be the force on m if the rod were a complete circle?

SOLUTION: Now every segment can be matched with another segment, diametrically opposite, which exerts a force of the same magnitude but opposite in direction. If you prefer to carry out the sum explicitly, change the upper limits on the integrals to 2π.

$$[\text{ans: } 0]$$

Some problems deal with the force of gravity at the Earth's surface or the variation of that force with altitude near the Earth's surface. Don't forget that the r in the gravitational force law for a sphere is a center-to-center distance. Other related problems deal with the apparent weight of an object. This differs from the force of gravity because the Earth is spinning and the object is spinning with it. Be sure to distinguish between the acceleration due to gravity and the free-fall acceleration.

PROBLEM 4 (19P).   You weigh 120 lb at the sidewalk level outside the World Trade Center in New York City. Suppose that you ride from this level to the top of one of its 1350-ft towers. Ignoring the Earth's rotation, how much less would you weigh there because you are slightly farther from the center of the Earth?

STRATEGY: Your weight at the bottom is given by $W_b = GMm/R^2$ and your weight at the top is given by $W_t = GMm/(R+h)^2$, where M is the mass of the Earth, m is your mass, R is the radius of the Earth, and h is the altitude of the tower. Notice that $W_t = W_b R^2/(R+h)^2$ and the change in your weight is given by $\Delta W = W_t - W_b = W_b[1 - R^2/(R+h)^2]$. Because R and $R+h$ differ only in the fifth significant figure there may be round-off error if you evaluate this expression as it is written. Put both terms in the brackets over the same denominator and cancel as much of the numerator as you can. You should obtain $W_t - W_b = W_b(2Rh + h^2)/(R+h)^2$. Evaluate this expression. Look up the radius of the Earth in the appendix and convert to feet.

SOLUTION:

[ans: 0.016 lb, less]

PROBLEM 5 (26P).   Certain neutron stars (extremely dense stars) are believed to be rotating at about 1 rev/s. If such a star has a radius of 20 km, what must be its minimum mass so that material on its surface remains in place during the rapid rotation?

STRATEGY: Assume material on the surface just barely goes around with the star. Then the normal force of the surface on the material is zero and gravity alone provides the centripetal force for uniform circular motion. Let m be the mass of the material, M be the mass of the star, and R be the radius of the star. Equate the gravitational force GmM/R^2 to the product of the mass and acceleration of the material $mR\omega^2$ and solve for M. You must convert 1 rev/s to rad/s.

SOLUTION:

[ans: 4.7×10^{24} kg]

PROBLEM 6 (28P).   A scientist is making a precise measurement of g at a certain point in the Indian Ocean (on the equator) by timing the swings of a pendulum of accurately known construction. To provide a stable base the measurements are conducted in a submerged submarine. It is observed that a slightly different result for g is obtained when the submarine is moving eastward than when it is moving westward, the speed in each case being 16 km/h. Account for this difference and calculate the fractional error $\Delta g/g$ in g for either travel direction.

STRATEGY: The free-fall acceleration g is related to the acceleration due to gravity a_g by $g = a_g - \omega^2 R$, where ω is the angular velocity and R is the radius of the Earth. For the submarine traveling eastward, in the direction of the Earth's rotation, $\omega = \omega_0 + v/R$, where v is the speed of the submarine relative to the Earth's surface and ω_0 is the angular velocity of the Earth. Thus $g = a_g - \omega_0^2 R - 2\omega_0 v - v^2/R$ and compared to the submarine at rest $\Delta g = -2\omega_0 v - v^2/R$. Calculate the value and divide by g to obtain the fractional change. For the submarine going westward $\omega = \omega_0 - v/R$ and $\Delta g = +2\omega_0 v - v^2/R$.

SOLUTION:

[ans: 6.6×10^{-5} for either direction]

Some problems ask you to use the shell theorems to find the gravitational force exerted on a point mass by a spherically symmetric distribution of charge. You use Newton's law, but for one of the masses you substitute the mass that is inside an imaginary spherical surface that passes through the point mass. You may need to carry out a side calculation to find that mass.

You can also find the force exerted on a point mass by an object with a cavity. First suppose the cavity is filled with material of the same mass density as the object and calculate the force exerted by the filled object. Then calculate the force exerted by the mass filling the cavity. Vectorially subtract the second force from the first.

PROBLEM 7 (30P).   Sensitive meters that measure the local free-fall acceleration **g** can be used to detect the presence of deposits of near-surface rocks of density significantly greater or less than that of the surroundings. Cavities such as caves and abandoned mine shafts can also be located. (*a*) Show that the vertical component of **g** a distance x from a point directly above the center of a spherical cavern (see Fig. 15–33) is less than what would be expected assuming a uniform distribution of rock of density ρ, by the amount

$$\Delta g = \frac{4\pi}{3} R^3 G \rho \frac{d}{(d^2 + x^2)^{3/2}} \, ,$$

where R is the radius of the cavern and d is the depth of its center.

STRATEGY: The difference in **g** that occurs because the cavern is there is the acceleration due to the gravitational force of matter that would fill the cavern. It has magnitude GM/r^2, where r is the distance from the cavern center to the observation point and M is the mass of rock that would fill the cavern, and it is directed *away* from the cavern center along the line that joins the center to the observation point. Since the volume of the cavern is $(4\pi/3)R^3$, $M = (4\pi/3)R^3\rho$. Furthermore $r = \sqrt{d^2 + x^2}$. If the line from the cavern center makes the angle θ with the vertical, you multiply the magnitude by $\cos\theta$ to obtain the vertical component. A little geometry shows that $\cos\theta = d/r = d/\sqrt{d^2 + x^2}$. Make these substitutions.

(*b*) These values of Δg, called *anomalies*, are usually very small and are expressed in milligals, where 1 gal = 1 cm/s². Suppose oil prospectors doing a gravity survey over a straight-line distance of 300 m find Δg varying from 10 milligals to a maximum of 14 milligals at the midpoint of the 300-m distance. Assuming that the larger anomaly value was recorded directly over the center of a spherical cavern known to be in the region, find its radius and the depth to its roof. Nearby rocks have a density of 2.8 g/cm³.

STRATEGY: Write the expression for Δg twice, one for $x = 0$ and once for $x = 150$ m (the distance from the center of the survey to one end. Solve the two equations simultaneously for d and R. The distance of the roof below the surface is given by $d - R$.

SOLUTION: The algebra here is a little complicated. Let $\Delta g_0 = (4\pi/3)R^3 G\rho/d^2$ be the anomaly value over the center ($x = 0$) and $\Delta g = (4\pi/3)R^3 G\rho d/(d^2 + x^2)^{3/2}$ be the anomaly value for $x = 300$ m. Then $\Delta g_0/\Delta g = (d^2 + x^2)^{3/2}/d^3$. This equation has one unknown, d. Raise both sides to the 2/3 rds power and show that $x^2/d^2 = (\Delta g_0/\Delta g)^{2/3} - 1$. Now you can solve for d. Finally, solve $\Delta g_0 = (4\pi/3)R^3 G\rho/d^2$ for R. When evaluating the result be sure to convert milligals to m/s^2 and ρ to kg/m^3.

$$[\text{ans: } 250\,\text{m}, 47\,\text{m}]$$

(c) Suppose that the cavern, instead of being empty, is completely flooded with water. What do the gravity readings in (b) now indicate for its radius and roof depth?

STRATEGY: The water attracts an object at the observation point but not as much as does rock. To convert the cavern filled with water (density = $1.0\,\text{g/cm}^3$) to one filled with rock (density = $2.8\,\text{g/cm}^3$) we must add material with a density of $1.8\,\text{g/cm}^3$. Carry out the same calculation with $\rho = 1.8\,\text{g/cm}^3$.

SOLUTION:

$$[\text{ans: } 290\,\text{m}, 7.0\,\text{m}]$$

PROBLEM 8 (31E). Two concentric shells of uniform density having masses M_1 and M_2 are situated as shown in Fig. 15–34. Find the force on a particle of mass m when the particle is located at (a) $r = a$, (b) $r = b$, and (c) $r = c$. The distance r is measured from the center of the shells.

STRATEGY: Sum the forces due to each shell individually. If the particle is outside a shell the force of that shell is as if the entire mass of the shell were at its center. If the particle is inside a shell the force of that shell is zero. In (a) the particle is outside both shells; in (b) it is outside the shell of mass M_1 and inside the shell of mass M_2; and in (c) it is inside both shells.

SOLUTION:

$$[\text{ans: } (a)\ G(M_1 + M_2)m/a^2;\ (b)\ GM_1m/b^2;\ (c)\ 0]$$

PROBLEM 9 (35P). (*a*) Figure 15–35 shows a planetary object of uniform density ρ and radius R. Show that the compressive stress S on the material near its center due to the weight pressing on that material is given by

$$S = \frac{2}{3}\pi G \rho^2 R^2 .$$

STRATEGY: Consider a column of material of small cross section A, reaching from the surface to the center. The compressive stress is the force that must be exerted by other material (below the bottom of the column in the figure) divided by A. Since the column is in equilibrium that force is equal to the total gravitational force on material in the column. Consider an infinitesimal portion of the column extending from r to $r + dr$. Its volume is $A\,dr$ and the mass it contains is $\rho A\,dr$. The gravitational force on it is the force of material within a sphere of radius r or $4\pi\rho r^3/3$. Thus the force on the infinitesimal portion of the column is $dF = (4\pi G/3)\rho^2 A r\,dr$. To find the total force on the column integrate from $r = 0$ to $r = R$. To find the stress divide the result by A.

SOLUTION:

(*b*) In our solar system, objects (for example, asteroids, small satellites, comets) with "diameters" less than 600 km can be very irregular in shape (see Fig. 15–36*b*, which shows Hyperion, a small satellite of Saturn), whereas those with larger diameters are spherical. Only if the rock has sufficient strength to resist gravitation can such an object maintain a non-spherical shape. Calculate the ultimate compressive strength of the rocks making up asteroids. Assume a density of $4000\,\text{kg/m}^3$.

STRATEGY: Evaluate the expression you derived for S. Use $R = 600\,\text{km}$ and $\rho = 4000\,\text{kg/m}^3$.

SOLUTION:

[ans: $2.0 \times 10^8\,\text{N/m}$]

(*c*) What is the largest possible "diameter" of a nonspherical satellite made of concrete with a density of $\rho = 3000\,\text{kg/m}^3$?

STRATEGY: Solve the expression for R and double the value to obtain the "diameter". According to Table 13–1 the yield strength for concrete in compression is $40 \times 10^6\,\text{N/m}^2$.

SOLUTION:

[ans: $3.6 \times 10^5\,\text{m}$]

    You might be asked for the initial speed that one of the masses must be given so it barely escapes from a mass distribution, while the rest of the mass remains at rest. After the mass is removed it is infinitely far from the other masses and its kinetic energy is zero. The initial energy is the potential energy of the system plus the kinetic energy of the mass to be removed. The final energy is simply the potential energy of the system with the mass removed. Use conservation of energy to find the initial kinetic energy of the mass that is removed.

PROBLEM 10 (43E). Show that the escape speed from the sun at the Earth's distance from the sun is $\sqrt{2}$ times the speed of the Earth in its orbit, assumed to be a circle. (This is a specific case of a general result for circular orbits: $v_{\text{esc}} = \sqrt{2}v_{\text{orb}}$.)

STRATEGY: If its orbit is a circle Newton's second law for the Earth is $GM_eM_s/r^2 = M_ev_{\text{orb}}^2/r$, where r is the radius of the orbit. When an object of mass m is a distance r from the sun, its gravitational potential energy is $U = -GmM_s/r$ and its potential energy is $\frac{1}{2}mv^2$. If $v = v_{\text{esc}}$ then the total mechanical energy is 0. Show that $v_{\text{orb}} = \sqrt{GM_s/r}$ and that $v_{\text{esc}} = \sqrt{2GM_s/r}$, so $v_{\text{esc}} = \sqrt{2}v_{\text{orb}}$.

SOLUTION:

Some problems might require you to find the potential energy of a collection of point masses. The problem might be phrased in terms of the work required to assemble the masses from infinite separation or to remove them to infinite separation. In any case identify all the possible *pairs* of masses and sum the potential energies of the pairs. If there are 4 masses, for example, the pairs are 1–2, 1–3, 1–4, 2–3, 2–4, and 3–4. For each pair find the separation r_{ij} of the two masses, then evaluate $-Gm_im_j/r_{ij}$ for the potential energy of the pair. Finally, sum all the contributions to find the total potential energy U. As the system is assembled gravity does work $-U$ and an external agent does work $+U$. As the system is disassembled gravity does work $+U$ and the agent does work $-U$. U itself is negative.

PROBLEM 11 (46P). The three spheres in Fig. 15–38, with masses $m_1 = 800$ g, $m_2 = 100$ g, and $m_3 = 200$ g, have their centers on a common line, with $L = 12$ cm and $d = 4.0$ cm. You move the middle sphere until its center-to-center separation from m_3 is $d = 4.0$ cm. How much work is done on m_2 (a) by you and (b) by the net gravitational force on m_2 due to m_1 and m_3?

STRATEGY: Calculate the gravitational potential energy of the three-sphere system before and after the move. The difference $U_f - U_i$ is the work done by you. The negative of this is the work done by the gravitational forces.

SOLUTION: The initial potential energy is given by

$$U_i = -G\left[\frac{m_1m_3}{d} + \frac{m_2m_3}{L-d} + \frac{m_1m_3}{L}\right]$$

and the final potential energy is given by

$$U_f = -G\left[\frac{m_1m_2}{L-d} + \frac{m_2m_3}{d} + \frac{m_1m_3}{L}\right]$$

[ans: (a) 5.0×10^{-11} J; (b) -5.0×10^{-11} J]

PROBLEM 12 (49P).   A projectile is fired vertically from the Earth's surface with an initial speed of 10 km/s. Neglecting air drag, how far above the surface of the Earth will it go?

STRATEGY: The projectile goes until its speed is 0. Use energy conservation. The initial potential energy is given by $U_i = -GM_em/r_i$, the initial kinetic energy is given by $K_i = \frac{1}{2}mv^2$, the final potential energy is given by $U_f = -GM_em/r_f$, and the final kinetic energy is 0. Solve $U_i + K_i = U_f + K_f = 0$ for r_f. Its highest altitude is given by $r_f - r_i$.

SOLUTION:

[ans: 2.6×10^7 m]

PROBLEM 13 (54P).   The gravitational force between two particles with masses m and M initially at rest at great separation, pulls them together. Show that at any instant the speed of either particle relative to the other is $\sqrt{2G(M + m)/d}$, where d is their separation at that instant.

STRATEGY: The particles move toward each other along the line that joins them. As they do the total mechanical energy and the total momentum of the two-particle system are both conserved. Take the zero of potential energy be at great separation. Then the initial energy is 0. The initial momentum is also 0. Let v be the velocity of m and V be the velocity of M when the particles have separation d. The energy can be written as $-GmM/d + \frac{1}{2}mv^2 + \frac{1}{2}MV^2$ and the momentum as $mv + MV$. Solve $-GmM/d + \frac{1}{2}mv^2 + \frac{1}{2}MV^2 = 0$ and $mv + MV = 0$ simultaneously for v and V. The relative speed of either particle is $|V - v|$.

SOLUTION: Solve $mv + MV = 0$ for v in terms of V and substitute the resulting expression for v into $-GmM/d + \frac{1}{2}mv^2 + \frac{1}{2}MV^2 = 0$. Solve for V. You should get $V = m\sqrt{2G/(m + M)d}$. Then use this expression, substituted into $mv + MV = 0$ to obtain v. You should get $v = -M\sqrt{2G/(m + M)d}$. Now find the expression for $|V - v|$.

Many planetary motion problems can be solved using the relationship between the energy E and the semimajor axis a: $E = -GmM/2a$, where m is the mass of the satellite and M is the mass of the central body. In terms of the distance r from the central body and the speed v, $E = -GmM/r + \frac{1}{2}mv^2$. If the energy and mass are known you can calculate the semimajor axis. If, in addition, the distance from the central body is known you can calculate the speed.

To solve some problems you will also need to know the relationship between the aphelion (or apogee) distance, the perihelion (or perigee) distance, and the semimajor axis: $a = (R_a + R_p)/2$. Sometimes the eccentricity e is given or requested. Then you will need $R_a = a(1 + e)$ and $R_p = a(1 - e)$. When the period is given or requested use $T^2 = (4\pi^2/GM_e)a^3$.

PROBLEM 14 (63E).   The sun's center is at one focus of the Earth's orbit. How far from this focus is the other focus? Express your answer in terms of the solar radius, 6.96×10^8 m. The eccentricity of the Earth's orbit is 0.0167 and the semimajor axis may be taken to be 1.50×10^{11} m. See Fig. 15–16.

STRATEGY: According to the figure the distance between foci is $2ea$, where e is the eccentricity and a is the semimajor axis. Compute this distance and divide by the solar radius.

SOLUTION:

$$\left[\text{ans: 7.20 solar radii}\right]$$

PROBLEM 15 (71P).   Show how, guided by Kepler's third law (Eq. 15–34), Newton could deduce that the force holding the moon in its orbit, assumed circular, depends on the inverse square of the distance from the center of the Earth.

STRATEGY: According to Newton's second law of motion the force of the Earth on the moon is related to the moon's speed by $F = mv^2/r$, where m is the mass of the moon and r is the radius of its orbit. The speed is given by $v = 2\pi r/T$, where T is the period. Thus $F = 4\pi^2 mr/T^2$. Use Kepler's third law of planetary motion to substitute for the period.

SOLUTION:

$$\left[\text{ans: } F = GMm/r^2, \text{ inversely proportional to } r^2\right]$$

PROBLEM 16 (72P).   A certain triple-star system consists of two stars, each of mass m, revolving about a central star of mass M in the same circular orbit of radius r. The two stars are always at opposite ends of a diameter of the circular orbit (see Fig. 15–43). Derive an expression for the period of revolution of the stars.

STRATEGY: Each of the stars of mass m is pulled toward the center of the orbit by the gravitational force of the other star of mass m and by the central star. Since the other star of mass m is $2r$ distant and the star of mass M is r distant the net force has magnitude $F = GMm/r^2 + Gm^2/4r^2$. This must equal mv^2/r. Furthermore the period and speed are related by $v = 2\pi r/T$. Thus $GMm/r^2 + Gm^2/4r^2 = m(2\pi r/T)^2/r$. Solve for T.

SOLUTION:

$$\left[\text{ans: } 2\pi r^{3/2}/\sqrt{G(M + m/4)}\right]$$

PROBLEM 17 (79P).   Use the conservation of mechanical energy and Eq. 15–47 to show that if an object is in an elliptical orbit about a planet, then its distance r from the planet and speed v are related by

$$v^2 = GM \left(\frac{2}{r} - \frac{1}{a}\right) .$$

STRATEGY: The total mechanical energy is given by $E = -GMm/2a$, the potential energy is given by $U = -GMm/r$, and the kinetic energy is given by $K = \frac{1}{2}mv^2$, where M is the mass of the planet, m is the mass of the object, and a is the semimajor axis of the orbit. Substitute these expressions into $E = U + K$ and solve for v^2.

SOLUTION:

PROBLEM 18 (80P).   Use the result of Problem 79 and data contained in Sample Problem 15–10 to calculate (a) the speed v_p of comet Halley at perihelion and (b) its speed v_a at aphelion.

STRATEGY: At perihelion $r = R_p = 8.9 \times 10^{10}$ m; at aphelion $r = R_a = 5.3 \times 10^{12}$ m. The semimajor axis is $a = 2.7 \times 10^{12}$ m and the mass of the sun is 1.99×10^{30} kg. Substitute these values into the expression derived in Problem 79.

SOLUTION:

$$[\text{ans: } (a)\ 5.4 \times 10^4 \text{ m/s}; (b)\ 9.6 \times 10^2 \text{ m/s}]$$

(c) Using conservation of angular momentum relative to the sun, find the ratio of the comet's perihelion distance R_p to it aphelion distance R_a, in terms of v_p and v_a.

STRATEGY: At both perihelion and aphelion the velocity of the comet is perpendicular to the position vector from the sun to the comet so the magnitude of its angular momentum is given by mrv. Conservation of angular momentum yields $R_p v_p = R_a v_a$, so $R_p/R_a = v_a/v_p$.

SOLUTION:

$$[\text{ans: } 1.8 \times 10^{-2}]$$

III. MATHEMATICAL SKILLS

Gravitational force near the Earth's surface.   The magnitude of the gravitational force exerted by the Earth on a body a distance h above its surface is given by $GmM/(R+h)^2$, where M is the mass of the Earth, m is the mass of the body, and R is the radius of the Earth. You may need to calculate the force for h small compared to R. Use the binomial expansion:

$$(R + h)^{-2} = R^{-2} - 2R^{-3}h + 3R^{-4}h^2 + \ldots .$$

This expression finds practical use in some calculations. Suppose you wished to find the difference in the gravitational field at opposite ends of a vertical rod. In principle, you could use Newton's law for gravity directly, substituting $R + h_1$ to find the force at one end and $R + h_2$ to find the field at the other end. R, however, is generally so much greater than h_1 or

h_2 that your calculator truncates both numbers to R. On the other hand, if you use the first two terms of the binomial expansion to write an expression for each force, then subtract the expressions, the first terms cancel and you are left with the terms that are proportional to h_1 and h_2. These can be calculated easily.

IV. NOTES

Chapter 16
FLUIDS

I. BASIC CONCEPTS

In this chapter you will study gases and liquids, first at rest and then in motion. The most important concepts for the study of static fluids are those of pressure and density. Learn their definitions well and learn to calculate the pressure in various situations, paying particular attention to the variation of pressure with depth in a fluid. Then use the concepts to understand two of the most basic principles of fluid statics: Archimedes' and Pascal's principles. Two equations are fundamental for understanding fluids in motion: the continuity and Bernoulli equations. They express the relationship between pressure, velocity, density, and height at points in a moving fluid. To understand these equations you must first understand the ideas of streamlines and tubes of flow.

Definitions. Describe some properties of solids, liquids, and gases that can be used to distinguish them from each other:

Solids: _____

_____

Liquids: _____

_____

Gases: _____

Both liquids and gases are fluids.

If a small volume ΔV of fluid has mass Δm then its <u>density</u> ρ at that place is given by

$$\rho =$$

The definition includes a limiting process in which the volume being considered shrinks to a point. Thus density is defined at each point in a fluid (or solid) and may vary from point to point. Note that density is a scalar. In many cases the density of a fluid is uniform and we may write $\rho = M/V$, where M is the mass and V is the volume of the fluid.

Normally a fluid exerts an outward force on the inner surface of the container holding it. The force on any small surface area ΔA is proportional to the area and is perpendicular to the surface. If ΔF is the magnitude of the force on the area ΔA then the <u>pressure</u> p exerted by the fluid at that place is defined by the scalar relationship

$$p =$$

You must take the limit as the area $\triangle A$ tends toward zero. Thus pressure is defined at each *point* and may vary from point to point on the surface. In many cases the surface being considered is a plane and the pressure is uniform over it. Then we may write $p = F/A$, where F is the force exerted by the fluid on the surface and A is the area of the surface.

Pressure ultimately arises from forces that must be exerted on particles of the fluid in order to contain them. Container walls must exert forces on particles that impinge on them in order to reverse the normal components of their velocities and keep the particles within the container. The particles, of course, exert forces with equal magnitude and opposite direction on the walls. If the container top is open, the atmosphere above exerts forces on fluid particles at the fluid surface.

Pressure also exists in the interior of a fluid. Fluid within any volume exerts an outward force on the fluid around it and fluid outside a volume exerts an inward force on the fluid it surrounds. The force is normal to the imaginary surface that bounds the volume and the pressure is defined in the same way as at the container walls.

The SI unit of pressure (N/m²) is called _____ and is abbreviated _____. Other units are: 1 atmosphere (atm) = _____ Pa, 1 bar = _____ Pa, 1 mm of Hg = _____ Pa, and 1 torr = _____ Pa. Pressure is a scalar quantity.

Table 16–1 of the text lists the densities of some materials and Table 16–2 lists various pressures that exist in nature. Note the wide range of values.

The density of a fluid depends on the pressure. If the pressure at any point is increased (by squeezing the container, for example) the density at that point increases. Most liquids are not readily compressible; great pressure is required to change their densities. Gases, on the other hand, are readily compressible.

Variation of pressure with depth in a fluid. Pressure varies with depth in a fluid subjected to gravitational forces. Consider an element of fluid at height y in a larger body of fluid. Suppose the upper and lower faces of the element each have area A and the element has thickness Δy, as shown. If the fluid has density ρ the mass of the element is _____ and the force of gravity on it is _____. The pressure at the upper face is $p(y + \Delta y)$ so the downward force of the fluid there is _____. The pressure at the lower face is $p(y)$ so the upward force of the fluid there is _____. Since the element is in equilibrium the net force must vanish. In the limit as Δy becomes infinitesimal this means

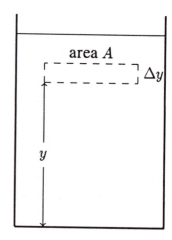

$$\frac{\mathrm{d}p}{\mathrm{d}y} =$$

The negative sign indicates that the pressure is _____ at points higher in the fluid than at lower points.

If the fluid is incompressible then the density is the same everywhere and the pressure difference between any two points in the fluid, at heights y_1 and y_2 respectively, is given by

$$p_2 - p_1 = \underline{\hspace{2cm}}$$

If p_0 is the pressure at the upper surface of an incompressible fluid then $p = \underline{\hspace{3cm}}.$
is the pressure a distance h below the surface. Notice that the pressure is same at any points that are at the same height in a homogeneous fluid.

To derive an expression for the pressure as a function of height in a fluid, no matter how the density depends on pressure, start with $\mathrm{d}p/\mathrm{d}y = -\rho g$. This leads to $g\mathrm{d}y = -\mathrm{d}p/\rho$ and then to

$$g(y_2 - y_1) = -\int_{p_1}^{p_2} \frac{\mathrm{d}p}{\rho}$$

where p_1 is the pressure at y_1 and p_2 is the pressure at y_2. If the density is a known function of pressure the integration can be carried out, at least in principle.

Assume the density does not depend on pressure, evaluate the integral, and show that $p_2 - p_1 = -\rho g(y_2 - y_1)$:

If the fluid is inhomogeneous, as it is if it consists of layers of immiscible fluids with different densities, you must apply the expression for $p(y)$ to each layer separately. For practice, consider the stack of two incompressible fluids shown on the right. The upper surface is open to the atmosphere and the pressure there is atmospheric pressure p_0. Each layer has thickness $2d$. Write an expression for the pressure at each of the labelled points:

$$p_A =$$
$$p_B =$$
$$p_C =$$
$$p_D =$$
$$p_E =$$

To work some problems you must also know when the pressure is the same at two points. Although atmospheric pressure varies with height, the variation is negligible over distances on the order of meters. The surfaces of two portions of a fluid that are both exposed to the atmosphere may be taken to be at the same pressure, atmospheric pressure. Thus the upper surface in each arm of an open U-tube may be taken to be at atmospheric pressure, even if the surfaces are at different heights.

If two points are at the same height *and* they can be joined by a line such that the fluid density does not change along the line, then the pressure is the same at the points. Thus the pressure is the same at any two points at the same height in the water shown in Fig. 16–6 of the text, even if the points are in different arms of the U-tube. A line can be drawn from one to the other through the liquid in the bottom of the tube. Note that the pressure is not the same at points at the same height if one point is in the water on the right and the other is in the oil on the left. A line joining them must pass through two liquids of different density.

Measurement of pressure. This section serves two purposes. It describes two instruments used to measure pressure: the barometer and the open-tube manometer. It also provides some excellent applications of ideas discussed earlier in the chapter. You should study it with both these purposes in mind.

In the space on the right draw a diagram of a mercury barometer. Clearly mark the mercury, the region of near vacuum, and the region where the pressure has the value indicated by the height of the mercury column.

You should be able to use what you know about the variation of pressure with height in an incompressible fluid to show that the pressure at the top of the mercury pool is given by $\rho g h$, where ρ is the density of mercury and h is the height of the top of the mercury column above the pool. Carry out the derivation in the space below:

Suppose a fluid with half the density of mercury is used in a barometer. The height of the column will be _____ the height of a mercury column.

You should understand the importance of the region above the mercury column. If this region is not a near vacuum but instead is filled with a gas, then the pressure at the surface of the mercury pool is NOT proportional to the height of the mercury column. In fact, if the pressure above the column is p_0 and the height of the column is h then the pressure above the pool is given by $p =$ _____ .

In the space to the right draw a diagram of an open-tube manometer. Mark the fluid. Use p_0 to label the fluid surface where the pressure is atmospheric and p to label the fluid surface where the pressure has the value being measured.

Describe exactly what characteristic of the fluid is measured and tell how this is related to the pressure to be found:

_____

_____

_____

Of these two instruments one measures <u>absolute pressure</u> and the other measures <u>gauge pressure</u>. Distinguish between absolute and gauge pressure and tell which instrument is used to measure each:_____

_____

_____

Pascal's principle. Suppose the external pressure applied to the surface of a fluid is changed by Δp_{ext}. Pascal's principle states that the pressure everywhere in the fluid changes by _____.

In the space to the right draw a diagram of a simple hydraulic system that can be used to lift a heavy object by applying a force that is much less than its weight. Clearly identify the input and output forces F_i and F_o. Suppose the input force is exerted over an area A_i and the output force is exerted over an area A_o. According to Pascal's principle these quantities are related by

$$\frac{F_i}{A_i} =$$

Suppose the fluid surface at the input force moves a distance d_i and the object moves a distance d_o. If the fluid is incompressible then the volume of fluid does not change and $d_i A_i = d_o A_o$. Substitute $A_o = d_i A_i / d_o$ into the equation for Pascal's principle to show that $F_i d_i = F_o d_o$:

This means that the work done by the external force on the fluid is the same as the work done by the fluid on the object. If the fluid is compressible the work done by _____ is larger than the work done by _____ . The difference results in an increase in the internal energy of the compressed fluid.

An automobile braking system uses hydraulics to transmit the force you apply at the brake pedal to the brake shoes. You may think of a small diameter hose containing brake fluid running from the pedal to a shoe. Since a small force applied by you results in a huge force applied to the shoe, you know that the surface area of fluid being pushed on by the pedal is much _____ than the area of fluid pushing on the shoe. Furthermore, since the fluid is essentially incompressible, the distance the pedal moves is much _____ than the distance the shoe moves.

Archimedes' principle. State Archimedes' principle in your own words: _____

_____

_____

The buoyant force on an object wholly or partially immersed in a fluid is a direct result of the pressure exerted on it by the fluid and depends on the variation of pressure with height. The pressure at the bottom of the object is greater than the pressure at the top and the net buoyant force is upward.

You know that Archimedes' principle is valid because, if the submerged portion of an object is replaced by an equal volume of fluid, the fluid would be in equilibrium and the net force associated with the pressure of surrounding fluid would equal the _____ of the fluid that replaced the object.

To calculate the buoyant force acting on a given object, first find the submerged volume. If the object is completely surrounded by fluid, this is the volume of the object. If the object is floating on the surface, it is the volume that is beneath the surface. Suppose the submerged volume is V_s. Then the magnitude of the buoyant force is given by $F_b =$ _____, where ρ_f is the density of _____ .

To predict if an object floats or sinks, first assume it is totally submerged and compare the buoyant force with the weight. If the weight is greater, the object _____ . If the buoyant force is greater, it _____ . For an object with uniform density in an incompressible fluid this procedure is the same as comparing the density of the object with the density of the fluid. If the density of the fluid is greater than the density of the object, the object _____ . If the density of the object is greater than the density of the fluid, the object _____ .

For purposes of calculating the torque on a floating object, the buoyant force of the fluid on the object can be treated as a single force acting at the center of mass of _____ . This point is called the center of _____ . The force of gravity, on the other hand, can be treated as a single force acting at the center of mass of _____ . If buoyancy and gravity together produce a net torque the object tilts.

Buoyant torque is used to test the stability of a boat. Assume the boat is tilted slightly and calculate the buoyant torque about the center of gravity. If it tends to right the boat, the boat is stable. If it tends to tilt the boat more, the boat is unstable. Three cases are shown below, with the centers of buoyancy marked B and the centers of gravity marked G. For each, draw vectors to indicate the directions of the force of gravity and the buoyant force, give the direction of the buoyant torque about the center of gravity (into or out of the page), and label the boats "stable" and "unstable", as appropriate.

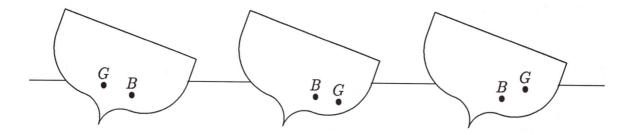

Moving fluids. Fluid flow can be categorized according to whether it is steady or nonsteady, compressible or incompressible, viscous or nonviscous, and rotational or irrotational. This chapter deals chiefly with steady, incompressible, nonviscous, irrotational flow.

The velocity and density of a fluid in <u>steady flow</u> do NOT depend on _____ _____ .

If we follow all the particles that eventually get to any selected point, we find they all have the same velocity as they pass that point, regardless of their velocities when they are elsewhere. In addition, particles flow into and out of any volume in such a way that the mass in the volume at any time is the same as at any other time. Be very careful here. Steady flow does NOT imply that the velocity of any fluid particle is constant nor does it imply that the density is everywhere the same.

If the flow is <u>incompressible</u> the density does not depend on either the _____ or the _____. It is the same everywhere in the fluid and retains the same value through time. Steady flow does NOT imply incompressible flow.

You should also know the meanings of the other terms. Qualitatively distinguish viscous from nonviscous flow: _____
_____

Qualitatively distinguish rotational from irrotational flow: _____
_____

<u>Streamlines</u> are a convenient and effective means of visualizing fluid flow. They are also important for several derivations carried out in this chapter. Define the term streamline:
_____
_____

In steady flow streamlines are fixed curves in space. Every fluid particle on the same streamline passes through the same sequence of points and at each point has the same velocity as other particles when they are at the point. At any point the fluid velocity is _____ to the streamline through the point.

In steady flow can the velocity of a particle be different when it is at different points on the same streamline? _____ Do different streamlines ever cross? _____ Give an argument to substantiate your claim: _____
_____

Define the term <u>tube of flow</u>: _____
_____

In steady flow do particles ever cross the boundary of a tube of flow? _____ Justify your answer: _____
_____

Does a streamline ever cross the boundary of a tube of flow? _____ Why or why not? _____
_____

We conclude from these statements that streamlines crowd closer together in narrow portions of tubes of flow than in wide portions. Consider a narrow tube of flow at a place where it has cross-sectional area A. If the fluid there has density ρ and speed v, then the mass of fluid that passes the cross section per unit time is given by _____ and the volume of fluid that passes the cross section per unit time is given by _____. The former quantity is called the <u>mass flow rate</u> and the later is called the <u>volume flow rate</u> and is denoted by R. The SI units of mass flow rate are_____ and the SI units of volume flow rate are _____. Other commonly used units for the volume flow rate are li/s, gal/s, and gal/min.

Equation of continuity. The equation of continuity is derived from the condition that ____ is conserved in fluid flow. If there are no sources or sinks of fluid a change in the mass contained in any given volume can come about only because the mass that flows into the volume differs from the mass that flows out at any time. In steady flow the mass in any volume does not change, so in any time interval the mass flowing in equals the mass flowing out.

For steady flow of an incompressible fluid the equation of continuity is a statement that the _____ is the same for every cross section along a tube of flow. Let the subscripts 1 and 2 label two places along a tube of flow with cross-sectional areas A_1 at 1 and A_2 at 2. Suppose the fluid speed is v_1 at 1 and v_2 at 2. Then for steady flow the equation of continuity is

$$Av_1 =$$

For incompressible steady flow the volume flow rate is uniform along a tube of flow. This means that fluid particles have greater speed in a _____ portion of a tube than in a _____ portion. Because streamlines crowd closer together in narrow portions of tubes of flow than in wider portions we conclude that a high concentration of streamlines corresponds to a _____ fluid speed and a low concentration corresponds to a _____ fluid speed.

Bernoulli's equation.   Bernoulli's equation for steady, incompressible, nonviscous, irrotational flow tells us that the quantity _____ has the same value at every point along a streamline. You should realize, however, that the value of the quantity may be different for different streamlines.

This equation arises directly from the work-energy theorem, applied to the fluid in a narrow tube of flow, in the limit as the tube becomes a streamline. The term p appears because the _____ exerted by neighboring portions of fluid do work, the term ρgy appears because _____ does work if the height of the tube varies, and the term $\frac{1}{2}\rho v^2$ appears because the _____ of the fluid changes if work is done on it.

If the tube does not change height, the gravitational term is not needed and Bernoulli's equation becomes _____ = constant along a streamline. This equation indicates that the pressure is large where the fluid speed is _____ and is small where the fluid speed is _____. Combining this result with the continuity equation, we can conclude that for steady, incompressible, horizontal flow the pressure is _____ where a tube of flow is narrow and _____ where it is wide.

Because the pressure is different in wide and narrow portions of a tube a net force acts on a fluid element as it moves from one region to another. This changes its velocity. On this basis we expect the pressure to be _____ where the fluid moves slowly and _____ where it moves faster.

To fix these ideas, consider the horizontal tube of flow shown on the right. Draw several streamlines within the tube, clearly illustrating where they crowd close together and where they spread apart. Label one end of the tube "high fluid velocity" and the other end "low fluid velocity", as appropriate. Label one end "high pressure" and the other end "low pressure", as appropriate.

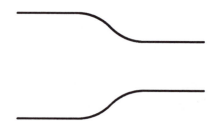

Applications   A number of applications of Bernoulli's equation depend on the lowering of pressure that accompanies an increase in fluid velocity in a horizontal tube of flow. Read the text carefully and for each of the two applications listed below tell what brings about the change in velocity. Draw a diagram. Indicate the direction of fluid flow, label regions of high and low velocity, and label regions of high and low pressure.

Venturi meter:
Mechanism of velocity change: _____   Diagram:
_____
_____
_____
_____

Airplane wing:
Mechanism of velocity change: _____   Diagram:
_____
_____
_____
_____

Water tank with a hole:
Mechanism of velocity change: _____   Diagram:
_____
_____
_____
_____

II. PROBLEM SOLVING

Some problems require you to know the definitions of pressure and density. Remember that if the pressure is uniform and the surface is a plane then $p = F/A$. If there several surfaces you may need to sum the forces vectorially to obtain the net force. Remember that each force is perpendicular to the surface on which it acts. In some cases you may need to use Newton's second law to find the force.

PROBLEM 1 (6E).   A fish maintains its depth in fresh water by adjusting the air content of porous bone or air sacs to make its density the same as that of the water. Suppose that with its air sacs collapsed a fish has a density of $1.08\,\mathrm{g/cm^3}$. To what fraction of its expanded body volume must the fish inflate the air sacs to reduce its density to that of water?

STRATEGY: Let V be the expanded body volume, V_a be the volume of the filled air sacs, ρ_f be density of the uninflated fish, ρ_w be the density of water, and ρ_a be the density of air. Then the mass of the fish with inflated air sacs is given by $\rho_f(V - V_a) + \rho_a V_a$. The first term represents the mass of the uninflated fish and the second represents the mass of air in the sacs. Set this equal to ρ_w and solve for V_a/V. Look up the densities of water and air in Table 16–1 of the text.

SOLUTION:

[ans: 0.074]

PROBLEM 2 (7P).   An airtight box having a lid with an area of 12 in.^{2} is partially evacuated. If a force of 108 lb is required to pull the lid off the box and the outside atmospheric pressure is 15 lb/in.^{2}, what is the air pressure in the box?

STRATEGY: Let F be the force required to pull the lid off, p_o be the outside pressure, and p_i be the inside pressure. When the minimum force is applied the lid is just on the verge of accelerating and the sides of box do not exert a force on the lid, so $F + p_i A - p_o A = 0$, where A is the area of the lid. The upward direction was taken to be positive and the weight of the lid was neglected. Solve for p_i.

SOLUTION:

$$[\text{ans: } 6.0 \, \text{lb/in}^2]$$

To calculate the variation of pressure with depth in a static incompressible fluid, use $p = p_0 + \rho g h$, where p is the pressure at depth h, p_0 is the pressure at the top of the fluid, and ρ is the density of the fluid. A variation on this type problem concerns a fluid that is accelerating. Then the pressure differential across any fluid element provides some or all of the force that accelerates the element.

PROBLEM 3 (12E).   Figure 16–35 displays the *phase diagram* of carbon, showing the ranges of temperature and pressure in which carbon will crystallize either as diamond or graphite. What is the minimum depth at which diamonds can form if the local temperature is 1000° C and the subsurface rocks have density 3.1 g/cm^3? Assume that, as in a fluid, the pressure is due to the weight of material lying above.

STRATEGY: According to the graph the minimum pressure for which diamond forms at 1000° C is about 4.0 GPa. Use $p = p_{\text{atm}} + \rho g d$, with $p = 4.0$ GPa, $\rho = 3.1$ g/cm^3 and $p_{\text{atm}} = 1.0 \times 10^5$ Pa. Notice that atmospheric pressure is negligible compared to the pressure of the rocks.

SOLUTION:

$$[\text{ans: } 1.3 \times 10^5 \, \text{m}]$$

PROBLEM 4 (14E).   A swimming pool has the dimensions 80 ft $\times$ 30 ft $\times$ 8.0 ft. (*a*) When it is filled with water, what is the force (resulting from the water alone) on the bottom, on the ends, and on the sides?

STRATEGY: The pressure on the bottom due to the water alone is given by $p_b = \rho g d$, where $\rho = 1.94$ slugs/ft^3, $g = 32$ ft/s^2, and $d = 8.0$ ft. The pressure at an end varies from top to bottom. To find the force you must add the forces on infinitesimal strips of length $L = 80$ ft. Let dy be the width of a strip at a depth y. Then the force on the strip is d$F = \rho g L y \, dy$ and the net force on the side is given by $F = \rho g L \int_0^d y \, dy$. Carry out the integration and evaluate the result. The same result is obtained for an end, but with L replaced by $W = 30$ ft.

SOLUTION:

$$[\text{ans: bottom: } 1.2 \times 10^6 \, \text{lb (590 tons); a side: } 1.6 \times 10^5 \, \text{lb (79 tons); an end: } 6.0 \times 10^4 \, \text{lb (30 tons)}]$$

(*b*) If you are concerned with whether or not the concrete walls and floor will collapse, is it appropriate to take the atmospheric pressure into account? Why?

SOLUTION:

Some problems involve immiscible, incompressible fluids in U-tubes. Usually you know the pressure at the top surface of the fluid in each arm of the tube. If the end is open to the atmosphere it is atmospheric pressure. If the end is closed and a vacuum exists above the fluid it is zero. You can find a relationship between values of the pressure at the various fluid interfaces. Start at the fluid surface in one arm and follow a line in the tube to the surface in the other arm. Calculate the pressure at each interface and finally the pressure at the top of the second arm in terms of the pressure at the top of the first arm. Lastly, set the expression you obtain equal to the known value of the pressure at the top of the second arm. This gives you an equation to solve for the unknown in the problem.

PROBLEM 5 (17E).   A simple, open U-tube contains mercury. When 11.2 cm of water is poured into the right arm of the tube, how high does the mercury rise in the left arm from its initial level?

STRATEGY: Let ℓ_m be the rise in the mercury column on the left side and let ℓ_w be the length of the water column on the right side. The mercury column, of course, drops by ℓ_m on the right side. The pressure is atmospheric at the top of both columns. If h is the original height of either column, then the pressure at the top of the left side after the water is poured in is given by $p_{\text{atm}} + \rho_w g \ell_w + \rho_m (h - \ell_m) - \rho_m (h + \ell_m)$. This must be p_{atm}, so $0 = \rho_w \ell_w - 2\rho_m \ell_m$. Solve for ℓ_m.

SOLUTION:

$$[\,\text{ans: } 0.411\,\text{cm}\,]$$

PROBLEM 6 (19P).   A cylindrical barrel has a narrow tube fixed to the top, as shown with dimensions in Fig. 16–35. The vessel is filled with water to the top of the tube. Calculate the ratio of the hydrostatic force exerted on the bottom of the barrel to the weight of the water contained inside. Why is the ratio not equal to one? (Ignore the presence of the atmosphere.)

STRATEGY: The hydrostatic pressure on the bottom, due to the weight of the water alone, is given by $p = \rho g d$, where d is the depth measured from the top of the tube. Divide this by the weight of the water $\rho g V$. The total volume V, of course, is the sum of the volume of the barrel ($\pi R^2 L$) and the volume of the tube ($A\ell$).

SOLUTION:

$$[\,\text{ans: } 2.0\,]$$

Because the tube is filled, the pressure at the bottom of the barrel is greater by an amount equal to the weight of the water in the tube divided by the cross-sectional area of the tube. This is the same as the weight of the water in the barrel divided by the cross-sectional area of the barrel, so filling the tube doubles the pressure and the force on the bottom. From another viewpoint, the top of the barrel pushes down on the water in the barrel.

PROBLEM 7 (20P). (*a*) Consider a container of fluid subject to a *vertical upward* acceleration *a*. Show that the pressure variation with depth in the fluid is given by

$$p = \rho h(g + a),$$

where h is the depth and ρ is the density.

STRATEGY: Consider an infinitesimally thin layer of fluid with its bottom surface at y and its upper surface at $y + dy$. It has mass $dm = \rho A\, dy$. The forces acting on it are the downward force of fluid above it $p(y + dy)A$ at the upper surface, the upward force of fluid below it $p(y)A$ at the lower surface, and the force of gravity $\rho g A\, dy$. This must be the product of its mass and acceleration, so $p(y)A - p(y + dy)A - \rho g A\, dy = \rho a A\, dy$ or $dp/dy = -\rho(g + a)$. Integrate this expression with respect to y.

SOLUTION:

(*b*) Show also that if the fluid as a whole undergoes a *vertical downward* acceleration *a*, the pressure at a depth h is given by
$$p = \rho h(g - a).$$

STRATEGY: The problem is the same but now the acceleration is $-a$, where a is the magnitude.

SOLUTION:

(*c*) What is the state of affairs in free fall?

STRATEGY: Set $a = g$ in the last expression.

SOLUTION:

$$[\text{ans: } p = 0]$$

PROBLEM 8 (25P). A U-tube is filled with a single homogeneous liquid. The liquid is temporarily depressed in one side by a piston. The piston is removed and the level of the liquid in each side oscillates. Show that the period of oscillation is $\pi\sqrt{2L/g}$, where L is the total length of the liquid in the tube.

STRATEGY: Suppose the original height of the liquid in each side is h but at some instant the liquid in the right side is higher by a distance x and the liquid in the left side is lower by the same distance. The mass of liquid on the right side is $\rho A(h + x)$. If p_t is the pressure at the top of each arm and p_b is the pressure at the bottom of the tube, then the forces acting on the liquid in the right side are p_t down, p_b up, and $\rho g A(h + x)$ down. The last is the force of gravity. Newton's second law for the liquid in the right side is $p_b A - p_t A - \rho g A(h + x) = \rho a A(h + x)$. Similarly, Newton's second law for the left side is $p_b A - p_t A - \rho g A(h - x) = -\rho a A(h - x)$. Note that the acceleration of the liquid in the left side is opposite the acceleration of the liquid in the right side. Eliminate $p_b - p_t$ by subtracting these two equations. You should obtain $a = -(2g/L)x$, where $L = 2h$ is the total length of the liquid. Recall from your study of oscillations that this equation implies the liquid is oscillating with an angular frequency of $\omega = \sqrt{2g/L}$. The period is given by $T = 2\pi/\omega$.

SOLUTION:

All Pascal's law problems are much the same. In every case a tube containing a fluid has a different cross section at different ends. A force F_1 is applied uniformly to an end with area A_1 and the fluid at the other end exerts a force F_2 on an object. If the area at the second end is A_2 then $F_1/A_1 = F_2/A_2$. Three of these quantities must be given, then the fourth can be evaluated.

Some problems ask you to calculate the work done by each of the forces. These are $W_1 = F_1 d_1$ and $W_2 = F_2 d_2$, where d_1 and d_2 are the distances moved. If the fluid is incompressible then $d_1 A_1 = d_2 A_2$ and $W_1 = W_2$.

PROBLEM 9 (29E).   A piston of small cross-sectional area a is used in a hydraulic press to exert a small force f on the enclosed liquid. A connecting pipe leads to a larger piston of cross-sectional area A (Fig. 16–41). (*a*) What force F will the larger piston sustain?

STRATEGY: Since the pressure is the same at the two pistons, $f/a = F/A$. Solve for F.

SOLUTION:

$$[\text{ans: } (A/a)f]$$

(*b*) If the small piston has a diameter of 1.5 in. and the large piston one of 21 in., what weight on the small piston will support 2.0 tons on the large piston?

SOLUTION:

$$[\text{ans: } 20\,\text{lb}]$$

PROBLEM 10 (30E).   In the hydraulic press of Exercise 29, through what distance must the large piston be moved to raise the small piston a distance of 3.5 ft?

STRATEGY: Use $FD = fd$, where D is the distance moved by the large piston and d is the distance moved by the small piston.

SOLUTION:

$$[\text{ans: } 0.0179\,\text{ft } (0.21\,\text{in.})]$$

Two fundamental Archimedes' principle problems have been covered in the Basic Concepts section. They involve finding the buoyant force on an object, either floating or completely submersed in an incompressible fluid, and deciding if an object floats or sinks. These and many other Archimedes' law problems start with the equations $F_g = \rho g V$ for the force of gravity and $F_b = \rho_f g V_s$ for the buoyancy, where ρ is the density of the object, ρ_f is the density of the fluid in which it is immersed, V is the volume of the object, and V_s is the submerged volume. If the object is floating with no other forces acting then $\rho V = \rho_f V_s$.

In some cases other forces are present. For example, a string may be tied to the object, either holding it up from above or pulling it down from below, or weights may be placed on the object. In these instances simply include the additional force F in the equation for equilibrium: $\rho_f g V_s - \rho g V + F = 0$, where F is positive if the force is directed upward. In other cases the object has an acceleration. Use Newton's second law: set the net force, including the buoyant force, equal to the product of the mass and acceleration.

PROBLEM 11 (32E). A boat floating in fresh water displaces 8000 lb of water. (a) How many pounds of water would this boat displace if it were floating in salt water of density 68.6 lb/ft³?

STRATEGY: The boat floats in salt water so it displaces water with weight equal to the weight of the boat just as it does in fresh water.

SOLUTION:

[ans: 8000 lb]

(b) Would the volume of water displaced change? If so, by how much?

STRATEGY: The weight of water is given by $W = \rho g V$, where ρ is the density (1.94 slug/ft³) for fresh water and 2.14 slug/ft³ for salt water) and V is the volume of water displaced. Calculate the volume for each case and find the difference.

SOLUTION:

[ans: decreases by 12.0 ft³]

PROBLEM 12 (47P). A car has a total mass of 1800 kg. The volume of air space in the passenger compartment is 5.00 m³. The volume of the motor and front wheels is 0.750 m³, and the volume of the rear wheels, gas tank, and luggage is 0.800 m³. Water cannot enter these areas. The car is parked on a hill; the handbrake cable snaps and the car rolls down the hill into a lake (see Fig. 16–44). (a) At first, no water enters the passenger compartment. How much of the car, in cubic meters, is below the water surface with the car floating as shown?

STRATEGY: Since the car is floating, its weight W equals the weight of the displaced water: $W = \rho_w g V_d$, where ρ_w is the density of the water and V_d is the volume of water displaced. Solve for V_d.

SOLUTION:

[ans: 1.80 m³]

(b) As water slowly enters, the car sinks. How many cubic meters of water are in the car as it disappears below the water surface? (The car remains horizontal, owing to a heavy load in the trunk.)

STRATEGY: Let V_c be the total volume of the car and V_w be the volume of water that enters. Then the volume of water displaced is given by $V_d = V_c - V_w$. If the acceleration of the car is very small (negligible) then $W = \rho_w g V_d = \rho_x g(V_c - V_w)$. Solve for V_w. Use $V_c = 5.00 + 0.750 + 0.800 = 6.55 \text{ m}^3$.

SOLUTION:

[ans: 4.75 m^3]

PROBLEM 13 (52P). The tension in a string holding a solid block below the surface of a liquid (of density greater than the solid) is T_0 when the containing vessel (Fig. 16–47) is at rest. Show that the tension T, when the vessel has an upward vertical acceleration a, is given by $T_0(1 + a/g)$.

STRATEGY: Consider the block when it has an upward acceleration a. The forces acting on it are the buoyancy force F_B, the force of gravity mg, and the tension T in the string. Newton's second law gives $F_B - T - mg = ma$.

Since the fluid is also accelerating the variation of pressure with depth is not the same as when the fluid is at rest and the buoyancy force must also be different. Consider a portion of the fluid (area A and thickness dy) with its lower face at y and its upper face at $y + dy$. Newton's second law for this portion of the fluid is $p(y)A - p(y + dy)A - \rho_f A g\, dy = \rho_f A a\, dy$, where ρ_f is the fluid density. Thus $dp/dy = -\rho_f(g + a)$ or $p = p_0 + \rho_f(g + a)h$, where p_0 is the pressure at the upper surface and h is the depth.

If the block has area A and thickness d the buoyancy force on it is given by $F_B = \rho_f(g + a)Ad = \rho_f(g + a)V$. Substitute this expression into Newton's second law for the block and solve for T. You should get $T = (\rho_f - \rho)(g + a)V$, where ρ is the density of the block. Set $a = 0$ to find an expression for T_0, the tension when the block is at rest. Use this expression to eliminate $(\rho_f - \rho)$ from the expression for T. You should obtain the desired result.

SOLUTION:

You should be able to relate the volume and mass flow rates to the dimensions of a tube of flow and to the fluid velocity. The volume flow rate gives the volume of fluid that passes a cross section per unit time and is given by Av, where A is the cross-sectional area of the tube and v is the fluid speed. The mass flow rate is the mass of fluid that passes a cross section per unit time and is given by $\rho A v$, where ρ is the fluid density. Notice that the mass flow rate is the product of the density and the volume flow rate.

For steady flow in a single tube of flow the mass flow rate is the same in all parts of the tube. If the fluid is incompressible the volume flow rate is also the same in all parts. If two tubes merge the sum of the two mass flow rates in the merging tubes must equal to the mass flow rate in the final tube. If the fluid is incompressible the same statement is true for the volume flow rates.

PROBLEM 14 (53E).   Figure 16–48 shows the merging of two streams to form a river. One stream has a width of 8.2 m, depth of 3.4 m, and current speed of 2.3 m/s. The other stream is 6.8 m wide, 3.2 m deep, and flows at 2.6 m/s. The width of the river is 10.5 m and the current speed is 2.9 m/s. What is its depth?

STRATEGY: The volume flow rate in the river must be the sum of the volume flow rates for the two streams: $A_1v_1 + A_2v_2 = A_rv_r$, where A_1 and A_2 are the cross-sectional areas (width times depth) for the streams and A_r is the cross-sectional area of the river; v_1 and v_2 are the current speeds in the streams and v_r is the current speed in the river. Let d_1, d_2, and d_r be the corresponding depths and w_1, w_2, and w_r be the corresponding widths. Then $d_1w_1v_1 + d_2w_2v_2 = d_rw_rv_r$. Solve for d_r.

SOLUTION:

[ans: 4.0 m]

PROBLEM 15 (57P).   A river 20 m wide and 4.0 m deep drains a 3000-km^2 land area in which the average precipitation is 48 cm/y. One-fourth of this rainfall returns to the atmosphere by evaporation, but the remainder ultimately drains into the river. What is the average speed of the river current?

STRATEGY: The volume of water that drains into the river per unit time is given by three-fourths the product of the land area and the precipitation rate. Be sure to change km^3 to m^3 and cm/y to m/s. Set this equal to Av, where A is the cross-sectional area of the river and v is its speed. Solve for v.

SOLUTION:

[ans: 43 cm/s]

Bernoulli's equation is used to solve some problems. It relates conditions (density, fluid speed, pressure, and height) at one point on a streamline to conditions at another point. For many problems you must solve the continuity and Bernoulli equations simultaneously. Sample Problem 16–9 in the text is a nice example. Study it carefully, then work the following problems.

PROBLEM 16 (59P).   Models of torpedoes are sometimes tested in a horizontal pipe of flowing water, much as a wind tunnel to used to test model airplanes. Consider a circular pipe of internal diameter 25.0 cm and a torpedo model, aligned along the axis of the pipe, with a diameter of 5.00 cm. The torpedo is to be tested with water flowing past it at 2.50 m/s. (*a*) With what speed must the water flow in the part of the pipe that is unconstricted by the model?

STRATEGY: Assume the water is incompressible and use the continuity equation in the form $A_1v_1 = A_2v_2$. If A_p is the cross-sectional area of the unconstricted pipe and A_t is the cross-sectional area of the torpedo, then $A_pv_p = (A_p - A_t)v_t$, where v_p in the water speed in the unconstricted portion of the pipe and v_t is the water speed past the torpedo. Use $A = \pi r^2$ to calculate the areas, then solve for v_p.

SOLUTION:

[ans: 2.40 m/s]

(*b*) What will the pressure difference be between the constricted and unconstricted parts of the pipe?

STRATEGY: Since the pipe is horizontal the Bernoulli equation becomes $\frac{1}{2}\rho v_1^2 + p_1 = \frac{1}{2}\rho v_2^2 + p_2$. Solve for the magnitude of $p_2 - p_1$. The density of water is $0.998 \times 10^3\,\text{kg/m}^3$.

SOLUTION:

$$\left[\,\text{ans: } 245\,\text{Pa}\,\right]$$

PROBLEM 17 (66E).   Air flows over the top of an airplane wing of area A with speed v_t and past the underside of the wing (also of area A) with speed v_u. Show that in this simplified situation Bernoulli's equation predicts that the magnitude of the upward lift force on the wing will be

$$L = \frac{1}{2}\rho A(v_t^2 - v_u^2),$$

where ρ is the density of the air.

STRATEGY: The lift force is given by $(p_u - p_t)A$, where p_u is the pressure at the underside of the wing and p_t is the pressure at the top of the wing. Neglect the gravitational terms in Bernoulli's equation and write $p_u + \frac{1}{2}\rho v_u^2 = p_t + \frac{1}{2}\rho v_t^2$. Solve for $p_u - p_t$ and substitute into the expression for the force.

SOLUTION:

PROBLEM 18 (72P).   In a hurricane, the air (density $1.2\,\text{kg/m}^3$) is blowing over the roof of a house at a speed of $110\,\text{km/h}$. (*a*) What is the pressure difference between inside and outside that tends to lift the roof?

STRATEGY: Inside the pressure is atmospheric pressure p_{atm} and the air speed is 0. Let the pressure outside be p and the air speed be v. Neglect gravitational terms in Bernoulli's equation. Then that equation becomes $p_{\text{atm}} = p + \frac{1}{2}\rho v^2$, where ρ is the density of the air. Calculate the pressure difference $p_{\text{atm}} - p$.

SOLUTION:

$$\left[\,\text{ans: } 560\,\text{Pa}\,\right]$$

(*b*) What would be the lifting force on a roof of area $90\,\text{m}^2$?

STRATEGY: The lifting force is given by the product of the pressure difference and the roof area: $F = (p_{\text{atm}} - p)A$.

SOLUTION:

$$\left[\,\text{ans: } 5.0 \times 10^4\,\text{N}\,\right]$$

PROBLEM 19 (75P). A hollow tube has a disk DD attached to its end (Fig. 16–52). When air is blown through the tube, the disk attracts the card CC. Let the area of the card be A, and let v be the average airspeed between the card and the disk. Calculate the resultant upward force on CC. Neglect the card's weight; assume that $v_0 \ll v$, where v_0 is the airspeed in the hollow tube.

STRATEGY: At the top of the tube the pressure is atmospheric pressure p_{atm} and the airspeed is v_0. Let the pressure between the card and the disk be p and the airspeed be v. Neglect the length of the tube. Then Bernoulli's equation becomes $p_{atm} + \frac{1}{2}\rho v_0^2 = p + \frac{1}{2}\rho v^2$, where ρ is the density of air. When the card lifts off the surface on which it rests, the pressure under it is atmospheric pressure so the net upward force on it is $(p_{atm} - p)A$. Use Bernoulli's equation to find an expression for $p_{atm} - p$. You should obtain $\frac{1}{2}\rho(v^2 - v_0^2)$. If $v_0 \ll v$ then $p_{atm} - p = \frac{1}{2}\rho v^2$.

SOLUTION:

$$\left[\text{ans: } \tfrac{1}{2}\rho v^2 A\right]$$

PROBLEM 20 (78P). A tank is filled with water to a height H. A hole is punched in one of the walls at a depth h below the water surface (Fig. 16–53). (*a*) Show that the distance x from the foot of the wall at which the stream strikes the floor is given by $x = 2\sqrt{h(H-h)}$.

STRATEGY: The stream starts at a height of $H - h$ from the floor. Let v be the initial speed, as the water leaves the tank. Use constant acceleration kinematics to show that $x = \sqrt{2v^2(H-h)/g}$. Now use Bernoulli's equation to find the speed v: take one point on a streamline to be at the water surface at the top of the tank and the other to be at the hole. The pressures at the top of the tank and at the hole are both atmospheric pressure. Assume the speed of the water at the top of the tank is negligible. Then Bernoulli's equation becomes $\rho g H = \rho g(H - h) + \frac{1}{2}\rho v^2$. Solve for v^2 and substitute into the expression for x.

SOLUTION:

(*b*) Could a hole be punched at another depth to produce a second stream that would have the same range? If so, at what depth?

STRATEGY: Solve $x = 2\sqrt{h(H-h)}$ for h. The equation is quadratic so there are two solutions. One is h. You want the other.

SOLUTION:

$$\left[\text{ans: } H - h\right]$$

(c) At what depth should the hole be placed to make the emerging stream strike the ground at the maximum distance from the base of the tank?

STRATEGY: Differentiate $x^2 = 4h(H - h)$ with respect to h, set the derivative equal to zero, and solve for h.

SOLUTION:

$$[\text{ans: } H/2]$$

PROBLEM 21 (80P). A *siphon* is a device for removing liquid from a container. It operates as shown in Fig. 16–55. Tube ABC must initially be filled, but once this has been done the liquid will flow through the tube until the level in the container drops below the tube opening at A. The liquid has density ρ and negligible viscosity. (*a*) With what speed does the liquid emerge from the tube at C?

STRATEGY: Use Bernoulli's equation with one point at the water surface in the container and the other at the free end of the tube. Take $y = 0$ at the water surface. Then the free end of the tube is at $y = -(h_2 + d)$. The pressure is atmospheric pressure at both points. Assume the fluid speed is negligible at the surface. Then the equation becomes $0 = -\rho g(h_2 + d) + \frac{1}{2}\rho v^2$. Solve for v.

SOLUTION:

$$\left[\text{ans: } \sqrt{2g(h_2 + d)}\right]$$

(*b*) What is the pressure in the liquid at the topmost point B?

STRATEGY: Assume the cross-sectional area of a tube of flow does not change from the top of the tube to the end. The fluid speed is then the same at those two points. The pressure is atmospheric pressure at the free end of the tube. Solve Bernoulli's equation for the pressure at the top.

SOLUTION:

$$[\text{ans: } p_{\text{atm}} - \rho g(h_1 + h_2 + d)]$$

(*c*) Theoretically what is the greatest possible height h_1 that a siphon can lift water?

STRATEGY: The pressure at the top of the tube cannot be negative, so $\rho g(h_1 + h_2 + d) < p_{\text{atm}}$. h_1 has its greatest value if $h_2 + d = 0$. Then $\rho g h_1 < p_{\text{atm}}$ or $h_1 < p_{\text{atm}}/\rho g$. Take $\rho = 0.998 \times 10^3 \text{ kg/m}^3$.

SOLUTION:

$$[\text{ans: } 10.3 \text{ m}]$$

III. NOTES

Chapter 17
WAVES — I

I. BASIC CONCEPTS

In this chapter you study wave motion, a mechanism by which a disturbance (or distortion) created at one place in a medium propagates to other places. A wave may carry energy and momentum with it, but it does not carry matter.

The general ideas are specialized to a mechanical wave on a taut string. Pay attention to those characteristics of a string that determine the speed of a wave. You will also learn about sinusoidal waves, for which the disturbance has the shape of a sine or cosine function. This type wave has a special terminology associated with it; you will need to know the meaning of the terms amplitude, frequency, period, angular frequency, wavelength, angular wave number, and wave number. You will also need to understand what determines the values of each of these quantities.

Two or more waves present at the same time and place give rise to what are called interference effects. Special combinations of traveling waves result in standing waves. Learn what these phenomena are and how to analyze them.

General characteristics of traveling waves. A distortion in one region of space causes a disturbance in neighboring regions and so the disturbance moves from place to place. After a taut string is distorted and released, for example, the distorted section pulls on neighboring sections to distort them and the neighboring sections pull the originally distorted portion back to its undistorted position.

Give some examples of mechanical waves: _____

_____

When a mechanical wave is present, both the waveform (the disturbance) and the medium (string, water, air) move. These motions are related to each other but they are NOT the same. While studying this chapter you should be continually aware of which of these motions is being discussed.

Waves are sometimes classified as transverse or longitudinal, according to the direction the medium moves relative to the direction the wave moves. In a <u>longitudinal</u> wave, the medium is displaced in a direction that is _____ to the direction the wave travels and in a <u>transverse</u> wave it is displaced in a direction that is _____ to the direction the wave travels. Many waves, water waves among them, are neither transverse nor longitudinal.

The <u>wave speed</u> is the speed with which the distortion moves. The graph below shows a distorted string at time $t = 0$. The string stretches along the x axis and its displacement is denoted by y. Suppose the distortion travels without change in shape to the right at 2.0 m/s. On the coordinates provided sketch the string at $t = 1\,$s and $t = 2\,$s.

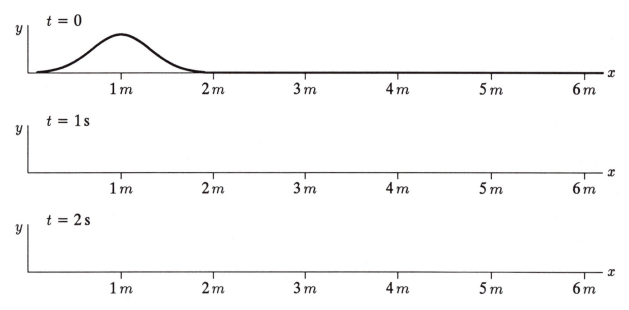

If the waveform moves with constant speed v and without change in shape then any particular point on the waveform (the maximum displacement, for example) moves so its coordinate x at time t is given by $x - vt = $ constant for waveforms traveling in the *positive* x direction and by $x + vt = $ constant for waveforms traveling in the *negative* x direction. Carefully note the signs in these expressions.

For the waveform shown on the three graphs above take x to be the coordinate at the maximum string displacement and fill in the following table.

t (s)	x (m)	$x - vt$ (m)
0		
1		
2		

If you did not get the same answer for $x - vt$ each time, you probably did not draw the graphs correctly.

A wave on a string is described by giving the string displacement $y(x, t)$ as a function of position and time. For example, the graphs above show $y(x, t)$ for three different values of t. If you know the function $y(x, t)$ you can find the displacement of any point on the string at any time. Simply substitute the value of the coordinate x of the point and the value of the time t into the function.

Sinusoidal waves. A sinusoidal wave is shown below for time t = 0.

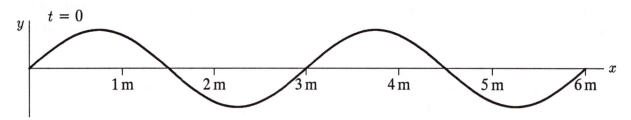

The function that describes the string displacement at $t = 0$ is $f(x) = y_m \sin(kx)$, where y_m and k are constants. y_m is called the _____ of the wave. Indicate it on the graph. k is called the _____ of the wave. It is related to the <u>wavelength</u> λ by $k =$ _____. Indicate the wavelength on the graph by marking an appropriate distance with the symbol λ. Any two points that are separated by a multiple of λ have identical displacements at every instant. The wavelength of the wave shown is about _____ m.

The wave number κ is related to the wavelength by $\kappa =$ _____ and to the angular wave number by $\kappa =$ _____. Physically, κ tells us how many waves there are per unit _____. Its SI unit is _____.

If the wave travels with speed v in the positive x direction, the displacement $y(x,t)$ at any time t is given by

$$y(x,t) = y_m \sin[k(x - vt)] = y_m \sin(kx - \omega t),$$

where $\omega = kv$.

Every point of the string moves in simple harmonic motion with <u>angular frequency</u> _____. In terms of the angular frequency the <u>frequency</u> of its motion is given by $f =$ _____ and the <u>period</u> of its motion is given by $T =$ _____. The period of an oscillation tells us _____

and the frequency tells us _____

The graph below shows the displacement of a point on the string as a function of time. On the time axis use the symbol T to label a time interval equal to a period.

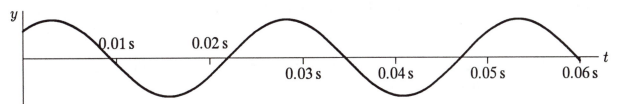

The period of the wave shown is about _____ s and the frequency is about _____ Hz.

When ω is written in terms of the period T and k is written in terms of the wavelength λ, $\omega = kv$ becomes $\lambda = vT$. That is, the wave moves a distance equal to one wavelength in a time equal to _____. This result should be obvious to you. Consider a point on the string that has maximum displacement at $t = 0$. There is another point with the same displacement a distance λ away. By the time this part of the wave gets to the first point that point has moved through one complete cycle.

If the wave shown moves in the *negative* x direction then

$$y(x,t) =$$

If you are given the wave speed and the function $f(x)$ that describes the string displacement at $t = 0$, then you can find the displacement $y(x,t)$ for later times. If the waveform is moving in the positive x direction you substitute _____ for x in the function $f(x)$. If the waveform is traveling in the negative x direction you substitute _____ for x. All traveling waves are functions of $x - vt$ or $x + vt$. The variables x and t cannot enter in any way but these two combinations.

Wave speed. Carefully read Section 17–6 of the text. There you are shown that Newton's second law leads directly to an expression for the wave speed in terms of the tension τ in the string and the linear density μ of the string. Specifically,

$$v =$$

The tension in the string is usually determined by the external forces applied at its ends to hold it taut.

You should carefully note that the wave speed does NOT depend on the frequency or wavelength. If the frequency is increased the wavelength must decrease so that $v = \lambda f$ has the same value. Suppose, for example, the wave speed on a certain string is 5.0 m/s. Then the wavelength of a 100 Hz sinusoidal wave is _____ m. If a 200 Hz sinusoidal wave travels on the same string with the same tension, its speed is _____ m/s and its wavelength is _____ m.

Particle velocity The transverse velocity $u(x, t)$ of the string at the position with coordinate x can be found as a function of time by differentiating _____ with respect to time. If the displacement is given by $y_m \sin(kx - \omega t)$ then

$$u(x, t) =$$

You should distinguish between wave and particle velocities. The wave velocity is associated with the motion of _____, while the particle velocity is associated with the motion of _____. Consider a transverse wave on a string and list some characteristics that are different for these two velocities: _____

_____

_____

The acceleration of a point on a string can be found by differentiating _____ with respect to time. For a sinusoidal wave the result is

$$a(x, t) =$$

Notice that it is proportional to the displacement $y(x, t)$ and that the constant of proportionality is negative, a result you should expect since the string at any point is in simple harmonic motion.

Energy and power. The mass in a segment of string of infinitesimal length dx is $dm =$ _____ if μ is the linear density. If u is the speed of the segment then its kinetic energy is $dK =$ _____ . This energy is transferred to a neighboring segment in time $dt = dx/v$, where v is the wave speed. Thus the rate at which kinetic energy is carried by the string is given by $dK/dt =$ _____ . For a sinusoidal wave, with $y(x, t) = y_m \sin(kx - \omega t)$,

$$dK/dt =$$

The average rate over an integer number of wavelengths is given by

$$\overline{\left(\frac{dK}{dt} \right)} =$$

since the average of $\cos^2(kx - \omega t)$ is $1/2$.

The string has potential energy because _____

_____

The average rate at which potential energy is transported is exactly the same as the rate at which kinetic energy is transported so, for a sinusoidal wave, the average rate of energy transport is given by

$$\overline{P} = $$

Notice that it depends on the square of the amplitude and on the square of the frequency.

Superposition and interference. To derive some results in this and the next chapter you should know the trigonometric identity

$$\sin \alpha + \sin \beta = 2 \sin \left(\frac{\alpha + \beta}{2} \right) \cos \left(\frac{\alpha - \beta}{2} \right) ,$$

valid for any angles α and β. It is proved in the Mathematical Skills section.

When two waves, one with displacement $y_1(x, t)$ and the other with displacement $y_2(x, t)$ are simultaneously on the same string, the displacement of the string is given by their sum: $y(x, t) = y_1(x, t) + y_2(x, t)$, provided the amplitudes are small. The addition of waveforms is called <u>superposition</u>. Carefully note that displacements, not transmitted powers, add. If the amplitudes are not small the presence of one wave might change the shape of a second wave. This situation is not considered here.

The addition of waveforms leads to some interesting and important phenomena, called <u>interference</u> phenomena. Consider two sinusoidal waves with the same frequency, traveling in the same direction on the same string. Suppose the waves have the same amplitude but different phase constants and let $y_1(x, t) = y_m \sin(kx - \omega t + \phi)$ and $y_2(x, t) = y_m \sin(kx - \omega t)$. The waves are identical except that at every instant the second is shifted along the x axis from the first by an amount that depends on the value of the phase constant ϕ. Eq. 17–44 gives the equation for the resultant string displacement. Copy it here:

$$y(x, t) = $$

Now use the trigonometric identity given above to prove the result:

Notice that the composite wave is sinusoidal with the same frequency and wavelength as either of the constituent waves.

Concentrate on the amplitude of the composite wave. It is a function of ϕ and, in particular, is given by _____. The maximum amplitude is given by _____ and the amplitude is a maximum when ϕ has any of the values _____. Maximum amplitude occurs when ϕ is adjusted so the crests of one of the constituent waves fall exactly on the _____ of the other. This condition is called _____ interference. The minimum amplitude is zero and the amplitude is a minimum when ϕ has any one of the values _____. Minimum amplitude occurs when ϕ is adjusted so the crests of one of the constituent waves fall exactly on the _____ of the other. This condition is called _____ interference. For other values of ϕ the interference is destructive but it is not complete.

Standing waves and resonance. In a standing wave each part of the string oscillates back and forth but the waveform does not move, as it does in a traveling wave. No energy is transmitted from place to place.

A standing wave can be constructed as the superposition of two traveling waves with the same amplitude and frequency, but moving in opposite directions. Let $y_1(x, t) = y_m \sin(kx - \omega t)$ and $y_2(x, t) = y_m \sin(kx + \omega t)$ represent the two traveling waves. The sum is given by Eq. 17–49 of the text. Copy it here:

$$y(x, t) =$$

Now use the trigonometric identity given above to prove the result:

Note that x and t do NOT enter the expression for $y(x, t)$ in either of the forms $x - vt$ or $x + vt$. The wave is NOT a traveling wave. The shape changes as the string moves but it does not travel along the string. Points of maximum displacement remain at the same points; so do points of minimum displacement. Also note that each point on the string vibrates in simple harmonic motion with an amplitude that varies with position along the string. In fact, the amplitude of the oscillation of the point with coordinate x is given by _____.

At certain points, called <u>nodes</u>, the amplitude is zero. Since their displacements are always zero, these points on the string do not vibrate at all. For the standing wave given above nodes occur at positions for which kx is a multiple of _____ or, what is the same thing, x in terms of the wavelength is a multiple of _____. Nodes are _____ wavelength apart.

At other points, called _____, the amplitude is a maximum. In terms of y_m it is _____. For the standing wave given above, these points occur at positions for which kx is an odd multiple of _____ or, what is the same thing, x is an odd multiple of _____.

The phase constant for both traveling waves was chosen to be 0 but that is not a necessity. If the phase constants of the constituent traveling waves are different, the nodes and antinodes are still the same distances apart but they are shifted in position from those that occur when the phase constants are the same.

One way to generate a standing wave is simply to allow a sinusoidal wave to be reflected from an end of the string. The reflection, of course, travels in the opposite direction. If such a wave is reflected from a *fixed* end the reflected and incident waves at that end have _____ signs and cancel each other there. The fixed end is a _____ of the standing wave pattern. On the other hand, if the same wave is reflected from a free end, the incident and reflected waves have _____ signs there and the free end of string is an _____ of the standing wave pattern.

For a string with both ends fixed, both ends are nodes of a standing wave pattern. This means that the traveling waves producing the pattern may have only certain wavelengths. Possible wavelengths are determined by the condition that the length of the string must be a multiple of _____, where λ is the wavelength. If v is the wave speed for traveling waves on the string, then the standing wave frequencies are given in terms of the string length L by $f =$ _____, where n is a positive integer, called the _____ number.

The lines below represent a string with fixed ends. Draw the amplitude as a function of position for the standing waves with the three lowest frequencies.

Standing wave frequencies are called the _____ frequencies of the string. If the string is driven at one of its standing wave frequencies by an applied sinusoidal force, the amplitude at the antinodes becomes large. The driving force is said to be in <u>resonance</u> with the string. Of course, the string can be driven at another frequency but then the amplitude remains small.

When an applied driving force is in resonance with a string the energy supplied by the applied force is dissipated in internal friction. In the absence of friction the standing wave amplitude would grow without bound. This is to be contrasted with the situation in which the driving force is not in resonance with the string. Then, in addition to frictional dissipation, the string loses energy by doing work on _____.

II. PROBLEM SOLVING

Many of the problems involving sinusoidal waves on a string deal with the relationships $v = \lambda f = \lambda/T = \omega/k$, where v is the wave speed, λ is the wavelength, f is the frequency, T is the period, ω is the angular frequency, and k is the angular wave number. Typical problems might give you the wavelength and frequency, then ask for the wave speed or might give you the wave speed and period, then ask for the wavelength or angular wave number.

Sometimes the quantities are given by describing the motion. For example, a problem might tell you that the string at one point takes a certain time to go from its equilibrium position to maximum displacement. This, of course, is one-fourth the period. In other problems the frequency of the source (a person's hand or a mechanical vibrator, for example) might be given. You must then recognize that the frequency of the wave is the same as the frequency of the source.

PROBLEM 1 (5E). A sinusoidal wave travels along a string. The time for a particular point to move from maximum displacement to zero is 0.170 s; what are the (a) period and (b) frequency?

STRATEGY: The time given is one-fourth the period T. The frequency is given by $f = 1/T$.

SOLUTION:

$$[\text{ans: } (a)\ 0.68\,\text{s; } (b)\ 1.47\,\text{Hz}]$$

(c) The wavelength is 1.40 m; what is the wave speed?

STRATEGY: Use $v = \lambda f$.

SOLUTION:

$$[\text{ans: } 2.06\,\text{m/s}]$$

For some problems you must be able to write an expression for the displacement as a function of position and time for a sinusoidal wave. Since it has the form $y(x, t) = y_m \sin(kx \pm \omega t + \phi)$, you must be able to determine k, ω and ϕ from data given in the problem statement. In many cases the initial conditions are not given. You may then select the time $t = 0$ so that $\phi = 0$ and write $y = y_m \sin(kx \pm \omega t)$. The sign in front of ωt is determined by the direction of travel. In other problems you may be given the displacement as a function of coordinate and time and asked to identify various quantities.

PROBLEM 2 (6E). Write the equation for a wave traveling in the negative direction along the x axis and having an amplitude of 0.010 m, a frequency of 550 Hz, and a speed of 330 m/s.

STRATEGY: Use $\omega = 2\pi f$ to find the angular frequency and $v = \omega/k$ to find the angular wave number. The position and time dependent terms must have the same sign since the wave is traveling in the negative x direction. Take the displacement to be parallel to the y axis.

SOLUTION:

$$[\text{ans: } 0.010 \sin(10.5x + 3.46 \times 10^3 t),\ \text{with } x \text{ and } y \text{ in meters and } t \text{ in seconds}]$$

PROBLEM 3 (13P). The equation of a transverse wave traveling along a very long string is given by $y = 6.0 \sin(0.020\pi x + 4.0\pi t)$, where x and y are expressed in centimeters and t is in seconds. Determine (a) the amplitude, (b) the wavelength, (c) the frequency, (d) the speed, (e) the direction of propagation of the wave, and (f) the maximum transverse speed of a particle in the string.

STRATEGY: The amplitude is 6.0 cm, the angular wave number is 0.020π rad/m, and the angular frequency is 4.0π rad/s. Use $k = 2\pi/\lambda$ to find the wavelength, $\omega = 2\pi f$ to find the frequency, and $v = \lambda f$ to find the wave speed. Look at the signs of the two terms in the argument of the sine function to determine the direction of travel. The transverse speed of the point at x, at time t, is given by $u(x,t) = \partial y/\partial t = 6.0 \times 4.0\pi \cos(0.020\pi x + 4.0\pi t)$ and its maximum value is $u_m = 6.0 \times 4.0\pi$. Evaluate this expression.

SOLUTION:

$\big[$ans: (a) 6.0 cm; (b) 100 cm; (c) 2.0 Hz; (d) 200 cm/s; (e) negative x direction; (f) 75 cm/s$\big]$

(g) What is the transverse displacement at $x = 3.5$ cm when $t = 0.26$ s?

STRATEGY: Evaluate $y = 6.0\sin(0.020\pi x + 4.0\pi t)$ for the given values of the coordinate and time.

SOLUTION:

$\big[$ans: -2.0 cm$\big]$

PROBLEM 4 (16P). A wave of frequency 500 Hz has a velocity of 350 m/s. (a) How far apart are two points that differ in phase by $\pi/3$ rad?

STRATEGY: The phase is given by $kx \pm \omega t$ and the difference in phase of two points separated by Δx is $k\Delta x$. Set this expression equal to $\pi/3$ and solve for Δx. You must first find the angular wave number k. Use $v = \lambda f$ to find the wavelength, then $k = 2\pi/\lambda$ to find k.

SOLUTION:

$\big[$ans: 0.117 m$\big]$

(b) What is the phase difference between two displacements at a certain point at times 1.00 ms apart?

STRATEGY: Now $\Delta x = 0$ and the phase difference is given by $\omega \Delta t$. Evaluate this expression. The angular frequency is given by $\omega = 2\pi f$.

SOLUTION:

$\big[$ans: 3.14 rad$\big]$

Some problems deal with the wave speed. The fundamental equation is $v = \sqrt{\tau/\mu}$, where τ is the tension in the string and μ is the linear density of the string. The tension and linear density may not be given directly but, if the problem asks for the wave speed, sufficient information will be given to calculate these quantities. In the first problem below the stress (force per unit area) and density (mass per unit volume) are given. In the second the tension is generated by mass hung on the string and can be computed from the value of the mass.

PROBLEM 5 (24E).   What is the fastest transverse wave that can be sent along a steel wire? Allowing for a reasonable safety factor, the maximum tensile stress to which steel wires should be subject is $7.0 \times 10^8 \, \text{N/m}^2$. The density of steel is $7800 \, \text{kg/m}^3$. Show that your answer does not depend on the diameter of the wire.

STRATEGY: The wave speed is given by $v = \sqrt{\tau/\mu}$, where τ is the tension in the wire and μ is the linear density. The stress S is the force per unit area on a cross section of the wire so the tension is given by $\tau = \pi R^2 S$, where R is the radius, and the linear density is the density times the cross-sectional area: $\mu = \pi R^2 \rho$. Substitute these expressions into the equation for the speed. Your answer should not depend on R. Evaluate the result.

SOLUTION:

[ans: 300 m/s]

PROBLEM 6 (29P).   In Fig. 17–25a, string 1 has a linear density of $3.00 \, \text{g/m}$, and string 2 has a linear density of $5.00 \, \text{g/m}$. They are under tension owing to the hanging block of mass $M = 500 \, \text{g}$. (a) Calculate the wave speed in each string.

STRATEGY: First find the tension in each string. Because the pulley is in rotational equilibrium the net torque on it is zero. If τ_1 is the tension in string 1 and τ_2 is the tension in string 2, then $(\tau_1 - \tau_2)R = 0$ and $\tau_1 = \tau_2$. Since the pulley is in translational equilibrium the net force on it is zero: $Mg - \tau_1 - \tau_2 = 0$. The solution to these equations is $\tau_1 = Mg/2$ and $\tau_2 = Mg/2$. Substitute into $v_1 = \sqrt{\tau_1/\mu_1}$ and $v_2 = \sqrt{\tau_2/\mu_2}$.

SOLUTION:

[ans: $v_1 = 28.6 \, \text{m/s}$, $v_2 = 22.1 \, \text{m/s}$]

(b) The block is now divided into two blocks (with $M_1 + M_2 = M$) and the apparatus rearranged as shown in Fig. 17–25b. Find M_1 and M_2 such that the wave speeds in the two strings are equal.

STRATEGY: The tension in string 1 is $M_1 g$ and the wave speed is $v_1 = \sqrt{M_1 g/\mu_1}$. The tension in string 2 is $M_2 g$ and the wave speed is $v_2 = \sqrt{M_2 g/\mu_2}$. Since the wave speeds are to be equal, $M_1/\mu_1 = M_2/\mu_2$. Solve this simultaneously with $M_1 + M_2 = M$ for M_1 and M_2.

SOLUTION:

[ans: $M_1 = 188 \, \text{g}$, $M_2 = 313 \, \text{g}$]

PROBLEM 7 (31P).   The type of rubber band used inside some baseballs and golf balls obeys Hooke's law over a wide range of elongation of the band. A segment of this material has an unstretched length ℓ and a mass m. When a force F is applied, the band stretches an additional length $\Delta\ell$. (a) What is the speed (in terms of m, $\Delta\ell$, and the force constant k) of transverse waves on this rubber band?

STRATEGY: The wave speed is given by $v = \sqrt{F/\mu}$, where μ is the linear density. Use $F = k\Delta\ell$ and $\mu = m/(\ell + \Delta\ell)$ to write v in terms of m, k, ℓ and $\Delta\ell$.

SOLUTION:

$$\left[\text{ans: } \sqrt{k\Delta\ell(\ell + \Delta\ell)/m}\,\right]$$

(b) Using your answer to (a), show that the time required for a transverse pulse to travel the length of the rubber band is proportional to $1/\sqrt{\Delta\ell}$ if $\Delta\ell \ll \ell$ and is constant if $\Delta\ell \gg \ell$.

STRATEGY: The time is given by $t = (\ell + \Delta\ell)/v = (\ell + \Delta\ell)/\sqrt{k\Delta\ell(\ell + \Delta\ell)/m}$. If $\Delta\ell \ll \ell$ then $\ell + \Delta\ell$ may be replaced by ℓ. You should obtain $t \approx \sqrt{\ell m/k\Delta\ell}$. If $\Delta\ell \gg \ell$ then you may replace $\ell + \Delta\ell$ with $\Delta\ell$. You should get $t \approx \sqrt{m/k}$.

SOLUTION:

The equation $\overline{P} = \frac{1}{2}\mu v \omega^2 y_m^2$ for the average power being transmitted at any time past any point is typically used in a straightforward manner to compute the $\overline{P}$, given the parameters of the wave. You may also be asked to find the value of one of the parameters (μ, ω, v, τ, y_m, or a related quantity) to achieve a given level of power transmission.

PROBLEM 8 (33E). Power P_1 is transmitted by a wave of frequency f_1 on a string with tension τ_1. What is the transmitted power P_2 in terms of P_1 (a) if the tension of the string is increased to $\tau_2 = 4\tau_1$, and (b) if, instead, the frequency is decreased to $f_2 = f_1/2$?

STRATEGY: Take the given power to the average over a cycle. Then the transmitted power is given by $\overline{P} = \frac{1}{2}\mu v \omega^2 y_m^2$. Since $\omega = 2\pi f$ and $v = \sqrt{\tau/\mu}$, this can be written $\overline{P} = 2\pi^2\sqrt{\tau\mu}f^2 y_m^2$. The only quantities that are changed are the tension and the frequency so $P_2/P_1 = \sqrt{\tau_2/\tau_1}(f_2/f_1)^2$. In (a) take $\tau_2 = 4\tau_1$ and $f_2 = f_1$. In (b) take $\tau_2 = \tau_1$ and $f_2 = f_1/2$.

SOLUTION:

$$\left[\text{ans: } (a)\ P_2 = 2P_1;\ (b)\ P_2 = P_1/4\right]$$

PROBLEM 9 (34E). A string 2.7 m long has a mass of 260 g. The tension in the string is 36 N. What must be the frequency of traveling waves of amplitude 7.7 mm in order that the average transmitted power be 85 W?

STRATEGY: Use $\overline{P} = \frac{1}{2}\mu v \omega^2 y_m^2$. Substitute $\mu = m/\ell$, $v = \sqrt{\tau/\mu}$, and $\omega = 2\pi f$, then solve for f.

SOLUTION:

$$\left[\text{ans: } 198\,\text{Hz}\right]$$

Interference problems may give the phase difference of two waves and ask for the resultant wave. In variations the resultant is given and you are asked for the phase difference.

PROBLEM 10 (37E).   What phase difference between two otherwise identical traveling waves, moving in the same direction along a stretched string, will result in the combined wave having an amplitude 1.50 times that of the common amplitude of the two combining waves? Express your answer both in degrees and radians.

STRATEGY: If the two waves are $y_1 = y_m \sin(kx - \omega t + \phi)$ and $y_2 = y_m \sin(kx - \omega t)$ then $y_1 + y_2 = 2y_m \sin(kx - \omega t + \phi/2)\cos(\phi/2)$. You want $2y_m \cos(\phi/2) = 1.50y_m$. Solve for $\cos(\phi/2)$, then ϕ.

SOLUTION:

$$[\text{ans: } 82.8° \ (1.45\,\text{rad})]$$

PROBLEM 11 (38P).   A source S and a detector D of radio waves are a distance d apart on level ground (Fig. 17–26). Radio waves of wavelength λ reach D either along a straight path or by reflecting (bouncing) from a certain layer in the atmosphere. When the layer is at height H, the two waves reaching D are exactly in phase. If the layer gradually rises, the phase difference between the two waves gradually shifts, until they are exactly out of phase when the layer is at height $H + h$. Express λ in terms of d, h, and H.

STRATEGY: The wave that reaches D along the straight path goes a distance d. Suppose the other wave goes a distance $d + \Delta x$. Then the composite wave at the detector is given by $y_1 + y_2 = y_m \sin(kd + \omega t) + y_m \sin(kd + \omega t + k\Delta x)$. The phase difference is $\phi = k\Delta x$, or $2\pi\Delta x/\lambda$. When the height of the layer is H an application of the pythagorean theorem tells us that $\Delta x = 2\sqrt{H^2 + d^2/4} - d$. Furthermore, $\phi = 2n\pi$, where n is an integer. Thus $2\sqrt{H^2 + d^2/4} - d = n\lambda$. When the layer is at height $H + h$, $\Delta x = 2\sqrt{(H + h)^2 + d^2/4} - d$ and $\phi = 2(n + \frac{1}{2})\pi$. Thus $2\sqrt{(H + h)^2 + d^2/4} - d = (n + \frac{1}{2})\lambda$. The altitude of the layer has changed so the indirect wave goes half a wavelength further than before. Use one of these equations to eliminate n from the other, then solve for λ.

SOLUTION:

$$\left[\text{ans: } \lambda = 4\sqrt{(H + h)^2 + d^d/4} - 4\sqrt{H^2 + d^2/4}\right]$$

PROBLEM 12 (41P).   Determine the amplitude of the resultant wave when two sinusoidal waves having the same frequency and traveling in the same direction are combined, if their amplitudes are 3.0 cm and 4.0 cm and they differ in phase by $\pi/2$ rad.

STRATEGY: The resultant wave is given by $y = y_{1m} \sin(kx - \omega t + \phi) + y_{2m} \sin(kx - \omega t)$. You want to write this expression in the form $y = y_m \sin(kx - \omega t + \alpha)$. Apply the trigonometric identity $\sin(A + B) = \sin A \cos B + \cos A \sin B$ to both $\sin(kx - \omega t + \phi)$ and $\sin(kx - \omega t + \alpha)$. Let $A = kx - \omega t$. In the first case let $B = \phi$ and in the second let $B = \alpha$. You should find $(y_{1m} \cos \phi + y_{2m}) \sin(kx - \omega t) + y_{1m} \cos(kx - \omega t) = y_m \sin(kx - \omega t) \cos \alpha + y_m \cos(kx - \omega t) \sin \alpha$. For this equality to hold for all values of x and t the coefficients of $\sin(kx - \omega t)$ on the two sides must be equal and the coefficients of $\cos(kx - \omega t)$ on the two sides must also be equal. That is, $y_m \cos \alpha = y_{1m} \cos \phi + y_{2m}$ and $y_m \sin \alpha = y_{1m} \sin \phi$. Solve for y_m. This is best done by adding the squares of the equations and using $\sin^2 \alpha + \cos^2 \alpha = 1$. You should get $y_m^2 = y_{1m}^2 + y_{2m}^2 + 2y_{1m}y_{2m} \cos \phi$. Evaluate this for $\phi = \pi/2$.

SOLUTION:

Some problems dealing with standing waves on a string give you the amplitude, frequency, and angular wave number (or related quantities) for the traveling waves and ask for the standing wave pattern. The inverse problem gives the standing wave in the form $y = A \sin(kx) \cos(\omega t)$ and asks for the component traveling waves. They have the form $y_m \sin(kx \pm \omega t)$ with $y_m = A/2$.

Instead of the wavelength or angular wave number of the traveling waves you might be told the distance between successive nodes or successive antinodes. Double it to find the wavelength. If you are told the distance between a node and a neighboring antinode, multiply it by 4 to find the wavelength.

If a standing wave is generated in a string with both ends fixed, the wave pattern must have a node at each end of the string. This means the length L of the string and the wavelength λ of the traveling waves must be related by $L = n\lambda/2$, where n is an integer. If one end is fixed and the other is free, the fixed end is a node and the free end is an antinode, so the length must be an odd multiple of $\lambda/4$.

PROBLEM 13 (46E). A nylon guitar string has a linear density of 7.2 g/m and is under a tension of 150 N. The fixed supports are 90 cm apart. The string is oscillating in the standing wave pattern shown in Fig. 17–27. Calculate the (*a*) speed, (*b*) wavelength, and (*c*) frequency of the waves whose superposition gives this standing wave.

STRATEGY: The speed is given by $v = \sqrt{\tau/\mu}$, where τ is the tension and μ is the linear density. According to the diagram there are three half-wavelengths in the length of the string, so $L = 3\lambda/2$. Solve for λ. Finally, use $v = f\lambda$ to find f.

SOLUTION:

[ans: (*a*) 140 m/s; (*b*) 60 cm; (*c*) 240 Hz]

PROBLEM 14 (52E). One end of a 120-cm string is held fixed. The other end is attached to a weightless ring that can slide along a frictionless rod as shown in Fig. 17–28. What are the three longest possible wavelengths for standing waves in this string?

STRATEGY: The length of the string must be an odd multiple of $\lambda/4$, so the four longest wavelengths are given by $L = \lambda/4$, $L = 3\lambda/4$, and $L = 5\lambda/4$, where L is the length of the string. Solve for λ in each case.

SOLUTION:

[ans: 480 cm, 160 cm, 96 cm]

Sketch the corresponding standing waves.

SOLUTION: On the axes below sketch the standing wave amplitude as a function of distance along the string. In each case one end is a node and the other is an antinode. For the longest wavelength there are no other nodes except the one at the end. For the next longest wavelength there is a single node between the two ends and for the third longest wavelength there are two nodes between the ends.

———————————   ———————————   ———————————

$\lambda = 4L$ $\lambda = 4L/3$ $\lambda = 4L/5$

PROBLEM 15 (60P). A 3.0-m-long string is oscillating as a three-loop standing wave whose amplitude is 1.0 cm. The wave speed is 100 m/s. (a) What is the frequency?

STRATEGY: Let L be the length of the string and λ be the wavelength of either of the traveling waves that combine to form the standing wave. Sketch the standing wave and convince yourself that since there are three loops, $1.5\lambda = L$. The frequency is given by $f = v/\lambda$.

SOLUTION:

[ans: 50 Hz]

(b) Write equations for two waves that, when combined, will result in this standing wave.

STRATEGY: Write the waves in form $y_m \sin(kx - \omega t)$ and $y_m \sin(kx + \omega T)$. The angular wave number is given by $k = 2\pi/\lambda$ and the angular frequency is given by $\omega = 2\pi f$. The amplitude of either wave is half the maximum amplitude in the standing wave.

SOLUTION:

[ans: $(0.05 \text{ cm}) \sin[(3.14 \text{ rad/s})x - (314 \text{ rad/s})t]$, $(0.05 \text{ cm}) \sin[(3.14 \text{ rad/m})x + 314 \text{ rad/s})t]$]

III. MATHEMATICAL SKILLS

Partial derivatives. Partial derivatives are important for understanding much of this chapter. The displacement $y(x, t)$ of a string carrying a wave is a function of two variables, the coordinate x of a point on the string and the time t. You may differentiate with respect to either variable.

The notation $\partial y(x, t)/\partial x$ stands for the derivative of y with respect to x, with t treated as a constant. Similarly, $\partial y(x, t)/\partial t$ means the derivative of y with respect to t, with x treated as a constant. The result of either differentiation may again be a function of x and t. For example, $\partial \sin(kx - \omega t)/\partial x = k\cos(kx - \omega t)$ and $\partial \sin(kx - \omega t)/\partial t = -\omega \cos(kx - \omega t)$.

You should understand the physical significance of a partial derivative as well as be able to evaluate it, given the function. The partial derivative $\partial y(x, t)/\partial x$ gives the slope of the string at the point x and time t; here you are evaluating the rate at which the displacement changes with *distance* along the string at some instant of time. For this to be meaningful the definition must make use of displacements for slightly separated points on the string *at the same time*, in the limit as the separation tends to zero. That is why t is treated as a constant.

The partial derivative $\partial y(x, t)/\partial t$ gives the string velocity at the point x and time t; here you are evaluating the rate at which the displacement changes with time at a given point. Clearly this is associated with the displacement *of the same point* but at slightly different times, in the limit as the time interval approaches zero. That is why x is treated as a constant.

Your instructor may require you to carry out partial differentiation of functions other than sine and cosine functions. For example, the string displacement associated with a bell-shaped pulse traveling in the positive x direction is given by

$$y(x, t) = Ae^{-(x-vt)^2/a},$$

where A and a are constants. The partial derivatives are

$$\frac{\partial y}{\partial x} = -\frac{2A(x - vt)}{a}e^{-(x-vt)^2/a}$$

and

$$\frac{\partial v}{\partial t} = \frac{2Av(x - vt)}{a}e^{-(x-vt)^2/a}$$

Check these yourself.

A trigonometric identity. The identity

$$\sin \alpha + \sin \beta = 2\sin\left(\frac{\alpha + \beta}{2}\right)\cos\left(\frac{\alpha - \beta}{2}\right)$$

plays an important role in this chapter. You should be able to show its validity. Write $\sin \alpha + \sin \beta = \sin\left(\frac{\alpha + \beta}{2} + \frac{\alpha - \beta}{2}\right) + \sin\left(\frac{\alpha + \beta}{2} - \frac{\alpha - \beta}{2}\right)$ and use the rules for expanding the sine of the sum and the sine of the difference of two angles: $\sin(\theta_1 + \theta_2) = \sin\theta_1 \cos\theta_2 + \cos\theta_1 \sin\theta_2$ and $\sin(\theta_1 - \theta_2) = \sin\theta_1 \cos\theta_2 - \cos\theta_1 \sin\theta_2$. These give

$$\sin \alpha + \sin \beta =$$

$$\sin \left(\frac{\alpha + \beta}{2} \right) \cos \left(\frac{\alpha - \beta}{2} \right) + \cos \left(\frac{\alpha + \beta}{2} \right) \sin \left(\frac{\alpha - \beta}{2} \right)$$

$$+ \sin \left(\frac{\alpha + \beta}{2} \right) \cos \left(\frac{\alpha - \beta}{2} \right) - \cos \left(\frac{\alpha + \beta}{2} \right) \sin \left(\frac{\alpha - \beta}{2} \right)$$

$$= 2 \sin \left(\frac{\alpha + \beta}{2} \right) \cos \left(\frac{\alpha - \beta}{2} \right) .$$

IV. NOTES

Chapter 18
WAVES — II

I. BASIC CONCEPTS

In this chapter the ideas of wave motion introduced in the last chapter are applied to sound waves. Pay particular attention to the dependence of the wave speed on the properties of the medium in which sound is propagating. Although the idea is the same, the properties are different from those that determine the speed of a wave on a taut string. Completely new concepts include beats and the Doppler shift. Be sure you understand what these phenomena are and how they originate.

General description. Sound waves are propagating distortions of a material medium. In fluids they are longitudinal: particles of the medium move back and forth along the line of _____. In solids they may be longitudinal, transverse, or neither.

Since particles at slightly different positions move different amounts the medium becomes compressed or rarefied as a sound wave passes. That is, we can consider a sound wave to be the propagation of a local increase or decrease in density. Since a change in pressure is associated with a change in density a sound wave may also be considered to be the propagation of a deviation in local pressure from the ambient pressure.

To understand this chapter you must know the relationships between density, volume, and pressure discussed in Chapter 16. Consider a tube of fluid with cross-sectional area A, density ρ, and bulk modulus B, at pressure p. Suppose an element of fluid originally of length ℓ is uniformly elongated by $\Delta\ell$ so its new length is $\ell + \Delta\ell$ and its new volume is $A(\ell + \Delta\ell)$. The mass of the fluid in the element is given by $m = \rho A \ell$, so the density after elongation is $\rho' = m/A(\ell + \Delta\ell) = \rho\ell/(\ell + \Delta\ell)$. This expression for ρ' can be written $\rho' = \rho + \Delta\rho$, where $\Delta\rho$ is the change in density. In the space below show that if $\Delta\ell$ is much smaller than ℓ then $\Delta\rho = -\rho\Delta\ell/\ell$:

Notice that $\Delta\rho$ is negative if $\Delta\ell$ is positive. The density decreases because the same mass occupies a larger volume after elongation. The change in pressure that accompanies the elongation is given by $\Delta p = -B\Delta V/V = -B\Delta\ell/\ell$. It is also proportional to $\Delta\ell$. The same expressions are also valid for a compression. Then $\Delta\ell$ is negative.

The relationships $\Delta\rho = -\rho\Delta\ell/\ell$ and $\Delta p = -B\Delta\ell/\ell$ are used to derive an expression for the speed of sound in terms of properties of the medium and to relate displacement, density, and pressure waves to each other.

Speed of sound.   The speed of sound in a fluid is determined by the density ρ and bulk modulus B of the fluid. Specifically, it is given by

$$v =$$

To understand this expression, apply Newton's second law to an element of fluid in which a sound pulse is traveling and make use of the relations between density, volume, and pressure given above. The derivation follows closely the derivation of the expression for the speed of a wave on a string, given in Section 17–6. You may wish to review that derivation.

We view a compressional pulse in a fluid from a reference frame that moves with the speed of sound v, so the pulse is stationary and the fluid moves with speed v into the right end and out of the left end of the pulse, as shown in Fig. 18–2 of the text. An element of fluid with cross-sectional area A and length ℓ (when just outside the pulse) takes time $\Delta t = \ell/v$ to completely enter the pulse. In terms of v and Δt the volume of the element when it is outside is _____ and if ρ is the fluid density outside then the mass of the element is $m =$ _____ .

Now consider the fluid element when it is partly inside and partly outside the pulse. If the pressure outside the pulse is p and the pressure inside the pulse is $p + \Delta p$ then the net force on the element is $F =$ _____ . Since the pressure inside the pulse is greater than the pressure outside, this force slows the element as it enters the pulse, so its velocity changes from v to $v + \Delta v$, where Δv is negative. The acceleration of the element is $a = \Delta v/\Delta t$. In $F = ma$ replace F with $-A\Delta p$, m with $\rho v A \Delta t$ and a with $\Delta v/\Delta t$, then solve for v:

Multiply the result by v to obtain $v^2 = -v\Delta p/\rho\Delta v$.

The key to the remainder of the derivation is to observe that the volume of the fluid element changes as it enters the pulse and that the fractional change in volume is the same as the fractional change in speed: that is, $\Delta V/V = \Delta v/v$. The leading edge of the element, inside the pulse, travels slower than the trailing edge, outside the pulse, by Δv. In time Δt it goes a distance $\Delta v \Delta t$ less, and this must be the amount by which the element is shortened. Thus $\Delta V = A\Delta v\Delta t$. Now $V = Av\Delta t$ is the volume V of the element when it is completely outside. Divide one of these results by the other to obtain $\Delta V/V = \Delta v/v$. Substitute $\Delta V/V$ for $\Delta v/v$ in the expression for ρv^2, then replace $-\Delta p/(\Delta V/V)$ with the bulk modulus B and solve for v:

You should have obtained $v = \sqrt{B/\rho}$. Thus the expression for the speed of sound can be derived straightforwardly from Newton's second law and is a direct result of the forces neighboring portions of the fluid exert on each other. You should know that the speed of sound is about _____ m/s in air, about _____ m/s in water, and about _____ m/s in solids.

Pressure and density waves. To describe a traveling sound wave you can use any of three quantities: the particle displacement, the deviation of the density from its ambient value, and the deviation of the pressure from its ambient value. All propagate as waves. They are related to each other and you should understand the relationships.

Suppose a sound wave is traveling along the x axis through a fluid and suppose further that the displacement of the fluid at coordinate x and time t is given by a known function $s(x,t)$. Consider an element of fluid that in the absence of the wave extends from x to $x + \Delta x$. Its length is Δx. In the presence of the wave the same fluid element extends from $x + s(x,t)$ to $x + \Delta x + s(x + \Delta x, t)$ so its change in length is $\Delta \ell = s(x + \Delta x, t) - s(x,t)$. In the limit as Δx becomes small $\Delta \ell$ becomes $[\partial s(x,t)/\partial x]\Delta x$. Thus the density in the presence of a wave is $\rho'(x,t) = \rho + \Delta\rho(x,t)$, where $\Delta\rho(x,t) = -\rho\Delta\ell/\Delta x = -\rho\partial s(x,t)/\partial x$.

Similarly, the pressure in the presence of a wave is $p'(x,t) = p + \Delta p(x,t)$ where $\Delta p = -B\Delta\ell/\Delta x = -B\partial s(x,t)/\partial x$ and B is the bulk modulus.

Suppose the sound wave is sinusoidal and the fluid displacement is given by $s(x,t) = s_m \cos(kx - \omega t)$, where k is the wave number and ω is the angular frequency. In terms of these quantities the deviation of the density from its ambient value is given by $\Delta\rho(x,t) = $ _____ and the deviation of the pressure from its ambient value is given by $\Delta p(x,t) = $ _____ .

These expressions can be written

$$\Delta\rho = \Delta\rho_m \sin(kx - \omega t)$$

and

$$\Delta p = \Delta p_m \sin(kx - \omega t),$$

where the density amplitude $\Delta\rho_m$ and the pressure amplitude Δp_m are given in terms of the displacement amplitude s_m, the wave speed v, and the wave number k by $\Delta\rho_m = $ _____ and $\Delta p_m = $ _____ .

Notice that the pressure and density waves are not in phase with the displacement wave. At points where the displacement is a maximum the deviation of the pressure from its ambient value is a _____ and the deviation of the density from its ambient value is a _____ . These results make physical sense because the fluid displacement is nearly uniform in the neighborhood of a displacement maximum. An element of fluid is neither compressed or elongated there. A _____ value of Δp corresponds to a compression, a _____ value corresponds to an expansion.

At points where the displacement is zero the deviation of the pressure from its ambient value is a _____ and the deviation of the density from its ambient value is a _____ . Here the rate of change of the displacement with distance has its greatest magnitude. The elongation or compression of the fluid is greater here than anywhere else. Thus we expect deviations of the pressure and density to be the greatest at these points.

Interference. To find the particle displacement when two or more sound waves are simultaneously present, add the _____ due to the individual waves. Let $s_1 = s_m \cos(kx - \omega T)$ represent one wave and $s_2 = s_m \cos(kx - \omega t + \phi)$ represent another. These waves interfere constructively if the phase difference ϕ has any of the values _____ rad, where n is an integer. The interference will be completely destructive if ϕ has any of the values _____ rad, where n is _____.

If the waves are generated by sources that are in phase but they travel to the detector along different paths they may have different phases at the detector. If one wave travels a distance x to the detector and the other travels a distance $x + \Delta d$, then the phase difference at the detector is given by $\phi =$ _____, where k is the angular wave number. Since $k = 2\pi/\lambda$, where λ is the wavelength, the phase difference can be written $\phi = 2\pi\Delta d/\lambda$. Constructive interference occurs if Δx is a multiple of _____; completely destructive interference occurs if Δx is an odd multiple of _____ .

Energy transport and intensity. If $s(x,t)$ is the particle displacement at coordinate x and time t then $v_s =$ _____ is the particle velocity. The mass contained in an infinitesimal length dx of fluid is given by $dm =$ _____, where ρ is the density and A is the cross-sectional area. This kinetic energy is transported from the segment of fluid to a neighboring segment in time $dt = dx/v$, where v is the speed of sound. Thus the rate at which kinetic energy flows in a sound wave is given by $dK/dt =$ _____. For a sinusoidal sound wave, for which $s = s_m \cos(kx - \omega t)$, this is $dK/dt =$ _____. The average over a cycle is $\overline{dK/dt} =$ _____, since the average of $\sin^2(kx - \omega t)$ is $1/2$. Potential energy is transported at the same rate, so the average rate of transport of mechanical energy is given by $\overline{P} =$ _____. This the also called the average <u>power</u> transported.

The <u>intensity</u> of a sound wave is the average rate of energy flow per unit _____ and, for the sinusoidal wave discussed above, is given by

$$I =$$

where s_m is the displacement amplitude. The SI units for intensity are _____ .

<u>Sound level</u> is often used instead of intensity. The sound level associated with intensity I is defined by $\beta =$ _____, where I_0 is the standard reference intensity (_____ W/m^2). Sound level is measured in units of _____, abbreviated _____ .

Carefully note that the sound level is defined in terms of the *logarithm to the base* 10 of I/I_0. If $I = I_0$ the sound level is _____ db. If the intensity is increased by a factor of 10 the sound level increases by _____ db.

The standard reference intensity is roughly the threshold of human hearing. The intensity at the upper end of the human hearing range (called the threshold of pain) is about _____ W/m^2 and the sound level is about _____ db. Look at Table 18–3 of the text for some other sound level values.

Standing sound waves. Two sinusoidal sound waves with the same frequency and amplitude but traveling in opposite directions combine to form a standing wave. At a displacement node the displacement is always _____ . At a displacement antinode the displacement oscillates between _____ and _____, where s_m is the displacement amplitude. At other points the displacement oscillates with an amplitude that is given by _____, where k is the angular wave number of either of the traveling waves and x is the coordinate of the point.

If there are no losses, standing waves are created in pipes by the superposition of a sinusoidal wave and its reflection from the end of the pipe. A displacement _____ exists at a closed end of a pipe. A displacement _____ exists at an open end.

If both ends of a pipe are open, the wavelengths associated with possible standing waves are such that the pipe length is a multiple of _____. In terms of L the standing wave wavelengths are given by $\lambda = $ _____ and the standing wave frequencies are given by $f = $ _____, where v is the speed of sound for the fluid that fills the pipe and n is _____.

For the lowest three standing wave frequencies use the coordinates below to plot the displacement amplitude A as a function of position along the pipe.

If one end of a pipe is open and the other is closed, the wavelengths associated with possible standing waves are such that the pipe length is an odd multiple of _____. In terms of L the standing wave wavelengths are given by $\lambda = $ _____ and the standing wave frequencies are given by $f = $ _____, where v is the speed of sound for the fluid that fills the pipe and n is _____.

For the lowest three standing wave frequencies use the coordinates below to plot the displacement amplitude A as a function of position along the pipe.

A string of a stringed instrument or the air in an organ pipe can vibrate at any one of its standing wave (or natural) frequencies. Which sound is produced depends on how the instrument is excited. Usually the lowest frequency dominates but higher frequency sound is mixed in. The admixture of higher frequencies gives any instrument the quality of sound peculiar to that instrument and, for example, allows us to distinguish a violin from a piano.

The lowest frequency is called the _____ frequency or first harmonic, while higher frequencies are higher harmonics and are numbered in order of increasing frequency.

Beats. Beats are created when two sound waves with nearly the same _____ are simultaneously present. We can view the resultant wave as one with a frequency that is the average of the frequencies of the constituent waves and an amplitude that varies with time, but much more slowly than either of the constituent waves.

Since we are interested in the time dependence of the wave, we can study the displacement at a single point in space and ignore variations with position. Let $s_1 = s_m \cos(\omega_1 t)$ represent the displacement at the point due to one of the waves and $s_2 = s_m \cos(\omega_2 t)$ represent the displacement at the same point due to the other wave. The resultant displacement is the sum of the two and, once the trigonometric identity given in the text is used, it can be written as Eq. 18–41 of the text:

$$s(t) =$$

Notice that there are two time dependent factors, both periodic. One has an angular frequency $\omega =$ _____ , the average of the two constituent angular frequencies. This is the greater of the angular frequencies associated with the two factors and if the two constituent frequencies are nearly the same it is essentially equal to either of them.

The angular frequency of the second time dependent factor is $\omega' =$ _____ . Note that it depends on the difference in the two constituent frequencies. If ω_1 and ω_2 are nearly the same the factor associated with this angular frequency is slowly varying. We may think of it as a slow variation in the amplitude of the faster vibration. The effect can be produced, for example, by blowing a note on a horn at the angular frequency ω, but modulating it so it is periodically loud and soft.

A <u>beat</u> is a maximum in the *intensity* and occurs each time $\cos(\omega't)$ goes from $+1$ to _____ . Thus the beat angular frequency ω_{beat} is NOT the same as ω'. In fact $\omega_{beat} =$ _____ ω', or in terms of ω_1 and ω_2, $\omega_{beat} =$ _____ . If one constituent wave has a frequency of 1000 Hz and the other has a frequency of 1005 Hz, then the beat frequency is $f_{beat} =$ _____ Hz and the beat angular frequency is $\omega_{beat} =$ _____ rad/s.

Doppler effect. Suppose a sustained note with a well-defined frequency f is played by a stationary trumpeter. If you move rapidly *toward* the trumpeter you will hear a note with a _____ frequency. If you move rapidly *away* from the trumpeter you will hear a note with a _____ frequency. Similar effects occur if you are stationary and the trumpeter is moving. The note has a higher frequency if the trumpeter is moving _____ and a lower frequency if the trumpeter is moving _____ . These are examples of the <u>Doppler</u> effect. The next time you hear a police or ambulance siren on the highway listen carefully as the vehicle approaches and then recedes from you. If it is going sufficiently fast you should hear the Doppler shift in frequency.

To understand how the Doppler effect comes about, suppose a sound detector is moving with speed v_D toward a source of sound with frequency f and wavelength λ ($= v/f$, where v is the speed of sound). If the detector were at rest it would receive _____ wave crests in time t. Because it is moving toward the source it receives _____ fewer crests in the same time. Thus the number of crests it receives in time t is _____ and the frequency it detects is this number divided by t, or $f' =$ _____ .

Write down the equations for the other possibilities:

> If the detector is moving with speed v_D *away from* a stationary source then the frequency it detects is $f' =$ _____ .
>
> If the detector is stationary and the source is moving with speed v_S *toward* it then the frequency it detects is $f' =$ _____ .
>
> If the detector is stationary and the source is moving *away from* it with speed v_S then the frequency it detects is $f' =$ _____ .

Eq. 18–51 in the text covers all possibilities. Write it here:

$$f' =$$

You can easily determine which signs to use in any particular situation by remembering that motion of the source toward the observer or the observer toward the source results in hearing a higher frequency while motion of the source away from the observer or the observer away from the source results in hearing a lower frequency than would be heard if both were stationary.

You should understand that the velocities in the Doppler effect equation are measured relative to the medium in which the wave is propagating (the air, for example). What counts is not the motion of the source relative to the observer but the motions of both the source and observer relative to the medium of propagation. You should also understand that the Doppler effect equations given above are valid only if the motion is along the line joining the source and detector. For motion in other directions v_D and v_S must be interpreted as components of the velocities along that line.

To obtain an understanding of the speeds involved, estimate the speed with which you would have to move toward a stationary source of sound in still air to hear a 10% increase in frequency: _____ m/s. Can this speed be achieved by walking slowly, walking fast, riding a bicycle, driving a car at a moderate speed, or driving a car at high speed?

The Doppler effect also occurs for electromagnetic radiation. Light from stars that are moving at high speeds away from the Earth is shifted toward the red and the extent of the shift is used to calculate the speeds of the stars. Doppler shifts of radar waves reflected from moving objects can be used to find their speeds. Police use the effect to detect speeders and TV technicians use it to find the speeds of thrown baseballs.

If a source of sound is moving through a medium faster than the speed of sound in the medium, a shock wave is produced. Then a wavefront has the shape of a _____, with the source at its _____, as shown in Fig. 18–21 of the text. The half angle θ of the cone is given by $\sin \theta =$ _____, where v is the speed of _____ and v_S is the speed of _____. Note that no shock wave is produced if $v < v_S$.

Shock waves are responsible for sonic booms. A listener hears a sonic boom when

_____

_____

In the space to the right draw a plane flying horizontally over level ground at an altitude h at the time when the envelope of the shock wave intersects the ground at point A. Show the envelope of the shock wave and label the half angle of the cone θ. Let d be the horizontal distance from A to the point on the ground under the plane. The relationship between θ, d, and h is $\tan \theta =$

_____ .

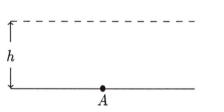

II. PROBLEM SOLVING

Since many of the problems are similar to those of the last chapter, you might want to review

Chapter 17 of this manual.

One difference is that the speed of sound in an isotropic medium is given by $\sqrt{B/\rho}$, where B is the bulk modulus and ρ is the density of the medium. If you are given the bulk modulus and density you can calculate the speed of sound. Alternatively you might use $v = \lambda\nu$ (or a related expression) to find the speed of sound, then use $v = \sqrt{B/\rho}$ to solve for B or ρ, given the other.

PROBLEM 1 (5E).   The average density of the Earth's crust 10 km beneath the continents is 2.7 g/cm^3. The speed of longitudinal seismic waves at that depth, found by timing their arrival from distant earthquakes, is 5.4 km/s. Use this information to find the bulk modulus of the Earth's crust at that depth. For comparison, the bulk modulus of steel is about 16×10^{10} Pa

STRATEGY: Solve $v = \sqrt{B/\rho}$ for B.

SOLUTION:

$$[\text{ans: } 7.9 \times 10^{10}\,\text{Pa}]$$

PROBLEM 2 (8P).   The speed of sound in a certain metal is V. One end of a long pipe of that metal of length L is struck a hard blow. A listener at the other end hears two sounds, one from the wave that has traveled along the pipe and the other from the wave that has traveled through the air. (a) If v is the speed of sound in air, what time interval t elapses between the arrival of the two sounds?

STRATEGY: The time taken by the wave that travels in air is $t_a = L/v$ and the time taken by the wave that travels along the pipe is $t_p = L/V$. Find an expression for the difference in the times: $t = t_a - t_p$.

SOLUTION:

$$[\text{ans: } L(V - v)/Vv]$$

(b) Suppose that $t = 1.00$ s and the metal is steel. Find the length L.

STRATEGY: Solve $\Delta t = L(V - v)/Vv$ for L. Use $v = 343$ m/s and $V = 5941$ m/s (see Table 18–1 of the text).

SOLUTION:

$$[\text{ans: } 364\,\text{m}]$$

Interference problems are quite similar to those of the last chapter. For complete destructive interference the two waves are out of phase by an odd multiple of π rad; for constructive interference their phases differ by zero or a multiple of 2π.

PROBLEM 3 (22P).   Two loudspeakers are located 11.0 ft apart on the stage of an auditorium. A listener is seated 60.0 ft from one and 64.0 ft from the other. A signal generator drives the two speakers in phase with the same amplitude and frequency. The frequency is swept through the audible range $(20 - 20,000\,\text{Hz})$. (*a*) What are the three lowest frequencies at which the listener will hear minimum signal because of destructive interference?

STRATEGY: The phase difference of the two waves at the listener is $\phi = k\Delta x$, where k is the angular wave number of the waves and Δx is the difference in the distance they travel from the speakers to the listener. For complete destructive interference ϕ must be an odd multiple of π or Δx must be an odd multiple of $\lambda/2$. Thus $\Delta x = n\lambda/2$ or $\lambda = 2\Delta x/n$, where n is an odd integer. The frequencies are given by $f = v/\lambda = nv/2\Delta x$. The lowest three occur when $n = 1$, 3, and 5. Use $v = 1125\,\text{ft/s}$ and $\Delta x = 4.0\,\text{ft}$.

SOLUTION:

[ans: 141 Hz, 422 Hz, 703 Hz]

(*b*) What are the three lowest frequencies at which the listener will hear maximum signal?

STRATEGY: Now the phase difference must be a multiple of 2π or Δx must be a multiple of λ. Show that the analysis leads to $f = nv/\Delta x$, where n is zero or an integer. The three lowest frequencies occur when $n = 0$, 1, and 2.

SOLUTION:

[ans: 281 Hz, 562 Hz, 844 Hz]

Sound waves can also form standing wave patterns. The requirements are the same as for waves on a string: two traveling waves with the same amplitude and frequency but traveling in opposite direction form a standing wave. Here's an example.

PROBLEM 4 (26P).   Two point sources, separated by a distance of 5.00 m, emit sound waves at the same amplitude and frequency (300 Hz), but the sources are exactly out of phase. At what points along the line between the sources do the sound waves result in maximum oscillations of the air molecules? (Hint: one such point is midway between the sources. Can you see why?)

STRATEGY: The two sources set up a standing wave pattern, with antinodes that are $\lambda/2$ apart. Let the right-traveling wave be given by $s_1 = s_m \sin(kx - \omega t)$ and the left-traveling wave be given by $s_2 = s_m \sin(kx + \omega t + \phi)$. If the sources are π rad out of phase then $s_1(0, t) = -s_2(L, t)$, where L is the distance between the sources. This means $\sin(-\omega t) = -\sin(kL - \omega t + \phi)$. Thus $kL + \phi = \pi$, or $\phi = \pi - kL$. The two waves are $s_1 = s_m \sin(kx - \omega t)$ and $s_2 = s_m \sin(kx + \omega t + \pi - kL)$. Using the trigonometric identity given in the last chapter their sum can be written $s = s_1 + s_2 = 2s_m \sin(kx + \pi/2 - kL/2) \cos(\omega t + \pi/2 - kL/2)$. Maximum displacement occurs where $kx + \pi/2 - kL/2 = n\pi/2$, where n is an odd integer. Find x for each odd integer value of n (both positive and negative) until you get values of x that are less than 0 or greater than L. Use $\lambda = v/f$, where v is the speed of sound, to find a value for λ.

SOLUTION:

[ans: 0.213, 0.785, 1.35, 1.93, 2.50, 3.07, 3.64, 4.22, 4.79 m from one end]

To solve power, intensity, and sound level problems use the equations $\overline{P} = \frac{1}{2}\rho A v \omega^2 s_m^2$ for the average power transmitted through area A, $I = \frac{1}{2}\rho v \omega^2 s_m^2$ for the intensity, and $\beta = 10\log(I/I_0)$ for the sound level. Remember that the intensity is the average power per unit area. If a source emits isotropically (the same in all directions) then energy crossing every sphere centered at the source is the same. Since the surface area of a sphere is $4\pi r^2$, the intensity a distance r from the source is given by $\overline{P}/4\pi r^2$.

PROBLEM 5 (33E).   A certain loudspeaker produces a sound with a frequency of 2000 Hz and an intensity of $0.960\,\text{mW/m}^2$ at a distance of 6.10 m. Assume that there are no refections and that the loudspeaker emits the same in all directions. (*a*) What is the intensity at 30.0 m?

STRATEGY: Assume there is no loss of energy as the spherical waves spread out. The intensity is then proportional to the inverse square of the distance from the source. Let I_1 be the intensity at $r_1 = 6.10$ m and I_2 be the intensity at $r_2 = 30$ m. Then $I_1/I_2 = r_2^2/r_1^2$. Solve for I_2.

SOLUTION:

[ans: $39.7\,\mu\text{W/m}^2$]

(*b*) What is the displacement amplitude at 6.10 m?

STRATEGY: Solve $I = \frac{1}{2}\rho v \omega^2 s_m^2$ for s_m. Use $v = 343$ m/s and $\rho = 1.21\,\text{kg/m}^3$.

SOLUTION:

[ans: 1.71×10^{-7} m]

(*c*) What is the pressure amplitude at 6.10 m?

STRATEGY: Use $\Delta p_m = v\rho\omega s_m$.

SOLUTION:

[ans: 0.893 Pa]

PROBLEM 6 (42P). A certain loudspeaker (assumed to be a point source) emits 30.0 W of sound power. A small microphone of effective cross-sectional area $0.750\,\text{cm}^2$ is located 200 m from the loudspeaker. Calculate (*a*) the sound intensity at the microphone and (*b*) the power intercepted by the microphone.

STRATEGY: The sphere centered at the loudspeaker and through the microphone has a surface area that is given by $A = 4\pi r^2$, where r is the distance from the loudspeaker to the microphone. The intensity at the microphone is $I = \overline{P}/A = \overline{P}/4\pi r^2$, where $\overline{P}$ is the average power output of the loudspeaker. The average power intercepted by the microphone is given by $\overline{P}_m = IA_m$, where A_m is the cross-sectional area of the microphone.

SOLUTION:

$$\left[\text{ans:} \; (a)\; 5.97 \times 10^{-5}\,\text{W/m}^2; \; (b)\; 4.48 \times 10^{-9}\,\text{W}\right]$$

PROBLEM 7 (43P). In a test, a subsonic jet flies overhead at an altitude of 100 m. The sound intensity on the ground as the jet passes overhead is 150 dB. At what altitude should the plane fly so that the ground noise is no greater than 120 dB, the threshold of pain? Ignore the finite time required for the sound to reach the ground.

STRATEGY: Use $\beta_2 - \beta_1 = (10\,\text{dB})\log(I_2/I_1)$. Since the intensity at the ground is inversely proportional to the square of the altitude, this becomes $\beta_2 - \beta_1 = (10\,\text{dB})\log(r_1^2/r_2^2)$. Let β_1 and r_1 refer to the lower altitude and β_2 and r_2 refer to the higher. Solve for r_2.

SOLUTION: You should obtain $r_1 = r_2 10^{-(\beta_2 - \beta_1)/20}$, for β_1 and β_2 in dB.

$$\left[\text{ans:} \; 3.16\,\text{km}\right]$$

Many problems deal with the production of sound by pipes with either both ends closed or one end open and one end closed. Remember that a displacement node and a pressure antinode occur at a closed end while a displacement antinode and a pressure node occur at an open end. Also remember that nodes are half a wavelength apart and antinodes occur halfway between adjacent nodes.

PROBLEM 8 (49E). In Fig. 18–29, a rod R is clamped at its center; a disk D at its end projects into a glass tube that has cork filings spread over its interior. A plunger P is provided at the other end of the tube. The rod is made to oscillate longitudinally at frequency f to produce sound waves inside the tube, and the location of the plunger is adjusted until a standing sound wave pattern is set up inside the tube. Once the standing wave is set up, the cork filings collect in a pattern of ridges at the displacement nodes. Show that if d is the average distance between the ridges, the speed of sound v in the gas within the tube is given by

$$v = 2fd.$$

This is *Kundt's method* for determining the speed of sound in various gases.

STRATEGY: The distance between nodes in a standing wave pattern is half the wavelength of either of the traveling waves that sum to form the standing wave. Thus $\lambda = 2d$. Use $v = \lambda f$ to find an expression for the speed of sound.

SOLUTION:

PROBLEM 9 (61P). A tube 1.20 m long is closed at one end. A stretched wire is placed near the open end. The wire is 0.330 m long and has a mass of 9.60 g. It is fixed at both ends and vibrates in its fundamental mode. It sets the air column in the tube into oscillation at its fundamental frequency by resonance. Find (a) the frequency of oscillation of the air column and (b) the tension in the wire.

STRATEGY: (a) Since one end of the tube is open and the other end is closed the tube length L_T is related to the wavelength λ by $L_T = n\lambda/4$, where n is an odd integer. Since the air column in the tube is vibrating in its fundamental mode, $n = 1$ and $\lambda = 4L_T$. The frequency is given by $f = v/\lambda = v/4L_T$, where v is the speed of sound in air (343 m/s). (b) The speed of waves on the wire is given by $v = \sqrt{\tau/\mu}$, where τ is the tension and μ is the linear density. It is also given by $v = \lambda f$, where λ is the wavelength and f is the frequency of a traveling sinusoidal wave on the wire. Thus $\lambda f = \sqrt{\tau/\mu}$ or $\tau = \lambda^2 f^2 \mu$. The given mass and length of the wire can be used to calculate μ. Use the standing wave condition to calculate λ. Since both ends of the wire are fixed the wavelength λ on the wire is related to the wire length L_W by $L_W = n\lambda/2$, where n is an integer. Since the wire is vibrating in its fundamental mode, $n = 1$ and $\lambda = 2L_W$.

SOLUTION:

[ans: (a) 71.5 Hz; (b) 64.8 N]

Some problems deal with the production of beats by two sinusoidal sound waves with nearly the same frequency. You may be given the frequency f_1 of one of the waves and the beat frequency f_{beat}, then asked for the frequency f_2 of the other wave. Since $f_{beat} = |f_1 - f_2|$, it is given by $f_2 = f_1 \pm f_{beat}$. You require more information to determine which sign to use in this equation. One way to give this information is to tell you what happens to the beat frequency if f_1 is increased (or decreased). If the beat frequency increases when f_1 increases, then f_1 must be greater than f_2 and $f_2 = f_1 - f_{beat}$. If the beat frequency decreases then f_1 must be less than f_2 and $f_2 = f_1 + f_{beat}$.

PROBLEM 10 (65E). The A string of a violin is a little too taut. Four beats per second are heard when the string is sounded together with a tuning fork that is oscillating accurately at concert A (440 Hz). What is the period of the violin string oscillation?

STRATEGY: Since there are four beats per second, the violin string is oscillating at either 444 Hz or at 436 Hz. Since the string is a little too taut the wave speed is a little too high. This means that either the wavelength or the frequency or both are a little too high. Since the wavelength is determined by the distance between the fixed ends and is not affected by the tension in the string, it must be that the frequency is too high. It must be 444 Hz. To find the period calculate the reciprocal of the frequency.

SOLUTION:

[ans: 2.25 ms]

PROBLEM 11 (66E). You given four tuning forks. The fork with the lowest frequency oscillates at 500 Hz. By striking two tuning forks at a time, the following beat frequencies are heard: 1, 2, 3, 5, 7, and 8 Hz. What are the possible frequencies of the other three tuning forks?

STRATEGY: Try to find 4 numbers, beginning with 500 and increasing, such that their differences are 1, 2, 3, 5, 7, and 8. 508 must be one of them since no beat frequency higher that 8 Hz occurs in the list. 507 might be one of them. Then beat frequencies of 1 and 7 Hz will occur. The addition of one other frequency to the list will produce the remaining beat frequencies of 2, 3, and 5 Hz.

SOLUTION:

[ans: 505, 507, and 508 Hz]

There is another solution. Try a list with 501 and 508 Hz in it.

SOLUTION:

[ans: 501, 503, 508 Hz]

Nearly all Doppler shift problems can be solved using

$$f' = f \left[\frac{v \pm v_D}{v \mp v_S} \right] ,$$

where v is speed of sound, v_S is the speed of the source, v_D is the speed of the observer, f is the frequency of the source, and f' is the frequency detected by the observer. The upper sign in the numerator refers to a situation in which the observer is moving toward the source; the lower sign to a situation in which the observer is moving away from the source. The upper sign in the denominator refers to a situation in which the source is moving toward the observer; the lower sign to a situation in which the source is moving away from the observer. Remember that all speeds are measured relative to the medium in which the sound is propagating. You might be given the velocities and one of the frequencies, then asked for the other frequency. In other situations you might be given the two frequencies and one of the velocities, then asked for the other velocity. In all cases, simple algebraic manipulation of the equation will produce the desired expression.

PROBLEM 12 (69E). A source S generates circular waves on the surface of a lake; the pattern of wave crests is shown in Fig. 18–32. The speed of the waves is 5.5 m/s, and the crest-to-crest separation is 2.3 m. You are in a small boat heading directly toward S at a constant speed of 3.3 m/s with respect to the shore. What frequency of the waves do you observe?

STRATEGY: You see the crests coming at you with a speed of $v + v_b$, where v is the wave speed and v_b is the speed of the boat, both relative to the water. If the crests are a distance Δx apart the time interval between them, from your point of view, is $\Delta t = \Delta x/(v + v_b)$ and the frequency is the reciprocal of this.

SOLUTION:

[ans: 3.8 Hz]

PROBLEM 13 (75E).   In 1845, Buys Ballot first tested the Doppler effect for sound. He put a trumpet player on a flatcar drawn by a locomotive and another player near the tracks. If each player blows a 440-Hz note, and if there are 4.0 beats/s as they approach each other, what is the speed of the flatcar?

STRATEGY: Think of the two waves joining to produce beats on the side of the stationary trumpeter opposite the flatcar. The note produced by the stationary trumpeter has a frequency of 440 Hz. The note produced by the moving trumpeter is higher and so must have a frequency of 444 Hz. If f is the frequency played by that trumpeter and f' is the frequency heard then $f' = fv/(v - v_S)$, where v is the speed of sound in air (343 m/s) and v_S is the speed of the flatcar. Solve for v_S.

SOLUTION:

[ans: 3.1 m/s]

PROBLEM 14 (79P).   Two identical tuning forks can oscillate at 440 Hz. A person is located somewhere on the line between them. Calculate the beat frequency as measured by this individual if (a) she is standing still and the tuning forks both move to the right, say, at 30.0 m/s, and (b) the tuning forks are stationary and the listener moves to the right at 30.0 m/s.

STRATEGY: (a) Tuning fork #1 is moving toward the listener, so she hears a frequency that is given by $f_1' = fv/(v - v_S)$, where f is the frequency of the turning fork, v is the speed of sound in air (343 m/s), and v_S is the speed of the tuning fork. The second tuning fork is moving away from the listener so she hears a frequency that is given by $f_2' = fv/(v + v_S)$. Calculate $f_1' - f_2'$. (b) The listener is moving away from tuning fork #1 so she hears a frequency that is given by $f_1' = f(v - v_D)/v$, where v_D is her speed. She is moving toward tuning fork #2 so she hears a frequency given by $f_2' = f(v + v_D)/v$. Calculate $f_2' - f_1'$.

SOLUTION:

[ans: (a) 77.6 Hz; (b) 77.0 Hz]

PROBLEM 15 (85P).   A siren emitting a sound of frequency 1000 Hz moves away from you toward the face of a cliff at a speed of 10 m/s. Take the speed of sound in air as 330 m/s. (a) What is the frequency of the sound you hear coming directly from the siren?

STRATEGY: You are stationary and the siren is moving away from you so the frequency you hear is given by $f' = fv/(v + v_S)$, where f is the emitted frequency, v is the speed of sound in air, and v_S is the speed of the siren. Notice that the frequency received is lower than the frequency of the siren, consistent with a source moving *away*.

SOLUTION:

[ans: 970 Hz]

(*b*) What is the frequency of the sound you hear reflected off the cliff?

STRATEGY: The siren is moving toward the cliff face, so the frequency received by the cliff is given by $f' = fv/(v - v_S)$. This is the same as the frequency of the reflected wave.

SOLUTION:

$$[\text{ans: } 1030\,\text{Hz}]$$

(*c*) What is the beat frequency between the 2 sounds? Is it perceptible (it must be less than 20 Hz)?

STRATEGY: The beat frequency is the difference of the two frequencies.

SOLUTION:

$$[\text{ans: } f_{\text{beat}} = 60\,\text{Hz; too high for the beats to be perceived}]$$

PROBLEM 16 (93P). Two trains are traveling toward each other at 100 ft/s relative to the ground. One train is blowing a whistle at 500 Hz. (*a*) What frequency will be heard on the other train in still air?

STRATEGY: Since the source is moving toward the listener and the listener is moving toward the source in still air, the frequency heard is given by $f' = f(v + v_D)/(v - v_S)$, where v_S is the speed of the train with the whistle, v_D is the speed of the other train, and v is the speed of sound in air (1125 ft/s).

SOLUTION:

$$[\text{ans: } 598\,\text{Hz}]$$

(*b*) What frequency will be heard on the other train if the wind is blowing at 100 ft/s toward the whistle and away from the listener?

STRATEGY: Use the same equation but take the train speeds to be those relative to the air. That is, the speed of the train with the whistle is $v_S = 100 + 100 = 200$ ft/s and the speed of the other train is $v_D = 0$.

SOLUTION:

$$[\text{ans: } 608\,\text{Hz}]$$

(*c*) What frequency will be heard if the wind direction is reversed?

STRATEGY: Now $v_S = 0$ and $v_D = 200$ ft/s.

SOLUTION:

$$[\text{ans: } 589\,\text{Hz}]$$

Solutions to most shock wave problems start with $\sin\theta = v/v_S$, where θ is the half angle of the shock envelope, v is the speed of sound, and v_S is the speed of the source. For an airplane flying horizontally at altitude h you may also need to make use of $\tan\theta = h/d$, where d is the horizontal distance from the intersection of the shock wave with the ground to the point on the ground under the plane. Some problems give the time interval t from when the plane is overhead to when the sonic boom is heard. Use $d = v_S t$.

PROBLEM 17 (80P).   A plane flies with 5/4 the speed of sound. The sonic boom reaches a man on the ground exactly 1 min after the plane passed directly overhead. What is the altitude of the plane? Assume the speed of sound to be 330 m/s.

STRATEGY: The Mach cone angle is given by $\sin\theta = v/v_S$, where v is the speed of sound and v_S is the speed of the plane. The altitude is given by $h = v_S t \tan\theta$. Find the value of θ, then calculate h.

SOLUTION:

[ans: 33.0 km]

III. NOTES

OVERVIEW II THERMODYNAMICS

Thermodynamic phenomena are familiar to everyone. During a long trip the tires of your car get hot and the air pressure in them increases; the car engine and coolant also warm but the fan blows cooler outside air over them and keeps them at appropriate operating temperatures; perhaps the car has an air conditioner to cool the air inside and you notice that you use fuel at a slightly greater rate when it is on; the brake shoe and disks become hot when you brake, but after the car has stopped for sufficient time they have the same temperature as the surrounding air.

Thermodynamics deals chiefly with the transfer of energy between large systems of particles and the changes that thereby occur in the systems. Simple gases are used to illustrate thermodynamic phenomena and, for them, the quantities considered are the volume, pressure, and temperature. The list is longer for other systems. Chapter 19 deals with the concept of temperature and when you study that chapter you will learn its precise definition and how it is measured.

An energy transfer can change both the internal energy of the system and the kinetic energy associated with the motion of the system as a whole. As you may recall from your study of Chapter 8, the internal energy is the sum of the kinetic energies of the particles, as measured in a reference frame that moves with the center of mass, and the potential energies of their interactions with each other. A kinetic energy is associated with the motion of the system as a whole but for simplic-ity we treat situations in which the center of mass remains at rest and only the internal energy changes. You will study ideal gases, for which the internal energy is simply the sum of the kinetic energies of the particles, including their rotational kinetic energies. These gases have no internal potential energy.

The study of energy transfers themselves form an important part of thermodynamics, apart from the changes they bring about in a system. You will learn that energy can be transferred to or from a system in two distinct ways, through work done on or by the system and through heat exchanged between the system and its environment. You are already familiar with work; heat is probably new to you. For processes involving a system with its center of mass at rest, work involves a change in volume. Heat, on the other hand, is energy that is transferred because the system and its environment are at different temperatures and does not involve a change in volume.

Work done by a system tends to decrease the internal energy and temperature of the system. Heat entering a system tends to increase those quantities. In fact, the principle of energy conservation, modified for a system with its center of mass at rest, tells us that the difference between the heat entering a system and the work done by the system is equal to the change in internal energy of the system. This principle, known as the first law of thermodynamics, is the central theme of Chapter 20. You will use it to calculate any of the three quantities, work, heat, and change in internal energy, in terms of the other two.

The pressure, volume, and temperature of a gas are related to each other: a change in one brings about changes in one or both of the others. The relationship is different for different systems but such a relationship, called an equation of state, exists for every system. Every equation of state contains at least three variables and so does not by itself determine the changes that accompany an exchange of energy. More information about the process must be known. You will study, for example, processes in which the volume, pressure, or temperature are held constant. Chapter 21 deals with the equation of state for an ideal gas.

To complete the study of changes in a system brought about by the transfer of energy you must know how the internal energy depends on the other thermodynamic variables. The internal energy of an ideal gas depends only on the temperature and changes in these quantities are directly proportional to each other. This important result is obtained in Chapter 21.

Thermodynamics, although it deals with macroscopic systems, is firmly based on the mechanics of particles. In Chapter 21 you will learn how the pressure of a gas is related to the speeds of the particles. You will also learn how the total kinetic energy of the particles is related to the temperature.

There are certain natural limitations on thermodynamic processes. For example, if no work is done on a system then heat flows out of the system if its temperature is higher than that of its environment and flows into the system if its temperature is lower. Only if work is done can heat flow be from the colder to the hotter body. Another limitation concerns processes in which heat enters a system and the system does work. No process can convert all the heat into work; at some point in the process heat must flow out of the system or else the internal energy must change. These limitations and others are codified in what is known as the second law of thermodynamics.

The limitations described by the second law have important practical consequences for the design of refrigerators, air conditioners, engines, electrical power generators, and many other devices. More importantly for our purposes, they also give us a glimpse at the way nature operates.

Chapter 22 is devoted to the second law of thermodynamics. You will learn several equivalent statements of the law and how to apply them. At the end of the chapter you will study another important thermodynamic property of systems, called the entropy. You will learn to calculate changes in entropy for various processes and will learn an important fact about it: the total entropy of a system and its environment never decreases. This is a direct consequence of the second law and is important for understanding many thermodynamic processes.

Chapter 19
TEMPERATURE

I. BASIC CONCEPTS

You now begin the study of thermodynamics. Temperature, the most important concept introduced in this chapter, is familiar to you but you have probably not thought about it in detail. Be sure you understand its definition and how it is measured. Pay particular attention to the zeroth law of thermodynamics, which codifies the characteristic of nature that allows temperature to be defined. You will also study the phenomenon of thermal expansion, the familiar change in the dimensions of an object that occurs when its temperature is changed.

Temperature. Temperature is intimately related to the idea of <u>thermal equilibrium</u>. If two objects that are not in thermal equilibrium with each other are allowed to exchange energy they will do so and, as a result, some or all of their macroscopic properties will change. When the properties stop changing there is no longer a net flow of energy and the objects are said to be in thermal equilibrium with each other. They are then at the same temperature.

Temperature is the property of an object that _____

_____

Note carefully that, when two systems are in thermal equilibrium with each other, all macroscopic quantities except temperature might have different values for the two systems. Suppose the systems are gases. When in thermal equilibrium with each other their pressures may be different, their volumes may be different, their particle numbers may be different, and their internal energies may be different. But their temperatures are the same.

The law of nature that legitimizes the temperature as a property of a body in thermal equilibrium with another is the <u>zeroth</u> <u>law</u> <u>of</u> <u>thermodynamics</u>. State the law here:

_____

_____

Suppose the law were not valid and suppose further that body A and body B are in thermal equilibrium and therefore at the same temperature. If body C is in equilibrium with A but not with B, then the temperature of C is not a well defined quantity and cannot be considered a property of C.

Another important consequence of the zeroth law is that it allows us to select some object for use as a thermometer. Suppose that when the thermometer is in thermal equilibrium with body A it is also in thermal equilibrium with body B. Then, because the law is valid, we know that A and B have the same temperature. We do not need to place A and B in thermal contact to test if they are equilibrium with each other.

Measurement of temperature. In general, temperature is measured by measuring some property of a system. You are familiar with an ordinary mercury thermometer, for which the _____ of the mercury column is measured. The text describes a constant-volume gas thermometer, for which the _____ of the gas in a bulb is measured. The gas must always have the same volume, no matter what its temperature. Look at Fig. 19–5 and tell how the gas is kept at the same volume. At high temperatures the gas in the bulb expands and pushes the top of the left-hand mercury column downward. The reservoir is then _____ to bring the top of the column back to _____. At low temperatures the gas contracts and the top of the left-hand column moves closer to the bulb. The reservoir is then _____ to bring the top of the column back to _____.

The temperature T in kelvins is taken to proportional to the pressure: $T = Cp$, where the constant of proportionality is chosen so $T =$ _____ at the _____ point of water. Thus $T = T_3(p/p_3)$, where T_3 is _____ K and p_3 is _____. Actually the ratio p/p_3 must be evaluated in the limit as the _____ of gas in the bulb approaches _____. In this limit the measured value of T does not depend on the type of gas used.

Three types of temperature scales are in common use: Celsius, Fahrenheit, and Kelvin (or absolute). A constant-volume gas thermometer gives the temperature on the Kelvin scale. The Celsius temperature T_C is defined in terms of the Kelvin-scale temperature T by $T_C =$_____. The Fahrenheit temperature T_F is given in terms of the Celsius temperature by $T_F =$ _____. A degree Celsius is the same as a kelvin and is _____ as large as a degree Fahrenheit.

Other properties of materials, such as length and electrical resistance, are often used to measure temperature. If X is such a property, then T is taken to be proportional to X and the constant of proportionality is chosen so $T = 273.16°$ at the triple point of water. If the value of the property is X_3 at the triple point and the value is X at temperature T, then $T = 273.16(T/T_3)$. This is not the Kelvin scale but rather it is a scale defined by the property X. To convert to the Kelvin scale the thermometer must be calibrated against a constant-volume gas thermometer or another carefully calibrated thermometer.

Thermal expansion. Most solids and liquids expand when the temperature is increased and contract when it is decreased. If the temperature is changed by ΔT, a rod originally of length L changes length by $\Delta L =$ _____, where α is the <u>coefficient</u> of <u>linear</u> <u>expansion</u>. Over small temperature ranges α may be considered to be independent of temperature. Be careful. For a situation in which the temperature is raised by ΔT, then lowered by the same amount, repeated application of the equation predicts the rod is shorter than its original length, whereas it actually returns to its original length.

Consult Table 19–3 of the text for values of some coefficients of linear expansion. You may need them to solve some problems. Carefully note that when values of α from the table are used ΔT must be in Celsius degrees or kelvins.

When the temperature changes, the length of every line in an isotropic material changes by the same *fractional* amount, given by $\Delta L/L = \alpha \Delta T$. The length of a scratch on the surface of an isotropic solid also changes by the same fractional amount. The fractional change in the diameter of a round hole in an object, for example, is given by $\Delta D/D =$ _____, where α is the coefficient of linear expansion of the object.

When the temperature changes by a small amount ΔT the area A of a face of a solid changes by $\Delta A = $ _____ and the volume V of the solid changes by $\Delta V = $ _____, where α is the coefficient of linear expansion. For a fluid the change in volume is given by the same expression. In the space below prove the first result for yourself. Consider a rectangular face with sides L_1 and L_2. Its area before a temperature change is $A = L_1 L_2$ and its area after the change is $A + \Delta A = (L_1 + \Delta L_1)(L_2 + \Delta L_2)$. Substitute $\Delta L_1 = \alpha L_1 \Delta T$ and $\Delta L_2 = \alpha L_2 \Delta T$ for the changes in length and retain only terms that are proportional to ΔT when you multiply the two factors in parentheses:

You should have obtained $\Delta A = 2\alpha A \Delta T$. The expression for the change in volume can be obtained in a similar manner.

Water near 4°C and a few other substances decrease in volume when the temperature is increased. These materials have negative coefficients of thermal expansion in the temperature range for which such a contraction occurs.

Qualitatively explain the microscopic basis of thermal expansion. As the temperature is raised the atoms of the material have greater vibrational energy and vibrate with larger

_____ .

In the space to the right draw a diagram showing the potential energy of interaction between two atoms as a function of their separation and show two possible energies. Mark the turning points and the average separation for each. For a given energy the average separation is halfway between the turning points. If you have drawn the curve correctly the average separation will be greater for greater vibrational energy than for smaller.

II. PROBLEM SOLVING

Many of the problems of this chapter deal with the definition of temperature. You will need to manipulate the relationship $T_1/T_2 = X_1/X_2$, where X_1 is the value of a thermodynamic property at temperature T_1 and X_2 is its value at temperature T_2. You must know three of the quantities that appear in the equation or else the ratio of two of them and the value of a third, then you can solve for the fourth. In some cases one of the temperatures is the triple point of water, 273.16° on the temperature scale defined by the thermodynamic property X. If the property is the limiting value of the pressure in a constant-volume gas thermometer then the temperature is in kelvins.

PROBLEM 1 (4E).   If the temperature of a gas at the steam point is 373.15 K, what is the limiting value of the ratio of the pressure of the gas at the steam point to its pressure at the triple point of water? (Assume the volume of the gas is the same at both temperatures.)

STRATEGY: Use $T = T_3(p/p_3)$, where T is the temperature at the steam point and T_3 is the temperature at the triple point of water (273.16 K). Solve for p/p_3.

SOLUTION:

[ans: 1.366]

PROBLEM 2 (5E).   A *resistance thermometer* is a thermometer whose electrical resistance changes with temperature. We are free to define temperatures measured by such a thermometer in kelvins (K) to be directly proportional to the resistance R, measured in ohms (Ω). A certain resistance thermometer is found to have a resistance R of 90.35 Ω when its bulb is placed in water at the triple-point temperature (273.16 K). What temperature is indicated by the thermometer if the bulb is placed in an environment such that its resistance is 96.28 Ω?

STRATEGY: Use $T = T_3(R/R_3)$, where R is the resistance at temperature T and R_3 is the resistance at the triple point.

SOLUTION:

[ans: 291.1°]

Strictly speaking, the value of the temperature you found in this problem is not on the Kelvin scale but is on the scale defined by the resistor. To convert to the Kelvin scale, for example, you need to know the resistance as a function of the absolute temperature.

PROBLEM 3 (7P).   A particular gas thermometer is constructed of two gas-containing bulbs, each of which is put into a water bath, as shown in Fig. 19–14. The pressure difference between the two bulbs is measured by a mercury manometer as shown. Appropriate reservoirs, not shown in the diagram, maintain constant gas volume in the two bulbs. There is no difference in pressure when both baths are at the triple point of water. The pressure difference is 120 mm Hg when one bath is at the triple point and the other is at the boiling point of water. Finally, the pressure difference is 90.0 mm Hg when one bath is at the triple point and the other is at an unknown temperature to be measured. What is the unknown temperature?

STRATEGY: Suppose the right hand bulb is held at the triple point of water. Let p_{3r} be its pressure. Let p_l be the pressure in the left bulb when the temperature is T and let p_{3l} be the pressure at the triple point. When both bulbs are at the triple point their pressures are the same, so $p_{3r} = p_{3l}$. Now suppose the left hand bulb is at the boiling point of water. This means $373.15 = 273.16(p_l/p_{3l})$. Substitute $p_l = p_{3l} + 120$ mm to obtain $373.15 = 273.16[(p_{3l} + 120)/p_{3l}]$. This can be solved for p_{3l}. Now suppose the left hand bulb is at the unknown temperature. Then $p_l = p_{3l} + 90.0$ mm. Evaluate $T = 273.16(p_l/p_{3l})$.

SOLUTION:

[ans: 348 K]

PROBLEM 4 (8P). A *thermistor* is a semiconductor device with a temperature-dependent electrical resistance. It is commonly used in medical thermometers and to sense overheating in electronic equipment. Over a limited range of temperature, the resistance is given by

$$R = R_a e^{B(1/T - 1/T_a)},$$

where R is the resistance of the thermistor at temperature T and R_a is the resistance at temperature T_a; B is a constant that depends on the particular semiconductor used. For one type of thermistor, $B = 4689\,\text{K}$ and the resistance at 273 K is $1.00 \times 10^4\,\Omega$ (ohms). What temperature is the thermistor measuring when its resistance is $100\,\Omega$?

STRATEGY: Take $T_a = 273\,\text{K}$, $R_a = 1.00 \times 10^4\,\Omega$, $R = 100\,\Omega$. Solve for T.

SOLUTION: Divide by R_a and take the natural logarithm of both sides to obtain $\ln(R/R_a) = B(1/T - 1/T_a)$. Now solve for $1/T$. You should obtain $1/T = (1/B)\ln(R/R_a) + (1/T_a)$. Finally, solve for T. The result is $T = T_a B/[T_a \ln(R/R_a) + B]$. Evaluate this expression.

[ans: 373 K]

You should know how to convert from one common temperature scale to another. The relevant equations are $T_C = T - 273.15$ (Kelvin to Celsius) and $T_F = (9/5)T_C + 32°$ (Celsius to Fahrenheit). The most straightforward problems give the temperature on one scale and ask for the temperature on another.

PROBLEM 5 (12E). (*a*) In 1964, the temperature in the Siberian village of Oymyakon reached a value of $-71°\,\text{C}$. What temperature is this on the Fahrenheit scale?

STRATEGY: Evaluate $T_F = (9/5)T_C + 32°$.

SOLUTION:

[ans: $-96°\,\text{F}$]

(*b*) The highest officially recorded temperature in the continental United States was $134°\,\text{F}$ in Death Valley, California. What is this temperature on the Celsius scale?

STRATEGY: Solve $T_F = (9/5)T_C + 32°$ for T_C.

SOLUTION:

[ans: $56.7°\,\text{C}$]

PROBLEM 6 (15P).   Suppose that on a temperature scale X, water boils at $-53.5°$ X and freezes at $-170°$ X. What would a temperature of 340 K be on the X scale?

STRATEGY: Assume a linear relationship between the Kelvin and X scales. Then $T_X = A + BT$, where T_X is the temperature on the X scale, T is the temperature on the Kelvin scale, and A and B are constants. Since water boils at 373.15 K and freezes at 273.15 K, $-53.5 = A + 373.15B$ and $-170 = A + 273.15B$. Solve for A and B, then use $T_X = A + BT$, with $T = 340$ K, to find T_X for the last temperature.

SOLUTION:

$$[\text{ans: } -92.1°\text{ X}]$$

Another set of problems is concerned with thermal expansion. The important relationship is $\Delta L = \alpha L\,\Delta T$. You might be given the original length L, the coefficient of linear expansion α, and the temperature change ΔT, then asked for the change in length ΔL or else the new length $L + \Delta L$. In other problems you might be asked for the temperature change required to achieve a given change in length or a given final length.

PROBLEM 7 (23E).   A circular hole in an aluminum plate is 2.725 cm in diameter at $0.000°$ C. What is its diameter when the temperature of the plate is raised to $100.0°$ C?

STRATEGY: The change in the diameter of the hole is given by $\Delta D = D_0\alpha\,\Delta T$, where D_0 is the original diameter, ΔT is the change in temperature, and α is the coefficient of linear expansion for aluminum ($23 \times 10^{-6}\,\text{K}^{-1}$). The new diameter is $D = D_0 + \Delta D$.

SOLUTION:

$$[\text{ans: } 2.731\,\text{cm}]$$

Sometimes the lengths for two different temperatures are given and you must calculate the coefficient, assumed to be independent of temperature. The fractional change $\Delta L/L$ might be given instead of the initial and final lengths.

PROBLEM 8 (24E).   An aluminum-alloy rod has a length of 10.000 cm at $20.000°$ C and a length of 10.015 cm at the boiling point of water. (*a*) What is the length of the rod at the freezing point of water?

STRATEGY: First find the coefficient of linear expansion. The change in temperature from $20.000°$ C to the boiling point of water is $\Delta T = 100.000 - 20.000 = 80.000°$ C and the change in length is $\Delta L = 10.015 - 10.000 = 0.015$ cm. Solve $\Delta L = L\alpha\,\Delta T$ for α. You should get $1.875 \times 10^{-5}\,\text{K}^{-1}$. The change in temperature from $20.000°$ C to the freezing point of water is $-20.000°$ C. Use $\Delta L = L\alpha\,\Delta T$ to find the change in length and $L + \Delta L$ to find the new length.

SOLUTION:

$$[\text{ans: } 9.996\,\text{cm}]$$

(*b*) What is the temperature if the length of the rod is 10.009 cm?

STRATEGY: The change in length from its length at 20.000° C is $\Delta L = +0.009$ cm. Solve $\Delta L = L\alpha\,\Delta T$, with $L = 10.000$ cm, for ΔT, then use $T + \Delta T$, with $T = 20.000°$ C, to calculate the new temperature.

SOLUTION:

$$[\text{ans: } 68°\text{ C}]$$

In some cases you are asked to compare the expansions of two objects.

PROBLEM 9 (27E).   A rod is measured to be exactly 20.05 cm long using a steel ruler at a room temperature of 20° C. Both the rod and the ruler are placed in an oven at 270° C, where the rod now measures 20.11 cm using the same ruler. What is the coefficient of thermal expansion for the material of which the rod is made?

STRATEGY: The change in the length of the rod is given by $\Delta L = L_0\alpha\,\Delta T$, where L_0 is its original length and α is its coefficient of linear expansion. The new length is $L = L_0 + \Delta L$. This is the value that would be measured by the ruler if the ruler had not also expanded. Because the ruler expands the reading is less by the factor $1 - \alpha_s\,\Delta T$, where α_s is the coefficient of linear expansion for steel. Thus the reading on the expanded ruler is $L' = L(1 - \alpha_s\,\Delta T) = (L_0 + L_0\alpha\,\Delta T)(1 - \alpha_s\,\Delta T) \approx L_0 + L_0(\alpha - \alpha_s)\,\Delta T$. Solve for α.

SOLUTION:

$$[\text{ans: } 23 \times 10^{-6}\,\text{K}^{-1}]$$

Some problems deal with areas and volumes. These are essentially the same as the problems dealing with linear dimensions but the coefficient of expansion is different. If α is the coefficient of linear expansion then 2α is the coefficient of area expansion and 3α is the coefficient of volume expansion. The fractional change in area is twice the fractional change in a linear dimension and the fractional change in volume is three times the fractional change in a linear dimension.

PROBLEM 10 (30E).   A brass cube has an edge of 30 cm. What is the increase in its surface area when it is heated from 20 to 75° C?

STRATEGY: The length of an edge of the cube increases by $\Delta L = L\alpha\,\Delta T$, where α is the coefficient of linear expansion. The area of a face increases by $\Delta A = (L + \Delta L)^2 - L^2 \approx 2L\,\Delta L = 2\alpha L^2\,\Delta T$. A cube has 6 faces so the increase in the total surface area is $12\alpha L^2\,\Delta T$.

SOLUTION:

$$[\text{ans: } 11\,\text{cm}^2]$$

PROBLEM 11 (42P).   When the temperature of a copper penny is raised by 100° C, its diameter increases by 0.18%. To two significant figures, give the percent increase in (a) the area of a face, (b) the thickness, (c) the volume, and (d) the mass of the penny.

STRATEGY: (a) The new diameter of the penny is given by $D = D_0(1 + 0.0018)$. The increase in the area is given by $\Delta A = \pi(D^2 - D_0^2)/4$. To two significant figures this is $2 \times 0.0018\pi D_0^2/4$ and the fractional change is $\Delta A/A = 2 \times \pi D_0^2/\pi D_0^2 = 2 \times 0.0018$. (b) The fractional change in the thickness is the same as the fractional change in the diameter. (c) The change in the volume is $\Delta V = (A + \Delta A)(L + \Delta L) - AL$, where L is the thickness. To two significant figures this is $\Delta V = 3 \times 0.0018AL$. (d) The mass of the penny does not depend on the temperature.

SOLUTION:

$$\left[\text{ans: } (a)\ 0.36\%;\ (b)\ 0.18\%;\ (c)\ 0.54\%;\ (d)\ 0\right]$$

(e) Calculate its coefficient of linear expansion.

STRATEGY: The fractional change in a linear dimension like the diameter is given by $\Delta D/D = \alpha\,\Delta T$, where α is the coefficient of linear expansion. Solve for α.

SOLUTION:

$$\left[\text{ans: } 1.8 \times 10^{-5}\,\text{K}^{-1}\right]$$

PROBLEM 12 (46P).   (a) Show that if the lengths of two rods of different solids are inversely proportional to their respective coefficients of linear expansion at the same initial temperature, the difference in length between them will be constant at all temperatures.

STRATEGY: After the temperature is changed by ΔT, the length of the first rod is given by $L_1 = L_{10} + L_{10}\alpha_1\,\Delta T$ and the length of the second is given by $L_2 = L_{20} + L_{20}\alpha_2\,\Delta T$. Substitute $\alpha_1 = B/L_{10}$ and $\alpha_2 = B/L_{20}$, where B is a constant, then calculate the difference in length $L_2 - L_1$. The result should be independent of ΔT.

SOLUTION:

(b) What should be the lengths of a steel and a brass rod at 0.00° C so that at all temperatures their difference in length is 0.30 m?

STRATEGY: You want $L_s = B/\alpha_s$, $L_b = B/\alpha_b$ and $L_s - L_b = 0.30$ m, where the subscript s refers to steel and the subscript b refers to brass. Solve these two equations simultaneously for L_s and L_b. The coefficients α_s and α_b can be found in Table 19–3.

SOLUTION:

[ans: 0.41 m and 0.71 m]

Here are some problems that deal with two objects in contact. Each individually obeys the law of thermal expansion and the sum of the two expansions is the total expansion of the composite system.

PROBLEM 13 (50P). A composite bar of length $L = L_1 + L_2$ is made from a bar of material 1 and length L_1 attached to a bar of material 2 and length L_2, as shown in Fig. 19–18. (*a*) Show that the effective coefficient of linear expansion α for this bar is given by $\alpha = (\alpha_1 L_1 + \alpha_2 L_2)/L$.

STRATEGY: The effective coefficient of linear expansion can be found from $\Delta L = L\alpha\,\Delta T$, where ΔL is the change in the length of the composite. ΔL is the sum of the expansions of the two parts of the composite rod: $\Delta L = \Delta L_1 + \Delta L_2 = L_1\alpha_1\,\Delta T + L_2\alpha_2\,\Delta T$. Equate these two expressions for ΔL and solve for α.

SOLUTION:

(*b*) Using steel and brass, design such a composite bar whose length is 52.4 cm and whose effective coefficient of linear expansion is $13.0 \times 10^{-6}/°\mathrm{C}$.

STRATEGY: Solve the two equations $(\alpha_s L_s + \alpha_b L_b)/L$ and $L_s + L_b = L$ for L_s and L_b.

SOLUTION:

[ans: $L_b = 13.1\,\mathrm{cm}$, $L_s = 39.3\,\mathrm{cm}$]

PROBLEM 14 (53P). Two rods of different materials but having the same lengths L and cross-sectional areas A are arranged end-to-end between fixed, rigid supports, as shown in Fig. 19–21a. The temperature is T and there is no initial stress. The rods are heated, so that their temperature increases by ΔT. (*a*) Show that the rod interface is displaced upon heating by an amount

$$\Delta L = \left(\frac{\alpha_1 E_1 - \alpha_2 E_2}{E_1 + E_2}\right) L\,\Delta T$$

(see Fig. 19-20*b*, where α_1, α_2 are the coefficients of linear expansion and E_1, E_2 are Young's moduli of the materials. Ignore changes in cross-sectional areas. (*b*) Find the stress at the interface after heating.

STRATEGY: Each rod might change length for two reasons: it tends to expand thermally and it tends to contract because the other rod pushes on it. The net change in length is the sum of the two. For the first rod $\Delta L_1 = L\alpha_1 \Delta T - LF/AE_1$, where F is the magnitude of the force of rod 2 on rod 1. Similarly, for the second rod $\Delta L_2 = L\alpha_2 \Delta T - LF/AE_2$. The magnitude of the force is, of course, the same. Since the rods are between rigid supports the total length does not change: $\Delta L_1 + \Delta L_2 = 0$. Solve these 3 equations for ΔL_1, ΔL_2, and F. The interface moves a distance equal to the magnitude of either ΔL_1 or ΔL_2. The stress is F/A.

SOLUTION:

$$[\text{ans: } (b)\ [(\alpha_1 + \alpha_2)E_1E_2/(E_1 + E_2)]\Delta T\,]$$

III. NOTES

Chapter 20
HEAT AND THE FIRST LAW
OF THERMODYNAMICS

I. BASIC CONCEPTS

The first law of thermodynamics, derived from the conservation of energy principle, governs the exchange of energy between a system and its environment. Be sure to distinguish between the two mechanisms of energy transfer, work and heat. Learn how to calculate the heat absorbed or rejected in terms of the heat capacity of the system and learn how to calculate the work done by a system, given the pressure as a function of volume.

Heat.   Recall from your study of Chapter 8 that the work done on an object by external forces may change its internal energy. There is another way in which the internal energy may change: the object may absorb or reject <u>heat</u>. Heat is the energy that flows between a system and its environment because _____

_____ .

Heat and work are alternate ways of transferring energy. Heat flows from the environment to the system when the temperature of the environment is _____ than the temperature of the system.

A sign convention is adopted for work and heat. Work W is taken to be positive when _____ and heat Q is taken to be positive when _____ .

The SI units of heat are _____ . Other units in common use and their SI equivalents are 1 cal = _____ J, 1 Btu (British thermal unit) = _____ J, and 1 Cal = _____ J. The Calorie unit (with a capital C) is commonly used in nutrition.

Heat capacity and specific heat.   The heat capacity relates the heat transferred into or out of the system to the change in temperature of the system. If during some process heat Q is absorbed and the temperature increases by a small increment ΔT, then the heat capacity of the system for that process is given by $C =$ _____ . The heat capacity depends not only on the kind of material in the system but also the amount of material. A related property that depends only on the kind of material and not the amount is the specific heat, defined in terms of C by $c =$ _____ , where m is the mass of the system. Another is the molar specific heat (or heat capacity per mole), defined by $C' =$ _____ , where n is the number of moles in the system. A mole is _____ elementary units (molecules for a gas). This number is called _____ number and is denoted by N_A. In this chapter of the study guide C' is used to denote a molar specific heat.

The heat capacity depends on the process during which heat is transferred. It is different for constant volume processes and constant pressure processes, for example. Thus there are many different heat capacities, one for each type process.

Suppose a certain system has molar specific heat C' for the process being considered. Then the heat capacity of n moles is $C =$ _____ and the heat capacity of N molecules is $C =$ _____ , where N_A is Avogadro's number. Suppose a certain system has specific heat c for a process and each of its molecules has mass m. Then the heat capacity of N molecules is $C =$ _____ and the heat capacity of n moles is $C =$ _____ . The molar specific heat for this system and process is $C' =$ _____ .

Consider an object of mass m, composed of material with specific heat c for some process. We assume c is independent of temperature. If the temperature of the material is changed from T_i to T_f by the process, then the heat exchanged is given by $Q =$ _____ . Strictly speaking, the heat capacity and specific heat are temperature dependent, but they may usually be approximated by constants if the temperature change is small.

Carefully note that $Q = C(T_f - T_i)$ is valid regardless of which temperature is higher. If $T_f > T_i$ then Q is positive, indicating that energy enters the object. If $T_f < T_i$ then Q is negative, indicating that energy leaves the object.

When two objects, initially at different temperatures, are placed in thermal contact with each other and isolated from their surroundings, the hotter substance cools, and the cooler substance warms until they reach the same temperature. The heat capacities of the objects can be used to find the final common temperature. The principle used is that the heat leaving the hotter object has the same magnitude as the heat entering the cooler object. The algebraic relationship is $Q_A + Q_B = 0$. If object A has mass m_A, specific heat c_A, and initial temperature T_A, then $Q_A =$ _____ , where T_f is the final temperature. If object B has mass m_B, specific heat c_B, and initial temperature T_B, then $Q_B =$ _____ . Note that the heat corresponding to the hotter object is negative, indicating that heat is actually transferred *from* it. Since $Q_A + Q_B = 0$, the final temperature is given by $T_f =$ _____ . You should use the specific heats (for constant volume, constant pressure, or some other process) appropriate to the actual process. If more than 2 objects are involved, use $\sum Q = 0$.

Both heat and work tend to change the internal energy of the system and thus tend to change the temperature. The change in temperature that accompanies an exchange of heat may be greater or less than it would be if no work were done and the same heat were exchanged. For example, if heat Q is absorbed by an object and positive work W is done on it, then the heat capacity is _____ than it would be if same amount of heat is absorbed with no work being done.

Heats of transformation. A substance accepts or rejects heat when it changes phase (melts, freezes, vaporizes, or condenses, for example), even though the temperature remains constant during a phase change. In words, a <u>heat</u> of <u>transformation</u> is _____

_____

If L is the heat of transformation for a certain phase change of a system with mass m then the magnitude of the heat accompanying a phase change is given by $|Q| =$ _____ . Q is positive (heat absorbed) for melting and vaporization; it is negative (heat rejected) for freezing and condensing. The <u>heat of fusion</u> L_F is the heat per unit mass transferred during _____ and _____ . The <u>heat of vaporization</u> L_V is the heat per unit mass transferred during _____ and _____ .

Work. The <u>thermodynamic state</u> of a gas can be described by giving the values of 3 thermo-dynamic variables: _____, _____, and _____. The state is changed by doing (positive or negative) work on the gas or by transferring energy as heat between the gas and its environment. Look at Fig. 20–3 of the text. Work is done on the gas by _____ and heat is transferred between the gas and _____ .

Thermodynamics deals with the work done *by* the system, rather than with the work done *on* the system. These are the negatives of each other. When the volume of a gas changes from V_i to V_f the work that is done *by* the gas is given by the integral of the pressure p:

$$W =$$

Processes exist for which work is done and the volume does not change but these necessarily involve an acceleration of the center of mass or an angular acceleration about the center of mass and we do not consider them. Rather, we take into account only changes in the *internal* energy of a system.

The integral for work is valid only if the process that changes the volume is carried out so the gas is nearly in thermal equilibrium at all times. Only then is the pressure well defined throughout the process. If $V_f > V_i$ then the sign of the work done by the system is _____; if $V_i > V_f$ then the sign of the work done by the system is _____ .

The process can be plotted as a curve on a graph with p as the vertical axis and V as the horizontal axis. Then the work W is the _____ under the curve.

The work done on the system is different for differ-ent processes, represented as different functional de-pendencies of p on V and plotted as different curves on a p-V diagram. To calculate the work you must know how the pressure varies as the volume changes.

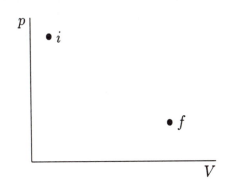

The diagram on the right shows an initial state, la-belled i and a final state, labelled f. Draw two different paths from i to f, illustrating two different processes. Label the curves 1 and 2, then tell for which process the gas does the greater work: _____

The first law of thermodynamics. The first law of thermodynamics postulates the existence of an internal energy and states that its change as the system goes from any initial thermal equilibrium state to any final thermal equilibrium state is independent of how the change is brought about. It then equates the change in the internal energy to $Q - W$, evaluated for any path between the initial and final equilibrium states. For an infinitesimal change in state the mathematical form of the first law of thermodynamics is

$$dE_{int} =$$

Carefully note the signs here. Positive work done by the system and heat leaving the system (dW positive and dQ negative) both tend to decrease the internal energy.

The internal energy is actually the sum of the kinetic energies of all the particles in the system and the potential energy of their interactions. We consider only systems for which the center of mass is at rest so the kinetic energy of translation of the system as a whole is zero.

Carefully contrast the internal energy with heat and work. When a system undergoes a change from one equilibrium state to another the work done by the system and the heat absorbed by the system depend on the actual process by which the state is changed and may be different for different processes. A change in internal energy, like a change in temperature, pressure, or volume, does NOT depend on the process but only on the initial and final states.

Because a change in the internal energy is a function of only the initial and final equilibrium states, the process used to change states is immaterial for its calculation. At intermediate stages of the process the system might not even be in thermal equilibrium.

You should understand some special cases. If a system undergoes a change of state at constant volume then the work done by it is _____ and the change in internal energy is related to the heat absorbed by ΔE_{int} = _____. When heat Q is absorbed at constant volume the internal energy increases by exactly that amount. If the change of state is adiabatic then the heat absorbed is _____ and the change in internal energy is related to the work by ΔE_{int} = _____. When work W is done by the system in an adiabatic process the internal energy decreases by exactly that amount. If the process is cyclic, so the initial and final states are the same, then the change in internal energy is _____ and the work done by the system is related to the heat absorbed by W = _____.

The free expansion of a gas is an important example of a process for which the system is not in thermal equilibrium during intermediate stages. Initially a partition confines the gas to one part of its container, which is thermally insulated. Then the partition is removed and the gas expands into the other part. For this process, W = _____ and Q = _____. Thus, according to the first law, ΔE_{int} = _____.

Transfer of heat. There are three mechanisms by which heat flows from one place to another. Briefly describe each of them in words, being sure the descriptions distinguish between them:

conduction: _____

_____

convection: _____

_____

radiation: _____

_____

The important quantity is the *rate* of heat transfer, with symbol H. If heat ΔQ is transferred in a short time Δt then the rate of transfer over that time is given by H = _____. The flow of energy is always from a _____ to a _____ region.

To discuss conduction we consider a homogeneous bar of material with one end held at temperature T_H (hot) and the opposite end held at temperature T_C (cold). Let the x axis be along the bar, with $x = 0$ at the hot face and $x = L$ at the cold face. The temperature in the bar varies from point to point along its length and so is a function of x.

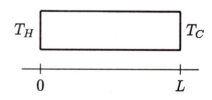

The rate of heat flow through the bar is proportional to the temperature difference of its ends and its cross-sectional area A. The constant of proportionality k is called the _____ _____ of the material. Mathematically the rate of heat flow through the bar is given by

$$H =$$

The thermal conductivity can be considered to be independent of the temperature over small temperature ranges but does, in fact, vary slightly with T.

Building materials are often characterized by their thermal resistances or R values rather than their thermal conductivities. The R value is related to the thermal conductivity k by $R =$ _____ . Table 20–4 gives some values.

II. PROBLEM SOLVING

The heat of transformation is the heat per unit mass transferred during a phase change. It is a property of the material that is changing phase. In the equation $Q = mL$, m is the mass of material that undergoes a phase change and is not necessarily the mass of all the material present.

PROBLEM 1 (5E).   Icebergs in the North Atlantic present hazards to shipping (see Fig. 20–15), causing the lengths of shipping routes to increase by about 30% during iceberg season. Attempts to destroy icebergs include planting explosives, bombing, torpedoing, shelling, ramming, and coating with black soot. Suppose that direct melting of the iceberg, by placing heat sources in the ice, is tried. How much heat is required to melt 10% of a $200,000$-metric ton iceberg?

STRATEGY: Use $Q = mL_F$, where m is the mass of the ice to be melted ($0.10 \times 200,000 \times 1000 = 2.0 \times 10^7$ kg) and L_F is the heat of fusion for water (6.01 kJ/kg). Note that the conversion 1 metric ton = 1000 kg was used.

SOLUTION:

[ans: 6.7×10^{12} J]

The heat Q absorbed or rejected by a substance is related to the change in temperature ΔT by $Q = C \, \Delta T$, where C is the heat capacity, or by $Q = mc \, \Delta T$, where m is the mass and c is the specific heat. You should recognize that if total energy ΔE in any form (work or heat) is added to an object the temperature increase is given by $\Delta T = \Delta E / C_V$, where C_V is the heat capacity for a constant volume process.

PROBLEM 2 (13E).   An object of mass 6.00 kg falls through a height of 50.0 m and, by means of a mechanical linkage, rotates a paddle wheel that stirs 0.600 kg of water. The water is initially at 15.0 °C. What is the maximum possible temperature rise of the water?

STRATEGY: The increase in the internal energy of the water equals the original potential energy of the object (Mgh, where M is the mass of the object and h is its initial height). Thus $Mgh = mc\,\Delta T$, where m is the mass of the water and c is its specific heat (4190 J/kg · K). Solve for ΔT. Remember that a kelvin is the same as a Celsius degree.

SOLUTION:

[ans: 1.17 °C]

You should know how to calculate the final temperature of two or more objects, initially at different temperatures, when they have been in contact long enough to achieve thermal equilibrium. Use the condition that the heat rejected by one object is absorbed by the others. If the final temperature is known this condition can be used to calculate the specific heat or mass of one of the objects.

PROBLEM 3 (18P).   Calculate the specific heat of a metal from the following data: A container made of the metal has a mass of 3.6 kg and contains 14 kg of water. A 1.8-kg piece of the metal initially at a temperature of 180 °C is dropped into the water. The container and water initially have a temperature of 16.0 °C and the final temperature of the entire system is 18.0 °C.

STRATEGY: Let c_m be the specific heat of the metal. If m_p is the mass of the piece that is dropped into the water then the energy it loses in the form of heat is $Q_p = m_p c(T - T_p)$, where T_p is its initial temperature and T is its final temperature. If m_w is the mass of the water and m_c is the mass of the container then the energy they gain in the form of heat is $Q_{cw} = (m_c c + m_w c_w)(T - T_{cw})$, where T_{cw} is the initial temperature of the water and container and c_w is the specific heat of water (4190 J/kg · K). The net heat into the entire system is zero, so $Q_p + Q_{cw} = m_p c(T - T_p) + (m_c c + m_w c_w)(T - T_{cw}) = 0$. Solve for c.

SOLUTION:

[ans: 413 J/kg · K]

PROBLEM 4 (25P).   The specific heat of a substance varies with temperature according to $c = 0.20 + 0.14T + 0.023T^2$, with T in °C and c in cal/g·K. Find the heat required to raise the temperature of 2.0 g of this substance from 5.0 to 15 °C.

STRATEGY: Since the specific heat depends on temperature you must evaluate the integral

$$Q = \int_{T_i}^{T_f} mc\,\mathrm{d}T,$$

where T_i is the initial temperature and T_f is the final temperature. You should obtain $Q = m[0.20(T_f - T_i) + 0.07(T_f^2 - T_i^2) + (0.023/3)(T_f^3 - T_i^3)]$.

SOLUTION:

[ans: 82 cal]

PROBLEM 5 (27P).   An insulated thermos contains $130\,cm^3$ of hot coffee, at a temperature of $80.0\,°C$. You put in a 12.0-g ice cube at its melting point to cool the coffee. By how many degrees will your coffee have cooled once the ice has melted? Treat the coffee as though it were pure water.

STRATEGY: Let L_F be the heat of fusion of water ($333\,kJ/kg$) and let m_i be the original mass of the ice. The ice absorbs heat $m_i L_F$. If T is the final temperature of the water (and coffee) then the water from the ice absorbs an additional heat $m_i cT$, where c is the specific heat of water ($4190\,J/kg \cdot K$). Let m_c be the mass of coffee originally in the thermos. Then the coffee gives up heat $m_c c(T-T_c)$, where T_c is the initial temperature of the coffee. Since no energy escapes the thermos, $m_i L_F + m_i cT + m_w c(T-T_c) = 0$. Solve for T, then calculate $\Delta T = T_c - T$. The density of water is $0.998 \times 10^3\,kg/m^3$ so $130\,cm^3$ has a mass of $130 \times 10^{-6} \times 0.998 \times 10^3 = 0.130\,kg$.

SOLUTION:

[ans: $13.5\,°C$]

PROBLEM 6 (29P).   A person makes a quantity of iced tea by mixing 500 g of the hot tea (essentially water) with an equal mass of ice at its melting point. If the initial hot tea is at a temperature of (a) $90°\,C$ and (b) $70°\,C$, what is the temperature and mass of the remaining ice when the tea and ice reach the same temperature?

STRATEGY: First suppose that not all the ice melts. Let M be the original mass and m be the mass that melts. Then the final temperature is $0\,°C$. The ice absorbs heat mL_F and the tea rejects heat $-McT_t$, where T_t is its initial temperature, c is the specific heat of water ($4190\,J/kg$), and L_F is the heat of fusion for water ($333\,kJ/kg$). Since no energy leaves the tea-ice system $mL_F - McT_t = 0$. Solve for m. If $m > M$ then the assumption is wrong; all the ice does melt. In that case, the ice and water formed from it absorb heat $ML_F + McT$, where T is the final temperature. The tea gives up heat $Mc(T - T_t)$. Since no energy leaves the tea-ice system $ML_F + McT + Mc(T - T_t) = 0$. Solve for T.

SOLUTION:

[ans: (a) no ice remains, $5.3\,°C$; (b) 60 g of ice remains, $0\,°C$]

PROBLEM 7 (33P).   By means of a heating coil, energy is transferred at a constant rate to a substance in a thermally insulated container. The temperature of the substance is measured as a function of time. (a) Show how we can deduce from this information the way in which the heat capacity of the body depends on the temperature.

STRATEGY: In any small time interval Δt the coil supplies heat $R\Delta t$, where R is the rate, a constant. During this interval the temperature of the substance changes by ΔT. Thus $R\,\Delta t = C\,\Delta T$, where C is the heat capacity. So $C = R/(\Delta T/\Delta t)$ and in the limit as the time interval becomes infinitesimal $C = R/(dT/dt)$. Use the data to calculate dT/dt as a function of temperature, then substitute the result into $C = R/(dT/dt)$ to find C as a function of temperature.

SOLUTION:

(b) Suppose that in a certain temperature range the temperature T is proportional to t^3, where t is the time. How does the heat capacity depend on T in this range?

STRATEGY: Let $T = At^3$, where A is a constant. Then $dT/dt = 3At^2$. Since $t = (T/A)^{1/3}$, $dT/dt = 3A(T/A)^{2/3}$. Substitute into the result of part a to find the functional dependence of C on T.

SOLUTION:

[ans: C is proportional to $T^{-2/3}$]

Some problems test your understanding of the first law of thermodynamics: $\Delta E_{int} = Q - W$. Be sure you understand the sign convention you must use with this equation.

PROBLEM 8 (37E).   Consider that 200 J of work are done on a system and 70.0 cal of heat are extracted from the system. In the sense of the first law of thermodynamics, what are the values (including algebraic signs) of (a) W, (b) Q, and (c) ΔE_{int}?

STRATEGY: In the sense of the first law W is the work done *by* the system. It is positive if the system does positive work and negative if the environment does positive work. Heat Q is positive if it enters the system and negative if it leaves the system. Use 1 cal = 4.186 J to convert to joules. Use $\Delta E_{int} = Q - W$ to calculate the change in the internal energy.

SOLUTION:

[ans: (a) −200 J; (b) −293 J; (c) −93 J]

PROBLEM 9 (38E).   A thermodynamic system is taken from an initial state A to another state B and back again to A, via state C, as shown by path $ABCA$ in the p-V diagram of Fig. 20–19a. (a) Complete the table in Fig. 20–19b by filling in either + or − for the sign of each thermodynamic quantity associated with each process.

STRATEGY: Over the path $A \rightarrow B$ the work done by the system is positive (the volume increases). Since $\Delta E_{int} = Q - W$ and ΔE_{int} is positive, Q must be positive. Over the path $B \rightarrow C$ the work done by the system is 0, so $\Delta E_{int} = Q$. Over the path $C \rightarrow A$ the work done by the system is negative (the volume decreases). Over the entire cycle the change in the internal energy is 0. Since it is positive over the other two portions of the cycle, it must be negative over this portion. Over the entire cycle $Q = W$. Now the area under the curve $A \rightarrow B$ is smaller than the area under the curve $C \rightarrow A$ so the total work done by the system over the cycle is negative.

SOLUTION:

$$\left[\text{ans: } A \text{ to } B\text{: } Q, W, \Delta E_{\text{int}} > 0; B \text{ to } C\text{: } Q, \Delta E_{\text{int}} > 0, W = 0; C \text{ to } A\text{: } Q, W, \Delta E_{\text{int}} < 0\right]$$

(*b*) Calculate the numerical value of the work done by the system for the complete cycle $ABCA$.

STRATEGY: The work done is the negative of the area of the triangle, or $\frac{1}{2} \Delta V \Delta p$.

SOLUTION:

$$\left[\text{ans: } -20\,\text{J}\right]$$

PROBLEM 10 (41P). Figure 20–22*a* shows a cylinder containing gas and closed by a movable piston. The cylinder is kept submerged in an ice-water mixture. The piston is *quickly* pushed down from position 1 to position 2. The piston is held at position 2 until the gas is again at the temperature of the ice-water mixture and then is *slowly* raised back to position 1. Figure 20–22*b* is a *p-V* diagram for the process. If 100 g of ice is melted during the cycle, how much work has been done *on* the gas?

STRATEGY: Over the complete cycle $\Delta E_{\text{int}} = 0$, so $W = Q$. Heat leaves the gas and melts the ice. Use $Q = mL_F$, where m is the mass of ice melted and L_F is the heat of fusion for water.

SOLUTION:

$$\left[\text{ans: } 33.3\,\text{kJ}\right]$$

Thermal conductivity problems all involve $H = kA(T_H - T_C)/L$. Usually all but one of the quantities that appear in this equation are given and you are asked for the other one. More complicated problems involve several rods, for example, welded either end to end or side by side. In the first case the rate of heat conduction must be the same in all rods, while in the second it is the sum of the rates for the individual rods. In other problems the heat accomplishes a purpose, such as the melting of ice at one end. Here are some examples.

PROBLEM 11 (44E). The average rate at which heat is conducted through the surface of the Earth in North America is 54.0 mW/m², and the average thermal conductivity of the near-surface rocks is 2.50 W/m · K As-suming a surface temperature of 10.0°C, what should be the temperature at a depth of 35.0 km (near the base of the crust)? Ignore the heat generated by the presence of radioactive elements.

STRATEGY: Use $H = kA(T_H - T_C)/L$, with $H/A = 54\,\text{mW/m}^2$, $T_C = 10°\text{C}$, $k = 2.50\,\text{W/m} \cdot \text{K}$, and $L = 35.0\,\text{km}$. Solve for T_H.

SOLUTION:

$$\left[\text{ans: } 766°\text{C}\right]$$

PROBLEM 12 (48E).   A cylindrical copper rod of length 1.2 m and cross-sectional area 4.8 cm² is insulated to prevent heat loss through its surface. The ends are maintained at a temperature difference of 100 °C by having one end in a water-ice mixture and the other in boiling water and steam. (*a*) Find the rate at which heat is conducted along the rod.

STRATEGY: Evaluate $H = kA(T_H - T_C)/L$. According to Table 20–4 the thermal conductivity of copper is 401 W/m · K.

SOLUTION:

$$[\text{ans: } 16.0\,\text{J/s}]$$

(*b*) Find the rate at which ice melts at the cold end.

STRATEGY: In a short time interval Δt the heat absorbed by the ice is $\Delta Q = H\,\Delta t$. If this melts a mass Δm of ice, then $\Delta Q = L_F\,\Delta m$, where L_F is the heat of fusion for water. Thus $H\,\Delta t = L_F\,\Delta m$ or in the limit as Δt becomes small, $dm/dt = H/L_F$. Evaluate this expression.

SOLUTION:

$$[\text{ans: } 4.8 \times 10^{-5}\,\text{kg/s}]$$

PROBLEM 13 (52P).   Two identical rectangular rods of metal are welded end to end as shown in Fig. 20–25*a* and 10 J of heat is conducted (in a steady state process) through the rods in 2.0 min. How long would it take for 10 J to be conducted through the rods if they are welded together as shown in Fig. 20–25*b*?

STRATEGY: Let L be the length of each rod, let A be cross-sectional area of each rod, and let ΔT be the temperature difference from one end of the combination to the other. For the arrangement of Fig. 20–25*a* the heat conducted in time Δt_1 is $Q = H\,\Delta t_1 = kA\,\Delta T\,\Delta t_1/2L$, since the total length is $2L$. For the arrangement of Fig. 20–25*b* the heat conducted in time Δt_2 is $Q = H\,\Delta t_2 = k2A\,\Delta T\,\Delta t_2/L$, since the cross-sectional area is $2A$. The heat is the same so $kA\,\Delta T\,\Delta t_1/2L = k2A\,\Delta T\,\Delta t_2/L$. Solve for Δt_2.

SOLUTION:

$$[\text{ans: } 0.50\,\text{min}]$$

PROBLEM 14 (55P).   A large cylindrical water tank with a bottom 1.7 m in diameter is made of iron boiler-plate 5.2 mm thick. As the water is being heated, the gas burner underneath is able to maintain a temperature difference of 2.3 °C between the top and bottom surfaces of the bottom plate. How much heat is conducted through that plate in 5.0 min? (Iron has a thermal conductivity of 67 W/m · K.)

STRATEGY: Use $H = kA(T_H - T_C)/L$. Here L is the thickness of the boilerplate and A is the area of the tank bottom, given by πR^2, where R is the radius.

SOLUTION:

[ans: 2.0×10^7 J]

PROBLEM 15 (58P). Ice has been formed on a shallow pond and a steady state has been reached, with the air above the ice at $-5.0\,°C$ and the bottom of the pond at $4.0\,°C$. If the total depth of ice + water is 1.4 m, how thick is the ice? (Assume that the thermal conductivities of ice and water are 0.40 and 0.12 cal/m $\cdot\,°C \cdot$ s, respectively.)

STRATEGY: Let x be the ice thickness and L be the total depth. Then the depth of the water under the ice is $L - x$. If T_1 is the temperature of the top surface of ice, T_2 is the temperature of the water-ice interface, and T_3 is the temperature at the pond bottom, then the rate of heat flow through the ice is $k_i A(T_2 - T_1)/x$ and the rate of heat flow through the water is $k_w A(T_3 - T_2)/(L - x)$. Here k_i is the thermal conductivity of ice and k_w is the thermal conductivity of water. At steady state the two rates of heat flow are equal, so $k_i(T_2 - T_1)/x = k_w(T_3 - T_2)/(L - x)$. Take $T_2 = 0.0\,°C$ and solve for x.

SOLUTION:

[ans: 1.1 m]

PROBLEM 16 (59P). Three metal rods, made of copper, aluminum, and brass, are each 6.00 cm long and 1.00 cm in diameter. These rods are placed end to end, with the aluminum between the other two. The free ends of the copper and brass rods are maintained at the boiling point and the freezing point of water, respectively. Find the steady-state temperatures of the copper-aluminum junction and the aluminum-brass junction. The thermal conductivity of brass is 109 W/m $\cdot$ K.

STRATEGY: Let T_C be the temperature at the free end of the brass rod, let T_1 be the temperature at the aluminum-brass junction, let T_2 be the temperature at the copper-aluminum junction, and let T_H the temperature at the free end of the copper rod. The rate of heat conduction through the brass rod is given by $H = k_B A(T_1 - T_C)/L$, the rate of heat conduction through the aluminum rod is given by $H = k_A A(T_2 - T_1)/L$, and the rate of heat flow through the copper rod is given by $H = k_C A(T_H - T_2)/L$. In steady state, these three rates of heat conduction are the same. Solve the three equations simultaneously for T_1 and T_2.

SOLUTION:

[ans: $T_1 = 57.6\,°C, T_2 = 84.3\,°C$]

III. NOTES

Chapter 21
THE KINETIC THEORY OF GASES

I. BASIC CONCEPTS

Kinetic theory is used to relate macroscopic quantities like pressure, temperature, and internal energy to the energies and momenta of the particles that comprise a gas. Here you will gain an understanding of the relationship between the microscopic and macroscopic descriptions of a gas. The example used throughout this chapter and the next is known as an ideal gas. Pay careful attention to its properties.

Ideal gas law. Since you will be dealing with ideal gases you should know what one is. Describe the properties that distinguish an ideal gas from a real gas: _____

_____

_____

   The quantity of matter in the system of interest is important for most of the discussions of this chapter. It might be given as the number of particles or as the number of moles. The number of molecules in a mole of molecules is _____. This is Avogadro's number, denoted by N_A. If a gas contains N molecules then it contains _____ moles of molecules. If the mass of one mole is M, then the mass of one molecule is $m =$ _____. If m is the molecular mass and M is the molar mass, then a gas containing N molecules has a total mass of _____ and a gas containing n moles has a total mass of _____.

   To understand the derivations of this chapter you must first understand the <u>ideal gas law</u> (or ideal gas equation of state), which expresses a relationship between the temperature, pressure, volume, and number of particles for an ideal gas. In terms of the number of moles n of gas and the universal gas constant R, it is

$$pV =$$

Here p is the _____, V is the _____, and T is the _____ of the gas. It is important to recognize that for this equation to be valid, the Kelvin or ideal gas temperature scale must be used. The SI value of R is $R =$ _____ J/mol·K.

   Real gases obey the ideal gas law only approximately; the approximation being better when _____ .
This is chiefly because the ideal gas law assumes the gas molecules are point particles that do not interact with each other except for elastic collisions of extremely short duration. In a real gas molecules have extent in space and interact continuously with each other. Molecular sizes and interactions play less of a role as the number of molecules per unit volume decreases.

Work. The work done by the system as its volume changes from V_i to V_f is given by the integral $W =$ _____ . Its value depends on the function $p(V)$. Different dependencies of p on V are represented by different paths on a p-V graph. You should know how to compute the work for the three special processes described below. The device depicted in Fig. 21–9 of the text allows the macroscopic properties of a gas to be varied individually. You should also understand how to manipulate the device to carry out each of the special processes.

For each of processes described below, all carried out on n moles of an ideal gas, write out the function $p(V)$, carry out the integration for the work done by the gas, and write the final expression for the work. Give expressions for the changes requested. On the p-V axes to the right, draw a curve that represents the process. Finally, describe how you would manipulate the device of Fig. 21–9 of the text to carry out the work.

1. Change in pressure from p_i to p_f at constant volume V
    $p(V)$ is immaterial
    $W =$

    Change in temperature: $\Delta T =$
    How would you manipulate the device?
    _____
    _____

2. Change in volume from V_i to V_f at constant pressure p
    $p(V) =$ constant
    $W =$

    Change in temperature: $\Delta T =$
    How would you manipulate the device?
    _____
    _____

3. Change in volume from V_i to V_f at constant temperature T
    $p(V) =$
    $W =$

    Change in pressure: $\Delta p =$
    How would you manipulate the device?
    _____
    _____

Pressure. The pressure in a gas is intimately related to the average of the squares of the speeds of the molecules. The relationship is

$$p =$$

where n is the number of moles, M is the molar mass, V is the volume, and $\overline{v^2}$ is the average value of the squares of the molecular speeds.

Pressure can be computed as the average force per unit area exerted by the molecules of a gas on the container walls. We assume that at each collision between a molecule and a wall the normal component of the molecule's _____ is reversed. By computing the change per unit time we can find the force exerted by the molecule on the wall. When this is divided by the area of the wall, the result is the contribution of the molecule to the pressure.

Consider an ideal gas in a cubical container with edge L. If m is the mass of a molecule, v_x is the component of its velocity normal to the wall, and Δt is the time from one collision to the next with the same wall, then the average force exerted by the molecule on the wall is given by _____. In terms of L and v_x, the time between collisions is $\Delta t =$ _____. Thus the average force of the molecule is _____ and its contribution to the pressure is _____. This is now summed over all molecules. Replace the sum of v_x^2 with the product of its average value and the number of molecules nN_A. You obtain $p =$ _____. Finally, replace the average of v_x^2 with one third the average of v^2, mN_A with M, and L^3 with V. These substitutions complete the derivation. You obtain $p =$ _____, which should be the expression you wrote above.

Explain why the average of v_x^2 is one third the average of v^2:

_____

_____

_____

The <u>root-mean-square</u> speed of the molecules in a gas is not quite the same as the average speed but it gives us a rough idea of the average. Its definition, in terms of the average of the squares of the molecular speeds, is $v_{\text{rms}} =$ _____ and, in terms of the pressure and density, it is given by

$$v_{\text{rms}} =$$

You can use this equation in conjunction with the ideal gas law to estimate the root-mean-square speed of a given gas with a known molar mass M, at a given temperature. Substitute $p = nRT/V$ into the expression for v_{rms} to obtain

$$v_{\text{rms}} =$$

Carefully note that v_{rms} depends only on the temperature and the molar mass. Thus v_{rms} remains the same if the temperature does not change as the pressure and volume are changed; that is, if the product pV does not change. v_{rms}^2 is directly proportional to the temperature and inversely proportional to the molar mass.

Suppose molecules in an ideal gas have a root-mean-square speed v_0. If the temperature is doubled the root-mean-square speed becomes _____.

Suppose two ideal gases are in thermal equilibrium with each other. Molecules in the first gas have mass M and root-mean-square speed v_0. If molecules in the second gas have mass $2M$ then their root-mean-square speed is _____ .

Some values of v_{rms} are listed in Table 21–1. For gases at room temperature they run from a low of _____ m/s (for _____) to a high of _____ m/s (for _____). You should know that the root-mean-square speed is roughly the same as the speed of sound for any given gas.

Kinetic energy and temperature. For an ideal gas the average translational kinetic energy per molecule is proportional to the temperature. In terms of the universal gas constant R and Avogadro's number N_A the exact relationship is

$$\overline{K} = \frac{1}{2}mv_{\mathrm{rms}}^2 =$$

where m is the molecular mass. The quantity $k = R/N_A$ is called the _____ constant and has the value $k =$ _____ J/K. In terms of k and T

$$\overline{K} =$$

Carefully note that the molecules of two ideal gases in thermal equilibrium with each other have exactly the same average translational kinetic energy. The root-mean-square speeds, however, are different if the molecular masses are different. Also note that the average translational kinetic energy depends on the temperature, but not separately on the pressure or volume of the gas.

Mean free path. The mean free path of a molecule in a gas is the average distance it travels between _____ with other molecules. It can be estimated theoretically by considering the number of molecules that are encountered by any given molecule as it moves through the gas. Assume the molecules each have a diameter d. Concentrate on one that is moving with speed v and assume all others are stationary. The moving molecule has a collision with another molecule if the distance between their centers is less than _____ . In terms of the diameter d and speed v, the number of collisions it has in time t equals the number of molecules in the volume of a cylinder with cross-sectional area $A =$ _____ and length $\ell =$ _____ . Since N/V is the number of molecules per unit volume in the gas, the number of collisions in time t is given by _____ . The mean free path λ is the distance vt divided by the number of collisions in time t or, in terms of d and N/V, $\lambda =$ _____ .

Because each molecule collides with moving rather than stationary molecules the mean free path is actually somewhat less. It is, in fact, given by

$$\lambda =$$

The mean free path of air molecules at sea level is about _____ ; the mean free path is about _____ at an altitude of 100 km and about _____ at an altitude of 300 km.

The distribution of molecular speeds. Not all molecules in a gas have the same speed. The distribution of speeds is described by the <u>Maxwell</u> <u>speed</u> <u>distribution</u> $P(v)$, defined so that $P(v)\,dv$ gives the fraction of molecules with speeds in the range from v to $v + dv$. If a gas at temperature T has molar mass M, then the Maxwell speed distribution is given by

$$P(v) =$$

Sketch the function $P(v)$ on the axes below:

To find the number of molecules with speed between v_1 and v_2 evaluate the definite integral

$$N(v_1, v_2) = \int_{v_1}^{v_2} P(v)\,dv \,.$$

If the interval $v_2 - v_1$ is small you can approximate this by $(v_2 - v_1)P(v_1)$. See Sample Problem 21–8 of the text.

The Maxwell speed distribution can be used to calculate the most probable speed, the average speed, and the root-mean-square speed. The <u>most</u> <u>probable</u> <u>speed</u> v_p is the one for which $P(v)$ is _____. In terms of the molar mass and the temperature, it is given by

$$v_p =$$

The <u>average</u> <u>speed</u> is given by

$$\bar{v} = \int_0^\infty v P(v)\,dv =$$

where the explicit expression for the Maxwell distribution was substituted for $P(v)$ and the integral was evaluated. The <u>mean-square</u> <u>speed</u> is given by

$$\overline{v^2} = \int_0^\infty v^2 P(v)\,dv =$$

where the explicit expression for the Maxwell distribution was again substituted for $P(v)$ and the integral was evaluated. The root-mean-square speed v_{rms} is the square root of this. When the square root is taken, the result is

$$v_{\text{rms}} =$$

Note that all three speeds (v_p, $\bar{v}$, and v_{rms}) depend on the temperature as well as on the molar mass. When the temperature increases, they all _____. For a given gas at a given temperature the greatest of the three characteristic speeds is _____ and the least is _____.

Internal energy and heat capacities.   For a *monatomic* ideal gas the internal energy is the total kinetic energy of the molecules. Since the average kinetic energy of a molecule is $\overline{K} = \frac{3}{2}kT$, the internal energy of n moles of an ideal gas is given in terms of R and T by $E_{int} = $ _____.

Note that the internal energy of an ideal gas depends only on the temperature and the amount of gas. This information will prove useful in discussing the thermodynamic temperature scale as well as in solving ideal gas problems. For other systems the internal energy may also depend on the pressure or other macroscopic quantities.

Suppose now that the temperature is changed by ΔT while the volume is held constant. Then the internal energy changes by $\Delta E_{int} = $ _____. Since no work was done the entire increase or decrease in internal energy is due to heat absorbed or rejected. The heat is $Q = $ _____. The heat capacity at constant volume is $Q/\Delta T = $ _____ and the molar specific heat at constant volume, in terms of R, is $C_V = $ _____. Its value is _____ J/mol·K. Notice that it does not depend on the temperature. Carefully note that in this chapter and the next C is used to represent a *molar* specific heat, not a heat capacity.

You should understand that $\Delta E_{int} = nC_V\Delta T$ for any process, whether it involves heat or not and whether the volume changes or not. However, $Q = nC_V\Delta T$ only for a constant volume process, during which no work is done.

The molar specific heat at constant pressure C_p is related to the molar specific heat at constant volume by $C_p = $ _____. More heat is absorbed at constant pressure than at constant volume for a given rise in temperature. Since the internal energy depends only on the temperature, its change is the same. What happens to the extra energy absorbed at constant pressure? _____
_____

The equipartition theorem tells us that, in thermal equilibrium, the internal energy is equally divided among the degrees of freedom of a system, with the energy associated with each degree of freedom being _____ kT per molecule or _____ RT per mole. One degree of freedom is associated with each independent energy term. For example, _____ degrees of freedom are associated with the translational kinetic energy of a molecule since it can move in any of three independent directions. Other degrees of freedom are associated with rotation of the molecules.

Consider three gases. The first consists of monatomic molecules, the second of diatomic molecules, and the third of polyatomic molecules. Fill in the following table giving the translational, rotational, and total internal energies per mole in terms of RT.

	MONATOMIC	DIATOMIC	POLYATOMIC
Number of translational degrees of freedom	_____	_____	_____
Translational energy	_____	_____	_____
Number of rotational degrees of freedom	_____	_____	_____
Total internal energy	_____	_____	_____

Fill in the following table for the molar specific heats for constant volume processes and for constant pressure processes.

	MONATOMIC	DIATOMIC	POLYATOMIC
C_V	_____	_____	_____
C_p	_____	_____	_____

Knowing how the internal energy depends on the temperature allows us to compute the change in temperature when work is done on the gas or heat is absorbed by the gas. If, for example, work W is done on n moles of a monatomic ideal gas in thermal isolation (no heat transfer) then the increase in internal energy is $\Delta E_{int} =$ _____ and the increase in temperature is $\Delta T =$ _____.

Adiabatic processes. As an ideal gas undergoes an adiabatic process the quantity pV^γ is _____. Here γ is the ratio _____. If p_i and V_i are the pressure and volume of the initial state and p_f and V_f are the pressure and volume of the final state then

$$p_i V_i^\gamma =$$

Use $pV = nRT$ to eliminate p from this expression. You should find that if T_i is the temperature of the initial state and T_f is the temperature of the final state then

$$T_i V_i^{\gamma - 1} =$$

If two of the three quantities p, V, and T are given for one state and one of them is given for another state, connected to the first by an adiabatic process, then you can use the equations given above and $pV = nRT$ to find values for the other thermodynamic variables.

II. PROBLEM SOLVING

You should know the relationships between the mass of a molecule, the mass of a mole of molecules, and the total mass of a collection of molecules. If a gas contains N molecules, each of mass m, the total mass is Nm. Since there are N_A molecules in a mole, the gas contains N/N_A moles. The molar mass M is the mass of N_A molecules: $M = N_A m$. If the gas contains n moles its total mass is nM. Note that this is exactly the same as Nm.

PROBLEM 1 (4P). Consider this sentence: A _____ of water contains about as many molecules as there are _____s of water in all the oceans. What single word best fits both blank spaces: drop, teaspoon, tablespoon, cup, quart, barrel, or ton? The oceans cover 75% of the Earth's surface and have an average depth of about 5 km. (After Edward M. Purcell.)

STRATEGY: Let V be the volume of the correct quantity to be placed in the blanks. The mass of water in V is $m = \rho V$, where ρ is the density of water ($0.998 \times 10^3 \text{ kg/m}^3$). The number of molecules in V is $N_m = N_A m/M = N_A \rho V/M$, where N_A is Avogadro's number and M is the molar mass of water. A water molecule consists of 2 hydrogen atoms (atomic mass 1) and 1 oxygen atom (atomic mass 16) so the molar mass is 18×10^{-3} kg.
    The area of the Earth's surface is $4\pi R^2$, where R is the radius (6.27×10^6 m), so the volume of water in all the oceans is $V_O = 0.75 \times 4\pi R^2 d$, where d is the average ocean depth. The number of entities of volume V in all the oceans is $V_O/V = 3\pi R^2 d/V$. Since this must be the same as the number of molecules in V,

$3\pi R^2 d/V = N_A \rho V/M$. Thus $V = \sqrt{3\pi R^2 dM/N_A\rho}$. Calculate V and compare with the possibilities listed in the problem statement.

SOLUTION:

[ans: $7.4\,\text{cm}^3$, about a tablespoon]

Many problems require you to use the ideal gas law $pV = nRT$ to compute one of the quantities that appear in it. Here is an example.

PROBLEM 2 (8E).   The best vacuum that can be attained in the laboratory corresponds to a pressure of about 1.00×10^{-18} atm, or 1.01×10^{-13} Pa. How many molecules are there per cubic centimeter in such a vacuum at $293\,\text{K}$?

STRATEGY: Use $pV = nRT$ to compute n/V. Multiply by Avogadro's number to obtain the number of molecules per unit volume.

SOLUTION:

[ans: 25 molecules/cm³]

You should understand how to compute the work done by a gas during a given process. Use $W = \int p\,dV$. First, you will need to find the functional form of the dependence of the pressure p on the volume V for the process. The ideal gas law is often helpful here. The next problem is an example of the calculation of work. The two following that are examples of the use of the ideal gas law to find the value of a thermodynamic quantity.

PROBLEM 3 (12E).   Calculate the work done by an external agent during an isothermal compression of $1.00\,\text{mol}$ of oxygen from a volume of $22.4\,\text{L}$ at $0\,^\circ\text{C}$ and $1.00\,\text{atm}$ to $16.8\,\text{L}$.

STRATEGY: The work done by the external agent is the negative of the work done by the gas and is given by

$$W_{\text{ext}} = -\int_{V_1}^{V_2} p\,dV\,.$$

Since $p = nRT/V$ and the compression is isothermal (T is constant), this is

$$W_{\text{ext}} = -nRT\int_{V_1}^{V_2} \frac{dV}{V} = -nRT\ln\frac{V_2}{V_1}\,.$$

SOLUTION:

[ans: $653\,\text{J}$]

PROBLEM 4 (18P).   A weather balloon is loosely inflated with helium at a pressure of 1.0 atm (= 76 cm Hg) and a temperature of 20°C. The gas volume is 2.2 m³. At an elevation of 20, 000 ft, the atmospheric pressure is down to 38 cm Hg and the helium has expanded, being under no restraint from the confining bag. At this elevation the gas temperature is −48°C. What is the gas volume?

STRATEGY: Since the amount of helium has not changed the ideal gas law gives $p_1V_1/T_1 = p_2V_2/T_2$, where the subscript 1 refers to the original conditions and the subscript 2 refers to conditions at 20, 000 ft. Solve for V_2.

SOLUTION: Notice that the pressures enter as a ratio, so you do not need to change units. Be sure to convert the temperatures to kelvins, however.

$$[\text{ans: } 3.4\,\text{m}^3]$$

PROBLEM 5 (22P).   A steel tank contains 300 g of ammonia gas (NH_3) at an absolute pressure of 1.35×10^6 Pa and temperature 77°C. (*a*) What is the volume of the tank?

STRATEGY: Use $pV = nRT$. The molar mass of NH_3 is 17 g so there are (300 g)/(17 g/mol) = 17.6 moles present.

SOLUTION:

$$[\text{ans: } 0.038\,\text{m}^3\ (38\,\text{L})]$$

(*b*) The tank is checked later when the temperature has dropped to 22° C and the absolute pressure has fallen to 8.7×10^5 Pa. How many grams of gas leaked out of the tank?

STRATEGY: Use $pV = nRT$ to calculate the number of moles of gas remaining in the tank. Subtract from 17.6 to find the number of moles that leaked out. Each mole has a mass of 17 g.

SOLUTION:

$$[\text{ans: } 71\,\text{g}]$$

You should know how the root-mean-square speed of molecules in a gas depends on the temperature: $v_{\text{rms}} = \sqrt{3RT/M}$, where M is the molar mass. You should also know how the average translational kinetic energy depends on the temperature: $\overline{K} = \frac{3}{2}RT/N_A$. The translational kinetic energy per mole is $\frac{3}{2}RT$, the energy of n moles is $\frac{3}{2}nRT$, and the translational kinetic energy for N molecules is $\frac{3}{2}NRT/N_A = \frac{3}{2}NkT$. For monatomic molecules the last two expressions give the internal energy. For diatomic and polyatomic molecules other terms, giving the rotational energy, must be added. Other problems deal with the mean free path.

PROBLEM 6 (28E).   (*a*) Compute the root-mean-square speed of a nitrogen molecule at 20.0 °C.

STRATEGY: Evaluate $v_{rms} = \sqrt{3RT/M}$, where M is the molar mass of N_2 (14 g/mol). Be sure to change the temperature value to the Kelvin scale.

SOLUTION:

$$[\text{ans: } 511 \text{ m/s}]$$

(*b*) At what temperatures will the root-mean-square speed be half that value and twice that value?

SOLUTION:

$$[\text{ans: } 73.25 \text{ K } (-200 \,^\circ\text{C}); 1172 \text{ K } (899 \,^\circ\text{C})]$$

PROBLEM 7 (30P).   At 273 K and 1.00×10^{-2} atm the density of a gas is 1.24×10^{-5} g/cm^3. (*a*) Find v_{rms} for the gas molecules.

STRATEGY: Since the total mass of the gas is nM, where n is the number of moles and M is the molar mass, the density is given by $\rho = nM/V$, where V is the volume. The pressure is given by $p = \rho v_{rms}^2/3$. Solve for v_{rms}. Recall that 1.00 atm $= 1.013 \times 10^5$ Pa.

SOLUTION:

$$[\text{ans: } 495 \text{ m/s}]$$

(*b*) Find the molar mass of the gas and identify it.

STRATEGY: The average kinetic energy per mole is given by $3RT/2$ and by $Mv_{rms}^2/2$. Equate these two expressions to obtain $M = 3RT/v_{rms}^2$. Evaluate this expression.

SOLUTION:

$$[\text{ans: } 28 \text{ g/mol}, N_2]$$

PROBLEM 8 (36P).   Show that the ideal gas equation, Eq. 21–4, can be written in the alternative forms: (*a*) $p = \rho RT/M$, where ρ is the mass density of the gas and M the molar mass; (*b*) $pV = NkT$, where N is the number of gas particles (atoms or molecules).

318   *Chapter 21: The Kinetic Theory of Gases*

STRATEGY: (a) Use $\rho = nM/V$ to obtain $n = \rho V/M$. Substitute this expression for n in $pV = nRT$. (b) Substitute $R = N_A k$ in $pV = nRT$, then replace $N_A n$ with N.

SOLUTION:

PROBLEM 9 (42E). Derive an expression, in terms of N/V, $\bar{v}$, and d, for the collision frequency of a gas atom or molecule.

STRATEGY: If λ is the mean free path then the average time between collisions is $\Delta t = \lambda/\bar{v}$ and the frequency of collisions is given by $f = 1/\Delta t = \bar{v}/\lambda$. Use Eq. 21–18 to substitute for λ.

SOLUTION:

$$\left[\text{ans: } \sqrt{2}\pi\bar{v}d^2 N/V\,\right]$$

PROBLEM 10 (45P). (a) What is the molar volume (the volume per mole) of an ideal gas at standard conditions (0.00°C, 1.00 atm)?

STRATEGY: Substitute $p = 1.013 \times 10^5$ Pa, $n = 1$, and $T = 273$ K into $pV = nRT$, then solve for V.

SOLUTION:

$$\left[\text{ans: } 0.0225\,\text{m}^3\ (22.5\,\text{L})\right]$$

(b) Calculate the ratio of the root-mean-square speed of helium atoms to that of neon atoms under these conditions.

STRATEGY: Since the root-mean-square speed is given by $v_{\text{rms}} = \sqrt{3RT/M}$, where M is the molar mass, the ratio is given by $\sqrt{M_{\text{Ne}}/M_{\text{He}}}$. The molar mass of helium is $M_{\text{He}} = 4.026 \times 10^{-3}$ kg and the molar mass of neon is $M_{\text{Ne}} = 20.183 \times 10^{-3}$ kg.

SOLUTION:

$$\left[\text{ans: } 2.25\right]$$

(c) What would be the mean free path of helium atoms under these conditions? Assume the atomic diameter d to be 1.00×10^{-8} cm.

STRATEGY: Evaluate $\lambda = 1/\sqrt{2}\pi d^2 N/V$. Take N to be Avogadro's number and V to be the molar volume.

SOLUTION:

$$\left[\text{ans: } 8.4 \times 10^{-7}\,\text{m}\right]$$

(d) What would be the mean free path of neon atoms under these conditions? Assume the same atomic diameter as for helium.

SOLUTION:

$$\left[\text{ans: } 8.4 \times 10^{-7}\,\text{m}\right]$$

(e) Comment on the results of parts (c) and (d) in view of the fact that the helium atoms are traveling faster than the neon atoms.

STRATEGY: Review the derivation of the Eq. 21–18 to see why the speed does not enter.

SOLUTION:

PROBLEM 11 (58E). (a) What is the internal energy of 1.0 mol of an ideal monatomic gas at 273 K?

STRATEGY: Take the internal energy to be 0 at temperature $T = 0$ K. Then the internal energy at any temperature is given by $E_{int} = nC_V T$, where C_V is the specific heat for a constant volume process. For an ideal monatomic gas $C_V = \frac{3}{2}R$.

SOLUTION:

$$\left[\text{ans: } 3400\,\text{J}\right]$$

(b) Does it depend on volume or pressure?

SOLUTION:

$$\left[\text{ans: no}\right]$$

You should be able to solve problems involving the first law of thermodynamics: $\Delta E_{int} = Q - W$. For some of these you need to know the specific heat or the molar specific heat. Remember that the change in the internal energy is given by $\Delta E_{int} = nC_V\,\Delta T$ and the heat input is given by $Q = nC\,\Delta T$, where C is the molar specific heat for the process.

PROBLEM 12 (61P).   Let 20.9 J of heat be added to a particular ideal gas. As a result, its volume changes from 50.0 to 100 cm^3 while the pressure remains constant at 1.00 atm. (*a*) By how much did the internal energy of the gas change?

STRATEGY: Use $\Delta E_{int} = Q - W$. Since the pressure is constant the work W done by the gas is $p \Delta V$. Since the volume increases W is positive.

SOLUTION:

$$[\,\text{ans: } 15.9\,\text{J}\,]$$

(*b*) If the quantity of gas present is 2.00×10^{-3} mol, find the molar specific heat at constant pressure.

STRATEGY: Since $pV = nRT$ the change in volume and change in temperature are related by $p\Delta V = nR\,\Delta T$, so $\Delta T = (p/nR)\,\Delta V$. The molar specific heat is determined by $C_p = Q/n\Delta T = (R/p)\,Q/\Delta V$.

SOLUTION:

$$[\,\text{ans: } 34.4\,\text{J/mol}\cdot\text{K}\,]$$

(*c*) Find the molar specific heat at constant volume.

STRATEGY: Use $C_V = \Delta E_{int}/n\Delta T = (R/p)\,\Delta E_{int}/\Delta V$.

SOLUTION:

$$[\,\text{ans: } 26.1\,\text{J/mol}\cdot\text{K}\,]$$

PROBLEM 13 (62P).   A quantity of ideal monatomic gas consists of n moles initially at temperature T_1. The pressure and volume are then slowly doubled in such a manner as to trace out a straight line on a p-V diagram. In terms of n, R, and T_1 what are (*a*) W, (*b*) ΔE_{int}, and (*c*) Q?

STRATEGY: (*a*) Suppose the volume increases from V_1 to V_2 and the pressure increases from p_1 to p_2. Then the straight line on the p-V diagram is described by $p - p_1 = (p_2 - p_1)(V - V_1)/(V_2 - V_1)$. Since $p_2 = 2p_1$ and $V_2 = 2V_1$ this is $p - p_1 = p_1(V - V_1)/V_1$. Carry out the integration for W:

$$W = \int_{V_1}^{V_2} p\,dV = \int_{V_1}^{2V_1} \left[p_1 + \frac{p_1(V - V_1)}{V_1} \right] dV\,.$$

You should get $W = \frac{3}{2}p_1V_1$. Use $p_1V_1 = nRT_1$ to write this in terms of T_1.

(*b*) Use $\Delta E_{int} = nC_V\,\Delta T$. The final temperature is given by $4p_1V_1 = nRT_2$ or $T_2 = 4T_1$. Thus $\Delta T = 3T_1$. The molar specific heat is $C_V = 3/2R$ for an ideal monatomic gas.

(*c*) Use the first law in the form $\Delta E_{int} = Q - W$ to find Q.

SOLUTION:

$$\left[\,\text{ans: } (a)\ \tfrac{3}{2}nRT_1;\ (b)\ \tfrac{9}{2}nRT_1;\ (c)\ 6nRT_1\,\right]$$

(*d*) If one were to define a molar specific heat for this process, what would be its value?

STRATEGY: Use $C = Q/\Delta T$.

SOLUTION:

$$\left[\,\text{ans: } 2R\,\right]$$

PROBLEM 14 (67E). Suppose 12.0 g of oxygen (O_2) is heated at constant atmospheric pressure from 25.0 to 125 °C. (*a*) How many moles of oxygen are present? (See Table 21–1.)

STRATEGY: According to Table 21–1 of the text the molar mass of oxygen is $M = 32.0\,\text{g/mol}$. If m is the mass of the sample then the number of moles present is $n = m/M$.

SOLUTION:

$$\left[\,\text{ans: } 0.375\,\text{mol}\,\right]$$

(*b*) How much heat is transferred to the oxygen? (The molecules rotate but do not oscillate.)

STRATEGY: Use $Q = nC_p\,\Delta T$. Since the molecules are diatomic $C_p = C_V + R = (5/2)R + R = 7R/2$.

SOLUTION:

$$\left[\,\text{ans: } 1090\,\text{J}\,\right]$$

(*c*) What fraction of the heat is used to raise the internal energy of the oxygen?

STRATEGY: The change in the internal energy is given by $\Delta E_{\text{int}} = nC_V\,\Delta T$, where $C_V = 5R/2$. The fraction of the heat that is used to increase the internal energy is given by $f = \Delta E_{\text{int}}/Q$.

SOLUTION:

$$\left[\,\text{ans: } 0.714\,\right]$$

Some problems deal with ideal gases as they undergo adiabatic processes. Remember that the combination pV^γ remains constant for such processes. Here γ is the ratio of the specific heats: $\gamma = C_p/C_V$. The ideal gas law is still valid and for some problems $pV = nRT$ must be solved simultaneously with $pV^\gamma = $ constant.

PROBLEM 15 (71E). (*a*) One liter of gas with $\gamma = 1.3$ is at 273 K and 1.0 atm pressure. It is suddenly (adiabatically) compressed to half its original volume. Find its final pressure and temperature.

STRATEGY: Since the initial and final states are on the same adiabat $p_1V_1^\gamma = p_2V_2^\gamma$ and $p_2 = (V_1/V_2)^\gamma p_1$. Substitute $V_2 = V_1/2$. The ideal gas law yields $p_1V_1/T_1 = p_2V_2/T_2$ so $T_2 = p_2T_1/2p_1$, once $V_2 = V_1/2$ is used. Notice that no changes of units are required since the pressures and volumes enter as ratios.

SOLUTION:

$$[\text{ans: } 2.5 \text{ atm}; 340 \text{ K}]$$

(*b*) The gas is now cooled back to 273 K at constant pressure. What is its final volume?

STRATEGY: Use the ideal gas law in the form $p_2V_2/T_2 = p_3V_3/T_3$. Substitute $p_3 = p_2$ and solve for V_3. Again no changes of units are required.

SOLUTION:

$$[\text{ans: } 0.40 \text{ L}]$$

PROBLEM 16 (80P). An ideal gas experiences an adiabatic compression from $p = 1.0$ atm, $V = 1.0 \times 10^6$ L, $T = 0.0\,^\circ$C to $p = 1.0 \times 10^5$ atm, $V = 1.0 \times 10^3$ L. (*a*) Is this a monatomic, a diatomic, or a polyatomic gas?

STRATEGY: For an adiabatic process $p_1V_1^\gamma = p_2V_2^\gamma$. Solve for γ. You should get $\gamma = \ln(P_1/p_2)/\ln(V_2/V_1)$. Compare your numerical result with 1.67 (monatomic), 1.40 (diatomic), and 1.29 (polyatomic).

SOLUTION:

$$[\text{ans: } \gamma = 1.67, \text{ monatomic}]$$

(*b*) What is the final temperature?

STRATEGY: Use the ideal gas law in the form $p_1V_1/T_1 = p_2V_2/T_2$. Solve for T_2.

SOLUTION:

$$[\text{ans: } 2.7 \times 10^4 \text{ K}]$$

(c) How many moles of gas are present?

STRATEGY: Use the ideal gas law in the form $p_1 V_1 = n R T_1$. Solve for n.

SOLUTION:

$$\left[\,\text{ans: } 4.5 \times 10^4 \text{ mol}\,\right]$$

(d) What is the total translational kinetic energy per mole before and after the compression?

STRATEGY: The translational kinetic energy per mole is given by $\frac{3}{2}RT$. For the energy before compression use $T_1 = 273\,\text{K}$ and for the energy after compression use $T_2 = 2.73 \times 10^4\,\text{K}$.

SOLUTION:

$$\left[\,\text{ans: } 3.40 \times 10^3 \text{ J}, 3.40 \times 10^5 \text{ J}\,\right]$$

(e) What is the ratio of the squares of the rms speeds before and after the compression?

STRATEGY: Since $K = \frac{1}{2} M v_{\text{rms}}^2$ the ratio of the rms speeds is the same as the ratio of the kinetic energies.

SOLUTION:

$$\left[\,\text{ans: } 0.010\,\right]$$

III. MATHEMATICAL SKILLS

Average and root-mean-square values. This chapter introduces some of the ideas of statistics, the most important of which are the average and root-mean-square of a collection of numbers. To find the average, add the numbers and divide the result by the number of terms in the sum. To find the root-mean-square, add the squares of the numbers, divide the sum by the number of terms, and take the square root of the result. The root-mean-square is the square root of the average of the squares of the numbers.

The speeds of 5 molecules are 350 m/s, 225 m/s, 432 m/s, 375 m/s, and 450 m/s. Find their average and root-mean-square speeds. The following table might help.

PARTICLE	SPEED (m/s)	SPEED SQUARED (m²/s²)
1	_____	_____
2	_____	_____
3	_____	_____
4	_____	_____
5	_____	_____
sum	_____	_____
average	_____	_____

$$\overline{v} = \underline{\hspace{2cm}} \text{ m/s} \qquad v_{\text{rms}} = \underline{\hspace{2cm}} \text{ m/s}$$

Your answers should be $\overline{v} = 366$ m/s and $v_{\text{rms}} = 375$ m/s. Notice that the average and root-mean-square values are different. In fact, the root-mean-square speed is greater than the average speed.

You should be able to show algebraically that if all the molecules have the same speed, their average and root-mean-square speeds are the same. If there are N molecules and each has speed v, then $\sum v = Nv$ and $\sum v^2 = Nv^2$. Complete the proof in the space below:

Integrals of the Maxwell distribution function. This chapter contains several integrals that are difficult to evaluate using only a knowledge of introductory calculus. They are associated with the average and mean-square speed for molecules with a Maxwellian speed distribution. The average speed is given by

$$\overline{v} = 4\pi \left[\frac{M}{2\pi RT} \right]^{3/2} \int_0^\infty v^3 e^{-Mv^2/2RT} \, dv$$

and the mean-square speed is given by

$$\overline{v^2} = 4\pi \left[\frac{M}{2\pi RT} \right]^{3/2} \int_0^\infty v^4 e^{-Mv^2/2RT} \, dv \,.$$

Most standard integral tables list the following integrals:

$$\int_0^\infty x^{2a} e^{-px^2} \, dx = \frac{(2a-1)!!}{2(2p)^a} \sqrt{\frac{\pi}{p}}$$

and

$$\int_0^\infty x^{2a+1} e^{-px^2} \, dx = \frac{a!}{2p^{a+1}} \,.$$

Here p is any positive number and a is any integer. $a!$ is the factorial of a; that is, $a! = 1 \cdot 2 \cdot 3 \cdot 4 \ldots a$. $(2a-1)!!$ is the product of all odd integers from 1 to $2a-1$; that is,

$(2a - 1)!! = 1 \cdot 3 \cdot 5 \cdot 7 \ldots (2a - 1)$. We use the first integral when x to an *even* power multiplies the exponential in the integrand and the second when x to an *odd* power multiplies the exponential.

To evaluate the integral for the average speed of a Maxwellian distribution substitute $v = x$ and $M/2RT = p$ into the equation for $\bar{v}$. You should obtain

$$\bar{v} = 4\pi \left[\frac{M}{2\pi RT}\right]^{3/2} \int_0^\infty x^3 e^{-px^2}\, \mathrm{d}x\,.$$

The integral is the same as the second integral taken from the tables, with $a = 1$. So

$$\bar{v} = 4\pi \left[\frac{M}{2\pi RT}\right]^{3/2} \frac{1}{2p^2} = 4\pi \left[\frac{M}{2\pi RT}\right]^{3/2} \frac{2R^2 T^2}{M^2} = \sqrt{\frac{8RT}{\pi M}}\,.$$

To evaluate the integral for the mean-square speed of a Maxwellian distribution make the same substitutions to obtain

$$\overline{v^2} = 4\pi \left[\frac{M}{2\pi RT}\right]^{3/2} \int_0^\infty x^4 e^{-px^2}\, \mathrm{d}x\,.$$

The integral is the same as the first one taken from the tables, with $a = 2$. So

$$\overline{v^2} = 4\pi \left[\frac{M}{2\pi RT}\right]^{3/2} \frac{3}{2(2p)^2} \sqrt{\frac{\pi}{p}} = 4\pi \left[\frac{M}{2\pi RT}\right]^{3/2} \frac{3R^2 T^2}{2M^2} \sqrt{\frac{2\pi RT}{M}} = \frac{3RT}{M}\,.$$

Thus the root-mean-square speed is $\sqrt{3RT/M}$.

IV. NOTES

Chapter 22
ENTROPY AND THE SECOND LAW
OF THERMODYNAMICS

I. BASIC CONCEPTS

Here you study the second law of thermodynamics and a closely associated property of any macroscopic system, its entropy. The law is formulated in three equivalent forms: in terms of heat engines, refrigerators, and entropy. Be sure you understand all three formulations and the far-reaching implications of the law. Pay careful attention to calculations of entropy changes during various process. Learn to distinguish between reversible and irreversible processes and understand how entropy behaves for each case.

The second law of thermodynamics. Quote the first form of the second law from the text:

_____

_____

There are many processes that convert heat to work. They do not violate the second law because they do not operate in cycles. Give an example: _____

_____

Although the second law is universally applicable, it is discussed in terms of <u>heat engines</u>. In terms of work and heat describe what a heat engine does: _____

_____

A heat engine makes use of a working substance whose thermodynamic state varies during the process but which periodically returns to the same state. The same sequence of steps is performed on it during every cycle. If a heat engine receives heat of magnitude $|Q_H|$ from a high temperature thermal reservoir and rejects heat of magnitude $|Q_C|$ to a low temperature reservoir, then according to the first law of thermodynamics it does work $W =$ _____ over a cycle. Note that because the engine works in cycles the change in the internal energy is

_____ .

The <u>thermal efficiency</u> of a heat engine is defined in terms of $|Q_H|$ and $|W|$ by

$$e =$$

In terms of $|Q_H|$ and $|Q_C|$, it is $e =$ _____ . .

A *perfect* heat engine takes in heat and does an identical amount of work. No heat is rejected. Since $|W| = |Q_H|$ for a perfect heat engine, its efficiency is _____ . Such an engine does not violate the first law of thermodynamics but it does violate the second law. According to the second law $|Q_C|$ cannot be zero, so $|W|$ cannot be the same as $|Q_H|$ and the efficiency of a heat engine cannot be exactly 1. The first law tells us that in a cycle heat input is turned to work and heat output. The second law tells us it cannot be turned completely to work; some must become heat output.

Quote the second form of the second law from the text: _____

_____

_____

This form of the second law is formulated in terms of refrigerators. In terms of work and heat tell what a refrigerator does: _____

_____

Suppose a refrigerator takes in heat of magnitude $|Q_C|$ at a low temperature reservoir and rejects heat of magnitude $|Q_H|$ at a high temperature reservoir. It does (negative) work W. Its coefficient of performance is defined in terms of $|Q_C|$ and $|W|$ by

$$K =$$

In terms of $|Q_C|$ and $|Q_H|$, the coefficient is $K =$ _____. For a *perfect* refrigerator $W =$ _____ and $K =$ _____. A perfect refrigerator does not violate the first law of thermodynamics but it does violate the second law and cannot be built. Specifically, $|W|$ cannot be zero. Work must be done to cyclically absorb heat from a low temperature reservoir and reject it to a high temperature reservoir. The statement does NOT preclude heat flowing from a high temperature reservoir to a low temperature reservoir without work being done. This is a natural occurrence.

The two statements of the second law are equivalent to each other. If the first form were not valid, then a perfect heat engine could be combined with an imperfect refrigerator in such a way that the combination would be a perfect refrigerator and thus would violate the second form. Similarly, if the second form were not valid then a perfect refrigerator could be combined with an imperfect heat engine in such a way that the combination would be a perfect heat engine and thus would violate the first form.

Reversible processes. The concepts of reversible and irreversible processes are important for understanding entropy and the third form of the second law. Explain what each is, being sure to distinguish between them.

reversible: _____

_____

irreversible: _____

_____

Reversible processes must be carried out quasi-statically; that is, extremely slowly. After a reversible process is carried out on a system, carrying out the steps in the reverse order brings the system back to its original state. This means, for example, that friction cannot be present.

If two different processes are used to take a system from the same initial equilibrium state to the same final equilibrium state, one reversible and one irreversible, the work and heat may be different but changes in the _____, _____, _____, and _____ will be the same. These quantities are called state variables. Their changes during any process that begins and ends at an equilibrium state depends only on the initial and final states and not on the process. Over a cycle the net change in any state variable is _____.

Carnot cycles.   A Carnot cycle consists of two isothermal processes at different temperatures, linked by two adiabatic processes. Here are the steps when the cycle is run as a heat engine. Tell how the device of Fig. 22–1 of the text is manipulated to carry out each step.

1. Isothermal expansion at high temperature: _____

_____

2. Adiabatic expansion: _____

_____

3. Isothermal compression at low temperature: _____

_____

4. Adiabatic compression: _____

_____

Diagram a Carnot cycle on the p-V axes to the right. Identify each of the isotherms and adiabats with one of the steps above and label it accordingly. Assume the cycle is run as a heat engine and mark the portion of the cycle during which heat flows from a reservoir to the working substance and the portion during which heat flows from the working substance to a reservoir.

In terms of the heat $|Q_H|$ absorbed at the higher temperature and the heat $|Q_C|$ rejected at the lower temperature the efficiency of a Carnot heat engine is given by $e =$ _____ . If an ideal gas is used as the working substance then, in terms of the reservoir temperatures T_H and T_C, $|Q_H|/|Q_C| =$ _____ and the efficiency is given by

$$e =$$

Notice the efficiency increases as the temperature of the hot reservoir increases but cannot reach 1 unless $T_C = 0\,\mathrm{K}$.

If a Carnot cycle is run as a refrigerator the coefficient of performance is given in terms of the temperatures by

$$K =$$

The efficiency of a Carnot cycle is an upper limit for the efficiency of all heat engines operating between the same temperatures. This is stated precisely by the theorem given at the beginning of Section 22–7: _____

_____

The theorem can be proven by assuming an engine exists with a greater efficiency than a Carnot engine. Work done by such an engine is used to drive the Carnot cycle as a refrigerator and the combination is shown to be a perfect refrigerator, in violation of the second form of the second law. Notice that the heat engine need not be reversible for the theorem to be valid.

Every *reversible* heat engine operating between two thermal reservoirs must have the same efficiency as a Carnot engine operating between the same reservoirs. If it did not we could combine the engine with a Carnot engine and build either an engine or a refrigerator that violates the second law.

These two theorems, taken together, tell us that a heat engine has the same efficiency as a Carnot engine operating between the same temperatures if the engine is reversible and a lower efficiency if it is not. Another important consequence is that the efficiency of a Carnot heat engine is given by $e = (T_H - T_C)/T_H$, regardless of the working substance.

Entropy. If heat dQ is received by the system when its state changes from an initial equilibrium state to a nearby equilibrium state, then the entropy of the system changes by $dS =$ _____, where T is the temperature. This defines entropy. If the state changes reversibly from some initial equilibrium state i to some final equilibrium state f, the change in entropy can be computed by evaluating the integral

$$\Delta S = S_f - S_i =$$

Recognize that the process connecting the initial and final states must be reversible for this equation to be valid. Notice that the integral $\int_i^f dQ$ depends on the process but the integral $\int_i^f dQ/T$ does not.

Entropy is another state variable. This means that whenever a system is in the same thermodynamic state it has the same entropy, just as it has the same temperature, pressure, and volume, and internal energy. The change in any of these variables over a cycle is _____.

To evaluate the integral for ΔS you must pick some reversible process that connects the initial and final states but, because entropy is a state variable, it need not be the process actually carried out on the system. To evaluate ΔS for an irreversible process pick a *reversible* process with the same initial and final states and use that process to find the value of the integral. Sample Problem 22–7 and examples in Section 22–9 of the text show how to calculate the entropy changes for various situations. Here are some examples.

1. Change in phase. Suppose mass m of a substance with heat of fusion L_F melts at temperature T. For this process T is constant and $\Delta S = Q/T$. Since $Q = mL_F$, $\Delta S =$ _____. The change in entropy on freezing is $\Delta S =$ _____. Note that the entropy of the substance increases on melting and decreases on freezing.

2. Constant volume process. Suppose the molar specific heat at constant volume C_V for a certain gas is independent of T, V, and p. If an n-mole sample of this gas undergoes an infinitesimal reversible change of state at constant volume, so its temperature changes by dT, then the heat absorbed is $dQ =$ _____ dT and the change in entropy is $dS = dQ/T =$ _____. For a finite change in temperature, from T_1 to T_2 say, the change in entropy is the integral of this expression. In the space below carry out the integration to show that $\Delta S = nC_V \ln(T_f/T_i)$:

3. **Constant pressure process.** Suppose the molar specific heat at constant pressure C_p for a certain gas is independent of T, V, and p. If an n-mole sample of this gas undergoes an infinitesimal reversible change of state at constant pressure, so its temperature changes by dT, then the heat absorbed is $dQ = $ _____ dT and the change in entropy is $dS = dQ/T = $ _____. For a finite change in temperature, from T_i to T_f, the change in entropy is the integral of this expression. In the space below carry out the integration to show that $\Delta S = nC_p \ln(T_f/T_i)$:

4. **Isothermal process.** If the gas undergoes an infinitesimal reversible change of state at constant temperature, then $dE_{int} = nC_V dT = 0$ and $dQ = dW = p\,dV$. Thus $dS = (p/T)\,dV$ and $\Delta S = \int (p/T)\,dV$, where the limits of integration are the initial volume V_i and the final volume V_f. To evaluate this integral p must be known as a function of volume and temperature. The equation of state gives this information. For n moles of an ideal gas $p/T = $ _____. In the space below show that $\Delta S = nR\ln(V_f/V_i)$:

As an example of an irreversible process consider the adiabatic free expansion of an ideal gas from volume V_i to volume V_f. The net heat transferred is $Q = $ _____, the net work done is $W = $ _____, the change in the internal energy is $\Delta E_{int} = $ _____, and the change in the temperature is $\Delta T = $ _____. The change in entropy is calculated using a reversible isothermal expansion from V_i to V_f. Explain why a reversible process is used: _____

Explain why an isothermal process is used: _____

The change in entropy is

$$\Delta S = $$

Notice that no heat is transferred but the entropy increases.

The second law of thermodynamics, in its third form, is expressed in terms of entropy changes. It is: _____

Notice you must consider the *total* entropy of the system *and* its environment. Many processes lower the entropy of a system but, according to the second law, they must then _____ the entropy of the environment.

If a change of state is reversible then the change in the total entropy of the system and its environment is _____. If the entropy of the system decreases in a reversible process then the entropy of the environment must _____. by exactly the same amount. If the system changes state through an irreversible process then the change in the total entropy of the system and its environment is _____.

Suppose two identical systems each go from the same initial state to the same final state. For system 1 the process is carried out reversibly while for system 2 it is carried out irreversibly. Suppose the change in the entropy for system 1 is ΔS_1. Then the change in the entropy of the environment of system 1 is _____. The change in entropy of system 2 is _____. The change in the magnitude of the entropy of the environment of system 2 is _____ than $|\Delta S_1|$.

You should be able to show that perfect heat engines and perfect refrigerators operating between non-zero temperatures violate this statement of the second law. For example, consider a heat engine that exchanges heat Q_H at temperature T_H and heat Q_C at temperature T_C (Q_C is negative). Over a cycle the change in entropy of the working substance is _____. The change in the entropy of the reservoir at T_H is _____ and the change in the entropy of the reservoir at T_C is _____. (Be careful about signs here: if heat is positive for the working substance it is negative for the reservoir.) The change in the total entropy of the working substance and the reservoirs is $\Delta S_{total} =$_____. Note that Q_C cannot be zero since that would make ΔS_{total} negative.

If the engine is reversible then over a cycle $\Delta S_{total} =$ _____ and, in terms of the temperatures, $|Q_C|/|Q_H| =$ _____, regardless of the working substance. If the engine is irreversible then ΔS is positive and $|Q_C| >$_____.

All real (non-ideal) processes are irreversible to some extent. We therefore expect the total entropy of every real system and its environment to _____ as time goes on. This has often been used to assign a "direction" to time.

At the microscopic level entropy is related to disorder. The relationship is $S =$ _____, where W is a measure of _____.

II. PROBLEM SOLVING

Many of the problems deal with heat engines and refrigerators. In each case heat is absorbed over part of the cycle, heat is rejected over another part, and work is done on or by the working substance. Given any one of these quantities and the efficiency or coefficient of performance, you should be able to calculate the others. The relevant equations are the first law of thermodynamics and the definition of either the efficiency or the coefficient of performance. Other problems give you two of the energy terms and ask for the efficiency or coefficient of performance.

You should also practice calculating the heat and work for various processes (isothermal, adiabatic, constant volume, and constant pressure). Use the heat capacity or specific heat to compute the heat, use $\int p\,dV$ to compute the work.

PROBLEM 1 (2E). A car engine delivers 8.2 kJ of work per cycle. (*a*) Before a tune up, the efficiency is 25%. Calculate, per cycle, the heat absorbed from the combustion of fuel and the heat exhausted to the atmosphere.

STRATEGY: The efficiency is given by $e = |W|/|Q_H|$, where $|Q_H|$ is the magnitude of the heat absorbed from fuel combustion. Thus $|Q_H| = e|W|$. According to the first law of thermodynamics $|Q_H| - |Q_C| - |W| = 0$, where $|Q_C|$ is the magnitude of the heat exhausted to the atmosphere. Solve for $|Q_C|$.

SOLUTION:

[ans: 33 kJ, 25 kJ]

(b) After a tune up, the efficiency is 31%. What are the new values of the quantities calculated in (a)?

SOLUTION:

[ans: 27 kJ, 18 kJ]

PROBLEM 2 (5P). One mole of a monatomic ideal gas initially at a volume of 10 L and a temperature of 300 K is heated at constant volume to a temperature of 600 K, allowed to expand isothermally to its initial pressure, and finally compressed isobarically (that is, at constant pressure) to its original volume, pressure, and temperature. (a) Compute the heat input to the system during one cycle.

STRATEGY: Heat is absorbed during the constant volume heating and during the isothermal expansion. It is rejected during the isobaric compression. Calculate the heat absorbed during the first two portions of the process.

During the constant volume heating $Q = nC_V \Delta T = n(3/2)R\Delta T$. During the isothermal expansion the change in the internal energy is 0 (it depends only on the temperature), so $Q = W = \int p\,dV = \int (nRT/V)\,dV = nRT\ln(V_f/V_i)$. For this process $T = 600$ K and $V_i = 10$ L. You must use $pV = nRT$ to find the final volume. The final pressure is the original pressure. Since the temperature has been doubled and the pressure is the same the volume must be twice as great, so $V_f = 20$ L.

SOLUTION:

[ans: 7200 J]

(b) What is the net work done by the gas during one cycle?

STRATEGY: During the constant volume heating the work done is 0. During the isothermal expansion the work done is $nRT\ln 2$, as found above. During the isobaric compression the work done is $p\,\Delta V$. Use $pV = nRT$, with $V = 10$ L and $T = 300$ K, to find the value of p. The initial volume is 20 L and the final volume is 10 L so $\Delta V = -10$ L.

SOLUTION:

(c) What is the efficiency of this cycle?

STRATEGY: Divide the work done by the heat absorbed.

SOLUTION:

PROBLEM 3 (7E).   To make ice, a freezer extracts 42 kcal of heat at $-12\,°C$ during each cycle. The freezer has a coefficient of performance of 5.7. The room temperature is $26\,°C$. (a) How much heat per cycle is delivered to the room?

STRATEGY: The coefficient of performance is given by $K = |Q_C|/(|Q_H| - |Q_C|)$. Solve for $|Q_H|$.

SOLUTION:

(b) How much work per cycle is required to run the freezer?

STRATEGY: According to the first law of thermodynamics $|Q_C| - |Q_H| - W = 0$. Solve for W.

SOLUTION:

Many problems deal with Carnot cycles, run either as heat engines or as refrigerators. You should know how to calculate the efficiency or coefficient of performance, given the temperatures of the two reservoirs. As before, you should know how to compute the heat absorbed, the heat rejected, and the work done.

PROBLEM 4 (10E).   In a Carnot cycle, the isothermal expansion of an ideal gas takes place at 400 K and the isothermal compression at 300 K. During the expansion, 500 cal of heat is transferred to the gas. Determine (*a*) the work performed by the gas during the isothermal expansion, (*b*) the heat rejected from the gas during the isothermal compression, and (*c*) the work done on the gas during the isothermal compression.

STRATEGY: (*a*) During the isothermal expansion the internal energy does not change (the internal energy of an ideal gas depends only on the temperature), so $Q_H - W = 0$.
   (*b*) The efficiency of a heat engine is given by $e = (|Q_H| - |Q_C|)/|Q_H|$. For a Carnot cycle $e_{\text{car}} = (T_H - T_C)/T_H$. Thus $(|Q_H| - |Q_C|)/|Q_H| = (T_H - T_C)/T_H$. Solve for $|Q_C|$.
   (*c*) During the isothermal compression the internal energy does not change, so $Q_C - W = 0$.

SOLUTION:

[ans: (*a*) 2090 J; (*b*) 1570 J; (*c*) 1570 J]

PROBLEM 5 (17E).   (*a*) A Carnot engine operates between a hot reservoir at 320 K and a cold reservoir at 260 K. If it absorbs 500 J of heat per cycle at the hot reservoir, how much work per cycle does it deliver?

STRATEGY: Since the engine operates as a Carnot cycle its efficiency can be calculated from the temperatures of the reservoirs: $e = (T_H - T_C)/T_H$. It is also given by $e = |W|/|Q_H|$. Solve $(T_H - T_C)/T_H = |W|/|Q_H|$ for $|W|$.

SOLUTION:

[ans: 94 J]

(*b*) If the same engine, working in reverse, functions as a refrigerator between the same two reservoirs, how much work per cycle must be supplied to remove 1000 J of heat from the cold reservoir?

STRATEGY: Since the refrigerator is operating as a Carnot cycle, the coefficient of performance can be calculated from the reservoir temperatures: $K = T_C/(T_H - T_C)$. It is also given by $K = |Q_C|/|W|$. Solve $T_C/(T_H - T_C) = |Q_C|/|W|$ for $|W|$.

SOLUTION:

[ans: 230 J]

Here are some problems that deal with real engines and refrigerators.

PROBLEM 6 (20P).   A heat pump is used to heat a building. The outside temperature is $-5.0\,°$C, and the temperature inside the building is to be maintained at $22\,°$C. The coefficient of performance is 3.8 and the heat pump delivers $1.8\,$Mcal of heat to the building each hour. At what rate must work be done to run the heat pump?

STRATEGY: The coefficient of performance is given by $K = -Q_C/W$. Q_C is positive and W is negative for a refrigerator. Since the heat pump runs in complete cycles the change in the internal energy is 0 and $Q_C + Q_H - W = 0$. Thus $Q_C = W - Q_H$ and $K = (Q_H - W)/W$. Consider the energy transfers that take place in $1\,$h and solve for W. Divide by $3600\,$s/h to obtain the rate in watts.

SOLUTION:

[ans: $440\,$W]

PROBLEM 7 (23P).   Find the relation between the efficiency of a reversible ideal heat engine and the coefficient of performance of the reversible refrigerator obtained by running the engine backward.

STRATEGY: The efficiency of the heat engine is given by $e = |W|/|Q_H|$ and the coefficient of performance is given by $K = |Q_C|/|W|$. The magnitudes of all the energy transfers are the same. Solve for $|Q_H|$ in terms of K: $|Q_C| = K|W|$. According to the first law of thermodynamics $|Q_C| - |Q_H| + |W| = 0$ so $|Q_H| = |Q_C| + |W| = (K + 1)|W|$. Substitute this result into the expression for the efficiency of the heat engine.

SOLUTION:

[ans: $e = 1/(K + 1)$]

PROBLEM 8 (28P).   An air conditioner operating between 93 and $70\,°$F is rated at $4000\,$Btu/h cooling capacity. It coefficient of performance is 27% of that of a Carnot refrigerator operating between the same two temperatures. What is the required horsepower of the air conditioner motor?

STRATEGY: The coefficient of performance of the Carnot refrigerator is given by $K_{\text{car}} = T_C/(T_H - T_C)$, so the coefficient of performance of the air conditioner is $K = 0.27T_C/(T_H - T_C)$. The coefficient of performance is also given by $K = |Q_C|/|W|$. Consider one hour of operation, for which $|Q_C| = 4000\,$Btu, and solve $0.27T_C/(T_H - T_C) = |Q_C|/|W|$ for $|W|$. This gives the power output of the motor in Btu/h. Use $1\,$Btu/h $= 3.929 \times 10^{-4}\,$hp to convert to horsepower.

SOLUTION:

[ans: $0.25\,$hp]

PROBLEM 9 (30P).   One mole of an ideal monatomic gas is used as the working substance of an engine that operates on the cycle shown in Fig. 22–21. Assume that $p = 2p_0$, $V = 2V_0$, $p_0 = 1.01 \times 10^5$ Pa, and $V_0 = 0.0225$ m^3. Calculate (a) the work done per cycle, (b) the heat added per cycle during the expansion stroke abc, and (c) the engine efficiency.

STRATEGY: (a) During the process bc the work done is $W_{bc} = p(V - V_0) = 2p_0V_0$. During the process da the work done is $W_{da} = p_0(V_0 - V) = -p_0V_0$. No work is done during the other two portions of the cycle, so the total work done per cycle is $W = 2p_0V_0 - p_0V_0 = p_0V_0$.

(b) The process ab is at constant volume so the heat added is $Q_{ab} = nC_V\,\Delta T = n(3/2)R\,\Delta T$. Use the ideal gas law to find the change in temperature: $\Delta T = (p_0V - p_0V_0)/nR = p_0V_0/nR$. Thus $Q_{ab} = (3/2)p_0V_0$. The process bc is at constant pressure so the heat added is $Q_{bc} = nC_p\,\Delta T = n(5/2)R\,\Delta T$. The change in temperature is $\Delta T = (pV - pV_0)/nR = 2p_0V_0/nR$, so $Q_{bc} = 5p_0V_0$. Heat is rejected during the other two portions of the cycle.

(c) The engine efficiency is given by $W/(Q_{ab} + Q_{bc})$.

SOLUTION:

$\big[$ ans: (a) 2270 J; (b) $14,800$ J; (c) $0.154\big]$

(d) What is the Carnot efficiency of an engine operating between the highest and lowest temperatures that occur in the cycle? How does this compare to the efficiency calculated in (c)?

STRATEGY: The highest temperature occurs when pV has it greatest value, at the upper right corner of the diagram. There $p = 2p_0$ and $V = 2V_0$, so $T_H = 4p_0V_0/nR$. The lowest temperature occurs when pV has its smallest value, at the lower left corner of the diagram. There $p = p_0$ and $V = V_0$, so $T_C = p_0V_0/nR$. The Carnot efficiency is given by $e_{car} = (T_H - T_C)/T_H$.

SOLUTION:

$\big[$ ans: $e_{car} = 0.750$; $e/e_{car} = 0.205\big]$

Here is a problem that asks you to test data for heat engines to see if they violate the first or second laws of thermodynamics. Remember that the second law places an upper limit on the efficiency of an engine.

PROBLEM 10 (36P). An inventor claims to have invented four engines, each of which operates between heat reservoirs at 400 and 300 K. Data on each engine, per cycle of operation, are as follows: Engine (a): $Q_H = 200 J$, $Q_C = -175 J$, $W = 40 J$; engine (b): $Q_H = 500 J$, $Q_C = -200 J$, $W = 400 J$; engine (c): $Q_H = 600 J$, $Q_C = -200 J$, $W = 400 J$; engine (d): $Q_H = 100 J$, $Q_C = -90 J$, $W = 10 J$. Of the first and second laws of thermodynamics, which (if either) does each engine violate?

STRATEGY: Check each engine to see if $Q_H + Q_C - W$ is 0. If it is not the engine violates the first law of thermodynamics. The efficiency of a Carnot heat engine operating between the given temperatures is $e_{car} = (T_H - T_C)/T_H = (400 - 300)/400 = 0.25$. Calculate the efficiency of each engine, using $e = |W|/|Q_H|$, and compare with e_{car}. If $e > e_{car}$ then the engine violates the second law of thermodynamics.

SOLUTION:

[ans: (a) first; (b) first and second; (c) second; (d) neither]

You should know how to calculate changes in entropy for various processes. Remember that $\Delta S = \int (1/T) \, dQ$ for a reversible process between the initial and final state.

PROBLEM 11 (37P). At very low temperatures, the molar specific heat C_V for many solids is (approximately) proportional to T^3; that is, $C_V = AT^3$, where A depends on the particular substance. For aluminum, $A = 7.53 \times 10^{-6}$ cal/mol $\cdot$ K^4. Find the entropy change of 4.00 mol of aluminum when its temperature is raised from 5.00 to 10.0 K.

STRATEGY: Since $dQ = C_V \, dT$, the change in entropy is given by

$$\Delta S = \int_{T_i}^{T_f} \frac{dQ}{T} = \int_{T_i}^{T_f} \frac{nC_V}{T} \, dT = nA \int_{T_i}^{T_f} T^2 \, dT,$$

where n is the number of moles, T_i is the initial temperature, and T_f is the final temperature. Carry out the integration and evaluate the result.

SOLUTION:

[ans: 8.79×10^{-3} cal/K]

PROBLEM 12 (41E).   Suppose that the same amount of heat, say 260 J, is transferred by conduction from a heat reservoir at a temperature of 400 K to another reservoir, the temperature of which is (*a*) 100 K, (*b*) 200 K, (*c*) 300 K, and (*d*) 360 K. Calculate the changes in entropy and discuss the trend.

STRATEGY: The high temperature reservoir loses entropy Q/T_H and the low temperature reservoir gains entropy Q/T_C, so the change in entropy is

$$\Delta S = \frac{Q}{T_C} - \frac{Q}{T_H}.$$

Evaluate this expression for the 4 values of T_C given.

SOLUTION:

[ans: (*a*) 1.95 J/K; (*b*) 0.650 J/K; (*c*) 0.217 J/K; (*d*) 0.072 J/K]

As the initial and final temperatures become closer in value the change in entropy becomes less. The process is becoming more nearly reversible.

PROBLEM 13 (45P).   An 8.0-g ice cube at $-10\,°C$ is dropped into a thermos flask containing 100 cm³ of water at 20 °C. What is the change in entropy of the system when a final equilibrium state is reached? The specific heat of ice is 0.50 cal/g · °C.

STRATEGY: 8.0 g of ice changes temperature from $T_1 = 263$ K to $T_2 = 273$ K. It absorbs heat $Q = mc_i(T_2 - T_1)$ and its entropy changes by $\Delta S = \int (1/T)\,dQ = \int (mc_i/T)\,dT = mc_i \ln(T_2/T_1)$. Here m_i is the mass of ice and c_i is its specific heat (2220 J/kg). The ice then melts. It absorbs heat $Q = m_i L_F$, where L_F is the heat of fusion for water (333 × 10³ J/kg), and its entropy increases by $\Delta S = m_i c_i/T_2$. Finally, the temperature of the resulting water increases from T_2 to the final equilibrium temperature T_3. It absorbs heat $Q = mc_w(T_3 - T_2)$, where c_w is the specific heat of water (4190 J/kg), and its entropy increases by $\Delta S = m_i c_w \ln(T_3/T_2)$. The original water in the thermos changes temperature from $T_4 = 293$ K to T_3. The heat is $Q = Mc_w(T_3 - T_4)$, where M is the mass of the water, and the entropy change is $\Delta S = Mc_w \ln(T_3/T_4)$.

   First find the final temperature T_3 of the system using $Q_{\text{total}} = mc_i(T_2 - T_1) + m_i L_F + mc_w(T_3 - T_2) + Mc_w(T_3 - T_4) = 0$ The mass of the water originally in the thermos can be found by multiplying its volume (100 cm³) by its density (0.998 × 10³ kg/m³), after changing units appropriately. Once a value is found for T_3, calculate and sum the entropy changes.

SOLUTION: Use the space below to find the final temperature.

Now calculate and sum the entropy changes.

$$\left[\,\text{ans: } 0.64\,\text{J/K} \ (0.15\,\text{cal/K})\,\right]$$

PROBLEM 14 (47P). A 50-g block of copper having a temperature of 400 K is placed in an insulating box with a 100-g block of lead having a temperature of 200 K. (*a*) What is the equilibrium temperature of this two-block system?

STRATEGY: The heat absorbed by the lead is given by $m_L c_L (T_F - T_L)$, where m_L is the mass of the lead, c_L is its specific heat (128 J/kg · K), T_L is its initial temperature, and T_F is the final temperature. The heat rejected by the copper is $m_C c_C (T_F - T_C)$, where m_C is the mass of the copper, c_C is its specific heat (386 J/kg · K), and T_C is its initial temperature. Since no energy entered or left the two-block system these heats must sum to 0. Solve $m_L c_L (T_F - T_L) + m_C c_C (T_F - T_C) = 0$ for T_F.

SOLUTION:

$$\left[\,\text{ans: } 320\,\text{K}\,\right]$$

(*b*) What is the change in the internal energy of the two-block system as it goes from the initial condition to the equilibrium condition?

STRATEGY: Since no work is done the energy rejected as heat by the copper comes from its store of internal energy, while the energy accepted as heat by the lead enters its store of internal energy. No energy is lost so these are equal in magnitude.

SOLUTION:

$$\left[\,\text{ans: } 0\,\right]$$

(*c*) What is the change in the entropy of the two-block system? (See Table 20–1.)

STRATEGY: The change in the entropy of the copper is given by $\Delta S_C = \int (1/T)\,dQ = m_C c_C \int (1/T)\,dT = m_C c_C \ln(T_F/T_C)$. Similarly, the change in the entropy of the lead is $\Delta S_L = m_L c_L \ln(T_F/T_L)$. Add the two values to obtain the change in the entropy of the system.

SOLUTION:

$$[\text{ans: } 1.7\,\text{J/K} \;(0.41\,\text{cal/K})]$$

PROBLEM 15 (49P). An ideal diatomic gas is caused to pass through the cycle shown on the *p-V* diagram in Fig. 22–23, where $V_2 = 3.00 V_1$. Determine, in terms of p_1, V_1, T_1, and R: (*a*) p_2, p_3, and T_3 and (*b*) W, Q, ΔE_{int}, and ΔS per mole for all three processes.

STRATEGY: (*a*) Points 1 and 2 lie on the same isothermal so according to the ideal gas law $p_1 V_1 = p_2 V_2$. Substitute $V_2 = 3.00 V_1$ and solve for p_2. Points 1 and 3 lie on the same adiabat so $p_1 V_1^{\gamma} = p_3 V_3^{\gamma}$. Substitute $V_3 = 3.00 V_1$ and solve for p_3. For a diatomic gas $\gamma = 1.4$. To find T_3 use the ideal gas law in the form $p_1 V_1 / T_1 = p_3 V_3 / T_3$.

(*b*) Over the isothermal portion of the cycle $W = \int p\,dV = \int (nRT_1/V)\,dV = nRT_1 \ln(V_2/V_1)$. Substitute for V_2. The change in internal energy is 0 since it depends only on the temperature for an ideal gas. The first law of thermodynamics then reduces to $Q = W$. The change in entropy is given by $\Delta S = Q/T_1$.

Over the constant volume portion of the cycle $W = 0$. The change in the internal energy is $\Delta E_{\text{int}} = nC_V \Delta T = (5/2)nR(T_3 - T_2)$. Substitute for the temperatures in terms of R and T_1. The first law reduces to $Q = \Delta E_{\text{int}}$. The change in entropy is given by $\Delta S = \int (1/T)\,dQ = nC_V \int (1/T)\,dT = (5/2)nR \ln(T_3/T_2)$. Substitute for the temperatures.

Over the adiabatic portion of the cycle $Q = 0$. The work is given by $W = \int p\,dV = \int (p_1 V_1^{\gamma}/V^{\gamma})\,dV = [p_1 V_1^{\gamma}/(1-\gamma)][V_1^{1-\gamma} - V_3^{1-\gamma}]$. Substitute for V_3. The first law reduces to $\Delta E_{\text{int}} = -W$. The change in entropy is 0.

SOLUTION:

$[\text{ans: } (a)\ p_1/3,\ p_1/3^{1.4},\ T_1/3^{0.4};\ (b)\ \text{isothermal: } W = 1.10 RT_1,\ Q = 1.10 RT_1,\ \Delta E_{\text{int}} = 0,\ \Delta S = 1.10 R;\ \text{constant}$
$\text{volume: } W = 0,\ Q = -0.889 RT_1,\ \Delta E_{\text{int}} = -0.889 RT_1,\ \Delta S = -1.10 R;\ \text{adiabatic: } W = -0.889 RT_1,\ Q = 0,$
$\Delta E_{\text{int}} = 0.889 RT_1,\ \Delta S = 0\ ]$

III. NOTES

OVERVIEW III
ELECTRICITY AND MAGNETISM

Electric and magnetic phenomena pervade our lives. We use electric power to heat and light our homes, schools, and places of work. We use it to carry out a vast number of industrial and manufacturing tasks. We use it, via our radios, television sets, and movie projectors, to obtain information and entertainment. Magnetic forces are used to run all electric motors and in every electric generating plant a coil of wire is rotated near a large magnet to produce electric power.

As important as all these uses are, electricity is even more vital to us in another area: electric forces bind particles together to form atoms and atoms together to form all materials. We literally cannot exist without electrical forces! Nor can any solid or liquid, or for that matter any atom.

Visible light, microwaves, x rays, and radio waves are all electromagnetic in nature, and differ only in their frequencies. Optics, x-ray imaging, radio transmission and reception, and microwave production and detection are all logically part of the field of electricity and magnetism.

Particles such as electrons and protons exert electric and magnetic forces on each other; other particles do not. The property of a particle that gives rise to these forces is called charge. Charged particles at rest interact only via electric forces; charged particles in motion exert both electric and magnetic forces on each other.

Electric forces between charged particles at rest are described in Chapter 23. You will learn there are two types of charge, called positive and negative, and that like charges repel each other while unlike charges attract each other. In each case the force is along the line that joins the charges and is proportional to the reciprocal of the square of the distance between them. Electric forces, of course, are vectors and, when two or more charges act on another, the resultant force is the vector sum of the individual forces.

We may think of a charge creating electric and magnetic fields in all of space; the fields then exert forces on other charges. The problem of finding the force exerted by a collection of charges is then broken down into two somewhat simpler problems: *what fields are produced by the collection?* and *what force do the fields exert on the other charge?* Electric and magnetic interactions can be described much more easily in terms of fields than in terms of forces. When charges are moving or produce electromagnetic radiation the field viewpoint is almost a necessity.

You start your study of electric fields, in Chapter 24, by learning about the force exerted by a field on a charge. If no counterbalancing forces act the charge accelerates. You will apply Newton's second law and your knowledge of kinematics to examine the motions of charges in electric fields.

The relationship between a collection of charges and the electric field it produces is discussed in Chapters 24 and 25. It can be formulated in two equivalent, but quite different, ways. In the first chapter it is given in terms of the charges and their positions while in the second it is given in terms of an integral of the field over a surface. You should understand both formulations. The first is useful when you are given the charges and their positions, then asked to find the electric field; the second, known as Gauss' law, is useful when you know the field and want to find the charge distribution. It can also be used to find the field in certain situations of high symmetry.

Electrostatic forces, those of charges at rest, are conservative, so a potential energy can be associated them. Chapter 26 deals with electric potential energy and a closely related quantity, electric potential. You also will learn to calculate the work that must be done to assemble a collection of charges.

In Chapter 27 the ideas of electric field and potential, as well as Gauss' law, are put to practical use in a study of capacitance. Capacitors are extremely useful devices for the storage of electrical energy, which can then be recovered in short intense bursts. Small capacitors are used in electronic flash units for cameras. Large capacitors are used to power the extremely intense lasers used in an attempt to produce controlled nuclear fusion. In a more fundamental vein you will learn here that you may think of the potential energy of a collection of charges as being stored in the electric field and you will learn how to calculate the energy per unit volume stored in any field.

An electric current is simply a collection of moving charges. Currents are studied for two reasons. First, electrical circuits carrying current have many important practical applications. Think of the miniature circuits that operate computers and the house and industrial circuits that operate lights, appliances, and machinery. Secondly, currents produce magnetic fields. Chapters 28 and 29, in which currents are studied, form a bridge to later chapters on the magnetic field.

In Chapter 28 you will learn that the current in any non-superconducting material is determined by the electric field. You will also learn about resistivity, the property of materials that determines the current for a given electric field. It is closely related to the transfer of energy from moving charges to the material through which they move. Chapter 29 deals with batteries and other devices used to produce currents. Here you will apply the ideas of this and the previous chapter to the analysis of simple electrical circuits carrying time-independent currents. Energy balance in electrical circuits, important because current is often used to carry energy from one place to another, is also studied in this chapter.

In Chapter 30 you start the study of magnetic fields by learning about the force exerted by a magnetic field on a moving charge. This is extended to a discussion of the force and torque exerted on a current-carrying circuit.

The relationship between a current and the magnetic field it produces is discussed in Chapter 31. Just as for electric fields, two equivalent views are presented. In the first, known as the Biot-Savart law, the field produced by an infinitesimal element of current is postulated and this expression is integrated over the entire current. In the second, known as Ampere's law, the integral of the magnetic field around a closed loop is related to the current passing through the loop. The first viewpoint is of value when the current circuit is known and the field is sought; the second is of value when the field is known and the current is sought. Ampere's law can also be used to find the magnetic fields of highly symmetric current distributions.

A time-varying current produces an electric as well as a magnetic field, even if the net charge is zero in every region of space.

The relationship between the fields is given by Faraday's law and is discussed in Chapter 32. The law also deals with situations in which an object moves in a constant magnetic field. The result is exactly as if an infinitesimal battery were placed at each point in the object. Current is induced if the object forms part of an electrical circuit. Faraday's law is applied to electrical circuits in Chapter 33. There you will learn about inductors: circuit elements that are used to control the rate of change of current in a circuit and that store energy in magnetic fields just as capacitors store energy in electric fields.

When an externally generated magnetic field is applied to a substance it produces a field of its own. You may think of currents circulating within the material. For paramagnetic materials the induced field is in the same direction as the applied field and the total field is greater than the applied field. For diamagnetic materials the induced field is in the opposite direction and the total field is smaller than the applied field. Certain materials, such as iron, are ferromagnetic and can be permanently magnetized. They produce a magnetic field even in the absence of an applied field. Magnetic properties of materials are examined in Chapter 34. Gauss' law for magnetism is also presented in this chapter. It is the mathematical equivalent of the statement that no particle exists that produces a magnetic field when at rest.

In an electrical circuit that contains both an inductor and a capacitor the current oscillates. Such circuits have great practical use in the tuning of radios and television sets. In Chapter 35 they are used to demonstrate the storage and flow of energy in a circuit. The material of this chapter will prove useful in your study of alternating currents.

Most electrical power is transmitted by means of alternating currents, for which the direction of motion of the charges changes periodically. Current in an AC circuit containing capacitors and inductors has a natural frequency of oscillation and this property is used, for example, to tune radios and television sets. You will study this type circuit in Chapter 36.

Electromagnetic theory is completed by an additional relationship, described in Section 37–4: a magnetic field is always associated with a time-dependent electric field, just as an electric field is always associated with a time-dependent magnetic field. You may think of a time-varying magnetic field as the source of an electric field and a time-varying electric field as the source of an magnetic field. You will learn later that the relationship is such that changes in the fields propagate through space as waves.

Gauss' law for electricity, Gauss' law for magnetism, Faraday's law, and Ampere's law provide a complete description of the electric and magnetic fields produced by any given charge and current distribution. Collectively they are known as Maxwell's equations and, taken with the equations for the electric and magnetic forces on a charge, they provide a complete classical description of electromagnetic phenomena. They are collected and reviewed in Chapter 37.

The force that one charge exerts on another, whether electric or magnetic, depends on the relative positions of the charges. If one charge moves the force it exerts on the other charge does not change immediately. On the contrary, the electric and magnetic fields first change in the near vicinity of the moving charge and these changes propagate as waves. The force on the second charge is not altered until the wave reaches it. Electromagnetic waves travel with the speed of light and, in fact, visible light is such a wave with frequency in a certain range. In Chapter 38 you will learn that the wave-like propagation of electric and magnetic fields is predicted by Maxwell's equations. Energy transport by electromagnetic radiation is also discussed in this chapter.

If the electric field of an electromagnetic wave is always parallel to the same line the wave is said to be linearly polarized. By way of contrast, the field in other waves, said to be circularly polarized, rotates as the wave travels and in still other waves, said to be unpolarized, jumps randomly. Some materials, such as are used in many sunglasses, transmit only light that is linearly polarized in a certain direction. Polarization is studied in Chapter 38

Once you understand the nature of electromagnetic waves you are ready for Chapters 39 through 41, the optics chapters. First you study two fundamental phenomena: reflection and refraction. Two new waves are generated when an electromagnetic wave impinges on the boundary between two regions with different optical properties. One, the reflected wave, propagates back into the region of the incident wave and the other, the refracted wave, continues into the other region. The propagation direction of the refracted wave is different from that of the incident wave and depends on the optical properties of both regions. In Chapter 39 you will learn how the reflecting properties of mirrors and the refracting properties of lenses are used to form images. The ideas are applied to microscopes and telescopes.

Electromagnetic waves obey a superposition principle like other waves: if two or more waves are simultaneously present the total electric field, for example, is the vector sum of the electric fields associated with the individual waves. Superposition gives rise to interference effects, the topic of Chapter 40. If two waves are in phase then they interfere constructively and the resultant amplitude is larger than either constituent amplitude. If their phases differ by 180° they interfere destructively and tend to cancel each other. If the difference in phase comes about because the waves travel different distances the effect is wavelength dependent; destructive interference of red light, for example, occurs at different places than the destructive interference of blue light. Interference is responsible for the colors on soap bubbles and oil films.

Waves passing by the edge of an object travel into the shadow region and produce interference-like fringes in that region: alternating dark and bright bands appear if only a single wavelength is present. Diffraction effects such as this are studied in Chapter 41. Diffraction gratings, which consist of closely spaced, extremely narrow, transparent lines ruled on glass or plastic, are used to analyze the spectrum of light from materials. X rays are diffracted from crystals and the pattern they produce can be used to find the atomic structure of the crystal. Diffraction by gratings and crystals is also discussed in Chapter 41.

Chapter 23
ELECTRIC CHARGE

I. BASIC CONCEPTS

You start your study of electromagnetism with Coulomb's law, which describes the force that one stationary charged particle exerts on another. Pay attention to both the magnitude and direction of the force and learn how to calculate the total force when more than one charge acts.

Electric charge. Electric charge is defined in terms of electric current, which in turn is defined in terms of the magnetic force exerted by two current carrying wires on each other. You must delay a precise understanding of charge until you study magnetism. For now you should know that charge is a property possessed by some particles and that two charged particles exert electric forces on each other. The SI unit for charge is _____ (abbreviated _____).

There are two types of charge, called <u>positive</u> and <u>negative</u>. In the equations of electromagnetism a positive number is substituted for the value of a positive charge and a negative number is substituted for the value of a negative charge.

The charge on an electron is _____ C; the charge on a proton is _____ C. Charge is quantized: the charge on all particles detected so far is a positive or negative multiple of a fundamental unit of charge e. The value of e is _____ C.

All macroscopic materials contain enormous numbers of electrons and protons but if the number of electrons in an object equals the number of protons then the net charge is zero and the object is said to be <u>neutral</u> (or uncharged). If there are more electrons than protons the object has a net negative charge; if there are more protons than electrons it has a net positive charge. In either case it said to be <u>charged</u>. Notice that the net charge is computed by algebraically adding the charges of all particles in the object, taking their signs into account. Normally, all macroscopic bodies are neutral.

Objects may be given net charges by rubbing them together. When a glass rod is rubbed with silk the rod becomes _____ charged. When a plastic rod is rubbed with fur the rod becomes _____ charged.

Charge is conserved. The net charge of a closed system (with no particles entering or leaving) is always the same. After rubbing a glass rod silk becomes _____ charged and, if no particles leak on or off, the magnitude of its charge is the same as the magnitude of the charge on the rod. Carefully note that it is the algebraic sum of the charges that remains the same. Before rubbing the net charge was zero. After rubbing it is still zero, the sum of the negative charge on one object and the positive charge on the other.

The force that one point charge exerts on another is along the line that joins the charges. If the charges have the same sign then the force is one of repulsion; if the charges have different signs then the force is one of attraction. The diagram below shows three possible pairings.

At each charge draw a vector that indicates the force exerted on it by the other charge in the pair.

$$\oplus \qquad \oplus \qquad\qquad\qquad \ominus \qquad \ominus \qquad\qquad\qquad \oplus \qquad \ominus$$

Conductors, insulators, and semiconductors. The *electrons* in materials move and are transferred to or from other objects by rubbing; the atomic nuclei (containing protons) are nearly immobile. Materials are often classified according the freedom with which electrons can move. Electrons in an insulator are not free to move far; any charge placed on an insulator remains where it is placed. On the other hand, conductors contain many electrons that are free to move throughout the conductor. When you touch a charged conductor to an uncharged conductor charge is transferred and, as a result, both conductors are charged. When you touch a charged insulator to a neutral conductor very little, if any, charge is transferred. Scraping or rubbing is required to transfer charge to or from an insulator.

Give some examples of conductors and insulators.

conductors: _____

insulators: _____

Your body is a moderately good conductor, particularly if there is moisture on it. Although you can charge a conductor by rubbing it, you cannot touch it while you do or else the charge will move to your body. Conductors in the laboratory are usually equipped with insulating handles.

Electrical force. Suppose two point particles, one with charge q_1 and the other with charge q_2, are a distance r apart. The magnitude of the force exerted by either of the charges on the other is given by

$$F =$$

You should write this equation with the factor $1/4\pi\epsilon_0$. The constant of nature ϵ_0 is called _____ and has the value $\epsilon_0 = $ _____ $C^2/N\cdot m^2$ (to three significant figures). The factor $1/4\pi\epsilon_0$ has the value _____ $N\cdot m^2/C^2$ (to three significant figures). When you use this force equation you must remember to enter the magnitudes of the charges. If you also enter their signs you might get a negative result, which is not correct for the magnitude of a vector.

Notice that the magnitude of the force is proportional to the inverse square of the distance between the charges and is also proportional to the product of the magnitudes of the charges. If the distance between the charges is doubled the force is reduced by a factor of _____; if one of the charges is doubled the force increases by a factor of _____; if both charges are doubled the force increases by a factor of _____.

Force is a vector. To find the direction of the electric force one charge exerts on another draw the line that joins the charges, then decide if the force is attractive (opposite sign charges) or repulsive (like sign charges). If it is attractive the force on either charge is along the line and toward the other charge. If it is repulsive it is along the line and away from the other charge. Electrostatic forces between two charges obey Newton's third law: they are equal in _____ and opposite in _____.

When more than two charges are present the force on any one of them is the *vector* sum of the forces due to the others. Each force is computed using Coulomb's law. This is the principle of <u>superposition</u> for electric forces.

To illustrate the principle of superposition, consider four identical charges q at the corners of the square with edge a, as shown. Develop an expression for the total force on the charge labelled 1 in the diagram. Fill in the following table, giving the magnitude of each force, the x component, and the y component, all in terms of q and a.

	<u>magnitude</u>	<u>x component</u>	<u>y component</u>
force of 2 on 1			
force of 3 on 1			
force of 4 on 1			

The x component of the total force on 1 is given by

$$F_x =$$

and the y component is given by

$$F_y =$$

Shell theorems. These are two theorems about electrical forces between a point charge and a special distribution of charge: a uniform spherical shell. You will find them extremely useful.

Describe what is meant by a uniform spherical shell of charge: _____
_____
_____

If a point charge q is outside a uniform shell of charge, a distance r from its center, the force exerted by the shell on the point charge has a magnitude that is given by $F =$ _____, where Q is the charge on the shell. The force is along the line joining the center of the shell and the point charge; it is away from the shell if q and Q have the same sign and toward the shell if they have opposite signs.

If a point charge q is anywhere inside the shell, the force exerted on it by the shell is _____ .

Forces on neutral objects. Charged objects attract neutral macroscopic objects. A neutral conductor is attracted to a charged rod because the electrons within the inductor are redistributed, making some regions positively charged and others negatively charged. The forces on the negative and positive regions are, of course, in opposite directions but the force is greater on the region nearer to the charged object.

The diagram on the right shows a positively charged insulating rod held near a neutral conductor. Indicate the distribution of charge on the conductor and draw arrows of different length to indicate the forces on the positive and negative regions of the conductor. The net force should be toward the rod.

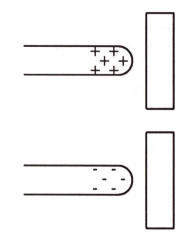

The diagram on the right shows a negatively charged insulating rod held near a neutral conductor. Indicate the distribution of charge on the conductor and draw arrows of different length to indicate the forces on the positive and negative regions of the conductor. The net force should be toward the rod.

Suppose you touch the conductor with the insulating rod nearby. Electrons flow from you to the conductor if the insulating rod is positive or from the conductor to you if the rod is negative. Remove your finger, then remove the rod. The conductor is now charged. It is _____ if the insulating rod was positive and _____ if the rod was negative. This process is called <u>charging</u> <u>by</u> <u>induction</u>.

Charged objects also attract neutral insulators but the force is much smaller than the force of attraction for a conductor. Under the influence of the external charge electrons in an insulator move slightly from their normal orbits so the center of the negative charge is slightly displaced from the center of the positive charge. This results in a net force. A charged rod can be used to pick up small pieces of paper and a rubbed balloon "sticks" to a wall.

II. PROBLEM SOLVING

The fundamental problems of this chapter deal with Coulomb's law. Given two charges and their separation you should be able to calculate the force (both magnitude and direction) each exerts on the other. You should know how to use vector addition to find the resultant force when more than one charge interacts with the charge of interest. Variations include problems that ask you to find one of the charges or the position of a charge, given the force.

To find the value of one or more of the charges in some problems a preliminary calculation must be performed prior to using Coulomb's law. For example, you may need to calculate the total positive charge in an object. The number of atoms in an object can be found from the mass of an atom and the mass of the object, the number of protons in an atom is given by the atomic number. Sometimes the atomic mass is given (or can be found in the periodic table of the chemical elements) and you will need to use that quantity to calculate the mass of an atom.

Other problems deal with conducting spheres and you will need to apply the shell theorems. In addition you should know that when two identical conducting spheres touch each other the total charge is shared equally between them.

Finally, you should know the relationship between charge and current: current is the charge that moves into or out of a region per unit time.

PROBLEM 1 (5E).   Two equally charged particles, held 3.2×10^{-3} m apart, are released from rest. The initial acceleration of the first particle is observed to be 7.0 m/s^2 and that of the second to be 9.0 m/s^2. If the mass of the first particle is 6.3×10^{-7} kg, what are (a) the mass of the second particle and (b) the magnitude of the common charge?

STRATEGY: (a) The magnitudes of the forces acting on the particles are the same, so $m_1 a_1 = m_2 a_2$, where m_1 and m_2 are the masses and a_1 an a_2 are the magnitudes of the accelerations. Solve for m_2. (b) The magnitude of the force is given by Coulomb's law: $F = (1/4\pi\epsilon_0)q^2/r^2$, where r is the separation of the particles. Solve $(1/4\pi\epsilon_0)q^2/r^2 = m_1 a_1$ for q.

SOLUTION:

[ans: 4.9×10^{-7} kg; 7.1×10^{-11} C]

PROBLEM 2 (7E).   Identical isolated conducting spheres 1 and 2 have equal amounts of charge and are separated by a distance large compared with their diameters (Fig. 23–13a). The electrostatic force acting on sphere 2 due to sphere 1 is F. Suppose now that a third identical sphere 3, having an insulating handle and initially neutral, is touched first to sphere 1 (Fig. 23–13b), then to sphere 2 (Fig. 23–13c), and finally removed (Fig. 23–13d). In terms of F, what is the electrostatic force F' that now acts on sphere 2?

STRATEGY: Let q be the charge originally on each of the spheres 1 and 2. The force on 2 has the magnitude $F = (1/4\pi\epsilon_0)q^2/r^2$, where r is the separation of the spheres. After 3 is touched to 1 they have equal charge. Since the total charge on 1 and 3 is q, each has charge $q/2$. After 3 is touched to 2 they have equal charge. Since the total charge on 2 and 3 is $q + q/2 = 3q/2$ they each have charge $3q/4$. Since the charge on 1 is $q/2$ and the charge on 2 is $3q/4$ the force between spheres 1 and 2 is now $F' = (1/4\pi\epsilon_0)(3q^2/8)/r^2$. Substitute F for $(1/4\pi\epsilon_0)q^2/r^2$. The force is in the same direction as before.

SOLUTION:

[ans: $3F/8$]

PROBLEM 3 (8P).   In Fig. 23–14, three charged particles lie on a straight line and are separated by a distance d. Charges q_1 and q_2 are held fixed. Charge q_3 is free to move but happens to be in equilibrium (no net electrostatic force acts on it). Find q_1 in terms of q_2.

STRATEGY: One of the fixed charges must attract and the other must repel q_3 so one is positive and the other is negative. The condition that the net force on q_3 is zero leads to $q_1/(2d)^2 + q_2/d^2 = 0$. Solve for q_1.

SOLUTION:

[ans: $q_1 = -4q_2$]

PROBLEM 4 (13P). Two fixed charges, $+1.0\,\mu C$ and $-3.0\,\mu C$, are $10\,cm$ apart. Where can a third charge be located so that no net electrostatic force acts on it?

STRATEGY: The third charge must be located on same line as the fixed charges for otherwise the two forces acting on it would not be antiparallel and so could not sum to zero. It cannot be between the fixed charges for then the two forces would be in the same direction. It must be nearer the smaller of the fixed charges for the magnitudes of the two forces to be the same. Place the $-3.0\,\mu C$ charge at the origin, the $1.0\,\mu C$ charge at $x = d$ ($= 10\,cm$), and the third charge at $x = d + \ell$, where ℓ is positive. If $q_1 = -3.0\,\mu C$ and $q_2 = 1.0\,\mu C$ the condition that the force vanish leads to $[q_1/(d + \ell)^2] + [q_2/\ell^2] = 0$. Solve for ℓ.

SOLUTION:

$$\left[\text{ans: } 24\,cm \text{ from } q_1 \text{ and } 14\,cm \text{ from } q_2 \right]$$

PROBLEM 5 (14P). The charges and coordinates of two charged particles held fixed in the xy plane are: $q_1 = +3.0\,\mu C$, $x_1 = 3.4\,cm$, $y_1 = 0.50\,cm$, and $q_2 = -4.0\,\mu C$, $x_2 = -2.0\,cm$, $y_2 = 1.5\,cm$. (a) Find the magnitude and direction of the electrostatic force on q_2.

STRATEGY: The relative displacement of q_2 from q_1 has components $\Delta x = x_2 - x_1$ and $\Delta y = y_2 - y_1$. The distance between the charges is $d = \sqrt{(\Delta x)^2 + (\Delta y)^2}$ and the magnitude of the force is $F = (1/4\pi\epsilon_)|q_1||q_2|/d^2$. The x component of the force is $F_x = F\,\Delta x/d$ and the y component is $F\,\Delta y/d$ so the angle between the force and the positive x axis is given by $\tan\theta = F_y/F_x$. Be sure to pick the proper value so the force is one of attraction.

SOLUTION:

$$\left[\text{ans: } 36\,N; \ 10° \text{ below the x axis}\right]$$

(b) Where could you locate a third charge $q_3 = +4.0\,\mu C$ such that the net electrostatic force on q_2 is zero?

STRATEGY: The third charge must be along the line that joins the first two. Since it is positive and attracts q_2 it must be on the side of q_2 opposite q_1. The magnitude of the force it exerts on q_2 must be the same as the magnitude of the force q_1 exerts on q_2. If it is a distance ℓ from q_2 then $|q_3|/\ell^2 = |q_1|/d^2$. Solve for ℓ. The coordinates of q_3 are given by $x_3 = x_2 - \ell\cos\theta$ and $y_3 = y_2 + \ell\sin\theta$.

SOLUTION:

$$\left[\text{ans: } x_3 = -8.3\,cm, \ y_3 = 2.7\,cm\right]$$

PROBLEM 6 (17P).   A charge Q is fixed at each of two opposite corners of a square. A charge q is placed at each of the other two corners. (*a*) If the net electrostatic force on each Q is zero, what is Q in terms of q?

STRATEGY: Since the two charges Q are separated by $\sqrt{2}a$, where a is the edge length of the square, the magnitude of the electrostatic force exerted by one on the other is $F = (1/4\pi\epsilon_0)Q^2/2a^2$. Each of the charges q exerts a force with magnitude $(1/4\pi\epsilon_0)qQ/a^2$, at 90° to each other and at 45° to the force of Q. For the net force on Q to be 0, $Q/2a^2 = 2q\cos 45°/a^2$. Use $\cos 45° = 1/\sqrt{2}$. Since the force of Q is repulsive, the force of each charge q must be attractive and Q and q must have opposite signs.

SOLUTION:

$$\left[\text{ans: } Q = -2\sqrt{2}q\right]$$

(*b*) Could q be chosen to make the net electrostatic force on each of the four charges zero? Explain your answer.

STRATEGY: To make the force on q vanish, the charges must be related by $q = -2\sqrt{2}Q$. This condition and $Q = -2\sqrt{2}q$ cannot be satisfied simultaneously.

SOLUTION:

$$\left[\text{ans: no}\right]$$

PROBLEM 7 (25E).   How many megacoulombs of positive (or negative) charge are in 1.00 mol of neutral molecular-hydrogen gas (H_2)?

STRATEGY: A mole contains 6.02×10^{23} molecules and each molecule contains 2 protons. The charge on a proton is 1.60×10^{-19} C. Divide by 1.00×10^6 to convert to megacoulombs.

SOLUTION:

$$\left[\text{ans: } 0.193\,\text{MC}\right]$$

PROBLEM 8 (32P).   The Earth's atmosphere is constantly bombarded by *cosmic ray protons* that originate somewhere in space. If the protons were all to pass through the atmosphere, each square meter of the Earth's surface would intercept protons at the average rate of 1500 protons per second. What would be the corresponding current intercepted by the total surface area of the Earth?

STRATEGY: The current is the charge per unit time intercepted by the Earth. If R is the radius of the Earth $(6.37 \times 10^6 \text{ m})$ then $A = 4\pi R^2$ is the area of its surface and $1500A$ is the number of photons that strike the Earth, on average, each second. Multiply by the charge on a proton $(1.60 \times 10^{-19}$ C) to obtain the current.

SOLUTION:

$$\left[\text{ans: } 0.122\,\text{A}\right]$$

PROBLEM 9 (35P).   We know that, within the limits of measurement, the magnitudes of the negative charge on the electron and the positive charge on the proton are equal. Suppose, however, that these magnitudes differ from each other by 0.00010%. With what force would two copper pennies, placed 1.0 m apart, repel each other? What do you conclude? (*Hint*: See Sample Problem 23–3.)

STRATEGY: According to Sample Problem 23–3 the total positive charge in a neutral copper penny is 1.37×10^5 C and the total negative charge is -1.37×10^5 C. If the positive charge were greater by a factor of 5.0×10^{-7} and the negative charge were less by the same factor, the net charge on the penny would be $1.37 \times 10^5 \times 1.0 \times 10^{-6} = 0.137$ C. Use Coulomb's law to find the force of one penny on another, separated by 1.0 m.

SOLUTION:

$$[\text{ans: } 1.7 \times 10^8 \text{ N}]$$

This is an enormous force! Since the penny has a mass of about 3 g, its acceleration would be about 6×10^{10} m/s. Since pennies are not observed to repel each other, we conclude that the magnitude of the electron and proton charges are much more nearly the same than was assumed in the problem.

PROBLEM 10 (37E).   In *beta decay* a massive fundamental particle changes to another massive particle, and an electron or a positron is emitted. (*a*) If a proton undergoes beta decay to become a neutron, which particle is emitted?

STRATEGY: Since charge is conserved the emitted particle must be positive.

SOLUTION:

$$[\text{ans: positron}]$$

(*b*) If a neutron undergoes beta decay to become a proton, which particle is emitted?

SOLUTION:

$$[\text{ans: electron}]$$

III. NOTES

Chapter 24
THE ELECTRIC FIELD

I. BASIC CONCEPTS

Here the idea of an electric field is introduced and used to describe electrical interactions between charges. It is fundamental to our understanding of electromagnetic phenomena and is used extensively throughout the rest of this course. Build a firm foundation by learning well the material of this chapter.

The electric field. A <u>field</u> is an important concept in physics. It gives some property of a region as a function of position. An example of a scalar field is _____ _____ ; an example of a vector field is _____.

To describe the electric force that one charge exerts on another, we associate an <u>electric field</u> with the first charge and say that this field exerts a force on the second charge. Consider two charges Q and q, for example. Charge Q creates an electric field and that field exerts a force on q. Charge q also creates an electric field and *that* field exerts a force on Q. The electric field of a charge exists throughout all space, so anywhere a second charge is placed it will experience a force.

The electric field at any point in space is defined as the electric force per unit charge on a stationary positive test charge placed at that point, in the limit as the test charge becomes vanishingly small. If the test charge is q_0 and $\mathbf{F}$ is the electric force on it then the electric field at the position of the test charge is

$$\mathbf{E} =$$

You should be aware that this field is *not* the field created by q_0; it is the total field created by all *other* charges. The test charge does not exert a force on itself. An electric field is associated with the charges that create it and exists whether or not a test charge is present. The SI units of an electric field are _____.

Explain why the limit as q_0 becomes vanishingly small must be included in the definition:

_____

_____

Suppose a collection of stationary charges creates an electric field $\mathbf{E}$ and a charge q is placed in this field. Then the force exerted on q is given by

$$\mathbf{F} =$$

If $\mathbf{E}$ varies from place to place, as it usually does, then you use the value at the position of q.

If an electric force is the only force acting on a charge then Newton's second law takes the form $q\mathbf{E} = m\mathbf{a}$ and gives the acceleration of the charge. If you know the initial position and velocity of the charge then you can, in principle, predict its subsequent motion.

Electric field lines.   Electric fields are often represented in diagrams by <u>electric field lines</u> (also called lines of force). These lines graphically depict both the direction and magnitude of the electric field throughout a region of space. Tell how the direction of the electric field at a point is related to the field line through that point: _____

_____

Arrows are placed on field lines to indicate the direction of the field but electric field lines themselves are *not* vectors.

   Tell how the magnitude of the electric field at a point is related to the field lines in the vicinity of the point: _____

_____

Electric field lines emanate from _____ charge and terminate on _____ charge.

   To see the electric field lines for some charge distributions look carefully at Figs. 24–2, 3, 4, and 5 of the text. Be aware that the configuration of lines is different for different charge distributions: the electric field of a point charge, for example, is *not* the same as the electric field of a dipole. For each figure notice the directions of the lines, where they are close together, and where they are far apart. You should recognize that the number of field lines is proportional to the charge. If all charge that creates the field is doubled then the number of lines doubles in every region.

   The diagram to the right shows some field lines. Indicate on it a region where the magnitude of the electric field is relatively large and a region where it is relatively small. Suppose an electron is at the point marked •. Draw an arrow to indicate the direction of the force on it and label the arrow **F**.

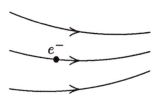

Point charges.   The magnitude of the electric field at a point a distance r from a single point charge q is given by

$$E = $$

The field is along the line that joins the charge q and the point. If q is positive it points away from q, if q is negative it points toward q.

   If more than one point charge is responsible for the electric field you calculate the total field by summing the fields due to the individual charges. Don't forget that electric fields are vectors and the sum is a vector sum. You can sum the magnitudes only if the fields are in the same direction.

   In any given situation you must distinguish between the total electric field and the field that acts on a given charge. Suppose, for example, there are three charges q_1, q_2, and q_3. The total field at any given point is the vector sum of the fields due to the three charges. The field that acts on q_1, however, is the vector sum of the fields due to q_2 and q_3 only.

Electric fields of continuous charge distributions.   A continuous charge distribution is characterized by a <u>charge density</u>. If the charge is on a line (straight or curved) the appropriate charge density is the _____ charge density, denoted by _____ and defined so that the charge dq in an infinitesimal segment ds of the line is given by dq = _____ .   If the charge is distributed on a surface the appropriate charge density is the _____ charge density, denoted by _____ and defined so that the charge dq in an infinitesimal element of area dA is given by $dq =$ _____. A uniform charge distribution means that the appropriate charge density is the same everywhere along the line or on the area.

Carefully study the derivations of expressions for the electric field on the axis of a uniform ring of charge and on the axis of a uniform disk of charge. You should understand why the field is parallel to the axis in each case.

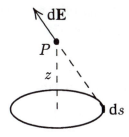

The diagram on the right shows an infinitesimal segment ds of the ring and the field it produces at point P on the axis, a distance z from the ring center. On the diagram show the location of another infinitesimal segment such that the sum of its field and the field shown is along the axis. Draw the electric field vector at P for this segment. Since every infinitesimal segment of the ring can be paired in this way with another segment, the total field must be along the axis.

If the ring has radius R and total charge q then the linear charge density is $\lambda =$ _____, the charge in an infinitesimal segment ds is $dq =$ _____, the magnitude of the field produced by the segment at P is

$$dE =$$

and the component along the axis is

$$dE_{\parallel} =$$

You must now integrate around the ring. The integration, of course, sums the components along the axis of the fields due to the ring segments. The result for the total field is

$$E =$$

To calculate the field produced by a uniform disk of charge, the disk is divided into rings, each having infinitesimal width dr. The area of a ring of radius r is $dA =$ _____ and if σ is the area charge density then $dq =$ _____ is the charge in a such a ring. The magnitude of the field it produces at a point on the axis a distance z from the ring center is

$$dE =$$

To find an expression for the total field this expression is integrated from $r = 0$ to $r = R$. The result is:

$$E =$$

By taking R to be much larger than z you can find an expression for the field produced by a uniform plane of charge. Notice that if $R \gg z$ then $z/\sqrt{x^2 + R^2}$ is much smaller than 1 and

$$E =$$

You will find this expression extremely useful in your study of later chapters. It is a good approximation to the field of a finite plane of charge at points that are close to the plane and far from the edges. Notice that the field does not depend on distance from the plane.

Note that at points near a ring, disk, or plane superposition produces total fields that are quite different from each other and quite different from the field of a point charge. At points far from a charge distribution, however, the electric field tends to become like that of a point particle with charge equal to the net charge in the distribution. For a ring of charge with uniform linear density λ the net charge is $q =$ _____ and the field far away ($z \gg R$) is given by $E =$ _____. For a disk with uniform area density σ the net charge is $q =$ _____ and the field far away ($z \gg R$) is given by $E =$ _____.

Electric dipoles. The geometry is shown in Fig. 24–8 of the text. An electric dipole consists of

_____

It is characterized by a dipole moment **p**, a vector with magnitude given by $p =$ _____, where q is _____

and d is _____.

The direction of **p** is _____.

Look carefully at Fig. 24–5 of the text to see the lines of force produced by a dipole.

Go over the derivation of Eq. 24–12 of the text for the electric field at point P in Fig. 24–8. Be sure you understand why the field there is in the positive z direction. In terms of q, z, and d the magnitude of the field produced at P by the positive charge is given by $E_+ =$ _____ and the magnitude of the field produced there by the negative charge is given by $E_- =$ _____. The results are summed to produce an expression for the total field. It is

$$E =$$

You will need to know about atomic dipoles, for which the separation of the charges is less than a nanometer and much less than z. Use the binomial theorem to expand $(z - d/2)^2$ and $(z + d/2)^2$. Retain only those terms that are proportional to d. In terms of the dipole moment the field is

$$E =$$

Notice that the field is inversely proportional to z^3, not z^2.

Consider a dipole in a uniform electric field **E** (created by other charges). The net force on the dipole is _____ because _____

_____.

The dipole does, however, experience a torque. If **p** is the dipole moment and **E** is the electric field then the torque is given by

$$\tau =$$

This torque tends to rotate the dipole so that **p** _____.

When an electric dipole rotates in an electric field the field does work on it. For a rotation during which the angle between the dipole and field changes from θ_0 to θ the work done by the field is $W =$ _____. The sign of the work is _____ if the angle increases and _____ if the angle decreases. During a rotation the potential energy of the field and dipole changes, the change being given by $\Delta U =$ _____. If we take the potential energy to be zero when the dipole moment is perpendicular to the field the potential energy for any orientation can be written as the scalar product $U =$ _____. The potential energy is a maximum when _____ and is a minimum when _____.

Suppose a dipole is initially oriented with its moment perpendicular to an electric field. It is then rotated through 90° so it is in the same direction as the field. During this process the sign of the work done by the field is _____ and the sign of the change in the potential energy is _____. Suppose, instead, that the dipole is rotated so its direction becomes opposite to that of the field. During this process the sign of the work done by the field is _____ and the sign of the change in the potential energy is _____.

II. PROBLEM SOLVING

Some problems deal with electric field lines. You should know that the electric field at any point is tangent to the line through that point, that the number of lines per unit area passing through an area perpendicular to the lines is proportional to the magnitude of the field, and that lines emanate from positive charge and end at negative charge. Here's an example that tests your knowledge of the relationship between the density of lines and the field strength. You will also need to know that the force on a charge in an electric field is given by the product of the charge and the field. For a positive charge it is in the direction of the field while for a negative charge it is directed opposite to the field.

PROBLEM 1 (1E). In Fig. 24–21 the electric field lines on the left have twice the separation as those on the right. (a) If the magnitude of the field at A is 40 N/C, what force acts on a proton at A?

STRATEGY: Use $F = eE$. A proton is positively charged so the force on it is in the direction of the field.

SOLUTION:

[ans: 6.4×10^{-18} N, to the left]

(b) What is the magnitude of the field at B?

STRATEGY: Assume that the separation along a line that is perpendicular to the page at B is the same as the separation along a line that is perpendicular to the page at A so that half as many lines pass through a given area perpendicular to the page at B as pass through the same area at A. Then the field is half as strong at B as at A.

SOLUTION:

[ans: 20 N/C]

You should know how to calculate the electric field of a point charge and how to calculate the total field of a collection of point charges.

PROBLEM 2 (7E).   What is the magnitude of a point charge that would create an electric field of $1.00\,\text{N/C}$ at points $1.00\,\text{m}$ away?

STRATEGY: Solve $E = (1/4\pi\epsilon_0)q/r^2$ for r.

SOLUTION:

$$\left[\text{ans: } 1.11 \times 10^{-10}\,\text{C}\right]$$

PROBLEM 3 (8E).   In Fig. 24–24 charges are placed at the vertices of an equilateral triangle. For what value of Q (both sign and magnitude) does the total electric field vanish at C, the center of the triangle?

STRATEGY: Draw the electric field vectors for the given charges. Note that their horizontal components cancel and their vertical components sum to give $E = 2(1/4\pi\epsilon_0)(q/\ell^2)\cos 60°$, where q is the charge and ℓ is the distance from a vertex to the triangle center. The field of Q must have this magnitude and point downward in the diagram.

SOLUTION:

$$\left[\text{ans: } 1.0\,\mu\text{C}\right]$$

You should be able to obtain this result by considering the symmetry of the situation.

PROBLEM 4 (18P).   In Fig. 24–29, what is the electric field at point P due to the four point charges shown?

STRATEGY: The fields due the two $5.0q$ charges have the same magnitudes at P but point in opposite directions, so they sum to 0. The field due to the $3.0q$ charge has magnitude $(1/4\pi\epsilon_0)\,3.0q/d^2$ and points away from the charge. The field due to the $-12q$ charge has magnitude $(1/4\pi\epsilon_0)\,12q/(2d)^2$ and points toward the charge. Sum these fields.

SOLUTION:

$$\left[\text{ans: } 0\right]$$

PROBLEM 5 (20P).   An electron is placed at each corner of an equilateral triangle having sides $20\,\text{cm}$ long. (*a*) What is the electric field at the midpoint of one of the sides?

STRATEGY: Draw a diagram and mark the midpoint of one of the sides. The field at that point due to the two charges at the ends of the side vanishes. You need calculate only the field of the charge at the opposite vertex. Use the Pythagorean theorem to show that the distance from the vertex to the midpoint of the opposite side is $\sqrt{3}\ell/2$, where ℓ is the length of the side. Then use Coulomb's law to calculate the field.

SOLUTION:

$$\left[\text{ans: } 4.8 \times 10^{-8}\,\text{N/C, toward the opposite vertex}\right]$$

(b) What force would act on another electron placed there?

STRATEGY: Use $\mathbf{F} = q\mathbf{E}$.

SOLUTION:

$\left[\text{ans: } 7.7 \times 10^{-27}\,\text{N, away from the opposite vertex}\right]$

PROBLEM 6 (26P). Calculate the electric field, in both magnitude and direction, due to an electric dipole, at a point P located at a distance $r \gg d$ along the perpendicular bisector of the line joining the charges (Fig. 24–32). Express your answer in terms of the magnitude and direction of the electric dipole moment $\mathbf{p}$.

STRATEGY: Draw a diagram, similar to Fig. 24–32, and show the electric field vector for the field at P of each charge. Each charge is a distance $\sqrt{r^2 + d^2/4}$ from P. The components of the two fields along the perpendicular bisector sum to zero and the components parallel to the line joining the charges sum to

$$E = \frac{qd}{4\pi\epsilon_0} \frac{1}{(r^2 + d^2/4)^{3/2}}.$$

For points that are far from the dipole $r \gg d/2$ and $(r^2 + d^2/4)^{-3/2} \approx 1/r^3$. Substitute the magnitude p of the dipole moment for qd. Notice from your drawing that the direction of the field is opposite that of the dipole moment.

SOLUTION:

$\left[\text{ans: } -(1/4\pi\epsilon_0)\mathbf{p}/r^3\right]$

You should know how to calculate the electric field of a continuous distribution of charge. Don't forget that the field is a vector and you must evaluate an integral for each component.

PROBLEM 7 (32P). A thin glass rod is bent into a semicircle of radius r. A charge $+Q$ is uniformly distributed along the upper half and a charge $-Q$ is uniformly distributed along the lower half, as shown in Fig. 24–35. Find the electric field $\mathbf{E}$ at P, the center of the semicircle.

STRATEGY: The diagram shows the upper half of the semicircle. The x component of the electric field is given by

$$E_x = \frac{1}{4\pi\epsilon_0} \int \frac{\lambda \sin\theta}{r^2}\,ds\,,$$

where ds is a line element of the semicircle and λ is the linear charge density. Since charge Q is distributed along a line of length $\pi r/2$, $\lambda = 2Q/\pi r$. Change the variable of integration to θ by making the substitution $ds = r\,d\theta$, integrate from $\theta = 0$ to $\theta = 90°$. You should get $E_x = Q/2\pi^2\epsilon_0 r^2$.

The y component of the electric field is given by

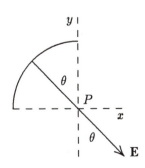

$$E_y = -\frac{1}{4\pi\epsilon_0} \int \frac{\lambda \cos\theta}{r^2}\,ds\,.$$

Make the same substitutions and obtain $E_y = -Q/2\pi^2 r^2$. Now carry out a similar derivation for the bottom half of the semicircle. You should obtain $E_x = -Q/2\pi^2\epsilon_0 r^2$ and $E_y = -Q/2\pi^2\epsilon_0 r^2$. The total field at P is the vector sum of these two fields.

SOLUTION:

$$\left[\text{ans: } Q/\pi^2\epsilon_0 r^2, \text{downward}\right]$$

PROBLEM 8 (34P). In Fig. 24–37, a nonconducting rod of length L has charge $-q$ uniformly distributed along its length. (a) What is the linear charge density of the rod?

STRATEGY: Since the charge is distributed uniformly the linear charge density is the total charge divided by the length of the rod.

SOLUTION:

$$\left[\text{ans: } -q/L\right]$$

(b) What is the electric field at point P, a distance a from the end of the rod?

STRATEGY: Put the origin at P and divide the rod into infinitesimal strips of width dx. The electric field due to the strip at x is given by $dE = (1/4\pi\epsilon_0)(\lambda/x^2)\,dx$, where λ is the linear charge density. The total field is

$$\frac{1}{4\pi\epsilon_0}\int_{-L-a}^{-a}\frac{\lambda}{x^2}\,dx\,.$$

Evaluate the integral.

SOLUTION:

$$\left[\text{ans: } q/4\pi\epsilon_0 a(L+a), \text{toward the rod}\right]$$

(c) If P were very far from the rod compared to L, the rod would look like a point charge. Show that your answer to (b) reduces to the electric field of a point charge for $a \gg L$.

STRATEGY: Replace $L + a$ in the denominator of the answer to (b) with a alone.

SOLUTION:

PROBLEM 9 (37P).   (*a*) What total charge q must the disk in Sample Problem 24–6 (Fig. 24–12) have in order that the electric field on the surface of the disk at its center equals the value at which air breaks down electrically, producing sparks? (See Table 24–1.)

STRATEGY: The electric field produced by a uniformly charged disk at a point on the axis a distance z away from the center is given by Eq. 27 of the text:

$$E = \frac{\sigma}{2\epsilon_0} \left(1 - \frac{z}{\sqrt{z^2 + R^2}} \right),$$

where R is the radius of the disk and σ is its area charge density. At the center of the disk $z = 0$ and $E = \sigma/2\epsilon_0$. Set $E = 3 \times 10^6$ N/C and solve for σ. Since σ is the charge per unit area, the total charge is given by $q = \pi R^2 \sigma$. According to Sample Problem 6, $R = 2.5$ cm.

SOLUTION:

[ans: 1.0×10^{-7} C]

(*b*) Suppose that each atom at the surface has an effective cross-sectional area of 0.015 nm^2. How many atoms are at the disk's surface?

STRATEGY: The area of the surface is given by $A = \pi R^2$. Divide A by the effective cross-sectional area of an atom.

SOLUTION:

[ans: 1.3×10^{17}]

(*c*) The charge in (*a*) results from some of the surface atoms having one excess electron. What fraction of the surface atoms must be so charged?

STRATEGY: The number of charged atoms is given by q/e and the ratio of this number to the number of atoms found in part (*b*) is the fraction of charged atoms.

SOLUTION:

[ans: 5.0×10^{-6}]

Some problems deal with the trajectories of charges in electric fields. If the field is uniform the acceleration of the charge is constant and you may use the kinematic equations of constant motion. The problems are quite similar to projectile motion problems; but the acceleration is now due to the force of electric field rather than to gravity.

PROBLEM 10 (47E).   An electron with a speed of 5.00×10^8 cm/s enters an electric field of magnitude 1.00×10^3 N/C, traveling along the field in the direction that retards its motion. (*a*) How far will the electron travel in the field before stopping momentarily and (*b*) how much time will elapse?

STRATEGY: If the electron is originally traveling in the positive direction its acceleration is $a = -eE/m$, where E is the magnitude of the field and m is the mass of the electron. The kinematic equations are $x = v_0 t - \frac{1}{2}(eE/m)t^2$ and $v = v_0 - (eE/m)t$, where v_0 is its initial velocity. Set $v = 0$ and solve for x and t.

SOLUTION:

$$\left[\text{ans: } (a)\ 7.12\,\text{cm};\ (b)\ 2.85 \times 10^{-8}\,\text{s}\right]$$

(c) If, instead, the region of electric field is only 8.00 mm wide (too small for the electron to stop), what fraction of the electron's initial kinetic energy will be lost that region?

STRATEGY: If ℓ is the width of the region then the work done by the electric field is $W = -eE\ell$. The magnitude of W is the kinetic energy lost by the electron. Calculate $|W|/K_0$, where $K_0 = \frac{1}{2}mv_0^2$ is the initial kinetic energy.

SOLUTION:

$$\left[\text{ans: } 0.112\right]$$

PROBLEM 11 (48E). A spherical water drop $1.20\,\mu\text{m}$ in diameter is suspended in calm air owing to a downward directed atmospheric electric field $E = 462\,\text{N/C}$. (a) What is the weight of the drop?

STRATEGY: The mass of the drop is given by $m = (4\pi/3)R^3\rho$, where R is its radius and ρ is the density of water ($0.998 \times 10^3\,\text{kg/m}^3$). Its weight is given by $W = mg$, where g is the acceleration due to gravity.

SOLUTION:

$$\left[\text{ans: } 8.87 \times 10^{-15}\,\text{N}\right]$$

(b) How many excess electrons does it have?

STRATEGY: Since the electric force balances the gravitational force the charge q on the drop is given by $qE = W$. The number of excess electrons is given by $n = q/e$.

SOLUTION:

$$\left[\text{ans: } 120\right]$$

PROBLEM 12 (52P). A uniform electric field exists in a region between two oppositely charged plates. An electron is released from rest at the surface of the negatively charged plate and strikes the surface of the opposite plate, 2.0 cm away, in a time $1.5 \times 10^{-8}\,\text{s}$. (a) What is the speed of the electron as it strikes the second plate? (b) What is the magnitude of the electric field E?

STRATEGY: Since the field is uniform the acceleration of the electron is constant. Its magnitude is given by $a = eE/m$, where E is the magnitude of the electric field, and m is the mass of the electron. If d is the separation of the plates the kinematic equations are $d = \frac{1}{2}(eE/m)t^2$ and $v = (eE/m)t$, where t is the time of flight. The origin was placed at the initial position of the electron. Solve these equations simultaneously for v and E.

SOLUTION:

[ans: (a) 2.7×10^6 m/s; (b) 1000 N/C]

PROBLEM 13 (56P).   In Fig. 24–41, a uniform, upward-pointing electric field E of magnitude 2.00×10^3 N/C has been set up between two horizontal plates by charging the lower plate positively and the upper plate negatively. The plates have length $L = 10.0$ cm and separation $d = 2.00$ cm. An electron is then shot between the plates from the left edge of the lower plate. The initial velocity v_0 of the electron makes an angle $\theta = 45.0°$ with the lower plate and has a magnitude of 6.00×10^6 m/s. (a) Will the electron strike one of the plates? (b) If so, which plate and how far horizontally from the left edge?

STRATEGY: The acceleration of the electron is downward in the diagram and its magnitude is given by $a = eE/m$. Take the x axis to be to the right and the y axis to be upward. Place the origin at the left edge of the lower plate. Then the kinematic equations are $x = v_{0x}t$ and $y = v_{0y}t - \frac{1}{2}(eE/m)t^2$, where $v_{0x} = v_0 \cos \theta_0$ and $v_{0y} = v_0 \sin \theta_0$. Here v_0 is the initial speed and θ_0 is the initial angle of the trajectory with the x axis.

To find if the electron strikes the upper plate, put $y = d$ and solve the second equation for t. You will obtain a quadratic equation. If the solution is complex (the quantity under the radical is negative) then the electron does not hit the upper plate. If the solution is real then substitute the value of t into the first equation to find the value of x. If $x > L$ the electron does not strike the upper plate. If $x < L$ it does.

If the electron does not strike the upper plate set $y = 0$ in the second kinematic equation and solve for t, then use the first kinematic equation to find the x coordinate for that value of t. If $x < L$ then the electron strikes the lower plate. If $x > L$ it strikes neither plate.

SOLUTION:

[ans: it strikes the upper plate 2.73 cm from the left edge]

You should know how to calculate the torque of a uniform electric field on an electric dipole and the potential energy of a dipole in an electric field. In some cases you are asked for the work that is done by the field on a dipole when the dipole turns. This is the negative of the change in the potential energy.

PROBLEM 14 (58E).   An electric dipole consists of charges $+2e$ and $-2e$ separated by 0.78 nm. It is in an electric field of strength 3.4×10^6 N/C. Calculate the magnitude of the torque on the dipole when the dipole moment is (a) parallel to, (b) perpendicular to, and (c) opposite the electric field.

STRATEGY: The torque is given by $\tau = \mathbf{p} \times \mathbf{E}$ and its magnitude is given by $\tau = pE \sin \theta$, where $\mathbf{p}$ is the dipole moment, $\mathbf{E}$ is the electric field, and θ is the angle between them when the vectors are drawn with their tails at the same point. The magnitude of the dipole moment is $2ed$, where d is the separation of the charges. In (a) $\theta = 0°$, in (b) $\theta = 90°$, and in (c) $\theta = 180°$.

SOLUTION:

$$[\text{ans: } (a)\ 0;\ (b)\ 8.5 \times 10^{-22}\,\text{N} \cdot \text{m};\ (c)\ 0]$$

PROBLEM 15 (59P).   Find the work required to turn an electric dipole end for end in a uniform electric field **E**, in terms of the magnitude p of the dipole moment, the magnitude E of the field, and the initial angle θ_0 between **p** and **E**.

STRATEGY: The work required of an external agent is the change in the potential energy of the dipole. The initial potential energy is $U_i = -pE\cos\theta_0$; the final potential energy is $U_f = -pE\cos(\theta_0 + 180°)$. Use the trigonometric identity $\cos(\theta_0 + 180°) = -\cos\theta_0$

SOLUTION:

$$[\text{ans: } 2pE\cos\theta_0]$$

III. NOTES

Chapter 25
GAUSS' LAW

I. BASIC CONCEPTS

Gauss' law is one of the four fundamental laws of electromagnetism. It relates an electric field to the charges that create it and is therefore closely akin to Coulomb's law. Unlike Coulomb's law, however, it is valid when the charges are moving, even at relativistic speeds. The central concept for an understanding of Gauss' law is that of electric flux, a quantity that is proportional to the number of field lines penetrating a given surface. You should pay careful attention to the definition of flux and learn how to compute it for various fields and surfaces. You should then learn how to use Gauss' law to compute the charge in any region if the electric field is known on the boundary and also how to use the law to compute the electric field in certain highly symmetric situations.

Electric flux.   Electric flux is associated with any surface (real or imaginary) through which electric field lines pass. It is defined by the integral

$$\Phi =$$

where $\mathbf{E}$ is the electric field on the surface and $d\mathbf{A}$ is an infinitesimal element of surface area. This integral tells us to evaluate the scalar product $\mathbf{E} \cdot d\mathbf{A}$ for each element of the surface and sum the results for all elements. Notice that the infinitesimal area element is written as a vector. What is its direction? _____
For an open surface, such as the top of a table, $d\mathbf{A}$ may be in either of the two directions that are perpendicular to the surface (up or down for a horizontal surface). Which one you pick determines the sign but not the magnitude of the flux. For a closed surface, one that completely surrounds a volume, $d\mathbf{A}$ is always chosen to be _____.

The scalar product that appears as the integrand may be interpreted as the product of the infinitesimal area and the component of the field normal to the surface. Thus only the field that pierces the surface contributes to the flux. Consider two identical surfaces and suppose a uniform electric field exists at each of them. If the fields have the same normal components then the flux is the same even if they have different tangential components. A field that is parallel to a surface does not contribute to the flux through that surface.

The flux through a surface is proportional to the number of field lines that penetrate the surface. Recall that the number of lines per unit area through an infinitesimal area dA perpendicular to the field is proportional to the magnitude of the field. Thus if $d\mathbf{A}$ is in the same direction as the field the number of lines that penetrate it is proportional to $E\,dA$. If the area is rotated so $d\mathbf{A}$ makes the angle ϕ with the field the number of lines that penetrate the area decreases to $E\,dA\cos\phi$. Carefully study Fig. 25–2 of the text.

When studying this and subsequent chapters you may think of the electric flux through a surface as giving the net number of lines crossing the surface. There are, however, two important distinctions you should make. First, Φ is not necessarily an integer. Second, Φ is negative if the field makes an angle of more than 90° with d**A** or, what is the same thing, the field lines cross the surface in a direction roughly opposite to d**A**. Field lines that cross a closed surface from inside to outside make a _____ contribution to Φ while lines that cross the surface from outside to inside make a _____ contribution. Since some parts of a closed surface may make positive contributions to the flux while other parts make negative contributions, the total flux through a surface may vanish even though an electric field exists at every point on the surface.

Gauss' law.   Gauss' law states that for any closed surface the total flux through the surface is proportional to the net charge enclosed by the surface. The constant of proportionality involves ϵ_0, the permittivity constant. Copy Eq. 25–8 of the text here:

When you use the law be sure to evaluate the integral for a *closed* surface. The small circle on the integral sign is a reminder. The surface is known as a <u>Gaussian surface</u>. When you use Gauss' law to solve a problem you must identify the Gaussian surface you are using. It might be the physical surface of an object or it might be a purely imaginary construction. Also be very careful that the charge you use in the Gauss' law equation is the net charge *enclosed* by the Gaussian surface. Charge outside does not contribute to the total flux through a closed surface, although it does contribute to the electric field at the surface.

The law should not surprise you since the magnitude of the electric field and the number of field lines associated with any charge are both proportional to the charge. All lines from a single positive charge within a closed surface penetrate the surface from inside to outside, so the total flux is positive and proportional to the charge. All lines associated with a single negative charge within a closed surface penetrate from outside to inside so the total flux is negative and again proportional to the charge. Some lines associated with a charge outside a surface penetrate the surface but those that do penetrate twice, once from outside to inside and once from inside to outside, so this charge does not contribute to the total flux through the surface. When more than one charge is present add the individual flux contributions, with their appropriate signs.

You should be aware that the electric field at points on a closed surface containing a net charge of zero is not necessarily zero. Each charge, whether in the interior or exterior, creates a field that does not vanish on the surface. Nevertheless, a net charge of zero inside means a total flux of zero. Zero total flux simply means there are just as many field lines entering the volume enclosed by the surface as there are leaving.

Gauss' law can be used to find an expression for the electric field of a point charge. Imagine a sphere of radius r with a positive point charge q at its center. Since the electric field is radially outward from the charge, the normal component of the field at any point on the surface of the sphere is the same as the magnitude E of the field. In terms of E and r the total flux through the sphere is $\Phi =$ _____. Equate this to q/ϵ_0 and solve for E. The result is: $E =$ _____, in agreement with Coulomb's law.

Gauss's law can be written in differential form and used to solve for the electric field of any distribution of charge. In its integral form it is useful in several ways. If the electric field is known at all points on a surface the law can be used to calculate the net charge enclosed by the surface. If the charge is known the law can be used to find the total electric flux through the surface. If the charge distribution is highly symmetric so a symmetry argument can be used to show that $E \cos \theta$ has the same value at all points on the surface then Gauss' law can be used to solve for the electric field at points on the surface.

Gauss' law and conductors. The electric field vanishes at all points in the interior of a conductor in electrostatic equilibrium, with all charge stationary. What would happen if this were not true? _____

_____

_____

Remember that $\mathbf{E} = 0$ inside a conductor even when excess charge is placed on it or when an external field is applied to it.

At electrostatic equilibrium the electric field at all points just outside a conductor is perpendicular to the surface. It has only a normal, not a tangential component. The charge on the surface can bring about a change in the normal component of the field so it is zero inside and non-zero outside but it cannot change the tangential component. Since the tangential component must be zero inside it must also be zero just outside.

The condition that the electric field is zero leads, via Gauss' law, to an interesting property of conductors: in an electrostatic situation any excess charge on a conductor must reside on its surface; there can be no net charge in its interior. Imagine a Gaussian surface that is completely within the conductor. The electric field is _____ at every point on the Gaussian surface, so the total flux through the surface is _____ and, according to Gauss' law, the net charge enclosed by the surface is _____. Any net charge in the conductor must lie outside the Gaussian surface. Since this result is true for *every* Gaussian surface that can be drawn completely within the conductor, no matter how close to its surface, we conclude that any excess charge must be on the surface of the conductor. Be careful! If the object is an insulator, excess charge may be distributed throughout its volume.

You should understand conductors with cavities. The diagram on the right shows the cross section of a conductor with a cavity containing a point charge q_1. The net charge within the Gaussian surface indicated by the dotted line is _____ so if q_2 is the charge on the inner surface of the conductor $q_1 + q_2 =$ _____ and $q_2 =$ _____. Furthermore, if Q is the total excess charge on the conductor and q_3 is the charge on its outer surface then $q_2 + q_3 =$ _____ and $q_3 =$ _____.

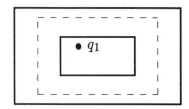

If the net charge on the conductor is zero the charge on its outer surface is _____. If $Q =$ _____ there is no net charge on the outer surface of the conductor. Notice that Gauss' law cannot tell us the distribution of charge on a surface, only the net charge there.

You should realize that the area charge density may not be the same at every point on the surface of a conductor. Whether it is or not the magnitude of the electric field at any point just outside is directly proportional to the area charge density at the corresponding point on the surface. The exact relationship is $E = \underline{\hspace{1cm}}$, where σ is the area charge density.

Gauss' law can be used to prove this result. Consider a small portion of the surface of a conductor and suppose its area is A and its area charge density is σ. The electric field is normal to the surface. Describe the Gaussian surface you will use: \underline{\hspace{4cm}}

\underline{\hspace{11cm}}

\underline{\hspace{11cm}}

One end of the Gaussian surface should be inside the conductor. The flux through this end is \underline{\hspace{1cm}}. Another end should be outside the conductor and perpendicular to the field. If the magnitude of the field is E the flux through this end is \underline{\hspace{1cm}}. The sides are parallel to the field so the flux through them is \underline{\hspace{1cm}}. In terms of E and A the total flux through the Gaussian surface is \underline{\hspace{1cm}}.

In terms of σ and A the net charge enclosed by the surface is \underline{\hspace{1cm}}. Gauss' law yields \underline{\hspace{1cm}} = \underline{\hspace{1cm}}. $E = \sigma/\epsilon_0$ follows immediately. You should realize that this field is produced by the charge on *all* parts of the surface and by external sources if they are present.

Calculating the electric field.   Successful use of Gauss' law to solve for the electric field depends greatly on choosing the right Gaussian surface. First, it must pass through the point where you want the value of the field. Second, either the electric field must have constant magnitude over the entire surface or else it must have constant magnitude over part of the surface and the flux through the other parts must be zero. A symmetry argument should be made to justify the use of the Gaussian surface you have chosen.

To understand how such an argument is made consider a long straight wire with positive charge distributed uniformly on it. Take the wire to be infinitely long and use symmetry to show that the electric field at any point is radially outward from the wire. The outline of the argument is: pretend that the field is *not* radially outward and show that this leads to a contradiction of the principle that identical charge distributions produce identical electric fields.

Assume the field at point P is in the direction shown in the upper diagram, a non-radial direction. Now turn the wire over end-for-end, through a 180° rotation about the dotted line through P. If the direction chosen for the field is correct, then after turning the wire the field will be in the direction shown in the lower diagram. This is impossible since the charge distribution is exactly the same in the two cases. A radial field is the only one that does not change when the wire is turned.

Note that if the wire is not infinite the argument works only if P is opposite the midpoint of the wire. Similarly the argument does not work (and the field is not radial) if the charge is not distributed uniformly on the wire. In those cases turning the wire changes the distribution of charge.

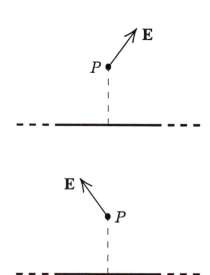

For an infinite wire with a uniform charge distribution the electric field must have the same magnitude at all points that are the same distance from the wire. The charge distribution looks exactly the same from point P as it does from a point 2 m (or 2 km) further along the wire, the same distance from the wire.

The points where the magnitude of the field are the same as at P form the rounded portion of the surface of a cylinder, so take a cylinder of length ℓ and radius r, with P on the surface, for your Gaussian surface. In the space to the right draw a diagram of the wire and the Gaussian surface. Label dA and **E** at point P.

The completion of the problem is now straightforward. First evaluate the integral $\int \mathbf{E} \cdot d\mathbf{A}$ over the rounded portion of the cylinder. Since **E** and d**A** are parallel at all points on the surface $\mathbf{E} \cdot d\mathbf{A} = $ _____ and since E is the same for all points on the surface $\int \mathbf{E} \cdot d\mathbf{A} = E \int dA = $ _____, in terms of r and ℓ. Since a Gaussian surface must be closed you must also consider the circles that form the ends of the cylinder. No flux passes through either end because _____.

Solve the Gauss' law equation. If q is the net charge enclosed by the cylinder, then the magnitude of the electric field is given by $E = $ _____. Usually the linear charge density λ of the wire is given. Since this is q/ℓ, $E = $ _____.

Now consider some of the other charge distributions discussed in the text. Think of these derivations in two ways. First, they will give you a chance to practice solving problems using Gauss' law. Second, in some cases the results are useful for later work. You should particularly remember the expression for the electric field of a large plane sheet of charge. You will use it several times when you study capacitors.

Consider an infinite sheet with uniform area charge density σ. In the space to the right sketch a portion of the sheet and the Gaussian surface used to find the electric field at a point away from the sheet. The field is perpendicular to the sheet and points away from it if it is positive. Draw a vector indicating the electric field at a point on the Gaussian surface. Also label the parts of the surface through which the flux is zero.

In terms of the magnitude E of the field and area A of one end of the Gaussian surface, the total flux through the surface is $\Phi = $ _____. In terms of σ and A the total charge enclosed by the surface is _____. Gauss' law thus gives the following expression for the electric field in terms of the area charge density: $E = $ _____.

You might think this result is inconsistent with the expression for the magnitude of the electric field just outside a conductor, $E = \sigma/\epsilon_0$. It is not. Consider an infinite plane conducting sheet with a uniform charge density σ on one surface. This charge produces an electric field with magnitude $\sigma/2\epsilon_0$, just like any other large uniform sheet of charge. The field exists on both sides of the sheet, in the interior of the conductor as well as in the exterior, and on each side it points away from the surface if σ is positive.

To obtain the layer of charge on the surface of the conductor another electric field must be present, produced perhaps by charge on another portion of the conductor or by external charge. In the diagram E_1 is the magnitude of the field due to the charge layer and E_2 is the magnitude of the second field. In the interior the second field exactly cancels the field due to the charge layer to produce a total field of zero and in the exterior it augments the field due to the charge layer to produce a total field with magnitude σ/ϵ_0.

interior	exterior
$E_1 = \sigma/2\epsilon_0$	$E_1 = \sigma/2\epsilon_0$
$\longleftarrow$	$\longrightarrow$
$\longrightarrow$	$\longrightarrow$
$E_2 = \sigma/2\epsilon_0$	$E_2 = \sigma/2\epsilon_0$

Another important charge configuration is a uniform spherical shell carrying total charge Q. Symmetry leads to us to conclude that the electric field must be radial so we use a Gaussian surface in the form of a sphere, concentric with the shell. If the radius of the Gaussian surface is r and the electric field has magnitude E at points on it then the flux through the surface is $\Phi =$ _____ . If the Gaussian surface is inside the shell the charge enclosed is _____ , so the electric field is $E =$ _____ at points inside the shell. If the Gaussian surface is outside the shell the charge enclosed is _____ , so the electric field is _____ at points outside the shell. Carefully study Sample Problem 25–7 to see how this result is applied to find the electric field inside a uniform spherical distribution of charge.

II. PROBLEM SOLVING

You should know how to calculate the electric flux through a given surface. A useful fact to remember is: if the field is uniform then the total flux through a closed surface is 0. Sometimes you can easily calculate the flux of a uniform field through part of a closed surface but you have been asked for the flux through the remaining part. Since the two contributions to the total flux sum to 0 they are the negatives of each other. Here are some examples.

PROBLEM 1 (3E). A cube with 1.40-m edges is oriented as shown in Fig. 25–25 in a region of uniform electric field. Find the electric flux through the right face if the electric field, in newtons per coulomb, is given by (a) $6.00\,\mathbf{i}$, (b) $-2.00\,\mathbf{j}$, and (c) $-3.00\,\mathbf{i} + 4.00\,\mathbf{k}$.

STRATEGY: Evaluate $\Phi = \int \mathbf{E} \cdot d\mathbf{A}$. For the right face $d\mathbf{A} = dA\,\mathbf{j}$. In (a) $\mathbf{E} \cdot d\mathbf{A} = 0$, in (b) $\mathbf{E} \cdot d\mathbf{A} = -2.00\,dA$, and in (c) $\mathbf{E} \cdot d\mathbf{A} = 0$.

SOLUTION:

$$[\,\text{ans: } 0;\ -3.92\,\text{N} \cdot \text{m}^2/\text{C};\ 0\,]$$

(d) What is the total flux through the cube for each of these fields?

STRATEGY: All the given fields are uniform. Whatever flux is through one face, flux of exactly the same magnitude but opposite sign is through the opposite face.

SOLUTION:

$$[\,\text{ans: } 0 \text{ for each field}\,]$$

PROBLEM 2 (4P).   Calculate Φ through (a) the flat base and (b) the curved surface of a hemisphere of radius R. The field $\mathbf{E}$ is uniform and perpendicular to the flat base of the hemisphere, and the field lines enter through the flat base.

STRATEGY: (a) Since the field is uniform and perpendicular to the flat base and the field lines go *into* the base, the flux through the base is given by $\Phi_b = \int \mathbf{E} \cdot d\mathbf{A} = -EA = -\pi R^2 E$. ($b$) Since the field is uniform the total flux through the whole hemisphere, including the base, is 0.

SOLUTION:

$$[\text{ans: } (a) -\pi R^2 E; \ (b) \ \pi R^2 E]$$

Some problems deal with Gauss' law in the form $\Phi = q/\epsilon_0$, where Φ is the electric flux through a closed surface and q is the charge enclosed by the surface. You might be given the charge and asked for the flux or you might be given the flux (or the information to calculate it) and asked for the enclosed charge. Here's some examples.

PROBLEM 3 (6E).   In Fig 25–27, the charge on a neutral isolated conductor is separated by a nearby positively charged rod. What is the flux through the five Gaussian surfaces shown in cross section? Assume that the charges enclosed by S_1, S_2, and S_3 are equal in magnitude.

STRATEGY: The charge enclosed by S_1 is q so the flux through the surface is $\Phi_1 = q/\epsilon_0$. The charge enclosed by S_2 is $-q$, the charge enclosed by S_3 is q, the charge enclosed by S_4 is $q - q = 0$, and the charge enclosed by S_5 is $q - q + q = q$.

SOLUTION:

$$[\text{ans: } \Phi_1 = q/\epsilon_0; \ \Phi_2 = -q/\epsilon_0; \ \Phi_3 = q/\epsilon_0; \ \Phi_4 = 0; \ \Phi_5 = q/\epsilon_0]$$

PROBLEM 4 (12P).   Find the net flux through the cube of Exercise 3 and Fig 25–25 if the electric field is given by (a) $\mathbf{E} = 3.00y\,\mathbf{j}$ and (b) $\mathbf{E} = -4.00\,\mathbf{i} + (6.00 + 3.00y)\,\mathbf{j}$. E is in newtons per coulomb, and y is in meters.

STRATEGY: Use $\Phi = \oint \mathbf{E} \cdot d\mathbf{A}$. ($a$) On the right face $\mathbf{E} \cdot d\mathbf{A} = (3.00y\,\mathbf{j}) \cdot (dx\,dz\,\mathbf{j})$, with $y = a$, the length of a cube edge. The contribution of this face to the electric flux is $(3.00a)(a^2) = 3.00a^3$. On the left face $\mathbf{E} \cdot d\mathbf{A} = (3.00y\,\mathbf{j}) \cdot (-dx\,dz\,\mathbf{j})$, with $y = 0$, so the contribution of this face is 0. On the other faces the field is parallel to the area (perpendicular to the area vector) so their contributions are 0.

   (b) On the right face $\mathbf{E} \cdot d\mathbf{A} = [-4.00\,\mathbf{i} + (6.00 + 3.00y)\,\mathbf{j}] \cdot (dx\,dz\,\mathbf{j})$, with $y = a$. The contribution of this face to the electric flux is $(6.00 + 3.00a)(a^2)$. On the left face $\mathbf{E} \cdot d\mathbf{A} = [-4.00\,\mathbf{i} + (6.00 + 3.00y)\,\mathbf{j}] \cdot (-dx\,dz\,\mathbf{j})$, with $y = 0$, so the contribution of this face is $-6.00a^2$. On the front face $\mathbf{E} \cdot d\mathbf{A} = [-4.00\,\mathbf{i} + (6.00 + 3.00y)\,\mathbf{j}] \cdot (dy\,dz\,\mathbf{i}) = -4.00\,dy\,dz$ so the contribution of this face is $-4.00a^2$. On the back face $\mathbf{E} \cdot d\mathbf{A} = [-4.00\,\mathbf{i} + (6.00 + 3.00y)\,\mathbf{j}] \cdot (-dy\,dz\,\mathbf{i}) = 4.00(dy\,dz)$ so the contribution of this face is $4.00a^2$. The top and bottom faces make no contributions.

SOLUTION:

$$[\text{ans: } (a) \ 8.23\,\text{N} \cdot \text{m}^2/\text{C}; \ (b) \ 8.23\,\text{N} \cdot \text{m}^2/\text{C}]$$

(c) In each case, how much charge is inside the cube?

STRATEGY: Use $q = \epsilon_0 \Phi$.

SOLUTION:

[ans: 72.8 pC in each case]

When a problem deals with a conductor you should remember that the electric field inside the conductor is 0, that the field just outside the conductor is perpendicular to the surface and has magnitude σ/ϵ_0, where σ is the area charge density, and that the total charge enclosed by any Gaussian surface completely inside the conductor is 0. Here are some problems that ask you to apply these results.

PROBLEM 5 (17E). Space vehicles traveling through the Earth's radiation belts can intercept a significant number of electrons. The resulting charge buildup can damage electronic components and disrupt operations. Suppose a spherical metallic satellite 1.3 m in diameter accumulates 2.4 μC of charge in one orbital revolution. (a) Find the resulting surface charge density.

STRATEGY: The surface charge density is the charge divided by the surface area: $\sigma = q/4\pi R^2$, where R is the radius of the satellite.

SOLUTION:

[ans: 4.5×10^{-7} C/m^2]

(b) Calculate the magnitude of the resulting electric field just outside the surface of the satellite due to the surface charge.

STRATEGY: Use $E = \sigma/\epsilon_0$

SOLUTION:

[ans: 5.1×10^4 N/C]

PROBLEM 6 (18E). A conducting sphere with charge Q is surrounded by a spherical conducting shell. (a) What is the net charge on the inner surface of the shell?

STRATEGY: Draw a diagram with a Gaussian surface in the form of a sphere completely inside the shell, close to its inner surface. The net charge inside the Gaussian surface is 0. If Q' is the charge on the inner surface of the shell then $Q + Q' = 0$.

SOLUTION:

[ans: $-Q$]

(*b*) Another charge q is placed outside the shell. Now what is the net charge on the inner surface of the shell?

SOLUTION:

[ans: $-Q$]

(*c*) If q is moved to a position between the shell and the sphere, what is the net charge on the inner surface of the shell?

SOLUTION:

[ans: $-(Q + q)$]

(*d*) Are your answers valid if the sphere and shell and not concentric?

STRATEGY: Review the arguments for a sphere and shell that are not concentric. You should see that none of them change.

SOLUTION:

[ans: yes]

Many problems ask you to use Gauss' law to solve for the electric field in given situations. Become adept in finding an expression for the total flux through a Gaussian surface for the three special cases discussed in the text: cylindrical, spherical and planar symmetry. For a cylindrical Gaussian surface, with a radial field that is uniform over the curved portion of the surface, $\int \mathbf{E} \cdot d\mathbf{A} = 2\pi r \ell E$, where r is the radius of the Gaussian cylinder and ℓ is its length. For a spherical Gaussian surface with a radial field that is uniform over the surface $\int \mathbf{E} \cdot d\mathbf{A} = 4\pi r^2 E$, where r is the radius of the Gaussian sphere. Also remember that the electric field due to a large plane with uniform area charge density σ is $\sigma / 2\epsilon_0$.

PROBLEM 7 (24P). Figure 25–31 shows a section through two long thin concentric cylinders of radii a and b with $a < b$. The cylinders have equal and opposite charges per unit length λ. Using Gauss' law prove (*a*) that $E = 0$ for $r < a$ and (*b*) that between the cylinders, for $a < r < b$,

$$E = \frac{1}{2\pi\epsilon_0} \frac{\lambda}{r}.$$

STRATEGY: (*a*) Draw a diagram with a Gaussian surface that is a cylinder of length ℓ and radius r, with $r < a$, concentric with the charge-carrying cylinders. The charge enclosed is 0. If an electric field existed in the region it would be radial and $\oint \mathbf{E} \cdot d\mathbf{A}$ would be $2\pi r \ell E$. Gauss' law becomes $2\pi r \ell E = 0$.

SOLUTION:

(*b*) Now draw a diagram with a cylindrical Gaussian surface that lies between the charge-carrying cylinders. Again $\oint \mathbf{E} \cdot d\mathbf{A} = 2\pi r \ell E$. The charge enclosed by the Gaussian cylinder is $\lambda \ell$, so Gauss' law becomes $2\pi r \ell = \lambda \ell / \epsilon_0$. Solve for E.

SOLUTION:

PROBLEM 8 (27P). A very long conducting cylindrical rod of length L with a total charge $+q$ is surrounded by a conducting cylindrical shell (also of length L) with total charge $-2q$, as shown in the section in Fig. 25–33. Use Gauss' law to find (a) the electric field at points outside the conducting shell, (b) the distribution of charge on the conducting shell, and (c) the electric field in the region between the shell and rod.

STRATEGY: (a) Draw a diagram with a Gaussian surface in the form of a cylinder, concentric with the rod and conducting shell and with radius r, greater than the outer radius of the shell. Take the length of the Gaussian cylinder to be L and neglect fringing effects. The electric field is then radial and since the net charge enclosed is negative, the field is inward. Thus $\oint \mathbf{E} \cdot d\mathbf{A} = -2\pi r L E$. The total charge enclosed is $q - 2q = -q$, so Gauss' law becomes $-2\pi r L E = -q/\epsilon_0$. Solve for E.

(b) To find the charge on the inner surface of the shell use a Gaussian surface in the form of a cylinder, similar to the one used in part (a) but within the shell. Draw a diagram. The electric field on this surface is 0 and the net charge enclosed is also 0. If q' is the charge on the inner surface then $q + q' = 0$. The charge on the outer surface of the shell is $-2q - q'$.

(c) Now draw a Gaussian cylinder that lies between the rod and shell. For this surface $\oint \mathbf{E} \cdot d\mathbf{A} = 2\pi r L E$ and the charge enclosed is q.

SOLUTION:

[ans: (a) $q/2\pi\epsilon_0 r L$, radially inward; (b) inner surface: $-q$, outer surface: $-q$; (c) $q/2\pi\epsilon_0 r L$, radially outward]

PROBLEM 9 (31E). Figure 25–34 shows two large parallel nonconducting sheets with identical distributions of positive charge. What is E at points (a) to the left of the sheets, (b) between them, and (c) to the right of the sheets?

STRATEGY: The magnitude of the electric field due to a large uniform sheet of charge is given by $E = \sigma/2\epsilon_0$, where σ is the area charge density. For each region vectorially add the individual fields due to the sheets. In (a) both fields point to the left, in (b) one field points to the left and other points to the right, and in (c) both fields point to the right.

SOLUTION:

[ans: (a) σ/ϵ_0, to the left; (b) 0; (c) σ/ϵ_0, to the right]

PROBLEM 10 (33E).   A large flat nonconducting surface has a uniform charge density σ. A small circular hole of radius R has been cut in the middle of the sheet, as shown in Fig. 25–35. Ignore fringing of the field lines around all edges, and calculate the electric field at point P, a distance z from the center of the hole along its axis. (*Hint*: See Eq. 24–27 and use superposition.)

STRATEGY: Let $\mathbf{E}_u$ be the electric field of a large uniform sheet with area charge density σ, $\mathbf{E}_d$ be the field of a uniform disk with the same charge density, and $\mathbf{E}$ be the field of a sheet with a hole. Then $\mathbf{E} = \mathbf{E}_u - \mathbf{E}_d$. $\mathbf{E}_u$ has magnitude $\sigma/2\epsilon_0$ and points away from the sheet if σ is positive. According to Eq. 24–27, $\mathbf{E}_d$ has magnitude $(\sigma/2\epsilon_0)[1 - z/\sqrt{z^2 + R^2}]$ and points away from the disk.

SOLUTION:

$$\left[\text{ans: } \frac{\sigma}{2\epsilon_0} \frac{z}{\sqrt{z^2 + R^2}}, \text{ away from the surface}\right]$$

PROBLEM 11 (37P).   Two large metal plates of area $1.0\,\text{m}^2$ face each other. They are $5.0\,\text{cm}$ apart and have equal but opposite charges on their inner surfaces. If E between the plates is $55\,\text{N/C}$, what is the magnitude of the charges on the plates? Neglect edge effects.

STRATEGY: Each plate produces a field of magnitude $\sigma/2\epsilon_0$, where σ is the area charge density on the plate. In the region between the plates the two fields are in the same direction so the magnitude of the total field is given by σ/ϵ_0. Solve for σ. The magnitude of the charge on either plate is $q = \sigma A$, where A is the area of the plate.

SOLUTION:

$$\left[\text{ans: } \pm 4.9 \times 10^{-10}\,\text{C}\right]$$

PROBLEM 12 (44E).   A thin, metallic, spherical shell of radius a has a charge q_a. Concentric with it is another thin, metallic, spherical shell of radius b (where $b > a$) and charge q_b. Find the electric field at radial points r where (*a*) $r < a$, (*b*) $a < r < b$, and (*c*) $r > b$.

STRATEGY: In each case the field, if it exists, is radial. Draw a diagram for each situation. Show a spherical Gaussian surface with radius r. Then $\oint \mathbf{E} \cdot d\mathbf{A} = 4\pi r^2 E$. (*a*) The charge enclosed is 0 so $4\pi r^2 E = 0$. (*b*) The charge enclosed is q_a so $4\pi r^2 E = q_a/\epsilon_0$. (*c*) The charge enclosed is $q_a + q_b$ so $4\pi r^2 E = (q_a + q_b)/\epsilon_0$. Solve for E.

SOLUTION:

$$\left[\text{ans: } (a)\ 0;\ (b)\ q_a/4\pi\epsilon_0 r^2;\ (c)\ (q_a + q_b)/4\pi\epsilon_0 r^2\right]$$

(*d*) Discuss the criterion one would use to determine how the charges are distributed on the inner and outer surfaces of the shells.

STRATEGY: Consider a Gaussian surface completely inside the inner shell. Since the electric field is 0 at every point on the surface the charge it encloses is 0 according to Gauss' law. The inner surface of the shell has charge 0 and the outer surface has charge q_a. Now consider a Gaussian surface that is completely within the outer shell. It also encloses a net charge of 0. The charge on its inner surface and the charge on the inner shell must sum to 0. Find the charge on the inner surface of the outer shell. Finally find the charge on the outer surface that makes the total charge on the outer shell q_b.

SOLUTION:

[ans: inner shell: inner surface has charge 0, outer surface has charge q_a; outer shell: inner surface has charge $-q_a$, outer surface has charge $q_a + q_b$]

PROBLEM 13 (48P). In Fig. 25–38 a sphere, of radius a and charge $+q$ uniformly distributed throughout its volume, is concentric with a spherical conducting shell of inner radius b and outer radius c. This shell has a net charge of $-q$. Find expressions for the electric field as a function of the radius r (*a*) within the sphere ($r < a$); (*b*) between the sphere and the shell ($a < r < b$); (*c*) inside the shell ($b < r < c$); and (*d*) outside the shell ($r > c$).

STRATEGY: In each case use a spherical Gaussian surface of radius r. For each situation draw a diagram that shows the Gaussian surface. The field is radial, so $\int E \cdot dA = 4\pi r^2 E$. In (*a*) the charge enclosed is qr^3/a^3, in (*b*) it is q, in (*c*) it is 0 (the electric field vanishes inside the shell), and in (*d*) it is also 0 (the sum of the charges on the sphere and shell is 0). Use Gauss' law to find E.

SOLUTION:

[ans: (*a*) $(q/4\pi\epsilon_0 a^3)r$; (*b*) $q/4\pi\epsilon_0 r$; (*c*) 0; (*d*) 0]

(*e*) What are the charges on the inner and outer surfaces of the shell?

STRATEGY: Consider a spherical Gaussian surface completely inside the shell. Since the electric field is 0 at all points on it, it encloses a net charge of 0. If q' is the charge on the inner surface then $q + q' = 0$. The total charge on the shell is $-q$. If q'' is the charge on the outer surface, then $q' + q'' = -q$. Solve for q' and q''.

SOLUTION:

[ans: $-q$ on inner surface, 0 on outer surface]

PROBLEM 14 (52P).   A nonconducting sphere of radius R has a nonuniform charge distribution of volume charge density $\rho = \rho_s r/R$, where ρ_s is a constant and r is the distance from the center of the sphere. Show that (a) the total charge on the sphere is $Q = \pi \rho_s R^3$ and (b) the electric field inside the sphere has a magnitude given by

$$E = \frac{1}{4\pi\epsilon_0} \frac{Q}{R^4} r^2 .$$

STRATEGY: (a) The charge is given by the volume integral $Q = \int \rho \, dV$ over the sphere. Since the charge density has spherical symmetry (it depends only on r) the infinitesimal volume may be replaced by the volume of a spherical shell with infinitesimal thickness dr: $dV = 4\pi r^2 \, dr$. Thus

$$Q = \frac{4\pi\rho_s}{R} \int_0^R r^3 \, dr .$$

Evaluate the integral.

(b) Draw a diagram that shows a spherical Gaussian surface with radius r, inside the sphere of charge. The electric field is radial, so $\int \mathbf{E} \cdot d\mathbf{A} = 4\pi r^2 E$. The charge enclosed is

$$q_{\text{enc}} = \frac{4\pi\rho_s}{R} \int_0^r r^3 \, dr .$$

Evaluate the integral, then substitute into Gauss' law and solve for E. Replace ρ_s with $Q/\pi R^3$ to obtain the result requested.

SOLUTION:

PROBLEM 15 (53P).   In Fig. 25–41, a nonconducting spherical shell, of inner radius a and outer radius b, has a volume charge density $\rho = A/r$, where A is a constant and r is the distance from the center of the shell. In addition, a point charge q is located at the center. What value should A have if the electric field in the shell ($a \leq r \leq b$) is to be uniform? (*Hint: A* depends on a but not on b).

STRATEGY: Draw a diagram showing a Gaussian sphere of radius r, completely within the shell. The field is radial, so $\int \mathbf{E} \cdot d\mathbf{A} = 4\pi r^2 E$. The charge enclosed is given by

$$Q = q + \int_a^r \frac{A}{r} 4\pi r^2 \, dr = q + 2\pi A(r^2 - a^2) .$$

Use Gauss' law and solve for E. You should obtain

$$E = \frac{q}{4\pi\epsilon_0 r^2} + \frac{A}{2\epsilon_0} \left(1 - \frac{a^2}{r^2} \right) .$$

Pick A so this expression does not depend on r.

SOLUTION:

[ans: $A = q/2\pi a^2$]

PROBLEM 16 (54P).   A nonconducting sphere has a uniform volume charge density ρ. Let $\mathbf{r}$ be the vector from the center of the sphere to a general point P within the sphere. (*a*) Show that the electric field at P is given by $\mathbf{E} = \rho\mathbf{r}/3\epsilon_0$. (Note that the result is independent of the radius of the sphere.)

STRATEGY: Draw a diagram with a Gaussian surface in the form a sphere with radius r, inside the charge distribution. The field is radial, so $\int \mathbf{E} \cdot d\mathbf{A} = 4\pi r^2 E$. Since the volume of Gaussian sphere is $4\pi r^3/3$ and the charge is uniformly distributed, the charge enclosed is $q = 4\pi\rho r^3/3$. Use Gauss' law to show that $E = \rho r/3\epsilon_0$. The field is in the direction of $\rho\mathbf{r}$, radially outward if ρ is positive and radially inward if ρ is negative.

SOLUTION:

(*b*) A spherical cavity is hollowed out of the sphere, as shown in Fig. 25–42. Using superposition concepts, show that the electric field at all points within the cavity is $\mathbf{E} = \rho\mathbf{a}/3\epsilon_0$ (uniform field), where $\mathbf{a}$ is the position vector pointing from the center of the sphere to the center of the cavity. (Note that the result is independent of the radii of the sphere and the cavity.)

STRATEGY: Let $\mathbf{E}_s$ be the field of the solid sphere, $\mathbf{E}_c$ be the field of a sphere that just fills the cavity and has the same volume charge density as the actual sphere, and $\mathbf{E}$ be the field of the sphere with the cavity. Then $\mathbf{E} = \mathbf{E}_s - \mathbf{E}_c$. Substitute $\mathbf{E}_s = \rho\mathbf{r}/3\epsilon_0$. Since $\mathbf{r} - \mathbf{a}$ is the vector from the center of the cavity to the point P, $\mathbf{E}_c = \rho(\mathbf{r} - \mathbf{a})/3\epsilon_0$. Also substitute this expression.

SOLUTION:

III. MATHEMATICAL SKILLS

Electric flux calculations.   To calculate the electric flux you should be familiar with the evaluation of simple area integrals. If the region of integration is a rectangle with sides a and b you will probably want to position the coordinate system so that the rectangle is in the x, y plane with two of its sides along the axes. The infinitesimal element of area can be taken to be a rectangle with sides dx and dy. The flux integral then becomes

$$\Phi = \int_{x=0}^{a} \int_{y=0}^{b} E_z(x,y) \, dx \, dy \, .$$

Integrations in x and y are carried out independently. If, for example, the z component of the electric field is given by $E_z = 9x^2y + 2y$ then

$$\Phi = \int_{x=0}^{a} \int_{y=0}^{b} (9x^2y^2 + 2y) \, dx \, dy = \int_{y=0}^{b} (3x^3y^2 + 2xy) \Big|_{x=0}^{a} \, dy$$

$$= \int_{y=0}^{b} (3a^3y^2 + 2ay) \, dy = (a^3y^3 + ay^2) \Big|_{y=0}^{b} = a^3b^3 + ab^2 \, ,$$

where the integration over x was carried out first, then the integration over y.

For many problems of this chapter a Gaussian surface can be chosen so that the normal component of the electric field has the same value at all points on a portion of it and is zero on the other portions. The integral for the flux then reduces to $\Phi = EA$, where A is the area of the region over which the normal component of the field is not zero and E is the magnitude of the normal component.

To evaluate the flux you must know how to calculate the areas of various surfaces. The area of a rectangle with sides of length a and b is ab; the area of a circle with radius R is πR^2; the surface area of a sphere with radius R is $4\pi R^2$; and the surface area of the rounded portion of a cylinder with radius R and length ℓ is $2\pi R\ell$.

If you have trouble remembering that $2\pi r\ell$ gives the area of a cylindrical surface, imagine a paper towel roll wrapped exactly once around with a towel. The surface area of the roll is the same as the area of the towel. Since the towel is a rectangle with sides of length $2\pi r$ and ℓ, its area is $2\pi r\ell$.

Calculations of charge. You must also be able to calculate the charge enclosed by a Gaussian surface when the volume, area, or linear charge density is given. If an object has a uniform volume charge density ρ, for example, the charge enclosed is given by ρV, where V is the volume of the part of the object that lies within the Gaussian surface. If the Gaussian surface is completely within the object and the object does not have any cavities then the region enclosed by the Gaussian surface is completely filled with charge and V is the volume enclosed by the Gaussian surface. If the object has a cavity that is wholly within the Gaussian surface then V is the volume enclosed by the Gaussian surface minus the volume of the cavity.

Suppose, for example, a sphere of radius R has a uniform volume charge density ρ and the Gaussian surface is a concentric sphere of radius r. If $r < R$ then the Gaussian sphere is filled with charge and the charge enclosed is $4\pi \rho r^3/3$. If, on the other hand, $R < r$ then the Gaussian surface is only partially filled with charge. The charge enclosed is the total charge, or $4\pi \rho R^3/3$.

If a spherical object has a spherical cavity with radius R_c and the Gaussian surface is within the object but outside the cavity then the charge enclosed is $\rho[(4\pi r^3/3) - (4\pi R_c^3/3)]$. The first term in the brackets is the volume enclosed by the Gaussian surface and the second is the volume of the cavity. If $R < r$ the charge enclosed is $\rho[(4\pi R^3/3) - (4\pi R_c^3/3)]$.

Now consider a solid cylinder with radius R and length ℓ, having a uniform volume charge density ρ. Suppose the Gaussian surface is a concentric cylinder with radius r and the same length as the cylinder of charge. If $r < R$ the charge enclosed is $2\pi \rho r\ell$ and if $R < r$ it is $2\pi \rho R\ell$. If the cylinder has a cavity that is inside the Gaussian surface its volume must be subtracted from the volumes in these expressions.

If the volume charge density varies from point to point in the object you must evaluate the integral $\int \rho \, dV$ over the volume of that part of the object lying within the Gaussian surface. The most common example is a sphere with a charge density that varies only with distance r from the center. Carry out the integration by dividing the sphere into spherical shells with infinitesimal thickness dr. A typical shell extends from r to $r + dr$ and has a volume of $4\pi r^2 \, dr$. Notice that this is the product of the surface area of the shell and its thickness. The charge in the shell is $4\pi\rho(r)\,r^2\,dr$. If the Gaussian surface is sphere of radius r, concentric with the sphere of charge and entirely within it, the charge enclosed is $\int_0^r 4\pi\rho(r)\,r^2\,dr$. If the Gaussian surface is entirely outside the charge distribution the charge enclosed is $\int_0^R 4\pi\rho(r)\,r^2\,dr$. Notice the upper limits of integration are different for these two cases.

For a cylinder with a charge density that depends only on the distance r from the axis, divide the cylinder into concentric cylindrical shells with thickness dr. The volume of a shell is $2\pi r\ell\,dr$, where ℓ is the length of the cylinder. If the Gaussian surface is a concentric cylinder with radius r and is inside the charge distribution then the charge enclosed is $\int_0^r 2\pi r\ell\rho(r)\,dr$. If the Gaussian surface is outside the distribution the charge enclosed is $\int_0^R 2\pi r\ell\rho(r)\,dr$.

IV. NOTES

Chapter 26
ELECTRIC POTENTIAL

I. BASIC CONCEPTS

Because charges exert forces on each other, work (perhaps negative) must be done to assemble a collection of charges and because the forces are conservative a potential energy is associated with the assembled collection. If the charges are released, potential energy is converted to kinetic energy as each charge moves in response to the forces of the other charges. In this chapter you will learn to calculate and use the potential energy of a system of charges.

Electric potential is closely related to potential energy and plays a vital role in most succeeding discussions of electricity. Play careful attention to its definition and learn how to compute it for collections of point charges and for continuous charge distributions.

Electric potential energy. Since the electric force is conservative a potential energy function can be defined. Consider a collection of interacting charged particles. If one of them, with charge q, moves through an infinitesimal displacement ds the electric field **E** acting on it does work $dW =$ _____ and the potential energy of the collection changes by $\Delta U = -dW =$ _____. If the displacement of q is not infinitesimal, but instead is from point i to point f, the work done by the field is given by the integral $W_{if} =$ _____ and the change in the potential energy is given by $U_f - U_i = -W_{if} =$ _____.

You should be familiar with the sign of the change in potential energy. If, for example, a positive charge moves in the direction of the electric field acting on it then the sign of the work done by the field is _____ and the sign of the change in the potential energy of the system is _____. If a positive charge moves in the direction opposite to the field acting on it then the sign of the work done by the field is _____ and the sign of the change in the potential energy is _____. These signs are reversed if the charge is negative.

Coulomb's law can be used to write the work and potential energy in terms of the charges and their separation. The simplest case is that of two point charges. Suppose charges q_1 and q_2 start a distance r_i apart and move so they end a distance r_f apart. Then the work done by the electric field is $W_{if} =$ _____ and the change in the potential energy of the two-charge system is $\Delta U =$ _____. The result is the same if either charge remains stationary while the other moves or if both move. All that matters is their initial and final separations.

If the total charge in the collection is finite the potential energy is usually taken to be zero for infinite charge separation. In the expression you wrote above for the change in the potential energy of two point charges, let the initial separation r_i become large without bound, replace r_f with r, and write the potential energy as a function of the final separation r:

$$U(r) =$$

This expression is valid no matter what the signs of the charges. If they have the same sign the potential energy is positive; it decreases if their separation becomes greater and increases if their separation becomes less. Note that the electrical force is repulsive so the field does positive work in the first case and negative work in the second. If the two charges have opposite signs the potential energy is negative; it increases (becomes less negative) if their separation becomes greater and decreases (becomes more negative) if their separation becomes less. The field now does negative work in the first case and positive work in the second.

To calculate the potential energy of a collection of point charges, sum the potential energies of all *pairs* of charges. If the system consists of four charges, for example, add the potential energies of q_1 and q_2, q_1 and q_3, q_1 and q_4, q_2 and q_3, q_2 and q_4, and q_3 and q_4. If r_{12} is the separation of q_1 and q_2, r_{13} is the separation of q_1 and q_3, etc., then the potential energy of this system is

$$U =$$

If one or more of the charges move then some or all of the separations change and a change in the potential energy results. To find the change calculate the initial and final potential energies, then subtract the initial potential energy from the final.

You may view the electric potential energy of the system as the work that must be done by an external agent to assemble the collection, bringing the charges from infinite separation to their final positions. The charges are at rest at the beginning and end of the process, so there is no change in kinetic energy. This work may be positive or negative.

You should recall how the principle of energy conservation is used. If the charges are released from some initial configuration their mutual attractions and repulsions might cause them to move. The potential energy of the system will decrease as potential energy is converted to kinetic energy. You write $\Delta U + \Delta K = 0$, where K is the total kinetic energy of the charges. Thus if you know the initial and final configurations of the charge you can compute the change in the total kinetic energy.

Electric potential. Electric potential and electric potential energy are closely related but they are not the same. Be careful to distinguish the two concepts. To find the electric potential of a collection of point charges, a reference point is chosen and the electric potential at that point is set equal to zero. Then a positive *test* charge, not one of the charges in the collection, is moved from the reference point to any point P and the work done by the electric field on the test charge is calculated. The electric potential at P is the negative of this work divided by the test charge. Note that this is also the change in the electric potential energy per unit test charge of the system consisting of the original collection of charges and the test charge. The charges of the collection must remain in fixed positions as the test charge is moved. If the total charge is finite the reference point is usually selected to be infinitely far removed from the charge collection. Note that a value for the electric potential is associated with each point in space and exists whether a test charge is present or not. To find the value for a particular point the test charge is moved from the reference point to that point.

Since electric potential is an energy divided by a charge its SI unit is J/C. This unit is called a _____ (abbreviation: _____). Since an electric potential is the product of an electric field and a distance, the unit of an electric field may be taken to V/m.

Equipotential surfaces. An equipotential surface is a surface (imaginary or real) such that the potential _____ at all points on it. An equipotential surface can be drawn by connecting neighboring points at which the potential has the same value and the surface can be labelled by giving the value of the potential on it. Equipotential surfaces do not cross each other. One and only one of them goes through any point in space.

The electric field line through any point is _____ to the equipotential surface through that point. If the field has a non-vanishing component tangent to a surface then a potential difference must exist between points on the surface and the surface cannot be an equipotential surface.

The equipotential surfaces of an isolated point charge are _____, centered on the charge. The equipotential surfaces of a uniform field are _____, perpendicular to the field. Equipotential surfaces associated with other charge distributions are more complicated but if the distribution has a net charge they are _____ far from the distribution. Look at Fig. 26–3 of the text to see the surfaces associated with a dipole, for which the net charge is zero. All conductors are equipotential volumes and their boundaries are equipotential surfaces.

You can approximate the magnitude of the electric field in any region by calculating $\Delta V/\Delta s$, where Δs is the perpendicular distance between two equipotential surfaces that differ in potential by ΔV. Equipotential surfaces are close together in regions of high field and far apart in regions of low field.

If the equipotential surfaces for some charge distribution are known they can be used to calculate changes in the potential energy and the work done by the electric field or an external agent as an additional charge is moved from one place to another. The diagram to the right shows a family of equipotential surfaces where they cut the plane of the page. Several points, labelled a, b, c, d, and e, are also shown.

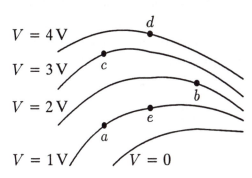

Calculate the change in potential energy as a particle is carried from one of these points to another, then calculate the work done by the electric field and the work done by the agent carrying the particle. Use the following table to record your answers.

PROCESS	CHANGE IN P.E.	WORK DONE BY FIELD	WORK DONE BY AGENT
electron from a to b	_____	_____	_____
electron from a to c	_____	_____	_____
electron from a to e	_____	_____	_____
proton from a to d	_____	_____	_____
proton from a to e	_____	_____	_____

The electric potential from the field. An expression for the potential difference between two points can be written as an integral involving the electric field at points along any path that connects the points. If **E** is the electric field then the potential difference between the points a and b is given by an integral:

$$V_b - V_a =$$

You should understand that this integral is a line integral: the path is divided into infinitesimal segments, the integrand is evaluated for each segment, and the results are summed. Because the electric field is conservative every path will give the same value for the potential difference of the given points.

You should also understand that the expression you wrote follows immediately from the definition of the potential difference as the negative of the work per unit test charge done by the field on a test charge as the test charge moves from a to b. The force of the electric field on the test charge q_0 is _____ and the work done by the field is _____ .

The integral expression for $V_b - V_a$ in terms of the electric field is valid for any situation, whether the field is produced by a single point charge or by a collection of charges. Specialize the expression for a situation in which the field is uniform, as it is outside a large sheet with uniform charge density, for example. If **E** is the electric field and $\Delta \mathbf{r}$ is the displacement from point a to point b then $V_b - V_a =$ _____ .

For some situations you must use different expressions for the electric field as the test charge passes through different regions. For example, the electric field of a uniform spherical distribution of charge is given by $E = Q/4\pi\epsilon_0 r^2$ for r outside the distribution and by $Qr/4\pi\epsilon_0 R^3$ for r inside. Here Q is the total charge and R is the radius of the distribution. The integral for the electric potential at a point inside must be broken into two parts. Use the first expression for the field as you integrate from far away to the surface of the charge distribution and the second expression as you integrate from the surface to a point inside.

Roughly speaking, if the potential varies from place to place, being high in one region and lower in a neighboring region, the electric field points from the region where the potential is _____ toward the region where the potential is _____ . A positive charge is accelerated toward the _____ potential region; a negative charge is accelerated toward the _____ potential region.

You have been considering the problem of finding the potential when the field is given. The inverse problem can also be solved: if the electric potential is a known function of position in a region of space then to find the component of **E** in any direction you calculate the rate of change of V with position along a line in that direction. Mathematically, if ds is the magnitude of an infinitesimal displacement in a given direction, then the component of **E** in that direction is given by the derivative $E_s =$ _____ . Specialize this relationship to each of the three coordinate directions. Suppose the potential is given as some function $V(x, y, z)$. Then the components of the electric field at the point with coordinates x, y, z are given by

$$E_x = \qquad\qquad E_y = \qquad\qquad E_z =$$

When you differentiate with respect to x to find E_x you treat y and z as constants. Similar statements hold for E_y and E_z. The partial derivative symbol ∂ is used to remind you. When you are finished differentiating, but not before, you evaluate the result for the coordinates of interest. As an example, suppose the potential in volts is given by $V(x, y, z) = 3x^2y - 5xy^2$, with the coordinates in meters. Then $E_x = -\partial V/\partial x = -6xy + 5y^2$ and $E_y = -\partial V/\partial y = -3x^2 + 10xy$. At $x = 2\,\text{m}$, $y = 3\,\text{m}$, $E_x = -6 \times 2 \times 3 + 5 \times 3^2 = 9\,\text{V/m}$ and $E_y = -3 \times 2^2 + 10 \times 2 \times 3 = 48\,\text{V/m}$.

Electric potentials of point charges. Suppose you wish to find the electric potential a distance r from a single isolated point charge q. Move a test charge q_0 from infinitely far away to a point r distant from q. As you do this the work done by the electric field of q is $W = $ _____ and the potential energy of the system consisting of q and q_0 changes from zero to $U = $ _____ . The electric potential at the final position of the test charge is given by

$$V = $$

This expression is also valid if q is negative. The potential then has a negative value. Be sure to note that the point charge q does not have a potential energy (only collections of more than one charge have electric potential energies) but it does create an electric potential in the space around it.

The electric potential of a collection of point charges is the sum of the individual potentials of the charges in the collection. Since the electric potential is a scalar the sum is algebraic. Suppose the system consists of charges q_1, q_2, and q_3. If you want to compute the potential at some point P you must know the distance from each charge to P. Suppose q_1 is a distance r_1 from P, q_2 is a distance r_2, and q_3 is a distance r_3. Then the potential at P is given by

$$V = $$

When you substitute values for the charges be sure you include the appropriate signs.

An electric dipole is another important example. It consists of a positive charge and a negative charge of the same magnitude, separated by a distance d. The magnitude of the dipole moment is given by $p = qd$, where q is the magnitude of either charge. The dipole moment is a vector, from the _____ charge toward the _____ charge. The potential at any point can be found by summing the potentials of the two charges. Let $\mathbf{r}$ be the vector from the midpoint of the line joining the charges to the point. The point is usually specified by giving the distance r and the angle θ between $\mathbf{r}$ and $\mathbf{p}$. If the charges are close together, compared to r, the potential is given by

$$V = $$

The electric potential of a collection of charges can be used to compute changes in the potential energy when another charge, not belonging to the collection, is moved from one point to another. If charge Q is placed at point a, where the electric potential due to charges in the collection is V_a, the potential energy of the system consisting of the collection and Q is $U = $ _____ and if Q is moved from that point to a point where the potential due to the collection is V_b the change in the potential energy of the system is $\Delta U = $ _____ .

Electric potentials of continuous charge distributions. Charge may be distributed continuously along a line, on a surface, or throughout a volume. If it is you divide the region into infinitesimal elements, each containing charge dq, and sum (integrate) the potential due to the regions. If an infinitesimal region is a distance r from point P then the contribution of that region to the potential at P is $dV = $ _____ and the potential due to the whole distribution is the integral

$$V = $$

In practice, dq is replaced by $\lambda \, ds$ for a line distribution and by $\sigma \, dA$ for a surface distribution. Here λ is a linear charge density and σ is an area charge density. To carry out the integration you will normally write the line element ds or area element dA in terms of variables that are appropriate to the geometry of the problem. Don't forget to write the distances in terms of these quantities also.

To see how the integration is done, study the calculations in the text of the potentials due to charge distributed uniformly along a straight line and uniformly over a disk.

Suppose you wish to find the potential due to a uniform line of charge, as shown in Fig. 13. If the linear charge density is λ then the charge in an infinitesimal segment dx is $dq = $ _____. If the coordinate of the segment is x and you are calculating the potential at a point that is a distance d from the line at $x = 0$, then the distance from the segment to the point is $r = $ _____. The potential at the point due to the segment is given by

$$dV = $$

This expression can be integrated. Suppose the line extends from $x = 0$ to $x = L$. Then

$$V = $$

Now calculate the potential produced by a uniform disk of charge at a point on the axis a distance z above the center of the disk. See Fig. 26–14. First consider a ring with radius R' and width dR'. The area of the ring is $dA = $ _____ and if σ is the area charge density the charge contained in the ring is $dq = $ _____. All charge on the ring is the same distance from the point; in particular, in terms of z and R', $r = $ _____. Thus the potential produced by the ring at z is $dV = $ _____ and the total potential is given by the integral _____. Since $\int (R'^2 + z^2)^{-1/2} R' \, dR' = (R'^2 + z^2)^{1/2}$ the integral can be evaluated easily. Carefully note the limits of integration. The result is

$$V = $$

Isolated conductors. All points in the interior and on the surface of an isolated conductor have the same electric potential, once equilibrium is established. This follows immediately from the definition of the potential difference between two points and the condition $\mathbf{E} = $ _____ inside a conductor in equilibrium.

Consider a conducting sphere of radius R, with charge Q uniformly distributed on its surface. If r is the distance from the sphere center to a point outside the sphere the electric potential there is given by $V = $ _____ . If r is the distance from the sphere center to a point inside the sphere, the potential there is given by $V = $ _____ . Notice that the potential inside does not depend on r and has the same value as the potential at the surface.

Suppose one conducting sphere has radius R_1 and charge Q_1 while another has radius R_2 and charge Q_2. They are far apart, which means the charge on one does not influence the distribution of charge on the other. Both have uniform distributions on their surfaces. The potential at the surface of the first sphere is $V_1 = $ _____ and the potential at the surface of the second is $V_2 = $ _____ . If the spheres are then connected by a long metal wire, these potentials must be the same. This condition can be used to solve for the ratio Q_1/Q_2. If this was not the ratio of the charges when the wire was attached, charge flows through the wire until this ratio is obtained.

II. PROBLEM SOLVING

To solve the problems of this chapter you should know how to calculate the electric potential of point charges and continuous distributions of charge. You should know how to calculate the potential energy of a collection of charges and the change in the potential energy when one or more of the charges moves. Finally, you should know how to interpret changes in potential energy as the work done by an external force that moves the charges (provided no kinetic energies change) or as the negative of the change in the total kinetic energy (when no external force acts).

PROBLEM 1 (6E).   Figure 26–25 shows, edge-on, an infinite nonconducting sheet with positive surface charge density σ on one side. (*a*) How much work is done by the electric field of the sheet as a small positive test charge q_0 is moved from an initial position on the sheet to a final position located a perpendicular distance z from the sheet?

STRATEGY: The electric field of an infinite sheet of charge is uniform, is perpendicular to the sheet, and has magnitude $E = \sigma/2\epsilon_0$. It points away from the sheet if the charge is positive and toward the sheet if the charge is negative. The work done by the electric field on the test charge q_0 is $W = q_0 \mathbf{E} \cdot \Delta \mathbf{s}$, where $\Delta \mathbf{s}$ is the displacement of the test charge.

SOLUTION:

$$\left[\text{ans: } q_0 \sigma z/2\epsilon_0 \right]$$

(*b*) Use Eq. 26–11 and the result from (*a*) to show that the electric potential of an infinite sheet of charge can be written

$$V = V_0 - (\sigma/2\epsilon_0)z ,$$

where V_0 is the potential at the surface of the sheet.

STRATEGY: According to Eq. 26–11, $V = V_0 - \int \mathbf{E} \cdot d\mathbf{s} = V_0 - W/q_0$. Substitute the result for W.

SOLUTION:

PROBLEM 2 (11P).   The electric field inside a nonconducting sphere of radius R, with charge spread uniformly throughout its volume, is radially directed and has magnitude

$$E(r) = \frac{qr}{4\pi\epsilon_0 R^3} .$$

Here q (positive or negative) is the total charge in the sphere, and r is distance from the sphere center. (*a*) Taking $V = 0$ at the center of the sphere, find the potential $V(r)$ inside the sphere.

STRATEGY: Take the infinitesimal displacement dr to be radially outward. The component of the field in this direction is $qr/4\pi\epsilon_0 R^3$, so the potential is given by

$$V(r) = -\int \mathbf{E} \cdot \mathbf{dr} = -\int_0^r \frac{qr}{4\pi\epsilon_0 R^3} \, dr .$$

Carry out the integration.

SOLUTION:

$$\left[\text{ans: } -qr^2/8\pi\epsilon_0 R^3 \right]$$

(*b*) What is the difference in electric potential between a point on the surface and the sphere's center? (*c*) If q is positive, which of these two points is at the higher potential?

STRATEGY: Use $\Delta V = V(R) - V(0)$.

SOLUTION:

$$\left[\text{ans: } (b) \ -q/8\pi\epsilon_0 R; \ (c) \text{ the sphere center is at the higher potential} \right]$$

PROBLEM 3 (15E).   Consider a point charge $q = +1.0\,\mu\text{C}$, point A (which is 2.0 m distant), and point B (which is 1.0 m distant). (*a*) If these points are diametrically opposite each other, as in Fig. 26–27a, what is the potential difference $V_A - V_B$?

STRATEGY: Use $V = q/4\pi\epsilon_0 r$ to calculate V_A and V_B, then evaluate the difference.

SOLUTION:

$$\left[\text{ans: } -4500\,\text{V} \right]$$

(b) What is that potential difference if points A and B are located as in Fig. 26–27b?

SOLUTION:

$$[\text{ans: } -4500 \text{ V}]$$

PROBLEM 4 (16E). Consider a point charge $q = 1.5 \times 10^{-8}$ C, and take $V = 0$ at infinity. (a) What are the shape and dimensions of an equipotential surface having a potential of 30 V due to q alone?

STRATEGY: Solve $V = q/4\pi\epsilon_0 r$ for r.

SOLUTION:

$$[\text{ans: a sphere with a radius of } 4.5 \text{ m}]$$

(b) Are surfaces whose potentials differ by a constant amount (1.0 V, say) evenly spaced?

STRATEGY: In any region the spacing is proportional to dV/dr.

SOLUTION:

$$[\text{ans: no; spacing is greater at larger } r]$$

PROBLEM 5 (23P). What are (a) the charge and (b) the charge density on the surface of a conducting sphere of radius 0.15 m whose potential is 200 V (with $V = 0$ at infinity)?

STRATEGY: The potential outside a conducting sphere is given by $V = Q/4\pi\epsilon_0 r$, where r is the distance from the center. At the surface $V = Q/4\pi\epsilon_0 R$, where R is the radius. Solve for Q. The charge density is given by $\sigma = Q/4\pi R^2$.

SOLUTION:

$$[\text{ans: } (a) \; 3.3 \times 10^{-9} \text{ C}; (b) \; 1.2 \times 10^{-8} \text{ C/m}^2]$$

PROBLEM 6 (27P).   A solid copper sphere whose radius is 1.0 cm has a very thin surface coating of nickel. Some of the nickel atoms are radioactive, each atom emitting an electron as it decays. Half of these electrons enter the copper sphere, each depositing 100 keV of energy there. The other half of the electrons escape, each carrying away a charge of $-e$. The nickel coating has an activity of 10 mCi (= 10 millicuries = 3.70×10^8 radioactive decays per second). The sphere is hung from a long, nonconducting string and isolated from its surroundings. (*a*) How long will it take for the potential of the sphere to increase by 1000 V?

STRATEGY: The potential of the sphere is given by $V = Q/4\pi\epsilon_0 R$, where R is its radius and Q is the charge on it. Solve for Q. Since, on average, each decay increases the charge on the sphere by $+e/2$ the rate at which the charge on the sphere is increasing is $eA/2$, where A is the activity of the nickel. Solve $Q = (eA/2)t$ for t.

SOLUTION:

[ans: 38 s]

(*b*) How long will it take for the temperature of the sphere to increase by 5.0 °C? The heat capacity of the sphere is 14.3 J/°C.

STRATEGY: The energy required to raise the temperature by ΔT is $\Delta E = C\,\Delta T$, where C is the heat capacity of the sphere. If E is the energy deposited by each absorbed electron then the rate at which the energy of the sphere is increasing is given by $EA/2$, where A is the activity of the nickel. Solve $C\,\Delta T = (EA/2)t$ for t.

SOLUTION:

[ans: 2.4×10^7 s (280 da)]

PROBLEM 7 (29E).   Two isolated charges of magnitudes Q_1 and Q_2 are separated by distance d. At an intermediate point $d/4$ from Q_1, the net electric field is zero. Setting $V = 0$ at infinity, locate a point (other than at infinity) at which the potential due to these charges is zero.

STRATEGY: Since the electric field is zero at a point between the charges, they have the same sign. Since the charges have the same sign their individual electric potentials cannot cancel anywhere.

SOLUTION:

[ans: no such points exist]

PROBLEM 8 (32P). A point charge $q_1 = +6.0e$ is fixed at the origin of a rectangular coordinate system, and a second point charge $q_2 = -10e$ is fixed at $x = 8.6$ nm, $y = 0$. The locus of all points in the xy plane with $V = 0$ (other than at infinity) is a circle centered on the x axis, as shown in Fig 26–31. Find (a) the location x_c of the center of the circle and (b) the radius R of the circle.

STRATEGY: Find the two points on the x axis where the potential vanishes. These points have coordinates x that satisfy $[(q_1/x) + (q_2/|x_2 - x|)] = 0$, where x_2 is the coordinate of q_2. First let $|x_2 - x| = x_2 - x$, then let $|x_2 - x| = -x_2 + x$. You should get 3.2 nm and -12.9 nm. The first point is at $x = x_c + R$ and the second is at $x = x_c - R$. Solve for x_c and R.

SOLUTION:

[ans: (a) -4.8 nm; (b) 8.1 nm]

(c) Is the cross section in the xy plane of the $V = 5$-V equipotential surface also a circle?

STRATEGY: Repeat the calculation for $[(q_1/x) + (q_2/|x_2 - x|)] = 5$ V. Then calculate the potential at $x = x_c$, $y = R$. If you do not get 5 V the equipotential is not a circle.

SOLUTION:

[ans: no, it is not a circle]

PROBLEM 9 (35P). In Fig. 26–34 P is at the center of the rectangle. With $V = 0$ at infinity, what is the net potential at P due to the six point charges?

STRATEGY: Sum the individual potentials. The distance from a rectangle corner to the rectangle center is $\sqrt{d^2 + (d/2)^2} = \sqrt{5}d/2$ so

$$V = \frac{1}{4\pi\epsilon_0}\left[\frac{5q}{\sqrt{5}d/2} - \frac{2q}{d/2} - \frac{3q}{\sqrt{5}d/2} - \frac{5q}{\sqrt{5}d/2} - \frac{2q}{d/2} + \frac{3q}{\sqrt{5}d/2}\right].$$

SOLUTION:

[ans: $-8q/4\pi\epsilon_0 d$]

PROBLEM 10 (36E).   (*a*) Figure 26–35*a* shows a positively charged plastic rod of length L and uniform linear charge density λ. Setting $V = 0$ at infinity and considering Fig. 26–13 and Eq. 26–28, find the electric potential at point P without written calculation.

STRATEGY: Each half of the rod produces a potential like that given by Eq. 26–28 except that each half is half as long as the rod of Fig. 26–13. Replace L with $L/2$ in the equation and double the result (since there are two halves).

SOLUTION:

$\left[\text{ans: } \dfrac{2\lambda}{4\pi\epsilon_0} \ln \dfrac{L + (L^2 + 4d^2)^{1/2}}{2d} \right]$

(*b*) Fig. 26–35*b* shows an identical rod, except that it is split in half and the right half is negatively charged; the left and right halves have the same magnitude λ of uniform linear charge density. What is the electric potential at point P in Fig. 26–35*b*?

SOLUTION:

[ans: 0]

PROBLEM 11 (38P).   (*a*) In Fig. 26–37*a*, what is the potential at point P due to charge Q at distance R from P? Set $V = 0$ at infinity.

STRATEGY: This is just the potential due to a point charge.

SOLUTION:

[ans: $Q/4\pi\epsilon_0 R$]

(*b*) In Fig. 26–37*b*, the same charge Q has been spread over a circular arc of radius R and central angle 40°. What is the potential at point P, the center of curvature of the arc?

STRATEGY: All the charge is the same distance from P so $(1/4\pi\epsilon_0) \int (1/R)\,dq = (1/4\pi\epsilon_0 R) \int dq$.

SOLUTION:

[ans: $Q/4\pi\epsilon_0 R$]

(c) In Fig. 26–37c, the same charge Q has been spread over a circle of radius R. What is the potential at point P, the center of the circle?

SOLUTION:

[ans: $Q/4\pi\epsilon_0 R$]

(d) Rank the three situations according to the magnitude of the electric field that is set up at P, from greatest to least.

STRATEGY: In each case, divide the total charge into infinitesimal elements dq. In (a) all elements are at the same place and the fields they produce at P are all in the same direction. In (b) the fields are not in the same direction and the vertical components cancel to produce a total field at P along the bisector of the arc. In (c) the field at P is 0.

SOLUTION:

[ans: from greatest to least: (a), (b), (c)]

PROBLEM 12 (41P).   What is the potential at point P in Fig. 26–40, a distance d from the right end of a plastic rod of length L and total charge $-Q$? The charge is uniformly distributed and $V = 0$ at infinity.

STRATEGY: The linear charge density is $\lambda = -Q/L$. Suppose the rod lies along the x axis with its left end at the origin. Consider an element of the rod at x. Its distance from P is $r = L + d - x$ and the potential it produces at P is $dV = (1/4\pi\epsilon_0)(\lambda/r)\,dx = (1/4\pi\epsilon_0)[(-Q/L)/(L + d - x)]\,dx$. Thus the total potential at P is given by the integral

$$V = -\frac{1}{4\pi\epsilon_0}\frac{Q}{L}\int_0^L \frac{1}{L + d - x}\,dx\,.$$

Carry out the integration.

SOLUTION:

[ans: $-\dfrac{1}{4\pi\epsilon_0}\dfrac{Q}{L}\ln\dfrac{L+d}{d}$]

PROBLEM 13 (45E).   In Section 26–8 the potential at a point on the central axis of a charged disk is shown to be

$$V = \frac{\sigma}{2\epsilon_0} \left(\sqrt{z^2 + R^2} - z \right).$$

Use Eq. 26–34 and symmetry to show that E for such a point is given by

$$E = \frac{\sigma}{2\epsilon_0} \left(1 - \frac{z}{\sqrt{R^2 + z^2}} \right).$$

STRATEGY: Symmetry tells us that the field is parallel to the axis. Its component along that axis is given by $E = -dV/dz$. Use

$$\frac{d}{dz}\sqrt{z^2 + R^2} = \frac{1}{2}\frac{2z}{\sqrt{z^2 + R^2}} = \frac{z}{\sqrt{z^2 + R^2}}$$

to evaluate the derivative.

SOLUTION:

PROBLEM 14 (50P).   In Fig. 26–43, a thin positively charged rod of length L, lying along the x axis with one end at the origin ($x = 0$), has a linear charge density given by $\lambda = kx$, where k is a constant. (*a*) Setting $V = 0$ at infinity, find V at point P on the y axis.

STRATEGY: Consider an infinitesimal element of the rod with coordinate x. Its distance from P is $\sqrt{x^2 + y^2}$ and the potential it produces at P is given by $dV = (1/4\pi\epsilon_0)(\lambda/\sqrt{x^2 + y^2})\,dx = (k/4\pi\epsilon_0)(x/\sqrt{x^2 + y^2})\,dx$. The total potential at P is given by

$$V = \frac{k}{4\pi\epsilon_0} \int_0^L \frac{x}{\sqrt{x^2 + y^2}}\,dx.$$

Use

$$\int \frac{x}{\sqrt{x^2 + y^2}}\,dx = \sqrt{x^2 + y^2}$$

to evaluate the integral.

SOLUTION:

$$\left[\text{ans: } (k/4\pi\epsilon_0)[(L^2 + y^2)^{1/2} - y] \right]$$

(*b*) Determine the vertical component E_y of the electric field intensity at P from the result of part (*a*) and also by integration of the differential fields due to differential charge elements.

STRATEGY: To find the field from the potential carry out the differentiation $E_y = -dV/dy$. The magnitude of the field due to the differential element at x is given by $dE = (1/4\pi\epsilon_0)kx(x^2 + y^2)^{-1} dx$ and its y component is this multiplied by $y/\sqrt{x^2 + y^2}$. The y component of the total field is thus given by the integral

$$E_y = \frac{1}{4\pi\epsilon_0} ky \int_0^L \frac{x}{(x^2 + y^2)^{3/2}} \, dx \,.$$

Use

$$\int \frac{x}{(x^2 + y^2)^{3/2}} \, dx = -(x^2 + y^2)^{-1/2}$$

to carry out the integration.

$$\left[\, \text{ans: } -\frac{k}{4\pi\epsilon_0} \left(\frac{y}{\sqrt{L^2 + y^2}} - 1 \right) \right]$$

(*c*) Why cannot E_x, the horizontal component of the electric field at P, be found using the result of part (*a*)?

STRATEGY: You found the potential only along the perpendicular line from the rod to P. Think about what information you need to evaluate the derivative $\partial V/\partial x$.

SOLUTION:

PROBLEM 15 (53E). The charges and coordinates of two point charges located in the xy plane are: $q_1 = +3.0 \times 10^{-6}$ C, $x = 3.5$ cm, $y = +0.50$ cm; and $q_2 = -4.0 \times 10^{-6}$ C, $x = -2.0$ cm, $y = +1.5$ cm. How much work must be done to locate these charges at their given positions, starting from infinite separation?

STRATEGY: The work that must be done is the same as the potential energy of the final configuration, calculated taking the potential energy to be 0 for infinite separation. This is given by

$$W = \frac{1}{4\pi\epsilon_0} \frac{q_1 q_2}{\sqrt{(x_1 - x_2)^2 + (y_1 - y_2)^2}} \,.$$

SOLUTION:

$$\left[\, \text{ans: } -1.9\,\text{J} \right]$$

PROBLEM 16 (59P). In the rectangle of Fig. 26–47, the sides have lengths 5.0 cm and 15 cm, $q_1 = -5.0\,\mu$C, and $q_2 = +2.0\,\mu$C. With $V = 0$ at infinity, what are the electric potentials (*a*) at corner A and (*b*) at corner B?

STRATEGY: Sum the individual potentials: $V = (1/4\pi\epsilon_0)[(q_1/r_1) + (q_2/r_2)]$. (*a*) For corner A, $r_1 = 15$ cm and $r_2 = 5.0$ cm. (*b*) For corner B, $r_1 = 5.0$ cm and $r_2 = 15$ cm.

SOLUTION:

$$\left[\text{ans: } (a)\ +6.0 \times 10^4\,\text{V};\ (b)\ -7.8 \times 10^5\,\text{V}\right]$$

(*c*) How much work is required to move a third charge $q_3 = +3.0\,\mu$C from B to A along a diagonal of the rectangle?

STRATEGY: When q_3 is at B its potential energy is $U_B = q_3 V_B$, where V_B is the electric potential at B due to q_1 and q_2. This was calculated in part (*b*). When q_3 is at A its potential energy is $U_A = q_3 V_A$, where V_A is the electric potential at A due to q_1 and q_2. This was calculated in part (*a*). The work required to move q_3 from B to A is given by $W = U_A - U_B$.

SOLUTION:

$$\left[\text{ans: } 2.5\,\text{J}\right]$$

(*d*) Does this work increase or decrease the electric energy of the three-charge system?

STRATEGY: Look at the sign of the work.

SOLUTION:

$$\left[\text{ans: increase}\right]$$

Is more, less, or the same work required if q_3 is moved along a path that is (*e*) inside the rectangle but not on a diagonal, and (*f*) outside the rectangle?

STRATEGY: The electric force is conservative and does not depend on the path.

SOLUTION:

$$\left[\text{ans: the same work is required for both } (e)\text{ and }(f)\right]$$

PROBLEM 17 (69P).   A thin, spherical, conducting shell of radius R is mounted on an isolating support and charged to a potential of $-V$. An electron is fired from point P at a distance r from the center of the shell ($r \gg R$) with an initial speed v_0, directed radially inward. What value of v_0 is needed for the electron to just reach the shell before reversing direction?

STRATEGY: Use conservation of energy: $K_i + U_i = K_f + U_f$, where the subscript i indicates the initial values, with the electron at r, and the subscript f indicates the final values, with the electron at R. The initial kinetic energy is given by $K_i = \frac{1}{2}mv_0^2$ and the final kinetic energy is $K_f = 0$. The final potential energy is $U_f = (-e)(-V) = eV$. Since the electric potential is inversely proportional to the distance from the center of the sphere, the potential at the initial position of the electron is $-VR/r$ and the initial potential energy is $U_i = (-e)(-VR/r) = eVR/r$. Since $r \gg R$ you may take U_i to be 0. Solve for v_0.

SOLUTION:

[ans: $\sqrt{2eV/m}$]

PROBLEM 18 (76E).   Consider two widely separated conducting spheres, 1 and 2, the second having twice the diameter of the first. The smaller sphere initially has a positive charge q, and the larger one is initially uncharged. You now connect the spheres with a long thin wire. (*a*) How are the final potentials V_1 and V_2 of the spheres related?

STRATEGY: The spheres and the wire form one large conductor and the electric potential has the same value throughout any conductor.

SOLUTION:

[ans: they are the same]

(*b*) Find the final charges q_1 and q_2 on the spheres in terms of q.

STRATEGY: The electric potentials are given by $V_1 = q_1/4\pi\epsilon_0 R_1$ and $V_2 = q_2/4\pi\epsilon_0 R_2$. Since these are equal and since $R_2 = 2R_1$, $q_1 = q_2/2$. Charge is conserved, so $q_1 + q_2 = q$. Solve these two equations for q_1 and q_2.

SOLUTION:

[ans: $q_1 = q/3$, $q_2 = 2q/3$]

(*c*) What is the ratio of the final surface charge density of sphere 1 to that of sphere 2?

STRATEGY: Evaluate $\sigma = q/4\pi R^2$ for each sphere.

[ans: 2]

III. NOTES

Chapter 27
CAPACITANCE

I. BASIC CONCEPTS

Capacitors are electrical devices that are used to store charge and electrical energy and, as you will see in a later chapter, are important for the generation of electromagnetic oscillations. This chapter is an excellent review of the principles you have learned in previous chapters: you will make extensive use of Gauss' law and the concepts of electric potential and electrical potential energy.

Capacitors and capacitance. A capacitor is simply two _____, isolated from each other. Each is called a <u>plate</u>. The symbol used to represent a capacitor in an electrical circuit is: _____ .

In normal use one plate holds positive charge and the other holds negative charge of the same magnitude. The charge creates an electric field, pointing roughly from the positive plate toward the negative plate, and because an electric field exists in the region between the plates, the plates are at different electric potentials. The positive plate, of course, is at a higher potential than the negative.

The potential difference of the plates and the charge on either one are proportional to each other. If the magnitude of the potential difference is V and the magnitude of the charge on either plate is q, then

$$q =$$

where _____ is the <u>capacitance</u> of the capacitor. Because q and V are proportional the capacitance is independent of both q and V. If q is doubled _____ doubles, but _____ remains the same. A capacitor can be charged by transferring charge from one plate to the other. The capacitance is a measure of how much is transferred for a given _____.

A battery connected to a capacitor transfers electrons from the plate at its positive terminal to the plate at its negative terminal. When fully charged, the potential difference of the plates is the same as the potential difference across the terminals of the battery. That is, for example, a 9 V battery will transfer charge until the potential difference of the plates is 9 V. Remember, however, that a potential difference exists between the plates of a charged capacitor whether or not a battery is connected.

The SI unit for capacitance is a coulomb/volt. This unit is called a _____ and is abbreviated _____ . Microfarad (abbreviated _____) and picofarad (abbreviated _____) capacitors are commonly used in electronic circuits. A 1-F parallel-plate capacitor must have a huge plate area or else an extremely small plate separation. Until recently the technology was not available to construct such a capacitor in a reasonable volume.

Calculation of capacitance. To calculate capacitance, first imagine charge q is placed on one plate and charge $-q$ is placed on the other. Use Gauss' law to find an expression for _____ in the region between the plates. You will find that the field at every point is proportional to q. Once an expression for the electric field is known, use _____ to compute the magnitude V of the potential difference of the plates. V is also proportional to q. Finally use _____ to find an expression for the capacitance. It will not depend on q or V, but in general terms it will depend on: _____ _____ .

Carefully study the examples given in the text. They clearly show that the electric field and potential difference used in the calculation are due to the charge on the plates and, in each case, they explicitly give the geometric quantities on which the capacitance depends.

Suppose a capacitor consists of two parallel metal plates, each of area A, separated by a distance d. If charge q is paced on one and charge $-q$ is placed on the other, the electric field between the plates is given by $E =$ _____ and the potential difference of the plates is given by $V =$ _____ . Thus the capacitance is given by $C =$ _____ . Carefully note that the capacitance depends only on the permittivity constant ϵ_0, the plate area A, and the plate separation d, not on q or V.

Remember the expression for the capacitance of a parallel plate capacitor. It will be used many times. Also remember that C is proportional to A and inversely proportional to d. If the plate separation is halved without changing the plate area, the capacitance is _____ . If the plate area is halved without changing the separation, the capacitance is _____ .

Suppose a capacitor consists of two coaxial cylinders of length L, the inner one with radius a and the outer one with radius b. Charge q is placed on the inner cylinder and charge $-q$ is placed on the outer cylinder. The electric field between the cylinders is given by $E =$ _____ , the potential difference of the cylinders is given by $V =$ _____ , and the capacitance is given by $C =$ _____ . Note that C depends only on ϵ_0, a, b, and L.

Suppose a capacitor consists of two concentric spherical shells, with radii a and b. Charge q is placed on the inner shell and charge $-q$ is placed on the outer shell. The electric field between the shells is given by $E =$ _____ , the potential difference of the shells is given by $V =$ _____ , and the capacitance is given by $C =$ _____ .

Capacitance is also defined for a single conductor. The other plate is assumed to be infinitely far away. To find an expression for the capacitance imagine charge q is on the conductor, find an expression for the electric field outside the conductor, calculate the potential V of the conductor relative to the potential at infinity, and use $q = CV$ to find the capacitance. For example, the capacitance of a spherical conductor of radius R is given by $C =$ _____ .

Capacitors in series and parallel. Be sure you can distinguish between <u>parallel</u> and <u>series</u> combinations of two or more capacitors. For a _____ combination the potential difference is the same for all the capacitors and for a _____ combination the charge is the same. If neither the potential difference nor the charge is the same then the capacitors do not form either a parallel or series combination.

In the space below draw two capacitors connected in parallel and two connected in series.

Assume a potential difference V is applied to each combination and indicate on the diagrams where it is applied. Label the charge on each plate to show which plates have the same charge and which have different charges.

PARALLEL	SERIES

In each case the capacitors can be replaced by a single capacitor with capacitance C_{eq} such that the charge transferred is the same as the total charge transferred for the original combination when the potential difference is the same as the potential difference across the original combination. If the capacitances are C_1 and C_2 then the value of C_{eq} for a parallel combination is given by $C_{eq} =$ _____ and the value of C_{eq} for a series combination is given by $C_{eq} =$ _____. For a parallel combination the equivalent capacitance is greater than the greatest capacitance in the combination; for a series combination the equivalent capacitance is less than the smallest capacitance in the combination.

Energy storage. As a capacitor is being charged, energy must be supplied by an external source, such as a battery, and energy is stored by the capacitor. You may think of the stored energy in either one of two ways: as the potential energy of the charge on the plates or as an energy associated with the electric field produced by the charges.

Suppose that, at one stage in the charging process, the positive plate has charge q' and the potential difference between the plates is V'. If an additional infinitesimal charge dq' is taken from the negative plate and placed on the positive plate, the energy is increased by $dU =$ _____ or, what is the same, by $dU = (q'/C)\,dq'$, where C is the capacitance. This expression is integrated from 0 to the final charge q. In terms of the final charge the total energy required to charge an originally uncharged capacitor is $U =$ _____. In terms of the final potential difference V it is $U =$ _____.

Instead of associating the energy of a charged capacitor with the mutual interactions of the charges on its plates, it may be associated with the electric field associated with those charges. The text shows that the energy density (or energy per unit volume) in a parallel plate capacitor is given by

$$u =$$

where E is the magnitude of the electric field between the plates.

This expression is quite generally valid for any electric field, whether in a capacitor or not. Most fields are functions of position so a volume integral must be evaluated to calculate the energy required to produce the field: $U = \frac{1}{2}\epsilon_0 \int E^2 \, dV$. Part c of Sample Problem 8 in the text shows how to evaluate this integral for a charged conducting sphere.

If, after charging a capacitor, the plates are connected by a conducting wire, then electrons will flow from the negative to the positive plate until both plates are neutral and the electric field vanishes. The stored energy is converted to kinetic energy of motion and eventually to internal energy in resistive elements of the circuit.

Dielectrics. If the space between the plates of a capacitor is filled with insulating material the capacitance is greater than if the space is a vacuum. In fact, if the space is completely filled the capacitance is given by $C = $ _____, where C_0 is the capacitance of the unfilled capacitor and κ is the _____ of the insulator. This last quantity is a property of the insulator, is unitless, and is always greater than 1. See Table 27–2 of the text for some values.

Consider two capacitors that are identical except capacitor A has a dielectric between its plates while capacitor B does not. If the capacitors have the same charge on their plates, the potential difference across capacitor _____ is greater than the potential difference across the other capacitor. If the capacitors have the same potential difference, the charge on the positive plate of capacitor _____ is greater than the charge on the positive plate of the other capacitor.

<u>Polarization</u> of the dielectric by the electric field brings about an increase in capacitance. Suppose the dielectric is composed of polar molecules and describe what happens when it is polarized: _____

_____

Now suppose the dielectric is composed of non-polar molecules and describe what happens:

_____

_____

The electric dipoles in a polarized dielectric produce an electric field that is directed opposite to the field produced by the charge on the conducting plates. Thus the total field is weaker than the field produced by charge on the plates alone. It is, in fact, weaker by the factor κ. That is, if $\mathbf{E}_0$ is the field produced by the charge on the plates then the total field is given by $\mathbf{E} = $ _____. If the electric field produced by charge on the plates is uniform, as it essentially is between the plates of a parallel plate capacitor, then the dipole field and the total field are also uniform.

Since the electric field for a given charge on the plates is weaker if the capacitor is filled with a dielectric than if it is not, the potential difference is _____ when the dielectric is present. Since $q = CV$, this means the capacitance is larger when the dielectric is present.

For a parallel plate capacitor the effect of the dielectric is exactly the same as a uniform distribution of positive charge q' on the surface nearest the negative plate and a uniform distribution of negative charge $-q'$ on the surface nearest the positive plate. If the dielectric constant of the dielectric is κ and the charge on the positive plate is q then $q' = $ _____. If the plates have area A then the electric field due to the charge on the plates is _____, the field due to the dipoles is _____, and the total field is _____. In terms of q, A, d, and κ the potential difference is given by _____.

The energy stored in a capacitor is still given by $U = \frac{1}{2}q^2/C = \frac{1}{2}CV^2$, where q is the magnitude of the charge on either plate and V is the potential difference. As a dielectric is inserted between the plates of a capacitor with the potential difference held constant (by a battery, say), the charge on the positive plate _____ and the stored energy _____. For the potential difference to remain the same the battery must transfer charge as the dielectric is inserted. As a dielectric is inserted with the capacitor in isolation, so the charge cannot change, the potential difference _____ and the stored energy _____.

If the dielectric is inserted by an external agent, the difference in energy is associated with _____ . If no external agent acts the difference in energy is associated with _____ .

Although a capacitor filled with a dielectric holds more charge than an identical one at the same potential difference, but without a dielectric, the dielectric places an upper limit on the charge and potential difference. The <u>dielectric strength</u> of an insulator is _____ _____ .

If a parallel plate capacitor has capacitance C and plate separation d and the insulator filling it has dielectric strength E_{max} then the maximum potential difference it can sustain is $V_{max} =$ _____ and the maximum charge than can be placed on the positive plate is $q_{max} =$ _____ .

II. PROBLEM SOLVING

To solve many problems of this chapter you should know the basic relationship between the magnitude q of the charge on either plate of a capacitor and the potential V difference across the plates: $q = CV$. You should also know how to calculate the capacitance of a parallel-plate capacitor, a cylindrical capacitor, and a spherical capacitor, all in terms of the geometry of the capacitor.

PROBLEM 1 (8E).   The plates of a spherical capacitor have radii 38.0 mm and 40.0 mm. (*a*) Calculate the capacitance.

STRATEGY: The capacitance of a spherical capacitor is given by $C = 4\pi\epsilon_0 ab/(b - a)$, where a is the radius of the inner plate and b is the radius of the outer plate.

SOLUTION:

[ans: 8.45×10^{-11} F]

(*b*) What must be the plate area of a parallel-plate capacitor with the same plate separation and capacitance?

STRATEGY: The capacitance of a parallel-plate capacitor is given by $C = \epsilon_0 A/d$, where A is the plate area and d is the plate separation. Set $d = 40.0 - 38.0 = 2.0$ mm and solve for A.

SOLUTION:

[ans: $0.0191\,\mathrm{m}^2$]

PROBLEM 2 (10E).   Two sheets of aluminum foil have a separation of 1.0 mm and a capacitance of 10 pF, and are charged to 12 V. (*a*) Calculate the plate area.

STRATEGY: Use $C = \epsilon_0 A/d$.

SOLUTION:

$[$ans: $0.0011\,\text{m}^2]$

The separation is now decreased by $0.10\,\text{mm}$ with the charge held constant. (*b*) What is the new capacitance?

STRATEGY: Evaluate $C = \epsilon_0 A/d'$, where d' is the new separation $(1.0 - 0.10 = 0.90\,\text{mm})$.

SOLUTION:

$[$ans: $1.1 \times 10^{-11}\,\text{F}]$

(*c*) By how much does the potential difference change?

STRATEGY: Since the charge does not charge $C_i V_i = C_f V_f$, where the subscript i refers to an initial value and the subscript f refers to a final value. Calculate V_f then the change $V_f - V_i$.

SOLUTION:

$[$ans: it decreases by $1.2\,\text{V}]$

PROBLEM 3 (12P). In Section 27–3 the capacitance of a cylindrical capacitor was calculated. Using the approximation (see Appendix G) that $\ln(1 + x) \approx x$ when $x \ll 1$, show that the capacitance approaches that of a parallel-plate capacitor when the spacing between the two cylinders is small.

STRATEGY: The capacitance of a cylindrical capacitor is given by

$$C = 2\pi\epsilon_0 \frac{L}{\ln(b/a)},$$

where a is the radius of the inner plate, b is the radius of the outer plate, and L is the length. Let $b = a + d$. Then $\ln(b/a) = \ln(1 + d/a) \approx d/a$ for $d/a \ll 1$ (that is, for a plate separation that is much smaller than the plate radius). Thus $C \approx 2\pi\epsilon_0 La/d$. The area of a plate is $A = 2\pi aL$. Make this substitution to obtain the formula for a parallel-plate capacitor.

SOLUTION:

Some problems deal with capacitors in series or parallel. In each case, you should know how to compute the equivalent capacitance, the charge on each capacitor, and the potential difference across each capacitor, given the potential difference across the combination.

PROBLEM 4 (19E). A capacitance $C_1 = 6.00\,\mu F$ is connected in series with a capacitance $C_2 = 4.00\,\mu F$, and a potential difference of 200 V is applied across the pair. (*a*) Calculate the equivalent capacitance.

STRATEGY: Since the two are connected in series the equivalent capacitance is given by $1/C_{eq} = (1/C_1) + (1/C_2)$.

SOLUTION:

$$[\text{ans: } 2.40\,\mu F]$$

(*b*) What is the charge on each capacitor?

STRATEGY: Since they are connected in series the charge is the same on both and it is the same as the charge on the equivalent capacitor for the same potential difference. Thus $q = C_{eq}V$.

SOLUTION:

$$[\text{ans: } 4.80 \times 10^{-4}\,C]$$

(*c*) What is the potential difference across each capacitor?

STRATEGY: Solve $q = C_1V_1$ and $q = C_2V_2$ for V_1 and V_2.

SOLUTION:

$$[\text{ans: } V_1 = 80\,V, V_2 = 120\,V]$$

PROBLEM 5 (20E). Work Exercise 19 for the same two capacitors connected in parallel.

STRATEGY: Since they are connected in parallel the equivalent capacitance is given by $C_{eq} = C_1 + C_1$. The potential difference across each of the capacitors is the same, 200 V. The charges are given by $q_1 = C_1V$ and $q_2 = C_2V$.

SOLUTION:

$$[\text{ans: } C_{eq} = 10.0\,\mu F, V_1 = 200\,V, V_2 = 200\,V, q_1 = 1.20 \times 10^{-3}\,C, q_2 = 8.00 \times 10^{-3}\,C]$$

PROBLEM 6 (22P).   A potential difference of 300 V is applied to a series connection of capacitance $C_1 = 2.0\,\mu F$ and capacitance $C_2 = 8.0\,\mu F$. (*a*) What are the charge and potential difference for each capacitor?

STRATEGY: The charges on the two capacitors are the same, so $q = C_1V_1$ and $q = C_2V_2$, where V_1 and V_2 are the potential differences. Now $V_1 + V_2 = V$, where V is the potential difference across the combination (300 V). Solve for V_1, V_2, and q. You should get $q = C_1C_2V/(C_1 + C_2)$, $V_1 = C_2V/(C_1 + C_2)$, and $V_2 = C_1V/(C_1 + C_2)$. This problem can also be worked using the idea of equivalent capacitance, as in Exercise 19.

SOLUTION:

$$\left[\text{ans: } q = 4.8 \times 10^{-4}\,\text{C}, V_1 = 240\,\text{V}, V_2 = 60\,\text{V}\right]$$

(*b*) The charged capacitors are disconnected from each other and from the battery. They are then reconnected, positive plate to positive plate and negative plate to negative plate, with no external voltage being applied. What are the charge and the potential difference for each now?

STRATEGY: The potential differences for the two capacitors are now the same. If q_1' is the charge on C_1 and q_2' is the charge on C_2, then $q_1'/C_1 = q_2'/C_2$. Since charge is conserved $q_1' + q_2' = 2q$. Solve for q_1' and q_2', then use $q_1' = C_1V$ to find the potential difference across either one of the capacitors.

SOLUTION:

$$\left[\text{ans: } q_1' = 1.9 \times 10^{-4}\,\text{C}, q_2' = 7.7 \times 10^{-4}\,\text{C}, V = 96\,\text{V}\right]$$

(*c*) Suppose the charged capacitors in (*a*) were reconnected with plates of *opposite* sign together. What then would be the steady-state charge and potential difference for each?

STRATEGY: Now $q_1' + q_2' = 0$.

SOLUTION:

$$\left[\text{ans: } q_1' = 0, q_2' = 0, V = 0\right]$$

PROBLEM 7 (29).   When switch S is thrown to the left in Fig. 27–30, the plates of the capacitor C_1 acquire a potential difference V_0. Capacitors C_2 and C_3 are initially uncharged. The switch is now thrown to the right. What are the final charges q_1, q_2, q_3 on the corresponding capacitors?

STRATEGY: When the switch is thrown to the left C_1 acquires charge $q = C_1V_0$. Let q_1', q_2', and q_3' be the charges on the capacitors after the switch is thrown to the right. Capacitors C_2 and C_3 must have the same charge, so $q_2' = q_3'$. Since charge is conserved $q_1' + q_2' = q$. Furthermore, the potential difference across C_2 and C_3 together must be the same as the potential difference across C_1, so $(q_2'/C_2) + (q_3'/C_3) = q_1'/C_1$. Solve for q_1', q_2', and q_3'.

SOLUTION:

$$\left[\, \text{ans: } q_1' = C_1^2(C_2 + C_3)V_0/(C_1C_2 + C_1C_3 + C_2C_3),\ q_2' = q_3' = C_1C_2C_3V_0/(C_1C_2 + C_1C_3 + C_2C_3) \,\right]$$

You should know how to compute the energy stored in a capacitor, given either its charge or potential difference. You should also know how to compute the energy density at points between the plates or, in general, at any point within an electric field.

PROBLEM 8 (35E).   A parallel-plate air-filled capacitor has a capacitance of 130 pF. (*a*) What is the stored energy if the applied potential difference is 56.0 V?

STRATEGY: Use $U = \frac{1}{2}CV^2$.

SOLUTION:

$$\left[\, \text{ans: } 2.04 \times 10^{-7}\,\text{J} \,\right]$$

(*b*) Can you calculate the energy density for points between the plates?

STRATEGY: The information given cannot be used to compute the electric field between the plates, the volume between the plates, the area of a plate, or the plate separation. If any of these quantities were given the energy density could be calculated.

SOLUTION:

$$\left[\, \text{ans: no} \,\right]$$

PROBLEM 9 (40P).   An isolated metal sphere whose diameter is 10 cm has a potential of 8000 V. Calculate the energy density in the electric field near the surface of the sphere.

STRATEGY: If the sphere has charge Q the electric potential at its surface is $V = Q/4\pi\epsilon_0 R$ and the electric field just outside its surface is $E = Q/4\pi\epsilon_0 R^2 = V/R$. The energy density just outside the surface is $u = \frac{1}{2}\epsilon_0 E^2 = \frac{1}{2}\epsilon_0 V^2/R^2$.

SOLUTION:

$$[\text{ans: } 0.11 \text{ J/m}^3]$$

PROBLEM 10 (42P).   For the capacitors of Problem 22, compute the energy stored for the three different connections of parts (a), (b), and (c). Compare these stored energies and explain any differences.

STRATEGY: For each connection you know the charge and potential difference for each capacitor. You may use either $U = \frac{1}{2}CV^2$ or $U = \frac{1}{2}q^2/C$. Calculate the energy in each capacitor and sum them to get the total energy stored.

SOLUTION:

$$[\text{ans: } (a)\ 7.2 \times 10^{-2}\text{ J}; (b)\ 4.6 \times 10^{-2}\text{ J}; (c)\ 0]$$

Energy was lost in going from the connection in (a) to the connection in (b) and in going from the connection in (a) to the connection in (c). The charge has less potential energy and so picks up kinetic energy. The kinetic energy is lost somehow in bringing the charges to rest in the final configuration, perhaps in the resistance of the wires connecting the capacitors, perhaps at the capacitor plates themselves, or perhaps in radiation.

You should know how to compute the capacitance and energy stored in a capacitor with dielectric material between its plates. You should also know how to compute the electric field between the plates and the charge induced on the surfaces of the dielectric. In addition, some problems deal with the dielectric strength of the material between the plates.

PROBLEM 11 (56E).   A coaxial cable used in a transmission line has an inner radius of 0.10 mm and an outer radius of 0.60 mm. Calculate the capacitance per meter for the cable. Assume that the space between the conductors is filled with polystyrene.

STRATEGY: The cable is cylindrical capacitor and its capacitance, if a vacuum exists between the conductors, is given by Eq. 27–14 of the text. Since the space is filled with polystyrene the capacitance is given by $C = 2\pi\kappa\epsilon_0 L/\ln(b/a)$, where κ is the dielectric constant for polystyrene (2.6 according to Table 27–2). The capacitance per unit length is given by C/L.

SOLUTION:

[ans: 8.1×10^{-11} F/m]

PROBLEM 12 (58P).   You are asked to construct a capacitor having a capacitance near 1 nF and a breakdown potential in excess of 10,000 V. You think of using the sides of a tall drinking glass (Pyrex), lining the inside and outside curved surfaces with aluminum foil. The glass is 15 cm tall with an inner radius of 3.6 cm and an outer radius of 3.8 cm. What are the (a) capacitance and (b) breakdown potential?

STRATEGY: The capacitance is given by Eq. 27–14 of the text, multiplied by the dielectric constant κ of Pyrex (4.7, according to Table 27–2 of the text): $C = 2\pi\kappa\epsilon_0 L / \ln(b/a)$, where a is the inner radius, b is the outer radius, and L is the height of the glass. If E_{max} is the dielectric strength of Pyrex and d is the thickness of the glass (the difference between the inner and out radii), then the breakdown potential is $V_b = E_{max}d$.

SOLUTION:

[ans: (a) 7.3×10^{-10} F; (b) 28 kV]

PROBLEM 13 (59P).   You have been assigned to design a transportable capacitor that can store 250 kJ of energy. You decide on a parallel-plate type with dielectric. (a) What is the minimum capacitor volume achievable using a dielectric selected from those whose dielectric strengths are listed in Table 27–2?

STRATEGY: The energy stored is given by $U = \frac{1}{2}\kappa\epsilon_0 E^2 V$, where E is the electric field, V is the volume, and κ is the dielectric constant of material between the plates. The field cannot exceed the dielectric strength of the material used between the plates. The minimum volume is obtained when the material with the largest value of κE_{max}^2 is used. Of those listed, this is strontium titanate, with $\kappa = 310$ and $E_{max} = 8$ kV/mm.

SOLUTION:

[ans: 2.85 m^3]

(b) Modern high-performance capacitors that can store 250 kJ have volumes of 0.0870 m^3. Assuming that the dielectric used has the same dielectric strength as in (a), what must be its dielectric constant?

STRATEGY: Solve $U = \frac{1}{2}\kappa\epsilon_0 E_{max}^2 V$ for κ.

SOLUTION:

[ans: 1.01×10^4]

PROBLEM 14 (61P).   A slab of copper of thickness b is thrust into a parallel-plate capacitor of plate area A, as shown in Fig. 27–33; it is exactly halfway between the plates. (*a*) What is the capacitance after the slab is introduced?

STRATEGY: The capacitor can be considered to be two parallel-plate capacitors in series, each with plate separation $(d-b)/2$. If A is the area of a plate the capacitance of either one is $2\epsilon_0 A/(d-b)$. Find an expression for the capacitance of the combination.

SOLUTION:

$$[\text{ans: } \epsilon_0 A/(d-b)]$$

(*b*) If a charge q is maintained on the plates, what is the ratio of the stored energy before to that after the slab is inserted?

STRATEGY: The energy stored before the slab is inserted is $U_b = \frac{1}{2}q^2/C_b$, where $C_b = \epsilon_0 A/d$ is the capacitance then. The energy stored after the slab is inserted is $U_a = \frac{1}{2}q^2/C_a$, where $C_a = \epsilon_0 A/(d-b)$ is the capacitance then. Find an expression for the ratio.

SOLUTION:

$$[\text{ans: } U_b/U_a = d/(d-b)]$$

(*c*) How much work is done on the slab as it is inserted? Is the slab sucked in or must it be pushed in?

STRATEGY: The work done on the slab is $W = U_a - U_b$.

SOLUTION:

$$[\text{ans: } -\tfrac{1}{2}q^2 b/\epsilon_0 A]$$

The minus sign indicates that the agent must pull back on the slab to keep it from accelerating inward. The slab is sucked in.

PROBLEM 15 (62P).   Rework Problem 61 assuming that the potential difference rather than the charge is held constant.

STRATEGY: Let V be the potential difference across the plates. The energy stored before the slab is inserted is $U_b = \frac{1}{2}C_b V^2$ and the energy stored afterwards is $U_a = \frac{1}{2}C_a V^2$. Since the potential difference is held constant the charge on the plates changes and the source of the potential difference does work as it moves charge from one plate to the other. If W_s is the work done by the source and W is the work done by the agent inserting the slab, then $U_a - U_b = W_s + W$ and $W = U_a - U_b - W_s$. W_s is $(q_a - q_b)V$, where $q_b = C_b V$ is the charge on the capacitor before the slab is inserted and $q_a = C_a V$ is the charge afterward.

412   *Chapter 27: Capacitance*

SOLUTION:

$$\left[\text{ans: }(a)\ \epsilon_0 A/(d-b);\ (b)\ U_b/U_a = (d-b)/d;\ (c)\ W = -\tfrac{1}{2}V^2\epsilon_0 Ab/d(d-b)\right]$$

The minus sign indicates that the slab is sucked in.

PROBLEM 16 (68P). Two parallel plates of area $100\,\text{cm}^2$ are given equal but opposite charges of $8.9 \times 10^{-7}\,\text{C}$. The electric field within the dielectric material filling the space between the plates is $1.4 \times 10^6\,\text{V/m}$. (a) Calculate the dielectric constant of the material.

STRATEGY: The electric field is reduced from its value in the absence of the dielectric material by the factor κ. Since the field when the dielectric is absent is $E_0 = \sigma/\epsilon_0$, where σ is the charge density on the positive plate, the field in the presence of the dielectric is $E = E_0/\kappa = \sigma/\kappa\epsilon_0$. Solve for κ.

SOLUTION:

$$\left[\text{ans: }7.1\right]$$

(b) Determine the magnitude of the charge induced on each dielectric surface.

STRATEGY: Let E' be the electric field produced by the induced charge. Since the total field is the sum of the field produced by the free and induced charges, $E = E_0 + E'$ or $E' = E - E_0 = (\sigma/\kappa\epsilon_0) - (\sigma/\epsilon_0) = (\sigma/\epsilon_0)(1-\kappa)/\kappa$. In terms of the induced charge density σ' this field is $E' = \sigma'/\epsilon_0$. Solve for σ'. You should get $\sigma' = (1-\kappa)\sigma/\kappa$. Use this to show that the magnitude of the induced charge on one surface is $q' = (\kappa-1)q/\kappa$.

SOLUTION:

$$\left[\text{ans: }7.7 \times 10^{-7}\,\text{C}\right]$$

III. NOTES

Chapter 28
CURRENT AND RESISTANCE

I. BASIC CONCEPTS

Here you begin the study of electric current, the flow of charge. Pay close attention to the definitions of current and current density and understand how they depend on the concentration and speed of the charged particles. For most materials a current is established and maintained only when an electric field is present. You will learn about the properties of materials that determine the magnitude and direction of the current for any given field.

Electric current and current density.   As precisely as you can, tell in words what it means for a conducting wire to contain an electric current: _____

_____

The current i through any area is given by

$$i = \frac{dq}{dt},$$

where dq is _____ that passes through _____ in the time interval _____.
Although current is a scalar, not a vector, it is assigned a direction. Positive charge moving to the right and negative charge moving to the left, for example, are both currents in the same direction, namely to the _____.   Positive charge moving to the right and negative charge moving in the same direction are currents in _____ directions: the current associated with the positive charge is to the _____ while the current associated with the negative charge is to the _____.

Explain why the flow of neutral atoms or molecules is not an electric current: _____

_____

Explain why electric current is not a vector: _____

_____

For the steady flow of charge through a conductor the current is the same for all cross sections along the conductor. Explain why: _____

_____

The SI unit for current is coulomb/second. This unit is called _____ and is abbreviated

_____.

Current is associated with an area, like the cross-sectional area of a conductor. On the other hand, <u>current density</u> **J** is a related quantity that can be associated with each point in a conductor. At any point it is the current per unit area through an area at the point, oriented perpendicularly to the charge flow, in the limit as the area shrinks to the point. Current density is a vector. If positive charge is flowing **J** is in the direction of the velocity; if negative charge is flowing **J** is in the direction opposite the velocity.

Consider any surface within a current-carrying conductor and let d**A** be an infinitesimal vector area, with direction perpendicular to the surface. If **J** is the current density, then the current i through the surface is given by the integral over the surface:

$$i =$$

The scalar product indicates that only the component of **J** normal to the surface contributes to the current. This is consistent with the statement that charge must flow through the surface for there to be a current through the surface. If the surface is a plane with area A and is perpendicular to the particle velocity and if the current density is uniform over the surface, then $i =$ _____ .

For a collection of particles with uniform concentration n, each with positive charge e and each moving with velocity $\mathbf{v}_d$, the current density is given in terms of e, n, and $\mathbf{v}_d$ by the vector relationship

$$\mathbf{J} =$$

This expression can be derived by considering the number of particles per unit time that pass through an area perpendicular to their velocity. Note that **J** and $\mathbf{v}_d$ are in the same direction for positively charged particles and in opposite directions for negatively charged particles. As Sample Problem 28–3 shows, the expression above can be used to compute the drift speed v_d for a given current in a given conductor.

All electrons in a current-carrying conductor do not actually move with the same velocity, but each of their velocities can be considered to be the sum of two velocities, one of which changes direction often as the electron collides with atoms. When all electrons are taken into account this component averages to zero: it does not contribute to the current. The second component, the drift velocity, is much smaller in magnitude than the first and is the same for all electrons if the current density is uniform. This is the velocity that enters the expression for the current density, not the total velocity. Be careful to distinguish between drift and total velocities.

In an ordinary conductor an electric field is required to produce a net drift and hence an electric current. Explain what a field does: _____

_____

For electrons how is the direction of the drift velocity related to the direction of the electric field? _____ This means the current is in the same direction as the field and it is directed from a region of high electric potential toward a region of lower electric potential.

That an electric field can be maintained in a conductor does not contradict the statement proved earlier that the electrostatic field in a conductor is zero. Describe how the situation considered in this chapter differs from that considered in Chapter 25: _____

_____

Resistance and resistivity. The current in any material (except a superconductor) is zero unless an electric field exists in the material. Resistance is a measure of the current generated in a given conductor by a given potential difference. To determine resistance a potential difference V is applied between two points in the material and the current i is measured. The resistance for that material and those points of application is defined by

$$R =$$

Remember that the resistance depends on where the current leads are attached, as well as on the material and its geometry. Resistance has an SI unit of volt/ampere, a unit that is called _____ and abbreviated _____. In drawing an electrical circuit, an element whose function is to provide resistance, called a resistor, is indicated by the symbol _____.

Resistance is intimately related to a property of the material called its resistivity. If at some point in the material the electric field is $\mathbf{E}$ and the current density is $\mathbf{J}$ then the resistivity ρ at that point is given by

$$\mathbf{E} =$$

We consider materials that are homogeneous and isotropic; the resistivity is the same at every point in the material and is the same for every orientation of the electric field.

The SI unit for resistivity is _____. Values for some materials are given in Table 28–1 of the text. Note that typical semiconductors have resistivities that are greater than those of metals by factors of 10^5 to 10^{11} and insulators have resistivities that are greater than those of metals by factors of 10^{18} to 10^{24}. The conductivity σ of any substance is related to the resistivity by $\sigma = $ _____.

Given the resistivity of the material and the points of application of a potential difference the resistance can, in principle, be calculated. The simplest case is that of a homogeneous wire with uniform cross section. Let L be the length of the wire, A be its cross-sectional area, and ρ be its resistivity. If one end of the wire is held at potential 0 and the other is held at potential V the electric field in the wire is given by $E = $ _____. If the current is i then, since the current density is uniform, it is given by $J = $ _____. Substitute $E = V/L$ and $J = i/A$ into $E = \rho J$ and solve for V/i ($= R$). The result is $R = $ _____. Note that the resistance depends on a property of the material (ρ) and on the geometry of the sample (L and A). Resistivity, on the other hand, is a property of the material alone.

Temperature dependence of the resistivity. The resistivity of a metal increases with increasing temperature, chiefly because electrons suffer more collisions per unit time at high temperatures than at low temperatures. The temperature dependence is characterized by a quantity α, called the temperature coefficient of resistivity, which is a measure of the deviation of the resistivity from its value at a reference temperature. Let ρ_0 be the resistivity at the reference temperature T_0 and let ρ be the resistivity at a nearby temperature T. Then, in terms of α,

$$\rho - \rho_0 =$$

Temperatures are given in K or °C.

Table 28–1 lists values for some materials. Notice that α is negative for semiconductors. For them the resistivity decreases as the temperature increases because the concentration of nearly free electrons increases rapidly with temperature. For metals, on the other hand, n is essentially independent of temperature.

Ohm's law.   Ohm's law describes an important characteristic of the resistance of certain samples, called ohmic samples (or devices). It is: _____

_____

_____

Carefully note that $V = iR$ defines the resistance R and so holds for every sample. For an ohmic sample V is a linear function of i or, what is the same, R does not depend on V or i. In addition, for ohmic samples a reversal of the potential difference simply reverses the direction of the current without changing its magnitude. For some non-ohmic samples, such as a pn junction diode, a reversal of the potential difference not only changes the direction of the current but also its magnitude.

In the space below sketch graphs of the current as a function of potential difference for an ohmic and for a non-ohmic sample. Include both positive and negative values of V.

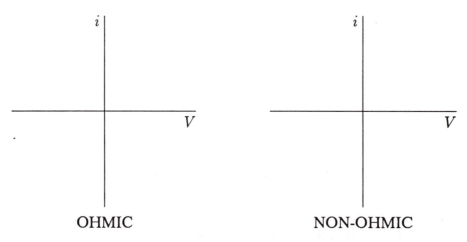

OHMIC   NON-OHMIC

For an ohmic substance the resistivity is independent of the electric field and the current density is therefore proportional to the electric field. This implies that the drift velocity is also proportional to the field. Since an electric field accelerates charges you should be surprised.

To show why Ohm's law is valid for some materials the text considers the collection of so-called free electrons in a conducting sample. These electrons are accelerated by the electric field applied to the sample and suffer collisions with atoms of the material. We suppose that the effect of a collision is to stop the drift of an electron so that, as far as drift is concerned, an electron starts from rest at a collision and is accelerated by the field until the next collision. The collision stops it and the process is repeated. The time from the last collision, averaged over all electrons, is designated by τ. Because the collisions occur at random times τ is independent of the time; its value is the same no matter when the averaging is done. In addition, τ gives the average time between collisions and is, in fact, called the <u>mean</u> <u>time</u> <u>between</u> <u>collisions</u> or the <u>mean</u> <u>free</u> <u>time</u>.

Since the magnitude of the acceleration is $a =$ _____, where E is the magnitude of the electric field and m is the mass of an electron, the average electron drift speed is given by $v_d =$ _____.   Although each electron accelerates between collisions v_d does not vary with time if the field is constant. Also note that v_d is proportional to E if τ does not depend on E. Explain why you expect τ to be essentially independent of E: _____

_____

In terms of E, m, e, and τ, the current density is given by $J = env_d =$ _____. This expression immediately gives $\rho =$ _____ for the resistivity. Because τ is independent of E the resistivity is also independent of E and the material obeys Ohm's law.

Give a qualitative argument to convince yourself that a long mean free time leads to a small resistivity. In particular, explain why a long mean free time leads to a larger current than a short mean free time if the electric field is the same: _____

_____

_____

Because the mean time between collisions is determined to a large extent by thermal vibrations of the atoms, it is temperature dependent. As the temperature increases we expect an electron to suffer more collisions per unit time so τ _____ as the temperature increases.

Energy transfers. If current i passes through a potential difference V, from high to low potential, the moving charges lose potential energy at a rate given by

$$P = dU/dt =$$

For a resistor the energy is transferred to atoms of the material in collisions. The result is an increase in the vibrational energy of the atoms and is usually accompanied by an increase in the temperature of the resistor. The phenomenon finds practical application in toasters and electrical heaters.

Since $V = iR$ for a resistor, the expression for the rate of energy loss can be written in terms of the current i and resistance R as $P =$ _____ or in terms of the potential difference V and resistance R as $P =$ _____.

Suppose the resistance of resistor A is greater than that of resistor B. If the same potential difference is applied to them then the internal energy of resistor _____ will increase at the greater rate. If they have the same current then the internal energy of resistor _____ will increase at the greater rate.

II. PROBLEM SOLVING

You should know the definitions of current and current density, their relationship and the relationship between current density and drift velocity. These examples will help you.

PROBLEM 1 (4P).   The belt of a Van de Graaf accelerator is 50 cm wide and travels at 30 m/s. The belt carries charge into the sphere at a rate corresponding to 100 μA. Compute the surface charge density on the belt. (See Section 26–11.)

STRATEGY: Let σ be the surface charge density and w be the width of the belt. A length L of the belt contains charge $q = \sigma w L$ and passes any point in time $t = L/v$, where v is its speed. Thus the rate at which charge is passing the point is $q/t = \sigma w v$. Solve for σ.

SOLUTION:

[ans: 6.7×10^{-6} C/m^2]

PROBLEM 2 (11P).   Near the Earth, the density of protons in the solar wind is $8.70\,\text{cm}^{-3}$ and their speed is $470\,\text{km/s}$. (*a*) Find the current density of these protons.

STRATEGY: Use $J = nev_d$.

SOLUTION:

$$\left[\text{ans: } 6.54 \times 10^{-7}\,\text{A/m}^2\right]$$

(*b*) If the Earth's magnetic field did not deflect them, the protons would strike the Earth. What total current would the Earth then receive?

STRATEGY: Imagine a circle with radius equal to the Earth's radius, with its plane tangent to the Earth and its center on the line from the Earth to the sun. All the protons that strike the Earth pass through this circle, so the current is given by $I = \pi R^2 J$, where $R = 6.37 \times 10^6\,\text{m}$.

SOLUTION:

$$\left[\text{ans: } 8.34 \times 10^7\,\text{A}\right]$$

PROBLEM 3 (13P).   How long does it take electrons to get from a car battery to the starting motor? Assume the current is $300\,\text{A}$ and the electrons travel through a copper wire with cross-sectional area $0.21\,\text{cm}^2$ and length $0.85\,\text{m}$. (See Sample Problem 28–3.)

STRATEGY: You need to know the drift velocity of the electrons. In terms of the current i and cross-sectional area A of the wire the current density is given by $J = i/A$ and in terms of the free-electron concentration n and the drift velocity v_d is given by $J = nev_d$. Thus $v_d = i/neA$. If L is the length of the wire the travel time is $t = L/v_d$. The free-electron concentration was found in Sample Problem 28–3 to be $n = 8.47 \times 10^{28}\,\text{electrons/m}^3$.

SOLUTION:

$$\left[\text{ans: } 810\,\text{s } (13\,\text{min})\right]$$

PROBLEM 4 (15P).   (*a*) The current density across a cylindrical conductor of radius R varies according to the equation

$$J = J_0(1 - r/R),$$

where r is the distance from the central axis. Thus the current density is a maximum J_0 at the axis $r = 0$ and decreases linearly to zero at the surface $r = R$. Calculate the current in terms of J_0 and the conductor's cross-sectional area $A = \pi R^2$.

STRATEGY: The current is normal to a cross-sectional area so $i = \int J \, dA = \int J_0(1 - r/R) \, dA$. Take dA to be the area of a ring of radius r and width dr. Its area is $dA = 2\pi r \, dr$, so $i = 2\pi J_0 \int (1 - r/R) r \, dr$. The limits of integration are 0 and R. Carry out the integration. When you are finished substitute A for πR^2.

SOLUTION:

[ans: $J_0 A/3$]

(b) Suppose that, instead, the current density is a maximum J_0 at the cylinder's surface and decreases linearly to zero at the axis, so that

$$J = J_0 r/R.$$

Calculate the current. Why is the result different from that in (a)?

SOLUTION:

[ans: $2J_0 A/3$]

Consider the current through rings of equal width dr. Those with small radii have small areas while those with large radii have larger areas. In (a) the current density is large near the axis, where the ring areas are small, and small near the surface, where the ring areas are large. In (b) the current density is small near the axis, where the ring areas are small, and large near the surface, where the ring areas are large. We therefore expect the current to be larger in (b) than in (a).

You should know the definition of resistance and be able to compute the resistance of a wire, given its resistivity, length, and cross-sectional area.

PROBLEM 5 (20E). A wire 4.00 m long and 6.00 mm in diameter has a resistance of 15.0 mΩ. If a potential of 23.0 V is applied between the ends, (a) what is the current in the wire?

STRATEGY: Solve $V = iR$ for i.

SOLUTION:

[ans: 1.53 kA]

(b) What is the current density?

STRATEGY: Evaluate $J = i/A$, where A is the cross-sectional area of the wire ($A = \pi R^2$, where R is the radius).

SOLUTION:

[ans: 5.42×10^7 A/m^2]

(*c*) Calculate the resistivity of the wire material. Identify the material. (Use Table 28–1.)

STRATEGY: The resistance is given by $R = \rho L/A$, where ρ is the resistivity of the material, L is the length of the wire, and A is the cross-sectional area. Solve for ρ.

SOLUTION:

$$\left[\text{ans: } 1.06 \times 10^{-7}\,\Omega \cdot \text{m}\right]$$

According to Table 28–1 this is the resistivity of platinum.

PROBLEM 6 (26E).   A cylindrical copper rod of length L and cross-sectional area A is reformed to twice its original length with no change in volume. (*a*) Find the new cross-sectional area.

STRATEGY: The volume does not change. Since the volume of a cylinder is given by AL, where A is the cross-sectional area and L is the length, $AL = A_f L_f$, where the subscript f refers to final values. Solve for A_f and substitute $L_f = 2L$.

SOLUTION:

$$\left[\text{ans: } A/2\right]$$

(*b*) if the resistance between its ends was R before the change, what is it after the change?

STRATEGY: The resistance after the change is $R_f = \rho L_f/A_f$. Substitute $L_f = 2L$ and $A_f = A/2$, then replace $\rho L/A$ with R.

SOLUTION:

$$\left[\text{ans: } 4R\right]$$

PROBLEM 7 (34P).   An electrical cable consists of 125 strands of fine wire, each having 2.65-$\mu\Omega$ resistance. The same potential difference is applied between the ends of all the strands and results in a total current of 0.750 A. (*a*) What is the current in each strand?

STRATEGY: Since the resistances of the strands are identical and the strands have the same end-to-end potential difference, each carries the same current. The total current is $i_{\text{total}} = Ni$, where N is the number of strands and i is the current in each strand.

SOLUTION:

$$\left[\text{ans: } 6.00\,\text{mA}\right]$$

(*b*) What is the applied potential difference?

STRATEGY: Use $V = iR$, where R is the resistance of a single strand and i is the current in each strand.

SOLUTION:

$$\left[\text{ans: } 1.59 \times 10^{-8}\,\text{V}\right]$$

(*c*) What is the resistance of the cable?

STRATEGY: Solve $V = i_{\text{total}} R_{\text{total}}$ for R_{total}.

SOLUTION:

$$\left[\text{ans: } 2.12 \times 10^{-8}\,\Omega\right]$$

PROBLEM 8 (37P).   A block in the shape of a rectangular solid has a cross-sectional area of $3.50\,\text{cm}^2$, a length of 15.8 cm, and a resistance of 935 Ω. The material of which the block is made has 5.33×10^{22} conduction electrons/m^3. A potential difference of 35.8 V is maintained between its ends. (*a*) What is the current in the block?

STRATEGY: Solve $V = iR$ for i.

SOLUTION:

$$\left[\text{ans: } 38.3\,\text{mA}\right]$$

(*b*) If the current density is uniform, what is its value?

STRATEGY: Evaluate $J = i/A$, where A is the cross-sectional area.

SOLUTION:

$$\left[\text{ans: } 109\,\text{A/m}^2\right]$$

(*c*) What is the drift velocity of the conduction electrons?

STRATEGY: Use $J = nev_d$, where n is the concentration of conduction electrons and v_d is the drift velocity. Solve for v_d.

SOLUTION:

$$\left[\text{ans: } 1.28\,\text{cm/s}\right]$$

(*d*) What is the electric field in the block?

STRATEGY: Since the current density is uniform the electric field is also uniform and its magnitude is given by $E = V/L$, where L is the length of the block.

SOLUTION:

[ans: 227 V/m]

You should understand that resistivity is determined to a large extent by the mean time between collisions and that this quantity is temperature dependent because the average velocity of the electrons is temperature dependent. The following problem illustrates the relationship.

PROBLEM 9 (43P). Show that, according to the free-electron model of electrical conduction in metals and classical physics, the resistivity of metals should be proportional to $\sqrt{T}$, where T is the temperature in kelvins. (See Eq. 21–23.)

STRATEGY: The resistivity is given by $\rho = m/e^2\pi\tau$, where m is the mass of an electron and τ is the mean time between collisions. The mean time between collisions is given by $\tau = \lambda/\bar{v}$, where λ is the mean free path and $\bar{v}$ is the average speed. According to classical physics the average speed is given by $\bar{v} = \sqrt{8kT/\pi m}$. Substitute for $\bar{v}$ in the expression for τ, then substitute for τ in the expression for ρ. Since the mean free path is independent of T you should get a result that is proportional to $\sqrt{T}$.

SOLUTION:

Carefully note that $\bar{v}$ is *not* the same as v_d. The drift velocity is the average velocity and is small because the velocities of electrons moving in opposite directions tend to cancel in the averaging process. On the other hand, $\bar{v}$ is the average of the *speeds*. No cancellation occurs in averaging.

Charge carriers lose kinetic energy in collisions with the atoms of a resistor. You should know how to compute the rate of energy dissipation, given the resistance and either the current in the resistor or the potential difference along the length of the resistor. A few problems deal with practical uses of the dissipated energy.

PROBLEM 10 (49E). An unknown resistor is connected between the terminals of a 3.00-V battery. The power dissipated in the resistor is 0.540 W. The same resistor is then connected between the terminals of a 1.50-V battery. What power is dissipated in this case?

STRATEGY: Let V_1 be the potential difference in the first case and V_2 be the potential difference in the second. The power dissipated in the first case is $P_1 = V_1^2/R$ and the poser dissipated in the second is $P_2 = V_2^2/R$, where R is the resistance. Use one of these equations to eliminate R from the other and show that $P_2 = (V_2^2/V_1^2)P_1$.

SOLUTION:

[ans: 0.135 W]

PROBLEM 11 (53P).   A potential difference V is applied to a wire of cross section A, length L, and resistivity ρ. You want to change the applied potential difference and draw out the wire so the power dissipated is increased by a factor of exactly 30 and the current is increased by a factor of exactly 4. What should be the new values of L and A?

STRATEGY: The power dissipated is given by $P = i^2 R = i^2 \rho L/A$, where i is the current and $\rho L/A$ was substituted for the resistance R. When the current is increased and the wire is drawn out the power dissipated is $P_{new} = 16 i^2 \rho L_{new}/A_{new}$. Since $P_{new} = 30P$, $L_{new}/A_{new} = (30/16)L/A$. Since the volume of the wire does not change $L_{new}A_{new} = LA$. Solve these two equations simultaneously for L_{new} and A_{new}.

SOLUTION:

[ans: $L_{new} = \sqrt{30}L/4$, $A_{new} = 4A/\sqrt{30}$]

The problem does not ask for the new potential difference, but if it did you would use $V_{new} = 4i\rho L_{new}/A_{new} = 4i\rho(\sqrt{30}/4)L(\sqrt{30}/4A) = (30/4)V$.

PROBLEM 12 (54P).   A cylindrical resistor of radius $5.0\,mm$ and length $2.0\,cm$ is made of material that has a resistivity of $3.5 \times 10^{-5}\,\Omega \cdot m$. What are (a) the current density and (b) the potential difference when the power dissipation in the resistor is $1.0\,W$?

STRATEGY: First find the resistance using $R = \rho L/A$, where L is the length, A is the cross-sectional area (πr^2, where r is the radius), and ρ is the resistivity. Then use $P = i^2 R$ to find the current and $J = i/A$ to find the current density. Finally, use $P = V^2/R$ to find the potential difference.

SOLUTION:

[ans: (a) $1.3 \times 10^5\,A/m^2$; (b) $94\,mV$]

PROBLEM 13 (61P).   A coil of current-carrying Nichrome wire is immersed in a liquid contained in a calorimeter. When the potential difference across the coil is 12 V and the current through the coil is 5.2 A, the liquid boils at a steady rate, evaporating at the rate of 21 mg/s. Calculate the heat of vaporization of the liquid, in cal/g.

STRATEGY: The coil supplies energy at the rate $P = iV$, where i is the current in the coil and V is the potential difference across the coil. If L_v is the heat of vaporization of the liquid and r is the rate of vaporization (as a mass per unit time) then the liquid absorbs energy at the rate $P = L_v r$. In steady state these are equal, so $iV = L_v r$. Solve for L_v and convert to cal/g.

SOLUTION:

[ans: 710 cal/g]

III. NOTES

Chapter 29
CIRCUITS

I. BASIC CONCEPTS

The concept of electromotive force (emf) is introduced in this chapter. Seats (or sources) of emf are used to maintain potential differences and to drive currents in electrical circuits. You will use this concept, along with those of electric potential, current, resistance, and capacitance, learned earlier, to solve both simple and complicated circuit problems. You will also learn about energy balance in electrical circuits.

Electromotive force. An emf device (or seat of emf) performs two functions: it maintains a potential difference and, to do that, it moves charge from one terminal to the other inside the device. An ideal emf device contains only a seat of emf; a real emf device, such as a battery, also contains internal resistance.

In electrical circuits an ideal emf device is indicated by the symbol _____. The arrow with the small circle at its tail points from the _____ terminal toward the _____ terminal. An emf tends to drive current in the direction of the arrow but you should remember that, for any circuit, the actual direction of the current through an emf device depends on other elements in the circuit.

An emf, denoted by $\mathcal{E}$, is defined in terms of the work it does as it transports charge. If it does work dW on charge dq then

$$\mathcal{E} =$$

Electromotive force has an SI unit of _____ .

An emf device does positive work if positive charge moves from the _____ to the _____ terminal and negative work if positive charge moves in the other direction. It does positive work if negative charge moves from the _____ to the _____ terminal and negative work if negative charge moves in the other direction. This can be summarized by saying the device does positive work on the charge if the current through it is from the negative to the positive terminal and negative work if the current is in the other direction.

Positive work done by an emf device results in a decrease in the store of energy of the device, chemical energy in the case of a battery. Negative work results in an increase in the energy of the device. If the device is a battery, then in the first case it is said to be discharging and in the second it is said to be charging.

The potential difference across an ideal emf device with emf $\mathcal{E}$ is _____ , with the positive terminal being at the higher potential. This statement is true no matter what the direction of the current through the device.

Single loop circuits. To solve single loop circuits you must be able to express the potential differences across emf devices and resistors in terms of the emf's, resistances, and currents. For each of the circuit elements shown below assume the potential is V_a at the left side and write an expression for the potential V_b at the right side in terms of V_a, $\mathcal{E}$, R, and i, as appropriate.

$V_b =$ $V_b =$ $V_b =$ $V_b =$

You will use <u>Kirchhoff's</u> <u>loop</u> <u>rule</u> to solve both single-loop and multiple-loop circuits. Write this rule in words: _____

_____

In general, the rules for solving a single loop circuit are: pick a direction for the current and draw a current arrow on the circuit diagram. For a single-loop circuit the current is the same in every circuit element. Go around the circuit and, as you do, add the changes in electric potential for the various circuit elements, with appropriate signs. For a resistor assume the current is in the direction of the arrow you drew. The diagrams above should help you. Then equate the sum to zero and solve for the unknown quantity.

Consider the single-loop circuit shown on the right and use Kirchhoff's loop rule to develop an equation that relates the emf $\mathcal{E}$, resistance R, and current i. In terms of these quantities the potential difference $V_b - V_a$ is _____ and the potential difference $V_c - V_b$ is _____. Now sum the potential differences around the loop. Since $V_c = V_a$, $(V_b - V_a) + (V_c - V_b) = 0$ or, in terms of $\mathcal{E}$, R, and i, $\mathcal{E} - iR = 0$. This equation can be solved for any one of the quantities appearing in it. If, for example, $\mathcal{E}$ and R are known, then $i = \mathcal{E}/R$ gives the current.

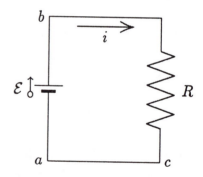

Notice that the value of i is positive, no matter what the values of $\mathcal{E}$ and R. This is because the current arrow was picked to be in the direction of the actual current. In other circuits the direction of the current might not be so obvious and you might not pick the proper direction for the current arrow. If you pick the wrong direction the value you obtain for i will be negative. Suppose, for example, the current arrow for the circuit above was picked in the opposite direction, upward through the resistor. Then the loop equation would be _____ and the expression for the current in terms of $\mathcal{E}$ and R would be $i =$ _____. The value for i is obviously negative now.

A real battery contains an internal resistance in addition to a seat of emf. Although the internal resistance is distributed inside the battery you may think of it as being in series with the seat. You must distinguish between the emf and terminal potential difference of a real battery. The latter includes the potential difference of the internal resistance. To give an example, suppose a battery is characterized by emf $\mathcal{E}$ and internal resistance r. If the current in the battery is zero then the terminal potential difference is $V =$ _____; if the battery is discharging, with current i in the direction of the emf, then $V =$ _____; and if the battery is charging, with current i opposite the emf, then $V =$ _____.

Suppose a single-loop circuit consists of a battery with emf $\mathcal{E}$ and internal resistance r, connected to an external resistance R. Then the current in the circuit is given by $i =$ _____. Study Sample Problems 29–1 and 2 of the text for an example of a circuit with two batteries. Note that the battery with the larger emf is discharging and determines the direction of the current while the battery with the smaller emf is charging.

Energy considerations. An emf with current in the direction of the emf supplies energy to the moving charges by increasing their potential energy. If the emf is $\mathcal{E}$ and the current is i then the rate at which energy is supplied is given by $P =$ _____. An emf with current opposite the direction of the emf removes energy from the moving charges by decreasing their potential energy. The rate at which energy is removed is _____. A resistor removes energy from moving charges in collisions between the charges and atoms of the resistor. If the resistance is R and the current is i then the rate at which energy is removed is given by _____. A discharging battery, being a combination of an emf and a resistor, both supplies and removes energy.

For any circuit in steady state the rate at which energy is supplied by discharging batteries exactly equals the rate at which energy is removed by resistors and charging batteries.

Multiloop circuits. To analyze multiloop circuits another rule, called <u>Kirchhoff's junction rule</u>, is needed in addition to the loop rule. State the junction rule in words: _____
_____

The diagrams below show two junctions and the current arrows associated with them. For each write the equation that follows from the junction rule.

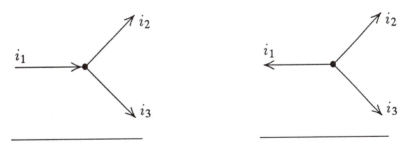

_____ _____

The form of the junction equation does not depend on the directions of the actual currents, only on the directions of the current arrows. Note that all of the currents in the second example above cannot have the same sign.

Successful analysis of a multiloop circuit depends on your ability to identify <u>junctions</u> and <u>branches</u>. A junction is a point where _____ or more wires are joined. A branch is a portion of a circuit with junctions at its ends and none between. You must place a current arrow in each branch and label the arrows in different branches with different current symbols.

Consider the circuit shown to the right and assume the battery has negligible internal resistance.
Identify the 4 junctions by writing the letters associated with them: _____

Identify the 6 branches by writing the letters corresponding to their end points: _____

_____

The currents in the various branches may be different. On the diagram draw a current arrow for each
branch and label them i_1 through i_6, with i_j being
the current in resistor R_j and i_6 being the current in
the battery. The directions of the arrows are immaterial. In general for any circuit with N junctions
there will be $N-1$ independent junction equations.
Pick any 3 junctions for the circuit shown and write
the junction equations here:

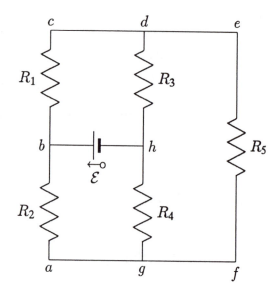

The total number of independent equations equals the number of branches so the number of
independent loop equations you will use equals the number of branches minus the number
of junction equations. For the circuit shown you will need $6 - 3 = 3$ loop equations. Pick
3 loops. You must be a little careful here. Every branch of the circuit must contribute to at
least one of the loop equations. You cannot pick $abhg$, $bcdh$, and $acdg$, for example, since
this set leaves out the branch deg. Substitute a loop containing R_5 for one of the 3 loops just
mentioned. Now write the corresponding loop equations:

The 6 equations can be solved for 6 unknowns. If the emf and resistances are known then the
equations can be solved for the currents, for example. If the numerical value of a current is
positive the actual current is in the direction of the corresponding arrow on the diagram and
electrons move opposite to the direction of the arrow. If the numerical value of a current is
negative the actual current is opposite the direction of the arrow and electrons move in the
direction of the arrow.

Resistors in series and parallel.   Be sure you can distinguish between parallel and series
combinations of two or more resistors. As for any circuit elements, the potential difference
is the same for all the resistors in a _____ combination and the current is the same

for all resistors in a _____ combination. If neither the potential difference nor the current is the same then the resistors do not form either a parallel or series combination.

In the space below draw two resistors connected in parallel and two connected in series. Assume a potential difference V is applied to each combination and indicate on the diagrams where it is applied. Draw a current arrow for each resistor and label them to show which resistors have the same current and which have different currents.

PARALLEL   SERIES

In each case the resistors can be replaced by a single resistor with resistance R_{eq} such that the current is the same as the total current into the original combination when the potential difference is the same as the potential difference across the original combination. If the resistances are R_1 and R_2 then the value of R_{eq} for a parallel combination is given by $R_{eq} =$ _____ and the value of R_{eq} for a series combination is given by $R_{eq} =$ _____. For a parallel combination the equivalent resistance is less than the least resistance in the combination; for a series combination the equivalent resistance is greater than the greatest resistance in the combination.

RC circuits. You will now study a circuit with a time dependent current: a single-loop containing a capacitor, a resistor, and an emf. Kirchhoff's loop rule is still valid and the circuit equation can be derived by summing the potential differences around the loop.

The diagram on the right shows the circuit; it might be used to charge the capacitor. $\mathcal{E}$ is the emf, R the resistance, and C the capacitance. Assume a switch (not shown) is closed at time $t = 0$, when the capacitor is uncharged. Let $i(t)$ be the current at time t, positive in the direction of the arrow; let $q(t)$ be the charge on the upper plate of the capacitor at time t. The potential difference from a to b is $V_b - V_a =$ _____; the potential difference from b to c is $V_c - V_b =$ _____; and the potential difference from c to d is $V_d - V_c =$ _____. Thus $\mathcal{E} - iR - q/C = 0$.

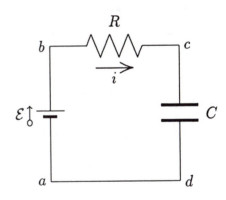

Both i and q are unknown, but since they are related you can eliminate one in favor of the other to obtain an equation in one unknown. Use $i = dq/dt$ to eliminate i and show that

$$R\frac{dq}{dt} + \frac{q}{C} - \mathcal{E} = 0.$$

This is a differential equation to be solved for $q(t)$.

The relationship between i and q depends in sign on the direction of the current arrow and the selection of q to represent the charge on the *upper* plate. For the choices made above positive i is equivalent to the statement that positive charge is flowing onto the upper plate and this, in turn, means dq/dt is positive. For a current arrow in the other direction $i = -dq/dt$.

The solution to the differential equation that obeys the initial condition $q(0) = 0$ is

$$q(t) =$$

This expression is differentiated with respect to time to obtain an expression for the current:

$$i(t) =$$

On the axes below sketch graphs of $q(t)$ and $i(t)$:

According to this result the current just after the switch is closed, at $t = 0$, is given by $i(0) =$ _____, just as if there were no capacitor. The results also predict that after a long time the charge on the capacitor is $q(\infty) =$ _____ and the current is $i(\infty) =$ _____. These conclusions make physical sense. At $t = 0$ the charge on the capacitor is zero so the potential difference across that element is _____. Since the loop equation reduces to $\mathcal{E} = iR$ we expect $i =$ _____. As charge builds up on the capacitor the potential difference across it increases and, as a consequence, the potential difference across the resistor _____ (they must always sum to $\mathcal{E}$). Since the potential difference across the resistor is iR this means the current _____ with time. When the capacitor is fully charged $dq/dt = 0$ and the loop equation becomes $\mathcal{E} = q/C$. Thus the current is _____ and the charge on the capacitor is _____.

The quantity $\tau = RC$ is called the <u>capacitive</u> <u>time</u> <u>constant</u> of the circuit. It has units of _____ and controls the time for the charge on the capacitor to reach any given value. Describe what happens to the curves you drew above if τ is made longer (by increasing C or R): _____

_____

Suppose now that a capacitor C, initially with charge q_0, is discharged by connecting it to a resistor R. In the space to the right draw a circuit diagram with a current arrow. In terms of the charge q and current i the loop equation is _____. Use the relationship between q and i to eliminate i and write the resulting differential equation for q:

The solution to the differential equation is

$$q(t) = $$

Note that $q(0) = q_0$ and that after a long time q becomes zero. The current as a function of time is

$$i(t) = $$

Initially the current is $i(0) = $ _____ ; after a long time the current is _____ . On the axes below sketch the behavior of $q(t)$ and $i(t)$:

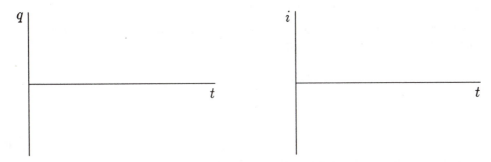

II. PROBLEM SOLVING

You should understand the definition of emf in terms of the work done on a charge as it moves through an emf device. You should also know that real emf devices have both emf's and internal resistances and because they do the terminal potential difference is not the same as the emf if charge is passing through the device. Be aware in each instance whether the current is in the direction of the emf or in the opposite direction. The terminal potential difference depends on the relative direction of the current. Here are some examples.

PROBLEM 1 (1E).   (a) How much work does an ideal battery with a 12.0-V emf do on an electron that passes through the battery from the positive to the negative terminal?

STRATEGY: The magnitude of the work done by the emf is given by $W = q\mathcal{E}$, where q is the magnitude of the charge. Since the electron passes from the positive to the negative terminal the emf does positive work.

SOLUTION:

$$[\text{ans: } 1.9 \times 10^{-18} \text{ J}]$$

(b) If 3.4×10^{18} electrons pass through each second, what is the power of the battery?

STRATEGY: The power is work done per unit time and is given by $P = NW$, where N is the number of electrons that pass through per unit time and W is the work done on each one of them.

SOLUTION:

$$[\text{ans: } 6.5 \text{ W}]$$

PROBLEM 2 (9E).   A car battery with a 12-V emf and an internal resistance of $0.040\,\Omega$ is being charged with a current of 50 A. (*a*) What is the potential difference across its terminals?

STRATEGY: Draw a diagram of the battery showing the emf and the internal resistance, with the current going through it from the positive to the negative terminal, as it is when the battery is being charged. Note that the potential difference across the terminals is given by $V = \mathcal{E} + iR$.

SOLUTION:

$$\left[\text{ans: }14\,\text{V}\right]$$

(*b*) At what rate is energy being dissipated as heat in the battery?

STRATEGY: Energy is dissipated in the internal resistance of the battery. The rate of dissipation is $P = i^2 r$, where r is the internal resistance and i is the current.

SOLUTION:

$$\left[\text{ans: }100\,\text{W}\right]$$

(*c*) At what rate is electric energy being converted to chemical energy?

STRATEGY: This rate is given by $P = i\mathcal{E}$, where $\mathcal{E}$ is the emf.

SOLUTION:

$$\left[\text{ans: }600\,\text{W}\right]$$

(*d*) What are the answers to (*a*) and (*b*) when the battery is used to supply 50 A to the starter motor?

STRATEGY: Now the battery is discharging and the terminal potential difference is $V = \mathcal{E} - ir$. The rate of dissipation is again $P = i^2 r$.

SOLUTION:

$$\left[\text{ans: }10\,\text{V},\,100\,\text{W}\right]$$

You should be able to write down Kirchhoff's loop equation for a single loop and solve it for an unknown (the current, a resistance, or an emf). If the current is known you should be able to compute the potential difference between any two points on the circuit. You should also be able to compute the rate with which an emf supplies energy and the rate with which a resistor dissipates energy. In one of the examples below a motor does mechanical work. In this case you should know that the rate with which energy is supplied to the motor (or any other electrical device) is the product of the current through it and the potential difference across it.

PROBLEM 3 (13E).   In Fig. 29–7a calculate the potential difference between a and c by considering a path that contains R, r_2, and $\mathcal{E}_2$. (See Sample Problem 29–2.)

STRATEGY: Take the current to be positive if it is from right to left across the top of the diagram. The loop equation is then $-\mathcal{E}_1 - ir_1 - iR - ir_2 + \mathcal{E}_2 = 0$, so $i = (\mathcal{E}_2 - \mathcal{E}_1)/(R + r_1 + r_2)$. The potential difference across points a and c is given by $V_c - V_a = -\mathcal{E}_2 + ir_2 + iR$. Values can be found in Sample Problem 29–1.

SOLUTION:

$$[\text{ans: } -2.5\,\text{V}]$$

PROBLEM 4 (15P).   (a) In Fig. 29–23 what value must R have if the current in the circuit is to be 1.0 mA? Take $\mathcal{E}_1 = 2.0\,\text{V}$, $\mathcal{E}_2 = 3.0\,\text{V}$, and $r_1 = r_2 = 3.0\,\Omega$.

STRATEGY: Take the current to be positive if it is counterclockwise around the circuit. The loop equation is then $\mathcal{E}_2 - \mathcal{E}_1 - ir_1 - iR - ir_2 = 0$. Solve for R.

SOLUTION:

$$[\text{ans: } 990\,\Omega]$$

(b) What is the rate at which thermal energy appears in R?

STRATEGY: Use $P = i^2 R$.

SOLUTION:

$$[\text{ans: } 9.9 \times 10^{-4}\,\text{W}]$$

PROBLEM 5 (19P).   The starting motor of an automobile is turning slowly, and the mechanic has to decide whether to replace the motor, the cable, or the battery. The manufacturer's manual says that the 12-V battery should have no more than $0.020\,\Omega$ internal resistance, the motor no more than $0.200\,\Omega$ resistance, and the cable no more than $0.040\,\Omega$ resistance. The mechanic turns on the motor and measures 11.4 V across the battery, 3.0 V across the cable, and a current of 50 A. Which part is defective?

STRATEGY: The potential difference across the battery is given by $V_b = \mathcal{E} - ir$, where r is the internal resistance, $\mathcal{E}$ is the emf, and i is the current. The potential difference across cable is given by $V_c = iR_c$, where R_c is the resistance of the cable. Because the battery, cable, and motor are in series, the loop equation is $\mathcal{E} - ir - iR_c - iR_m = 0$, where R_m is the resistance of the motor. Calculate the values of r, R_c, and R_m and compare them with the maximum values given by the manual.

SOLUTION:

[ans: the cable, with a resistance of 0.060 Ω, is defective]

PROBLEM 6 (21P).   A solar cell generates a potential difference of 0.10 V when a 500-Ω resistor is connected across it and a potential difference of 0.15 V when a 1000-Ω resistor is substituted. What are (*a*) the internal resistance and (*b*) the emf of the solar cell?

STRATEGY: Let $\mathcal{E}$ be the emf of the cell and r be its internal resistance. When resistance R is connected across it the loop equation is $\mathcal{E} - ir - iR = 0$, so the current is $i = \mathcal{E}/(r + R)$. The potential difference across its terminals is $V = \mathcal{E} - ir = \mathcal{E} - \mathcal{E}r/(r + R) = R\mathcal{E}/(r + R)$. Write this equation twice, once for $R = 500\,\Omega$ and once for $R = 1000\,\Omega$. Solve the two equations simultaneously for r and $\mathcal{E}$.

SOLUTION:

[ans: (*a*) 1000 Ω; (*b*) 0.30 V]

(*c*) The area of the cell is $5.0\,\text{cm}^2$ and the rate per unit area at which it receives light energy is $2.0\,\text{mW/cm}^2$. What is the efficiency of the cell for converting light energy to thermal energy in the 1000-Ω external resistor?

STRATEGY: The rate at which energy is dissipated in R is $P = i^2R$, where the current is given by $i = \mathcal{E}/(r + R)$. The efficiency is $e = P/LA$, where L is the rate per unit area with which light energy is received and A is the area of the cell.

SOLUTION:

[ans: 2.3×10^{-3}]

PROBLEM 7 (23P).   Conductors A and B, having equal lengths of 40.0 m and equal diameters of 2.60 mm, are connected in series. A potential difference of 60.0 V is applied between the ends of the composite wire. The resistances of the wires are 0.127 and 0.729 Ω, respectively. Determine (*a*) the current density in each wire and (*b*) the potential difference across each wire.

STRATEGY: The resistance of the composite wire is $R = R_A + R_B$, where R_A and R_B are the resistances of the individual conductors. The current in each part of the composite wire is $i = V/R$, where V is the potential difference. The current density in A is $J_A = i/A_A$ and the current density in B is $J_B = i/A_B$, where A_A and A_B are the cross-sectional areas. The potential difference across A is $V_A = iR_A$ and the potential difference across B is $V_B = iR_B$.

SOLUTION:

$$[\text{ans:} \ (a) \ J_A = J_B = 1.32 \times 10^7 \, \text{A/m}^2; \ (b) \ V_A = 8.90 \, \text{V}, V_B = 51.1 \, \text{V}]$$

(c) Identify the wire materials. See Table 28–1.

STRATEGY: Use $R = \rho L/A$ to compute the resistivity of the material in each wire, then compare the results with the values listed in the table.

SOLUTION:

$$[\text{ans:} \ A \ \text{is copper}, B \ \text{is iron}]$$

PROBLEM 8 (24P). A battery of emf $\mathcal{E} = 2.00$ V and internal resistance $r = 0.500 \, \Omega$ is driving a motor. The motor is lifting a 2.00-N mass at constant speed $v = 0.500$ m/s. Assuming no energy losses, find (a) the current i in the circuit and (b) the potential difference V across the terminals of the motor.

STRATEGY: Let i be the current. Then the emf of the battery supplies energy at the rate $\mathcal{E}i$ and energy is dissipated in the internal resistance at the rate i^2r. The motor does work at the rate mgv. Conservation of energy yields $\mathcal{E}i = i^2r + mgv$. Solve this equation for i. It is a quadratic equation and you will get two solutions.

SOLUTION:

$$[\text{ans:} \ (a) \ 3.41 \, \text{A or } 0.586 \, \text{A}; \ (b) \ 0.293 \, \text{V or } 1.71 \, \text{V}]$$

(c) Discuss the fact that there are two solutions to this problem.

STRATEGY: The motor does work at the same rate in the two cases. What is different?

SOLUTION:

Some problems ask you to calculate the equivalent resistance when two or more resistors are connected in parallel or series. You should also know how to use equivalent resistances to solve circuit problems. Remember that for a series connection the current is the same in all the resistors and is the same as the current in the equivalent resistor. The potential difference across the equivalent resistor is the sum of the potential differences across the resistors of the combination. For a parallel connection the potential difference across all the resistors is the same and is the same as the potential difference across the equivalent resistor. The current in the equivalent resistor is the sum of the currents in the resistors of the combination.

PROBLEM 9 (27E).   A total resistance of $3.00\,\Omega$ is to be produced by connecting an unknown resistance to a $12.0\,\Omega$ resistance. What must be the value of the unknown resistance and how should it be connected?

STRATEGY: Since the total resistance is less than one of the resistances in the combination the combination must be a parallel connection. Let $R = 3.00\,\Omega$, $R_1 = 12.0\,\Omega$, and solve $1/R = (1/R_1) + (1/R_2)$ for R_2.

SOLUTION:

[ans: $4.00\,\Omega$]

PROBLEM 10 (32E).   In Fig. 29–27, find the equivalent resistance between points D and E.

STRATEGY: The two 4.00-Ω resistors are in parallel. Find their equivalent resistance and redraw the diagram with that equivalent resistance in series with the 2.50-Ω resistance. Find the equivalent resistance for this series combination.

SOLUTION:

[ans: $4.50\,\Omega$]

PROBLEM 11 (36E).   A 120-V power line is protected by a 15-A fuse. What is the maximum number of 500-W lamps that can be simultaneously operated in parallel on this line without "blowing" the fuse?

STRATEGY: The potential difference is the same across each lamp, so the resistance R of each lamp can be found using $P = V^2/R$ and the current in each lamp can be found using $V = iR$. Here $V = 120\,\text{V}$. The total current in the power line is the sum of the currents in the lamps. If N is the number of lamps then the maximum value for N is given by $Ni = 15\,\text{A}$. Solve for N and round down to the nearest integer.

SOLUTION:

[ans: 3]

PROBLEM 12 (41P). You are given two batteries of emf $\mathcal{E}$ and internal resistance r. They may be connected either in parallel (Fig. 29–29a) or in series (Fig. 29–29b) and are used to establish a current in a resistor R. (a) Derive expressions for the current in R for both methods of connection.

STRATEGY: When the batteries are in series the loop equation is $\mathcal{E} + \mathcal{E} - ir - ir - iR = 0$. Solve for i. For the parallel connection let i_b be the current in either battery (the two currents are the same since the batteries are identical) and let i_R be the current in R. The junction equation is $2i_b = i_R$. For either loop containing a single battery and R, the loop equation is $\mathcal{E} - i_b r - i_R R = 0$. Solve the junction and loop equations for i_R.

SOLUTION:

$$\left[\text{ans: series: } 2\mathcal{E}/(2r + R), \text{ parallel: } 2\mathcal{E}/(r + 2R)\right]$$

Which configuration will yield the larger current (b) when $R > r$ and (c) when $R < r$?

SOLUTION:

$$\left[\text{ans: series connection for } R > r, \text{ parallel connection for } R < r\right]$$

You should know how to write and solve Kirchhoff's junction and loop equations for multiloop circuits. Here are some examples.

PROBLEM 13 (43P). (a) Calculate the current through each ideal battery in Fig. 29–31. Assume that $R_1 = 1.0\,\Omega$, $R_2 = 2.0\,\Omega$, $\mathcal{E}_1 = 2.0\,\text{V}$, and $\mathcal{E}_2 = \mathcal{E}_3 = 4.0\,\text{V}$.

STRATEGY: Let i_1 be the current in battery 1, i_2 be the current in battery 2, i_3 be the current in battery 3. Take each current to be positive if it is upward through the corresponding emf. The junction equation for the junction at a is $i_1 + i_2 + i_3 = 0$. The loop equation for the left-hand loop is $\mathcal{E}_1 - 2i_1 R_1 + i_2 R_2 - \mathcal{E}_2 = 0$ and the loop equation for the right-hand loop is $\mathcal{E}_2 - i_2 R_2 + 2i_3 R_1 - \mathcal{E}_3 = 0$. Solve these equations for i_1, i_2, and i_3.

SOLUTION:

$$\left[\text{ans: } i_1 = -0.67\,\text{A}, i_2 = i_3 = 0.33\,\text{A}\right]$$

(b) Calculate $V_a - V_b$.

STRATEGY: Follow the path from b to a through the center branch of the circuit to obtain $V_a = V_b + \mathcal{E}_2 - i_2 R_2$.

SOLUTION:

[ans: 3.3 V, with a at the higher potential]

PROBLEM 14 (47P). In the circuit of Fig. 29–34, for what value of R will the ideal battery transfer energy to the resistors *(a)* at a rate of 60.0 W, *(b)* at the maximum possible rate, and *(c)* at the minimum possible rate?

STRATEGY: The rate of energy transfer is given by $P = i\mathcal{E}$, where i is the current in the battery and $\mathcal{E}$ is the emf of the battery. Find the current by first finding the equivalent resistance. Three of the resistors are in parallel and their equivalent resistance is given by $1/R_{eq} = (1/12.0) + (1/4.00) + (1/R) = (1/3.00) + (1/R)$. Thus $R_{eq} = 3.00R/(3.00 + R)$. The fourth resistor is in series with the combination of other three so the total equivalent resistance is $R_{eq\ total} = 7.00 + 3.00R/(3.00 + R) = (10.0R + 21.0)/(3.00 + R)$. The current is $i = \mathcal{E}/R_{eq\ total} = (3.00 + R)\mathcal{E}/(10.0R + 21.0)$ and the rate of energy transfer is $P = (3.00 + R)\mathcal{E}^2/(10.0R + 21.0)$. Solve for R. You may want to sketch a graph of P vs. R see the maximum and minimum values. Remember that R cannot be negative.

SOLUTION:

[ans: *(a)* 19.5 Ω; *(b)* 0; *(c)* infinite]

(d) For *(b)* and *(c)*, what are those rates?

STRATEGY: For the values of R found above, calculate P.

SOLUTION:

[ans: the maximum rate is 82.3 W, the minimum rate is 57.6 W]

PROBLEM 15 (60P). A voltmeter (resistance R_V) and an ammeter (resistance R_A) are connected to measure a resistance R, as in Fig. 29–43a. The resistance is given by $R = V/i$, where V is the voltmeter reading and i is the current in the resistor R. Some of the current (i') registered by the ammeter goes through the voltmeter so that the ratio of the meter readings (= V/i') gives only an *apparent* resistance reading R'. Show that R and R' are related by

$$\frac{1}{R} = \frac{1}{R'} - \frac{1}{R_V} .$$

Note that as $R_V \to \infty$, $R' \to R$.

STRATEGY: Let i_V be the current through the voltmeter. The junction rule gives $i + i_V = i'$. Divide by V to obtain $(i/V) + (i_V/V) = i'/V$. Identify i/V with $1/R$, i_V/V with $1/R_V$, and i'/V with $1/R'$. Solve for $1/R$.

SOLUTION:

Some problems deal with a resistor and capacitor in a series circuit, with the capacitor either charging or discharging. You should know how to solve problems using the expression for the charge on the capacitor as a function of time: $q(t) = C\mathcal{E}(1 - E^{-t/RC})$ for a charging capacitor and $q(t) = q_0 e^{-t/RC}$ for a discharging capacitor. Also know how to compute the current in the circuit and the potential difference across the capacitor.

PROBLEM 16 (68E). A 15.0-kΩ resistor and a capacitor are connected in series and then a 12.0-V potential difference is suddenly applied across them. The potential difference across the capacitor rises to 5.00 V in 1.30 μs. (*a*) Calculate the time constant of the circuit.

STRATEGY: The charge on the capacitor is given as a function of time t by $q(t) = CV_0(1 - e^{-t/\tau})$ and the potential difference across it is given by $V(t) = q/C = V_0(1 - e^{-t/\tau})$, where V_0 is the applied potential difference and τ is the time constant of the circuit. Solve for τ.

SOLUTION: A little algebra yields $e^{-t/\tau} = 1 - (V/V_0)$. Now take the natural logarithm of both sides to obtain $-t/\tau = \ln(1 - V/V_0)$. Finally, $\tau = -t/\ln(1 - V/V_0)$.

[ans: 2.41 μs]

(*b*) Find the capacitance of the capacitor.

STRATEGY: The time constant is the product of the resistance and capacitance: $\tau = RC$. Solve for C.

SOLUTION:

[ans: 1.61×10^{-10} F]

PROBLEM 17 (75P). An initially uncharged capacitor C is fully charged by a device of constant emf $\mathcal{E}$, in series with a resistor R. (*a*) Show that the final energy stored in the capacitor is half the energy supplied by the emf device.

STRATEGY: The final potential difference across the capacitor is $V_f = \mathcal{E}$ and the energy that is then stored in it is $U = \frac{1}{2}CV^2 = \frac{1}{2}C\mathcal{E}^2$. The rate at which the emf supplies energy is $P = i\mathcal{E}$ and the total energy supplied is $E = \int_0^\infty P\,dt$. Since the charge on the capacitor during charging is given by $q(t) = C\mathcal{E}(1 - e^{-t/RC})$, the current is given by $i = dq/dt = (\mathcal{E}/R)e^{-t/RC}$, and $E = (\mathcal{E}^2/R)\int_0^\infty e^{-t/RC}\,dt$. Evaluate the integral. You should get $E = C\mathcal{E}^2$, which is $2U$.

SOLUTION:

(b) By direct integration of i^2R over the charging time, show that the thermal energy dissipated by the resistor is also half the energy supplied by the emf device.

STRATEGY: The rate at which energy is dissipated in the resistor is $P = i^2R = (\mathcal{E}^2/R)\,e^{-2t/RC}$ and the energy dissipated over the charging times is $E = \int P\,dt = (\mathcal{E}^2/R)\int_0^\infty e^{-2t/RC}\,dt$. Evaluate the integral. You should get $E = \frac{1}{2}C\mathcal{E}^2$, which is the same as the energy stored in the capacitor and half the energy supplied by the emf.

SOLUTION:

III. NOTES

Chapter 30
THE MAGNETIC FIELD

I. BASIC CONCEPTS

Charged particles, when moving, exert magnetic as well as electric forces on each other. Magnetic fields are used to describe magnetic forces and, although the geometry is a little more complicated, the idea is much the same as the idea of using electric fields to describe electric forces. In the next chapter you will learn how a magnetic field is related to the motion of the charges that produce it. Here you will learn about the force exerted by a magnetic field on a moving charge and on a wire carrying an electric current. In each case pay particular attention to what determines the magnitude and direction of the force. As you read the chapter note similarities and differences in electric and magnetic forces.

Magnetic force on a moving charge. _____ charges exert magnetic forces on other _____ charges. The view taken is that a moving charge creates a <u>magnetic</u> <u>field</u> in all of space and this magnetic field exerts a (necessarily magnetic) force on any other moving charge. Charges at rest do not produce magnetic fields nor do magnetic fields exert forces on them. A moving charge does not exert a magnetic force on itself.

The force exerted by the magnetic field $\mathbf{B}$ on a charge q moving with velocity $\mathbf{v}$ is given by

$$\mathbf{F}_B =$$

Notice that the force is written in terms of the vector product $\mathbf{v} \times \mathbf{B}$. You might review Section 3–7 of the text if you do not remember how to determine the magnitude and direction of a vector product. The magnitude of the magnetic force on the moving charge is given by $F_B =$ _____, where θ is the angle between $\mathbf{v}$ and $\mathbf{B}$ when they are drawn with their tails at the same point. The direction of $\mathbf{v} \times \mathbf{B}$ is determined by the right-hand rule: _____
_____
_____

Carefully note that the magnetic force is always perpendicular to both the magnetic field and to the velocity of the charge. Because the force is perpendicular to the velocity a magnetic field cannot change the speed or _____ energy of a charge. It can, however, change the direction of motion.

Also note that the sign of the charge is important for determining the direction of the force. A positive charge and a negative charge moving with the same velocity in the same magnetic field experience magnetic forces in _____ directions.

If a proton is traveling in the positive x direction in a region in which the magnetic field is in the positive z direction, the magnetic force on it is in the _____ direction. If an electron is traveling in the same direction in the same region, the force on it is in the

_____ direction. If either particle is traveling parallel to the magnetic field the force on it is _____ .

The magnetic field can be defined in terms of the force on a moving positive test charge. First, the electric force is found by measuring the force on the test charge when it is _____ . This force is subtracted vectorially from the force on the test charge when it is moving in order to find the magnetic force. Second, the direction of the magnetic field is found by causing the test charge to move in various directions and seeking the direction for which the magnetic force is _____ . Lastly, the test charge is given a velocity perpendicular to the field and the magnetic force on it is found. If its charge is q_0, its speed is v, and the magnitude of the force on it is F_B, then the magnitude of the magnetic field is given by $B =$ _____ .

The SI unit for a magnetic field is _____ and is abbreviated _____ . In terms of the units coulomb, kilogram, and second, $1\,\mathrm{T} = 1$ _____ . Another unit in common use is the gauss. $1\,\mathrm{T} =$ _____ gauss.

Magnetic field lines. Magnetic field lines are drawn so that at any point the field is tangent to the line through that point and so that the magnitude of the field is proportional to the number of lines through a small area perpendicular to the field. Recall that electric field lines are drawn in the same manner.

The diagram on the right shows the magnetic field lines in a certain region of space. On the diagram label a region of high magnetic field and a region of low magnetic field. Suppose an electron is moving out of the page at the point marked with the symbol • and draw an arrow to indicate the direction of the magnetic force on it.

Magnetic field lines form closed loops. They continue through the interior of a magnet, for example. The end of a magnet from which they emerge is called a _____ pole; the end they enter is called a _____ pole. If the magnet is free to rotate in the earth's magnetic field, the north pole of the magnet will tend to point toward the _____ geographic pole.

Crossed electric and magnetic fields. If both a magnetic field **B** and an electric field **E** act simultaneously on a charge q moving with velocity **v**, the total force on the charge is given by

$$\mathbf{F} =$$

An electric force can be used to balance a magnetic force on a charge, but the magnitude and direction used depend on the velocity of the charge. In particular, if a charge is in a magnetic field **B** and has velocity **v** then the total force is zero if $\mathbf{E} =$ _____ . Notice that this result does not depend on either the sign or magnitude of the charge. Also notice that the balancing electric field is perpendicular to both the magnetic field and the velocity of the charge. An important special case occurs if the velocity of the charge is perpendicular to the magnetic field. Then the magnitude of the electric field is given by $E =$ _____ .

Perpendicular electric and magnetic fields form the basis for a velocity selector, used to select charges with a given speed from among a group of charges with a variety of speeds.

Uniform fields **E** and **B**, with known values and perpendicular to each other, are established in a region of space and the charges are incident in a direction that is perpendicular to both fields. Those with a speed given by $v =$ _____ continue straight through the region without being deflected while those with other speeds are deflected. By changing the ratio of the field strengths different sets of charges can be selected.

Crossed electric and magnetic fields can be used to measure the mass-to-charge ratio m/q of an electron (or other charged particle). A beam of fast electrons, emitted from a hot filament, is formed by accelerating them in an electric field and passing them through a slit. They then enter a region of crossed electric and magnetic fields. The beam hits a fluorescent screen and produces a spot where it hits. The steps of the measurement are:

1. The fields are turned off and the position of the spot is noted.

2. Only the electric field is turned on and the deflection y of the beam is computed. Review Sample Problem 24–8 to see how this is done. In terms of the length L of the plates, the mass m, charge q, and speed v of the electron, and the electric field E, it is given by

$$y =$$

Notice that y is the deflection of the beam as it exits the region between the plates. It is not the deflection of the spot but it can be calculated from the spot deflection and knowledge of the geometry. The expression for y cannot be used directly to calculate m/q because the speed v is not known.

3. The magnetic field is now adjusted until the spot returns to its original position. Then the magnetic and electric forces cancel and, in terms of E and B, the speed is $v =$ _____ . Substitute this expression into the equation for y and solve for m/q. The result is

$$m/q =$$

All of the quantities on the right side can be determined experimentally.

The Hall effect. A Hall effect experiment is one of the few experiments that can be carried out to determine the sign of the charge carriers in a current-carrying sample. The diagram shows a rectangular sample of width d and thickness ℓ, carrying a current i, in a uniform magnetic field that points out of the page. Assume the current consists of *positive* charges moving in the direction of the current arrow. The magnetic field pushes them to one side of the sample — mark that side with a series of + signs and the opposite side with a series of − signs.

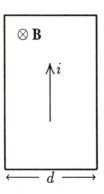

As charge accumulates at the sample sides it creates an electric field in the sample, transverse to the current. For the situation described above the electric field points from the _____ side to the _____ side. Charge continues to accumulate until the electric force on charges in the current is exactly the right strength to balance the magnetic force. In terms of the drift speed v_d of the particles and the magnetic field B, the final electric field strength is given by $E = $ _____ . Once this condition is met, usually in times of about 10^{-13} or 10^{-14} s, charge in the current is no longer deflected and no additional charge accumulates at the sample sides.

A potential difference is associated with the transverse electric field. Since the charge carriers are positive, a point on the _____ side of the sample is at a higher potential than the point directly opposite on the other side.

If the current consists of *negative* charges going in the direction opposite to the current arrow they are forced to the _____ side by the magnetic field. The final electric field now points from the _____ side toward the _____ side and the _____ side is at the higher potential. Thus the sign of the charge carriers can be found by noting the sign of the deflection of a voltmeter attached across the sample width.

In the usual experiment the transverse potential difference, the current, and the magnetic field are measured and the data is used to calculate the carrier concentration n. No matter what the sign of the carriers the magnitude of the transverse potential difference is given by $V = Ed$, where E is the magnitude of the transverse electric field. If only one type of carrier is present in the current the current is given by $i = qnAv_d$, where $A (= \ell d)$ is the cross-sectional area of the sample (see Eq. 28–7). In the space below substitute $E = V/d$ and $v_d = i/qn\ell d$ into $E = v_d B$ and solve for n.

You should have obtained $n = iB/qV\ell$. If i, B, V, and ℓ are measured, n can be calculated.

Cyclotron motion. Since a magnetic force is always perpendicular to the velocity of the charge on which it acts it can be used to hold a charge in a circular orbit. All that is required is to fire the charge perpendicularly into a uniform field. Suppose a region of space contains a uniform magnetic field with magnitude B and a particle with charge of magnitude q is given a velocity $\mathbf{v}$ perpendicular to the field. Equate the magnitude of the magnetic force (qvB) to the product of the mass and acceleration (mv^2/r) and solve for the radius r of the orbit. In terms of q, m, v, and B the result is

$r = $

The time taken by the charge to go once around its orbit does not depend on its speed because the radius of the orbit and hence the distance traveled are directly proportional to _____ . You can see how this comes about by developing an expression for the period. In terms of the radius r and speed v, the period is given by $T = $ _____ . Now substitute the expression for r in terms of the magnetic field, mass, charge, and speed. The result, $T = m/qB$, is independent of v. The number of times the charge goes around per unit time, or the frequency of the motion, is given by $f = 1/T = $ _____ . The angular frequency is given by $\omega = 2\pi f = $ _____ .

If the particle velocity is not perpendicular to the field, but has both parallel and perpendicular components its trajectory is a _____ . The radius is determined by the

_____ component and the pitch is determined by the _____ component of the velocity.

These results are important for the operation of a cyclotron, a type of particle accelerator. The two dees of a cyclotron are diagramed to the right. A uniform magnetic field is everywhere out of the page and an electric field is in the region _____.
A positive charge enters the cyclotron near the center. The _____ field causes the charge to travel in a circular orbit. The _____ field causes the speed of the charge to increase and when it does the charge moves to an orbit of larger radius. Assume the charge goes around four times before leaving the accelerator and draw the path on the diagram. In reality charges circulate many thousands of times.

$\odot$ **B**

For the electric field to increase the speed of a positive charge each time the charge encounters it, it must be in the direction of motion. For it to increase the speed of a negative charge it must be opposite the direction of motion. In either case its direction must be reversed twice each orbit. This is fairly easy to do since the time between reversals does not change as long as the particle speed is significantly less than the speed of light. At speeds near the speed of light the period does depend on the particle speed and the interval between reversals of the electric field must be _____ as the charge speeds up. Accelerators that do this are called _____.

Magnetic force on a current. A current is just a collection of moving charge and so experiences a force when a magnetic field is applied. If the current is in a wire the force is transmitted to the wire itself.

To calculate the force on a wire carrying current i, divide the wire into infinitesimal segments, calculate the force on each segment, then vectorially sum the forces. Let dL be the length of an infinitesimal segment of wire with cross-sectional area A. If n is the concentration of charge carriers and each carrier has charge q then the total charge in the segment is given by $qnA\,dL$. The magnetic force on the segment is the product of this and the force on a single charge. That is, $d\mathbf{F}_B = qnA\mathbf{v}_d \times \mathbf{B}\,dL$, where $\mathbf{v}_d$ is the drift velocity. Notice that the combination $qnAv_d$ appears in the expression for the magnitude of the force. This is the current i. If we take the vector $d\mathbf{L}$ to be in the direction of $q\mathbf{v}_d$ (i.e. in the direction of the current arrow) then the force can be written in terms of i, $d\mathbf{L}$, and $\mathbf{B}$: $d\mathbf{F}_B = $ _____ .

For a finite wire the resultant force is given by the integral

$$\mathbf{F}_B = $$

Check to be sure you have the correct order for the factors in the vector product. For a straight wire of length L in a uniform field $\mathbf{B}$, perpendicular to the wire, the magnitude of the force is $F_B = $ _____ . Explain how the direction of the force is found: _____

For a wire of any shape in a *uniform* magnetic field the magnetic force is given by $\mathbf{F}_B = i\mathbf{L} \times \mathbf{B}$, where $\mathbf{L}$ is the vector from the end where the current enters to the end where the current exits the wire. Since the field is uniform it can be factored from the integral for the force, with the result $\mathbf{F}_B = i\left(\int d\mathbf{L}\right) \times \mathbf{B}$. The integral $\int d\mathbf{L}$ is just $\mathbf{L}$. For a closed loop $\mathbf{L} = 0$ and the total force of a uniform field is zero. The force of a non-uniform field is not zero.

Torque on a current loop. Although a uniform field does not exert a net force on a closed loop carrying current, it may exert a torque. Thus the center of mass of the loop does not accelerate in a uniform field but if the magnetic torque is not balanced the loop has an angular acceleration about the center of mass.

Consider the rectangular loop of wire shown, in the plane of the page and carrying current i. A uniform magnetic field points from left to right. The force on the upper segment is zero since the current there is parallel to the field. The torque exerted on it is also zero. Both the force and torque on the lower segment are zero for the same reason. The force on the left segment has magnitude iaB and is directed into the page. The torque on this segment about the center of the loop has magnitude $iabB/2$ and is directed toward the bottom of the page. The force on the right segment has magnitude _____ and is directed toward _____. The torque on this segment about the center of the loop has magnitude _____ and is directed toward _____. The total force is _____ and the total torque has magnitude _____ and is directed toward _____. The loop tends to turn so its _____ side comes out of the page and its _____ side goes into the page. Use a dotted line to show the axis of rotation on the diagram. Label it. If the wire forms a coil with N turns the magnitude of the torque on it is _____.

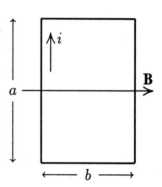

For the orientation shown above the magnitude of the torque is a maximum. It decreases as the loop turns. The diagram on the right shows the view looking down on the loop from the top of the page after it has turned through the angle ϕ. The magnitude of the torque on it is now given by _____. The torque always tends to orient the loop so it is _____ to the magnetic field and when it has this orientation the torque is _____.

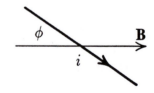

The torque exerted by a uniform field on any planar current loop is easily expressed in terms of the <u>magnetic dipole moment</u> μ of the loop. For a loop with area A, having N turns and carrying current i, the magnitude of the dipole moment is given by $\mu =$ _____. The direction of the dipole moment is determined by the right hand rule: _____

A magnetic dipole moment is not the same as an electric dipole moment. Do not confuse them.

The torque exerted by a field **B** on a loop with magnetic dipole moment μ is given by

$$\tau = \underline{\hspace{3cm}}$$

Its magnitude is given by _____, where θ is the angle between μ and **B** when they are drawn with their tails at the same point.

The magnitude of the dipole moment of the loop diagramed above is $\mu = $ _____. On the diagram above draw a vector representing the dipole moment and label the angle θ. The vector product $\mu \times $ **B** has magnitude _____ and direction _____. Your answers should be in agreement with the results of the direct calculation of the torque.

Although magnetic fields are not conservative and a potential energy cannot be associated with a charge in a magnetic field, a potential energy can be associated with a magnetic dipole μ in a magnetic field **B**. It is given by

$$U = \underline{\hspace{3cm}}$$

The potential energy is a minimum when the dipole moment is _____ to the magnetic field and is in the _____ direction. It is a maximum when the dipole moment is _____ to the magnetic field and is in the _____ direction. In both these cases the plane of the loop is _____ to the field. If the loop is in the plane of the page and the field is pointing toward you, then the direction of the current is _____ at a minimum of potential energy and is _____ at a maximum.

Suppose a dipole initially makes the angle θ_i with the magnetic field, but then rotates so it makes the angle θ_f. During this rotation the work done on the dipole by the field is given by $W = $ _____.

II. PROBLEM SOLVING

You should know how to compute the magnetic force on a moving charge. Use $F = qvB \sin \theta$ to compute the magnitude and use the right hand rule to find the direction. Draw a diagram to help find the angle θ between the vectors **v** and **B**. In other cases, the velocity and field might be given in unit vector notation. Use $\mathbf{i} \times \mathbf{j} = \mathbf{k}$, $\mathbf{j} \times \mathbf{k} = \mathbf{i}$, and $\mathbf{k} \times \mathbf{i} = \mathbf{j}$. You also need to know that the sign of the vector product reverses if the factors are interchanged: $\mathbf{j} \times \mathbf{i} = -\mathbf{k}$, for example. Remember that the vector product of two parallel vectors is 0: $\mathbf{i} \times \mathbf{i} = 0$, for example.

PROBLEM 1 (2E). Four particles follow the paths shown in Fig. 30–28 as they pass through the magnetic field there. What can one conclude about the charge of each particle?

STRATEGY: The magnetic force on a charge q moving with velocity **v** in a magnetic field **B** is given by $\mathbf{F}_B = $ **v** $\times$ **B** and this must be toward the center of the circular path. The arrows on the paths indicate the directions of the particle velocities. In each case find the direction of **v** $\times$ **B**. If it is toward the center of the path the particle is positive; if it is away from the center of the path the particle is negative. If the path is a straight line the particle is uncharged.

SOLUTION:

[ans: 1: positive; 2: negative; 3: neutral; 4: negative]

PROBLEM 2 (3E).   An electron in a TV camera tube is moving at 7.20×10^6 m/s in a magnetic field of strength 83.0 mT. (*a*) Without knowing the direction of the field, what can you say about the greatest and least magnitudes of the force the electron could feel due to the field?

STRATEGY: Since the magnitude of the magnetic force is given by $F = evB \sin \theta$, where θ is the angle between **v** and **B**, the greatest force occurs when **v** and **B** are perpendicular ($\theta = 90°$) and its magnitude is evB. The least force occurs when **v** and **B** are parallel ($\theta = 0$ or 180°) and its magnitude is 0.

SOLUTION:

$$[\text{ans: } 9.56 \times 10^{-14}\,\text{N}, 0]$$

(*b*) At one point the acceleration of the electron is 4.90×10^{14} m/s². What is the angle between the electron's velocity and the magnetic field?

STRATEGY: Solve $a = F_B/m = (evB/m) \sin \theta$ for θ.

SOLUTION:

$$[\text{ans: } 0.267°]$$

PROBLEM 3 (8P).   An electron has an initial velocity $(12.0\,\text{km/s})\,\mathbf{j} + (15.0\,\text{km/s})\,\mathbf{k}$ and a constant acceleration of $(2.00 \times 10^{12}\,\text{m/s}^2)\,\mathbf{i}$ in a region in which uniform electric and magnetic fields are present. If $\mathbf{B} = (400\,\mu\text{T})\,\mathbf{i}$, find the electric field **E**.

STRATEGY: The acceleration of the electron is given by $\mathbf{a} = \mathbf{F}/m = -(e/m)(\mathbf{E} + \mathbf{v} \times \mathbf{B})$, where m is its mass. Thus $\mathbf{E} = -(m/e)\mathbf{a} - \mathbf{v} \times \mathbf{B}$. You should be able to show that $\mathbf{v} \times \mathbf{B} = (6.00\,\text{m} \cdot \text{T/s})\,\mathbf{j} - (4.80\,\text{m} \cdot \text{T/s})\,\mathbf{k}$. Then evaluate the expression for **E**.

SOLUTION:

$$[\text{ans: } -(11.4\,\text{V/m})\,\mathbf{i} - (6.00\,\text{V/m})\,\mathbf{j} + (4.80\,\text{V/m})\,\mathbf{k}]$$

Some problems deal with the simultaneous influence of electric and magnetic fields. Use $\mathbf{F} = q(\mathbf{E} + \mathbf{v} \times \mathbf{B})$. In many cases the charge is not accelerated but continues in a straight line with constant speed. Then $\mathbf{E} + \mathbf{v} \times \mathbf{B} = 0$. This equation is often used to find the value of one of three quantities that appear in it.

PROBLEM 4 (13P).   An ion source is producing ions of ^{6}Li (mass = 6.0 u), each with a charge of +e. The ions are accelerated by a potential difference of 10 kV and pass horizontally into a region in which there is a uniform vertical magnetic field $B = 1.2$ T. Calculate the strength of the smallest electric field, to be set up over the same region, that will allow the ^{6}Li ions to pass through undetected.

STRATEGY: Since the net force on an ion is 0, the electric field is given by $\mathbf{E} = -\mathbf{v} \times \mathbf{B}$, where $\mathbf{v}$ is the velocity of the ion. Since $\mathbf{v}$ and $\mathbf{B}$ are perpendicular the magnitude of the electric field is $E = vB$. Conservation of energy can be used to find the speed of an ion. As an ion crosses the accelerating potential it loses potential energy eV and gains kinetic energy $\frac{1}{2}mv^2$, where m is its mass and V is the potential difference. Thus $eV = \frac{1}{2}mv^2$ and $v = \sqrt{2eV/m}$. Convert the given mass to kilograms: $1\,\mathrm{u} = 1.661 \times 10^{-27}\,\mathrm{kg}$.

SOLUTION:

$$[\text{ans: } 6.80 \times 10^5 \,\mathrm{V/m}]$$

Several problems deal with the Hall effect. Here the electric and magnetic forces cancel as in the last problem but now you might need to relate the electric field to the transverse potential difference ($E = V/d$, where d is the width of the sample) and the carrier speed to the current density ($v_d = J/ne$, where n is the carrier concentration) or current ($v_d = i/neA$, where A is the cross-sectional area).

PROBLEM 5 (14E). Figure 30–30 shows the cross section of a conductor carrying a current perpendicular to the page. (a) Which pair of the four terminals (a, b, c, d) should be used to measure the Hall voltage if the magnetic field is in the $+x$ direction, the charge carriers are negative, and they move out of the page? Which terminal of the pair is at the higher potential?

STRATEGY: Since the carriers are negative and move out of the page the magnetic force on them is toward c. Negative charges collect on the lower surface of the sample and the electric field they produce points from the upper surface (at a) toward the lower surface (at c).

SOLUTION:

$$[\text{ans: terminals } a \text{ and } c, a \text{ is at the higher potential}]$$

(b) Repeat for a magnetic field in the $-y$ direction and positive charge carriers moving out of the page.

SOLUTION:

$$[\text{ans: terminals } b \text{ and } d, b \text{ is at the higher potential}]$$

(c) Discuss the situation if the magnetic field is in the $+z$ direction.

STRATEGY: The magnetic field is now parallel to the velocity of the carriers. The magnetic force is 0.

SOLUTION:

$$[\text{ans: the Hall voltage is } 0]$$

PROBLEM 6 (16P).   In a Hall-effect experiment, a current of 3.0 A length-wise in a conductor 1.0 cm wide, 4.0 cm long, and 10 μm thick produces a transverse (across the width) Hall voltage of 10 μV when a magnetic field of 1.5 T is passed perpendicularly through the thin conductor. From these data, find (*a*) the drift velocity of the charge carriers and (*b*) the number density of charge carriers.

STRATEGY: Since the forces of the magnetic field **B** and the transverse electric field **E** cancel, the drift velocity is given by $v_d = E/B$. The transverse field and the Hall voltage V are related by $E = V/d$, where d is the width of the sample. Thus $v_d = V/Bd$. The current density is given by $J = nev_d$ and this is $i/\ell d$, where i is the current and ℓ is the thickness of the sample. Thus $n = i/ev_d\ell d.$

SOLUTION:

$$\left[\text{ans: } (a)\ 6.7 \times 10^{-4}\,\text{m/s}; \ (b)\ 2.8 \times 10^{29}\,\text{m}^{-3}\right]$$

(*c*) Show on a diagram the polarity of the Hall voltage with a given current and magnetic field direction, assuming the charge carriers are electrons.

STRATEGY: The diagram shows the sample and the direction of the current. Take the magnetic field to be out of the page. Find the direction of the magnetic force on the electrons in the current. They will be forced to the side in the direction of the force and the potential on that side of the sample will be negative compared to the other side.

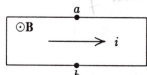

SOLUTION:

$$\left[\text{ans: } a \text{ is at a higher potential than } b\right]$$

Some problems deal with a charge that moves around a circular orbit, under the influence of a uniform magnetic field. You should know the relationship between the particle speed, the orbit radius, and the magnetic field.

PROBLEM 7 (22E).   An electron with kinetic energy 1.20 keV circles in a plane perpendicular to a uniform magnetic field. The orbit radius is 25.0 cm. Calculate (*a*) the speed of the electron, (*b*) the magnetic field, (*c*) the frequency of revolution, and (*d*) the period of the motion.

STRATEGY: (*a*) The kinetic energy K is related to the speed v by $K = \frac{1}{2}mv^2$, where m is the mass. Solve for v. Use $1\,\text{eV} = 1.60 \times 10^{-19}\,\text{J}$ to convert the kinetic energy from electron volts to joules.

   (*b*) The magnitude of the magnetic force on the electron is evB, where v is its speed and B is the magnitude of the magnetic field. Since the electron moves in uniform circular motion the magnitude of its acceleration is v^2/r, where r is the radius of its orbit. Newton's second law yields $evB = mv^2/r$. Solve for B.

   (*c*) The frequency of the motion is given by $f = v/2\pi r$ and (*d*) the period is given by $T = 1/f$.

SOLUTION:

$$\left[\text{ans: } (a)\ 2.05 \times 10^7\,\text{m/s}; \ (b)\ 4.67 \times 10^{-4}\,\text{T}; \ (c)\ 13.1\,\text{MHz}; \ (d)\ 7.65 \times 10^{-8}\,\text{s}\right]$$

PROBLEM 8 (29P).   A proton, a deuteron, and an alpha particle, accelerated through the same potential difference, enter a region of uniform magnetic field **B**, moving perpendicular to **B**. (*a*) Compare their kinetic energies.

STRATEGY: If q is the charge on the particle then $K = qV$, where V is the accelerating potential. A proton has charge e, a deuteron has charge e, and an alpha particle has charge $2e$.

SOLUTION:

[ans: the kinetic energies of the proton and deuteron are equal, the kinetic energy of the alpha is twice as much: $K_\alpha = 2K_p = 2K_d$]

If the radius of the proton's circular path is 10 cm, what are the radii of (*b*) the deuteron and (*c*) the alpha-particle paths?

STRATEGY: The orbit radius is given by $r = mv/qB$ and since $v = \sqrt{2K/m} = \sqrt{2qV/m}$, this can be written $r = (1/B)\sqrt{2mV/q}$. The radius is proportional to $\sqrt{m/q}$. A deuteron has the same charge as a proton but twice the mass, an alpha particle has twice the charge and 4 times the mass.

SOLUTION:

[ans: (*b*) 14 cm; (*c*) 14 cm]

PROBLEM 9 (34P).   Bainbridge's mass spectrometer, shown in Fig. 30–34, separates ions having the same velocity. The ions, after entering through slits S_1 and S_2, pass through a velocity selector composed of an electric field produced by the charged plates P and P', and a magnetic field **B** perpendicular to the electric field and the ion path. Those ions that pass undeviated through the crossed **E** and **B** fields enter into a region where a second magnetic field **B**′ exists, and are bent into circular paths. A photographic plate registers their arrival. Show that for the ions, $q/m = E/(rBB')$, where r is the radius of the circular orbit.

STRATEGY: In the velocity selector portion of the instrument the electric and magnetic forces cancel for those ions that pass through, so their speed is $v = E/B$. After leaving the velocity selector the radius of the orbit is given by $r = mv/qB'$. Substitute the expression for v and solve for q/m.

SOLUTION:

PROBLEM 10 (41P).   A deuteron in a cyclotron is moving in a magnetic field with $B = 1.5\,\text{T}$ and an orbit radius of 50 cm. Because of a grazing collision with a target, the deuteron breaks up, with a negligible loss of kinetic energy, into a proton and a neutron. Discuss the subsequent motions of each. Assume that the deuteron energy is shared equally by the proton and neutron at breakup.

STRATEGY: The neutron, being neutral, moves off in a straight line that is tangent to the deuteron orbit at the point of the breakup. The proton, being charged, moves in a circle. The kinetic energy of the proton is half the kinetic energy of the deuteron, so $m_p v_p^2 = \frac{1}{2}m_d v_d^2$, where the subscript d refers to the deuteron and the subscript p refers to the proton. Since $m_p = m_d/2$, $v_p = v_d$. The radius of an orbit in a magnetic field is given by $r = mv/qB$, so the ratio of the proton orbit to the radius of the deuteron orbit is $r_p/r_d = m_p v_p/m_d v_d = 1/2$.

SOLUTION:

[ans: neutron moves tangent to the original path, proton moves in a circle of radius 25 cm]

You should know how to calculate the force of a magnetic field on a current-carrying wire and the torque of a uniform magnetic field on a current-carrying loop. Often the easiest way to compute the torque is by using $\tau = \mu \times \mathbf{B}$. You will need to know how to compute the dipole moment μ, both magnitude and direction, of a current loop.

PROBLEM 11 (43E).   Figure 30–35 shows four views of a magnet and a straight wire in which electrons are flowing out of the page, perpendicular to the plane of the magnet. In which case will the force on the wire point toward the top of the page?

STRATEGY: For each situation find the direction of $i\mathbf{L} \times \mathbf{B}$. L is into the page and B points from the north pole toward the south pole of the magnet. Pick the situation for which the direction is toward the top of the page.

SOLUTION:

[ans: (b)]

PROBLEM 12 (50P).   A long, rigid conductor, lying along the x axis, carries a current of 5.0 A in the $-x$ direction. A magnetic field **B** is present, given by $\mathbf{B} = 3.0\mathbf{i} + 8.0x^2\mathbf{j}$, with x in meters and B in milliteslas. Calculate the force on the 2.0-m segment of the conductor that lies between $x = 1.0$ m and $x = 3.0$ m.

STRATEGY: Consider an infinitesimal segment of the conductor, with coordinate x and length dx. In general, the magnetic force on an infinitesimal segment of wire is given by Eq. 30–26: $d\mathbf{F}_B = i\,d\mathbf{L} \times \mathbf{B}$. In this case $d\mathbf{L} = -dx\,\mathbf{i}$ (remember that dL is in the direction of the current). Thus $d\mathbf{F}_B = -i\,dx\,\mathbf{i} \times (3.0 \times 10^{-3}\mathbf{i} + 8.0 \times 10^{-3}x^2\mathbf{j}) = -8.0 \times 10^{-3}x^2 i\,dx\,\mathbf{k}$, where $\mathbf{i} \times \mathbf{i} = 0$ and $\mathbf{i} \times \mathbf{j} = \mathbf{k}$ was used. The total force on the segment is the integral

$$\mathbf{F}_B = -8.0 \times 10^{-3}i \int_{1.0}^{3.0} x^2\,dx\,\mathbf{k}\,.$$

Evaluate the integral.

SOLUTION:

[ans: $-(0.35\,\mathrm{N})\,\mathbf{k}$]

PROBLEM 13 (59P).   Figure 30–40 shows a wire ring of radius a that is perpendicular to the general direction of a radially-symmetric diverging magnetic field. The magnetic field at the ring is everywhere of the same magnitude B, and its direction at the ring everywhere makes an angle θ with a normal to the plane of the ring. The twisted lead wires have no effect on the problem. Find the magnitude and direction of the force the field exerts on the ring if the ring carries a current i.

STRATEGY: Consider an infinitesimal segment of the ring, with length dL. Since the angle between the wire and the field is 90°, the magnitude of the magnetic force on the segment is d$F_B = iB$ dL. Since the force is normal to both the wire and the field it makes the angle θ with the horizontal. The horizontal component of the force is d$F_{Bh} = iB \cos\theta$ dL and the vertical component is d$F_{Bv} = iB \sin\theta$ dL. These must be summed over the ring. All the horizontal components point toward the center of the ring and sum to 0. All vertical components due to segments of equal lengths are the same so the sum over vertical components is easy to carry out.

SOLUTION:

[ans: $2\pi aiB \sin\theta$, vertically upward]

PROBLEM 14 (64E). A circular wire loop whose radius is 15.0 cm carries a current of 2.60 A. It is placed so that the normal to its plane makes an angle of 41.0° with a uniform magnetic field of 12.0 T. (*a*) Calculate the magnetic dipole moment of the loop.

STRATEGY: The magnitude of the dipole moment is given by $\mu = iA$, where A is the area of the loop (πR^2, where R is the radius). It is directed along the normal.

SOLUTION:

[ans: 0.184 A · m^2]

(*b*) What torque acts on the loop?

STRATEGY: Use $\tau = \mu B \sin\theta$, where θ is the angle between the dipole moment and the field.

SOLUTION:

[ans: 1.45 N · m]

PROBLEM 15 (67P). A circular loop of wire having a radius of 8.0 cm caries a current of 0.20 A. A unit vector parallel to the dipole moment μ of the loop is given by 0.60 **i** − 0.80 **j**. If the loop is located in a magnetic field given by **B** = (0.25 T) **i** + (0.30 T) **k**, find (*a*) the torque on the loop (in unit vector notation) and (*b*) the magnetic potential energy of the loop.

STRATEGY: The magnitude of the dipole moment is given by $\mu = iA$, where A is the area of the loop (πR^2, where R is the radius). The vector dipole moment is $\mu = \mu(0.60\,\mathbf{i} - 0.80\,\mathbf{j})$. The torque on the loop is given by $\tau = \mu \times \mathbf{B}$. Use $\mathbf{i} \times \mathbf{i} = 0$, $\mathbf{i} \times \mathbf{k} = -\mathbf{j}$, $\mathbf{j} \times \mathbf{i} = -\mathbf{k}$, and $\mathbf{j} \times \mathbf{k} = \mathbf{i}$ to evaluate the vector product. The potential energy of the loop is given by $U = -\mu \cdot \mathbf{B}$. Use $\mathbf{i} \cdot \mathbf{i} = 1$, $\mathbf{i} \cdot \mathbf{k} = 0$, $\mathbf{j} \cdot \mathbf{i} = 0$, and $\mathbf{j} \cdot \mathbf{k} = 0$ to evaluate the scalar product.

SOLUTION:

[ans: (*a*) −(9.7 × 10^{-4} N · m) **i** − (7.2 × 10^{-4} N · m) **j** + (8.0 × 10^{-4} N · m) **k**; (*b*) 6.0 × 10^{-4} J]

PROBLEM 16 (68P). Figure 30–43 shows a current loop $ABCDEFA$ carrying a current $i = 5.00\,\text{A}$. The sides of the loop are parallel to the coordinate axes, with $AB = 20.0\,\text{cm}$, $BC = 30.0\,\text{cm}$, and $FA = 10.0\,\text{cm}$. Calculate the magnitude and direction of the magnetic dipole moment of this loop. (*Hint*: Imagine equal and opposite currents i in the line segment AD; then treat the two rectangular loops $ABCDA$ and $ADEFA$.)

STRATEGY: A current i from D to A completes the horizontal loop and a current i from A to D completes the vertical loop. The two currents sum to zero, so their effect is as if no current were present along the line AD. The dipole moment of the vertical loop is in the negative y direction and has magnitude given by $\mu_v = iA_v$, where A_v is the area of the loop. The dipole moment of the horizontal loop is in the negative z direction and its magnitude is given by $\mu_h = iA_h$, where A_h is the area of the loop. The magnitude of the total dipole moment is given by $\mu = \sqrt{\mu_v^2 + \mu_h^2}$. The moment is in the yz plane and the angle θ it makes with the z axis is given by $\tan\theta = \mu_v/\mu_h$.

SOLUTION:

$$[\text{ans: } 0.335\,\text{A} \cdot \text{m}^2;\ 26.6°\ \text{away from the} -z\ \text{axis, toward the} +y\ \text{axis}]$$

III. NOTES

Chapter 31
AMPERE'S LAW

I. BASIC CONCEPTS

You will learn to calculate the magnetic field produced by a current using two techniques, one based on the Biot-Savart law and the other based on Ampere's law. When using the first you will sum the fields produced by infinitesimal current elements. Carefully note how the directions of the current and the position vector from a current element to the field point influence the direction of the field. Ampere's law relates the integral of the tangential component of the field around a closed path to the net current through the path. Pay attention to the role played by symmetry when you use this law to find the field.

The Biot-Savart law. The diagram on the right shows a portion of a current-carrying wire. Suppose you wish to calculate the magnetic field it produces at point P. First, mark an infinitesimal element **ds** of the wire, in the same direction as _____. Now draw the displacement vector **r** from the selected current element to P. Draw an arrow head ($\odot$) or tail ($\otimes$) near P to represent the (infinitesimal) field produced there by the element. Be careful about the direction. Finally, write the Biot-Savart law in vector form for the (infinitesimal) magnetic field produced at P by the current element:

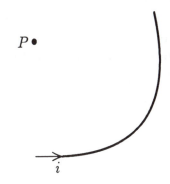

$$\mathbf{dB} =$$

Notice that the field of the element is perpendicular to both ds and **r**. To find the total field at P sum the contributions from all elements of the wire. The result is the vector integral

$$\mathbf{B} =$$

You should recognize that the integrand is a vector and you must actually carry out three integrations, one for each component. In many cases, however, you can choose the coordinates so only one component does not vanish.

The Biot-Savart law contains the constant μ_0. It is called the _____ constant and its value is _____ T·m/A. Do not confuse the symbol with that for the magnitude of a magnetic dipole moment (μ without the subscript 0).

In Section 31–2 of the text the Biot-Savart law is used to find an expression for the magnitude of the magnetic field produced by a long straight wire carrying current i. Each infinitesimal element of the wire produces a field in the same direction so the magnitude of the total field is the sum of the magnitudes of the fields produced by all the elements. Go over the calculation carefully. The diagram shows an infinitesimal element of the wire at x, a distance r from point P. In terms of i and r the magnitude of the field it produces at P is $dB = $ _____.
The direction of the field is _____ the page. In terms of x and R, $r = $ _____ and $\sin\theta = $ _____, so $dB = $ _____. Integrate in x from $-\ell$ to $+\ell$, then take the limit as $\ell \to \infty$. The result is

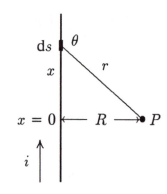

$$B = $$

You should recognize that the field lines around a long straight wire are circles centered on the wire. Three are shown in the diagram. The field has the same magnitude, given by the equation you wrote above, at all points on any given circle. The direction is determined by a right-hand rule: if the thumb of the right hand points in the direction of the current then the fingers will curl around the wire in the direction of the magnetic field lines. The current in the wire shown is out of the page. Put arrows on the field lines to show the direction of the field.

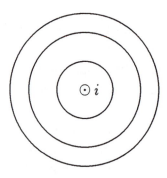

For current carrying wires that are not straight the magnitude of the field is not uniform along a field line. You should be able to make a symmetry argument to show that it is for a straight wire.

The expression for the field of a long straight wire can be combined with the expression for the magnetic force on a wire, developed in the last chapter, to find an expression for the force exerted by one wire on another, parallel to the first. The magnetic field produced by one wire at a point on the other wire is _____ to the second wire. If the wires are separated by a distance d and carry currents i_1 and i_2 then the magnitude of the force per unit length of one on the other is given by

$$F/L = $$

If the currents are in the same direction the wires _____ each other; if the currents are in opposite directions they _____ each other. Notice that the forces of the wires on each other obey Newton's third law: they are equal in magnitude and opposite in direction.

Ampere's law.   Ampere's law is given by Eq. 31–16. Copy it here:

The integral on the left side is a path integral around any *closed* path. The current i on the right side is the net current through a surface that is bounded by the path. For example, if the path is formed by the edges of this page then i is the net current through the page. You should recognize that the Amperian path need not be the boundary of any physical surface and, in fact, may be purely imaginary. You should also recognize that the surface need not be a plane. The sides and bottom of a wastepaper basket form a valid surface with the rim as the boundary.

You should know how to evaluate the right side of the Ampere's law equation when the current is distributed among several wires. First, choose a direction, clockwise or counter-clockwise, to be used in evaluating the integral on the left side of the Ampere's law equation. The choice is immaterial but it must be made since it determines the direction of ds. If the tangential component of the field is in the direction of ds then the integral is positive, if it is in the opposite direction the integral is negative.

Now curl the fingers of your right hand around the loop in the direction chosen. Your thumb will point in the direction of positive current. Examine each current through the surface and algebraically sum them. If a current arrow is in the direction your thumb pointed it enters the sum with a positive sign; if it is in the opposite direction it enters with a negative sign. If the net current through the path is positive then the average tangential component of **B** is in the direction chosen for ds. If the net current is negative then the average tangential component of **B** is in the opposite direction.

A current outside the path does not contribute to the right side of the Ampere's law equation. You should be aware that a current outside the path produces a magnetic field at every point on the path but the integral $\oint \mathbf{B} \cdot d\mathbf{s}$ of the field it produces vanishes.

Only the tangential component of the magnetic field enters. When Ampere's law is used to calculate the magnetic field of a current distribution the path is taken, if possible, to be either parallel or perpendicular to field lines at every point. Then the integral reduces to $\int B\, ds$, along those parts of the path that are parallel to field lines. If, in addition, the magnitude of the field is constant along the path, then the integral is Bs, where s is the total length of those parts of the path that are parallel to field lines.

Ampere's law can be used to find an expression for the magnetic field produced by a long straight wire carrying current i. The Amperian path used is a circle in a plane perpendicular to the wire and centered at the wire. Why? _____

_____

Since the magnetic field is tangent to the circle and has the same magnitude at all points around the circle, the integral $\oint \mathbf{B} \cdot d\mathbf{s}$ is _____, where r is the radius of the circle. In the space below equate this expression to $\mu_0 i$ and solve for B:

Your result should be $B = \mu_0 i / 2\pi r$, as before.

Ampere's law can also be used to find the magnetic field *inside* a long straight wire. The diagram shows the cross section of a cylindrical wire of radius R carrying current i, uniformly distributed throughout its cross section. Take the Amperian path to be the dotted circle of radius r. The magnetic field is tangent to the circle and has constant magnitude around the circle, so the integral $\oint \mathbf{B} \cdot d\mathbf{s}$ is _____. Not all the current in the wire goes through the dotted circle. In fact, the fraction through the circle is the ratio of the circle area to the wire area: r^2/R^2. Thus the right side of the Ampere's law equation is _____. Equate the two sides and solve for B. The result is $B =$ _____.

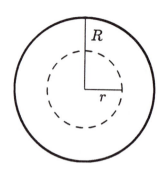

The magnetic field at the center of the wire is _____. The field has it largest value at _____ and this value is given by _____.

Ampere's law can be applied to a solenoid, a cylinder tightly wrapped with a thin wire. For an ideal solenoid (long, with the current approximated by a cylindrical sheet of current rather than a wrapped wire) the magnetic field outside is negligible and the field inside is uniform and is parallel to _____.

The diagram shows the cross section of a solenoid along its length, with the current in each wire coming out of the page at the top and going into the page at the bottom. Use dotted lines to draw the rectangular path you will use to evaluate the integral on the left side of the Ampere's law equation. The integral is $\oint \mathbf{B} \cdot d\mathbf{s} =$ _____, where h is the length of the rectangle. Label it on the diagram.

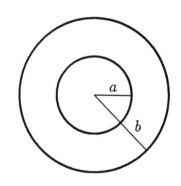

If the solenoid has n turns of wire per unit length then the number of turns that pass through the surface bounded by the Amperian path is _____. Each carries current i so the right side of the Ampere's law equation is _____. The equation can be solved for the magnitude of the field, with the result $B =$ _____.

Ampere's law can also be applied to a toroid, with a core shaped like a doughnut and wrapped with a wire, like a solenoid bent so its ends join.

The diagram to the right shows a cross section with the wire omitted for clarity. The magnetic field is confined to the interior of the core ($a < r < b$) and the field lines are concentric circles centered at the center of the hole. Draw the Amperian path you will use to find an expression for the field a distance r from the center. The integral $\oint \mathbf{B} \cdot d\mathbf{s}$ is _____ and if there are N turns of wire, each carrying current i, the total current through the surface bounded by the path is _____. Thus the Ampere's law equation is _____ = _____ and $B =$ _____.

Consider an Amperian path in the form of a circle, centered at the center of the hole and inside the hole. The net current through the path is _____ and the field in the "doughnut hole" is _____. Consider a similar path outside the toroid. The net current through this path is _____ and the field outside the toroid is _____.

The magnetic field of a circular current loop. This calculation is an excellent illustration of how the Biot-Savart law is used. Pay careful attention to it and notice particularly that the vector integral for the field is evaluated one component at a time. First write the expression for the magnitude dB of the field produced by an infinitesimal element of the current. This involves the distance r from the current element to the field point and the sine of the angle θ between the element **ds** and the vector **dr**. Next determine the direction of the field associated with **ds** and let α be the angle between the field and one of the coordinate axes. $dB \cos \alpha$ is the component of the infinitesimal field along the chosen axis. Integrate $dB \cos \alpha$ over the current to find the component of the total field along the axis. The quantities r, θ, and α may be different for different segments of the current. If they are you must use geometry write them in terms of a single variable of integration. Finally, repeat the process for the other coordinate axes.

Two angles enter the calculation: the angle θ between **ds** and **r** and the angle α between the field and a coordinate axis. The first is used to find an expression for the magnitude of the field produced by an infinitesimal element of current, the second is used to find a component, once the expression for the magnitude is known. Be careful to keep them straight.

Now use the Biot-Savart law to develop an expression for the field on the axis of a circular loop, a distance z from its center. Consider the element **ds** shown. The angle between **r** and **ds** is _____ so the magnitude of the infinitesimal field at P reduces to

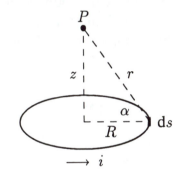

$$dB = $$

Draw a vector at P to indicate the direction of d**B** and verify that it makes the angle α with the z axis. The expression for the component of d**B** along the axis is $dB_{\parallel} = $ _____. This must be integrated around the loop.

All quantities in the integral are the same for all segments of the loop, so the integral can be evaluated using $\oint ds = 2\pi R$. The result is $B_{\parallel} = $ _____. In terms of z and R, $r = $ _____ and $\cos \alpha = $ _____. Make these substitutions to obtain

$$B_{\parallel} = $$

A symmetry argument can be made to show all other components vanish.

The field of a magnetic dipole. An important result can be derived from this equation. For points far away from the loop ($z \gg R$) the expression for the field can be written in terms of the magnetic dipole moment μ of the loop. **B** is in the direction of μ for points above the loop, in the direction of the dipole moment and is in the opposite direction for points below the loop. Write the vector expression for the magnetic field:

B =

This expression is valid for the magnetic field of *any* plane loop, regardless of its shape, for points far away along the axis defined by the direction of the dipole moment.

You should be able to estimate the direction of the field produced in its plane by a current carrying loop of wire. Near any segment of the wire you may treat the segment as a long straight wire and use the right-hand rule given above. Since the field decreases with distance take the direction at any point to be roughly in the same direction as the field produced by the nearest segment of the loop, perhaps adjusted for other nearby segments.

For example, the circle on the right represents a current loop, with current i as shown. Points A, B, C, and D are in the plane of the loop. The magnetic field is directed _____ the page at A, _____ the page at B, _____ the page at C, and _____ the page at D. At points far away from the loop the field closely resembles that of a bar magnet (both are dipole fields). The _____ pole is at the upper surface of the loop, where magnetic field lines leave the "magnet" and the _____ pole is at the lower surface of the loop, where field lines enter the "magnet".

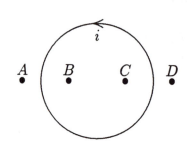

II. PROBLEM SOLVING

A few of the early problems deal with the field of a long straight wire. You should be able to find the magnitude and direction of the field at any point in space, given the current in the wire. The first problem below tests your knowledge by asking you to superpose two fields, one of which is the field of a long straight wire.

You should know how to calculate the magnetic force on a moving charge using $q\mathbf{v} \times \mathbf{B}$. Don't forget to include the sine of the angle between $\mathbf{v}$ and $\mathbf{B}$ when you calculate the magnitude. Know how to use the right-hand rule to find the direction of the force. Be sure to take into account the sign of the charge. The second problem below asks you to calculate the force on a charge moving near a long straight wire.

PROBLEM 1 (6E).   A long wire carrying a current of 100 A is placed in a uniform external magnetic field of 5.0 mT. The wire is perpendicular to this magnetic field. Locate the points at which the resultant magnetic field is zero.

STRATEGY: Draw a diagram with the current coming out of the page. Draw a field line for the field of the wire. It circles the wire in the counterclockwise direction. Show two field lines for the external field, both tangent to the circle. Suppose the lines run from the bottom toward the top of the page and draw one on the left side of the wire and one on the right. Notice that the fields tend to cancel on the left. They will cancel exactly if $B_{\text{wire}} = B_{\text{ext}}$. Since $B_{\text{wire}} = \mu_0 i/2\pi r$, where i is the current in the wire and r is the distance from the wire, the point of zero resultant field is given by $\mu_0 i/2\pi r = B_{\text{ext}}$. Solve for r.

SOLUTION:

[ans: on a line through the wire and perpendicular to the external field, 1.0×10^{-7} m from the wire on the side where the fields are in opposite directions]

PROBLEM 2 (8E).   A positive point charge of magnitude q is a distance d from a long straight wire carrying a current i and is traveling with speed v perpendicular to the wire. What are the direction and magnitude of the force acting on it if the charge is moving (a) toward or (b) away from the wire?

STRATEGY: Draw a diagram with the current coming out of the page. Show the charge and its velocity vector. The force on the charge is given by $\mathbf{F} = q\mathbf{v} \times \mathbf{B}$. Since the particle velocity is perpendicular to the field the magnitude of the force is given by $F = qvB$. The magnitude of the magnetic field is given by $B = \mu_0 i / 2\pi d$. Use the right-hand rule for vector products to find the direction of the force

SOLUTION:

[ans: (a) $\mu_0 q v i / 2\pi d$, opposite the direction of the current; (b) same magnitude but in the direction of the current]

Some problems ask you to use the Biot-Savart law to compute the magnetic field of a current. Divide the current into infinitesimal elements, write the expression for the field of an element, then integrate each component over the current. You will need to write the integrand in terms of single variable. Here are some examples.

PROBLEM 3 (13P).   Use the Biot-Savart law to calculate the magnetic field $\mathbf{B}$ at C, the common center of the semicircular arcs AD and HJ in Fig. 31–33. The two arcs, of radii R_2 and R_1, respectively, form part of the circuit $ADJHA$ carrying current i.

STRATEGY: First develop an expression for the magnetic field at the center of a current-carrying wire in the form of an arc. All infinitesimal segments of the wire produce fields in the same direction, normal to the plane of the arc. All segments are the same distance from the center and for all of them the angle between the line element $\mathbf{ds}$ and the vector $\mathbf{r}$ from the segment to the arc center is $90°$. The magnetic field is $B = (\mu_0 i / 4\pi r^2) \int ds$. The arcs of this problem are semicircles and $\int ds = \pi r$, so for each of them $B = \mu_0 i / 4r$. The smaller arc produces a field that is into the page at C and the larger arc produces a field that is out of the page. Take the field to positive if it is into the page and sum the fields. The two straight segments (AH and JD) do not produce fields at C.

SOLUTION:

$$\left[\,\text{ans: }\ \frac{\mu_0 i}{4}\left(\frac{1}{R_1}-\frac{1}{R_2}\right)\right]$$

PROBLEM 4 (15P). A wire carrying current i has the configuration shown in Fig. 31–35. Two semi-infinite straight sections, both tangent to the same circle, are connected by a circular arc, of central angle θ, along the circumference of the circle, with all sections lying in the same plane. What must θ be in order for B to be zero at the center of the circle?

STRATEGY: The field due to each of the straight wires is half the field of an infinite wire, or $\mu_0 i/4\pi R$. Both of these fields are out of the page. The length of the circular arc is $R\theta$, for θ in radians, so the field it produces is $\mu_0 i\theta/4\pi R$. It is into the page. Set the expression for the total field equal to 0 and solve for θ.

SOLUTION:

$$\left[\,\text{ans: } 2\,\text{rad}\right]$$

In many cases you may think of a circuit as composed of segments, each of which produces a magnetic field. You can then calculate the field produced by each segment and vectorially sum the individual fields to find the total field. Here are some examples.

PROBLEM 5 (19P). Show that the magnitude of the magnetic field produced at the center of a rectangular loop of wire of length L and width W, carrying a current i is

$$B = \frac{2\mu_0 i}{\pi}\,\frac{(L^2 + W^2)^{1/2}}{LW}.$$

Show that for $L \gg W$, this reduces to a result consistent with the result of Sample Problem 31–3.

STRATEGY: The center of the rectangle is on the perpendicular bisectors of the sides, so the result of Problem 17P can be used. Use it four times, once for each side of the rectangle. The four fields are in the same direction so simply add the magnitudes. For a side of length L replace R with $W/2$; for a side of length W replace L with W and R with $L/2$.

SOLUTION:

In the limit as L becomes much greater than W, $(L^2 + W^2)^{1/2}$ can be replaced by L and the expression for the field becomes $B = 2\mu_0 i/\pi W$. This is identical to the result of Sample Problem 31–3 for $x = 0$ and $d = W/2$.

PROBLEM 6 (22P).   A straight section of wire of length L carries current i. Show that the magnetic field associated with this segment at P, a perpendicular distance D from one end of the wire (see Fig. 31-38), is given by

$$B = \frac{\mu_0 i}{4\pi D} \frac{L}{(L^2 + D^2)^{1/2}} .$$

STRATEGY: Take the wire segment to be along the x axis from $x = 0$ to $x = L$. Consider a line element with coordinate x and length dx. The distance from this element to P is $r = \sqrt{(L - x)^2 + D^2}$ and the sine of the angle between the vector from the element to P and the positive x axis is $\sin\theta = D/\sqrt{(L - x)^2 + D^2}$. The magnetic field produced by the element is $dB = (\mu_0 i/4\pi)D[(L - x)^2 + D^2]^{-3/2}\,dx$. All the elements produce fields in the same direction, out of the page, so the total field at P is given by

$$B = \frac{\mu_0 i D}{4\pi} \int_0^L [(L - x)^2 + D^2]^{-3/2}\,dx .$$

Use $\int [(L - x)^2 + D^2]^{-3/2}\,dx = -(L - x)/D^2\sqrt{(L - x)^2 + D^2}$ to evaluate the integral.

SOLUTION:

PROBLEM 7 (25P).   Calculate the magnetic field **B** at point P in Fig. 31–41. Assume that $i = 10\,\text{A}$ and $a = 8.0\,\text{cm}$.

STRATEGY: Consider the circuit to be composed of 8 segments, all with the point P on a line that is perpendicular to the segment and through one of its ends. The result of the last problem can then be used. All contributions are in the same direction, into the page, so you may add the magnitudes of the individual fields. For the upper portion of the left wire and the left portion of the upper wire replace both L and D with $a/4$. For the lower portion of the left wire and the right portion of the upper wire replace L with $3a/4$ and D with $a/4$. For the upper portion of the right wire and the left portion of the lower wire replace L with $a/4$ and D with $3a/4$. For the right portion of the lower wire and the lower portion of the right wire replace both L and D with $3a/4$.

SOLUTION:

[ans: 2.0×10^{-4} T]

Here are some more problems that deal with the magnetic fields of long straight wires. Some of them ask you to use what you learned in the last chapter to calculate the force that one wire exerts on another.

PROBLEM 8 (28E). Two long straight parallel wires, separated by 0.75 cm, are perpendicular to the plane of the page as shown in Fig. 31–43. Wire 1 carries a current of 6.5 A into the page. What must be the current (magnitude and direction) in wire 2 for the resultant magnetic field at point P to be zero?

STRATEGY: The fields must be in opposite directions so the current in wire 2 must be out of the page. The field due to wire 1 is given by $B_1 = \mu_0 i_1 / 4\pi r_1$, where r_1 is the distance from the wire to P and i_1 is the current in the wire. The field due to wire 2 is given by $B_2 = \mu_0 i_2 / 4\pi r_2$, where r_2 is distance from the wire to P and i_2 is the current in the wire. Equate the two magnitudes and solve for i_2. You should obtain $i_2 = (r_2/r_1) i_1$.

SOLUTION:

[ans: 4.3 A]

PROBLEM 9 (37P). Two long wires a distance d apart carry equal antiparallel currents i, as in Fig. 31-47. (a) Show that the magnitude of the magnetic field at point P, which is equidistant from the wires, is given by

$$B = \frac{2\mu_0 i d}{\pi(4R^2 + d^2)}.$$

(b) In what direction does **B** point?

STRATEGY: Draw a diagram and include the perpendicular lines from the wires to P. Draw the magnetic field vector for the field at P of each wire. Each field is perpendicular to the line from the wire to P. Use the right-hand rule to find the directions. The distance from either wire to P is $r = \sqrt{R^2 + d^2/4} = \sqrt{4R^2 + d^2}/2$ and the magnitude of the field due to either wire is $\mu_0 i / 2\pi r = \mu_0 i / \pi \sqrt{4R^2 + d^2}$. Observe that the angle θ between either of the fields and the horizontal axis is the same as the angle between line from the wire to P and the line joining the wires, so that $\cos\theta = (d/2)/\sqrt{R^2 + d^2/4} = d/\sqrt{4R^2 + d^2}$. The vertical components of the fields cancel. Sum the horizontal components.

SOLUTION:

[ans: (b) to the right]

PROBLEM 10 (38P). In Fig. 31–48, the long straight wire carries a current of 30 A and the rectangular loop carries a current of 20 A. Calculate the resultant force acting on the loop. Assume that $a = 1.0$ cm, $b = 8.0$ cm, and $L = 30$ cm.

STRATEGY: The long wire produces a magnetic field with magnitude $B = \mu_0 i_w / 2\pi r$, where r is the distance from the wire and i_w is the current in the wire. The field is into the page at all points below the wire. The forces it exerts on the left and right sides of the loop are equal in magnitude and opposite in direction, so they cancel when the resultant force is calculated. The force on the upper segment of the loop has magnitude $F_u = i_\ell L B = \mu_0 i_w i_\ell L / 2\pi a$ and the force on the lower segment of the loop has magnitude $B_\ell = \mu_0 i_w i_\ell L / 2\pi (a + b)$. The first is upward and the second is downward. Sum the forces.

SOLUTION:

[ans: 3.2×10^{-3} N, up]

Notice that the field of the long straight wire is not uniform, so the force on the loop does not vanish.

Ampere's law problems take three forms. The most straightforward give the currents and ask you to find the value of $\oint \mathbf{B} \cdot d\mathbf{s}$ around a given path. You must pay attention to the directions of the currents as they pierce the plane of the path. Other problems ask you to use Ampere's law to calculate the magnetic field. Carefully choose the Amperian path you will use and pay attention to the evaluation of the Ampere's law integral in terms of the unknown field. Still other problems give you the magnetic field as a function of position and ask for the current through a given region. Carry out the line integral of the tangential component of the field around the boundary of the region and equate the result to $\mu_0 i$, then solve for i.

PROBLEM 11 (41E). Eight wires cut the page perpendicularly at the points shown in Fig. 31–51. A wire labeled with the integer k ($k = 1, 2, \ldots, 8$) carries the current $k i_0$. For those with odd k, the current is out of the page; for those with even k, it is into the page. Evaluate $\oint \mathbf{B} \cdot d\mathbf{s}$ along the closed green path in the direction shown.

STRATEGY: Use Ampere's law: $\oint \mathbf{B} \cdot d\mathbf{s} = \mu_0 i$, where i is the net current piercing the loop. First decide which currents are inside the loop and which are outside. Shading the region inside with a pencil helps. Start with the region inside the long tail that curves around the lower left portion of the diagram. You should find that currents labeled 1, 3, 6, and 7 are inside while the rest are outside. Use the right-hand rule to find that currents coming out of the page (odd labels) are positive while currents going into the page (even labels) are negative. Sum the currents with their signs.

SOLUTION:

$\left[\text{ans: } +5\mu_0 i_0\right]$

PROBLEM 12 (46P).   Figure 31–55 shows a cross section of a hollow cylindrical conductor of radii a and b, carrying a uniformly distributed current i. (*a*) Show that $B(r)$ for the range $b < r < a$ is given by

$$B = \frac{\mu_0 i}{2\pi(a^2 - b^2)} \left(\frac{r^2 - b^2}{r}\right).$$

STRATEGY: Use Ampere's law with the dotted circle shown in the figure as an Amperian path. The magnetic field lines are circles centered on the axis of the cylinder. If the current is out of the page they are counterclockwise. Perform the line integral on the left side of the Ampere's law equation in that direction. Then $\oint \mathbf{B} \cdot d\mathbf{s} = 2\pi r B$. You must now find the net current through the Amperian path: the current between $r = b$ and the path. Since the current density (current per unit area) is uniform the fraction of the current inside the path is the ratio of the cross-sectional area of the portion of the cylinder inside the path to the total cross-sectional area of the cylinder. The area inside the path is $\pi(r^2 - b^2)$ while the total area is $\pi(a^2 - b^2)$. The net current is $(r^2 - b^2)i/(a^2 - b^2)$. Thus $2\pi r B = \mu_0(r^2 - b^2)i/(a^2 - b^2)$. Solve for B.

SOLUTION:

(*b*) Show that when $r = a$, this equation gives the magnetic field B for a long straight wire; when $r = b$, it gives zero magnetic field, and when $b = 0$, it gives the magnetic field inside a solid conductor.

STRATEGY: When $r = a$ the equation reduces to $B = \mu_0 i/2\pi a$. When $r = b$ it reduces to $B = 0$. When $b = 0$ it reduces to $B = \mu_0 i r/2\pi a^2$. Check to be sure the first is the field of a long straight wire and the last is the field inside a solid conducting cylinder.

SOLUTION:

(*c*) Assume $a = 2.0\,\mathrm{cm}$, $b = 1.8\,\mathrm{cm}$, and $i = 100\,\mathrm{A}$ and plot $B(r)$ for the range $0 < r < 6\,\mathrm{cm}$.

STRATEGY: For $r < b$ the field is 0; for $a < r < b$ the field is given by the expression derived in (*a*); for $r > a$ the field is that of a long straight wire carrying current i: $B = \mu_0 i/2\pi r$. Use the axes below to plot the graph.

468   *Chapter 31: Ampere's Law*

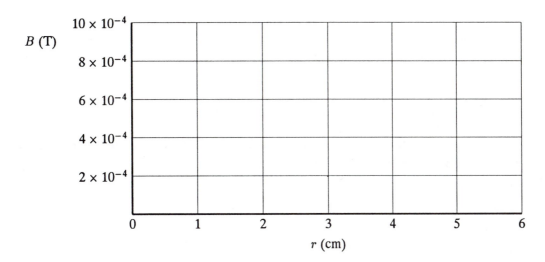

B (T)

r (cm)

PROBLEM 13 (48P).   The current density inside a long, solid, cylindrical wire of radius a is in the direction of the central axis and varies linearly with radial distance r from the axis according to $J = J_0 r/a$. Find the magnetic field inside the wire.

STRATEGY: Use Ampere's law with an Amperian path that is a circle of radius r ($r < a$) in the plane of a cross section of the wire. The magnetic field is tangent to the path at all points on it and has the same magnitude at all points on it. Thus $\oint \mathbf{B} \cdot d\mathbf{s} = 2\pi r B$. The current through the region bounded by the path is given by $i = \int \mathbf{J} \cdot d\mathbf{A}$, where $d\mathbf{A}$ is an infinitesimal area element, normal to the cross section. Take dA to be the area of a ring of radius r and width dr: $dA = 2\pi r \, dr$. Then $i = (2\pi J_0/a) \int_0^r r^2 \, dr$. Carry out the integration, then solve the Ampere's law equation for B.

SOLUTION:

$$\left[\, \text{ans: } \mu_0 J_0 r^2/3a \,\right]$$

PROBLEM 14 (52P).   In a certain region there is a magnetic field given in milliteslas by $\mathbf{B} = 3.0\mathbf{i} + 8.0(x^2/d^2)\mathbf{j}$, where x is the x-coordinate distance in meters and d is a constant with unit of length. Some current must be flowing through the region to cause the specified $\mathbf{B}$ field. (a) Evaluate the integral $\int \mathbf{B} \cdot d\mathbf{s}$ along the straight path from $(d, 0, 0)$ to $(d, d, 0)$.

STRATEGY: Along the line $y = d$ the field is given by $\mathbf{B} = 3.0 \times 10^{-3}\mathbf{i} + 8.0 \times 10^{-3}\mathbf{j}$. The line element $d\mathbf{s}$ is $dy\,\mathbf{j}$, so $\int \mathbf{B} \cdot d\mathbf{s} = 8.0 \times 10^{-3} \int_0^d dy$. Carry out the integration.

SOLUTION:

$$\left[\, \text{ans: } 8.0 \times 10^{-3}d, \text{ with } d \text{ in meters} \,\right]$$

(b) Let $d = 0.50\,$m in the expression for **B** and apply Ampere's law to determine what current flows perpendicularly through the plane of a square of side length 0.5 m that lies in the first quadrant of the xy plane, with one corner at the origin. (c) Is this current in the **k** or $-$**k** direction?

STRATEGY: First calculate the contribution of the other sides of the square to the integral in Ampere's law. Along the line from $(d, d, 0)$ to $(0, d, 0)$ the x component of the field is $3.0 \times 10^{-3}\,$T and the contribution to the integral is $-3.0 \times 10^{-3}d$. The result is negative because **B** and d**s** are in opposite directions. Along the line from $(0, d, 0)$ to $(0, 0, 0)$ the y component of the field is 0 so this side does not contribute. Along the line $(0, 0, 0)$ to $(d, 0, 0)$ the x component of the field is $3.0 \times 10^{-3}\,$T and this side contributes $+3.0 \times 10^{-3}d$ to the integral. Sum the contributions and equate the sum to $\mu_0 i$, then solve for i.

SOLUTION:

$$\left[\text{ans: } 3.2 \times 10^3\,\text{A}\right]$$

You should know how to compute the magnetic fields of some special current configurations (in addition to a long straight wire): a solenoid, a toroid, and a magnetic dipole. In the last case you studied only the field along the dipole axis. You should also recall from the last chapter how to find the dipole moment of a current loop (both magnitude and direction) and how to compute the torque of a uniform magnetic field on a dipole.

PROBLEM 15 (60P).   A long solenoid with 10.0 turns/cm and a radius of 7.00 cm carries a current of 20.0 mA. A current of 6.00 A flows in a straight conductor located along the axis of the solenoid. (a) At what radial distance from the axis will the direction of the resulting magnetic field be at 45.0° to the axial direction?

STRATEGY: The field produced by the solenoid and the field produced by the wire along its axis are perpendicular to each other. The resultant field is at 45.0° to the axial direction where the magnitudes of the two fields are the same. The field of the solenoid is given by $\mu_0 n i_s$, where i_s is the current in the solenoid and n is the number of turns per unit length. The field of the wire is given by $\mu_0 i_w / 2\pi r$, where i_w is the current in the wire and r is the radial distance from the wire. Equate the magnitudes of the two fields and solve for r.

SOLUTION:

$$\left[\text{ans: } 4.77\,\text{cm}\right]$$

(b) What is the magnitude of the magnetic field there?

STRATEGY: Evaluate either $B_s = \mu_0 i_s n$ or $B_w = \mu_0 i_w / 2\pi r$ for the magnitudes of both fields. The magnitude of the total field is given by $B = \sqrt{B_s^2 + B_w^2}$.

SOLUTION:

$$\left[\text{ans: } 3.55 \times 10^{-5}\,\text{T}\right]$$

PROBLEM 16 (67E).   The magnitude $B(x)$ of the magnetic field at points on the axis of a square current loop of side a is given in Problem 20. (*a*) Show that the axial magnetic field for this loop, for $x \gg a$, is that of a magnetic dipole (see Eq. 31–25).

STRATEGY: According to Problem 20 the axial field of the loop is given by

$$B(x) = \frac{4\mu_0 i a^2}{\pi (4x^2 + a^2)(4x^2 + 2a^2)^{1/2}} .$$

For $x \gg a$ the terms a^2 and $4x^2$ in the denominator can be neglected. When this is done the result is $B(x) = \mu_0 i a^2 / 2\pi x^3$. Compare this with Eq. 31–25 and note that the field is inversely proportional to the cube of the distance from the loop.

SOLUTION:

(*b*) What is the magnetic dipole moment of this loop?

STRATEGY: Compare the result of part (*a*) and Eq. 31–25.

SOLUTION:

[ans: ia^2]

Notice that the result is iA, where A is the area of the loop.

PROBLEM 17 (71P).   A circular loop of radius 12 cm carries a current of 15 A. A second loop of radius 0.82 cm, having 50 turns and a current of 1.3 A, is at the center of the first loop. (*a*) What magnetic field **B** does the large loop set up at its center?

STRATEGY: Each infinitesimal element of the loop produces a magnetic field at the center in the same direction, perpendicular to the plane of the loop. Furthermore, each element is the same distance from the center and the vector from any element to the center makes an angle of 90° with the current element. The field at the center is therefore $B = (\mu_0 i_\ell / 4\pi R^2) \int ds = \mu_0 i_\ell / 4\pi R^2)(2\pi R) = \mu_0 i_\ell / 2R$, where R is the radius of the large loop and i_ℓ is the current in it.

SOLUTION:

[ans: 7.9×10^{-5} T]

(*b*) What torque acts on the small loop? Assume that the planes of the two loops are perpendicular and that the magnetic field due to the large loop is essentially uniform throughout the volume occupied by the small loop.

STRATEGY: Use $\tau = \mu \times \mathbf{B}$, where μ is the dipole moment of the small loop. Since the loops are perpendicular to each other the dipole moment of the small loop lies in the plane of the large loop and is perpendicular to the magnetic field produced by the large loop. Therefore the magnitude of the torque is given by $\tau = \mu B$. The magnitude of the dipole moment is found using $\mu = Ni_s A = N\pi r^2 i_s$, where A is the area of the small loop, r is its radius, i_s is its current, and N is the number of turns.

SOLUTION:

$$\left[\text{ans: } 1.1 \times 10^{-6}\,\text{N} \cdot \text{m}\right]$$

III. NOTES

Chapter 32
FARADAY'S LAW OF INDUCTION

I. BASIC CONCEPTS

As the magnetic flux through any area changes, an emf is generated around the boundary and, if the boundary is conducting, a current is induced. Faraday's law tells us the relationship between the emf and the rate of change of the flux. You will learn that an emf can be generated by changing the magnetic field or by moving a physical object through a magnetic field. You should concentrate on how to calculate the magnetic flux and the emf in each case. In addition, you will learn that a non-conservative electric field is always associated with a changing magnetic field and it is this field that is responsible for the emf when the magnetic field is changing.

Faraday's law. To understand Faraday's law you must first understand <u>magnetic flux</u>. The magnetic flux through any area is proportional to the number of magnetic field lines through that area. Recall that the number of field lines through a small area perpendicular to the field is proportional to the magnitude of the field and that the number of lines through *any* small area is proportional to the component B_n along a normal to the area.

The diagram shows an area A bounded by the curve C. If a magnetic field pierces the plane of the page then there is magnetic flux through the area. It is the integral over the area of the normal component of the magnetic field. Divide the area into a large number of small elements, each with area ΔA, and for each element calculate the quantity $B_n \Delta A$, then sum the results. The magnetic flux is the limit of this sum as the area elements become infinitesimal. That is, the flux through the area is given by the integral $\Phi_B = \int B_n \, dA$.

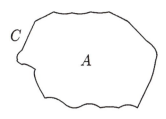

Usually the infinitesimal area is assigned a direction normal to the surface and written as the vector dA. Then $B_n dA = \mathbf{B} \cdot d\mathbf{A}$ and the integral for the flux becomes

$$\Phi_B = $$

If the field is uniform over the entire area the flux is given by

$$\Phi_B = $$

where θ is the angle between _____ and _____. The SI unit of magnetic flux is _____ and is abbreviated _____.

Faraday's law tells us that whenever the flux through any area is changing with time then an emf is generated around the boundary of that area. Symbolically, the law is

$$\mathcal{E} = $$

where $\mathcal{E}$ is the induced emf. The induced emf is distributed around the boundary and is not localized as is the emf of a battery. Its direction is specified as clockwise or counterclockwise around a loop in the plane of the page. The minus sign that appears in Faraday's law is closely related to the direction of the emf.

Notice that the direction of d$\mathbf{A}$ is ambiguous since two directions are normal to any surface. For the area shown above the normal directions are into and out of the page. Φ_B has the same magnitude for both choices but is positive for one and negative for the other. You may choose either but you may find thinking about Faraday's law a bit easier if you pick the one for which Φ_B is positive. If you point the thumb of your right hand in the direction chosen for d$\mathbf{A}$ then your fingers will curl around the boundary in the direction of positive emf. If $\mathcal{E}$, calculated using Faraday's law including the minus sign, turns out to be positive then the emf is in the direction of your fingers. If it turns out to be negative then it is in the opposite direction. If the boundary of the region is conducting and no other emf's are present, the induced current will be in the direction of the induced emf.

Some problems deal with a coil of wire consisting of several identical turns, like a solenoid, with the same magnetic field through each turn. If Φ_B is the flux through each turn and there are N turns, then the total emf induced around the coil is given by

$$\mathcal{E} =$$

The flux through the interior of a loop can be changed by changing the magnitude or direction of the magnetic field, by changing the area of the loop, or by changing the orientation of the loop. No matter how the emf is generated it represents work done on a unit test charge as the test charge is carried around the loop. If the magnetic field is changing in either magnitude or direction, an electric field is generated and the emf is the work per unit charge done by that electric field. If the magnetic field is constant and the loop changes area or orientation, work must be done by an external agent, the agent that is changing the loop.

Lenz' law. Lenz' law provides another way to determine the direction of an induced emf. Imagine that the bounding curve is a conducting wire so current flows in it when an emf is induced, the current being in the direction of the emf. The induced current produces a magnetic field everywhere but you should concentrate on the field it produces inside the loop and the flux associated with that field. According to Lenz' law when the flux of the externally applied field changes current flows in the loop in such a way that the flux it produces in the interior of the loop counteracts the change.

Suppose a bar magnet is held above the loop shown above, with its north pole toward the loop. Then the magnetic field points into the page through the loop. If the magnet is moved toward the loop, the magnitude of its field in the plane of the loop increases (the field of a magnet is stronger near the magnet). According to Lenz' law, current will flow in the loop in such a way that its field in the interior of the loop is directed _____ the page. If the magnet is moved away from the loop, the magnitude of the applied field decreases and, as it does, current flows in the loop in such a way that its magnetic field in the interior of the loop is directed _____ the page. Thus the law is used to determine the direction of the field produced by the induced current.

Once you have determined the direction of the magnetic field produced by the induced current you can determine the direction of the current itself and hence that of the emf. Very near any segment of the loop the field is quite similar to the field of a long straight wire; the lines are nearly circles around the segment. You can use the right-hand rule explained in the last chapter: curl your fingers around the segment so they point in the direction of the field in the interior of the loop. That is, they should curl upward through the loop if the field is upward through the loop and they should curl downward if the field is downward. Your thumb will then point along the segment in the direction of the current.

For the situation being considered, the current and the emf are clockwise around the loop if the induced field inside the loop is _____ the page and are counterclockwise if the induced field inside the loop is _____ the page.

Motional emf. An emf is also be generated if all or part of the loop moves in a manner that changes the flux through it. You should realize that if the loop above is moved through a uniform field the flux through it does not change and no emf is generated. If, however, the field is not uniform then the flux changes as the loop moves and an emf is generated. You must find an expression for the flux as a function of the position of the loop, then differentiate it with respect to time to obtain $d\Phi_B/dt$. The emf clearly depends on the velocity of the loop. Either Lenz' law or the sign convention associated with Faraday's law can be used to find the direction of the emf.

If part of the loop moves in such a way that the area changes, an emf is generated even if the field is uniform. The flux through the loop is a function of time because the area is a function of time. In terms of the rate dA/dt with which the area changes in a uniform field B, the emf is given by $\mathcal{E} = $ _____. It is generated only along the moving part, not around the whole loop.

Suppose a wire forms three sides of a rectangle and a rod, free to move on the wire, forms the fourth side, as shown. If W is the fixed dimension of the rectangle, x is the variable dimension, and the magnetic field is uniform and perpendicular to the rectangle, then the flux through it is $\Phi_B = $ _____ and the emf generated around it is _____. Note that the emf is proportional to the speed.

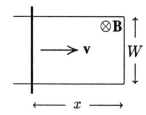

The loop need not consist entirely of physical objects, such as wires. Part of it may be imaginary. An emf is generated along an isolated rod moving in a magnetic field, for example. To calculate it you may complete the loop with imaginary lines. If the rod is conducting, electrons move in the direction opposite to the emf when the emf first appears. They collect at one end of the rod, leaving the other end positively charged, and as a result an electric field exists along the rod. Very quickly the field prevents further build-up of charge and the current becomes zero. A steady current cannot flow in an isolated rod.

Do not confuse the electric field created by charge at the ends of a moving rod with an electric field associated with a time-varying magnetic field. In the first case the magnetic field is not changing and no electric field is associated with it.

An emf is also generated around a loop that is changing orientation so the angle between its normal and the magnetic field is changing with time. Suppose a circular loop with radius R is rotated about a diameter in a uniform magnetic field, perpendicular to the axis of rotation. The flux through the loop is given by $\Phi_B = BA\cos\theta = \pi R^2 B \cos\theta$, where θ is the angle between the normal to the plane of the loop and the magnetic field. If θ is changing according to $\theta = \omega t$, then $d\Phi_B/dt =$ _____.

You should be able to find the direction of the induced emf (and current if the loop is conducting). When ωt is between 0 and $\pi/2$ the flux is positive and so is the emf. Positive flux tells us that the direction we picked for the normal makes an angle of less than 90° with the magnetic field. If we look from a point where the magnetic field is pointing toward us from the loop, the emf is counterclockwise. When ωt is between 90° and 180° the flux is _____ and the emf is _____. When viewed from the same point the emf is _____. When ωt is between 180° and 270° the flux is _____ and the emf is _____. When viewed from the same point the emf is _____. When ωt is between 270° and 360° the flux is _____ and the emf is _____. When viewed from the same point the emf is _____.

Energy considerations. When a current is induced energy is transferred, via the emf, to the moving charges. Recall that the rate at which a seat of emf $\mathcal{E}$ does work on a current i is given by $P =$ _____. When the emf is associated with a changing magnetic field the energy comes from the agent changing the field; when the emf is motional the energy comes from the agent moving the loop or from the kinetic energy of the loop. In either case the energy is dissipated in the resistance of the loop. You will learn more about the mechanism of energy transfer by a changing magnetic field in the next chapter. Here you learn about energy transfer by a motional emf.

Consider the example above in which one side of rectangular loop moves in a uniform magnetic field and recall that the emf generated is given by $\mathcal{E} = BWv$. If the resistance of the loop is R then the current is given by $i =$ _____, so in terms of B, W, v, and R the emf transfers energy at the rate $P =$ _____. This is precisely the rate at which the external agent does work on the rod to keep it moving at a constant speed.

In terms of i, B, and W the external field **B** exerts a force of magnitude $F =$ _____ on the rod. The direction of the current is _____ in the diagram and the magnetic field is into the page so the magnetic force on the moving rod is directed to the _____. If the rod is to move with constant velocity an external agent must apply a force of equal magnitude but in the opposite direction. Thus the force of the agent is directed to the _____. Since the rod is moving to the right with speed v the rate at which the agent is doing work is $P = Fv$. In terms of B, v, W, and R, this is $P =$ _____.

The rate at which energy is dissipated in the resistance of the loop is given by $P = i^2 R$ and, in terms of B, v, W, and R, this is $P =$ _____. All of the energy supplied by the agent is dissipated. The magnetic field provides the mechanism of energy transfer but it supplies no energy.

If the agent stops pushing, the magnetic field exerts a force that slows the moving rod and eventually stops it. The kinetic energy of the rod is then dissipated in the resistance of the loop. This is the basis of magnetic braking.

Induced electric fields. An electric field is always associated with a changing magnetic field and this electric field is responsible for the induced emf. The relationship between the electric field **E** and the emf around a closed loop is given by the integral

$$\mathcal{E} = $$

Note that the integral is zero for a conservative field, such as the electrostatic field produced by charges at rest. The electric field induced by a changing magnetic field, however, is non-conservative and the integral is not zero.

Suppose a cylindrical region of space contains a uniform magnetic field, directed along the axis of the cylinder, as shown. The field is zero outside the region. If the magnetic field changes with time the lines of the electric field it produces form circles, concentric with the cylinder. You should be able to derive an expression for the magnitude of the electric field at points inside and outside the cylinder.

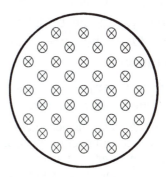

First consider a point inside the cylinder, a distance r from the center. Draw a circle through the point, concentric with the cylinder cross section. The magnetic flux through the circle is given by _____ and in terms of dB/dt its rate of change is given by _____.

The circle coincides with an electric field line and the electric field has uniform magnitude around the circle. In terms of the magnitude of the electric field and the radius of the circle the emf is given by $\mathcal{E} =$ _____. Solve the Faraday's law equation for the magnitude of the electric field as a function of r. It is $E =$ _____.

Now repeat the calculation for a point outside the cylinder. Draw a circle of radius r. The magnetic flux through this circle is given by _____, where R is the radius of the cylinder. The magnitude of the emf around the circle is given by _____ and the magnitude of the electric field is given as a function of r by $E =$ _____.

If the magnitude of the magnetic field is increasing, the emf (either inside or outside the cylinder) is in the _____ direction and so is the electric field. If the magnitude of the magnetic field is decreasing the emf is in the _____ direction and so is the electric field.

II. PROBLEM SOLVING

You should know how to compute the flux through a given area. In some cases the magnetic field is uniform and you simply multiply the perpendicular component by the area. In other cases the field is not uniform and you must carry out an integration of the perpendicular component over the area. In some problems the magnetic field is given while in others the current that produces the field is given and you must use what you learned in the last chapter to write an expression for it as a function of position. For example, the field may be produced by current in a long straight wire.

Once an expression for the magnetic flux is found you differentiate it with respect to time to find the emf around the boundary of the region. Be very careful: you integrate with respect to *position* to find the flux and differentiate the flux with respect to *time* to find the emf. Before you do look carefully at the expression for the flux and decide which quantity that enters it is changing with time. Sometimes it is the magnitude of the magnetic field, sometimes it is the magnitude of the area of the region, and sometimes it is the orientation of the area with respect to the field.

In some problems the boundary of the region is a conductor and you are asked to find the current in it. First note all the emfs around the boundary. One of them is the emf associated with the changing magnetic flux but there may be others, produced by batteries or generators, for example. Add them with their correct signs and divide the total by the resistance in the loop. Obviously, you must be very careful about the signs of the various emfs here.

Finally, you may be asked about the force on a wire carrying current or perhaps its acceleration. Once the current is found use $\mathbf{F} = i\mathbf{L} \times \mathbf{B}$ to find the force and $\mathbf{F} = m\mathbf{a}$ to find the acceleration.

PROBLEM 1 (13P). A toroid having a 5.00-cm square cross section and an inside radius of 15.0 cm has 500 turns of wire and carries a current of 0.800 A. What is the magnetic flux through the cross section?

STRATEGY: The magnetic field inside a toroid is given by $B = \mu_0 N i / 2\pi r$, where r is the distance from the center, N is the number of turns, and i is the current. The field is perpendicular to any cross section. If a is the length of an edge of a cross section the outer radius of the toroid is $R + a$, where R is the inner radius. Take a strip of the cross section with length a and width dr. Its area is $dA = a\,dr$ and the flux through it is $d\Phi = B\,a\,dr = (\mu_0 N i a / 2\pi)(1/r)\,dr$. Integrate this from R to $R + a$ to find the total flux. You should get

$$\Phi = \frac{\mu_0 N i a}{2\pi} \ln \frac{R + a}{R}.$$

Evaluate this expression.

SOLUTION:

[ans: 1.15×10^{-6} Wb]

PROBLEM 2 (15P). A closed loop of wire consists of a pair of equal semicircles, of radius 3.7 cm, lying in mutually perpendicular planes. The loop was formed by folding a plane circular loop along a diameter until the two halves became perpendicular. A uniform magnetic field **B** of magnitude 76 mT is directed perpendicular to the fold diameter and makes equal angles (= 45°) with the planes of the semicircles as shown in Fig. 32–36. The magnetic field is reduced to zero at a uniform rate during a time interval of 4.5 ms. Determine the magnitude of the induced emf and the direction of the induced current in the loop during this interval.

STRATEGY: The magnetic flux through each semicircle is $BA\cos 45°$, where A is the area of the semicircle, so the total flux through the loop is $\pi R^2 B \cos 45°$. The magnitude of the induced emf is $\mathcal{E} = d\Phi/dt = \pi R^2 |dB/dt| \cos 45°$. Since the field is reduced uniformly to zero the magnitude of the rate at which it is changing is given by $|dB/dt| = (76 \times 10^{-3}\,\text{T})/(4.5 \times 10^{-3}\,\text{s}) = 16.9\,\text{T/s}$.

SOLUTION:

[ans: 51 Mv]

The current induced in the loop produces a magnetic field in the same direction as the external field. The current and the induced emf are clockwise when viewed from the left side of the diagram.

PROBLEM 3 (18P).   One hundred turns of insulated copper wire are wrapped around a wooden cylindrical core of cross-sectional area $1.20 \times 10^{-3}\,\text{m}^2$. The two terminals are connected to a resistor. The total resistance in the circuit is $13.0\,\Omega$. If an externally applied uniform longitudinal magnetic field in the core changes from $1.60\,\text{T}$ in one direction to $1.60\,\text{T}$ in the opposite direction, how much charge flows through the circuit? (*Hint:* See Problem 17.)

STRATEGY: The induced emf is given by $\mathcal{E} = -N\,d\Phi/dt = -NA\,dB/dt$, where N is the number of turns, Φ is the magnetic flux through one turn, A is the cross-sectional area of the cylinder, and B is the magnitude of the magnetic field. The current in the circuit is $i = \mathcal{E}/R$, where R is the resistance. The total charge that flows through the circuit is $q = \int i\,dt = (1/R)\int \mathcal{E}\,dt = -(A/R)\int (dB/dt)\,dt = -(A/R)\,\Delta B$, where ΔB is the change in the magnetic field from the beginning to the end of the interval. Since the field simply changes direction $\Delta B = B_f - B_i = -2B_i$, where B_i is the initial field.

SOLUTION:

[ans: $2.95 \times 10^{-4}\,\text{C}$]

Notice that the charge does *not* depend on the time taken to change the field. However, the current does.

PROBLEM 4 (22E).   A circular loop of wire 10 cm in diameter is placed with its normal making an angle of 30° with the direction of a uniform 0.50-T magnetic field. The loop is "wobbled" so that its normal rotates in a cone about the field direction at the constant rate of 100 rev/min; the angle (= 30°) between the normal and the field direction remains unchanged during the process. What emf appears in the loop?

STRATEGY: The magnetic flux is given by $\Phi = BA\cos 30°$, where A is the area of the loop. This does not change as the loop wobbles.

SOLUTION:

[ans: 0]

PROBLEM 5 (25E).   In Fig. 32–41 a conducting rod of mass m and length L slides without friction on two long horizontal rails. A uniform vertical magnetic field **B** fills the region in which the rod is free to move. The generator G supplies a constant current i that flows down one rail, across the rod, and back to the generator along the other rail. Find the velocity of the rod as a function of time, assuming it to be at rest at $t = 0$.

STRATEGY: As the rod slides an emf is induced in the circuit but the potential difference across the generator changes to keep the current constant. You do not need to calculate the induced emf. The magnetic field is perpendicular to the sliding rod, so the force on it is $F = iBL$. The acceleration of the rod is $a = F/m$ and the velocity is $v = at = (F/m)t$. Substitute for F in terms of the magnetic field. Use the right-hand rule to find the direction of the force on the rod, and hence the direction of its velocity.

SOLUTION:

[ans: $(iBL/m)t$, away from the generator]

PROBLEM 6 (27P).   Two straight conducting rails form a right angle where their ends are joined. A conducting bar in contact with the rails starts at the vertex at time $t = 0$ and moves with a constant velocity of 5.20 m/s to the right, as shown in Fig. 32–42. A 0.350-T magnetic field points out of the page. Calculate (a) the flux through the triangle formed by the rails and bar at $t = 3.00$ s and (b) the emf around the triangle at that time. (c) In what manner does the emf around the triangle vary with time?

STRATEGY: Let x be the perpendicular distance from the vertex to the sliding bar at time t. The distance along the sliding bar from rail to rail is then $2x$ and the area of the triangle is $A = x^2$. Since $x = vt$, where v is the velocity of the bar, $A = v^2t^2$, and the flux through the triangle is $\Phi = BA = Bv^2t^2$. The magnitude of the emf is $\mathcal{E} = d\Phi/dt = 2v^2t$. Evaluate these expressions for $t = 3.00$ s.

SOLUTION:

[ans: (a) 85.2 Wb; (b) 56.8 V; $\mathcal{E}$ is proportional to t]

PROBLEM 7 (32P).   In Exercise 25 (see Fig. 32–41) the constant-current generator G is replaced by a battery that supplies a constant emf $\mathcal{E}$. (a) Show that the velocity of the rod now approaches a constant terminal value **v** and give its magnitude and direction.

STRATEGY: Assume the battery is connected so the current is in the direction shown on the diagram. The magnetic force on the rod is away from the battery and the rod moves in that direction. As it moves the magnetic flux through the loop increases and an additional emf is induced in the circuit; its magnitude is BLv. It is clockwise and tends to decrease the total emf. Thus the total emf is $\mathcal{E} - BLv$, the current is $(\mathcal{E} - BLv)/R$, where R is the resistance of the circuit, and the force on the rod is $F = iBL = (\mathcal{E} - BLv)BL/R$. This vanishes when terminal velocity is reached. Solve for v.

SOLUTION:

[ans: $\mathcal{E}/BL$]

(b) What is the current in the rod when this terminal velocity is reached?

STRATEGY: Evaluate $i = (\mathcal{E} - BLv)/R$ for $v = \mathcal{E}/BL$

SOLUTION:

[ans: 0]

(c) Analyze both this situation and that of Exercise 25 from the point of view of energy transfers.

STRATEGY: Obtain expressions for the rate at which the battery is supplying energy and the rate at which energy is being dissipated in the resistance of the loop.

SOLUTION:

PROBLEM 8 (35P).   For the situation shown in Fig. 32–46, $a = 12.0$ cm and $b = 16.0$ cm. The current in the long straight wire is given by $i = 4.50t^2 - 10.0t$, where i is in amperes and t is in seconds. (*a*) Find the emf in the square loop at $t = 3.00$ s.

STRATEGY: The magnetic field of the long straight wire a distance r from the wire is given by $B = \mu_0 i/2\pi r$, where i is the current in the wire. The field is into the page at points below the wire and out of the page at points above it. Take the flux to be positive where the field is into the page. Divide the region into strips of length b and width dr. Each has area $dA = b\, dr$. Since the flux through the region above the wire is canceled by the flux through a region of equal width below the net flux is given by $\Phi = (\mu_0 ib/2\pi) \int_{b-a}^{a}(1/r)\, dr = (\mu_0 ib/2\pi) \ln[(b - a)/a]$. The emf is given by $\mathcal{E} = -d\Phi/dt = (\mu_0 b/2\pi)(di/dt) \ln[a/(b - a)]$. The rate of change of the current is $di/dt = 9.00t - 10.0$.

SOLUTION:

[ans: $0.598\,\mu$V]

(*b*) What is the direction of the induced current in the loop?

STRATEGY: The flux is increasing, so the field of the induced current must be out of the page inside the loop.

SOLUTION:

[ans: counterclockwise]

PROBLEM 9 (37P).   A rectangular loop of wire with length a, width b, and resistance R is placed near an infinitely long wire carrying current i, as shown in Fig. 32–48. The distance from the long wire to the center of the loop is r. Find (*a*) the magnitude of the magnetic flux through the loop and (*b*) the current in the loop as it moves away from the long wire with speed v.

STRATEGY: (*a*) The magnetic field of the wire a distance r' from the wire is given by $B = \mu_0 i/2\pi r'$, where i is the current in the wire. The field is into the page in the diagram. Divide the region of the loop into strips of length a and width dr'. The magnetic flux through the loop is $\Phi = (\mu_0 ia/2\pi) \int_{r-b/2}^{r+b/2}(1/r')\, dr'$. Carry out the integration.

(*b*) First find the emf: differentiate Φ with respect to time and replace dr/dt with v. Use $d\ln[(2r + b)/(2r - b)]/dt = -4vb/(4r^2 - b^2)$. The current is given by $\mathcal{E}/R$.

SOLUTION:

[ans: (*a*) $\dfrac{\mu_0 ia}{2\pi} \ln \dfrac{2r + b}{2r - b}$; (*b*) $2\mu_0 iabv/\pi R(4r^2 - b^2)$]

PROBLEM 10 (39P).   A wire whose cross-sectional area is $1.2\,\text{mm}^2$ and whose resistivity is $1.7 \times 10^{-8}\,\Omega \cdot \text{m}$ is bent into a circular arc of radius $r = 24\,\text{cm}$ as shown in Fig. 32–50. An additional straight length of this wire, OP, is free to pivot about O and makes sliding contact with the arc at P. Finally, another straight length of this wire, OQ, completes a loop. The entire arrangement is located in a magnetic field $B = 0.15\,\text{T}$ directed out of the plane of the figure. The straight wire OP starts from rest with $\theta = 0$ and has a constant angular acceleration of $12\,\text{rad/s}^2$. (a) Find the resistance of the loop $OPQO$ as a function of θ.

STRATEGY: Use $R = \rho L/A$, where ρ is the resistivity, L is the length, and A is the cross-sectional area. The length of each straight portion of the loop is r and, if θ is measured in radians, the length of the arc is $r\theta$.

SOLUTION:

$$\left[\,\text{ans:}\ 3.4 \times 10^{-3}(2.0 + \theta),\ \text{in ohms}\,\right]$$

(b) Find the magnetic flux through the loop as a function of θ.

STRATEGY: Use $\Phi = BA$, where A is the area enclosed by the loop. The ratio of the area enclosed by the loop to the area of a circle is $\theta/2\pi$, where θ is in radians.

SOLUTION: .

$$\left[\,\text{ans:}\ 4.3 \times 10^{-3}\theta,\ \text{in webers}\,\right]$$

(c) For what value of θ is the induced current in the loop a maximum?

STRATEGY: The emf is $\mathcal{E} = -d\Phi/dt = -4.3 \times 10^{-3}\omega$, where ω is the angular velocity. Since the wire starts from rest and has constant angular acceleration, $\omega = \alpha t$ and $\theta = \frac{1}{2}\alpha t^2$. You need to find ω as a function of θ. The second kinematic equation gives $t = \sqrt{2\theta/\alpha}$ and the first gives $\omega = \alpha t = \alpha\sqrt{2\theta/\alpha} = \sqrt{2\alpha\theta}$. Thus $\mathcal{E} = 4.3 \times 10^{-3}\sqrt{2\alpha\theta}$. The current is $i = \mathcal{E}/R = 4.3 \times 10^{-3}\sqrt{2\alpha\theta}/3.4 \times 10^{-3}(2.0 + \theta)$. Find the maximum by setting $di/d\theta$ equal to 0.

SOLUTION:

$$\left[\,\text{ans:}\ 2.0\,\text{rad}\,\right]$$

(*d*) What is the maximum value of the induced current in the loop?

STRATEGY: Evaluate $i = \mathcal{E}/R$ for $\theta = 2.0\,\text{rad}$.

SOLUTION:

[ans: 2.2 A]

Some problems ask for the electric field associated with a changing magnetic field. Use Faraday's law in the form $\oint \mathbf{E} \cdot d\mathbf{s} = -d\Phi/dt$. All of the problems have cylindrical symmetry, with the electric field lines forming circles around the cylinder axis. For them, you integrate the tangential component of the electric field around one of the field lines, with result $\oint \mathbf{E} \cdot d\mathbf{s} = 2\pi r E$, where r is the radius of the circle. Find the rate of change of the magnetic flux through the circle and solve the Faraday's law equation for E.

PROBLEM 11 (42P). Early in 1981 the Francis Bitter National Magnet Laboratory at M.I.T. commenced operation of a 3.3-cm diameter cylindrical magnet, which produces a 30-T field, then the world's largest steady state field. The field can be varied sinusoidally between the limits of 29.6 and 30.0 T at a frequency of 15 Hz. When this is done, what is the maximum value of the induced electric field at a radial distance of 1.6 cm from the axis? (*Hint*: See Sample Problem 32–4.)

STRATEGY: Note that the point is closer to the axis than the outer rim of the cylinder. Start with Faraday's law in the form $\oint \mathbf{E} \cdot d\mathbf{s} = -d\Phi/dt$. Take the path for the integral on the left side to be a circle of radius r (= 1.6 cm), which is an electric field line and along which the electric field has uniform magnitude. The integral is $2\pi r E$. The magnetic field is perpendicular to the plane of the circle, so the magnetic flux through it is $\pi r^2 B$. Thus $2\pi r E = -\pi r^2\, dB/dt$ and $E = -\frac{1}{2} r\, dB/dt$. Take $B = B_0 + B_m \sin(2\pi f t)$, where $B_0 = 29.8\,\text{T}$ and $B_m = 0.2\,\text{T}$. This field varies sinusoidally between 29.6 T and 30.0 T and has frequency f. The rate of change of B is given by $dB/dt = 2\pi f B_m \cos(2\pi f t)$ and its maximum value is $(dB/dt)_{\text{max}} = 2\pi f B_m$. Thus the maximum of the induced electric field is $E_{\text{max}} = \pi r f B_m$.

SOLUTION:

[ans: 0.15 V/m]

PROBLEM 12 (43P). Figure 32–52 shows a uniform magnetic field **B** confined to a cylindrical volume of radius R. **B** is decreasing in magnitude at a constant rate of 10 mT/s. What is the instantaneous acceleration (direction and magnitude) experienced by an electron placed at a, at b, and at c? Assume $r = 5.0\,\text{cm}$.

STRATEGY: The electron is at rest so its acceleration is due to the electric field acting on it. The field is induced by the changing magnetic flux through the cylinder. Use Faraday's law in the form $\oint \mathbf{E} \cdot d\mathbf{s} = -d\Phi/dt$ to show that the magnitude of the electric field is given by $E = \frac{1}{2} r\, dB/dt$. Also show that at a the electric field is to the right, at b it is zero, and at c it is to the left. The acceleration of the electron is given by $\mathbf{a} = \mathbf{F}/m = -(e/m)\mathbf{E}$.

SOLUTION:

[ans: at a; $4.4 \times 10^7 \, \text{m/s}^2$, to the right; at b: 0; at c: $4.4 \times 10^7 \, \text{m/s}^2$, to the left]

III. NOTES

Chapter 33
INDUCTANCE

I. BASIC CONCEPTS

Here you learn about inductive circuit elements, which produce emfs via Faraday's law when their currents are changing and which store energy in magnetic fields. Pay attention to the definition of inductance and learn to calculate its value for solenoids and toroids. Learn about the influence of inductance on the current in a circuit and study the calculations of current for series circuits containing inductors. Also learn what factors determine the energy stored in the magnetic field of an inductor and learn how to calculate the stored energy.

Inductance. Current in a circuit produces a magnetic field and magnetic flux through the circuit. Both are proportional to the current. If the circuit consists of N turns and the flux is the same through all of them, then its inductance L is defined by

$$L =$$

where i is the current that produces flux Φ through each turn. L does not depend on the current or its rate of change. In general terms it does depend on

_____

_____

The SI unit of inductance is called the _____ (abbreviated _____). The quantity $N\Phi$ is called the _____.

When the current changes, the flux changes and an emf in induced in the circuit. If the circuit has inductance L the induced emf is given in terms of the rate of change of the current by

$$\mathcal{E}_L =$$

Any circuit has an inductance, usually small, but there are electrical devices, called inductors, that are used expressly for the purpose of adding given amount of inductance to a circuit. They usually consist of a coil of wire, like a solenoid. The symbol for an inductor is

_____ .

A solenoid is an excellent example of an inductor because it concentrates the field in its interior and the magnetic flux through each turn is large. For all practical purposes, the emf induced by a changing current in a circuit containing a solenoid appears in its entirety along the wire of the solenoid.

The diagram shows an inductor carrying current i (the rest of the circuit is not shown). The direction of positive current is from a toward b. If the current is increasing then the emf induced in the inductor is from _____ toward _____ ; if it is decreasing then the emf is from _____ toward _____ .

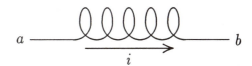

In either case the potential V_b at point b is given by $V_b = V_a - L\,di/dt$, where V_a is the potential at point a and L is the inductance.

For an ideal solenoid of length ℓ and cross-sectional area A, with n turns per unit length and carrying current i, the magnetic field in the interior is given by $B = $ _____, the flux through each turn is given by $\Phi_B = $ _____, and the inductance is given by $L = $ _____. Notice that the inductance depends on the geometry (the length, area, and number of turns per unit length) but not on the current.

Toroids are also often used as inductors. Consider a toroid with N turns, a square cross section, inner radius a, and outer radius b. If the current in the toroid is i the magnetic field a distance r from the center, within the toroid, is $B = $ _____ and the flux through a cross section is $\Phi = $ _____. The inductance is therefore $L = $ _____.

An LR circuit. The diagram on the right shows a series LR circuit. Take the current to be positive in the direction of the arrow and develop the loop equation in terms of L, R, $\mathcal{E}$, i, and di/dt. Take the electric potential to be 0 at point a. Then the potential at point b is _____ and the potential at point c is given by _____. The potential at point a is this value minus iR and the result must, of course, be zero. Thus the loop equation is

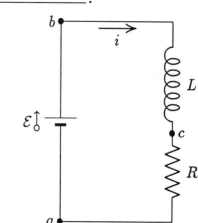

Assume the source of emf $\mathcal{E}$ is connected to the circuit at time t = 0, at which time the current is 0. The solution to the loop equation is then

$$i(t) = $$

Note that this equation predicts $i = 0$ at $t = 0$. It also predicts that long after the emf $\mathcal{E}$ is connected the current is $i = $ _____. The rate at which the current increases to its final value is controlled by the inductive time constant τ_L, which in terms of L and R, is $\tau_L = $ _____. The larger the time constant the slower the rate. A large time constant can be obtained by making _____ large and _____ small. When $t = \tau_L$ the current is about _____ per cent of its final value.

The potential difference across the inductor is given by

$$V_L(t) = L\frac{di}{dt} = $$

and the potential difference across the resistor is given by

$$V_R(t) = iR = $$

On the axes to the left below draw a graph of the current as a function of time. On the time axis mark the approximate position of $t = \tau_L$. On the axes to the right draw graphs of the potential differences across the resistor and the inductor, all as functions of time. Here's some information that might help you: $V_R = V_L$ at $t \approx 0.7\tau_L$ and each of these potential differences have the value $\mathcal{E}/2$ at that time. Label the graphs V_R and V_L, as appropriate.

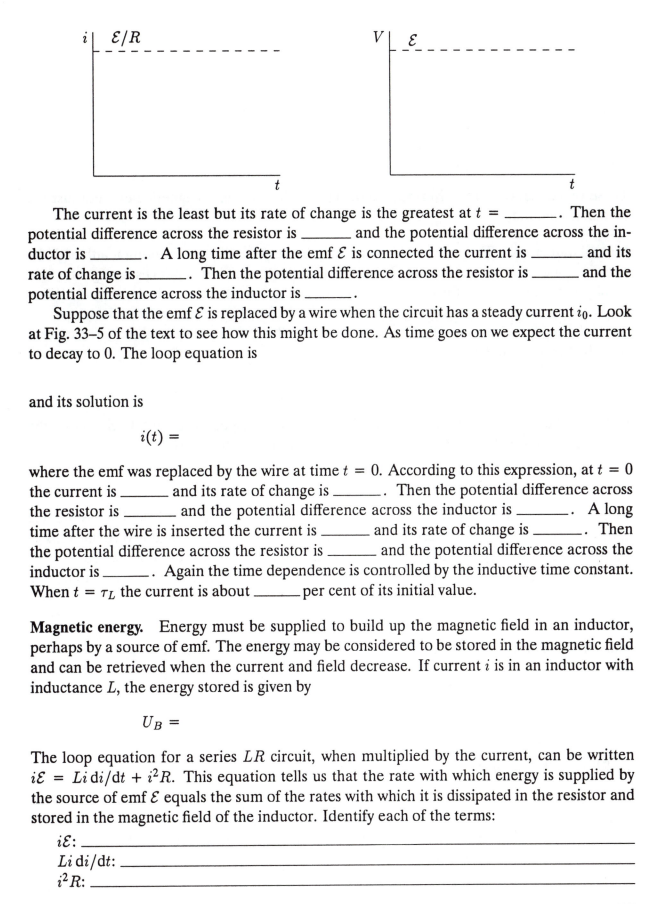

The current is the least but its rate of change is the greatest at $t =$ _____ . Then the potential difference across the resistor is _____ and the potential difference across the inductor is _____ . A long time after the emf $\mathcal{E}$ is connected the current is _____ and its rate of change is _____ . Then the potential difference across the resistor is _____ and the potential difference across the inductor is _____ .

Suppose that the emf $\mathcal{E}$ is replaced by a wire when the circuit has a steady current i_0. Look at Fig. 33–5 of the text to see how this might be done. As time goes on we expect the current to decay to 0. The loop equation is

and its solution is

$$i(t) =$$

where the emf was replaced by the wire at time $t = 0$. According to this expression, at $t = 0$ the current is _____ and its rate of change is _____ . Then the potential difference across the resistor is _____ and the potential difference across the inductor is _____ . A long time after the wire is inserted the current is _____ and its rate of change is _____ . Then the potential difference across the resistor is _____ and the potential difference across the inductor is _____ . Again the time dependence is controlled by the inductive time constant. When $t = \tau_L$ the current is about _____ per cent of its initial value.

Magnetic energy. Energy must be supplied to build up the magnetic field in an inductor, perhaps by a source of emf. The energy may be considered to be stored in the magnetic field and can be retrieved when the current and field decrease. If current i is in an inductor with inductance L, the energy stored is given by

$$U_B =$$

The loop equation for a series LR circuit, when multiplied by the current, can be written $i\mathcal{E} = Li\,di/dt + i^2 R$. This equation tells us that the rate with which energy is supplied by the source of emf $\mathcal{E}$ equals the sum of the rates with which it is dissipated in the resistor and stored in the magnetic field of the inductor. Identify each of the terms:

$i\mathcal{E}$: _____

$Li\,di/dt$: _____

$i^2 R$: _____

The energy density (energy per unit volume) stored in a magnetic field **B** is given by

$$u_B =$$

The total energy stored in a field is given by the volume integral

$$u_B = \frac{1}{2\mu_0} \int B^2 \, dV \, .$$

These two equations are universally valid. They hold for every magnetic field, not just the field of a solenoid.

Mutual induction. The current in one circuit influences the current in another, although the two are not connected electrically. The current in the first circuit produces a magnetic field at all points in space and thus is responsible for a magnetic flux through the second. When the current changes, the flux changes and an emf is induced in the second circuit.

If the second circuit has N_2 turns and Φ_{21} is the flux produced through it by the current i_1 in the first circuit, then

$$M_{21} =$$

is the coefficient of mutual inductance of circuit 2 with respect to circuit 1. Similarly, the coefficient of mutual inductance of circuit 1 with respect to circuit 2 is

$$M_{12} =$$

where N_1 is the number of turns in circuit 1, i_2 is the current in circuit 2, and Φ_{12} is the flux through circuit _____ produced by the current in circuit _____. The two coefficients of mutual inductance M_{12} and M_{21} are always equal and the subscripts are not needed.

When the current i_1 in circuit 1 changes at the rate di_1/dt the emf induced in circuit 2 is

$$\mathcal{E}_2 =$$

and when the current i_2 in circuit 2 changes the emf induced in circuit 1 is

$$\mathcal{E}_1 =$$

II. PROBLEM SOLVING

Some problems ask you to use the definition of self-inductance ($L = N\Phi/i$) to compute one of the quantities that appear in it. To carry out this task you may need to compute the magnetic flux Φ of the magnetic field, perhaps by carrying out an integration. Other problems deal with the emf generated by an inductor when the current changes. Use $\mathcal{E} = -L \, di/dt$.

PROBLEM 1 (2E).   A circular coil has a 10.0-cm radius and consists of 30.0 closely wound turns of wire. An externally produced magnetic field of 2.60 mT is perpendicular to the coil. (*a*) If no current is in the coil, what is the flux linkage?

STRATEGY: The flux linkage is given by $N\Phi = NBA = \pi NBR^2$, where N is the number of turns, B is the magnetic field, and R is the radius of the coil.

SOLUTION:

$$[\text{ans: } 2.45 \times 10^{-3}\,\text{Wb}]$$

(*b*) When the current in the coil is 3.80 A in a certain direction, the net flux through the coil is found to vanish. What is the inductance of the coil?

STRATEGY: The flux linkage due to the current in the coil is given by Li, where L in the inductance and i is the current. This must be the same as the flux linkage due to the external field.

SOLUTION:

$$[\text{ans: } 0.641\,\text{mH}]$$

PROBLEM 2 (5P).   *Inductors in series.* Two inductors L_1 and L_2 are connected in series and are separated by a large distance. (*a*) Show that the equivalent inductance is given by

$$L_{eq} = L_1 + L_2.$$

STRATEGY: The equivalent inductor has the same induced emf as the total emf in the two original inductors when the current is changing at the same rate. The emf induced in inductor 1 is $\mathcal{E}_1 = -L_1\,di/dt$ and the emf induced in inductor 2 is $\mathcal{E}_2 = -L_2\,di/dt$ so the total emf is $\mathcal{E} = \mathcal{E}_1 + \mathcal{E}_2 = -(L_1 + L_2)\,di/dt$. The equivalent inductance is given by $\mathcal{E} = -L_{eq}\,di/dt$. Substitute for $\mathcal{E}$ and calculate L_{eq} in terms of L_1 and L_2.

SOLUTION:

(*b*) Why must their separation be large for this relationship to hold?

STRATEGY: The flux through each inductor was assumed to be due to the current in that inductor only. What has been neglected?

SOLUTION:

$$[\text{ans: expression neglects mutual induction of the coils}]$$

(*c*) What is the generalization of (*a*) for N inductors in series?

SOLUTION:

$$\left[\text{ans: } L_{eq} = \sum_{i=1}^{N} L_i\right]$$

PROBLEM 3 (6P). *Inductors in parallel.* Two inductors L_1 and L_2 are connected in parallel and separated by a large distance. (*a*) Show that the equivalent inductance is given by

$$\frac{1}{L_{eq}} = \frac{1}{L_1} + \frac{1}{L_2}.$$

STRATEGY: The emfs induced in the inductors are the same. The emf induced in the equivalent inductor is the same as that induced in either of the original inductors when the current in the equivalent inductor is the same as the total current in the original inductors. The total current is $i = i_1 + i_2$ and its rate of change is $di/dt = (di_1/dt) + (di_2/dt)$. Since $\mathcal{E} = -L_1 di_1/dt$ and $\mathcal{E} = -L_2 di_2/dt$, $di/dt = -\mathcal{E}[(1/L_1) + (1/L_2)]$. Solve $\mathcal{E} = -L_{eq} di/dt$ for L_{eq}.

SOLUTION:

(*b*) Why must their separation be large for this relationship to hold?

SOLUTION:

(*c*) What is the generalization of (*a*) for N inductors in parallel?

SOLUTION:

$$\left[\text{ans: } \frac{1}{L_{eq}} = \sum_{i=1}^{N} \frac{1}{L_i}\right]$$

PROBLEM 4 (7P). A wide copper strip of width W is bent to form a tube of radius R with two planar extensions, as shown in Fig. 33–14. A current i flows through the strip, distributed uniformly over its width. In this way a "one-turn solenoid" has been formed. (a) Derive an expression for the magnitude of the magnetic field **B** in the tubular part (far away from the edges). (*Hint*: Assume that the magnetic field outside this one-turn solenoid is negligibly small.)

STRATEGY: The derivation follows closely the derivation of the expression for the field inside a solenoid. Draw a cross section parallel to the axis of the tube. Draw an Amperian path in the form of a rectangle, with one side parallel to the tube axis and inside the tube, another side parallel to the tube axis and outside the tube, and the remaining two sides perpendicular to the tube axis. Take the width of the rectangle, as measured along the tube axis, to be h. The integral on the left side of the Ampere's law equation is $\oint \mathbf{B} \cdot d\mathbf{s} = Bh$. The current through the rectangle is ih/W.

SOLUTION:

$$[\text{ans: } \mu_0 i/W]$$

(b) Find the inductance of this one-turn solenoid, neglecting the two planar extensions.

STRATEGY: The flux through the tube is $\Phi = BA = \pi R^2 B$. Substitute for B and use $L = \Phi/i$ to find an expression for L.

SOLUTION:

$$[\text{ans: } \mu_0 \pi R^2/W]$$

PROBLEM 5 (9E).   At a given instant the current and the induced emf in an inductor are as indicated in Fig. 33–15. (a) Is the current increasing or decreasing?

STRATEGY: Notice that the emf is in the same direction as the current. That is, it is tending to increase the current. Remember that an induced emf tends to counteract whatever the current is doing.

SOLUTION:

$$[\text{ans: decreasing}]$$

(b) The emf is 17 V and the rate of change of the current is 25 kA/s; what is the value of the inductance?

STRATEGY: Solve $\mathcal{E} = -L\,di/dt$ for L.

SOLUTION:

$$[\text{ans: } 0.68\,\text{mH}]$$

PROBLEM 6 (12E).   The inductance of a closely wound coil is such that an emf of 3.0 mV is induced when the current changes at the rate of 5.0 A/s. A steady current of 8.0 A produces a magnetic flux of 40 μWb through each turn. (*a*) Calculate the inductance of the coil.

STRATEGY: Solve $\mathcal{E} = -L\,di/dt$ for L.

SOLUTION:

[ans: 0.60 mH]

(*b*) How many turns does the coil have?

STRATEGY: Solve $L = N\Phi/i$ for N.

SOLUTION:

[ans: 120]

Some problems deal with RL series circuits. If a source of emf, such as a battery, is in the circuit and the current is 0 at time $t = 0$ (because, for example, a switch is closed then), the current is given by $i(t) = (\mathcal{E}/R)(1 - e^{-t/\tau_L})$, where the inductive time constant is $\tau_L = L/R$. If the current is i_0 at $t = 0$ and no source of emf is in the circuit then $i = i_0 e^{-t/\tau_L}$.

Other circuits may be considered, in which case you must use Kirchhoff's loop and junction rules to find the differential equation obeyed by the current. In most situations the equation can be solved by comparing it to the equation for a series connection, Eq. 33–15 of the text.

You may be asked to compute the potential difference across the resistor (iR) or across the inductor ($L\,di/dt$). Be sure you can tell which end is at the higher potential. You may also be asked for the rate at which the source of emf is supplying energy ($i\mathcal{E}$), the rate at which the resistor is dissipating energy (i^2R), or the rate at which the inductor is storing energy ($Li\,di/dt$).

PROBLEM 7 (14E).   The current in an RL circuit builds up to one-third of its steady-state value in 5.00 s. Calculate the inductive time constant.

STRATEGY: The current is given by $i(t) = i_m(1 - e^{-t/\tau_L})$, where i_m is the steady-state value and τ_L is the inductive time constant. Solve for τ_L.

SOLUTION: A little rearrangement gives $e^{-t/\tau_L} = 1 - (i/i_m)$. Take the natural logarithm of both sides to obtain $-t/\tau_L = \ln[1 - (i/i_m)]$. Set $t = 5.00$ s and $i/i_m = 1/2$, then solve for τ_L.

[ans: 12.3 s]

PROBLEM 8 (21P).   Suppose the emf of the battery in the circuit of Fig. 33–6 varies with time t so that the current is given by $i(t) = 3.0 + 5.0t$, where i is in amperes and t is in seconds. Take $R = 4.0\,\Omega$, $L = 6.0\,\mathrm{H}$, and find an expression for the battery emf as function of time. (*Hint*: Apply the loop theorem.)

STRATEGY: The loop theorem gives $\mathcal{E} - iR - L\,\mathrm{d}i/\mathrm{d}t = 0$. Substitute $i = 3.0 + 5.0t$ and $\mathrm{d}i/\mathrm{d}t = 5.0$, in SI units.

SOLUTION:

$$\left[\,\text{ans: } 42 + 20t, \text{ in volts for } t \text{ in seconds}\,\right]$$

PROBLEM 9 (26P).   In the circuit shown in Fig. 33–18, $\mathcal{E} = 10\,\mathrm{V}$, $R_1 = 5.0\,\Omega$, $R_2 = 10\,\Omega$, and $L = 5.0\,\mathrm{H}$. For the two separate conditions (I) switch S just closed and (II) switch S closed for a long time, calculate (*a*) the current i_1 through R_1, (*b*) the current i_2 through R_2, (*c*) the current i through the switch, (*d*) the potential difference across R_2, (*e*) the potential difference across L, and (*f*) $\mathrm{d}i_2/\mathrm{d}t$.

STRATEGY: The inductor causes the current i_2 to increase slowly toward its steady-state value, so just after the switch is closed $i_2 = 0$. This means the potential difference across R_2 is zero and the current i through the switch is the same as the current i_1 through R_1. This current is given by $i_1 = \mathcal{E}/R_1$. Since the potential difference across R_2 is zero the potential difference across the inductor must be $\mathcal{E}$, the emf of the battery. Use $\mathcal{E} = -L\,\mathrm{d}i_2/\mathrm{d}t$ to find the rate of change of the current in the inductor.

After the switch has been closed for a long time the currents are at their steady-state values. Since i_2 is not changing with time the potential difference across the inductor is zero. The potential difference across each resistor is $\mathcal{E}$, the emf of the battery. Use $\mathcal{E} = i_1 R_1$ and $\mathcal{E} = i_2 R_2$ to find the currents in the resistors and $i = i_1 + i_2$ to find the current in the switch.

SOLUTION:

$$\left[\,\text{ans: I: } (a)\ 2.0\,\mathrm{A};\ (b)\ 0;\ (c)\ 2.0\,\mathrm{A};\ (d)\ 0;\ (e)\ 10\,\mathrm{V};\ (f)\ 2.0\,\mathrm{A/s}\right.$$
$$\left.\text{II: } (a)\ 2.0\,\mathrm{A};\ (b)\ 1.0\,\mathrm{A};\ (c)\ 3.0\,\mathrm{A};\ (d)\ 10\,\mathrm{V};\ (e)\ 0;\ (f)\ 0\,\right]$$

PROBLEM 10 (28P).   In the circuit shown in Fig. 33–20, switch S is closed at time $t = 0$. Thereafter the constant current source, by varying its emf, maintains a constant current i out of its upper terminal. (*a*) Derive an expression for the current through the inductor as a function of time.

STRATEGY: Let $\mathcal{E}$ be the emf of the constant current source, i_1 be the current in the resistor, and i_2 be the current in the inductor. Take each current to positive if it is downward in the diagram. The loop rule applied to the loop containing the resistor and inductor gives $i_1 R - L\, di_2/dt = 0$. The junction rule gives $i = 1_1 + i_2$. Use the first of these equations to eliminate i_1 from the second, with the result $i_2 R + L\, di_2/dt = iR$. This equation has the same form as Eq. 33–15 of the text: to obtain the equation you just derived simply replace i in the text equation with i_2 and $\mathcal{E}$ with iR. Do the same with the solution (Eq. 33–18 of the text) to find an expression for i_2 as a function of time.

SOLUTION:

$$\left[\text{ans: } i(1 - e^{-Rt/L})\right]$$

(*b*) Show that the current through the resistor equals the current through the inductor at time $t = (L/R)\ln 2$.

STRATEGY: At that time $i_2 = i/2$. Solve $i/2 = i(1 - e^{-Rt/L})$ for t.

SOLUTION: First show that $e^{-Rt/L} = 1/2$, then take the natural logarithm of both sides to obtain $Rt/L = \ln 2$. Solve for t.

PROBLEM 11 (32E).   A coil with an inductance of 2.0 H and a resistance of 10 Ω is suddenly connected to a resistanceless battery with $\mathcal{E} = 100$ V. At 0.10 s after the connection is made, what are the rates at which (*a*) energy is being stored in the magnetic field, (*b*) thermal energy is appearing, and (*c*) energy is being delivered by the battery?

STRATEGY: You may consider the circuit to be a series connection of an ideal resistor and an ideal inductor. If R is the resistance, L is the inductance, and $\mathcal{E}$ is the battery emf, then energy is stored in the magnetic field at the rate $P_L = Li\, di/dt$, thermal energy is produced at the rate $P_R = i^2 R$, and energy is delivered by the battery at the rate $P_{\mathcal{E}} = i\mathcal{E}$. The current is given by $i = (\mathcal{E}/R)(1 - e^{-Rt/L})$.

494   *Chapter 33: Inductance*

SOLUTION:

$[$ ans: (a) 240 W; (b) 150 W; (c) 390 W $]$

Notice that $P_\mathcal{E} = P_R + P_L$.

PROBLEM 12 (33P).   Suppose that the inductive time constant for the circuit of Fig. 33–6 is 37.0 ms and the current in the circuit is zero at time $t = 0$. At what time does the rate at which energy is dissipated in the resistor equal the rate at which energy is being stored in the inductor?

STRATEGY: The rate of energy dissipation in the resistor is i^2R and the rate of energy storage in the inductor is $Li\,di/dt$. You want the time for which $i^2R = Li\,di/dt$, or if i is canceled, when $iR = L\,di/dt$. When $i = (\mathcal{E}/R)(1 - e^{-t/\tau})$ is substituted and L/R is replaced by τ, the result is $e^{-t/\tau} = 1/2$. Take the natural logarithm of both sides to obtain $-t/\tau = \ln 1/2$. Solve for t.

SOLUTION:

$[$ ans: 25.6 ms $]$

PROBLEM 13 (35P).   For the circuit of Fig. 33–6, assume that $\mathcal{E} = 10.0$ V, $R = 6.70\,\Omega$, and $L = 5.50$ H. The battery is connected at time $t = 0$. (a) How much energy is delivered by the battery during the first 2.00 s?

STRATEGY: The rate at which the battery delivers energy is given by $P_\mathcal{E} = i\mathcal{E}$, so the energy delivered is given by $U_\mathcal{E} = \int i\mathcal{E}\,dt$, where the limits of integration are 0 and 2.0 s. The current is given by $i = (\mathcal{E}/R)(1 - e^{-Rt/L})$, so $U_\mathcal{E} = (\mathcal{E}^2/R) \int (1 - e^{-Rt/L})\,dt$. Evaluate the integral.

SOLUTION:

$[$ ans: 18.7 J $]$

(b) How much of this energy is stored in the magnetic field of the inductor?

STRATEGY: Use $U_L = \frac{1}{2}Li^2$.

SOLUTION:

[ans: 5.10 J]

(c) How much of this energy has been dissipated in the resistor?

STRATEGY: The energy supplied by the battery is either dissipated in the resistor or else is stored in the magnetic filed of the inductor: $U_{\mathcal{E}} = U_L + E_R$, where E_R is the energy that has been dissipated in the resistor. Solve for E_R. As a check you might evaluate the integral $E_R = R \int i^2 \, dt = (\mathcal{E}^2/R) \int (1 - 2e^{-Rt/L} + e^{-2Rt/L}) \, dt$.

SOLUTION:

[ans: 13.6 J]

Here is an example of the calculation of the magnetic and electric energy densities.

PROBLEM 14 (45P). A length of copper wire carries a current of 10 A, uniformly distributed. Calculate (a) the energy density of the magnetic field and (b) the energy density of the electric field at the surface of the wire. The wire diameter is 2.5 mm and its resistance per unit length is 3.3 Ω/km.

STRATEGY: The magnetic field at the surface of the wire is given by $B = \mu_0 i/2\pi r$, where r is the radius of the wire and i is the current in the wire. The magnetic energy density is given by $u_B = B^2/2\mu_0$. Consider a length L of wire. The potential difference from end to end is $V = iR$, where R is the resistance. The electric field is $E = V/L = iR/L$. Use $u_E = \epsilon_0 E^2/2$ to compute the electric energy density.

SOLUTION:

[ans: (a) 1.0 J/m^3; (b) 4.8 × 10^{-15} J/m^3]

Some problems deal with the calculation of the coefficient of mutual inductance. Here you assume a current in one circuit and compute the magnetic flux it produces in another circuit.

PROBLEM 15 (50P).   A coil C of N turns is placed around a long solenoid S of radius R and n turns per unit length, as in Fig. 33–22. Show that the coefficient of mutual inductance for the coil-solenoid combination is given by

$$M = \mu_0 \pi R^2 n N .$$

Explain why M does not depend on the shape, size, or possible lack of close-packing of the coil.

STRATEGY: The magnetic field inside the solenoid is uniform, along the axis, and has magnitude $B = \mu_0 in$. The magnetic field outside the solenoid is essentially 0. The flux through the coil is $\Phi = BA = \pi R^2 B$. Notice that the radius of the solenoid is used here since the field is 0 outside. The coefficient of mutual inductance is given by $M = N\Phi/i$. Substitute for Φ.

SOLUTION:

Notice that since the coil is outside the solenoid the geometry of the solenoid, not the coil, determines the coefficient of mutual inductance.

PROBLEM 16 (51P).   Figure 33–23 shows a coil of N_2 turns linked as shown to a toroid of N_1 turns. The toroid's inner radius is a, its outer radius is b and its height is h. Show that the coefficient of mutual inductance M for the toroid-coil combination is

$$M = \frac{\mu_0 N_1 N_2 h}{2\pi} \ln \frac{b}{a} .$$

STRATEGY: The magnetic field a distance r from the center of the toroid is given by $B = \mu_0 N_1 i/2\pi r$. The field is perpendicular to each turn of the coil and the flux through each turn is given by $\Phi = \int B\, dA = (\mu_0 N_1 ih/2\pi) \int_a^b (1/r)\, dr$. Here the infinitesimal area was taken to be $dA = h\, dr$. Notice that the toroid determines the limits of integration because the field is 0 outside the toroid. Carry out the integration, then use $M = N_2\Phi/i$ to find an expression for the mutual inductance.

SOLUTION:

III. NOTES

Chapter 34
MAGNETISM AND MATTER

I. BASIC CONCEPTS

No particles have been observed that produce magnetic fields while at rest. This statement is codified in Gauss' law for magnetism, one of the fundamental laws of electromagnetic theory. You should understand both the mathematical statement and the important ramifications of the law. Magnetic dipoles, not monopoles, are the fundamental magnetic entities responsible for the magnetic properties of matter. You should understand that the motions of electrons in matter produce atomic dipole moments and that the dipole moment of an atom is intimately related to its angular momentum. You should be able to describe the magnetic properties of paramagnetic, diamagnetic, and ferromagnetic materials and you should understand how the behavior of atomic dipoles leads to these properties.

Gauss' law for magnetism. Suppose the normal component of the magnetic field is integrated over a *closed* surface (one that completely surrounds a volume). The result is _____ no matter what closed surface is considered and no matter what the distribution of current. Write the mathematical statement of the law, Eq. 34–13 of the text, here:

The direction of the infinitesimal element of area dA is _____ . and $\mathbf{B} \cdot \mathbf{dA} = B_n \, dA$, where B_n is the normal component of the magnetic field.

The law does not necessarily mean that the magnetic field is zero at any point on the surface, only that the total magnetic flux through any closed surface is zero. This usually happens because the field is essentially outward over some portions and essentially inward over others. No field lines start or stop in the interior of any surface: every line that enters any volume also leaves the volume.

The law can be interpreted to mean that magnetic monopoles do not exist, or at any rate are so rare that their influence has not be detected. Explain what a magnetic monopole is:

_____

_____

If monopoles existed and if a closed surface surrounded one of them the right side of the equation would not be 0. Magnetic field lines would start and stop at monopoles so a net magnetic flux would pass through the surface. On the other hand, magnetic dipoles do exist: a charge circulating around a small loop is an example. They do not violate Gauss' law for magnetism and are, in fact, the fundamental sources of magnetism in matter.

The magnetic field in the exterior of a permanent bar magnet can be closely approximated by the field that would be produced by a positive monopole (a north pole) at one end and a negative monopole (a south pole) at the other but the field does not actually arise from single monopoles but rather from magnetic dipoles associated with electron motion. As proof that the field is not due to monopoles, the magnet can be cut in half with the result that _____ _____.

This process can be continued to the atomic level, with the same result.

Dipole moments and angular momentum. Review some properties of magnetic dipoles:

1. If current i is in a loop that bounds area A the magnitude of its dipole moment is given by $\mu = $ _____. The direction of the moment is given by a right hand rule. Curl the fingers of your right hand in the direction of the _____, then your thumb points in the direction of the _____.

2. On the line defined by the moment the magnitude of the magnetic field produced by a magnetic dipole is given by $B = $ _____, where r is the distance from _____ to _____. If a dipole is in the page with its moment normal to the plane of the page, pointing upward out of the page, then the direction of the magnetic field at points above the dipole is roughly _____ and the direction of the magnetic field at points below the dipole is roughly _____.

3. An external magnetic field **B** exerts a torque on a magnetic dipole. The torque is given by $\tau = $ _____. Its magnitude is given by $\tau = $ _____, where θ is the angle between _____ and _____.

The magnetic dipole moment of a circulating charge is closely related to its angular momentum. Suppose a charge q is traveling with speed v around a circle of radius r. The period of its motion is given by $T = $ _____, the current is given by $i = q/T = $ _____, and the magnitude of its dipole moment is given by $\mu_{orb} = iA = $ _____. If its mass is m then the magnitude of its angular momentum is $L_{orb} = $ _____. In terms of L_{orb}, $\mu_{orb} = $ _____. The vector relationship between μ_{orb} and L_{orb} is:

$$\mu_{orb} = $$

If q is positive then μ_{orb} and L_{orb} are in the same direction; if q is negative they are in opposite directions. Suppose an electron is traveling in a counterclockwise direction around a circular orbit in the plane of the page. The direction of its orbital angular momentum vector is _____ the page and the direction of its magnetic dipole moment is _____ the page.

Every electron has an intrinsic angular momentum, often called its spin angular momentum or simply its spin. It is denoted by **S**. Its magnitude is given by $S = $ _____, where h is the Planck constant. The numerical value of S is _____ J·s. An intrinsic magnetic dipole moment is associated with spin and its magnitude is given in terms of S by

$$\mu_S = $$

Note that the relationship differs by a factor of 2 from the analogous relationship for orbital angular momentum. The dipole moment and spin angular momentum are in opposite directions for electrons.

For a collection of electrons, as in an atom, the individual electron dipole moments sum vectorially to produce the total dipole moment of the atom. For most atoms the individual moments cancel and the total moment is zero; some atoms, however, have a net moment. Whether or not the atoms of a material have net dipole moments or not is important for the determination of the magnetic properties of the material.

Particle and atomic magnetic moments are often measured in units of what is called a Bohr magneton and denoted μ_B. In terms of the Planck constant h and the mass m of an electron, a Bohr magneton is given by μ_B = _____ and its SI value is _____ J/T. In units of a Bohr magneton the magnitude of the intrinsic dipole moment of an electron is _____.

Magnetic dipole moments are also associated with the spins of protons and neutrons. The moments, however, are extremely small compared with that of an electron because _____ _____.

In spite of the smallness of the magnetic moments of protons and neutrons, nuclear magnetism has found important applications in medicine and elsewhere.

The magnetic field of the Earth. The magnetic field of the Earth can be approximated by the field of a magnetic dipole with moment μ = _____ J/T, located at the Earth's center. The dipole is not along the axis of rotation but is tilted slightly from that direction. It points from _____ to _____, so magnetic field lines leave the Earth in the _____ hemisphere and enter it in the _____ hemisphere. Thus the north geomagnetic pole is actually a _____ pole of the Earth's magnetic dipole.

The direction of the Earth's magnetic field is often specified in terms of its declination and inclination. The declination is _____ _____.

The inclination is _____ _____.

If the field has horizontal component B_h and vertical component B_v then the inclination ϕ_i is given by $\tan\phi_i$ = _____ .

The origin of the Earth's magnetic field cannot be explained in detail but the field probably arises from _____ _____ _____.

Magnetization. In words, the magnetization of a uniformly magnetized object is _____ _____ per unit _____ . If the object is not uniformly magnetized you may need the definition of the magnetization at a point. It is the limiting value of the expression you wrote as the volume shrinks to zero around the point.

Some materials, called paramagnetic and diamagnetic, become magnetized only when an external magnetic field is applied; others, called ferromagnetic, can be permanently magnetized. In any event, the dipoles of a magnetized object produce a magnetic field of their own and, to find the total magnetic field at any point, this field must be added vectorially to any applied field that is present. If a magnetic field $\mathbf{B}_0$ is applied to magnetic material the total field is given by $\mathbf{B} = \mathbf{B}_0 + \mathbf{B}_M$, where $\mathbf{B}_M$ is the field produced by the magnetic dipoles of the material. Since the field $\mathbf{B}_M$ depends on the geometry of the sample and the uniformity of the magnetization it is often difficult to calculate.

Paramagnetism. Atoms of paramagnetic materials have permanent dipole moments. When no external field is applied, however, the magnetization is zero because _____
_____ .

When an external field is applied the material becomes magnetized because _____
_____ .

When the applied field is removed the magnetization quickly reduces to zero because _____
_____ .

When an external field is applied the direction of the field produced by dipoles of the material is _____ the direction of the applied field, so the total field in the material is _____ in magnitude than the applied field.

According to Curie's law (Eq. 34–15) the magnetization at any point is directly proportional to the _____ at that point and inversely proportional to the _____ . Symbolically,

$$M =$$

where C is called the _____ constant for the material. The relationship is valid only for small values of B/T. Explain qualitatively why the magnetization increases if the local magnetic field is increased without changing the temperature: _____
_____

Explain qualitatively why the magnetization decreases if the temperature is increased without changing the applied field: _____
_____

For large magnetic fields the magnetization is no longer proportional to the field and, in fact, is less in magnitude than a proportional relationship predicts. For sufficiently high fields the magnetization is independent of the field. The magnetization is then said to be <u>saturated</u>. This occurs when all the dipoles are _____ .

On the axes to the right draw a graph of the magnetization as a function of the magnetic field for a typical paramagnetic material. Label the saturation magnetization M_{max} and indicate the linear region where Curie's law is valid.

M

If the sample contains N dipoles per unit volume, each with dipole moment μ, then the saturation value of the magnetization is given by

B

$$M_{max} =$$

When the orientation of a magnetic dipole changes with respect to a magnetic field its energy changes. The potential energy of a dipole μ in a magnetic field $\mathbf{B}$ is given by $U =$

_____. Suppose a dipole originally aligned with a field is turned end for end. Its initial potential energy is _____, its final potential energy is _____, and the energy required to turn the dipole is _____. To see if the thermal energy of the material is adequate to inhibit paramagnetism this energy is compared with the mean kinetic energy of the atoms of the material. For an ideal gas of atoms this is _____ at absolute temperature T.

Diamagnetism. In the absence of an applied field the atoms of a diamagnetic substance have no magnetic dipole moments, but moments are induced when a field is applied. As an external field is turned on the orbits of the electrons change so the electrons produce an opposing field, in accordance with _____ law. The direction of the magnetization (and the induced dipole moments, on average) is _____ that of the local magnetic field. The total magnetic field in a diamagnetic material is _____ in magnitude than the applied field.

The effect occurs for all materials, but if the atoms have permanent dipole moments the effect of their alignment with the field dominates and the material is _____ rather than _____.

Ferromagnetism. The atoms of ferromagnetic materials have permanent dipole moments but unlike the atomic moments of paramagnetic materials they *spontaneously* align with each other. List some ferromagnetic materials: _____

_____

You should realize that the spontaneous alignment of dipoles in a ferromagnet is *not* due to the magnetic torque exerted by one magnetic dipole on another. These torques are not sufficiently strong to overcome thermal agitation that tends to randomize the dipole directions. The torques that align the dipoles have their source in the quantum mechanics of electrons in solids. Above a certain temperature, called its _____ temperature, a ferromagnetic object becomes paramagnetic. For iron this temperature is _____ K.

Ferromagnetic materials exhibit <u>hysteresis</u>. On the axes to the right draw a curve that shows the field B_M due to atomic dipoles for a typical ferromagnet as an applied field B_0 increases from 0 to B_{0f}. Assume the material is unmagnetized to start. Next draw the curve as the applied field decreases from B_{0f} to 0. The magnetization does not retrace the original curve and it does not become 0 when the applied field is 0. The ferromagnet is magnetized although there is no external field.

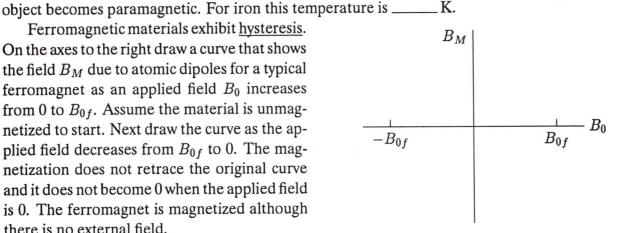

Now suppose the applied field is reversed (has negative values on the graph) and its magnitude is increased to B_{0f}. Draw the curve. Note that a field must be applied opposite to the magnetization in order to reduce the magnetization to 0. Now draw the curve that shows what happens if the applied field is reduced to 0. Finally, draw the curve that shows what happens if the applied field is reversed again and increased to B_{0f}.

Hysteresis is a direct result of the existence of ferromagnetic <u>domains</u>. In a domain the dipole moments are _____ to each other, while dipoles in a neighboring domain are in _____ direction. The field B_M you plotted above is the vector sum of the fields due to all domains. If a ferromagnet is unmagnetized the vector sum of the dipole moments of the domains is _____. When an external field is applied domains with dipoles parallel to the field tend to _____ in size while domains in other directions tend to _____. This amounts to changes in the orientations of dipoles near domain boundaries. Some re-orientation of dipoles within a domain may also take place.

Hysteresis comes about because the growth and shrinkage of domains is not reversible. When an external field is applied to an unmagnetized sample and then turned off the domains do not spontaneously revert to their original sizes.

II. PROBLEM SOLVING

The first problem deals with the alignment of magnetic dipoles as a result of the torques they exert on each other. You should consider this as practice in using the properties of magnetic dipoles, not as an example of how dipoles behave in magnetic materials. In diamagnetic and paramagnetic materials the dipoles are induced or aligned chiefly by an external field. In ferromagnetic materials other torques, not due to the magnetic fields of the dipoles, align them.

PROBLEM 1 (2P).   Figure 34–18 shows four arrangements of a pair of small compass needles, set up in a space in which there is no external magnetic field. Identify the equilibrium in each case as stable or unstable. For each pair consider only the torque acting on one needle due to the magnetic field set up by the other. Explain your answers.

STRATEGY: Take the needles to be magnetic dipoles and look at Fig. 34–3b to see the field lines. In (a) the magnetic field of the left-hand needle points downward at the position of the right-hand needle. If that needle is rotated counterclockwise slightly the direction of the magnetic torque on it, found using $\tau = \mu \times \mathbf{B}$, is into the page, which means the torque tends to turn the dipole clockwise. If the right-hand dipole is rotated clockwise slightly, the magnetic torque tends to turn it counterclockwise. Thus the torque is always a restoring torque and the equilibrium is stable. Carry out similar analyses for the other configurations.

SOLUTION:

[ans: (a) stable; (b) unstable; (c) stable; (d) unstable]

Many problems that deal with Gauss' law for magnetism give you a closed surface such that you can easily calculate the magnetic flux through part of it. You are then asked for the flux through the rest of the surface. You must know that the flux through the complete surface is zero. Here are some an examples.

PROBLEM 2 (8P). A Gaussian surface in the shape of a right circular cylinder has a radius of 12.0 cm and a length of 80.0 cm. Through one end there is an inward magnetic flux of 25.0 μWb. At the other end there is a uniform magnetic field of 1.60 mT, normal to the surface and directed outward. What is the net magnetic flux through the curved surface?

STRATEGY: The two ends and the curved surface together form a closed surface, with a net magnetic flux of zero through it. Write $\Phi_c + \Phi_1 + \Phi_2 = 0$, where Φ_c is the magnetic flux through the curved surface, Φ_1 is the magnetic flux through the first end, and Φ_2 is the magnetic flux through the second end. Φ_1 is given in the problem statement. It is negative since it is inward. At the second end the field is uniform, normal to the end, and outward. This means $\Phi = BA = \pi R^2 B$, where B is the magnitude of the field, A is the area of the end, and R is the radius of the end. Solve for Φ_c.

SOLUTION:

$$[\text{ans: } -4.74 \times 10^{-5}\,\text{Wb}]$$

PROBLEM 3 (9P). Two wires, parallel to the z axis and a distance $4r$ apart, carry equal currents i in opposite directions, as shown in Fig. 34–21. A circular cylinder of radius r and length L has its axis on the z axis, midway between the wires. Use Gauss' law for magnetism to calculate the net outward magnetic flux through the half of the cylindrical surface above the x axis. (*Hint:* Find the flux through that portion of the xz plane that is within the cylinder.)

STRATEGY: According to Gauss' law for magnetism the total flux through the surface of the half-cylinder above the x axis, including the flat portion in the xz plane, is zero. Thus the flux through the rounded surface has the same magnitude as the flux through the flat surface. Consider the magnetic field of the wire on the right side. Along the flat surface it is inwardly normal to the surface and has magnitude $\mu_0 i/2\pi d$, where d is the distance from the wire at a point on the surface. If x is the coordinate of a point on the surface $d = 2r - x$. Divide the surface into strips parallel to the z axis, with length L and width dx. The area of a strip is $L\,dx$ and the flux through the strip at coordinate x is

$$d\Phi = -\frac{\mu_0 i L}{2\pi} \frac{1}{2r - x}\,dx\,.$$

The minus sign appears because the field point inward at the surface. The total flux through the flat surface, due to a single wire is given by the integral

$$\Phi = -\frac{\mu_0 i L}{2\pi} \int_{-r}^{+r} \frac{1}{2r - x}\,dx\,.$$

Carry out the integration. The flux due to the other wire is exactly the same, so you double the result to obtain the total flux.

SOLUTION:

$$[\text{ans: } (\mu_0 i L/\pi)\ln 3]$$

The Earth's magnetic field can be approximated by the field of magnetic dipole located at the center of the Earth. Some problems ask you to use this approximation to calculate the magnitude of the Earth's field or else its inclination angle.

PROBLEM 4 (13P).   The magnetic field of the Earth can be approximated as a dipole magnetic field, with horizontal and vertical components, at a point a distance r from the Earth's center, given by

$$B_h = \frac{\mu_0 \mu}{4\pi r^3} \cos \lambda_m , \qquad B_v = \frac{\mu_0 \mu}{2\pi r^3} \sin \lambda_m ,$$

where λ_m is the *magnetic latitude* (latitude measured from the magnetic equator toward the north or south magnetic pole). Assume that the magnetic dipole moment is $\mu = 8.00 \times 10^{22} \, \text{A} \cdot \text{m}^2$. (*a*) Show that the strength at latitude λ_m is given by

$$B = \frac{\mu_0 \mu}{4\pi r^3} \sqrt{1 + 3 \sin^2 \lambda_m} .$$

STRATEGY: The magnitude of the field is given by $B = \sqrt{B_h^2 + B_v^2}$. After substituting the expressions for B_h and B_v, use $\cos^2 \lambda_m + \sin^2 \lambda_m = 1$ to obtain the result given in the problem statement.

SOLUTION:

(*b*) Show that the inclination ϕ_i of the magnetic field is related to the magnetic latitude λ_m by

$$\tan \phi_i = 2 \tan \lambda_m .$$

STRATEGY: The inclination is given by $\tan \phi_i = B_v / B_h$. Substitute the expressions for B_h and B_v and use $\sin \lambda_m / \cos \lambda_m = \tan \lambda_m$.

SOLUTION:

PROBLEM 5 (16P).   Using the dipole field approximation to the Earth's magnetic field given in Problem 13, calculate the maximum strength of the magnetic field at the core-mantle boundary, which is 2900 km below the Earth's surface.

STRATEGY: According the results of Problem 13 the magnitude of the field is given by

$$B = \frac{\mu_0 \mu}{4\pi r^3} \sqrt{1 + 3 \sin^2 \lambda_m} ,$$

where μ is the magnetic moment ($8.00 \times 10^{22} \, \text{A} \cdot \text{m}^2$), λ_m is the magnetic latitude, and r is the distance from the Earth's center. The maximum magnitude occurs at the magnetic poles, where $\lambda_m = 90°$ and $\sin \lambda_m = 1$. There $B = \mu_0 \mu / 2\pi r^3$. Since the radius of the Earth is $6.37 \times 10^6 \, \text{m}$, $r = 6.37 \times 10^6 - 2.90 \times 10^6 = 3.47 \times 10^6 \, \text{m}$.

SOLUTION:

$$[\text{ans: } 3.83 \times 10^{-4} \, \text{T}]$$

Here are some examples of problems that deal with paramagnetic, diamagnetic, and ferromagnetic materials. To solve them you should understand the concept of magnetization.

PROBLEM 6 (18E).   A 0.50-T magnetic field is applied to a paramagnetic gas whose atoms have an intrinsic magnetic dipole moment of 1.0×10^{-23} J/T. At what temperature will the mean kinetic energy of translation of the gas atoms be equal to the energy required to reverse such a dipole end for end in this magnetic field?

STRATEGY: The mean kinetic energy of the gas atoms is given by $\frac{3}{2}kT$ and the energy needed to reverse a dipole is $2\mu B$. Here k is the Boltzmann constant (1.38×10^{-23} J/K), T is the temperature in kelvins, μ is the atomic dipole moment, and B is the magnetic field. Solve $\frac{3}{2}kT = 2\mu B$ for T.

SOLUTION:

[ans: 0.48 K]

PROBLEM 7 (23P).   A sample of the paramagnetic salt to which the magnetization curve of Fig. 34–11 applies is immersed in a magnetic field of 2.0 T. At what temperature would the degree of magnetic saturation of the sample be (*a*) 50%? (*b*) 90%?

STRATEGY: According to the graph (Fig. 11) $B/T \approx 0.5$ for $M/M_{\text{max}} = 0.5$ and $B/T \approx 1.5$ for $M/M_{\text{max}} = 0.9$. Solve each of these for T.

SOLUTION:

[ans: (*a*) 4.0 K; (*b*) 1.3 K]

PROBLEM 8 (25P).   Consider a solid containing N atoms per unit volume, each atom having a magnetic dipole moment μ. Suppose the direction of μ can be only parallel or antiparallel to an externally applied magnetic field **B** (this will be the case if μ is due to the spin of a single electron). According to statistical mechanics, it can be shown that the probability of an atom being in a state with energy U is proportional to $e^{-U/kT}$, where T is the temperature and k is Boltzmann's constant. Thus, since $U = -\mu \cdot \mathbf{B}$, the fraction of atoms whose dipole moment is parallel to **B** is proportional to $e^{\mu B/kT}$ and the fraction of atoms whose dipole moment is antiparallel to **B** is proportional to $e^{-\mu B/kT}$. (*a*) Show that the magnetization of this solid is $M = N\mu \tanh(\mu B/kT)$. Here tanh is the hyperbolic tangent function: $\tanh(x) = (e^x - e^{-x})/(e^x + e^{-x})$.

STRATEGY: Let N_p be the number of atoms per unit volume with their dipole moments parallel to the magnetic field and N_a be the number per unit volume with their dipole moments antiparallel to the magnetic field. The magnetization (net dipole moment per unit volume) is $M = \mu(N_p - N_a)$. Now $N_p = Ae^{\mu B/kT}$ and $N_a = Ae^{-\mu B/kT}$, where A is a constant of proportionality. The condition $N_p + N_a = N$ can be used to find A. It is $A = N/(e^{\mu B/kT} + e^{-\mu B/kT})$. Make these substitutions in the expression for M and note that the combination of exponentials that occurs can be replaced by $\tanh(\mu B/kT)$.

SOLUTION:

(*b*) Show that the result given in (*a*) reduces to $M = N\mu^2 B/kT$ for $\mu B \ll kT$.

STRATEGY: Let $x = \mu B/kT$. For x very small e^x may be replaced by $1 + x$ and e^{-x} may be replaced by $1 - x$. Then $\tanh(x) \approx (1 + x - 1 + x)/(1 + x + 1 - x) = x$. Replace $\tanh(\mu B/kT)$ with $\mu B/kT$ in the expression for M.

SOLUTION:

(c) Show that the result of (a) reduces to $M = N\mu$ for $\mu B \gg kT$.

STRATEGY: Let $x = \mu B/kT$. For x large e^x is large and e^{-x} is small. The second type exponential can be neglected in the definition of the hyperbolic tangent. Thus $\tan(x) \approx e^x/e^x = 1$. Replace $\tanh(\mu B/kT)$ with 1 in the expression for M.

SOLUTION:

(d) Show that (b) and (c) agree qualitatively with Fig. 34–11.

STRATEGY: Notice that Curie's law is obeyed if $\mu B \ll kT$. The magnetization is proportional to the applied field and inversely proportional to the temperature. If $\mu B \gg kT$ the magnetization is independent of both the applied field and the temperature. All the dipoles are aligned and the magnetization is saturated.

SOLUTION:

PROBLEM 9 (27P).   An electron of mass m and charge magnitude e moves in a circular orbit of radius r about a nucleus. A magnetic field **B** is then established perpendicular to the plane of the orbit. Assuming that the radius of the orbit does not change and that the change in the speed of the electron due to the field **B** is small, find an expression for the change in the orbital magnetic moment of the electron.

STRATEGY: The orbital magnetic moment is given by $\mu = \pi r^2 e/T = evr/2$, where T is the period of the motion and v is the electron's speed. The substitution $T = 2\pi r/v$ was made. Suppose that when the magnetic field is turned on the speed changes from v_0 to $v_0 + \Delta v$. The dipole moment changes from $ev_0 r/2$ to $e(v_0 + \Delta v)r/2$, so the change in the dipole moment is $\Delta\mu = \frac{1}{2}er\,\Delta v$.

Before the field is turned on the electron is held in its orbit by the electrical force of the nucleus. Let F represent the magnitude of this force (given by Coulomb's law). Newton's second law becomes $F = mv_0^2/r$. When the magnetic field is turned on an additional force, of magnitude evB, acts on the electron and $F + evB = mv^2/r$. In addition, $v = v_0 + \Delta v$. Make the substitution for v. Since Δv is small compared to v_0, $(v_0 + \Delta v)^2$ can be approximated by $v_0^2 + 2v\,\Delta v$ and the Newton's second law equation becomes $F + e(v_0 + \Delta v))B = (mv_0^2/r) + (2mv_0/r)\,\Delta v$. Notice that F and mv_0^2/r cancel and the equation becomes $e(v_0 + \Delta v)B = (2mv_0/r)\,\Delta v$. Δv can also be neglected on the left side and you may write $ev_0 B = (2mv_0/r)\,\Delta v$. Solve for Δv and substitute the expression you find into the equation for $\Delta\mu$.

SOLUTION:

[ans: $r^2 e^2 B/4m$]

You should be able to show that $\Delta\mu$ is directed opposite to **B**. The magnetization of a diamagnetic material can be found by multiplying $\Delta\mu$ by the number of diamagnetic atoms per unit volume.

PROBLEM 10 (29E). The exchange coupling mentioned in Section 34–8 as being responsible for ferromagnetism is *not* the mutual magnetic interaction between two elementary magnetic dipoles. To show this calculate (*a*) the magnetic field a distance 10 nm away along the dipole axis from an atom with magnetic dipole moment 1.5×10^{-23} J/T (cobalt), and (*b*) the minimum energy required to turn a second identical dipole end for end in this field.

STRATEGY: The magnetic field of a dipole at a point along its axis is given by $B = \mu_0 \mu/2\pi r^3$, where μ is the dipole moment and r is the distance from the dipole to the field point. Substitute the given values. The energy required to turn the dipole end for end, starting with the dipole aligned with the field, is $\Delta U = 2\mu B$.

SOLUTION:

[ans: (*a*) 3.0×10^{-6} T; (*b*) 9.0×10^{-29} J (5.6×10^{-10} eV)]

Compare with the results of Sample Problem 34–4. What do you conclude?

STRATEGY: According to the results of Sample Problem 34–4 the average translational energy of atoms in an ideal gas at room temperature is 6.2×10^{-21} J.

SOLUTION:

[ans: there would be no possibility of ferromagnetism at room temperature]

PROBLEM 11 (31E). The dipole moment associated with an atom of iron in an iron bar is 2.1×10^{-23} J/T. Assume that all the atoms in the bar, which is 5.0 cm long and has a cross-sectional area of 1.0 cm^2, have their dipole moments aligned. (*a*) What is the dipole moment of the bar? (*b*) What torque must be exerted to hold this magnet at right angles to an external field of 1.5 T? The density of iron is 7.9 g/cm^3.

STRATEGY: First find the number of atoms in the bar. The atomic mass of iron is 55.847 g/mol (see Appendix D), so the mass of a single atom is $55.847/6.02 \times 10^{23} = 9.27 \times 10^{-23}$ g. The number of atoms per unit volume is the density of iron divided by the mass of a single iron atom and the number of atoms in the bar is the number per unit volume multiplied by the volume of the bar. You should get 4.26×10^{23} atoms. The dipole moment of the bar is the product of the dipole moment of a single atom and the number of atoms in the bar. The magnitude of the torque exerted on the bar by a magnetic field that is perpendicular to it is given by $\tau = \mu B$, where μ is the dipole moment of the bar.

SOLUTION:

$$[\,\text{ans:}\ \ (a)\ 8.9\,\text{A}\cdot\text{m}^2;\ (b)\ 13\,\text{N}\cdot\text{m}\,]$$

III. NOTES

Chapter 35
ELECTROMAGNETIC OSCILLATIONS

I. BASIC CONCEPTS

You will study three important electrical circuits, all closely related to each other. In each case the combination of a capacitor and an inductor connected in series produces a sinusoidally varying current. The basic circuit consists of a capacitor-inductor combination only and for it the current amplitude remains constant. Learn what factors determine the frequency of oscillation. When a resistor is added the amplitude decays exponentially with time. When a sinusoidally varying emf is added the amplitude is again constant but it is small unless the frequency of the emf matches the natural frequency of the circuit.

Energy is an important theme of this chapter. Pay careful attention to the form it takes, sometimes as energy stored in the electric field of the capacitor and sometimes as energy stored in the magnetic field of the inductor. In the first circuit energy is conserved but when a resistor is added it is dissipated as thermal energy. At resonance a sinusoidal emf replaces the energy dissipated in the resistor.

An LC circuit. The circuit shown in the diagram to the right consists of an inductor with inductance L and a capacitor with capacitance C, connected in series. As you will see, the charge on either plate of the capacitor oscillates sinusoidally, being positive for half a cycle and negative for the other half. The current also oscillates, being in one direction for half a cycle and in the other direction for the other half cycle. Finally the energy also oscillates between the capacitor and the inductor. You should understand the behavior of the charge, current, and energy.

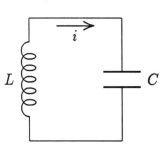

The first step is to develop and solve the loop equation for the circuit. Let q represent the charge on the upper plate of the capacitor and take the current to be positive in the direction of the arrow. The potential difference across the capacitor is given by _____, with the _____ plate at the higher potential if q is positive. The potential difference across the inductor is given by _____, with the _____ end at the higher potential if i is increasing. In terms of q and di/dt the loop equation is _____. Substitute $i = dq/dt$ to obtain the loop equation in terms of q:

Suppose that at time $t = 0$ the current is zero and charge Q is on the capacitor, with the upper plate positive. Then the solution to the loop equation is

$$q(t) =$$

and the current is given by

$$i(t) = \frac{dq}{dt} = $$

Both q and i oscillate with an angular frequency that is determined by the values of L and C: $\omega =$ _____ . Solutions with other phase constants, of the form $q(t) = Q\cos(\omega t + \phi)$, are valid for other initial conditions. Initial conditions, however, do not influence ω.

No matter what the value of the phase constant, the maximum magnitude I of the current is _____ and $i = \pm I$ when $q =$ _____ . When q is either $+Q$ or $-Q$ the current is _____ .

Energy is stored in the electric field of the capacitor and in the magnetic field of the inductor. As functions of time the energy in the capacitor is given by

$$U_E(t) = \frac{1}{2}\frac{q^2}{C} = $$

and the energy in the inductor is

$$U_B(t) = \frac{1}{2}Li^2 = $$

Note that the energy oscillates back and forth between the inductor and the capacitor. The energy is entirely magnetic when the _____ is a maximum and the _____ is zero; it is entirely electric when the _____ is a maximum and the _____ is zero. The sum, $U = U_E + U_B$, is constant and can be written in terms of the current amplitude I as $U =$ _____ or in terms of the charge amplitude Q as $U =$ _____ .

Study Fig. 35–1 carefully. In (a) the sign of the charge on the upper plate of the capacitor is _____ and the current through the inductor is _____ . At this stage all the energy is stored in the _____ field of the _____ . As soon as this situation is set up charge begins to flow. Why? _____

_____

In (b) the charge on the upper plate of the capacitor is _____ (the current is away from that plate). A magnetic field is building up in the inductor and the energy is shared between the inductor and capacitor. In (c) the discharge of the capacitor has been completed. The current and the magnetic field in the inductor are now at their maximum values and all the energy is stored in the _____ field of the _____ .

Since the current continues as before positive charge collects on the _____ plate of the capacitor. Because the charge on the plate repels other charge of the same sign the current in (d) is _____ . This means the magnetic field and the energy stored in the inductor are _____ . In (e) the capacitor is maximally charged but now the _____ plate is positive. The current is _____ and all energy is stored in the _____ . Now the capacitor begins discharging. In (f) charge on the capacitor is _____ , the current is _____ , and its direction is _____ through the inductor. The energy is again shared by the capacitor and inductor. In (g) the discharge is completed, the current is a maximum, and the energy is all stored in the _____ . The current continues and in (h) the capacitor is again charging, this time with the _____ plate positive. The current is _____ and the energy is shared. In (a) the capacitor is fully charged, the current is zero, and the cycle begins again.

Electromagnetic oscillations of an LC circuit are quite similar to the mechanical oscillations of a block-spring system. If a block of mass m moves in one dimension on a frictionless horizontal surface, acted on by a spring with spring constant k, its coordinate obeys the differential equation

$$m\frac{d^2}{dt^2} + kx = 0.$$

This can be changed into the differential equation for the charge on a capacitor plate in an LC series circuit by making the following replacements: $x \rightarrow$ _____, $m \rightarrow$ _____, and $k \rightarrow$ _____. Although the velocity of the block does not appear in the differential equation, in other equations it is replaced by _____.

The potential energy stored in the spring is given by $U = \frac{1}{2}kx^2$. When the replacements are made this becomes _____, which is the energy stored in the _____. The kinetic energy of the block is given by $K = \frac{1}{2}mv^2$. When the replacements are made this becomes _____, which is the energy stored in the _____.

The coordinate and velocity of the block oscillate with a phase difference of 90° just like the charge and current in the electrical circuit. The energy of the block-spring system remains constant but changes from potential to kinetic and back again. The energy of the LC circuit also remains constant. It changes from electrical energy stored in the capacitor to magnetic energy stored in the inductor and back again.

Such analogies between mechanical and electrical systems help us think about the systems. Since you have studied the block-spring system in detail you can use your knowledge to help answer questions about an LC series circuit. Engineers who need to study complicated mechanical systems often build the analogous electrical circuit (at a savings in cost) and study it experimentally.

RLC circuits. If a resistor with resistance R is added in series to the LC circuit shown above, the charge and current amplitudes decay exponentially with time. The loop equation becomes

and its general solution is

$$q(t) =$$

where the angular frequency of oscillation is now $\omega' =$ _____. Each time the charge on the capacitor reaches a maximum it is less than the previous time. Does this violate the principle of charge conservation? _____ Explain: _____

If the resistance is small, so it may be neglected in the expression for the angular frequency, then $\omega' =$ _____.

The sum of the electric and magnetic energies is not constant but decreases exponentially because _____

_____.

In terms of the current i and resistance R the rate of change of the energy is given by

$$\frac{dU}{dt} =$$

Suppose a sinusoidal emf is added in series to the RLC circuit. Take the emf to be $\mathcal{E} = \mathcal{E}_m \sin(\omega t)$. After transients die out the current is given as a function of time by

$$i(t) =$$

where I is the magnitude of the maximum current. Notice that the angular frequency of the current is the same as that of the source and is not the natural angular frequency ($\omega_0 = 1/\sqrt{LC}$). The current amplitude I depends on the difference between the imposed angular frequency ω and the natural angular frequency ω_0. See Fig. 35–6. As the imposed frequency approaches the natural frequency from either side the current amplitude _____ and reaches its maximum value when ω = _____. This is the condition for _____. Notice that the resonance peak can be made sharper by reducing the _____ in the circuit.

You should realize that at resonance energy is being dissipated in the resistor but it is being supplied by _____ at the same rate.

II. PROBLEM SOLVING

You should know the sequence of events in an LC circuit oscillation:

1. capacitor maximally charged (plate A positive, say), current zero

2. capacitor uncharged, maximum current in one direction

3. capacitor maximally charged (plate B positive), current zero

4. capacitor uncharged, maximum current in the other direction

The cycle then repeats. These events are one-quarter cycle apart ($T/4$, where T is the period of the oscillation). You may need to use $T = 2\pi/\omega$, where ω ($= 1/\sqrt{LC}$) is the angular frequency of the oscillation. Also remember that the electrical energy stored in the capacitor is a maximum when the charge on the capacitor is a maximum and the magnetic energy stored in the inductor is a maximum when the current is a maximum.

PROBLEM 1 (5E).   For a certain LC circuit the total energy is converted from electrical energy in the capacitor to magnetic energy in the inductor in $1.50\,\mu s$. (a) What is the period of oscillation?

STRATEGY: As Fig. 35–1 shows, the given time interval is one-fourth of a cycle, so the period is 4 times as long as the interval.

SOLUTION:

[ans: $6.00\,\mu s$]

(b) What is the frequency of oscillation?

STRATEGY: Use $f = 1/T$, where T is the period.

SOLUTION:

[ans: 1.67×10^5 Hz]

(c) How long after the magnetic energy is a maximum will it be a maximum again?

STRATEGY: The magnetic energy is a maximum every half cycle. At the two times each cycle when the magnetic energy is maximum the current is in opposite directions.

SOLUTION:

[ans: $3.00\,\mu$s]

PROBLEM 2 (6P).   The frequency of oscillation of a certain LC circuit is 200 kHz. At time $t = 0$, plate A of the capacitor has maximum positive charge. At what times $t > 0$ will (a) plate A again have maximum positive charge, (b) the other plate of the capacitor have maximum positive charge, and (c) the inductor have maximum magnetic field?

STRATEGY: Look at Fig. 35–1. (a) Plate A has maximum positive charge once every cycle. Use $T = 1/f$, where f is the frequency of oscillation, to find the period. (b) The other plate has maximum positive charge a half cycle later. (c) The magnetic field of the inductor is a maximum one quarter cycle later.

SOLUTION:

[ans: (a) $5.00\,\mu$s; (b) $2.50\,\mu$s; (c) $1.25\,\mu$s]

Some problems deal with the analogy between an LC electrical circuit and a block-spring mechanical system. Remember that $q \leftrightarrow x$, $L \leftrightarrow m$, $C \leftrightarrow 1/k$, and $i \leftrightarrow v$.

PROBLEM 3 (8P).   The energy in an LC circuit containing a 1.25-H inductor is $5.70\,\mu$J. The maximum charge on the capacitor is $175\,\mu$C. Find (a) the mass, (b) the spring constant, (c) the maximum displacement, and (d) the maximum speed for the analogous mechanical system.

STRATEGY: (a) The mass m is analogous to the inductance. (b) The spring constant is analogous to the reciprocal of the capacitance. Use $U = Q^2/2C$ to find the capacitance C. Here U is the energy and Q is the maximum charge on the capacitor. (c) The maximum displacement is analogous to the maximum charge. (d) The maximum speed is analogous to the maximum current. Use $U = LI^2/2$ to find the maximum current I.

SOLUTION:

[ans: (a) 1.25 kg; (b) 372 N/m; (c) $175\,\mu$m; (d) 3.02 m/s]

You should know the relationship between the maximum charge on the capacitor and the maximum current: $I = \omega Q$. You should also know that the total energy of the circuit can be written in terms of either of these: $U = Q^2/2C = LI^2/2$. Some problems ask you to compute the maximum charge on the capacitor, given the maximum potential difference across the plates: $Q = CV_{max}$.

PROBLEM 4 (15P). An oscillating LC circuit consisting of a 1.0-nF capacitor and a 3.0-mH coil has a peak voltage of 3.0 V. (*a*) What is the maximum charge on the capacitor?

STRATEGY: Evaluate $Q = CV$, where V is the peak voltage and C is the capacitance.

SOLUTION:

$$[\text{ans: } 3.0\,\text{nC}]$$

(*b*) What is the peak current through the circuit?

STRATEGY: Use $I = \omega Q$, where $\omega\ (= 1/\sqrt{LC})$ is the angular frequency of oscillation and Q is the maximum charge on the capacitor.

SOLUTION:

$$[\text{ans: } 1.7\,\text{mA}]$$

(*c*) What is the maximum energy stored in the magnetic field of the coil?

STRATEGY: Evaluate $U_B = LI^2/2$.

SOLUTION:

$$[\text{ans: } 4.5 \times 10^{-9}\,\text{J}]$$

PROBLEM 5 (17P). In an LC circuit in which $C = 4.00\,\mu\text{F}$ the maximum potential difference across the capacitor during the oscillations is 1.50 V and the maximum current through the inductor is 50.0 mA. (*a*) What is the inductance L?

STRATEGY: $Q = CV$, where C is the capacitance and V is the maximum potential difference across the capacitor, gives the maximum charge on the capacitor. The maximum value of the current is given by $I = \omega Q$, where $\omega\ (= 1/\sqrt{LC})$ is the angular frequency of oscillation. Thus $I = CV/\sqrt{LC}$. Solve for L.

SOLUTION:

$$[\text{ans: } 3.60\,\text{mH}]$$

(*b*) What is the frequency of the oscillation?

STRATEGY: Use $f = \omega/2\pi = 1/2\pi\sqrt{LC}$.

SOLUTION:

$$[\text{ans: } 1.33 \times 10^3\,\text{Hz}]$$

(*c*) How much time does it take for the charge on the capacitor to rise from zero to its maximum value?

STRATEGY: This is one-fourth of a cycle. Use $T = 1/f$ to calculate the period.

SOLUTION:

$$[\text{ans: } 1.88 \times 10^{-4}\,\text{s}]$$

PROBLEM 6 (18P). In the circuit shown in Fig. 35–12 the switch has been in position *a* for a long time. It is now thrown to position *b*. (*a*) Calculate the frequency of the resulting oscillating current.

STRATEGY: Once the switch is thrown the circuit becomes an *LC* series circuit. Use $\omega = 1/\sqrt{LC}$ and $f = \omega/2\pi$.

SOLUTION:

$$[\text{ans: } 275\,\text{Hz}]$$

(*b*) What is the amplitude of the current oscillations?

STRATEGY: Use $I = \omega Q$, where Q is the maximum charge on the capacitor. This is the charge just after the switch is thrown since the current is zero then. Q is given by $C\mathcal{E}$, where $\mathcal{E}$ is the emf of the battery used to charge the capacitor.

SOLUTION:

$$[\text{ans: } 364\,\text{mA}]$$

PROBLEM 7 (20P). An *LC* circuit oscillates at a frequency of 10.4 kHz. (*a*) If the capacitance is 340 μF, what is the inductance?

STRATEGY: Solve $f = \omega/2\pi = 1/2\pi\sqrt{LC}$ for L.

SOLUTION:

$$[\text{ans: } 0.689\,\mu\text{H}]$$

(*b*) If the maximum current is 7.20 mA, what is the total energy in the circuit?

STRATEGY: Evaluate $U = LI^2/2$.

SOLUTION:

$$[\text{ans: } 1.79 \times 10^{-11}\,\text{J}]$$

(*c*) Calculate the maximum charge on the capacitor.

STRATEGY: Solve $U = Q^2/2C$ for Q.

$$[\text{ans: } 1.10 \times 10^{-7}\,\text{C}]$$

An *LC* circuit may be started (at $t = 0$) with the capacitor less than fully charged and with current in the circuit. If the expression for the charge on the capacitor as a function of time is written $q = Q\cos(\omega t + \phi)$, the phase angle ϕ is not zero. The phase angle is determined by the initial values of the charge and current. The initial charge is given by $q_0 = Q\cos\phi$ and the initial current is given by $i_0 = -I\sin\phi$, where I is the maximum current ($I = \omega Q$). Thus $\tan\phi = -i_0 Q/Iq_0$. Remember that this equation has two solutions for ϕ, the one your calculator gives you and one that is 180° greater. You must be sure that the one you pick gives the correct signs for q_0 and i_0.

PROBLEM 8 (25P).   In an *LC* circuit $L = 25.0\,\text{mH}$ and $C = 7.80\,\mu\text{F}$. At time $t = 0$ the current is 9.20 mA, the charge on the capacitor is $3.80\,\mu\text{C}$, and the capacitor is charging. (*a*) What is the total energy in the circuit?

STRATEGY: There is energy in both the capacitor and the inductor. Use $U = (q^2/2C) + (Li^2/2)$ to calculate the total.

SOLUTION:

$$[\text{ans: } 1.98\,\mu\text{J}]$$

(*b*) What is the maximum charge on the capacitor?

STRATEGY: Solve $U = Q^2/2C$ for Q.

SOLUTION:

$$[\text{ans: } 5.56\,\mu\text{C}]$$

(*c*) What is the maximum current?

STRATEGY: Solve $U = LI^2/2$ for I.

SOLUTION:

$$[\text{ans: } 12.6\,\text{mA}]$$

(*d*) If the charge on the capacitor is given by $q = Q\cos(\omega t + \phi)$, what is the phase angle ϕ?

STRATEGY: The initial charge (at $t = 0$) is given by $q_0 = Q\cos\phi$. The current at any time t is given by $i = dq/dt = -\omega Q\sin(\omega t + \phi) = -I\sin(\omega t + \phi)$ and the current at $t = 0$ is $i_0 = -I\sin\phi$, where $I\ (= \omega Q)$ is the maximum current. Solve the first equation for $\cos\phi$ and second for $\sin\phi$, then divide one by the other to obtain $\tan\phi = -i_0 Q/Iq_0$. Evaluate this expression, then find the inverse tangent. You substitute $+9.20 \times 10^{-3}\,\text{A}$ for i_0. Check to be sure that $\cos\phi$ is positive and $\sin\phi$ is negative. This means the charge is initially positive and increasing in magnitude.

SOLUTION:

$$[\text{ans: } -46.9°\ (-0.819\,\text{rad})]$$

(e) Suppose the data are the same, except that the capacitor is discharging at $t = 0$. What then is the phase angle ϕ?

STRATEGY: The calculation is the same and you obtain $\tan\phi = -i_0 Q/Iq_0$, as before. Now you must substitute -9.20×10^{-3} A for i_0 and you must pick the value for ϕ such that both $\cos\phi$ and $\sin\phi$ are positive.

SOLUTION:

$$[\text{ans: } +46.9° \ (+0.819\,\text{rad})]$$

Some problems deal with damped circuits. You should know that the frequency of oscillation changes with the addition of resistance and that in succeeding cycles the maximum charge on the capacitor decreases exponentially, the relevant factor being $e^{-Rt/2L}$.

PROBLEM 9 (34P). A single-loop circuit consists of a 7.20-Ω resistor, a 12.0-H inductor, and a 3.20-μF capacitor. Initially the capacitor has a charge of 6.20 μC and the current is zero. Calculate the charge on the capacitor N complete cycles later for $N = 5$, 10, and 100.

STRATEGY: The charge on the capacitor at the end of each cycle is given by $q = Q e^{-Rt/2L}$, where R is the resistance, L is the inductance, and Q is the maximum charge, which in this case is the same as the initial charge. The time t is given by $t = NT$, where T is the period of the oscillations. Use $T = 2\pi/\omega$, with $\omega = \sqrt{(1/LC) - (R/2L)^2}$.

SOLUTION:

$$[\text{ans: } N = 5: 5.85\,\mu\text{C}; \ N = 10: 5.52\,\mu\text{C}; \ N = 100: 1.93\,\mu\text{C}]$$

PROBLEM 10 (35P). *(a)* By direct substitution of Eq. 35–18 into Eq. 35–17, show that

$$\omega' = \sqrt{(1/LC) - (R/2L)^2}.$$

STRATEGY: You need to differentiate $q = Q e^{-RT/2L} \cos(\omega' + \phi)$ twice with respect to t. You should obtain

$$\frac{dq}{dt} = -\frac{RQ}{2L} e^{-Rt/2L} \cos(\omega't + \phi) - Q\omega' e^{-Rt/2L} \sin(\omega't + \phi)$$

and

$$\frac{d^2q}{dt^2} = \frac{R^2Q}{4l^2} e^{-Rt/2L} \cos(\omega't + \phi) + \frac{2RQ\omega'}{2L} e^{-Rt/2L} \sin(\omega't + \phi) - Q(\omega')^2 e^{-Rt/2L} \cos(\omega't + \phi).$$

Substitute into

$$L\frac{d^2q}{dt^2} + R\frac{dq}{dt} + \frac{1}{C}q = 0.$$

Notice that all the terms containing $\sin(\omega't + \phi)$ cancel. Solve for ω'.

SOLUTION:

(b) By what fraction does the frequency of oscillation shift when the resistance is increased from 0 to $100\,\Omega$ in a circuit with $L = 4.40\,\mathrm{H}$ and $C = 7.30\,\mu\mathrm{F}$?

STRATEGY: Calculate $\omega = \sqrt{1/LC}$ and $\omega' = \sqrt{(1/LC) - (R/2L)^2}$. The fractional shift is $(\omega' - \omega)/\omega$.

SOLUTION:

$\left[\text{ans: } 2.10 \times 10^{-3}\right]$

III. NOTES

Chapter 36
ALTERNATING CURRENTS

I. BASIC CONCEPTS

Here you study in detail a series circuit consisting of a resistor, an inductor, and a capacitor, driven by a sinusoidal emf. The current is an alternating current (ac): it is sinusoidal and periodically reverses direction. Pay close attention to the relationship between the potential difference across each circuit element and the current in the element. Learn how to compute the current amplitude and phase in terms of the generator emf. You will be able to apply what you learn to many other circuits.

An AC circuit. The circuit you consider is diagramed on the right. The ac generator is symbolized by _____ and the emf it produces is given by $\mathcal{E} = \mathcal{E}_m \sin(\omega t)$, where ω is its angular frequency. Its frequency is given by $f =$ _____. $\mathcal{E}$ is taken to be positive if the upper terminal of the generator is positive and negative if the upper terminal is negative. The arrow on the diagram shows the direction of positive current.

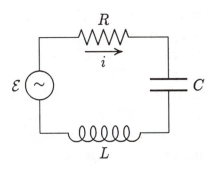

After transients die out the current is given by $i(t) = I \sin(\omega t - \phi)$, where I is the current amplitude and ϕ is a phase constant. Note that the current has the same frequency as the generator emf but because the circuit contains a capacitor and an inductor it may not be in phase with the generator emf. If ϕ is between 0 and 90° the current is said to _____ the emf. If it is between 0 and −90° the current _____ the emf.

Quantities with sinusoidal time dependence, such as the generator emf and the current, can be represented by <u>phasors</u>. Tell what a phasor is: _____

_____

The diagram on the right shows the phasor associated with the generator emf at some instant of time. Its length is $\mathcal{E}_m$ and it rotates in the counterclockwise direction. Its projection on the vertical axis is $\mathcal{E}(t) = \mathcal{E}_m \sin(\omega t)$. Label the appropriate angle ωt. Assume ϕ is about 40° and draw the phasor associated with the current. Label its length I. Label the angle $\omega t - \phi$ and the angle ϕ between the two phasors.

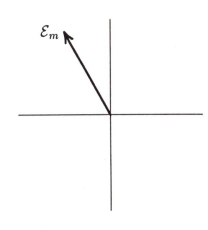

The potential difference v_R across the resistor is given by $v_R = iR = IR \sin(\omega t - \phi)$; v_R and i are in phase. The phasor associated with v_R is on the same line as the phasor associated with _____ and its length is $V_R =$ _____.

The potential difference v_L across the inductor is given by $v_L = L\,di/dt = \omega LI\cos(\omega t - \phi)$. Use the trigonometric identity $\sin(A + \pi/2) = \cos(A)$ to write this $v_L = \omega LI\sin(\omega t - \phi + \pi/2)$. v_L is said to lead the current by $\pi/2$ radians. Its amplitude is given by $V_L = $ _____. When i is positive and increasing the _____ end of the inductor in the circuit diagram is at a higher potential than the _____ end.

The phasor associated with v_L is $\pi/2$ radians ahead of (more counterclockwise than) the phasor associated with the current. On the axes to the right draw the phasors associated with the current and with the potential difference across the inductor at the instant for which the previous phasor diagram was drawn. Label their lengths I and V_L, respectively. Also label the right angle between them.

The potential difference v_C across the capacitor is given by $v_C = q/C$. Now $q = \int i\,dt = -(I/\omega)\cos(\omega t - \phi)$, so $v_C = -(I/\omega C)\cos(\omega t - \phi)$. Use the trigonometric identity $\sin(A - \pi/2) = -\cos(A)$ to write this $v_C = (I/\omega C)\sin(\omega t - \phi - \pi/2)$. v_C is said to lag the current by $\pi/2$ radians. Its amplitude is given by $V_C = $ _____. When q is positive the _____ plate of the capacitor in the circuit diagram is at a higher potential than the _____ plate.

The phasor associated with v_C is $\pi/2$ radians behind (less counterclockwise than) the phasor associated with the current. On the axes to the right draw the phasors associated with the current and with the potential difference across the capacitor at the instant for which the previous phasor diagrams were drawn. Label their lengths I and V_C, respectively. Also label the right angle between them.

The relationship between the current and potential amplitude is often written

$$V_C = IX_C$$

for a capacitor and

$$V_L = IX_L$$

for an inductor. Here $X_C = $ _____ and $X_L = $ _____. X_C is called the <u>capacitive reactance</u> and X_L is called the <u>inductive reactance</u>. These relations and $V_R = IR$ for a resistor hold no matter what circuit the elements are in. In addition, the potential difference across an inductor always leads the current in the inductor by $\pi/2$ radians and the potential difference across a capacitor always lags the current into the capacitor by $\pi/2$ radians.

Reactances are not resistances but they play a similar role: they relate the amplitude of the current through a circuit element to the amplitude of the potential difference across the element. X_L, X_C, and R all have SI units of _____.

Carefully note that the reactances depend on the generator frequency. As the frequency increases X_C _____ and X_L _____. If the frequency is somehow increased without changing I the maximum potential difference across the capacitor and the maximum charge on the capacitor both decrease. Explain why this makes sense physically: _____

_____

If the frequency is increased without changing I the maximum potential difference across the inductor increases. Explain why this makes sense physically: _____

_____

Now you are ready to put the pieces together and solve for the current amplitude I and phase constant ϕ, given the generator emf. On the axes to the left below draw the phasors for the potential differences across the resistor, capacitor and inductor at the time for which the previous phasor diagrams were drawn.

According to the loop equation for the circuit $\mathcal{E}(t) = v_R(t) + v_L(t) + v_C(t)$. This means the phasors associated with v_R, v_L, and v_C must add like vectors to produce the phasor associated with $\mathcal{E}$. Since the phasors for v_L and v_C are parallel to each other they can be summed easily. Assume $V_L > V_C$ and on the axes to the right below draw phasors associated with v_R and $v_L + v_C$. The first has length V_R and the second has length $(V_L - V_C)$. Now draw the "vector" sum of the two phasors with its tail at the origin. This must be the emf phasor. Label the angle ϕ between it and the phasor associated with v_R. This is also the angle between the emf and current phasors.

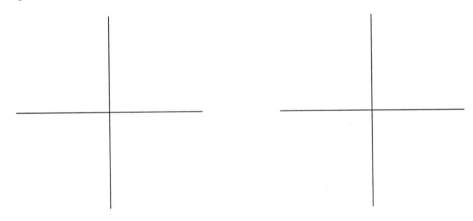

The phasors associated with v_R and $v_L + v_C$ form two sides of a right triangle with a hypotenuse of length $\mathcal{E}_m$, so $\mathcal{E}_m^2 = V_R^2 + (V_L - V_C)^2$. Substitute $V_R = IR$, $V_L = IX_L$, and $V_C = IX_C$, then solve for I. The result can be written

$$I = \mathcal{E}_m/Z ,$$

where

$$Z =$$

Z is called the <u>impedance</u> of the circuit.

If the impedance of a circuit is increased without changing the generator emf, the current amplitude _____. The impedance can be increased by increasing the resistance or by increasing the difference between the inductive and capacitive reactances. Suppose $X_L >$ X_C. Then as the frequency increases, the impedance _____. If $X_C > X_L$ and the frequency increases, then the impedance initially _____.

According to the phasor diagram the phase constant for the current is given in terms of the potential differences by

$$\tan \phi =$$

and in terms of the resistance and the reactances by

$$\tan \phi =$$

Notice from your phasor diagram that ϕ must be between $-90°$ and $+90°$. Since R is positive the value given for ϕ by your calculator is correct. You will never need to add $180°$.

Give the relation between X_L and X_C for which each of the following occurs:

the current lags the generator emf ($0 < \phi < 90°$) _____

the current leads the generator emf ($-90° < \phi < 0$) _____

the current is in phase with the generator emf ($\phi = 0$) _____

Note that the angular frequency for which the current is in phase with the generator emf is the resonance angular frequency $\omega = \sqrt{1/LC}$. For a given emf, the current amplitude has its greatest value for this angular frequency.

Power. The rate at which energy is supplied by the generator is given by $P_\mathcal{E} = i\mathcal{E}$, the rate at which energy is stored in the capacitor is given by $P_C = iv_C$, the rate at which energy is stored in the inductor is given by $P_L = iv_L$, and the rate at which energy is dissipated in the resistor is given by $P_R = iv_R$. All of these vary with time, being periodic with a period equal to the period of the generator.

Usually the time variations are of no interest. We consider instead averages over a cycle of oscillation. To compute averages you should know that the average of $\sin^2(\omega t - \phi)$ is $1/2$ and the average of $\sin(\omega t - \phi)\cos(\omega t - \phi)$ is 0. See the Mathematical Skills section.

You should be able to show that the averages of P_C and P_L are zero. For example, use $i = I\sin(\omega t - \phi)$ and $v_C = -(I/\omega C)\cos(\omega t - \phi)$ to obtain $P_C = -(I^2/\omega C)\sin(\omega t - \phi)\cos(\omega t - \phi)$. This averages to zero because it contains the product of the sine and cosine of $\omega t - \phi$. In the space below show that the average of P_L is also zero:

The power supplied by the generator is $P_\mathcal{E} = i\mathcal{E} = I\mathcal{E}_m \sin(\omega t - \phi)\sin(\omega t)$. The first sine function can be expanded as $\sin(\omega t)\cos\phi - \cos(\omega t)\sin\phi$. When the first term is multiplied by $\sin(\omega t)$ and averaged, the result is $\frac{1}{2}\cos\phi$. When the second term is multiplied by $\sin(\omega t)$ and averaged, the result is 0. Thus the average power supplied is

$$P_{av} = \frac{1}{2}I\mathcal{E}_m \cos\phi.$$

This can also be written $P_{av} = \frac{1}{2}(\mathcal{E}_m^2/Z)\cos\phi$ or $P_{av} = \frac{1}{2}I^2 Z \cos\phi$, where $\mathcal{E}_m = IZ$ was used. The quantity $\cos\phi$ is called the _____ .

In the space below use $\tan\phi = (X_L - X_C)/R$ and the trigonometric identity $\cos^2 A = 1/(1 + \tan^2 A)$ to show that $\cos\phi = R/Z$:

Thus the average power supplied by the generator is also given by $P_{av} = \frac{1}{2}I^2 R$.

The rate at which energy is dissipated in the resistor is $P_R = i^2 R = I^2 R \sin^2(\omega t - \phi)$ and its average value is $P_{av} = \frac{1}{2}I^2 R$, the same as the power supplied by the generator. Once transients have died out all energy supplied by the generator is dissipated in the resistor.

For a given emf the greatest power dissipation occurs when the power factor has the value $\cos\phi = $ _____ . For this to occur the impedance must be $Z = $ _____ , which means the inductive and capacitive reactances must be related by $X_L = $ _____ . This is the condition for resonance. If L and C are fixed it occurs if the angular frequency has the value given by $\omega = $ _____ . Often a circuit is designed to deliver as much power as possible to a resistive load. L and C are then adjusted so $X_L = X_C$ for the frequency of the generator.

Average power is often expressed in terms of root-mean-square quantities instead of amplitudes. The meaning of the term "root-mean-square" (rms) is _____

_____ .

Because the average over a cycle of $\sin^2(\omega t + \phi)$ is $1/2$ the rms value of a sinusoidal function of time is the amplitude divided by $\sqrt{2}$. For example, the rms value of $\mathcal{E}(t) = \mathcal{E}_m \sin(\omega t)$ is $\mathcal{E}_{rms} = $ _____ . In terms of $\mathcal{E}_{rms}$, $P_{av} = $ _____ and in terms of i_{rms}, $P_{av} = $ _____ .

The transformer.   Power lines transport electrical energy at low current, with a high potential difference across the ends. A low current is advantageous because it reduces _____ . On the other hand, a low potential difference is advantageous at both the generator and the consumer ends. A transformer is used to change the potential difference.

A transformer consists of two coils wrapped around the same iron core. Changing magnetic flux produced by changing currents in the coils induces emfs in the coils. Consider an ideal transformer (negligible resistance and hysteresis loss) with N_p turns in the primary coil and N_s turns in the secondary coil. The potential difference V_p across the primary coil is related to the potential difference V_s across the secondary coil by

$$\frac{V_p}{N_p} =$$

because, according to Faraday's law the emf generated in each turn of primary coil is the same as the emf generated in each turn of secondary coil.

If $N_s > N_p$ then the potential differences are related by _____ > _____ and the transformer is called a _____ transformer. If $N_s < N_p$ then the potential differences are related by _____ > _____ and the transformer is called a _____ transformer.

Since there are no losses in the transformer, the rate at which energy enters through the primary coil equals the rate at which energy leaves through the secondary coil and

$$I_p V_p =$$

These are rms values.

If the secondary circuit has resistance R and no inductance, the current in the secondary is given by $I_s = V_s/R$. In terms of the potential difference across the primary coil the current in the secondary circuit is

$$I_s =$$

and the current in the primary circuit is

$$I_p =$$

The current in the primary circuit does not change if the transformer and the secondary circuit are replaced by a resistor with resistance $R_{eq} =$ _____ . Transformers are often used to match the impedances of generators (or amplifiers) and their loads. That is, they are used to make R_{eq} the same as the generator resistance. Maximum energy transfer occurs if this condition is met.

II. PROBLEM SOLVING

You should know how to calculate the reactances and the impedance for an RLC series circuit: $X_L = \omega L$, $X_C = 1/\omega C$, and $Z = \sqrt{R^2 + (X_L - X_C)^2}$, where ω is the angular frequency. Some problems give the frequency f and you must use $\omega = 2\pi f$.

You should know how to compute the current given the generator emf. If $\mathcal{E} = \mathcal{E}_m \sin \omega t$, then $i = I \sin(\omega t - \phi)$, where $I = \mathcal{E}_m/Z$ and $\tan \phi = (X_L - X_C)/R$. Think of ϕ as a phase *difference*. That is, is $\mathcal{E} = \mathcal{E}_m \sin(\omega t + \alpha)$, then $i = I \sin(\omega t + \alpha - \phi)$.

Given the current you should know how to compute the potential differences across the individual circuit elements: $v_R = iR$, $v_C = IX_C \sin(\omega t - \pi/2)$, and $v_L = IX_L \sin(\omega t + \pi/2)$.

Finally, you should know that the generator supplies energy at the rate $P_{\mathcal{E}} = i\mathcal{E}$, the inductor stores energy at the rate $P_L = iv_L$, the capacitor stores energy at the rate $P_C = iv_C$, and the resistor dissipates energy at the rate $P_R = i^2 R$. These are all time dependent quantities.

PROBLEM 1 (8P). The output of an ac generator is $\mathcal{E} = \mathcal{E}_m \sin \omega t$, with $\mathcal{E}_m = 25.0$ V and $\omega = 377$ rad/s. It is connected to a 12.7-H inductor. (*a*) What is the maximum value of the current?

STRATEGY: The current through the inductor is given by $i(t) = (\mathcal{E}_m/X_L)\sin(\omega t - \pi/2)$ and its maximum value is $I = \mathcal{E}_m/X_L$, where X_L $(= \omega L)$ is the inductive reactance.

SOLUTION:

[ans: 5.22 mA]

(b) When the current is a maximum, what is the emf of the generator?

STRATEGY: The current is a maximum, for example, when $\sin(\omega t - \pi/2) = 1$ or $\omega t - \pi/2 = \pi/2$. Thus $\omega t = \pi$. Evaluate $\mathcal{E}_m \sin \omega t$ for the generator emf.

SOLUTION:

[ans: 0]

(c) When the emf of the generator is -12.5 V and increasing in magnitude, what is the current?

STRATEGY: When the generator emf is -12.5 V, $\sin \omega t = -1/2$. This means $\omega t = -0.524$ rad or 3.665 rad. In the first case the magnitude of the emf is decreasing (the emf is becoming less negative) while in the second case it is increasing. The current is given by $i = I\sin(\omega t - \pi/2)$.

SOLUTION:

[ans: 4.51 mA]

(d) For the conditions of part (c), is the generator supplying energy to or taking energy from the rest of the circuit?

STRATEGY: The rate at which energy is supplied by the source of emf is given by $P = i\mathcal{E}$. If the result is negative the source of emf is taking energy from the rest of the circuit.

SOLUTION:

[ans: $P = -56.5$ mW, the generator is taking energy]

PROBLEM 2 (9P). The ac generator of Problem 8 is connected to a 4.15-μF capacitor. (*a*) What is the maximum value of the current?

STRATEGY: The current into the capacitor is given by $i(t) = (\mathcal{E}_m/X_C)\sin(\omega t + \pi/2)$ and its maximum value is $I = \mathcal{E}_m/X_C$, where X_C ($= 1/\omega C$) is the capacitive reactance.

SOLUTION:

$$[\text{ans: } 39.1\,\text{mA}]$$

(*b*) When the current is a maximum, what is the emf of the generator?

STRATEGY: The current is a maximum, for example, when $\sin(\omega t + \pi/2) = 1$ or $\omega t + \pi/2 = \pi/2$. Thus $\omega t = 0$. Evaluate $\mathcal{E}_m \sin \omega t$ for the generator emf.

SOLUTION:

$$[\text{ans: } 0]$$

(*c*) When the emf of the generator is -12.5 V and increasing in magnitude, what is the current?

STRATEGY: As for Problem 8, $\omega t = 3.665$ rad. The current is given by $i = I \sin(\omega t + \pi/2)$.

SOLUTION:

$$[\text{ans: } -33.9\,\text{mA}]$$

(*d*) For the conditions of part (*c*), is the generator supplying energy to or taking energy from the rest of the circuit?

STRATEGY: Evaluate $P = i\mathcal{E}$. If the result is negative the source of emf is taking energy from the rest of the circuit.

SOLUTION:

$$[\text{ans: } P = 423\,\text{mW, the generator is supplying energy}]$$

PROBLEM 3 (11P).   The output of an ac generator is given by $\mathcal{E} = \mathcal{E}_m \sin(\omega t - \pi/4)$, where $\mathcal{E}_m = 30.0\,\text{V}$ and $\omega = 350\,\text{rad/s}$. The current is given by $i(t) = I \sin(\omega t + \pi/4)$, where $I = 620\,\text{mA}$. (a) At what time, after $t = 0$, does the generator emf first reach a maximum?

STRATEGY: The first maximum after $t = 0$ of the function $\sin(\omega t - \pi/4)$ occurs for $\omega t = \pi/2$. Solve for t.

SOLUTION:

[ans: 6.73 ms]

(b) At what time, after $t = 0$, does the current first reach a maximum?

STRATEGY: The first maximum after $t = 0$ of the function $\sin(\omega t + \pi/4)$ occurs for $\omega t + \pi/4 = \pi/2$. Solve for t.

SOLUTION:

[ans: 2.24 ms]

(c) The circuit contains a single element other than the generator. Is it a capacitor, an inductor, or a resistor? Justify your answer.

STRATEGY: The current leads the potential difference by $\pi/2$ rad. According to Eqs. 36–1 and 10 of the text the current leads the potential difference by $\pi/2$ for a capacitor, according to Eqs. 36–1 and 17 the current lags the potential difference by $\pi/2$ for an inductor, and according to Eqs. 36–1 and 4 the current and potential difference are in phase for a resistor.

SOLUTION:

[ans: a capacitor]

(d) What is the value of the capacitance, inductance, or resistance, as the case may be?

STRATEGY: The emf and current amplitudes are related by $\mathcal{E}_m = IX_C$, where $X_C = 1/\omega C$. Solve $\mathcal{E}_m = I/\omega C$ for C.

SOLUTION:

[ans: 59.0 μF]

PROBLEM 4 (19P).   A coil of inductance 88 mH and unknown resistance and a 0.94-μF capacitor are connected in series with an alternating emf of frequency 930 Hz. If the phase constant between the applied voltage and current is 75°, what is the resistance of the coil?

STRATEGY: The phase constant is given by $\tan\phi = (X_L - X_C)/R$, where X_L is the inductive reactance (ωL) and X_C is the capacitive reactance $(1/\omega C)$. Solve for R. Don't forget to use the *angular* frequency ($\omega = 2\pi f$, where f is the frequency).

SOLUTION:

$$[\text{ans: } 89\,\Omega]$$

PROBLEM 5 (22P).   For a certain RLC circuit the maximum generator emf is 125 V and the maximum current is 3.20 A. If the current leads the generator emf by 0.982 rad, what are (*a*) the impedance and (*b*) the resistance of the circuit?

STRATEGY: (*a*) Solve $\mathcal{E}_m = IZ$ for Z.

(*b*) The phase constant ϕ is given by $\tan\phi = X/R$, where $X = X_L - X_C$, and the impedance is given by $Z = \sqrt{R^2 + X^2}$. Solve these equations for R. The second gives $X^2 = Z^2 - R^2$. Substitute this expression into the square of the first equation to obtain $\tan^2\phi = (Z^2 - R^2)/R^2$. Complete the solution.

SOLUTION:

$$[\text{ans: } (a)\ 39.1\,\Omega; \ (b)\ 21.7\,\Omega]$$

(*c*) Is the circuit predominantly capacitive or inductive?

STRATEGY: Notice that the current leads the generator emf.

SOLUTION:

$$[\text{ans: capacitive}]$$

PROBLEM 6 (28P). The ac generator in Fig. 36–12 supplies 120 V (rms) at 60.0 Hz. With the switch open as in the diagram, the current leads the generator emf by 20.0°. With the switch in position 1 the current lags the generator emf by 10.0°. When the switch is in position 2 the rms current is 2.00 A. Find the values of R, L, and C.

STRATEGY: Let ϕ_1 be the current phase constant when the switch is open ($-20°$), ϕ_2 be the current phase constant when the switch is in position 1 ($+10°$), X_L be the reactance of the inductor (ωL), and X_C be the reactance of either one of the capacitors ($1/\omega C$). When the switch is open the circuit is a series RLC circuit and $\tan\phi_1 = (X_L - X_C)/R$. When the switch is in position 1 the two capacitors are in parallel, the total capacitance of the circuit is $2C$, and the capacitive reactance is $X_C/2$. Then $\tan\phi_2 = (X_L - X_C/2)/R$. When the switch is in position 2 the resistance of the circuit is 0, the impedance of the circuit is $Z = |X_L - X_C|$, and $\mathcal{E}_{rms} = i_{rms}Z$. These three equations are to be solved simultaneously for R, L, and C.

The first equation gives $|X_L - X_C| = R|\tan\phi_1|$ and when this is used in the third equation the result is $\mathcal{E} = i_{rms}R|\tan\phi_1|$. Solve for R. Then solve $X_L - X_C = R\tan\phi_1$ and $X_L - X_C/2 = R\tan\phi_2$ for X_L and X_C. Finally use $X_L = \omega L$ and $X_C = 1/\omega C$ to calculate L and C. Don't forget to use $\omega = 2\pi f$ to calculate the angular frequency.

SOLUTION:

$$\left[\text{ans:}\ R = 165\,\Omega;\ L = 313\,\text{mH};\ C = 14.9\,\mu\text{F}\right]$$

PROBLEM 7 (38P). In an RLC circuit, $R = 16.0\,\Omega$, $C = 31.2\,\mu\text{F}$, $L = 9.20\,\text{mH}$, and $\mathcal{E} = \mathcal{E}_m \sin\omega t$ with $\mathcal{E}_m = 45.0\,\text{V}$ and $\omega = 3000\,\text{rad/s}$. For time $t = 0.442\,\text{ms}$ find (a) the rate at which energy is being supplied by the generator, (b) the rate at which energy is being stored in the capacitor, (c) the rate at which energy is being stored in the inductor, and (d) the rate at which energy is being dissipated in the resistor.

STRATEGY: (a) The rate at which energy is being supplied by the generator is given by $P_\mathcal{E} = i\mathcal{E}$, where i is the instantaneous value of the current and $\mathcal{E}$ is the instantaneous value of the generator emf. The current is given by $i(t) = (\mathcal{E}/Z)\sin(\omega t - \phi)$, where Z is the impedance ($\sqrt{R^2 + (X_L - X_C)^2}$) and ϕ is the phase constant ($\tan\phi = (X_L - X_C)/R$). The reactances are given by $X_L = \omega L$ and $X_C = 1/\omega C$.

(b) The rate at which energy is being stored in the capacitor is given by $P_C = iv_C$, where v_C is the instantaneous potential difference across the capacitor. It is given by $v_C = X_C I \sin(\omega t - \phi - \pi/2) = (\mathcal{E}X_C/Z)\sin(\omega t - \phi - \pi/2)$. Notice that this expression includes the proper phase difference between the current and the potential difference across the capacitor.

(c) The rate at which energy is being stored in the inductor is given by $P_L = iv_L$, where v_L is the instantaneous potential difference across the inductor. It is given by $v_L = X_L I \sin(\omega t - \phi + \pi/2) = (\mathcal{E}X_L/Z)\sin(\omega t - \phi + \pi/2)$. Notice the addition of $\pi/2$ to the phase.

(d) The rate at which energy is being dissipated in the resistor is given by $P_R = i^2 R$.

SOLUTION:

$$\big[\text{ans: } (a)\ 41.4\,\mu\text{W}; (b)\ -17.1\,\mu\text{W}; (c)\ 44.1\,\mu\text{W}; (d)\ 14.4\,\mu\text{W}\big]$$

(e) What is the meaning of a negative result for any of parts (a), (b), and (c)?

SOLUTION:

[ans: in (a) it means energy is being transferred from the circuit to the generator; in (b) and (c) it means energy is being transferred from the circuit element to the rest of the circuit]

(f) Show that the results of parts (b), (c), and (d) sum to the result of part (a).

SOLUTION:

Sometimes the rms values of current and potential differences are given or requested. You should remember that if $i = I\sin(\omega t - \phi)$, then $i_{\text{rms}} = I/\sqrt{2}$. Similar expressions hold for the potential differences. You should also recognize that equations such as $v_L = IX_L\sin(\omega t + \pi/2)$ lead to $V_{L\,\text{rms}} = i_{\text{rms}}X_L$.

PROBLEM 8 (43P). In Fig. 36–16, $R = 15.0\,\Omega$, $C = 4.70\,\mu\text{F}$, and $L = 25.0\,\text{mH}$. The generator provides a sinusoidal voltage of 75.0 V (rms) and frequency $f = 550\,\text{Hz}$. (a) Calculate the rms current.

STRATEGY: Evaluate $i_{\text{rms}} = \mathcal{E}_{\text{rms}}/Z$, where Z is the impedance ($\sqrt{R^2 + (X_L - X_C)^2}$). The reactances are $X_L = \omega L$ and $X_C = 1/\omega C$. Use $\omega = 2\pi f$ to find the angular frequency.

SOLUTION:

$$\big[\text{ans: } 2.59\,\text{A}\big]$$

(*b*) Find the rms voltages V_{ab}, V_{bc}, V_{cd}, V_{bd}, V_{ad}.

STRATEGY: The potential difference across the resistor is $v_R = iR$ and the rms value is $V_{ab} = i_{\text{rms}}R$. The potential difference across the capacitor is $v_C = IX_C \sin(\omega t - \pi/2)$ and the rms value is $V_{bc} = i_{\text{rms}}X_C$. The potential difference across the inductor is $v_L = IX_L \sin(\omega t + \pi/2)$ and the rms value is $V_{cd} = i_{\text{rms}}X_L$. Here X_L is the inductive reactance (ωL) and X_C is the inductive reactance ($1/\omega C$). Use $\omega = 2\pi f$, where f is the frequency. The potential difference V_{ad} is the same as the potential difference across the generator.

SOLUTION:

$$\left[\, \text{ans: } V_{ab} = 38.8\,\text{V}, \; V_{bc} = 159\,\text{V}, \; V_{cd} = 224\,\text{V}, \; V_{bd} = 64.2\,\text{V}, \; V_{ad} = 75.0\,\text{V} \,\right]$$

(*c*) At what average rate is energy dissipated by each of the three circuit elements?

STRATEGY: Use $P_{\text{av}} = i_{\text{rms}}^2 R$ to compute the average rate of energy dissipation in the resistor.

SOLUTION:

$$\left[\, \text{ans: } 100\,\text{W in the resistor, 0 in each of the other elements} \,\right]$$

III. MATHEMATICAL SKILLS

You should be able to prove that the average over a cycle of $\sin^2(\omega t)$ is $1/2$ and the average over a cycle of $\sin(\omega t)\cos(\omega t)$ is 0. First consider the average of $\sin^2(\omega t)$, given by the integral

$$I = \frac{1}{T} \int_0^T \sin^2(\omega t)\, dt \,,$$

where T is the period of oscillation. Use the trigonometric identity $\cos(2\omega t) = \cos^2(\omega t) - \sin^2(\omega t) = 1 - 2\sin^2(\omega t)$, where $\cos^2(\omega t) + \sin^2(\omega t)$ was used. Thus $\sin^2(\omega t) = \frac{1}{2}[1 - \cos(2\omega t)]$ and

$$I = \frac{1}{2T} \int_0^T [1 - \cos(2\omega t)]\, dt = \frac{1}{2} - \frac{1}{4\omega T}\sin(2\omega T)\,.$$

Now $2\omega T = 4\pi f T = 4\pi$ and $\sin(2\omega T) = \sin(4\pi) = 0$, so $I = 1/2$.

The average of $\sin(\omega t)\cos(\omega t)$ is somewhat easier to find. You must evaluate the integral

$$I = \frac{1}{T} \int_0^T \sin(\omega t)\cos(\omega t)\, dt$$

and this is

$$I = \frac{1}{2\omega T} \sin^2(\omega t)\big|_0^T = 0.$$

The last result follows from $\sin(\omega T) = \sin(4\pi) = 0$.

IV. NOTES

Chapter 37
MAXWELL'S EQUATIONS

I. BASIC CONCEPTS

The four equations that tell us about the electric and magnetic fields produced by any source are collectively known as Maxwell's equations and are presented together in this chapter so you can see how they complement each other. Most of this material should be a review for you. The new topic deals with displacement current. You should pay close attention to its definition and learn how to calculate it and the magnetic field it produces.

Maxwell's equations. Complete the following statements of Maxwell's equations (see Table 37–2 of the text):

$$\text{I.} \qquad \oint \mathbf{E} \cdot d\mathbf{A} =$$

$$\text{II.} \qquad \oint \mathbf{B} \cdot d\mathbf{A} =$$

$$\text{III.} \qquad \oint \mathbf{E} \cdot d\mathbf{s} =$$

$$\text{IV.} \qquad \oint \mathbf{B} \cdot d\mathbf{s} =$$

Additional terms must be added to some of these equations if magnetic or dielectric materials are present but otherwise they are completely general.

You should understand what each of the Maxwell equations says. Equations I and II contain surface integrals on their left sides and the surface must be specified to evaluate the integral. In each case, d$\mathbf{A}$ is a vector whose magnitude is an infinitesimal element of _____ and whose direction is _____ to the surface. In equation I the _____ component of the electric field is integrated over a _____ surface. The symbol q on the right side represents the net charge _____ by the surface. In equation II the right side is zero because there are no _____ .

Earlier in this course you used equation I (Gauss' law for electricity) to find the electric field of a point charge, a large plate of charge, a line of charge, and other charge configurations. You also used it to show that static charge on a conductor resides on its surface. The equation tells us that electric field lines can be drawn so they start on positive charge and end on negative charge. Equation II (Gauss' law for magnetism) tells us that magnetic field lines form closed loops.

Equations III and IV contain path integrals on their left sides. In each case, ds is a vector whose magnitude is an infinitesimal element of _____ and whose direction is _____ to the path. On the left side of equation III the _____ component of the electric field is integrated around a _____ path. This integral gives the _____ around the path. On the left side of equation IV the _____ component of the magnetic field is integrated around a _____ path. On the right side of equation III, Φ_B is the magnetic flux through the _____ bounded by the path. On the right side of equation IV i is the net _____ through the area and Φ_E is the _____ through the area.

Equation III is Faraday's law. You used it to compute the emf in a loop that is rotating in a magnetic field, the emf generated in a loop by the changing field of a nearby long straight wire, and other emf's generated by changing magnetic fluxes.

Equation IV is the Ampere-Maxwell law. It has been used to find the magnetic field of a long straight wire and the magnetic fields inside solenoids and toroids. For these situations the fields are produced by currents alone and $d\Phi_E/dt = 0$. $\mu_0\epsilon_0\, d\Phi_E/dt$ is the term introduced in this chapter.

Although these equations have been used in specific examples, chosen in most cases for ease in computation, they are generally valid. The first two are true for *any* closed surface, the second two are true for *any* closed path.

Take special care to distinguish between the various equations. Remember that a changing magnetic field produces an electric field and that *both* a changing electric field and a current produce a magnetic field. Remember that the fields produced appear on the left sides of Maxwell's equations, the sources of the field appear on the right sides. Look at the equations and explain to yourself what each describes.

How can you generate an electric field in a region without having a net charge anywhere?

_____

_____

_____

For this situation which of the Maxwell equations have zero for the value of their right sides? _____ Which do not? _____

How can you generate an electric field in a region without having a current anywhere?

_____

_____

_____

For this situation which of the Maxwell equations have zero for the value of their right sides? _____ Which do not? _____

How can you generate a magnetic field in a region without having a net charge anywhere?

_____

_____

_____

For this situation which of the Maxwell equations have zero for the value of their right sides? _____ Which do not? _____

The displacement current. Equation IV above tells us that there must be a magnetic field around the boundary of any area through which there is a changing electric flux, just as if a current passed through the area. In fact, the quantity i_d = _____ is called the <u>displacement current</u> through the area. You should realize that a displacement current is emphatically NOT a true current, which consists of moving charge. A displacement current exists in any region of space that contains a changing electric field, whether or not charge is moving in the region. True currents exist only in regions that contain moving charge.

Suppose a cylindrical region with radius R contains a uniform electric field **E** parallel to its axis. A cross section is shown to the right, with the field pointing into the page. First consider the whole cross section. If d**A** is also into the page the electric flux through the circle of radius R is given by Φ_E = _____ , where E is the magnitude of the field. If the magnitude is changing at the rate dE/dt then the displacement current through the entire cross section is given by i_d = _____ .

Now consider the circle of radius r $(< R)$. The electric flux through this circle is given by Φ_E = _____ and the displacement current through it is given by i_d = _____ .

Finally consider a circle of radius r $(> R)$. The electric flux through this circle is given by Φ_E = _____ and the displacement current through it is given by i_d = _____ .

A magnetic field is associated with a displacement current just as a magnetic field is associated with moving charge. For the cylindrical region considered above, the displacement current is uniformly distributed over every cross section, so the cylinder acts like a long straight wire of radius R with uniform current density. The magnetic field lines are _____ centered at the _____ of the cylinder. Take the left side of the Ampere-Maxwell law to be an integral around a circle in the plane of a cross section and centered at the cylinder axis. Then, in terms of the magnitude B of the magnetic field and the radius r of the circle, $\oint \mathbf{B} \cdot d\mathbf{s}$ = _____ . Let i_d be the total displacement current through a cross section of the cylinder. Then, if $r < R$ the displacement current through the amperian circle is given by _____ . In terms of i_d the magnitude of the magnetic field is given by B = _____ and in terms of dE/dt it is given by B = _____ . If $r > R$ the displacement current through the amperian circle is the same as that through the entire cross section of the cylinder; that is, it is i_d. For this case the magnitude of the field is given in terms of i_d by B = _____ and in terms of dE/dt by B = _____ .

You should be able to find the direction of the magnetic field lines. First chose a direction for d**A**. It is in one of the two directions that are perpendicular to the surface. This choice determines the signs of Φ_E and $d\Phi_E/dt$. Point the thumb of your right hand in the direction chosen. Your fingers will curl in the direction you *must* use for d**s** when you evaluate $\oint \mathbf{B} \cdot d\mathbf{s}$. Now evaluate $d\Phi_E/dt$. If it is positive the magnetic field lines follow your fingers; if it is negative they go the opposite way.

A charging (or discharging) capacitor provides an excellent example of a displacement current. The charge on the plates produces an electric field in the interior and since the charge is changing so is the field. Consider a parallel plate capacitor with plate area A and let $q(t)$ be the charge on the positive plate at time t. Then the magnitude of the electric field between the plates is given by $E(t) = $ _____. Consider a cross section between the plates, parallel to them, and having the same area as a plate. The electric flux through this cross section is _____ and the total displacement current in the interior of the capacitor is $i_d = $ _____, where i ($= dq/dt$) is the current into the capacitor. You should have shown that $i_d = i$. The total displacement current in the interior is the same as the current in the capacitor wires. The total current (true current plus displacement current) is continuous. Although the true current stops at the plates, the displacement current continues into the interior.

II. PROBLEM SOLVING

You should be able to calculate the magnetic field induced by a uniform time-varying electric field in a cylindrical region. Remember that although the electric field exists only inside the region a magnetic field is induced at all points, both inside and outside.

You should know how the displacement current and the rate of change of the electric flux are related: $i_d = \epsilon_0 \, d\Phi_E/dt$. You should also know how to calculate the electric flux and the displacement current when the electric field is uniform. If the field is uniform and normal to the surface then $\Phi_E = EA$, where A is the area of the surface. In this case $i_d = \epsilon_0 A \, dE/dt$. In some cases you must find the electric field from a given electric potential function. Recall, for example, that the electric field between the plates of a capacitor is related to the potential difference across the plates by $E = V/d$, where d is the plate separation.

Also remember that the right side of the Ampere-Maxwell law contains both true current and displacement current. In most cases one or the other of the currents is likely to be zero but you must think about both when you solve a problem.

PROBLEM 1 (4P). Suppose that a circular-plate capacitor has a radius R of 30 mm and a plate separation of 5.0 mm. A sinusoidal potential difference with a maximum value of 150 V and a frequency of 60 Hz is applied between the plates. Find $B_m(R)$, the maximum value of the induced magnetic field at $r = R$.

STRATEGY: Use $\oint \mathbf{B} \cdot d\mathbf{s} = \mu_0 \epsilon_0 \, d\Phi_E/dt$. The integral on the left is around a circle of radius R, between the plates, parallel to them, and in the direction of the magnetic field lines. Integration yields $2\pi RB$. The electric flux through the circle is $\pi R^2 E$, so $2\pi RB = \mu_0 \epsilon_0 \pi R^2 \, dE/dt$ and $B = (\mu_0 \epsilon_0 R/2) \, dE/dt$. The magnitude E of the electric field and the potential difference V across the plates are related by $E = V/d$, where d is the plate separation, so $dE/dt = (1/d) \, dV/dt$. Assume $V = V_m \sin \omega t$, then $dE/dt = (\omega V_m/d) \cos \omega t$. The maximum induced magnetic field occurs when dE/dt is a maximum. This means $\cos \omega t = 1$ and the maximum value of dE/dt is $\omega V_m/d$. Substitute this expression into the equation for B.

SOLUTION:

[ans: 1.9×10^{-12} T]

PROBLEM 2 (9E). Figure 37–5 shows the plates P_1 and P_2 of a circular parallel-plate capacitor of radius R. They are connected as shown to long straight wires in which a constant conduction current i exists. A_1, A_2, and A_3 are hypothetical circles of radius r, two of them outside the capacitor and one between the plates. Show that the magnetic field at the circumference of each of these circles is given by

$$B = \frac{\mu_0 i}{2\pi r} \, .$$

STRATEGY: Use the Ampere-Maxwell law in the form $\oint \mathbf{B} \cdot d\mathbf{s} = \mu_0 (i + i_d)$, where the path of integration is the circumference of any one of the circles, i is the true current through the path, and i_d is the displacement current through the path. For A_1 and A_3 the displacement current is zero while for A_2 the true current is zero. In addition, the circumference of A_2 is outside the plates so i_d is the total displacement current between the plates. Use $i_d = i$.

SOLUTION:

PROBLEM 3 (10P). In Sample Problem 37–1 show that the expressions derived for $B(r)$ can be written as

$$B(r) = \frac{\mu_0 i_d}{2\pi r} \quad (\text{for } r \geq R)$$

and

$$B(r) = \frac{\mu_0 i_d r}{2\pi R^2} \quad (\text{for } r \leq R) \, .$$

Note that these expressions are of the same form as those derived in Chapter 31, except that the conduction current i has been replaced by the displacement current i_d.

STRATEGY: For $r \geq R$ the displacement current is the total displacement current between the capacitor plates. The total electric flux is $\pi R^2 E$ and the total displacement current is $i_d = \epsilon_0 \, d\Phi_E/dt = \epsilon_0 \pi R^2 \, dE/dt$. Substitute $dE/dt = i_d/\epsilon_0 \pi R^2$ into $B(r) = (\mu_0 \epsilon_0 R^2/2r) \, dE/dt$, the expression given in the sample problem for the field at $r \geq R$.

For $r \leq R$ you must consider only the electric flux through a circle of radius r: $\Phi_E = \pi r^2 E$. The displacement current through the circle is $\epsilon_0 \pi r^2 \, dE/dt = (r^2/R^2) i_d$, where i_d is the total displacement current. Substitute $dE/dt = (1/\epsilon_0 \pi R^2) i_d$ into $B = (\mu_0 \epsilon_0 r/2) \, dE/dt$, the expression given in the sample problem for the field at $r \leq R$.

SOLUTION:

PROBLEM 4 (11P).   As a parallel-plate capacitor with circular plates 20 cm in diameter is being charged the displacement current density throughout the region between the plates is uniform and has a magnitude of $20 \, \text{A/m}^2$. (*a*) Calculate the magnitude B of the magnetic field at a distance $r = 50 \, \text{mm}$ from the axis of symmetry of the region.

STRATEGY: Use the Ampere-Maxwell law in the form $\oint \mathbf{B} \cdot \mathbf{ds} = \mu_0 i_d$. The integral on the left side is around a circle of radius r, parallel to the plates and centered on the symmetry axis. The circumference is a magnetic field line, so $\oint \mathbf{B} \cdot \mathbf{ds} = 2\pi r B$. The displacement current through the circle is $i_d = \pi r^2 J$, where J is the displacement current density. Notice that this expression for i_d is valid because $r < R$, where R is the radius of a plate, and because the displacement current density is uniform. Solve $2\pi r B = \mu_0 \pi r^2 J$ for B.

SOLUTION:

$$[\text{ans: } 6.3 \times 10^{-7} \, \text{T}]$$

(*b*) Calculate dE/dt in this region.

STRATEGY: Consider the total displacement current between the plates. In terms of the displacement current density it is $i_d = \pi R^2 J$. In terms of the rate of change of the electric field it is $\epsilon_0 \pi R^2 \, dE/dt$. Solve $\pi R^2 J = \epsilon_0 \pi R^2 \, dE/dt$ for dE/dt.

SOLUTION:

$$[\text{ans: } 2.3 \times 10^{12} \, \text{V/m} \cdot \text{s}]$$

PROBLEM 5 (13P).   A parallel-plate capacitor has square plates 1.0 m on a side as in Fig. 37–7. There is a charging current of 2.0 A flowing into (and out of) the capacitor. (*a*) What is the displacement current through the region between the plates?

STRATEGY: The true current into a capacitor has the same value as the total displacement current between the plates.

SOLUTION:

$$[\text{ans: } 2.0 \, \text{A}]$$

(*b*) What is dE/dt in this region?

STRATEGY: Solve $i_d = \epsilon_0 A \, dE/dt$, where A is the area of a plate.

SOLUTION:

$$[\text{ans: } 2.3 \times 10^{11} \, \text{V/m} \cdot \text{s}]$$

(c) What is the displacement current through the square dashed path between the plates?

STRATEGY: Evaluate $i'_d = \epsilon_0 A' \, dE/dt$, where A' is the area of the region bounded by the dashed path.

SOLUTION:

[ans: 0.50 A]

(d) What is $\oint \mathbf{B} \cdot d\mathbf{s}$ around this square dashed path?

STRATEGY: Use $\oint \mathbf{B} \cdot d\mathbf{s} = \mu_0 i_d$.

SOLUTION:

[ans: $6.3 \times 10^{-7} \, \text{T} \cdot \text{m}$]

Some problems of this chapter deal with Maxwell's equations. Many are designed to show you similarities and differences. Here are two examples.

PROBLEM 6 (17P). *A self-consistency property of two of the Maxwell equations* (Eqs. III and IV in Table 37–2). Two adjacent closed paths $abefa$ and $bcdeb$ share the common edge be as shown in Fig. 37–9. (*a*) We may apply $\oint \mathbf{E} \cdot d\mathbf{s} = -d\Phi_B/dt$ (Eq. III) to each of the two closed paths separately. Show that, from this alone, Eq. III is *automatically* satisfied for the composite closed path $abcdefa$.

STRATEGY: For the loop on the left side write

$$\int_a^b \mathbf{E} \cdot d\mathbf{s} + \int_b^e \mathbf{E} \cdot d\mathbf{s} + \int_e^f \mathbf{E} \cdot d\mathbf{s} + \int_f^a \mathbf{E} \cdot d\mathbf{s} = -d\Phi_{BL}/dt$$

and for the loop on the right side write

$$\int_b^c \mathbf{E} \cdot d\mathbf{s} + \int_c^d \mathbf{E} \cdot d\mathbf{s} + \int_d^e \mathbf{E} \cdot d\mathbf{s} + \int_e^b \mathbf{E} \cdot d\mathbf{s} = -d\Phi_{BR}/dt .$$

Notice that both loops are traversed in the clockwise direction. Now add the two equations. The common line segment is traversed twice in the sum of the left sides, once from b to e and once from e to b. The magnitudes of the two integrals are the same but their signs are different (**B** is the same for both of them but d**s** is oppositely directed). The two integrals cancel. What remains is the integral around the outside of the composite rectangle. Furthermore the total electric flux through the composite is the sum of the fluxes through the two smaller rectangles: $\Phi_B = \Phi_{BL} + \Phi_{BR}$.

SOLUTION:

(b) Repeat (a) for Eq. IV.

SOLUTION:

PROBLEM 7 (18P).   *A self-consistent property of two of the Maxwell equations* (Eqs. I and II in Table 2). Two adjacent closed parallelepipeds share a common face $abcd$ as shown in Fig. 37-10. (*a*) We may apply $\oint \mathbf{E} \cdot d\mathbf{A} = q/\epsilon_0$ (Eq. I) to each of these two closed surfaces separately. Show that, from this alone, Eq. I is *automatically* satisfied for the composite closed surface.

STRATEGY: For each parallelepiped write the integral on the left side of Gauss' law as the sum of six integrals, one over each face. Add the two equations. The sum of integrals on the left side contains two integrals over the common face. They are equal in magnitude but opposite in sign ($\mathbf{E}$ is the same but $d\mathbf{A}$ is oppositely directed). The two integrals cancel and what is left is the integral over the surface of the composite. The sum of the charges in the two parts is, of course, the net charge in the composite.

SOLUTION:

(*b*) Repeat (*a*) for Eq. II.

SOLUTION:

III. NOTES

Chapter 38
ELECTROMAGNETIC WAVES

I. BASIC CONCEPTS

Maxwell's equations predict the possibility of electric and magnetic fields that propagate in free space, with the changing magnetic field producing changes in the electric field, via Faraday's law, and the changing electric field producing changes in the magnetic field, via the displacement current term in the Ampere-Maxwell law. Pay close attention to the relationship between the amplitudes of the fields, to the relationship between their phases, and to the relationship between their directions. One important consequence of the equations is that they produce an expression for the wave speed, the speed of light, in terms of ϵ_0 and μ_0.

The electromagnetic spectrum. Electromagnetic waves exist for all wavelengths and frequencies, from very large to very small. Various ranges of wavelengths have been named and you should be familiar the ranges and their names.

1. Gamma radiation extends from the very shortest wavelengths (highest frequencies) to a wavelength of about _____ m (a frequency of about _____ Hz).

2. X-ray radiation extends from a wavelength of about _____ m (a frequency of about _____ Hz) to a wavelength of about _____ m (a frequency of about _____ Hz).

3. Ultraviolet radiation extends from a wavelength of about _____ m (a frequency of about _____ Hz) to a wavelength of about _____ m (a frequency of about _____ Hz).

4. Visible radiation extends from a wavelength of about _____ m (a frequency of about _____ Hz) to a wavelength of about _____ m (a frequency of about _____ Hz).

5. Infrared radiation extends from a wavelength of about _____ m (a frequency of about _____ Hz) to a wavelength of about _____ m (a frequency of about _____ Hz).

6. Microwave radiation extends from a wavelength of about _____ m (a frequency of about _____ Hz) to a wavelength of about _____ m (a frequency of about _____ Hz).

7. Above this in wavelength lies the TV and radio portion of the spectrum, which extends from a wavelength of about _____ m (a frequency of about _____ Hz) to the longest wavelengths (lowest frequencies).

All these electromagnetic radiations are exactly alike except for their frequencies and wavelengths. They are all electromagnetic. That is, they all consist of traveling electric and magnetic fields. They all travel in free space with the same speed, the speed of light $c =$ _____ m/s.

Electromagnetic waves   Classically, electromagnetic waves are produced by charges that are accelerating. Charges at rest or moving with constant velocity do not radiate. In the quantum mechanical picture electromagnetic radiation is produced when a charge (perhaps an electron in an atom or an atomic nucleus) changes its state or in reactions of fundamental particles.

Figs. 38–3, 4, and 5 show the electric and magnetic fields of a radiating antenna consisting of straight wires in which the current varies sinusoidally. Look at these figures and identify the important characteristics of electromagnetic radiation (take the x axis to be positive to the right and the z axis to be positive out of the page):

1. The electric and magnetic fields are perpendicular to each other. At the point on the x axis of Fig. 38–4 where the most advanced electric field line shown crosses the axis the electric field is in the _____ direction and the magnetic field is in the _____ direction.

2. The wave travels in a direction that is perpendicular to both the electric and magnetic fields. At the point you just considered the wave is traveling in the _____ direction.

3. The electric and magnetic fields are in phase. Explain what this means: _____
_____

4. The wave is linearly polarized. Explain what this means: _____
_____

Very far from the source the wave becomes a plane wave: the electric and magnetic field lines are very close to straight lines at the detector. Suppose the detector is on the x axis so that waves reaching it are traveling in the positive x direction. The diagram shows the fields in the wave near the detector. The electric field is parallel to the y axis and the magnetic field is parallel to the z axis. Mathematically the fields are given by

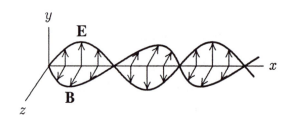

$$E(x,t) =$$

$$B(x,t) =$$

where $k = 2\pi/\lambda$, λ is the wavelength, and ω is the angular frequency. These quantities are related by $\omega/k =$ _____, where c is the speed of the wave. The phase constants are the same for the two fields.

The electric field is different at a point with coordinate x and a point with coordinate $x + dx$, an infinitesimal distance away, because a changing magnetic flux penetrates the region between these points. The magnetic field is different at x and $x + dx$ because a changing electric flux penetrates the region between these points. Differences in the fields at different points can be computed using the Faraday and Ampere-Maxwell laws.

When these laws are applied you find that the amplitudes of the fields are related by

$$\frac{E_m}{B_m} =$$

and that the wave speed is determined by the constants ϵ_0 and μ_0:

$$c =$$

You should understand how the laws of electromagnetism lead to these relationships. The diagram on the right shows a small region of space in which a wave is traveling toward the right. The electric fields at two infinitesimally separated points, x and $x + \Delta x$, are shown, as is the magnetic field in the region between. Apply Faraday's law to this situation. To calculate the magnetic flux take d**A** to be out of the page and to calculate the emf traverse the path shown in the counterclockwise direction. Clearly, $\Phi_B = Bh\Delta x$ and $\oint \mathbf{E} \cdot \mathbf{ds} = h[E(x + \Delta x) - E(x)]$. Why does $E(x + \Delta x)$ enter with a plus sign and $E(x)$ enter with a minus sign? _____

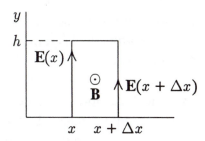

_____

As Δx becomes small $[E(x + \Delta x) - E(x)]/\Delta x$ becomes $\partial E/\partial x$ and Faraday's law yields

$$\frac{\partial E}{\partial x} =$$

The diagram on the right shows the view looking up from underneath the previous diagram. The magnetic field at the two points and the electric field between are shown. Apply the Ampere-Maxwell law to this situation. Take d**A** to be into the page and traverse the path in the clockwise direction. Clearly, $\Phi_E = Eh\Delta x$ and $\oint \mathbf{B} \cdot \mathbf{ds} = h[B(x) - B(x + \Delta x)]$. Why does $B(x)$ enter with a plus sign and $B(x + \Delta x)$ enter with a minus sign?

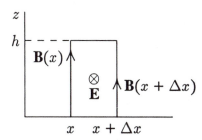

_____

As Δx becomes small $[B(x) - B(x + \Delta x)]/\Delta x$ becomes $-\partial B/\partial x$ and the Ampere-Maxwell law yields

$$\frac{\partial B}{\partial x} =$$

Substitute $E = E_m \sin(kx - \omega t)$ and $B = B_m \sin(kx - \omega t)$ into these two equations and show that $kE_m = \omega B_m$ and that $kB_m = \mu_0\epsilon_0\omega E_m$:

Now use $\omega/k = c$ to show that $E_m = cB_m$ and $c^2 = 1/\mu_0\epsilon_0$:

Energy transport.   Electromagnetic waves carry energy. The electrical energy per unit volume associated with an electric field $\mathbf{E}$ is given by $u_E =$ _____ and the magnetic energy per unit volume associated with a magnetic field $\mathbf{B}$ is given by $u_B =$ _____. You can use $B = E/c$ and $c = 1/\sqrt{\epsilon_0\mu_0}$ to show that for a plane wave $u_E = u_B = EB/2\mu_0 c$ and that the total energy density is $u = EB/\mu_0 c$.

This energy moves with the wave, with speed c. Consider a region of space with infinitesimal width dx and cross-sectional area A, perpendicular to the direction of travel of a plane electromagnetic wave. The volume of the region is $A\,dx$ so, in terms of the energy density u, the electromagnetic energy in the region is $dU =$ _____. All this energy will pass through the area A in time dt, given by $dt = dx/c$. Substitute $dx = c\,dt$ and divide by dt. The energy passing through the area per unit time is $dU/dt =$ _____ and the energy per unit area passing through per unit time is $dU/A\,dt =$ _____. In terms of the field magnitudes E and B, this is $dU/A\,dt =$ _____.

The transport of energy is described in terms of the <u>Poynting vector</u> $\mathbf{S}$, defined by the vector product

$$\mathbf{S} =$$

Since $\mathbf{E}$ and $\mathbf{B}$ are perpendicular to each other the magnitude of the Poynting vector is given by $S =$ _____. In terms of the magnitude S of the Poynting vector the energy passing through a surface of area A, perpendicular to the direction of travel, per unit time, is given by $dU/dt =$ _____ and the energy passing through per unit area per unit time is given by $dU/A\,dt =$ _____.

For most sinusoidal electromagnetic waves of interest the fields oscillate so rapidly that their instantaneous values cannot be detected or else are not of interest. Energies and energy densities are then characterized by their average over a period of oscillation. The average value of E^2, for example, is $\frac{1}{2}E_m^2$, where E_m is the amplitude.

The <u>intensity</u> of a wave is the magnitude of the Poynting vector averaged over a period of oscillation. In terms of the electric field amplitude E_m it is $I =$ _____ and in terms of the magnetic field amplitude B_m it is $I =$ _____. For sinusoidal waves the intensity is often written in term of the rms value E_{rms} of the electric field: $I =$ _____.

The direction of $\mathbf{S}$ is the same as the direction in which the wave is traveling. Recall that this direction is intimately connected with the relative signs of the terms kx and ωt in the argument of the trigonometric function that describes the fields. If the wave travels in the negative x direction you write $E_y(x,t) = E_m \sin(kx + \omega t)$ and $B_z(x,t) =$ _____. Be sure you have selected the appropriate signs so $B_z(x,t)$ represents a wave traveling in the negative x direction and so that $\mathbf{E} \times \mathbf{B}$ is also in the negative x direction. Carefully note that $B_z = (E_m/c)\sin(kx + \omega t)$ is *wrong* since this means $\mathbf{E} \times \mathbf{B}$ is in the positive x direction.

Electromagnetic waves also carry momentum. If ΔU is the energy in any infinitesimal portion of a wave then

$$\Delta p =$$

is the magnitude of the momentum in that portion. The momentum is in the direction of **S**, the direction of propagation. If u is the energy per unit volume then _____ is the momentum per unit volume and if I is the energy transported through an area per unit area per unit time then _____ is the momentum transported through the area per unit area per unit time.

When an electromagnetic wave interacts with a charge the electric field component of the wave does work on the charge and energy is transferred from the wave to the charge. Since the fields exert forces on the charge momentum is also transferred. When electromagnetic radiation is absorbed by a material object its energy and momentum are transferred to the object. Consider a plane wave with time averaged energy density u, incident normally on the plane surface of an object. If the area of the surface is A and the wave is completely absorbed then averaged over a period the energy transferred in time Δt is given by _____ and the momentum transferred is given by _____. The momentum transferred per unit time is the force of the radiation on the object and the force per unit area is the radiation pressure. In terms of u the force is given by _____ and the radiation pressure is given by _____.

If the wave is reflected without loss back along the original path, the energy transferred is _____ and the momentum transferred is _____. The radiation pressure is now _____, twice what it would be if the radiation were absorbed. Why? _____

Polarization. The electric field of a <u>linearly</u> <u>polarized</u> electromagnetic wave is always parallel to the same _____. The <u>direction of polarization</u> is parallel to the _____ field and the <u>plane of polarization</u> is determined by the directions of _____ and _____. For maximum signal, the wires in an electric dipole antenna used to detect polarized waves must be aligned with the _____ direction.

You should be able to contrast polarized and unpolarized electromagnetic radiation. Describe the orientation of the electric field of unpolarized radiation: _____

Light from an incandescent bulb or from the sun is not polarized. The direction of its electric field changes rapidly in a random fashion. Explain why this comes about: _____
_____.

You should recognize that any linearly polarized wave can be treated as the sum of two waves, polarized in any two mutually orthogonal directions that are perpendicular to the direction of propagation. Suppose a wave with an electric field amplitude E_m is traveling out of the page and is polarized with its electric field along a line that makes an angle θ with the x axis, as shown. You can consider the electric field to be the vector sum of two fields, one with amplitude _____, polarized along the x axis and one with amplitude _____, polarized along the y axis.

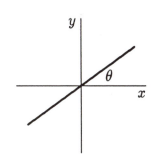

Polarized radiation can be produced by shining unpolarized radiation through a sheet of Polaroid. These sheets contain certain long-chain molecules that are aligned in the manufacturing process. Radiation with its electric field vector parallel to the chains is preferentially _____ while radiation with its electric field perpendicular to the chains is preferentially _____ . A line perpendicular to the molecules is said to be along the <u>polarizing</u> <u>direction</u> of the sheet.

Suppose that linearly polarized radiation with amplitude E_m and intensity I_m is incident on an ideal polarizing sheet and that its electric field is along a line that makes an angle θ with the polarizing direction. Then the amplitude of the transmitted radiation is given by $E =$ ___ and its intensity is given by

$$I =$$

This is called the law of Malus. Note that the transmitted amplitude is the component of the incident amplitude along the polarization direction of the sheet. You simply resolved the incident electric field into components parallel and perpendicular to the polarization direction. The parallel component is transmitted while the perpendicular component is absorbed.

If unpolarized radiation with intensity I_i is incident on an ideal polarizing sheet the intensity of the transmitted radiation is given by

$$I =$$

Be sure you understand the conditions for which each of these expressions for the transmitted intensity applies. In both cases the transmitted radiation is linearly polarized with its electric field parallel to _____ .

II. PROBLEM SOLVING

Many problems deal with the relationships between various characteristics of plane electromagnetic waves. You should know that the angular wave number k and the wavelength λ are related by $k = 2\pi/\lambda$, that the angular frequency ω and the frequency f are related by $\omega = 2\pi f$, and that the wavelength and frequency are related by $\lambda f = c$, where c is the speed of light. This last relationship leads to $\omega/k = c$.

You should also know that the electric and magnetic fields are in phase, that their amplitudes are related by $B_m = cE_m$, that **E** and **B** are perpendicular to each other, and that **E** × **B** is in the direction of propagation. The last statement implies that both **E** and **B** are perpendicular to the direction of propagation. You should also remember that the direction of propagation determines the argument of the trigonometric function in the expressions for **E** and **B**. That is, for example, $E_m \sin(kx - \omega t)$ is propagating in the positive x direction and $E_m \sin(kx + \omega t)$ is propagating in the negative x direction.

PROBLEM 1 (3E).   (a) The wavelength of the most energetic x rays produced when electrons accelerated to a kinetic energy of 18 GeV in the Stanford Linear Accelerator slam into a solid target is 0.067 fm. What is the frequency of these x rays? (b) A VLF (very low frequency) radio wave has a frequency of only 30 Hz. What is its wavelength?

STRATEGY: The frequency f and wavelength λ of a sinusoidal electromagnetic wave are related by $c = f\lambda$. In (a) the wavelength is given and you calculate the frequency. In (b) the frequency is given and you calculate the wavelength.

SOLUTION:

$$\left[\, \text{ans:} \ (a) \ 4.5 \times 10^{24} \, \text{Hz}; \ (b) \ 1.0 \times 10^{7} \, \text{m} \,\right]$$

PROBLEM 2 (12E). The electric field associated with a certain plane electromagnetic wave is given by $E_x = 0$, $E_y = 0$, $E_z = 2.0 \cos[\pi \times 10^{15}(t - x/c)]$, with $c = 3.0 \times 10^8$ m/s and all quantities in SI units. The wave is propagating in the positive x direction. Write expressions for the components of the magnetic field of the wave.

STRATEGY: Consider an instant of time when the electric field is in the positive z direction and find the direction of the magnetic field so that $\mathbf{E} \times \mathbf{B}$ is in the positive x direction. Choose the trigonometric function that multiplies the field amplitude so that the electric and magnetic fields are in phase. The magnetic field amplitude is given by $B_m = E_m/c$.

SOLUTION:

$$\left[\, \text{ans:} \ B_x = 0, \ B_y = -6.7 \times 10^{-9} \cos[\pi \times 10^{15}(t - x/c)], \ B_z = 0, \text{ in SI units} \,\right]$$

B_y can also be written $+6.7 \times 10^{-9} \cos[\pi \times 10^{15}(t - x/c) + \pi]$ in SI units.

Eqs. 38–15 and 21 of the text relate the electric and magnetic field components of an electromagnetic wave. Each field also obeys a wave equation, a differential equation that contains only a single field. The general wave equation is one of the most important in physics. Any quantity that obeys it propagates as a wave. The next example shows how the wave equation for electromagnetic waves is derived. The one following shows that any function of $kx - \omega t$ or $kx + \omega t$ propagates as a wave, sinusoidal or not, in the positive or negative x direction. This means, for example, that any oscillating electric or magnetic disturbance created in some region propagates to other regions.

PROBLEM 3 (13P). Start from Eqs. 38–15 and 38–21 and show that $E(x,t)$ and $B(x,t)$, the electric and magnetic field components of a plane traveling electromagnetic wave, must satisfy the "wave equations"

$$\frac{\partial^2 E}{\partial t^2} = c^2 \frac{\partial^2 E}{\partial x^2}$$

and

$$\frac{\partial^2 B}{\partial t^2} = c^2 \frac{\partial^2 B}{\partial x^2}$$

STRATEGY: To find the equation for E you must eliminate B from the two given equations. Notice that B enters Eq. 38–15 in a derivative with respect to time and enters Eq. 38–21 in a derivative with respect to x. Differentiate Eq. 38–15 with respect to x to obtain $\partial^2 E/\partial x^2 = -\partial^2 B/\partial t \partial x$ and differentiate Eq. 38–21 with respect to t to obtain $-\partial^2 B/\partial t \partial x = \epsilon_0 \mu_0 \partial^2 E/\partial t^2$. Use the first of these to eliminate $\partial^2 B/\partial t \partial x$ from the second. To obtain the equation for B eliminate E from the two given equations in a similar manner.

SOLUTION:

PROBLEM 4 (14P).   (*a*) Show that Eqs. 38–1 and 38–2 satisfy the wave equations displayed in Problem 13.

STRATEGY: Differentiate Eq. 38–1 twice with respect to t to obtain $\partial^2 E/\partial t^2 = -\omega^2 E_m \sin(kx - \omega t)$. Now differentiate it twice with respect to x to obtain $\partial^2 E/\partial x^2 = -k^2 E_m \sin(kx - \omega t)$. Multiply the last equation by c^2 and replace c^2 on the right side with ω^2/k^2. You should get $c^2 \partial^2 E/\partial x^2 = -\omega^2 E_m \sin(kx - \omega t)$. This is the same as $\partial^2 E/\partial t^2$. Now do the same for Eq. 38–2.

SOLUTION:

(*b*) Show that any expressions of the form

$$E = E_m f(kx \pm \omega t)$$

and

$$B = B_m f(kx \pm \omega t),$$

where $f(kx \pm \omega t)$ denotes an arbitrary function, also satisfy these wave equations.

STRATEGY: Let $u = kx \pm \omega t$. Then $E = E_m f(u)$, $\partial E/\partial x = (dE/du)(\partial u/\partial x) = k\, dE/du$, and $\partial^2 E/\partial x^2 = k^2\, d^2E/du^2$. Similarly, $\partial^2 E/\partial t^2 = \omega^2\, d^2E/du^2$. Multiply the first equation by c^2 and replace c^2 with ω^2/k^2 on the right side. You should get $c^2 \partial^2 E/\partial x^2 = \omega^2\, d^2E/du^2$, which is identical to $\partial^2 E/\partial t^2$. Do the same for B.

SOLUTION:

You should know that the magnitude of the Poynting vector, given for a plane wave by $S = E^2/c\mu_0$, is the energy per unit area per unit time that crosses an area that is perpendicular to the direction of propagation. Its average over a cycle is called the intensity and is given by $I = E_m^2/2c\mu_0$ or by $I = E_{rms}^2/c\mu_0$. For a point source that radiates uniformly in all directions the intensity at a point is proportional to the inverse square of the distance from the source to the point.

You should also know that when an object absorbs electromagnetic radiation it receives momentum: $\Delta p = \Delta U/c$, where ΔU is the energy absorbed. When the radiation is reflected back along the path of incidence the momentum received is $\Delta p = 2\Delta U/c$, where ΔU is the reflected energy. For total absorption the radiation pressure is $p_r = I/c$ and the force on the object is $F = IA/c$, where A is the area struck by radiation. If the radiation is reflected back on its path of incidence, the radiation pressure and force are both twice as great.

PROBLEM 5 (21E).   The intensity of direct solar radiation that is not absorbed by the atmosphere on a particular summer day is $100 \, \text{W/m}^2$. How close would you have to stand to a 1.0-kW electric heater to feel the same intensity? Assume that the heater radiates uniformly in all directions.

STRATEGY: The intensity at a distance r from the heater is $I = P/4\pi r^2$, where P is the power output of the heater. Solve for r.

SOLUTION:

$$\big[\,\text{ans: } 89 \, \text{cm}\,\big]$$

PROBLEM 6 (27P).   Sunlight just outside the Earth's atmosphere has an intensity of $1.40 \, \text{kW/m}^2$. Calculate E_m and B_m for sunlight, assuming it to be a plane wave.

STRATEGY: In terms of E_m the intensity is given by $I = E_m^2/c\mu_0$. Solve for E_m. Use $B_m = E_m/c$ to calculate the magnetic field amplitude.

SOLUTION:

$$\big[\,\text{ans: } E_m = 1.03 \times 10^3 \, \text{V/m}, \ B_m = 3.43 \times 10^{-6} \, \text{T}\,\big]$$

PROBLEM 7 (40P).   A helium-neon laser of the type often found in physics laboratories has a beam power of $5.00 \, \text{mW}$ at a wavelength of 633 nm. The beam is focused by a lens to a circular spot whose effective diameter may be taken to be 2.00 wavelengths. Calculate (a) the intensity of the focused beam, (b) the radiation pressure exerted on a tiny perfectly absorbing sphere whose diameter is that of the focal spot, (c) the force exerted on this sphere, and (d) the acceleration imparted to it. Assume a sphere density of $5.00 \times 10^3 \, \text{kg/m}^3$.

STRATEGY: (a) The intensity is given by $I = P/\pi R^2$, where R is the radius of the spot (633 nm). (b) The radiation pressure is given by $p_r = I/c$. (c) The net force is given by $F = \pi R^2 p_r$. (d) The acceleration of the sphere is given by $a = F/m$, where m is the mass of the sphere, the product of its density and volume.

SOLUTION:

$$\big[\,\text{ans: } (a) \ 3.97 \times 10^9 \, \text{W/m}^2; \ (b) \ 13.2 \, \text{Pa}; \ (c) \ 1.67 \times 10^{-11} \, \text{N}; \ (d) \ 3.14 \times 10^3 \, \text{m/s}^2\,\big]$$

PROBLEM 8 (42P).   Radiation of intensity I is normally incident on an object that absorbs a fraction *frac* of it and reflects the rest back along the original path. What is the radiation pressure on the object?

STRATEGY: For the fraction that is absorbed the radiation pressure is I/c and for the fraction that is reflected the radiation pressure is $2I/c$. Compute the total.

SOLUTION:

$$\big[\,\text{ans: } (2 - frac)I/c\,\big]$$

PROBLEM 9 (46P).   It has been proposed that a spaceship might be propelled in the solar system by radiation pressure, using a large sail made of foil. How large must the sail be if the radiation force is to be equal in magnitude to the sun's gravitational attraction? Assume that the mass of the ship + sail is 1500 kg, that the sail is perfectly reflecting, and that the sail is oriented perpendicularly to the sun's rays. See Appendix C for needed data. (With a larger sail, the ship is continually driven away from the sun.)

STRATEGY: The magnitude of the radiation force on the sail is $F_r = p_r A = 2IA/c$, where p_r is the radiation pressure, A is the sail area, and I is the radiation intensity. If P is the rate at which the sun radiates energy, then $I = P/4\pi r^2$, where r is the distance from the sun to the ship. The magnitude of the gravitational force of the sun on the ship is $F_g = GmM/r^2$, where m is the mass of the ship and sail together and M is the mass of the sun. Equate the magnitudes of the two forces and solve for A. According to the appendices $P = 3.90 \times 10^{26}$ W, $M = 1.99 \times 10^{30}$ kg, and $G = 6.67 \times 10^{-11}$ m^3/s$^2 \cdot$ kg.

SOLUTION:

[ans: 4.8×10^5 m^2]

Some problems deal with polarization. You should know how to identify the direction and plane of polarization. You should also know how to compute the intensity of radiation transmitted by a polarizing sheet. Remember than the exiting radiation is polarized in the polarizing direction of the sheet.

PROBLEM 10 (48E). The magnetic field equations for an electromagnetic wave in vacuum are $B_x = B\sin(ky + \omega t)$, $B_y = B_z = 0$. (a) What is the direction of propagation?

STRATEGY: Look at the argument of the sine function.

SOLUTION:

[ans: negative y direction]

(b) Write the electric field equations.

STRATEGY: The electric field amplitude is $E = cB$. The direction of the electric field must such that $\mathbf{E} \times \mathbf{B}$ is in the negative y direction and the electric field is in phase with the magnetic field.

SOLUTION:

[ans: $E_x = 0$, $E_y = 0$, $E_z = -cB\sin(ky + \omega t)$]

(c) Is the wave polarized? If so, in what direction?

STRATEGY: Is the electric field always parallel to the same line? If so, what is the direction of the line?

SOLUTION:

[ans: yes, parallel to the z axis]

PROBLEM 11 (50E).   A beam of unpolarized light is sent through two polarizing sheets placed one on top of the other. What must be the angle between the polarizing directions of the sheets if the intensity of the transmitted light is one-third the intensity of the incident light?

STRATEGY: After passing through the first sheet the intensity is one-half the incident intensity and the wave is polarized in the polarizing direction of the sheet. After passing through the second sheet it is $\frac{1}{2}\cos^2\theta$ times the incident intensity. Here θ is the angle between the polarizing directions. Solve $\frac{1}{2}\cos^2\theta = 1/3$ for θ.

SOLUTION:

$$[\text{ans: } 35°]$$

PROBLEM 12 (53P).   In Fig. 38–23, initially unpolarized light is sent through three polarizing sheets whose polarizing directions make angles of $\theta_1 = 40°$, $\theta_2 = 50°$, and $\theta_3 = 40°$ with the direction of the y axis. What percentage of the initial intensity is transmitted by the system?

STRATEGY: After passing through the first sheet the intensity is one-half the incident intensity and the wave is polarized in the polarizing direction of the sheet. The polarizing directions of the first and second sheets differ by 10° so the intensity after the second sheet is $\frac{1}{2}\cos^2 10°$ times the incident intensity. The polarizing directions of the second and third sheets also differ by 10° so the final intensity is $\frac{1}{2}\cos^4 10°$ times the incident intensity.

SOLUTION:

$$[\text{ans: } 47\%]$$

PROBLEM 13 (58P).   A beam of partially polarized light can be considered as a mixture of polarized and unpolarized light. Suppose we send such a beam through a polarizing filter and then rotate the filter through 360° while keeping it perpendicular to the beam. If the transmitted intensity varies by a factor of 5.0 during the rotation, what fraction of the intensity of the original beam is associated with the beam's polarized light?

STRATEGY: Let f be the fraction of the incident intensity that is associated with polarized light. If I is the total incident intensity then the intensity associated with unpolarized light is $(1-f)I$ and the intensity of this light after transmission is $(1-f)I/2$. The incident intensity associated with polarized light is fI and the intensity of this light after transmission is $fI\cos^2\theta$, where θ is the angle between the direction of polarization of the light and the polarizing direction of the sheet. The total intensity after transmission is $(1-f)I/2 + fI\cos^2\theta$. As the filter is rotated the maximum intensity is $I_{\max} = (1-f)I/2 + fI = (1+f)I/2$ and minimum is $I_{\min} = (1-f)I/2$. Solve $I_{\max} = 5.0I_{\min}$ for f.

SOLUTION:

$$[\text{ans: } 2/3]$$

PROBLEM 14 (59P).   We want to rotate the direction of polarization of a beam of polarized light through 90° by sending the beam through one or more polarizing sheets. (*a*) What is the minimum number of sheets required?

STRATEGY: The beam will be polarized along the polarizing direction of the final sheet. One sheet will not do since it will not transmit any light if it is placed with its polarizing direction at 90° to the direction of polarization of the incident light. Explain how to do it with 2.

SOLUTION:

[ans: 2]

(*b*) What is the minimum number of sheets required if the transmitted intensity is to be more than 60% of the original intensity?

STRATEGY: Suppose there are n sheets, with the angle $\theta = 90°/n$ between the polarizing directions of successive sheets. The ratio of the transmitted intensity I to the incident intensity I_i is given by $(I/I_i) = \cos^{2n}\theta$. Solve $\cos^{2n}(90°/n) = 0.60$ for n. Use trial and error. Round down to the nearest smaller integer.

SOLUTION:

[ans: 5]

III. NOTES

Chapter 39
GEOMETRICAL OPTICS

I. BASIC CONCEPTS

With this chapter you begin the study of optics. In particular, you will learn what happens when light is incident on the boundary between two materials and learn to determine the propagation directions of the reflected and refracted light. These are given by the laws of reflection and refraction. First, you should understand what geometrical optics is and when it is valid.

The law of reflection is applied to spherical mirrors and the law of refraction is applied to spherical refracting surfaces. Each type surface forms images of objects placed in front of it and you will learn the relationship between the position of the object, the position of the image, and the radius of curvature of the surface. You will also apply what you learn to lenses with spherical faces and to systems of lenses, such as those used in telescopes, microscopes, and cameras.

Pay careful attention to ray tracing techniques. They will help you to visualize image formation and to understand the position and size of the image. When using the equations relating object and image distances be sure to concentrate on the signs of the variables. Knowing how to choose the signs of given quantities and how to interpret the signs of results is vital for obtaining correct answers.

Geometrical and wave optics. In physical optics (or wave optics) the wave nature of light is used to describe and understand optical phenomena. Wave optics is close to the fundamental principles, Maxwell's equations, and is valid for all optical phenomena outside the quantum realm. However, when any obstacles to light or any openings through which light passes are large compared to the wavelength of the light then details of its wave nature are not important. What is important is the direction of travel of the light. This is the realm of geometrical optics (or ray optics).

Geometrical optics uses rays to describe the path of light. A ray is a line in the direction of _____ of a light wave. The propagation direction changes when light is reflected or enters a region where the wave speed is different.

The laws of reflection and refraction. When light in one medium encounters a boundary with another region, in which the wave speed is different, some light is reflected back into the region of incidence and some is transmitted into the second region. Two laws, the law of reflection and the law of refraction, describe the directions of propagation of the reflected and transmitted light.

The diagram shows a ray in medium 1 incident on the boundary with medium 2. It makes the angle θ_1 with the normal to the boundary. Reflected and refracted rays all lie in the plane of incidence, defined by the incident ray and the normal to the surface. The reflected ray makes the angle θ_1' with the normal and the transmitted ray makes the angle θ_2 with the normal. The law of reflection is

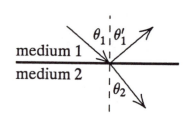

$$\theta_1' =$$

and the law of refraction is

$$n_2 \sin \theta_2 =$$

where n_1 is the index of refraction for medium 1 and n_2 is the index of refraction for medium 2. The angle θ_1 is called the angle of _____, θ_1' is called the angle of _____, and θ_2 is called the angle of _____. The law of refraction is also known as Snell's law.

The index of refraction of a medium is the ratio of the speed of light in vacuum to the speed of light in the medium. Specifically, if v is the speed of light in the medium then

$$n =$$

Look at Table 39–1 of the text for some values for yellow sodium light. Of the materials listed the one with the largest index of refraction is _____, with $n =$ _____. The speed of light is _____ in this material than in any other listed. The index of refraction depends on the wavelength; light of different wavelengths, incident at the same angle, is refracted through different angles. See Fig. 39–2 of the text.

You should understand Snell's law: $n_1 \sin \theta_1 = n_2 \sin \theta_2$. Recall that $\sin \theta$ increases as θ increases from 0 to 90°. If n_1 is greater than n_2 then θ_1 is less than θ_2. Light incident from a medium with a small index of refraction bends toward the normal as it enters a medium with a higher index of refraction. Light incident from a medium with a large index of refraction bends away from the normal as it enters a medium with a smaller index of refraction. When you shine light into water from air it bends _____ the normal. Light from an underwater source bends _____ the normal as it enters the air.

<u>Total internal reflection</u> can occur when light travels in an optically dense medium (large index of refraction) toward a less optically dense medium (smaller index of refraction). If the angle of incidence is greater than a certain critical value denoted by θ_c no light is transmitted into the less optically dense medium. All the light incident on the boundary is reflected. Suppose $n_1 > n_2$ and light is incident from medium 1 on the boundary with medium 2. Then $\theta_1 = \theta_c$ if $\theta_2 =$ _____ and $\sin \theta_2 =$ _____. In terms of n_1 and n_2, $\theta_c =$ _____.

Polarization by reflection. If unpolarized light is incident on a boundary between two different materials, both the reflected and refracted waves are partially polarized for most directions of incidence. There is, however, a special angle of incidence, called Brewster's angle and denoted by θ_B, for which the reflected wave contains only one polarization component.

If θ_r is the angle of refraction for incidence at Brewster's angle, then $\theta_B + \theta_r =$ _____ . If n_1 is the index of refraction for the medium of incidence and n_2 is the index of refraction for the medium of refraction then Snell's law leads to

$$\tan \theta_B =$$

This is Brewster's law for the polarizing angle.

The direction of polarization for light reflected at the polarizing angle is normal to the plane determined by the _____ and the _____ . Carefully note that even at Brewster's angle the *refracted* light is partially, not completely, polarized. It contains light with all polarization directions but is deficient in light with the polarization of the reflected light.

Images in mirrors. When you view light from an object after it has been reflected by a plane mirror the light appears to come from points behind the mirror and, in fact, you see an image that appears to be behind the mirror.

The diagram shows a point source P in front of a mirror and two rays emanating from it. The reflected rays are drawn according to the law of reflection: the angles of incidence and reflection are the same. Draw dotted lines to extend the reflected rays to behind the mirror and use P' to label the point where they intersect. This is the image of the source. When your eyes view the light reflected from the mirror it appears to come from a point source at P'.

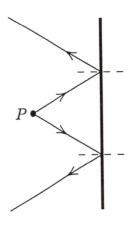

You should realize that many other rays from the source to the mirror could have been drawn. All reflected rays, extended backward to the other side of the mirror, intersect at P'. When you look at mirror only a very narrow bundle of light enters your eyes, but all rays in the bundle diverge from the image.

The image point P' is on the line through the source P that is _____ to the plane of the mirror. It is the same distance behind the mirror as _____ is in front. If you want to know if you can see the image when your eye is in any given position, draw a line from the image to your eye. If the line intersects the mirror then light reflected from the mirror gets to your eye and you see the image. If the line does not intersect the mirror you do not see the image.

If the source is extended you may think of each point on it as a point source, emitting light in all directions. Only the light that reaches the mirror is reflected, of course. For each point that sends light to the mirror an image point is formed. In many examples an extended source is made up of one or more straight lines. To find the image of a line simply find the images of each end and connect them with a straight line. Clearly the image of an extended source is the same size as the source.

You should be familiar with some of the terms used. A source of light is often called an object. An image may be real or virtual. In both cases light diverges from the image but if the image is real the light actually _____ , while if it is virtual it does not. Images formed by plane mirrors are _____ .

The distance from a point source to a mirror is denoted by p. It is positive. The image position is denoted by i. It is negative (for virtual images) and its magnitude is the distance from the image to the mirror. In terms of these quantities the statement that the image is as far behind the mirror as the object is in front is written

$$i =$$

Spherical mirrors. The diagrams below show two spherical mirrors, with their centers labelled c and their centers of curvature labelled C. A point source of light S is in front of each mirror. The line through the center of curvature and the center of the mirror is called the _____. Usually angles are measured with respect to this line and distances are measured along it. Light from the source is reflected by the mirror.

On each diagram draw a ray from the source to the mirror, about halfway to its upper edge. Use a dotted line to draw the normal to the mirror at the point where the ray strikes, then draw the reflected ray. The rays should obey the law of reflection at the mirror, but you will probably have some drawing error.

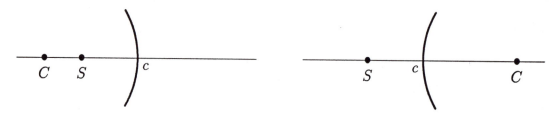

In some cases reflected rays converge to a point on the source side of the mirror. Reflected light then forms a _____ image. In other cases the reflected rays diverge as they leave the mirror but they follow lines that pass through a single point behind the mirror. That is, they form a _____ image. The region in front of the mirror, where the source is located and where real images are formed, is called the R-side of the mirror. The region behind, where virtual images are located, is called the V-side of the mirror. Indicate the R-side and V-side on each of the diagrams above. Strictly speaking, sharp images are formed only by light whose rays make small angles with the central axis.

The object distance p is the distance from the source to the mirror, measured along the central axis. It is positive for a source in front of a single mirror. Indicate p on each of the diagrams above. The image distance i is the distance from the mirror to the image position, measured along the central axis. A sign is associated with it. It is positive for a _____ image, located _____ the mirror and negative for a _____ image, located _____ the mirror.

The mirror equation relates the object distance p, the image distance i, and the radius of curvature r. It is

$$\frac{1}{p} + \frac{1}{i} =$$

This is often written in terms of the <u>focal length</u>, defined by $f =$ _____. Then

$$\frac{1}{p} + \frac{1}{i} =$$

Signs are associated with the radius of curvature and focal length as well as with object and image distances. Both r and f are positive if the center of curvature is on the _____-side of the mirror and negative if the center of curvature is on the _____-side. A mirror that is <u>concave</u> with respect to the source has a _____ radius of curvature and focal length; a mirror that is <u>convex</u> with respect to the source has a _____ radius of curvature and focal length. On each diagram above indicate whether the mirror is concave or convex with respect to S and give the sign of r and f.

The <u>focal point</u> of a mirror is important for tracing rays and graphically finding the position of an image when the object is not on the central axis. It is on the central axis, a distance $|f|$ from the mirror. If the mirror is concave it is on the _____-side; if the mirror is convex it is on the _____-side. For each of the mirrors above place a dot at the focal point and label it F.

Incident light with rays that are parallel to the central axis is reflected so its rays are along lines that pass through _____. If the focal point is on the R-side the rays actually pass through it. If the focal point is on the V-side they do not but their extensions backward into the V-region do. Incident light with rays along lines that pass through the focal point is reflected so its rays are along lines that are _____.

The image is located at the intersection of all small-angle reflected rays. The two special rays mentioned above are easy to locate and draw, so they are often used to locate an image. Another that is used for the same reason is a ray along a line through the center of curvature. Its reflection is along a line that passes through _____.

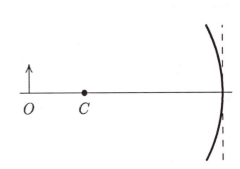

The diagram shows a spherical mirror with its center of curvature at C. Locate the focal point and mark it F. Locate the image of the arrow at O by drawing two rays both before and after reflection. The first is from the head of the arrow to the mirror and is parallel to the axis. The reflected ray goes through _____. The second is from the head of the arrow through C to the mirror. The reflected ray goes through _____. Draw the image of the arrow with its head at the point where these rays intersect and its tail on the axis.

You might also draw a ray from the arrow head through the focal point to the mirror. The reflected ray is parallel to the axis and passes through the intersection of the rays you have already drawn. All of the rays you draw may not intersect at precisely the same point and none of the intersections may agree exactly with the image position as calculated by the mirror equation. Discrepancies arise because the rays you have drawn do not make sufficiently small angles with the central axis. You can obtain a single intersection point and agreement with the mirror equation if you draw the rays from the object to the dotted line shown, rather than to the actual mirror surface.

Note that the image is real. Light actually passes through points of the image. It is also inverted. If you move the object arrow to the left, away from the mirror, the image moves _____ the mirror. To verify this imagine the object after it is moved to the left and note what happened to the ray through the center of curvature and its intersection with the ray that is parallel to the axis. The latter does not change as the object moves.

Now suppose the object is placed between the focal point and the mirror, as shown. Draw a line from the center of curvature through the head of the arrow to the mirror. The portion between the arrow head and the mirror is a ray. It is reflected through _____. Draw a ray from the arrow head to the mirror, parallel to the axis. It is reflected through _____. Use dotted lines to extend the reflected rays backward into the region behind the mirror, then locate the image of the arrow head at their intersection and draw the image of the arrow.

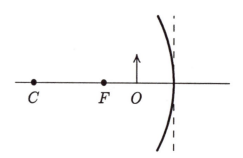

You might also use a line from the focal point through the arrow head to the mirror. The portion between the head and the mirror is a ray and its reflection is parallel to the axis. Its extension passes through the intersection of the dotted lines you have drawn.

Notice that the image is virtual and erect. It is behind the mirror and no light reaches it but reflected light in front of the mirror appears to come from it. If the object arrow is moved closer to the mirror, its image moves _____ the mirror.

The results you found graphically are predicted by the mirror equation. When it is solved for i the result is $i = fp/(p - f)$. If $p > f$ then i is positive: the image is real and is on the R-side. If $p < f$ then i is negative: the image is virtual and is on the V-side. If $p = f$ all small-angle reflected rays are parallel to each other and the image is said to be at infinity.

Now consider a convex mirror. Locate the focal point on the diagram and label it F. Draw a ray from the head of the arrow to the mirror, parallel to the axis. Draw the reflected ray. It is along a line that passes through _____. Draw a ray from the arrow head to the mirror, along the line that passes through C. Draw the reflected ray. It is along a line that passes through _____. Use dotted lines to extend the reflected rays to behind the mirror and draw the image of the arrow at their intersection. It is virtual and erect.

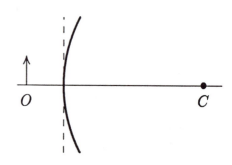

You might also use a line from the head of the arrow through the focal point. The reflected ray is parallel to the axis and its extension behind the mirror should pass through the intersection of the other lines.

The mirror equation also predicts a virtual image. Since f is negative $i = fp/(p - f)$ is negative for any positive value of p.

The lateral magnification m of a mirror is the ratio of the lateral size of the _____ to the lateral size of the _____. The term *lateral size* means the dimension perpendicular to the central axis. In terms of the object distance p and image distance i the lateral magnification of a spherical mirror is given by

$$m =$$

The negative sign here has special meaning. Values for p and i are substituted with their signs. If m is positive then the orientations of the object and image are _____; if m is negative then the orientations of the object and image are _____. Notice that virtual images of an erect object are erect and real images of an erect object are inverted.

Spherical refracting surfaces. The diagrams below show two spherical refracting surfaces, with their centers of curvature labelled C. Each separates a medium with an index of refraction n_1 from a medium with index of refraction n_2. Light from point source S in front of a surface is refracted at the surface and continues into the medium on the other side. Suppose n_2 is greater than n_1, so the rays are bent toward the normal, and for each diagram draw a ray from the source that strikes the surface about halfway up. Use a dotted line to draw the normal to the surface at the point where the ray strikes and show the ray after refraction.

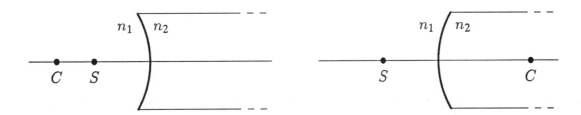

If the rays make small angles with the central axis, the light forms a sharp image of the source. If the image is virtual it is formed in the region containing the source and this is called the V-side of the surface. If the image is real it is formed on the side into which light is transmitted. This region is called the R-side of the surface. Label the two sides on each diagram above.

For a single refracting surface the object distance p is positive. Indicate it on each diagram above. The image distance is positive if the image is real and negative if it is virtual. For the surfaces and sources of the diagrams above images with positive values of i are to the _____ of the surface and images with negative values of i are to the _____.

The law of refraction yields a relationship between the object distance, the image distance, the radius of curvature of the surface, and the indices of refraction for the two sides. It is

$$\frac{n_1}{p} + \frac{n_2}{i} =$$

For this equation to be valid the appropriate sign must be associated with the radius of curvature. If the surface is concave with respect to the source, the center of curvature is on the _____-side and r is _____. If the surface is convex with respect to the source, the center of curvature is on the _____-side and r is _____. In terms of the labels R and V used to designate the regions the sign convention for r and i is the same as the convention for spherical mirrors. Remember that the R-side is the side the light goes to after striking the surface. For a mirror this is the side of the source; for a refracting surface it is the other side.

If $n_2 > n_1$ then a convex surface forms a real image of an object that is far from it and a virtual image of an object that is near it. Details depend on the values of the indices of refraction and the radius of curvature. A concave surface, on the other hand, always forms a _____ image. If $n_1 > n_2$ then a convex surface always forms a _____ image and a concave surface forms a _____ image for an object that is far from it and a _____ image for an object that is near it.

Thin lenses. A lens consists of two refracting surfaces. The image formed by the first may be considered the object for the second. If the surrounding medium is a vacuum and the thickness of the lens can be neglected, then the object distance p, image distance i, and focal length f are related by

$$\frac{1}{p} + \frac{1}{i} =$$

where the focal length is given in terms of the radii of curvature by

$$\frac{1}{f} =$$

Here r_1 is the radius of curvature of the surface nearer the object, r_2 is the radius of curvature of the surface farther away, and n is the index of refraction of the lens material. The equation is a good approximation if the surrounding medium is air.

Again you should be familiar with the sign convention. Real images are formed on the opposite side of the lens from the object and this side is labelled the _____ -side. For these images i is _____. Virtual images are formed on the same side of the lens as the object and this side is labelled the _____ -side. For a virtual image i is _____. A surface radius is positive if the center of curvature is on the _____ -side and negative if the center of curvature is on the _____ -side. You should know that the focal length of a lens does not depend on which side faces the object. Lenses with positive focal lengths are said to be _____, while lenses with negative focal lengths are said to be _____.

For each of the lenses shown below give the sign of each radius, the sign of the focal length, and say if the lens is converging or diverging. Assume the object is to the left of the lens.

sign of r_1:   _____   _____   _____   _____
sign of r_2:   _____   _____   _____   _____
sign of f: _____   _____   _____   _____

                   _____   _____   _____   _____

A lens has two focal points, located equal distances $|f|$ on opposite sides of the lens. The first focal point, denoted by F_1, is on the side of the incident light (the V-side) for a converging lens (f positive) and on the side of the refracted light (the R-side) for a diverging lens (f negative). Light rays that are along lines that pass through F_1 are bent by the lens to become parallel to the axis. For a converging lens the rays before refraction actually pass through F_1. For a diverging lens the rays strike the surface, so only their extensions into the R-side pass through F_1. For each of the lenses shown below draw an incident ray along a line that passes through F_1. Also draw the refracted ray, parallel to the axis. For the diverging lens use a dotted line to continue the line of the incident ray to the focal point. Since we have neglected the thickness of the lenses you may assume refraction takes place at the plane through the lens center, perpendicular to the axis. These planes are shown as dotted lines on the diagrams.

The second focal point, denoted by F_2, is on the side of the refracted light (the R-side) for a converging lens (f positive) and on the side of the incident light (the V-side) for a diverging lens (f negative). Light rays that are parallel to the central axis are bent by the lens to lie along lines that pass through F_2. For a converging lens the rays after refraction actually do pass through F_2. For a diverging lens the backward extension of the rays into the V-side pass through F_2. For each of the lenses shown below draw an incident ray parallel to the central axis. Also draw the refracted ray, along a line through F_2. For the diverging lens use a dotted line to continue the line of the refracted ray backward to the focal point.

The position of an off-axis image can be found graphically by tracing two or more rays originating at the same point on the object. Two of these might be the rays you drew above: one along a line through F_1 and then parallel to the axis and the other parallel to the axis and then along a line through F_2. The third ray you might use goes through _____ of the lens. It is not refracted.

In terms of the object and image distances the lateral magnification associated with a lens is given by

$$m =$$

an expression that is identical to the expression for a spherical mirror. If m is negative the image of an upright object is _____; if m is positive the image of an upright object is _____. A virtual image formed by single thin lens is always _____; a real image is always _____.

Angular magnification. Because it takes into account the apparent diminishing of the size of an object with distance, the <u>angular magnification</u> is often a better measure of the usefulness of a lens used for viewing than is the lateral magnification.

If an object has a lateral dimension (perpendicular to the central axis) of h and is a distance d from the eye, as in Fig. 39–25(a) of the text, its angular size is given in radians by $\theta =$ _____. If the object is viewed through a lens and the image is a distance d' from the eye and has a lateral dimension of h' then the angular size of the image is $\theta' =$ _____. The angular magnification of the lens is the ratio of these, or $m_\theta = \theta'/\theta =$ _____, where the last expression gives the angular magnification in terms of h, h', d, and d'.

A single converging lens used as a magnifying glass is usually positioned so the object is just inside the focal point F_1. See Fig. 39–25(c) of the text. The image is then virtual and far away. Its lateral size is $h' = mh = -ih/p$, where m is the lateral magnification. Thus $m_\theta = -(i/p)(d/d')$. The eye is close to the lens so $d' \approx |i|$. Furthermore $p \approx f$. Once these substitutions are made the result is $m_\theta = d/f$. Usually d is taken to be the distance to the near point of the eye, about 15 cm, so $m_\theta = 15/f$, where f must be measured in centimeters.

The near point is used since the object is in focus and has its largest angular size when it is this distance from an unaided eye. The angular magnification then tells us how much better the lens is than the best the unaided eye can do.

Lens systems. Telescopes, microscopes, and other optical instruments usually consist of a series of lenses. They can be analyzed graphically by tracing a few rays as they pass through each lens in succession. The lens equation can also be applied to each lens in succession to find the position of the image. Consider a system of two lenses, a distance ℓ apart on the same central axis, as shown below. Let p_1 be the distance from the object O to the first lens struck. This lens forms an image I_1 at i_1, given by $1/p_1 + 1/i_1 = 1/f_1$, where f_1 is the focal length of the lens. You may think of this image as the object for the second lens. The object distance is $p_2 = \ell - i_1$ and the image distance i_2 is given by $1/p_2 + 1/i_2 = 1/f_2$, where f_2 is the focal length of the second lens. On the diagram identify and label p_1, i_1, p_2, and i_2. Verify that $i_1 + p_2 = \ell$.

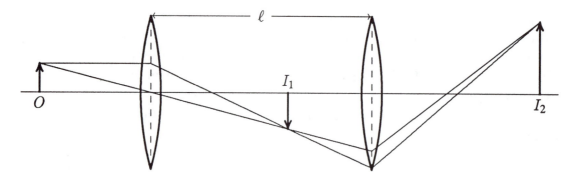

For some systems the object distance for the second lens may be negative. This occurs if the image formed by the first lens is behind the second lens, so the light exiting the first lens is converging as it strikes the second lens. Such objects are called *virtual* objects. The equation relating object and image distances is still valid — just substitute a negative value for p_2.

A simple microscope consists of an objective lens, near the object, and an eyepiece (or ocular). The focal length of the objective lens is small and the lens is positioned so the object lies just outside the _____ of the lens. The image produced by the objective is real, large, inverted, and far from the lens. If the object has a lateral dimension h then the image has a lateral dimension $h' = mh = ih/p$. The magnification is great since i is large and p is small.

The distance s between the second focal point of the objective lens and the first focal point of the eyepiece is called the _____ of the microscope. In terms of this quantity $h' = $ _____, where θ is the angle the ray through the focal point makes with the central axis. Since $h = f_{\text{ob}} \tan \theta$, the magnification of the objective lens is given by $m = $ _____.

The eyepiece is positioned so the image produced by the objective lens falls at its _____ focal point. The eye is placed close to the eyepiece, which then acts as a simple magnifying glass with an angular magnification of $15\,\text{cm}/f_{\text{ey}}$, where f_{ey} is the focal length of the eyepiece. The overall angular magnification is given by $M = $ _____. The expression again compares the angular size of the image produced by the microscope with the angular size of the object when it is at the near point of an unaided eye.

Telescopes are used to view objects that are far away. Rays entering the objective lens are essentially parallel to the central axis and that lens produces an image that is close to the _____ focal point. To compute the angular magnification of a telescope we compare the angular size of the object at its far-away position (not the near point) to the angular size of the image produced by the telescope.

Since the length of the telescope is much less than the object distance the angle subtended by the object at the eye is the same as the angle it subtends at the objective lens. This is the same as the angle subtended at the objective lens by the image produced by that lens, so $\theta_{\text{ob}} = h'/f_{\text{ob}}$, where h' is the lateral dimension of the image. The angle subtended by the final image at the eye is the same as the angle subtended by the intermediate image at the eyepiece, or $\theta_{\text{ey}} = h'/f_{\text{ey}}$. The angular magnification of the telescope is given by $m_\theta = \theta_{\text{ob}}/\theta_{\text{ey}} = $ _____. To obtain a large angular magnification, a telescope should have an objective lens with a _____ focal length and an eyepiece with a _____ focal length.

II. PROBLEM SOLVING

Many problems deal with refraction at a single plane surface and with total internal reflection. For refraction problems use Snell's law: $n_1 \sin \theta_1 = n_2 \sin \theta_2$. Remember to measure the angle from the surface normal. For total internal reflection remember that $n_1 \sin \theta > n_2$, where n_1 is the index of refraction for the medium of incidence and n_2 is the index of refraction for the medium beyond the surface. Other problems ask you calculate the Brewster angle. Use $\tan \theta_B = n_2/n_1$, where n_1 is the index of refraction for the medium of incidence and n_2 is the index of refraction for the medium of the refracted light. In some cases you will also need to use geometry to trace rays. Here are some examples.

PROBLEM 1 (8P). A penny lies at the bottom of a pool with depth d and index of refraction n, as shown in Fig. 39–35. Show that light rays that are close to the normal appear to come from a point that is a distance $d_a = d/n$ below the surface. This distance is the apparent depth of the pool.

STRATEGY: Suppose one of the rays shown makes the angle θ_1 with the vertical for points in the water and the angle θ_2 with the vertical for points outside the water. Snell's law gives $n_1 \sin \theta_1 = \sin \theta_2$. Let ℓ be half the distance between the rays at the points where they emerge from the water. Then $\tan \theta_1 = \ell/d$. Extend the emerging rays backward to the point where they meet, a distance d_a (the apparent depth) below the water surface. Then $\tan \theta_2 = \ell/d_a$. Use one of these equations to eliminate ℓ from the other, then solve for d_a. You should obtain $d_a = (\tan \theta_1/ \tan \theta_2)d$. Since the angles are small each tangent may be replaced by the sine of the same angle: $d_a = (\sin \theta_1/ \sin \theta_2)d$. Use Snell's law to eliminate $\sin \theta_1/ \sin \theta_2$.

SOLUTION:

PROBLEM 2 (13P).   The index of refraction of the Earth's atmosphere decreases monotonically with height from its surface value (about 1.00029) to the value in space (1.00000) at the top of the atmosphere. This continuous (or graded) variation can be approximated by considering the atmosphere to be composed of three (or more) plane parallel layers in each of which the index of refraction is constant. Thus, in Fig. 39–39, $n_3 > n_2 > n_1 > 1.00000$. Consider a ray of light from a star S that strikes the top of the atmosphere at an angle θ with the vertical. (a) Show that the apparent direction θ_3 of the star relative to the vertical as seen by an observer at the Earth's surface is given by

$$\sin \theta_3 = \left(\frac{1}{n_3} \right) \sin \theta .$$

(*Hint:* Apply the law of refraction to successive pairs of layers of the atmosphere; ignore the curvature of the Earth.)

STRATEGY: The successive application of the law of refractions yields $\sin \theta = n_1 \sin \theta_1$, $n_1 \sin \theta_1 = n_2 \sin \theta_2$, and $n_2 \sin \theta_2 = n_3 \sin \theta_3$. Solve for $\sin \theta_3$ in terms of $\sin \theta$.

SOLUTION:

(b) Calculate the angular shift $\theta - \theta_3$ of a star observed to be 20.0° from the vertical. (The very small effects due to the atmospheric refraction can be most important; for example, they must be taken into account in using navigation satellites to obtain accurate fixes of position on Earth.)

STRATEGY: Take $\theta_3 = 20.0°$ and calculate θ, then $\theta - \theta_3$. Use $n_3 = 1.00029$.

SOLUTION:

[ans: 0.00605°]

PROBLEM 3 (14P).   In Fig. 39–40, a ray is incident on one face of a glass prism in air. The angle of incidence θ is chosen so that the emerging ray also makes the same angle θ with the normal to the other face. Show that the index of refraction n of the glass prism is given by

$$n = \frac{\sin \frac{1}{2}(\psi + \phi)}{\sin \frac{1}{2}\phi} ,$$

where ϕ is the vertex angle of the prism and ψ is the *deviation angle*, the total angle through which the beam is turned in passing through the prism. (Under these conditions the deviation angle ψ has the smallest possible value, which is called the *angle of minimum deviation*.)

STRATEGY: Apply Snell's law to the refraction that takes place at the left face of the prism to obtain $n = \sin\theta / \sin\theta'$, where θ' is the angle of refraction. You want to find expressions for θ and θ' in terms of ϕ and ψ. Consider the triangle formed by the ray inside the prism, the incident ray extended forward to inside the prism, and the emerging ray extended backward to inside the prism. Two of the angles are equal. Label them β. The third angle is $180° - \psi$. Since the sum of the interior angles of a triangle is $180°$, $2\beta + 180° - \psi = 180°$. Solve for β. It is $\psi/2$. Now consider the triangle formed by the ray in the prism and the prism faces. Again two of the angles are equal. Label them γ. Since the interior angles sum to $180°$, $\gamma = 90° - \phi/2$. Now $\theta' + \gamma = 90°$ so $\theta' = \phi/2$. Furthermore $\theta = \theta' + \beta$ so $\theta = \frac{1}{2}(\phi + \psi)$. Substitute these expressions into the equation for n.

SOLUTION:

PROBLEM 4 (18E). In Fig. 39–42, a ray of light is perpendicular to the face ab of a glass prism ($n = 1.52$). Find the largest value for the angle ϕ so that the ray is totally reflected at face ac if the prism is immersed (*a*) in air and (*b*) in water.

STRATEGY: The angle of incidence θ at the right face of the prism is given by $\theta = 90° - \phi$. If ϕ is the largest angle for which total internal reflection occurs then the angle of refraction is $90°$. Snell's law gives $n_1 \sin\theta = n_2$, where n_1 is the index of refraction for the prism and n_2 is the index of refraction for the surrounding medium. Substitute $90° - \phi$ for θ and solve for ϕ.

SOLUTION:

$$[\text{ans: } (a)\ 49°;\ (b)\ 29°]$$

PROBLEM 5 (22P). A ray of white light traveling in fused quartz strikes a plane surface of the quartz, with an angle of incidence θ. Is it possible for the internally reflected beam to appear (*a*) bluish or (*b*) reddish? (*c*) If so, what value of θ is required? (*Hint:* White light will appear bluish if wavelengths corresponding to red are removed from the spectrum, and vice-versa.)

STRATEGY: For the reflected light to appear bluish we want blue to be totally internally reflected and red to be partially transmitted. That is, we want $\sin\theta > 1/n$ for blue and $\sin\theta < 1/n$ for red. Look at Fig. 39–2 of the text and notice that n decreases with increasing wavelength. If $\sin\theta > 1/n$ for blue then $\sin\theta > 1/n$ for red. For the reflected light to appear reddish we want $\sin\theta > 1/n$ for red and $\sin\theta < 1/n$ for blue. This is possible. Pick a wavelength between blue and red (500 nm, say), obtain the value for n from Fig. 39–2, and calculate θ.

SOLUTION:

$$[\text{ans: } (a)\ \text{no; } (b)\ \text{yes; } (c)\ \text{in the neighborhood of } 42°]$$

PROBLEM 6 (26P). An optical fiber consists of a glass core (index of refraction n_1) surrounded by a coating (index of refraction $n_2 < n_1$). Suppose a beam of light enters the fiber from air at an angle θ with the fiber axis as shown in Fig. 39–44. (a) Show that the greatest possible value of θ for which a ray can travel down the fiber is given by $\theta = \sin^{-1} \sqrt{n_1^2 - n_2^2}$.

STRATEGY: Suppose θ' is the angle of refraction at the fiber-air interface. Then $90° - \theta'$ is the angle of incidence at the fiber-coating interface. For total internal reflection to occur there we must have $n_1 \sin(90° - \theta') > n_2$. This means $n_1 \cos \theta' > n_2$. We want to write this in terms of θ instead of θ'. In preparation, square the equation and replace $\cos^2 \theta'$ with $1 - \sin^2 \theta'$. You should obtain $n_1^2(1 - \sin^2 \theta') > n_2^2$. At the air-fiber interface $\sin \theta = n_1 \sin \theta'$, so $\sin \theta' = (1/n_1) \sin \theta$. Substitute for $\sin^2 \theta'$ in the inequality and solve for the largest possible value of θ for which the inequality holds.

SOLUTION:

(b) If the indices of refraction of the glass and coating are 1.58 and 1.53, respectively, what is this value of θ?

SOLUTION:

[ans: 23.2°]

PROBLEM 7 (30E). Calculate the upper and lower limits of the Brewster angles for white light incident on fused quartz. Assume that the wavelength limits of the light are 400 and 700 nm, and use the curve of Fig. 39–2.

STRATEGY: The Brewster angle θ_B is given by $\tan \theta_B = n_2/n_1$, where n_1 is the index of refraction for the surrounding medium and n_2 is the index of refraction for the quartz. Take $n_1 = 1$, $n_2 = 1.47$ for 400 nm, and $n_2 = 1.455$ for 700 nm.

SOLUTION:

[ans: 55.8°, 55.5°]

You should be able to calculate the image position for an object in front of a plane or spherical mirror. Use the mirror equations: $i = -p$ for a plane mirror and $(1/p) + (1/i) = (1/f)$, where $f = 2/r$ for a spherical mirror. Be careful about the signs of the various quantities. Be sure you know the location of the image formed by a plane mirror and how to tell if the image is visible to an observer at a given location. Recall that the lateral magnification is given by $m = -i/p$ and that the lateral size h' of the image is related to the lateral size h of the object by $h' = mh$. Know how to interpret signs.

PROBLEM 8 (35E).   You look through a camera toward an image of a hummingbird in a plane mirror. The camera is 4.30 m in front of the mirror. The bird is at camera level, 5.00 m to your right, and 3.30 m from the mirror. For what distance must you focus your camera lens to get a clear image of the bird; that is, what is the distance between the lens and the apparent position of the image?

STRATEGY: The image is 3.30 m behind the mirror, directly in front of the bird (assumed to be facing the mirror). Thus the image is 5.00 m to the right of the camera and $3.30 + 4.30 = 7.60$ m in front of it. The distance for which the camera must be focused is the square root of the sum of the squares of these distances.

SOLUTION:

$$[\text{ans: } 9.10\,\text{m}]$$

PROBLEM 9 (40P).   Figure 39–49 shows an overhead view of a corridor with a plane mirror mounted at one end. A burglar B sneaks along the corridor directly toward the center of the mirror. If $d = 3.0$ m, how far from the mirror will she be when the security guard S can first see her in the mirror?

STRATEGY: On the diagram continue the path of the burglar to the region behind the mirror. Her image will be on that line. Draw the line from the guard to the right edge of the mirror and continue it until it intersects the path of the image. That is the position of the image when the guard first sees it. Calculate the distance from the mirror to the image. That is the same as the distance from the burglar to the mirror when the guard first sees the image.

SOLUTION:

$$[\text{ans: } 1.5\,\text{m}]$$

PROBLEM 10 (42P).   A point object is 10 cm away from a plane mirror while the eye of an observer (with pupil diameter 5.0 mm) is 20 cm away. Assuming both the eye and the point to be on the same line perpendicular to the mirror surface, find the area of the mirror used in observing the reflection of the point.

STRATEGY: Draw a diagram showing the eye 20 cm in front of the mirror and the image of the point 10 cm behind the mirror. Draw rays from the image point to the circumference of the eye. Use similar triangles to show that the radius R of the circular region on the mirror defined by the rays is given by $R = r/3.0$, where r is the radius of the pupil. The area is πR^2.

SOLUTION:

$$[\text{ans: } 2.2\,\text{mm}^2]$$

PROBLEM 11 (50P).   (a) A luminous point is moving at speed v_o toward a spherical mirror, along its axis. Show that the speed at which the image of this point is moving is given by

$$v_I = -\left(\frac{r}{2p - r}\right)^2 v_o\,,$$

where p is the distance of the luminous point from the mirror at any given time. (*Hint*: Start with Eq. 39–10.)

STRATEGY: Solve

$$\frac{1}{p} + \frac{1}{i} = \frac{2}{r}$$

for i: $i = (rp/(2p - r))$. Differentiate with respect to time. Replace di/dt with v_I and dp/dt with v_o.

SOLUTION:

Now assume that the mirror is concave, with $r = 15\,\text{cm}$, and let $v_o = 5.0\,\text{cm/s}$. Find the speed of the image when (b) $p = 30\,\text{cm}$ (far outside the focal point), (c) $p = 8.0\,\text{cm}$ (just outside the focal point), and (d) $p = 10\,\text{mm}$ (very near the mirror).

SOLUTION:

$$[\,\text{ans:}\ (b)\ -0.56\,\text{cm/s};\ (c)\ -1100\,\text{cm/s};\ (d)\ -5.1\,\text{cm/s}\,]$$

You should know how to find the image formed by a spherical refracting surface and by a lens. If the incoming rays make small angles with the central axis use $(n_1/p) + (n_2/i) = (n_2 - n_1)/r$ for a single surface and $(1/p) + (1/i) = (1/f)$ for a thin lens. You should also know how to compute the focal length, given the radii of the two surfaces of a lens. See Eq. 16.

PROBLEM 12 (51P). A parallel beam of light from a laser is incident on a solid transparent sphere of index of refraction n, as shown in Fig. 39–52. (a) If the beam is brought to a focus at the back of the sphere, what is the index of refraction of the sphere?

STRATEGY: Draw a diagram showing the top ray of the figure. Also draw the normal to the surface at the point where the ray strikes the sphere. The normal, of course, goes through the center of the sphere. Draw the refracted ray through the sphere to the point at the back. Suppose the angle of incidence is θ. Then the angle of refraction θ' satisfies $n \sin \theta' = \sin \theta$. Now consider the triangle formed by the normal to the surface (extended to the sphere center), the refracted ray through the sphere and the line from the sphere center to the focus point at the back of the sphere. Two of the angles are equal to the angle of refraction θ' and the third, at the sphere center, is equal to $180° - \theta$. Since the interior angles of a triangle sum to $180°$, $\theta' = \theta/2$. This is the condition for the beam to be focused at the back of the sphere. Snell's law gives $\sin \theta = n \sin(\theta/2)$. Use the trigonometric identity $\sin \theta = 2 \sin(\theta/2) \cos(\theta/2)$ to write the law $n = 2 \cos(\theta/2)$. Since n depends on θ the beam is not focused by a uniform sphere. If the beam is narrow, however, $\cos(\theta/2) \approx 1$ and the beam is focused.

SOLUTION:

$$[\,\text{ans:}\ n = 2.00\ \text{for narrow beams}\,]$$

(b) What index of refraction, if any, will focus the beam at the center of the sphere?

STRATEGY: To reach the center of the sphere a ray must be refracted along the normal to the surface at the point where it strikes the surface. The angle of refraction must be 0 for all angles of incidence.

SOLUTION:

[ans: not possible]

PROBLEM 13 (58E). You focus an image of the sun on a screen, using a thin lens whose focal length is 20.0 cm. What is the diameter of the image? (See Appendix C for needed data on the sun.)

STRATEGY: The object distance is the distance from the Earth to the sun: $p = 1.5 \times 10^{11}$ m (see the appendix). Use $(1/p) + (1/i) = (1/f)$ to find the image distance. The magnitude of the magnification is given by $m = i/p$ and the diameter of the image is given by $d' = md$, where d is the diameter of the sun (1.4×10^9 m).

SOLUTION:

[ans: 1.85 mm]

PROBLEM 14 (66P). A converging lens with a focal length of $+20$ cm is located 10 cm to the left of a diverging lens having a focal length of -15 cm. If an object is located 40 cm to the left of the converging lens, locate and describe completely the final image formed by the diverging lens.

STRATEGY: Let $p_1 = 40$ cm be the object distance for the formation of an image by the converging lens. Let $f_1 = +20$ cm be the focal length of that lens. Solve $(1/p_1) + (1/i_1) = (1/f_1)$ for the position i_1 of the intermediate image. The image produced by the converging lens becomes the object for the diverging lens. The object distance is given by $p_2 = \ell - i_1$, where $\ell = 10$ cm is the distance between the lenses. Let $f_2 = -15$ cm be the focal length of the diverging lens and solve $(1/p_2) + (1/i_2) = (1/f_2)$ for the final image position (relative to the diverging lens). Use $m = m_1 m_2 = (-i_1/p_1)(-i_2/p_2)$ to find the overall magnification. If it is positive the image has the same orientation as the object; if it is negative the image is inverted. The image is real if it is to the right of the diverging lens. Otherwise it is virtual.

SOLUTION:

[ans: the final image is virtual, has the same size and orientation as the object, and is located 30 cm to the left of the diverging lens]

PROBLEM 15 (73P). An illuminated slide is held 44 cm from a screen. How far from the slide must a lens of focal length 11 cm be placed in order to focus an image on the screen?

STRATEGY: The lens must be between the slide and the screen. The distance from the slide to the lens is the object distance p. The image distance is $\ell - p$, where ℓ is the distance from the slide to the screen. Solve

$$\frac{1}{p} + \frac{1}{\ell - p} = \frac{1}{f}$$

for p.

SOLUTION:

[ans: 22 cm]

You should understand compound lens systems, such as microscopes and telescopes. Be sure you can calculate the angular magnification for each of these instruments. Here are a few examples.

PROBLEM 16 (77E). If the magnifying power of an astronomical telescope is 36 and the diameter of the objective lens is 75 mm, what is the minimum diameter of the eyepiece required to collect all the light entering the objective from a distant point source on the axis of the instrument?

STRATEGY: In an astronomical telescope the second focal point of the objective lens is at the first focal point of the eyepiece and all the light that enters the objective from a distance source goes through this point. Draw a diagram showing the two lenses. Now draw a ray from the outer rim of the objective lens through the focal point to the eyepiece. Consideration of the similar triangles so formed leads to $d_{ob}/f_{ob} = d_{ey}/f_{ey}$, where d_{ob} is the diameter of the objective lens and d_{ey} is the diameter of the eyepiece. Use $m = -f_{ob}/f_{ey}$ to eliminate the focal lengths.

SOLUTION:

[ans: 2.1 mm]

PROBLEM 17 (78P). (a) Show that if the object o in Fig. 39–25(c) is moved from focal point F_1 toward the eye, the image moves in from infinity and the angle θ' (and thus the angular magnification m_θ) increases.

STRATEGY: Solve $(1/p)+(1/i) = (1/f)$ for i. Notice that i is negative for $p < f$. Differentiate the expression to show that

$$\mathrm{d}i = -\frac{f^2}{(p-f)^2}\,\mathrm{d}p.$$

$\mathrm{d}i$ is positive for $\mathrm{d}p$ negative, indicating that i becomes less negative as p becomes smaller.

SOLUTION:

(b) If you continue this process, at what image location will m_θ have its maximum usable value? (You can then still increase m_θ, but the image will no longer be clear.)

STRATEGY: The maximum usable value occurs when the image is at the near point.

SOLUTION:

[ans: $i = -15$ cm]

(c) Show that the maximum usable value of m_θ is $1 + (15\,\text{cm})/f$.

STRATEGY: Let h be the length of the object. The length of the image is then $h' = -(i/p)h$ and the image subtends the angle $\theta' = -h'/i = h/p$. The angular magnification is $m_\theta = (h/p)(15/h) = 15/p$, where p is in centimeters. Solve $(1/p) + (1/i) = (1/f)$ for p when $i = -15\,\text{cm}$. You should get $p = 15f/(15 + f)$, for f in centimeters. Substitute for p in the expression for m_θ.

SOLUTION:

(d) Show that in this situation the angular magnification is equal to the linear magnification.

STRATEGY: The linear magnification is $m = -i/p$. Substitute for p.

SOLUTION:

PROBLEM 18 (83P). In a compound microscope, the object is 10 mm from the objective lens. The lenses are 300 mm apart and the intermediate image is 50 mm from the eyepiece. What magnification is produced?

STRATEGY: Use $m = -15s/f_{ey}f_{ob}$, where s is the tube length, f_{ey} is the focal length of the eyepiece, f_{ob} is the focal length of the objective, and all distances are in centimeters. For the objective lens the object distance is $p = 0.10\,\text{cm}$ and the image distance is $i = 30\,\text{cm} - 5\,\text{cm} = 25\,\text{cm}$. Solve $(1/p) + (1/i) = (1/f_{ob})$ for f_{ob}. The focal length of the eyepiece is 5.0 cm and the tube length is $30\,\text{cm} - f_{ob} - f_{ey}$.

SOLUTION:

[ans: −75]

III. MATHEMATICAL SKILLS

Geometrical optics makes extensive use of the properties of similar triangles. Similar triangles are triangles:

1. such that every angle of one equals an angle of the other, and
2. every side of one is proportional to a side of the other

Two triangles are similar if any of the following are true:

1. any two angles of one are equal to angles of the other
2. an angle of one is equal to an angle of the other and the sides that form that angle in one triangle are proportional to the sides that form the equal angle in the other
3. every side of one is proportional to a side of the other
4. every side of one is either parallel or perpendicular to a side of the other

Use any of these conditions to prove that two triangles are similar. They then are guaranteed to have equal angles and proportional sides.

IV. NOTES

Chapter 40
INTERFERENCE

I. BASIC CONCEPTS

You studied the fundamentals of interference in Chapter 17. Now the results are specialized to light waves and applied to double-slit interference, thin-film interference, and the Michelson interferometer, an important instrument for measuring distances. Pay special attention to the role played by the distances traveled by interfering waves in determining their relative phase.

Huygens' principle. Maxwell's equations lead to a geometrical construction, called Huygens' principle, that shows us how to construct the wavefront for an electromagnetic wave at some time $t + \Delta t$, given the wavefront at an earlier time t. According to the principle you may think of each point on a wavefront as a point source of spherical waves, called Huygens wavelets. After time Δt the radius of the wavelets will be $v\Delta t$, where v is the wave speed, and the wavefront will be tangent to the wavelets.

The principle is illustrated by the diagram to the right, which shows a plane wavefront at some instant of time. Use the time t to label it. Three spherical Huygens wavelets are shown. Each is centered on the plane wavefront and each has a radius of $c\Delta t$. The position of the plane wavefront at time $t + \Delta t$ is the common tangent to the spherical wavefronts. Draw it on the diagram and label it with the time $t + \Delta t$.

You will use Huygens' principle in this chapter and the next to understand what happens to light when it passes through one or more slits or passes by the edge of an obstacle. In each case the light spreads out to enter the region you might think is in shadow. You may think of the spreading Huygens wavelets as being responsible for the phenomenon. To obtain some practice with the principle first use it to prove the law of refraction.

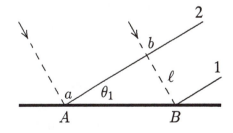

In the diagram to the right two wavefronts impinging on a plane boundary between two media are drawn with solid lines; two rays are drawn with dashed lines. Wavefront 1 has reached the boundary at B and wavefront 2 has reached the boundary at A. Point b on wavefront 2 has not yet reached the boundary but it will arrive at B in time $t = \ell/v_1$, where ℓ is the distance between the wavefronts and v_1 is the wave speed in medium of incidence.

The next diagram shows the two wavefronts in the region on the other side of the boundary, when b has reached B. If the wave speed in this region is v_2, the Huygens wavelet emanating from A has a radius of $v_2 t$ so $\ell' = $ _____. Draw an arc of the wavelet through a. Notice that the line AB is the hypotenuse of a right triangle for which ℓ' is one side. Thus the length of AB is given by $v_2 t / \sin \theta_2$.

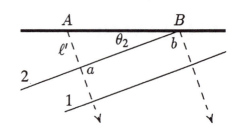

According to the diagram showing the *incident* wavefronts, the length of AB is also given by $v_1 t / \sin \theta_1$. In the space below set the two expressions for AB equal to each other, replace v_1 with c/n_1, replace v_2 with c/n_2, and obtain the law of refraction:

Phasors. To find the total disturbance when two or more waves are present you add the waves. Here you consider two sinusoidal waves with the same frequency and traveling in the same direction. The resultant is again a sinusoidal wave. Its amplitude may be as much as the sum of the amplitudes of the individual waves or as small as their difference, depending on their relative phase.

Phasors can be used to sum waves. The diagram on the right shows a phasor used to represent the electric field of a light wave at some point in space. Suppose the field is given by $E = E_0 \sin(\omega t)$, where ω is the angular frequency. The length of the phasor is proportional to the amplitude E_0 and its projection on the vertical axis is proportional to the wave itself. Label the angle ωt and indicate the direction of rotation of the phasor. Its angular speed is _____. Indicate the projection on the vertical axis.

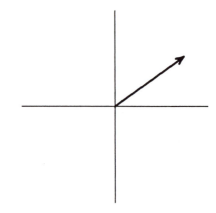

Now draw a second phasor, associated with the field $E = E_0 \sin(\omega t + \phi)$, where ϕ is 75°. Draw it with its tail at the head of the first phasor. Finally draw the phasor that represents the sum of the two waves. Label it E.

The phasors are redrawn to the right. Since two sides of the triangle are the same the two angles marked β are equal. Furthermore, since the third angle of the triangle is $180° - \phi$ and the angles of any triangle must sum to 180°, $\beta = $ _____. The dotted line is the perpendicular bisector of the phasor E. It is one side of a small right triangle with hypotenuse E_0 and one interior angle β. The other side is $E/2$. Thus, $E/2 = E_0 \cos \beta$ and, in terms of E_0 and ϕ,

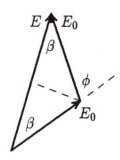

$$E =$$

If $\cos(\phi/2) = $ _____ constructive interference is complete and $E = 2E_0$ This means ϕ is a multiple of _____ rad. Both waves have their maximum values at the same time. If $\cos(\phi/2) = $ _____ destructive interference is complete and $E = 0$ This means ϕ is an odd multiple of _____ rad. At all times one wave is the negative of the other.

You should recognize that ϕ in the equations above is the phase of one wave *relative* to the other. Any constant can be added to the phase of both waves without changing the amplitude of the resultant. Said another way, if the waves are given by $E_0 \sin(\omega t + \phi_1)$ and $E_0 \sin(\omega t + \phi_2)$ then $\phi = \phi_1 - \phi_2$. Notice that ϕ can be replaced by $-\phi$ without changing the expression for the amplitude. This is because $\cos(\phi/2) = \cos(-\phi/2)$. Thus you may also take ϕ to be $\phi_2 - \phi_1$. In addition, any multiple of _____ can be added to or subtracted from ϕ without changing the resultant.

Phase differences can come about because two waves, starting in phase, travel different distances to get to the same point. If two waves of the same frequency travel in the same medium, so their speeds and wavelengths are the same, and one goes a distance ΔL further than the other the phase difference is $\phi = $ _____, where λ is the wavelength.

If a wave with wavelength λ in vacuum enters a medium with index of refraction n the frequency does not change but the wave speed does. The wavelength in the medium is $\lambda' = $ _____. Suppose two waves have the same frequency and start in phase. One travels a distance L_1 in a medium with index of refraction n_1 and the other travels a distance L_2 in a medium with index of refraction n_2. At the end their phase difference is _____.

The electric fields of the two waves are vectors and strictly speaking the waves should be added vectorially. We assume, however, that the waves are plane polarized and the fields are along the same line. Then scalar addition can be used. This is a good approximation for most of the situations considered in this chapter.

Double-slit interference. A monochromatic plane wave is incident normally on a barrier with two slits, as diagramed in Fig. 9 of the text. You can find the intensity at any point on the screen by summing the Huygens wavelets emanating from points within the slits. Note that wavelets arrive at every point on the screen, not just those directly behind the slits. For now assume the slits are so narrow that only one wavelet from each slit is required. Furthermore assume the amplitude of each wavelet is the same at the screen. At the point P in the diagram the wavelet from the upper slit can be written $E_0 \sin(\omega t - k r_1)$ and the wavelet from the lower slit can be written $E_0 \sin(\omega t - k r_2)$, where $k = 2\pi/\lambda$ and λ is the wavelength. The phase difference is $\phi = k(r_2 - r_1) = 2\pi(r_2 - r_1)/\lambda$.

Carefully note that the phase difference arises because the wavelets travel different distances from the slits to the same point on the screen. Also note that because a plane wave is incident normally on the slits the phases of the wavelets are the same at the slits. This is changed if the wave is incident at some other angle or if the slits are covered by transparent materials with different indices of refraction. The expression for ϕ must then be modified.

If the screen is far from the slits then rays from the slits to any point on the screen are nearly parallel to each other and the difference in the distances can easily be written in terms of the angle θ between a ray and a line normal to the barrier. The geometry is shown in the diagram to the right. Since the screen is far away the distance from the dotted line between the rays to P is the same along each ray. One of the interior angles of the right triangle formed by the dotted line, the barrier, and the lower ray is also θ. Label it. The lower ray is longer than the upper by $d\sin\theta$, where d is the slit separation. Indicate this distance on the diagram and label it with the expression for its length.

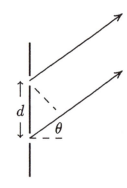

In terms of d, θ, and λ, the phase difference at the screen is

$$\phi =$$

The amplitude of the resultant wave at P is denoted by E. In terms of E_0 and ϕ it is given by

$$E =$$

The intensity I at P is proportional to the square of the amplitude and in terms of ϕ is given by

$$I =$$

where I_0 is the intensity associated with a single wave. The same result is obtained for a nearby screen if a lens is used to focus light on the screen.

You should recognize that light emanates from each slit in all forward directions but only the portions of wavefronts that follow the rays shown get to point P. Other portions of the wavefronts get to other places on the screen and for them θ has a different value. For different values of θ the phase difference is different and, as a result, so are the resultant amplitude and intensity. Alternating bright and dark regions (fringes) are seen on the screen. Centers of bright fringes (maxima of intensity) occur at points where the phases of the two wavelets differ by _____, where m is an integer. To reach these places one wave travels further than the other by a multiple of _____. That is,

$$d\sin\theta =$$

Centers of dark fringes (minima of intensity) occur at points where the phases of the two wavelets differ by _____, where m is an integer. To reach these places one wave travels further than the other by an odd multiple of _____. That is,

$$d\sin\theta =$$

Notice that complete constructive interference occurs at the point on the screen directly back of the slits, for which $\theta = 0$. The angular separation of the first minima on either side of the central maximum is a measure of the extent to which the intensity pattern is spread on the screen. This is given by $2\theta_0$, where $\sin\theta_0 = \lambda/2d$. As d _____ or λ _____ the angular separation increases and the pattern spreads. If $d =$ _____ then $\theta_0 = 90°$ and no bright fringes appear beyond the central maximum. For any given values of d and λ the number of maxima that are seen can be found by setting θ equal to _____ and solving $d \sin \theta = m\lambda$ for _____ .

Coherence. Two sinusoidal waves are said to be coherent if the difference in their phases is constant. Light is emitted from atoms in bursts lasting on the order of _____ s and each burst may have a different phase constant associated with it. Thus light from atoms emitting independently of each other is not coherent.

If an extended incoherent source, such as an incandescent lamp, is used to produce light that is incident on a double slit barrier the light from each atom goes through each slit and combines on the other side to form an interference pattern. Light from different atoms, however, form patterns that are shifted with respect to each other, the amount of the shift depending on the separation of the atoms in the source. Describe a technique that can be used to insure that only light from a small region of the source reaches the slits: _____

_____

_____

All the light from another type light source, a _____ , is coherent, even though many different atoms are emitting simultaneously. When this light is incident on a double slit it produces an interference pattern without any additional apparatus.

Thin-film interference. If light is incident normally on a thin film, some is reflected from the front surface and some from the back. These two waves interfere and the resultant intensity may be quite large or quite small, depending on the thickness of the film.

To calculate the intensity for a given wavelength light you must be able to find the relative phases of the two waves. For normal incidence the wave reflected from the far surface travels _____ further than the wave reflected from the near surface. Here L is the thickness of the film. In addition, for one or both of the waves the medium beyond the surface of reflection may have a higher index of refraction than the medium of incidence. On reflection the wave then suffers a phase change of _____ radians.

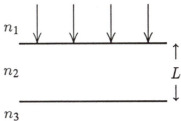

The diagram shows a film of thickness L with a plane wave incident normally. You should be able to show that the phase difference for waves reflected from the two surfaces is given by $\phi = 4\pi L n_2/\lambda$ if $n_1 > n_2 > n_3$ or $n_1 < n_2 < n_3$. In the first case neither wave suffers a phase change on reflection. In the second they both suffer a phase change of π rad.

If $n_1 < n_2 > n_3$ or $n_1 > n_2 < n_3$ the phase difference is given by $(4\pi L n_2/\lambda) \pm \pi$. In the first case the wave reflected from the front surface suffers a phase change of π rad but the wave reflected from the back surface does not. In the second case the wave reflected from the back surface suffers a phase change of π but the wave reflected from the front surface does not. Notice that the wavelength for the film must be used to calculate the difference in phases. It is given by λ/n_2, where n_2 is the index of refraction of the film and λ is the wavelength in vacuum. Since you may always add 2π to any phase, it is immaterial whether the sign in front of π is plus or minus.

When white light (a combination of all wavelengths of the visible spectrum) is incident on a thin film the reflected light is colored. It consists chiefly of those wavelengths for which interference of the two reflected waves produces a maximum or nearly a maximum of intensity. Those wavelengths for which interference produces a minimum are missing. This phenomena accounts for the colors of oil films and soap bubbles, for example.

Suppose a film with thickness L has an index of refraction n and is in air (with a smaller index of refraction). Monochromatic light is incident normally on a surface. There is a change in phase on reflection at the _____ surface but not at the _____ surface. The difference in phase of the two reflected waves is $\phi = $ _____. Interference maxima occur for wavelengths given by $\lambda = $ _____ and interference minima occur for wavelengths given by $\lambda = $ _____, where m is an integer. How are these results changed if the film is deposited on glass with a higher index of refraction? _____

_____

_____

When monochromatic light is incident on a thin film with a varying thickness, like a wedge, interference produces bright and dark bands. Bright bands appear in regions for which the film thickness is such that the phase difference between the two reflected waves is close to _____ rad; dark bands appear in regions for which the film thickness is such that the phase difference is close to _____ rad. Here m is an integer.

The Michelson interferometer. The diagram on the right is a schematic drawing of the instrument. Light from a source S is incident on a half-silvered mirror M, where the beam is split. Half is reflected to mirror M_2 where it is reflected back through M to the eye at E. The other half is transmitted to mirror M_1 where it is reflected back to M. There half is reflected to the eye. The two beams reaching the eye interfere. To measure a distance, one of the mirrors (M_2 say) is moved, thereby changing the interference pattern at the eye. As the mirror moves alternately bright and dark fringes are seen. The number of fringes that appear are counted and related to the distance moved by the mirror.

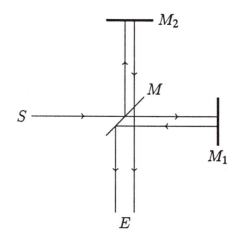

Suppose the distance traveled by the light that strikes mirror M_1 is d_1 and the distance traveled by the light that strikes mirror M_2 is d_2. If the wavelength of the light is λ then the difference in phase of the two waves at the eye is $\phi = $ _____. Suppose d_1 and d_2 happen to have values so a maximum of intensity is produced at the eye. Then mirror M_2 is moved until the intensity is again a maximum. In terms of the wavelength the distance it moved must have been _____. Interferometers have been used to measure distances with an accuracy of about _____ the wavelength of the light used.

II. PROBLEM SOLVING

You should know the relationship between the wave speed in a material medium and the wave speed in vacuum. You should also know that when a wave enters a medium the frequency does not change but the wave speed and wavelength do.

PROBLEM 1 (1E).   The wavelength of yellow sodium light in air is 589 nm. (*a*) What is its frequency?

STRATEGY: Use $\lambda f = c$. Take the speed of light in air to be the same as in vacuum.

SOLUTION:

$$[\text{ans: } 5.09 \times 10^{14} \text{ Hz}]$$

(*b*) What is its wavelength in glass whose index of refraction is 1.52?

STRATEGY: Use $\lambda' = \lambda/n$.

SOLUTION:

$$[\text{ans: } 388 \text{ nm}]$$

(*c*) From the results of (*a*) and (*b*) find its speed in this glass.

STRATEGY: Use $v = \lambda' f$.

SOLUTION:

$$[\text{ans: } 1.97 \times 10^8 \text{ m/s}]$$

You should know how to calculate the phase difference of two waves. If wave 1 travels a distance r_1 in a medium with refractive index n_1 and wave 2 travels a distance r_2 in a medium with refractive index n_2, the difference in phase is $\phi = (2\pi/\lambda)(n_2 r_2 - n_1 r_1)$.

PROBLEM 2 (12P).   The two waves in Fig. 40–3 have wavelength 500 nm in air. In wavelengths, what is their phase difference after traversing media 1 and 2 if (*a*) $n_1 = 1.50$, $n_2 = 1.60$, and $L = 8.50\,\mu$m; (*b*) $n_1 = 1.62$, $n_2 = 1.72$, and $L = 8.50\,\mu$m; and (*c*) $n_1 = 1.59$, $n_2 = 1.79$, and $L = 3.25\,\mu$m?

STRATEGY: The phase difference in wavelengths is given by $(L/\lambda)(n_2 - n_1)$.

SOLUTION:

$$[\text{ans: } (a) \text{ 1.70 (or 0.70); } (b) \text{ 1.70 (or 0.70); } (c) \text{ 1.30 (or 0.30)}]$$

(d) Suppose that in each of these three situations the waves arrive at a common point. Rank the situations according to the brightness the waves produce at the common point.

STRATEGY: The brightest spot is produced by those waves with a phase difference closest to 1 wavelength. The least bright spot is produced by those waves with a phase difference closest to half a wavelength. Note that a phase difference of 0.3 wavelengths is equivalent to a phase difference of 0.7 wavelengths because we can always add or subtract an integer number of wavelengths and we can always reverse the sign of the phase difference without changing the resultant amplitude.

SOLUTION:

[ans: all produce the same brightness]

PROBLEM 3 (13P). In Fig. 40–3, two light waves of wavelength 620 nm are initially out of phase by π rad. The indices of refraction of the media are $n_1 = 1.45$ and $n_2 = 1.65$. (a) What is the smallest thickness L that will put the waves exactly in phase once they pass through the media?

STRATEGY: After traversing the media the phase difference is $\pi + (2\pi L/\lambda)(n_2 - n_1)$. You want this to be $2\pi m$, where m is an integer. Solve for L, then select m so L has the smallest possible value.

SOLUTION:

[ans: $1.55\,\mu$m]

(b) What is the next smallest L that will do this?

SOLUTION:

[ans: $4.65\,\mu$m]

You should understand the conditions for the formation of bright and dark fringes by a double-slit arrangement. Find an expression for the phase difference between the waves from the two slits and set it equal to $2\pi m$ for a bright fringe and $2\pi(m + \frac{1}{2})$ for a dark fringe. In many cases the phase difference is due to the difference in distance traveled by the waves. This is $d \sin \theta$, where d is the slit separation. If a medium with thickness L and refraction index n is placed in front of a slit you must add $2\pi nL/\lambda$ to the phase of the wave through that slit.

You should also be able to calculate the intensity at points on the interference pattern, relative to the intensity at the center of the central maximum. If the waves have equal amplitudes at the screen then $I = I_m \cos^2(\phi/2)$. One of the problems below asks to find the intensity when the amplitudes are not equal. You must then sum the phasors.

PROBLEM 4 (15E). Monochromatic green light, of wavelength 550 nm, illuminates two parallel narrow slits 7.70 μm apart. Calculate the angular deviation (θ in Fig. 40–9) of the third order (for $m = 3$) bright fringe (*a*) in radians, and (*b*) in degrees.

STRATEGY: Solve $d \sin \theta = m\lambda$ for θ.

SOLUTION:

$$[\,\text{ans:}\ (a)\ 0.216\,\text{rad};\ (b)\ 12.4°\,]$$

PROBLEM 5 (21E). In a double-slit arrangement the slits are separated by a distance equal to 100 times the wavelength of the light passing through the slits. (*a*) What is the angular separation in radians between the central maximum and an adjacent maximum?

STRATEGY: Solve $d \sin \theta = \lambda$ for θ. This gives the angular position of the first maximum outside the central maximum, so it also gives the angular separation between the central maximum and an adjacent maximum.

SOLUTION:

$$[\,\text{ans:}\ 0.010\,\text{rad}\,]$$

(*b*) What is the distance between these maxima on a screen 50.0 cm from the slits?

STRATEGY: The separation is $D \tan \theta$, where D is the distance from the slits to the screen.

SOLUTION:

$$[\,\text{ans:}\ 0.50\,\text{cm}\,]$$

PROBLEM 6 (22E). In a double-slit experiment, $\lambda = 546$ nm, $d = 0.10$ mm, and $D = 20$ cm. On a viewing screen, what is the distance between the fifth maximum and seventh minimum from the central maximum?

STRATEGY: The path difference for the two waves is $d \sin \theta$. For the fifth maximum this must be 5λ. For the seventh minimum it must be $13\lambda/2$. Solve for θ in each case. The distance from the center of the pattern on the screen is given by $y = D \tan \theta$. Calculate y for each case, then calculate the difference.

SOLUTION:

$$[\,\text{ans:}\ 1.6\,\text{mm}\,]$$

PROBLEM 7 (23E). A double-slit arrangement produces interference fringes for sodium light ($\lambda = 589$ nm) that are 0.20° apart. What is the angular fringe separation if the entire arrangement is immersed in water ($n = 1.33$)?

STRATEGY: In vacuum the angular position of the first maximum outside the central maximum is given by $d \sin \theta = \lambda$ and in water it is given by $d \sin \theta' = \lambda/n$, where d is the slit separation and n is the index of refraction for water. Use the first equation to eliminate d from the second, then solve for θ'.

SOLUTION:

$$[\text{ans: } 0.15°]$$

PROBLEM 8 (27P).   In Fig. 40–22, A and B are identical radiators of waves that are in phase and of the same wavelength λ. The radiators are separated by distance $d = 3.00\lambda$. Find the largest distance from A, along the x axis, for which fully destructive interference occurs. Express this in terms of λ.

STRATEGY: Let x be the coordinate of a point on the x axis. The phase difference of the two waves there is $\phi = (2\pi/\lambda)(\sqrt{x^2 + d^2} - x)$. You want this to be $(m + \frac{1}{2})\lambda$, where m is an integer. Solve for x. You should get

$$x = \frac{d^2 - (m + \frac{1}{2})^2\lambda^2}{2(m + \frac{1}{2})\lambda} \; .$$

Since x clearly decreases as m increases, the largest value of x is associated with $m = 0$.

SOLUTION:

$$[\text{ans: } 8.75\lambda]$$

PROBLEM 9 (33P).   One slit of a double-slit arrangement is covered by a thin glass plate of refractive index 1.4 and the other by a thin glass plate of refractive index 1.7. The point on the screen where the central maximum fell before the glass plates were inserted is now occupied by what had been the $m = 5$ bright fringe before. Assuming that $\lambda = 480$ nm and the plates have the same thickness t, find t.

STRATEGY: The point on the screen is directly in front of the slits, so $\theta = 0$. The two waves travel the same distance to this point so their phase difference is $(2\pi t/\lambda)(n_2 - n_1)$. This must be $2\pi m$, where $m = 5$. Solve for t.

SOLUTION:

$$[\text{ans: } 8.0 \, \mu\text{m}]$$

PROBLEM 10 (39P).   Two waves of the same frequency have amplitudes 1.00 and 2.00, respectively. They interfere at a point where their phase difference is 60.0°. What is the resultant amplitude?

STRATEGY: Draw the phasor diagram and use the law of cosines (see the Mathematical Skills section at the end of this chapter of the manual). If E_1 and E_2 are the amplitudes and ϕ is the phase difference, the resultant amplitude E is given by $E^2 = E_1^2 + E_2^2 + 2E_1E_2 \cos \phi$.

SOLUTION:

[ans: 2.65]

PROBLEM 11 (42P).   The horizontal arrow in Fig. 40–10 marks the points on the intensity curve where the intensity of the central fringe is half the maximum intensity. Show that the angular separation $\Delta\theta$ between the corresponding points on the screen is

$$\Delta\theta = \frac{\lambda}{2d}$$

if θ in Fig. 40–9 is small enough so that $\sin \theta \approx \theta$.

STRATEGY: Since the intensity is given by $I = 4I_0 \cos^2(\phi/2)$ you want the value of θ for which $\cos^2(\phi/2) = 1/2$. This means $\phi/2 = \pi/4$ rad or $\phi = \pi/2$ rad. Since $\phi = (2\pi/\lambda)d \sin \theta$, $\sin \theta = \lambda/4d$. Make the small angle approximation and solve for θ. The two half-intensity points are positioned symmetrically on either side of the central maximum so their angular separation is given by $\Delta\theta = 2\theta$.

SOLUTION:

Some problems deal with the interference of waves reflected from the front and back surfaces of a thin film. The difference in phase comes from the difference is distances traveled and, in some cases, from the change in phase by π on reflection from a medium with a higher index of refraction. Don't forget to think about both of these sources of phase difference. Also don't forget to use the wavelength of the wave in the film, not the wavelength in vacuum, to compute the phase difference.

PROBLEM 12 (47E).   Bright light of wavelength 585 nm is incident perpendicularly on a soap film ($n = 1.33$) of thickness 1.21 μm, suspended in air. Is the light reflected by the two surfaces of the film closer to interfering fully destructively or fully constructively?

STRATEGY: Calculate the phase difference of the two waves (don't forget the change of phase by π for the wave reflected from the outer surface): $\phi = \pi + 4\pi nL/\lambda$, where L is the film thickness. One wave traverses the film twice. Subtract multiples of 2π until the result is between 0 and 2π. If the result is less than $\pi/2$ or greater than $3\pi/2$ the reflected wave is closer to fully constructive interference. If the result is between $\pi/2$ and $3\pi/2$ the wave is closer to fully destructive interference.

SOLUTION:

[ans: $\phi = 0.008\pi$, closer to fully constructive interference]

PROBLEM 13 (49E). A lens with index of refraction greater than 1.30 is coated with a thin transparent film of index of refraction 1.30 to eliminate by interference the reflection of red light at wavelength 680 nm that is incident perpendicularly on the lens. What minimum film thickness is needed?

STRATEGY: There is a change of phase by π for both reflected waves so the phase difference is $4\pi nL/\lambda$, where L is the film thickness and n is its refractive index. You want this to be $2\pi(m + \frac{1}{2})$, where m is an integer. Solve for L and pick m so L has the smallest possible value.

SOLUTION:

[ans: 131 nm]

PROBLEM 14 (57P). A plane wave of monochromatic light is incident normally on a uniformly thin film of oil that covers a glass plate. The wavelength of the source can be varied continuously. Fully destructive interference of the reflected light is observed for wavelengths of 500 and 700 nm and for no wavelengths between them. If the index of refraction of the oil is 1.30 and that of the glass is 1.50, find the thickness of the oil film.

STRATEGY: The wave reflected from the front surface of the film suffers a change of phase of π but the wave reflected from the back surface does not, so the phase difference for the two waves is $\pi + 4\pi nL/\lambda$, where L is the thickness of the film and n is its index of refraction. You want this to be $2\pi(m + \frac{1}{2})$ for the two given values of λ and for no value between. The last condition means that the two values of m differ by an integer. Let m_1 be the value for the longer wavelength (λ_2). Then $m_1 + 1$ is the value for the shorter wavelength (λ_1). Solve $\pi + 4\pi nL/\lambda_1 = 2\pi(m + \frac{3}{2})$ and $\pi + 4\pi nL/\lambda_2 = 2\pi(m + \frac{1}{2})$ for L. You should get $L = \lambda_1\lambda_2/2n(\lambda_2 - \lambda_1)$.

SOLUTION:

[ans: 673 nm]

PROBLEM 15 (63P). From a medium of index of refraction n_1, monochromatic light of wavelength λ is incident normally on a thin film of uniform thickness L (where $L > 0.1\lambda$) and index of refraction n_2. The transmitted light travels in a medium with index of refraction n_3. Find expressions for the minimum film thickness (in terms of λ and the indices of refraction) for the following cases: (a) minimum light is reflected (maximum light is transmitted) with $n_1 < n_2 > n_3$; (b) minimum light is reflected (maximum light is transmitted) with $n_1 < n_2 < n_3$; and (c) maximum light is reflected (minimum light is transmitted) with $n_1 < n_2 < n_3$.

STRATEGY: (a) The wave reflected from the first surface suffers a change of phase of π while the wave reflected from the second does not. The phase difference is $\pi + 4\pi n_2 L/\lambda$. You want this to be $2\pi(m + \frac{1}{2})$. Solve for L and pick m to have the integer value that makes L a minimum (but not 0). (b) Each wave suffers a change of phase of π so the phase difference is $4\pi n_2 L/\lambda$. You want this to be $2\pi(m + \frac{1}{2})$. Solve for L and pick the integer value of m that makes L a minimum. (c) Each wave suffers a change of phase of π, so the phase difference is $4\pi n_2 L/\lambda$. You want this to be $2\pi m$. Solve for L and pick the integer value of m that makes L a minimum.

SOLUTION:

[ans: (a) $\lambda/2n_2$; (b) $\lambda/4n_2$; (c) $\lambda/2n_2$]

PROBLEM 16 (70P). Two pieces of plate glass are held together in such a way that the air space between them forms a very thin wedge. Light of wavelength 480 nm strikes the upper surface perpendicularly and is reflected from the lower surface of the top glass and the upper surface of the bottom glass, thereby producing a series of interference fringes. How much thicker is the air wedge at the sixteenth fringe than it is at the sixth?

STRATEGY: The wave reflected from the upper surface of the bottom glass suffers a change of phase of π. The wave reflected from the lower surface of the upper glass does not. At a place where the thickness of the air wedge is L the difference in phase of the two waves is $\pi + 4\pi L/\lambda$. For fully constructive interference this must be $2\pi m$, where m is an integer. Solve $\pi + 4\pi L/\lambda = 2\pi m$ for L, first with $m = 6$ and then with $m = 16$. Calculate the difference in the two values of L.

SOLUTION:

[ans: $2.4\,\mu m$]

Some problems deal with the Michelson interferometer. Again you calculate the phase difference for the two waves and set it equal to $2\pi m$ for a bright fringe and $2\pi(m + \frac{1}{2})$ for a dark fringe. The phase difference may be due to the different distances traveled by the waves and due to materials with different indices of refraction in the two arms of the interferometer.

PROBLEM 17 (79E). A thin film with index of refraction $n = 1.40$ is placed in one arm of a Michelson interferometer, perpendicular to the optical path. If this causes a shift of 7.0 fringes of the pattern produced by light of wavelength 589 nm, what is the film thickness?

STRATEGY: In the absence of the film the phase difference of the two waves is $4\pi(d_1 - d_2)/\lambda$, where d_1 and d_2 are the distances defined in Fig. 40–18 of the text. To produce a maximum of intensity this is $2\pi m$, where m is an integer. Now put the film in the arm containing the movable mirror, say. If L is its thickness the wave in that arm goes a distance L with a wavelength of λ/n and a distance $d_2 - L$ in air (or vacuum). The phase difference is now $4\pi(d_2 - d_1)/\lambda + 4\pi n L/\lambda - 4\pi L/\lambda = 4\pi(d_2 - d_1)/\lambda + 4\pi L(n - 1)/\lambda$. Since there is a shift of 7 fringes this must be $2\pi(m + 7)$. Use the first expression to eliminate $4\pi(d_2 - d_1)/\lambda$ from the second, then solve for L.

SOLUTION:

[ans: $5.2\,\mu m$]

PROBLEM 18 (81P). Write an expression for the intensity observed in Michelson's interferometer (Fig. 40–18) as a function of the position of the movable mirror. Measure the position of the mirror from the point at which $d_1 = d_2$.

STRATEGY: Let one arm be x longer than the other. The path difference is $2x$. Neglect the thickness of mirror M. Each wave is reflected the same number of times, so the path difference is the only source of a phase difference. The phase difference is $\phi = (2\pi/\lambda)2x$ and the intensity is given by $I = I_m \cos^2(\phi/2)$, where I_m is the intensity when the waves are in phase ($x = 0$).

SOLUTION:

[ans: $I_m \cos^2(2\pi x/\lambda)$]

III. MATHEMATICAL SKILLS

A few problems deal with the addition of two waves with different amplitudes. Suppose one wave is given by $E_1 \sin(\omega t)$ and the second by $E_2 \sin(\omega t + \phi)$. The phasor diagram is shown to the right. According to the law of cosines, $E^2 = E_1^2 + E_2^2 - 2E_1 E_2 \cos \alpha$. Since $\alpha = 180° - \phi$ and $\cos(180° - \phi) = -\cos \phi$, this can be written

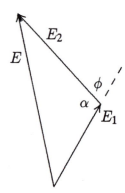

$$E^2 = E_1^2 + E_2^2 + 2E_1 E_2 \cos \phi$$

Use this equation to calculate the amplitude of the resultant.

If the waves are in phase then $\phi = 0$, $\cos \phi = 1$, and $E^2 = E_1^2 + E_2^2 + 2E_1 E_2 = (E_1 + E_2)^2$. If the waves are 180° out of phase then $\cos \phi = -1$ and $E^2 = E_1^2 + E_2^2 - 2E_1 E_2 = (E_1 - E_2)^2$. These results should agree with what you expect.

If $E_1 = E_2$ then the general expression reduces to $E^2 = 2E_1^2(1 + \cos \phi)$. Now $\cos \phi = \cos(\phi/2 + \phi/2) = \cos^2(\phi/2) - \sin^2(\phi/2) = 2\cos^2(\phi/2) - 1$, where $\cos^2(\phi/2) + \sin^2(\phi/2) = 1$ was used. Thus $E^2 = 4E_1^2 \cos^2(\phi/2)$ and $E = 2E_1 \cos(\phi/2)$, in agreement with the result given in the text.

IV. NOTES

Chapter 41
DIFFRACTION

I. BASIC CONCEPTS

Diffraction is the flaring out of light as it passes by the edge of an object or through an opening in a barrier. Pay attention to the description of this phenomenon in terms of Huygen wavelets. You should also understand how interference of the wavelets produces bright and dark fringes in the diffraction pattern of an object or opening. When you study diffraction by a double slit pay close attention to the relationship between the single-slit diffraction pattern and the double-slit interference pattern. In particular, know what characteristics of the slits determine the maxima and minima in each pattern.

Two important applications are discussed. Diffraction patterns produced by diffraction gratings are used to study spectra and patterns produced by x rays incident on crystals are used to study crystalline structure. Your goals should be to learn what the patterns look like and to understand the details of wave interference that leads to these patterns. In addition, learn how to calculate the angles for maximum and minimum intensity.

Diffraction. The phenomenon is described in terms of Huygen wavelets. If there is no barrier the wavelets combine to produce a wave that continues moving in its original direction. If a barrier blocks some of the wavelets then those that are not blocked combine to produce a wave that moves into the geometric shadow. In addition, if the light is coherent, the Huygen wavelets interfere to form a series of bright and dark bands, called the diffraction pattern of the object. Figs. 41-1, 2, and 3 show some diffraction patterns.

Single-slit diffraction. Consider plane waves of monochromatic light incident normally on a barrier with a single slit of width a, as shown in Fig. 41–4 of the text. To find the intensity at a point P on a screen add the Huygen wavelets emanating from the slit and take the limit as the number of wavelets becomes infinite. We suppose the viewing screen is far away from the slit and consider parallel rays as shown in Fig. 41–5b.

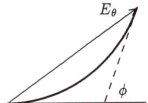

Each wavelet has an infinitesimal amplitude and a phase that differs from that of a neighboring wavelet by an infinitesimal amount. The phasors form an arc of a circle, as shown to the right. The angle ϕ is the difference in phase of wavelets from the edges of the slit. It is also the angle subtended by the arc at its center. Look at Fig. 41–8 of the text. If the arc has radius R then its length is given by $E_m = R\phi$, for ϕ in radians. E_m is the sum of the amplitudes of all the wavelets and thus is the amplitude if they all have the same phase.

E_θ, the amplitude at P, is the chord of the arc. A little geometry shows that $E_\theta = 2R\sin(\phi/2)$. Eliminating R between these two expressions yields an expression for E_θ in terms of E_m and ϕ:

$$E_\theta =$$

Differences in the phases come about because the wavelets travel different distances to P. If the screen is far away and the slit width is a then the lower wavelet travels $a \sin \theta$ further than the upper wavelet. Label the slit width and mark the distance $a \sin \theta$ on the diagram to the right. If the wavelength of the light is λ then the difference in the phases of these two wavelets is given by

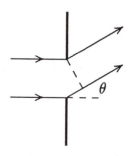

$$\phi =$$

The expression for the amplitude is often written in terms of $\alpha = \phi/2$, rather than in terms of ϕ. Then the amplitude at P is given by

$$E_\theta =$$

where, in terms of a and θ,

$$\alpha =$$

The intensity at P is proportional to the square of the amplitude and is given by

$$I_\theta =$$

where I_m is the intensity when all wavelets are in phase.

Carefully study Fig. 41–7 of the text. It shows the intensity as a function of θ for several values of the slit width a. The most prominent feature is a broad central maximum, centered at $\theta = 0$. Note particularly that $\theta = 0$ corresponds to an intensity maximum, not a minimum. In the limit as $\alpha \to 0$, $(\sin \alpha)/\alpha \to 1$. If the slit width is small the central maximum spreads to cover the entire screen and no zeros of intensity occur. For a wide slit the central maximum is narrow and is followed on both sides by secondary maxima. These are narrower and of considerably less intensity than the central maximum. They are roughly midway between zeros of intensity. The number that appear depends on the slit width.

In the space below draw phasor diagrams corresponding to the peak of the central maximum, the first zero, and the second zero to one side of the central maximum. Remember that the total arc length is the same in all cases.

CENTRAL MAXIMUM   FIRST ZERO   SECOND ZERO

The first zero on either side of the central maximum corresponds to $\alpha = \pm\pi$. The angles θ for which these occur are given by

$$\sin\theta =$$

For these angles every wavelet from the upper half of the slit can be paired with a wavelet from the bottom half, emanating from a point $a/2$ away. The phase difference of two wavelets in a pair is _____ and these wavelets sum to _____ .

Notice that as a decreases the angle θ for the first minimum increases. The central maximum is broader for narrow slits than for wide slits and diffraction is more pronounced. If a is greater than λ no zeros of intensity occur ($\sin\theta$ cannot be greater than 1) and the entire screen is within the central maximum.

In the space to the right sketch the phasor diagram corresponding to the first secondary maximum to one side of the central maximum. In terms of the diagram explain why the intensity at a secondary maximum is less than that at the central peak: _____

_____

_____

_____

Double-slit diffraction.   Consider two identical slits, each of width a, with a center-to-center separation d. Any point P on a screen is reached by a wave from each slit, the resultant of the Huygen wavelets from that slit, and we may think of the pattern as being formed by the interference of these two waves. If light is incident normally on the slits and the observation point is far away the wave that reaches it from each slit has amplitude $E_m(\sin\alpha)/\alpha$, where $\alpha = (\pi a/\lambda)\sin\theta$, and the two waves differ in phase by $(2\pi d/\lambda)\sin\theta$. When they are combined the resultant amplitude is

$$E_\theta =$$

where $\beta = (\pi d/\lambda)\sin\theta$.

As you learned in the last chapter double-slit interference minima occur for $\beta = (2m + 1)\pi/2$ or $\sin\theta = (2m + 1)\lambda/2d$. Single-slit diffraction minima occur for $\alpha = m\pi$ or $\sin\theta = m\lambda/a$. In each case m is an integer but it may have different values in the two expressions. Since d must be larger than a, the interference minima must be closer together than the diffraction minima. The single-slit diffraction pattern forms an envelope, with the interference pattern inside. Study Fig. 41–13 of the text.

For any double-slit situation there are 3 parameters you must be aware of: the wavelength λ, the slit width a, and the center-to-center slit separation d. The ratio a/λ controls the width of the central diffraction maximum, which extends from $-\sin^{-1}(\lambda/a)$ to $+\sin^{-1}(\lambda/a)$. The ratio also controls the positions of the secondary diffraction maxima, if any. The ratio d/λ controls the angular positions of the interference maxima and minima.

Finally, the ratio d/a controls how many interference maxima fit within the central diffraction maximum or any of the secondary diffraction maxima. This number is independent of the wavelength. As the wavelength increases the central diffraction maximum widens but the interference pattern also spreads, with the result that just as many interference fringes are within the central diffraction maximum.

Diffraction from a circular aperture. When plane waves pass through a circular aperture and onto a screen a diffraction pattern is formed there. The pattern consists of a bright central disk, followed by a series of alternating dark and bright rings. The first minimum occurs at an angle θ, measured from the normal to the aperture and given by

$$\sin \theta =$$

where d is the diameter of the aperture. This angle can be used as a measure of the angular size of the central disk.

Stars are effectively point sources of light and lenses act like circular apertures. The image of a star formed by a lens is not a point but is broadened by diffraction to a disk and rings. Two stars do not form distinct images if the central disks of their diffraction patterns overlap too much. Describe the Rayleigh criterion for the resolution of two far-away point sources: _

Multiple slits. Consider a barrier in which N parallel slits have been cut, with distance d between adjacent slits. Monochromatic plane waves are incident normally on the barrier and the intensity pattern formed by waves passing through the slits is viewed on a screen far away. The slits are so narrow that single-slit diffraction can be ignored. That is, the interference pattern is well within the central maximum of the single-slit diffraction pattern.

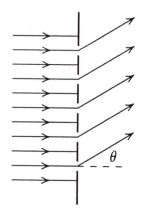

The pattern on the screen consists a series of intense, narrow bands, called _____. Secondary maxima lie between but they are much less intense and not important for our purposes. The pattern is usually described in terms of the angle θ made with the normal by a ray from the slit system to a point on the screen. On the axes below sketch a graph of the intensity as a function of θ for a small number of slits, 4 or 5, say.

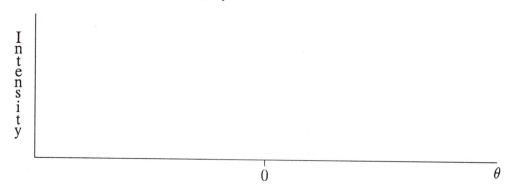

A diffraction line occurs when the phases at the screen of waves from any two adjacent slits are either _____ or differ by a multiple of _____ rad. Since waves from two adjacent slits travel distances that differ by $d \sin \theta$, where d is the slit separation, the condition for a line is

$$d \sin \theta =$$

where λ is the wavelength. The integer m in this equation is called _____.
High order lines occur at _____ angles than low order lines.

Notice that the locations of the lines are determined by the ratio d/λ and are independent of the number of slits. Also notice that the lines occur at different angles for different wavelengths of light. For the same order line the angle for red light is _____ than that for violet light. If white light is incident on the barrier, the color of an observed band continuously changes from red at one end to violet at the other.

The width of a line is indicated by the angular position $\Delta\theta$ of an adjacent minimum. For the line that occurs at angle θ,

$$\Delta\theta =$$

Notice that it depends on the number of slits. In fact, as the number of slits increases without change in their separation the width of every line _____. Also notice that maxima near the normal (small θ) are _____ than maxima farther away from the normal (larger θ).

Diffraction gratings. A diffraction grating consists of many thousands of closely spaced rulings on either a transparent or highly reflecting surface. When light is incident on a grating a diffraction pattern is formed, just like the multiple-slit pattern you studied above.

Because light with different wavelengths produces lines at different angles, diffraction gratings are often used to analyze the spectra of light sources. Two parameters are used to measure the quality of a diffraction grating. The <u>dispersion</u> of a grating measures the angular separation of lines of the same order for wavelengths differing by $\Delta\lambda$. It is defined by

$$D =$$

and for order m, occurring at angle θ, it is given by

$$D =$$

where d is the slit separation. Notice that dispersion does not depend on the number of rulings. Large dispersion means _____ angular separation.

The second parameter is the <u>resolving power</u>. It measures the difference in wavelength for waves with the angular separation of their lines equal to half the angular width of a line; that is, for two lines that obey the Raleigh criterion for resolution. Mathematically it is defined by $R = \lambda/\Delta\lambda$, where $\Delta\lambda$ is the difference in wavelength. For a system of N slits the resolving power at the line of order m is given by

$$R =$$

Be sure you understand the difference between dispersion and resolving power. Describe in words the pattern produced by a grating with large dispersion and small resolving power:

_____

_____

Consider a principal maximum near $\theta = 0$ and tell which quantities (of m, d, and N) should be large and which should be small to produce such a pattern: _____

_____

Describe the intensity pattern produced by a grating with small dispersion and large resolving power: _____

_____

Tell which quantities should be large and which should be small to produce such a pattern:

_____

X-ray diffraction.   Atoms in a crystal form a periodic array in three dimensions and x-ray radiation scattered by their electrons produces a diffraction pattern. Why are x rays with wavelengths on the order of 0.1 nm, rather than visible light with wavelengths on the order of 500 nm, used to form the pattern? _____

_____

Diffraction occurs only when the x rays are incident at certain angles to <u>crystal planes</u>. A crystal plane is a plane that _____

_____.

Suppose monochromatic plane-wave x rays with wavelength λ are incident at an angle θ to a set of crystal planes with separation d. An intense spot is produced if the angle of incidence satisfies

$$2d \sin \theta =$$

where m is an integer. When this condition is met a high intensity beam is radiated at the angle θ to the planes. Carefully note that θ is *not* the angle between the incident rays and the normal to the planes. Rather, it is the angle between the incident rays and the planes themselves.

The diagram on the right shows the edges of two crystal planes as horizontal dotted lines and two rays reflected from them. If d is the separation of the planes, the lower ray travels a distance $2d \sin \theta$ further than the upper ray. On the diagram draw a normal to the rays through the point of reflection of the upper ray and point out the distance $d \sin \theta$. For constructive interference to occur the difference in the distance traveled must be a multiple of _____, so $2d \sin \theta = m\lambda$. There are many more crystal planes parallel to the ones shown. If x rays from two adjacent planes interfere constructively then x rays from all of them interfere constructively and an intense beam is formed.

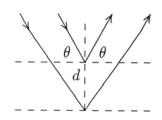

X rays are used to experimentally determine the atomic arrangements in crystals. The idea is to find the orientations and separations of a great many sets of crystal planes and use these to reconstruct the crystal. The symmetry of the diffraction spots is indicative of the symmetry of the crystal and can often be used to great advantage.

Crystals with known atomic arrangements are used as filters to separate x rays of a given wavelength from an incident beam containing a mixture of wavelengths. The crystal is oriented so that only the waves with the desired wavelength form an intense scattered beam.

II. PROBLEM SOLVING

The single-slit diffraction pattern is fundamental to this chapter. You should know how to find minima: $a \sin \theta = m\lambda$, and how to calculate the intensity: $I = I_m (\sin \alpha)^2 / \alpha^2$, where $\alpha = (\pi a / \lambda) \sin \theta$.

PROBLEM 1 (2E). Monochromatic light of wavelength 441 nm is incident on a narrow slit. On a screen 2.00 m away, the distance between the second diffraction minimum and the central maximum is 1.50 cm. (*a*) Calculate the angle of diffraction θ of the second minimum.

STRATEGY: Use $\tan \theta = y/D$, where y is the displacement on the screen of the second minimum from the central maximum and D is the slit-to-screen distance.

SOLUTION:

[ans: 0.430°]

(*b*) Find the width of the slit.

STRATEGY: Use $a \sin \theta = m\lambda$, with $m = 2$.

SOLUTION:

[ans: 0.118 mm]

PROBLEM 2 (10P). Manufacturers of wire (and other objects of small dimensions) sometimes use a laser to continually monitor the thickness of the product. The wire intercepts the laser beam, producing a diffraction pattern like that of a single slit of the same width as the wire diameter (see Fig. 41–25). Suppose a helium-neon laser, of wavelength 632.8 nm, illuminates a wire , and the diffraction pattern appears on a screen 2.60 m away. If the desired wire diameter is 1.37 mm, what is the observed distance between the two tenth-order minima (one on each side of the central maximum)?

STRATEGY: Use $a \sin \theta = m\lambda$, with $m = 10$, to calculate θ, then $y = D \tan \theta$ to calculate the displacement y of one minimum from the central maximum. The distance between the minima is $2y$.

SOLUTION:

[ans: 2.40 cm]

PROBLEM 3 (15P).   The full width at half maximum (FWHM) of the central diffraction maximum is defined as the angle between the two points in the pattern where the intensity is one-half that at the center of the pattern. (See Fig. 41–7b.) (*a*) Show that the intensity drops to one-half of the maximum value when $\sin^2 \alpha = \alpha^2/2$.

STRATEGY: The intensity is given by

$$I = I_m \left(\frac{\sin \alpha}{\alpha} \right)^2 ,$$

where I_m is the maximum intensity. Set $I = I_m/2$ and solve for $\sin^2 \alpha$.

SOLUTION:

(*b*) Verify that $\alpha = 1.39$ radians (about $80°$) is a solution to the transcendental equation of part (*a*).

STRATEGY: Substitute $\alpha = 1.39$ rad into $\sin^2 \alpha$ and into $\alpha^2/2$. Compare the results. You might also try $\alpha = 1.38$ rad and $\alpha = 1.40$ rad to convince yourself that the given result is the closest to the correct result of any 3 significant digit number.

SOLUTION:

(*c*) Show that the FWHM is $\Delta \theta = 2 \sin^{-1}(0.443\lambda/a)$.

STRATEGY: Solve $\alpha = (\pi a/\lambda) \sin \theta$ for θ. The result must be doubled to obtain $\Delta \theta$.

SOLUTION:

(*d*) Calculate the FWHM of the central maximum for slits whose widths are 1.0, 5.0, and 10 wavelengths.

STRATEGY: Use the expression you just developed, with $a = 1.0\lambda, 5.0\lambda$, and 10λ..

SOLUTION:

$$\left[\text{ans: } 53°, 10°, 5.1° \right]$$

You should know that the diffraction angle for the first minimum of a circular aperture is given by $\theta_R = \sin^{-1}(1.22\lambda/d)$, where d is the diameter of the aperture. You should also know that this expression also gives the Rayleigh criterion for the resolution of two far-away objects. If $d \gg \lambda$ then θ_R is given in radians by $\theta_R = 1.22\lambda/d$.

PROBLEM 4 (27P).   (a) How far from grains of red sand must you be so that you are just at the limit of resolving the grains if your pupil diameter is 1.5 mm, the grains are spherical with radius 50 μm, and the light from the grains has wavelength 650 nm?

STRATEGY: The minimum angle between lines from adjacent grains to your eye is given in radians by $\theta_R = 1.22\lambda/d$, where λ is the wavelength of the light and d is the diameter of the eye pupil. The separation of two adjacent grains is twice the radius of one of them. If r is the radius and D is the distance from the grains to the eye, then $\theta_R = 2r/D$. Thus $D = 2rd/1.22\lambda$.

SOLUTION:

[ans: 19 cm]

(b) If the grains were blue and the light from them had wavelength 400 nm, would the answer to (a) be larger or smaller?

SOLUTION:

[ans: larger]

A double-slit pattern combines double-slit interference and single-slit diffraction. The interference minima are given by $d\sin\theta = (2m+1)\lambda/2$ and the maxima are given by $d\sin\theta = m\lambda$. The diffraction minima are given by $a\sin\theta = m\lambda$. When an interference maxima coincides with a diffraction minima the order is said to be missing. You should also be able to calculate the intensity: $I = I_m(\cos\beta)^2(\sin\alpha)^2/\alpha^2$, where $\beta = (\pi d/\lambda)\sin\theta$ and $\alpha = (\pi a/\lambda)\sin\theta$. Don't confuse the slit width a and slit separation d.

PROBLEM 5 (36E).   For $d = 2a$ in Fig. 41–27, how many bright interference fringes lie in the central diffraction envelope?

STRATEGY: The first diffraction minimum occurs at the angle θ such that $a\sin\theta = \lambda$. Interference maxima occur for $d\sin\theta = m\lambda$, where m is an integer. Solve for the largest value of m such that $m\lambda/d$ is less than λ/a. In this case one of the interference maxima coincides with the diffraction minimum and so cannot be counted (the intensity is zero there). Remember that you have found the interference maxima on only one side of the central diffraction peak. Double the number you found and add 1 (for the central peak) to find the total number of interference maxima.

SOLUTION:

[ans: 3]

PROBLEM 6 (37P).   If we put $d = a$ in Fig. 41–27, the two slits coalesce into a single slit of width $2a$. Show that Eq. 41–15 reduces to the diffraction pattern for such a slit.

STRATEGY: For $d = a$, $\beta = (\pi a/\lambda)\sin\theta$ and this is the same as α. Thus the intensity is given by $I = I_m(\cos\alpha\sin\alpha)^2/\alpha^2$. Use the trigonometric identity $\cos\alpha\sin\alpha = \frac{1}{2}\sin(2\alpha)$ to write this $I = I_m(\sin 2\alpha)^2/(2\alpha)^2$. Since $\alpha = (\pi a/\lambda)\sin\theta$ this is the same as the intensity for a slit of width $2a$.

SOLUTION:

PROBLEM 7 (41P). Light of wavelength 440 nm passes through a double slit, yielding a diffraction pattern whose graph of intensity I versus deflection angle θ is shown in Fig. 41–28. Calculate (a) the slit width and (b) the slit separation.

STRATEGY: (a) According to the graph, the first diffraction minimum is at $\theta = 5.0°$. Use $a\sin\theta = \lambda$ to calculate the slit width a. (b) The second interference maximum beyond the central maximum is at $\theta = 2.5°$. Use $d\sin\theta = 2\lambda$ to calculate the slit separation.

SOLUTION:

$$[\text{ans: } (a)\ 5.05\,\mu\text{m; } (b)\ 20.2\,\mu\text{m}]$$

(c) Verify the displayed intensities of the $m = 1$ and $m = 2$ interference fringes.

STRATEGY: Use $I = I_m(\cos\beta)^2(\sin\alpha)^2/\alpha^2$, where $\beta = (\pi d/\lambda)\sin\theta$ and $\alpha = (\pi a/\lambda)\sin\theta$. Read the values of θ from the graph.

SOLUTION:

Diffraction gratings produce intense lines. The condition is $d\sin\theta = m\lambda$, where d is the slit separation. Use this expression, for example, to find the angular positions of the lines. You should also know how to compute the angular width of a line: $\Delta\theta = \lambda/Nd\cos\theta$, where N is the number of lines. You should know the meaning of the dispersion and know how to calculate it in terms of the angular separation of lines of different wavelength ($D = \Delta\theta/\Delta\lambda$) and in terms of the grating properties ($D = m/d\cos\theta$). You should know the meaning the resolving power and how to calculate it in terms of the difference in wavelength ($R = \lambda/\Delta\lambda$) and in terms of grating properties ($R = Nm$).

PROBLEM 8 (43E). A diffraction grating 20.0 mm wide has 6000 rulings. (a) Calculate the distance d between adjacent rulings.

STRATEGY: Divide the width by the number of rulings.

SOLUTION:

$$[\text{ans: } 3.33\,\mu\text{m}]$$

(b) At what angles will intensity maxima occur if the incident radiation has a wavelength of 589 nm?

STRATEGY: Use $\sin\theta = m\lambda/d$. Start with $m = 0$ and increase m by 1 for each successive calculation until $m\lambda/d > 1$ is reached.

SOLUTION:

$$[\text{ans: } 0, \pm 10.2°, \pm 20.7°, \pm 32.0°, \pm 45.0°, \pm 62.2°]$$

PROBLEM 9 (48E). A diffraction grating 1.0 cm wide has 10,000 parallel slits. Monochromatic light that is incident normally is deviated through 30° in first order. What is the wavelength of the light?

STRATEGY: Use $d\sin\theta = m\lambda$. Calculate d by dividing the width by the number of slits. Take $m = 1$.

SOLUTION:

$$[\text{ans: } 500\,\text{nm}]$$

PROBLEM 10 (50P). A diffraction grating is made up of slits of width 300 nm with separation 900 nm. The grating is illuminated by monochromatic plane waves of wavelength $\lambda = 600$ nm at normal incidence. (a) How many diffraction maxima are there in the full pattern?

STRATEGY: Solve $d\sin\theta = m\lambda$, with $\theta = 90°$, for m. Round down to the nearest integer. Double the result to include both sides of the central maximum and add 1 to include the central maximum itself.

SOLUTION:

$$[\text{ans: } 3]$$

(b) What is the width of the spectral lines observed in first order if the grating has 1000 slits?

STRATEGY: The angular width in radians is given by $\Delta\theta = \lambda/Nd\cos\theta$. Use $d\sin\theta = m\lambda$ to find θ for the first order line.

SOLUTION:

$$[\text{ans: } 8.94 \times 10^{-4}\,\text{rad} \ (0.051°)]$$

PROBLEM 11 (55P).   Two spectral lines have wavelengths λ and $\lambda + \Delta\lambda$, respectively, where $\Delta\lambda \ll \lambda$. Show that their angular separation $\Delta\theta$ in a grating spectrometer is given approximately by

$$\Delta\theta = \frac{\Delta\lambda}{\sqrt{(d/m)^2 - \lambda^2}},$$

where d is the slit separation and m is the order at which the lines are observed. Note that the angular separation is greater in the higher orders than in lower orders.

STRATEGY: Differentiate $d\sin\theta = m\lambda$ to obtain $d\cos\theta\,\Delta\theta = m\,\Delta\lambda$. Since $\sin\theta = m\lambda/d$ and $\sin^2\theta + \cos^2\theta = 1$, $\cos\theta = \sqrt{1 - (m\lambda/d)^2}$. Substitute for $\cos\theta$ in first equation and solve for $\Delta\theta$.

SOLUTION:

PROBLEM 12 (61E).   A grating has 600 rulings/mm and is 5.0 mm wide. (*a*) What is the smallest wavelength interval that can be resolved in the third order at $\lambda = 500$ nm?

STRATEGY: Use $R = Nm$ to find the resolving power and $R = \lambda/\Delta\lambda$ to find the interval $\Delta\lambda$.

SOLUTION:

[ans: 56 pm]

(*b*) How many higher orders of maxima can be seen?

STRATEGY: Solve $d\sin\theta = m\lambda$, with $\theta = 90°$, for m. Round down to the nearest integer and subtract 3.

SOLUTION:

[ans: none, only 3 orders can be seen]

PROBLEM 13 (67P).   Light containing a mixture of two wavelengths, 500 nm and 600 nm, is incident normally on a diffraction grating. It is desired (1) that the first and second maxima for each wavelength appear at $\theta \leq 30°$, (2) that the dispersion be as high as possible, and (3) that the third order for 600 nm be a missing order. (*a*) What should be the slit separation?

STRATEGY: The dispersion is given by $D = m/d\cos\theta$. To make it as large as possible you want to select d to be as small as possible and θ to be as large as possible, still staying within the other constraints. Take θ to be 30° for the second order 600 nm line and solve $d\sin\theta = m\lambda$ for d.

SOLUTION:

[ans: 2.4 μm]

(b) What is the smallest possible individual slit width?

STRATEGY: Select a so the third order line for 600 nm is missing. The third order line occurs for $d \sin \theta = 3\lambda$ and the first minimum of the diffraction pattern occurs for $a \sin \theta = \lambda$. Eliminate $\sin \theta$ to find that $d/a = 3$.

SOLUTION:

[ans: $0.80 \, \mu m$]

(c) For the 600-nm wavelength, which orders of intensity maxima are produced by the grating, assuming the values derived in (a) and (b)?

STRATEGY: Put $\theta = 90°$ in $d \sin \theta = m\lambda$, then solve for m. Round down to the nearest integer. Recall that the $m = 3$ line is missing.

SOLUTION:

[ans: 0, 1, and 2]

You should be able to use the Bragg condition for diffraction from a crystal: $2d \sin \theta = m\lambda$. Remember that θ is measured from the reflecting planes, not their normal. One of the problems below shows how the Bragg condition can be used to find the atomic separation in a crystal.

PROBLEM 14 (72E). If first-order reflection occurs in a crystal at Bragg angle 3.4°, at what Bragg angle does second-order reflection occur from the same family of reflecting planes?

STRATEGY: For first-order reflection $2d \sin \theta = \lambda$, where d is the interplanar spacing. Solve for λ/d. For second-order reflection $2d \sin \theta = 2\lambda$. Solve for θ using the value of λ/d you just found.

SOLUTION:

[ans: 6.8°]

PROBLEM 15 (78P). In Fig. 41–33, first-order reflection from the reflection planes shown occurs when an x-ray beam of wavelength 0.260 nm has an angle of 63.8° to the top face of the crystal. What is the unit cell size a_0?

STRATEGY: A little geometry shows that the x-ray beam makes the angle $\theta = 18.8°$ with the reflecting planes. Use $2d \sin \theta = m\lambda$ to calculate the spacing of the planes. Observe that the diagonal of a square with side a_0 is $2d$, so $a_0 = \sqrt{2}d$.

SOLUTION:

[ans: 0.570 nm]

PROBLEM 16 (80P).   In Fig. 41–32, a beam of x rays of wavelength 0.125 nm is incident on a NaCl crystal at an angle of 45° to the top face of the crystal. The reflecting planes have separation $d = 0.252$ nm. Through what angles must the crystal be turned about an axis that is perpendicular to the plane of the page for these reflecting planes to give intensity maxima in their reflections?

STRATEGY: Use $2d \sin \theta = m\lambda$ to find all possible Bragg angles. To find the angle that the crystal must be turned subtract θ from 45° if θ is less than 45° or subtract 45° from θ if θ is greater than 45°. In the first case the crystal must be turned clockwise; in the second case it must be turned counterclockwise.

SOLUTION:

$$\left[\text{ans: } 30.6°, 15.3° \text{ clockwise; } 3.08°, 37.8° \text{ counterclockwise} \right]$$

III. NOTES

OVERVIEW IV
MODERN PHYSICS

This section opens with a chapter on special relativity, which tells us how measurements of the same phenomenon made by observers who are moving with respect to each other are related. The theory has forced us to redefine our concepts of time, energy, and momentum. The results are important and interesting in their own right. They are absolutely essential for the discussions of nuclear and particle physics in Chapters 47, 48, and 49.

You have learned that electric and magnetic fields propagate as waves. Waves are predicted by Maxwell's equations and phenomena such as reflection, refraction, interference, diffraction, and polarization are all explained in terms of electromagnetic waves. However, experiments that test the nature of electromagnetic radiation at the atomic level show that in some aspects it behaves like particles, exchanging energy and momentum in localized collisions.

You have also learned that electrons are particles. They are highly localized in space and exchange energy and momentum in collisions with other particles. However, experiments show that in some aspects matter behaves like waves, exhibiting interference and diffraction effects.

A highly successful theory, the quantum theory, has been developed to explain in a comprehensive manner the behavior of nature at the atomic level. The fundamentals are discussed in Chapters 43 and 44. Many of the results will mystify you but the ideas of the quantum theory will not be difficult for

you to understand if you keep an open mind as you study these chapters.

Some of the experiments that force us to treat electromagnetic radiation as particles are described in Chapter 43. Here you will be introduced to relationships between particle properties such as energy and momentum and wave properties such as frequency and wavelength. The experiments that force us to associate waves with matter are discussed in Chapter 44. Here you will learn that the same relationships hold for particles and the waves associated with them.

In Chapter 44 you will also get a glimmer of how the wave and particle natures of both light and particles are made compatible. In each case the wave is not a manifestation of the motion of the medium in which it propagates. Instead it is closely associated with the probability that the particle is in a given region of space. It is the probability that shows interference effects, for example.

Interesting and important consequences arise because the probability of finding the particle in a certain given region propagates as a wave. The more closely the position of the particle is defined, for example, the less closely its momentum is defined. By necessity, a measurement of the position destroys information about the momentum. Another important consequence is that a particle initially on one side of a barrier may appear on the other side. It is said to have tunneled through. Still another consequence is that the energy of a bound particle is quantized. It may have only certain discrete values.

Remaining chapters of this section deal with applications of the quantum principles. In Chapter 45 quantum mechanics is applied to the electron in a hydrogen atom. This is the simplest atom to understand. Here you will see in detail how the quantization of energy comes about because restrictions must be placed on the wave function. You will also see graphs of the probability density for an electron in an atom. The ideas are then extended to atoms with more than 1 electron.

Not only is the energy of an electron in an atom quantized but so is its angular momentum. You will learn to categorize the quantum mechanical state of an electron by giving its energy, the magnitude of its angular momentum, and one component of its angular momentum. You will be introduced to the idea of spin angular momentum, the intrinsic angular momentum of a particle.

The various states available to electrons in atoms along with the knowledge of how they are filled lead to an understanding of the periodic table of the chemical elements. You will see why atoms in the same column have similar chemical properties. You will also learn how x rays can be used to distinguish between atoms with the same chemical properties.

Quantum mechanics has enormously increased our understanding of the properties of materials. You learned in Chapter 28 that the resistivity of a material is determined by the density of conduction electrons and by the mean free time between their collisions with atoms. In Chapter 46 quantum mechanics is used to provide a deeper understanding. You will learn how metals, insulators, and semiconductors differ in their conducting properties and about the role played by quantum mechanics in determining the dif-

ferences. You will also learn about the basis for the temperature dependence of the resistivity. Chapter 46 also includes an introduction to semiconductor devices, so pervasive in modern technology.

The next two chapters deal with the important topic of nuclear physics, the fundamentals in Chapter 47 and applications to the harnessing of nuclear energy in Chapter 48. In the first of these chapters you will learn what particles are emitted by radioactive nuclei and about the rate of emission for a collection of such nuclei. You will learn to use conservation of energy to predict the energies of the emission products.

Energy is released when a heavy nucleus fragments. The basics of this phenomenon (nuclear fission) and its utilization in modern nuclear power plants are discussed in Chapter 48. Energy is also released when two light nuclei fuse together. This phenomenon (nuclear fusion) powers the sun (and other stars) and is a candidate for power plants of the future. It is also discussed in this chapter.

Chapter 49 is devoted to the very small (fundamental particles, the building blocks of matter) and the very large (the universe itself). You will learn that some particles, such as protons and neutrons, are actually made of more fundamental entities called quarks. Other particles, such as electrons, seem to be indivisible. You will learn a little about the basic interactions of the fundamental particles with each other. You will see that the basic forces and the properties of the fundamental particles played pivotal roles in the evolution of the universe from the big bang to its present state. Evidence for the big bang is discussed and a chronological description of events in the early universe is given.

Chapter 42
RELATIVITY

I. BASIC CONCEPTS

When two observers who are moving relative to each other measure the same physical quantity, they may obtain different values. The theory of special relativity tells how the values are related to each other when both observers are at rest in different inertial frames. Although the complete theory deals with all physical quantities, the ones you consider here are the coordinates and time of an event and the velocity, momentum, and energy of a particle.

All the equations you will use are linear algebraic equations, so the mathematics is quite simple. The concepts, however, are difficult for some students because they run counter to experience. You must get used to the idea, for example, that the time interval and spatial distance between two events depend on the velocity of the observer carrying out the measurements. Relativity has significantly altered our concepts of length and time.

The basis of special relativity. The theory is based on two postulates. The first deals with the laws of physics. It is: _____
_____

Keep in mind that the laws of physics are relationships between physical quantities, not the quantities themselves. Newton's second law and the conservation principles are examples of laws. The momentum of a system has a different value for different reference frames but if it is conserved for one inertial frame it is conserved for all inertial frames, according to the postulate.

The second postulate deals with the speed of light. It is: _____
_____

Suppose a light source sends a pulse of light toward you. If you are stationary with respect to the source, the speed of the pulse relative to you is _____ . If you are moving at $c/2$ toward the source the speed of the pulse relative to you is _____ . If you are moving away from the source the speed of the pulse relative to you is _____ . This postulate is consistent with the notion that electromagnetic radiation does not require a medium for its propagation.

Special relativity deals with measurements made in <u>inertial</u> <u>reference</u> <u>frames</u>. Tell what an inertial frame is: _____
_____

As you study this chapter you will be concerned with <u>events</u>. An event has four numbers associated with it: three of them are _____ that designate the location of the event; the fourth designates the _____ of the event. To measure the coordinates of an event meter sticks must be laid out in an inertial reference frame, at rest with respect to the frame. To measure the time of an event a clock must be present at the location of the event and it must be a rest with respect to the reference frame. In addition, it must be synchronized with other clocks at rest in the frame. Synchronization is accomplished by _____

_____

_____.

The coordinates and time of any event may be measured by meter sticks and clocks at rest in any inertial frame. In general, different results are obtained for difference frames. Relativity tells how the measurements made in one frame are related to those made in another.

Simultaneity. Two events, separated in space and simultaneous to one observer, are NOT simultaneous to another observer, moving with respect to the first along the line joining the positions of the events. Carefully study Fig. 42–4 of the text. It shows two events, labeled Red and Blue. The events are simultaneous according to Sam. He knows they are simultaneous because the events occurred at the ends of his spaceship and electromagnetic waves from the events met at _____. Since the waves travel at the same speed and go the same distance they must have started at the same time.

The events also occur at the ends of Sally's space ship but she is moving away from the position of the Blue event and toward to the position of the Red event. Waves from the _____ event reach the midpoint of her spaceship before waves from the _____ event. In her reference frame the waves move with the speed of light, just as they do in Sam's reference frame. Therefore she knows that the _____ event occurred before the _____ event.

Time dilation. The time interval between two events is different when measured with clocks at rest in two inertial frames that are moving relative to each other. The digram to the right shows a clock. The flash unit F emits a light pulse that travels to mirror M and is reflected back to the flash unit. It is detected there and immediately triggers the next flash. If the flash unit and mirror are separated by a distance D then the time interval between flashes is given by $\Delta t_0 = $ _____ .

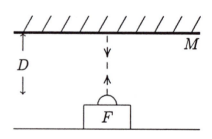

Sally carries a clock like this at speed v past Sam. Complete the diagram on the right by drawing the path of one pulse as seen by Sam. If Δt is the time interval between emission and detection of the pulse, as measured by Sam, then during this interval the flash unit moves a distance $\ell = $ _____ and the light pulse moves a distance $2L = $ _____ . This follows because L is the hypotenuse of a right triangle with sides of length D and $v\Delta t/2$. Substitute $2L = c\Delta t$ and solve for Δt:

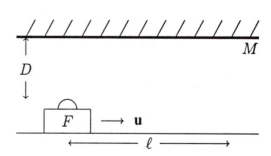

You should obtain $\Delta t = \gamma \Delta t_0$, where the Lorentz factor γ is $1/\sqrt{1-(v/c)^2}$. The speed is often given as a fraction of the speed of light. In terms of $\beta\ (= v/c)$, $\gamma =$ _____ . Notice that the derivation of the result for Δt depends strongly on the second postulate of relativity. An observer comparing a moving clock with his clocks concludes that the moving clock ticks at a slower rate. It is not important that the clock utilize light as does the clock you used to derive the relationship above. Any clock will do, even a heartbeat.

You should realize that there is perfect symmetry between the two reference frames. If Sally watches a clock at rest with respect to Sam she sees it tick at a slower rate.

The concept of <u>proper time</u> is important for understanding the relativity of time. Suppose two events, such as the emission and detection of a light pulse, occur at the same coordinate in one frame. The time interval between them, as measured in that frame, is the proper time between the events. The time interval between the same two events, as measured in a frame that is moving relative to the first, is longer by the factor γ. Since the speed of a reference frame is always less than the speed of light, the factor γ is always greater than 1.

Length contraction. If you measure the length of a rod that is moving past you at high speed the result is less than if you measure it when it is at rest with respect to you. Which of these measurements gives the <u>proper length</u> of the rod? _____

_____

Suppose you know the speed of the rod to be v. You place a marker on your coordinate system and measure the interval from time the front end of the rod is at the marker to the time the back end of the rod is at the marker. If that interval is Δt_0 then the length of the moving rod is $L =$ _____ . Notice that Δt_0 is the proper time interval between the two events. A single clock at the position of the marker can be used. L is NOT the proper length because the rod is moving relative to the frame used to measure L.

Now consider the same events from the point of view of someone moving with the rod. He sees the marker move from the front to the back of the rod in time Δt and gives the length of rod as $L_0 = v\,\Delta t$. This is the proper length because the rod is rest relative to the observer. Since Δt_0 is the proper time interval between the events, Δt and Δt_0 are related by $\Delta t =$ _____ , where $\gamma = 1/\sqrt{1-(v/c)^2}$. Thus, in terms of its proper length, the length of the moving rod is $L =$ _____ .

The Lorentz transformation. You should understand the phenomena of relativity not only qualitatively but also from the mathematical viewpoint of the <u>Lorentz transformation</u>. Suppose an event is viewed by two observers, one at rest in inertial frame S and another at rest in inertial frame S'. From the viewpoint of S, S' is moving in the positive x direction with velocity v. The coordinate systems and clocks are arranged so that the origins coincide at time $t = 0$ and the clocks of S' are synchronized to read 0 when the origins coincide. The coordinates of the event are x, y, z and time of the event is t, all as measured in S. The same event has coordinates x', y', z' and occurs at time t', as measured in S'. Then, according to the Lorentz transformation, the coordinates and times are related by

$$x' =$$

$$y' =$$

$$z' =$$

$$t' =$$

where $\gamma =$ _____ .

You will often deal with the time and spatial intervals between two events. Suppose the spatial interval has components Δx, Δy, and Δz in S and has components $\Delta x'$, $\Delta y'$, and $\Delta z'$ in S'. The time interval is Δt in S and $\Delta t'$ in S'. Then the intervals are related by

$$\Delta x' =$$

$$\Delta y' =$$

$$\Delta z' =$$

$$\Delta t' =$$

You should recognize that the same equations can be used if S' is moving in the negative x direction relative to S. The velocity v is then negative.

You might be given the intervals in S' and asked for those in S. You then use

$$\Delta x =$$

$$\Delta y =$$

$$\Delta z =$$

$$\Delta t =$$

Carefully note that distance measurements along the y or z axis produce the same values in all inertial frames that move relative to each other along their x axes. Different values are obtained only when the measurement is along the direction of relative motion.

Whether you must take special relativity into account or not is controlled by the value of γ. If γ is close to 1 you can safely ignore relativity. But if γ is much greater than 1 relativity is important. If $v = 0.1c$, $\gamma =$ _____; if $v = 0.5c$, $\gamma =$ _____; if $v = 0.9c$, $\gamma =$ _____; if $v = 0.95c$, $\gamma =$ _____; and if $v = 0.99c$, $\gamma =$ _____ .

Use the Lorentz transformation equations to obtain the time dilation equation. Take $\Delta x' = 0$ (the events occur at the same coordinate in S') and solve for Δt:

You should obtain $\Delta t = \gamma \Delta t'$. The proper time interval between the events is measured in the _____ frame.

Now suppose the two events occur at the same coordinate in the S frame. Take $\Delta x = 0$ and solve for $\Delta t'$:

You should obtain $\Delta t' = \gamma \Delta t$. The proper time interval between the events is now measured in the _____ frame.

Notice that two events might occur at different coordinates in both frames. Then the clocks in neither frame measure the proper time interval between the events.

The length contraction equation can also be derived by means of the Lorentz transformation. Suppose an object of proper length L_0 is at rest along the x axis of frame S and its length is measured in S', a frame that is moving with speed v in positive x direction relative to S. One way of measuring the length is to place two marks on the x' axis of S', one at the front of the object and one at the back, then measure the distance between the marks. The marks, of course, must be made simultaneously. Set $\Delta x' = L$, $\Delta t' = 0$, and $\Delta x = L_0$. Use the transformation equations to solve for $\Delta x'$ in terms of Δx:

You should obtain $\Delta x' = \Delta x / \gamma$.

Now suppose the object is at rest on the x' axis of S' and its length is measured in S. Set $\Delta x' = L_0$, $\Delta x = L$, and $\Delta t = 0$. Solve for Δx in terms of $\Delta x'$:

You should obtain $\Delta x = \Delta x' / \gamma$.

You can also use the Lorentz transformation equations to investigate simultaneity. Suppose two events are simultaneous in S and are spatially separated by Δx. Set $\Delta t = 0$. The time interval between them, as measured in S', is given by $\Delta t' =$ _____ if S' is moving with speed v in the positive x direction relative to S.

Now suppose two events are simultaneous in S' and are spatially separated by $\Delta x'$. Give an expression for the time interval between them, as measured in S: $\Delta t =$ _____ .

Relativistic velocities. Here you deal with the relationship between the velocity of an object as measured in one frame and its velocity as measured in another, moving with speed u in the positive x direction. Carefully note that the symbol u is used for the velocity of one reference frame relative to another, and the symbol v is used for the velocity of an object.

Suppose a particle is moving with velocity v' along the x' axis of S'. Then its velocity as measured in S is given by

$$v =$$

This expression can be derived easily from the Lorentz transformation equations. Divide $\Delta x = \gamma(\Delta x' + u\Delta t')$ by $\Delta t = \gamma(\Delta t' + u\Delta x'/c^2)$ to obtain $\Delta x/\Delta t =$ _____ . Now divide both the numerator and the denominator by $\Delta t'$, replace $\Delta x'/\Delta t'$ with v', and replace $\Delta x/\Delta t$ with v:

When $u \ll c$ the quantity $1 + uv'/c^2$ in the denominators can be approximated by 1 and the Galilean velocity transformation equation is obtained. It is $v = $ _____ .

One of the most important consequences of the relativistic velocity transformation equations is: if the speed of an object, as measured in one frame, equals the speed of light then its speed, as measured in any other frame, also equals the speed of light. No matter how fast you travel away from an approaching light pulse its speed relative to you will be c. The same is true no matter how fast you travel toward it. Prove this for an object moving in the positive x direction: replace v' with c in the expression for v and show that the result is $v = c$.

A corollary is: if the speed of an object, as measured in one frame, is less than the speed of light, then its speed, as measured in any frame, is less than the speed of light.

The Doppler Effect.   When a source of light is moving toward an observer or an observer is moving toward a source the observed frequency is greater than the frequency f_0 measured in the rest frame of the source (the proper frequency). When the relative motion of the source and observer is one of increasing separation the observed frequency is less than the proper frequency. Relativity predicts that the observed frequency is given by

$$ f = $$

where $\beta = v/c$ and v is the relative velocity of the source and observer. If the distance between the source and observer is increasing β is _____ ; if the distance is decreasing β is _____ . Remember that the equation is valid only if the motion is along the line joining the source and observer.

A Doppler shift in frequency also occurs if the motion is transverse to the line joining the source and observer. Then the observed frequency is given by

$$ f = $$

This expression is a direct result of applying the relativistic time-dilation equation to the period of oscillation.

Relativistic momentum and energy.   Relativity theory requires that the definitions of momentum and kinetic energy be revised if these quantities are to obey the familiar conservation laws. More precisely, if measurements taken in one inertial frame show that momentum is conserved then measurements taken in any other inertial frame should also show that momentum is conserved. Similarly if energy is conserved in one inertial frame then it should be conserved in all inertial frames.

If a particle has mass m and travels with velocity $\mathbf{v}$ then its momentum $\mathbf{p}$ is given by

$$ \mathbf{p} = $$

If $v \ll c$ this expression reduces to $\mathbf{p} = $ _____, the non-relativistic definition.

If a system consists of several particles the total momentum is the vector sum of the individual momenta. If the net external force on particles of a system vanishes then the total momentum of the system is conserved.

Suppose a particle with mass m is moving in the positive x direction with velocity v. Its momentum, of course, is in the positive x direction and the magnitude is given by $p = mv/\sqrt{1 - (v/c)^2}$. The same particle is observed in a frame S' that moves with velocity u in the positive x direction. The magnitude of the particle's momentum, as measured in this frame, is given by $p' = mv'/\sqrt{1 - (v'/c)^2}$, where $v' = $ _____.

The kinetic energy of a particle with mass m moving with speed v is given by

$$K = $$

This expression is valid no matter what the speed v. If $v \ll c$ it reduces to the familiar non-relativistic expression $K = \frac{1}{2}mv^2$. The total kinetic energy of a system of particles is the scalar sum of the individual kinetic energies. This quantity is conserved in collisions provided the masses of the particles do not change. Work must be done to change the kinetic energy of a particle and the change in kinetic energy equals the net work done.

In many nuclear scattering and decay processes the masses do change. You must then take into the account the <u>rest energies</u> of the particles. The rest energy of a particle of mass m is given by $E_0 = $ _____. The total energy, the sum of the rest and kinetic energies, of a particle is given in terms of the mass and speed of the particle by

$$E = $$

The total energy and the magnitude of the momentum of any particle are related by

$$E^2 = $$

This expression replaces $E = p^2/2m = mv^2/2$, the non-relativistic relationships between E and p and between E and v. Note that the energy E in the relativistic relationship includes both rest and kinetic energies.

II. PROBLEM SOLVING

In several problems you are asked to find the time interval between two events in one reference frame, given the interval in another. If the two events occur at the same coordinate in one of the frames, so the interval is measured by a single clock, then that clock measures the proper time interval Δt_0 and the interval in the other frame is given by $\Delta t = \gamma \Delta t_0$, where $\gamma = 1/\sqrt{1 - (v/c)^2}$ and v is the velocity of the second frame relative to the first. You must be able to identify which interval is the proper time interval. You must also know that the proper length of an object is measured in the frame in which the object is at rest. Its length in a frame for which it is moving is given by $L = L_0/\gamma$.

PROBLEM 1 (6P).   An unstable high-energy particle enters a detector and leaves a track 1.05 mm long before it decays. Its speed relative to the detector was $0.992c$. What is its proper lifetime? That is, how long would it have lasted before decay had it been at rest with respect to the detector?

STRATEGY: The proper lifetime is measured by a clock at rest with respect to the particle, not by clocks at rest with respect to the detector. Evaluate $\Delta t_0 = \Delta t/\gamma$, where $\gamma = 1/\sqrt{1-(v/c)^2}$ and Δt is the lifetime as measured in the lab. Use $\Delta t = \ell/v$, where ℓ is the length of the track, to find that time. Δt_0 would also be the measured lifetime if the particle were at rest in the lab.

SOLUTION:

$$\left[\text{ans: } 4.45 \times 10^{-13}\,\text{s}\right]$$

PROBLEM 2 (7P).   A pion is created in the higher reaches of the Earth's atmosphere when an incoming high-energy cosmic-ray particle collides with an atomic nucleus. A pion so formed descends toward the earth with a speed of $0.99c$. In a reference frame in which they are at rest, pions decay with a mean life of 26 ns. As measured in a frame fixed with respect to the Earth, how far (on the average) will such a pion move through the atmosphere before it decays?

STRATEGY: It goes a distance $\ell = v\,\Delta t$, where Δt is its lifetime as measured in the rest frame of the Earth. The proper lifetime is measured in the rest frame of the pion, so $\Delta t = \gamma\,\Delta t_0$, where $\Delta t_0 = 26$ ns.

SOLUTION:

$$\left[\text{ans: } 55\,\text{m}\right]$$

PROBLEM 3 (10E).   The length of a spaceship is measured to be exactly half its rest length. (*a*) What is the speed of the spaceship relative to the observer's frame?

STRATEGY: Solve $L = L_0/\gamma$ for γ, then $\gamma = 1/\sqrt{1-(v/c)^2}$ for v.

SOLUTION:

$$\left[\text{ans: } 2.560 \times 10^8\,\text{m/s }(0.866c)\right]$$

(*b*) By what factor do the spaceship's clocks run slow, compared to clocks in the observer's frame?

STRATEGY: We think of a single clock on the spaceship and compare its reading with two different clocks on the earth, at the beginning and end of some trip. In this case the proper time is measured on the spaceship's clock. The time interval as measured on the earth clocks is $\gamma\,\Delta t_0$. The spaceship's clock runs slow by the factor $1/\gamma$.

SOLUTION:

$$\left[\text{ans: } 1/2\right]$$

PROBLEM 4 (15P). An airplane whose rest length is 40.0 m is moving at uniform velocity with respect to the Earth, at a speed of 630 m/s. (*a*) By what fraction of its rest length will it appear to be shortened to an observer on Earth?

STRATEGY: The observer on Earth measures its length to be $L = L_0/\gamma$, where L_0 is the rest length. The fractional shortening is $(L_0 - L)/L_0 = 1 - \sqrt{1 - (v/c)^2}$. The quantity v/c is so small that your calculator probably cannot calculate the fractional change in length if you substitute directly into this equation. Use the first two terms of a binomial expansion: $(1 - v^2/c^2)^{1/2} \approx 1 - v^2/2c^2$.

SOLUTION:

$$\left[\, \text{ans: } 2.21 \times 10^{-12} \,\right]$$

(*b*) How long would it take, according to Earth clocks, for the airplane's clock to fall behind by $1.00 \, \mu s$? (Use special relativity in your calculations.)

STRATEGY: Let Δt be the time as measured by earth clocks and Δt_0 be the time as measured by the airplane clock. This the proper time and the two intervals are related by $\Delta t = \gamma \Delta t_0$. You want to find the value of Δt so that $\Delta t - \Delta t_0 = 1.00 \, \mu s$. Solve $\Delta t - \Delta t/\gamma = 1.00 \, \mu s$ for Δt.

SOLUTION:

$$\left[\, \text{ans: } 4.53 \times 10^5 \, \text{s} \; (5.25 \, \text{d}) \,\right]$$

PROBLEM 5 (16P). (*a*) Can a person, in principle, travel from Earth to the galactic center (which is about 23,000 ly distant) in a normal lifetime? Explain, using either time-dilation or length-contraction arguments.

STRATEGY: In the rest frame of the Earth the time for the trip is ℓ/v, where ℓ is the distance to the galactic center and v is the speed of the spaceship. In the frame of the space ship (where the proper time interval is measured) the time is $t = \ell/v\gamma$. You want t to be a normal lifetime. You might also consider length contraction. The distance as measured in the frame of the spaceship is ℓ/γ since the rest distance ℓ is measured in the frame of the Earth. The time, as measured in the spaceship, is again $t = \ell/v\gamma$.

SOLUTION:

$$\left[\, \text{ans: yes, if the spaceship goes fast enough} \,\right]$$

(*b*) What constant speed would be needed to make the trip in 30 y (proper time)?

STRATEGY: Solve $t = \ell/v\gamma$ for v. The first step is to square both sides of the equation and multiply by v^2 to obtain $v^2 t^2 = \ell^2(1 - v^2/c^2)$. You should find $v/c = \ell/\sqrt{\ell^2 + c^2 t^2}$. Use the binomial theorem to evaluate this expression.

$$\left[\, \text{ans: } 0.99999915c \,\right]$$

In some cases two events do not occur at the same place in either frame. Then the proper time interval between the events is not measured in either frame and you cannot use the time-dilation equation. Similarly the two events may not occur at the same time in either frame. Then the proper distance between the events is not measured in either frame and you cannot use the length-contraction equation. You must instead use the full Lorentz transformation equations: $\Delta x' = \gamma(\Delta x - c\Delta t)$ and $\Delta t' = \gamma(\Delta t - v\,\Delta x/c^2)$.

PROBLEM 6 (19E).   Inertial frame S' moves at a speed of $0.60c$ with respect to frame S. Two events are recorded. In frame S, event 1 occurs at the origin at $t = 0$ and event 2 occurs on the x axis at $x = 3.0\,\text{km}$ at $t = 4.0\,\mu\text{s}$. What times of occurrence does observer S' record for these same events? Explain the difference in the time order.

STRATEGY: Assume the origins of the two frames coincide at $t = 0$ and the clocks in S' are synchronized to read $t' = 0$ at that time. Evaluate the Lorentz transformation equation $t' = \gamma(t - vx/c^2)$. for each event.

SOLUTION:

$$\left[\text{ans: event 1: } t' = 0,\ \text{event 2: } t' = -2.5\,\mu\text{s}\right]$$

Suppose the events are simultaneous in S. Then in S' event 2 would be earlier than event 1. Event 2 is actually later than event 1 in S but it is not sufficiently later to cause event 2 to also be later in S'.

PROBLEM 7 (23P).   An observer S sees a flash of red light 1200 m from his position, and a flash of blue light 720 m closer to him directly in line with the red flash. He measures the time interval between the flashes to be $5.00\,\mu\text{s}$, the red flash occurring first. (*a*) What is the relative velocity **v** (give both magnitude and direction) of a second observer S' who records these flashes as occurring at the same place?

STRATEGY: Let Δx be the distance between the events and Δt be the time interval between the events, as measured by S. Then the distance as measured by S' is given by $\Delta x' = \gamma(\Delta x - v\,\Delta t)$. Δx is 720 m and Δt is $-5.00\,\mu\text{s}$. Solve for v so that $\Delta x' = 0$.

SOLUTION:

$$\left[\text{ans: } -1.44 \times 10^8\ \text{m/s } (-0.480c),\ \text{in the negative } x \text{ direction}\right]$$

(*b*) From the point of view of S', which flash occurs first? (*c*) What time interval between them would S' measure?

STRATEGY: Evaluate $\Delta t' = \gamma(\Delta t - v\,\Delta x/c^2)$. A negative result means the red flash occurs first.

SOLUTION:

$$\left[\text{ans: } (b) \text{ the red flash; } (c) -4.39\,\mu\text{s}\right]$$

Here are some examples that deal with the transformation of velocities. Typically you are given the velocity of a particle as measured in one reference frame and asked for its velocity as measured in another. Use $v = (v' + u)/(1 + uv'/c^2)$. Be careful to distinguish between the particle velocity v (and v') and the velocity u of the primed frame relative to the unprimed frame.

PROBLEM 8 (26E). Frame S' moves relative to frame S at $0.62c$ in the direction of increasing x. In frame S' a particle is measured to have a velocity of $0.47c$ in the direction of increasing x'. (a) What is the velocity of the particle with respect to frame S? (b) What would be the velocity of the particle with respect to S if it moved (at $0.47c$) in the direction of *decreasing* x' in the S' frame? In each case, compare your answers with the predictions of the classical velocity transformation equation.

STRATEGY: Evaluate $v = (v' + u)/(1 + uv'/c^2)$. The classical result is given by $v = v' + u$. In (a) $v' = 0.47c$ and $u = 0.62c$; in (b) $v' = -0.47c$ and $u = 0.62c$.

SOLUTION:

[ans: $0.84c$ (2.5×10^8 m/s), classical: $1.1c$; (b) $0.21c$ (6.3×10^7 m/s), classical: $0.15c$]

PROBLEM 9 (32P). A spaceship, at rest in a certain reference frame S, is given a speed increment of $0.50c$. Relative to its new frame of rest, it is then given a further $0.50c$ increment. This process is continued until its speed with respect to its original frame S exceeds $0.999c$. How many increments does this process require?

STRATEGY: At the beginning of each increment take S' to be the rest frame of the spaceship. It is moving with speed u relative to S. At the end of the increment the speed of the spaceship in S' is given by $v = (v' + u)/(1 + vu'/c^2)$. This is also the speed of S' for the next increment. For the first increment, $u = 0$ and $v' = 0.50c$. You should find $v = 0.80c$. For the second increment, $u = 0.80c$ and $v' = 0.50c$. Continue until v is greater than $0.999c$.

SOLUTION:

[ans: 7]

Some problems deal with the Doppler effect. Use $f = f_0\sqrt{(1-\beta)/(1+\beta)}$ when the motion is along the line joining the source and observer and $f = f_0/\gamma$ if the motion is transverse to that line. In the first case pay careful attention to the sign of β.

PROBLEM 10 (37P). A radar transmitter T is fixed to a reference frame S' that is moving to the right with speed v relative to reference frame S (see Fig. 42–14). A mechanical timer (essentially a clock) in frame S', having a period τ_0 (measured in S') causes transmitter T to emit timed radar pulses, which travel at the speed of light and are received by R, a receiver fixed in frame S. (a) What is the period τ of the timer as detected by observer A, who is fixed in frame S?

STRATEGY: Take the two events to be the sending of two successive pulses. The proper time interval between these events is measured by the mechanical timer. The time interval in S is given by $\tau = \gamma \tau_0$.

SOLUTION:

[ans: $\tau_0/\sqrt{1-(v/c)^2}$]

(b) Show that at the receiver R the time interval between pulses arriving from T is not τ or τ_0, but

$$\tau_R = \tau_0 \sqrt{\frac{c+v}{c-v}}.$$

STRATEGY: Use the Doppler shift equation $f = f_0\sqrt{(1-v/c)/(1+v/c)}$. The period is the reciprocal of the frequency.

SOLUTION:

(c) Explain why the observer at R measures a different period for the transmitter than observer A, who is in the same reference frame. (*Hint*: A clock and a radar pulse are not the same.)

STRATEGY: The period of a wave (τ_R) is the time from the beginning to the end of a cycle, measured at the same place. τ is the time from the beginning to the end of the timer cycle. During that time the timer moves.

SOLUTION:

You should know the relativistic definition of kinetic energy: $K = mc^2(\gamma-1)$; rest energy: mc^2; total energy: $E = mc^2\gamma$; and momentum: $mv\gamma$, as well as the relationship between total energy and momentum: $E^2 = (mc^2)^2 + (pc)^2$.

PROBLEM 11 (43E). A particle has a speed of $0.990c$ in a laboratory reference frame. What are its kinetic energy, its total energy, and its momentum if the particle is (*a*) a proton or (*b*) an electron?

STRATEGY: The kinetic energy is given by $K = mc^2(\gamma - 1)$, the total energy is given by $E = mc^2\gamma$ or by $E = K + mc^2$, and the momentum can be calculated from $E^2 = (mc^2)^2 + (pc)^2$. Use $m = 1.673 \times 10^{-27}$ kg for a proton and $m = 9.109 \times 10^{-31}$ kg for an electron.

SOLUTION:

[ans: (*a*) $K = 9.16 \times 10^{-10}$ J (5.71 GeV), $E = 1.07 \times 10^{-9}$ J (6.65 GeV), $p = 3.52 \times 10^{-18}$ kg·m/s (6.59 GeV/c);
(*b*) $K = 4.98 \times 10^{-13}$ J (3.11 MeV), $E = 5.80 \times 10^{-13}$ J (3.62 MeV), $p = 1.92 \times 10^{-21}$ kg·m/s (3.59 MeV/c)]

PROBLEM 12 (46P).   How much work must be done to increase the speed of an electron from (*a*) $0.18c$ to $0.19c$ and (*b*) $0.98c$ to $0.99c$? Note that the speed increase ($= 0.01c$) is the same in each case.

STRATEGY: The work done equals the change in kinetic energy: $W = K_f - K_i$, where K_f is the final kinetic energy and K_i is the initial kinetic energy. Use $K = mc^2(\gamma - 1)$ to compute the kinetic energy.

SOLUTION:

$$\left[\text{ans: } (a)\ 1.0\,\text{keV}\ (1.6 \times 10^{-16}\,\text{J});\ (b)\ 1.1\,\text{MeV}\ (1.7 \times 10^{-13}\,\text{J})\right]$$

PROBLEM 13 (53P).   (*a*) If the kinetic energy K and the momentum p of a particle can be measured, it should be possible to find its mass m and thus identify the particle. Show that

$$m = \frac{(pc)^2 - K^2}{2Kc^2}.$$

STRATEGY: Start with $E^2 = (mc^2)^2 + (pc)^2$ and substitute $E = K + mc^2$, then solve for m.

SOLUTION:

(*b*) Show that this expression reduces to an expected result as $u/c \to 0$, in which u is the speed of the particle.

STRATEGY: Substitute $p = mu$ and $K = \frac{1}{2}mu^2$.

SOLUTION:

(*c*) Find the mass of a particle whose kinetic energy is $55.0\,\text{MeV}$ and whose momentum is $121\,\text{MeV}/c$. Express your answer in terms of the mass of the electron.

SOLUTION:

$$\left[\text{ans: } 1.88 \times 10^{-28}\,\text{kg}\ (207m_e)\right]$$

PROBLEM 14 (59P).   A 10-GeV proton in cosmic radiation approaches the Earth in the plane of the Earth's geomagnetic equator (**v** is perpendicular to **B**), in a region over which the Earth's average magnetic field is 55 μT. What is the radius of the proton's curved path in that region?

STRATEGY: Use $r = mv\gamma/qB$ (see Problem 57). Since the kinetic energy of the proton is given by $K = mc^2(\gamma - 1)$, $\gamma = (K/mc^2) + 1$ and since $\gamma = 1/\sqrt{1 - v^2/c^2}$, $v/c = 1 - (1/\gamma)$.

SOLUTION:

$$\left[\text{ans: } 630\,\text{km}\right]$$

III. NOTES

Chapter 43
QUANTUM PHYSICS — I

I. BASIC CONCEPTS

With this chapter you begin your study of quantum mechanics. You will learn that electromagnetic radiation, which is described classically as a wave-like propagation of electric and magnetic fields, also has particle-like properties. In the next chapter you will learn that particles, such as electrons, also have waves associated with them. Pay particular attention to the relationship between the energy of a particle of light and the frequency of the wave and to the relationship between the momentum of the particle and the wavelength of the wave. Learn to use conservation of energy and momentum to explain the experiments that provide evidence for the particle nature of electromagnetic radiation.

Photons. The quantum hypothesis, applied to electromagnetic radiation, says that the energy in a beam is concentrated in discrete bundles, called _____ . The energy of a single photon associated with a wave of frequency f is given by

$$E =$$

where h is the Planck constant. Its value is $h = $ _____ J·s = _____ eV·s. In terms of photons the rate with which energy is carried by a beam is given by Rhf, where R is the number of photons per unit time that cross a plane perpendicular to their direction of travel. The intensity of a uniform beam is given by $I = Rhf/A$, where A is the cross-sectional area of the beam.

Photons carry momentum in addition to energy, the momentum carried by a single photon associated with a wave of wavelength λ being given by

$$p =$$

If we compare photons associated with waves of different frequency, the one with the higher frequency has the _____ energy and the _____ momentum. The energy and momentum are related by $E = $ _____ , where c is the speed of light. The energy of a single photon of visible light is about _____ eV (or _____ J) and its momentum is about _____ kg·m/s.

In some respects electromagnetic radiation behaves as a wave while in other respects it behaves as a collection of particles. Wavelength and frequency are characteristic of a wave; energy and momentum quanta are characteristic of particles. Carefully note that each of the equations displayed above relate a wave characteristic to a particle characteristic. In this chapter you study some of the experiments that justify the quantum hypothesis.

The photoelectric effect. When monochromatic light shines on a sample, energy is trans-ferred to electrons of the material. As a result, some electrons overcome the potential energy barrier that normally keeps them inside the sample and they leave the sample with various values of kinetic energy. Data taken in a photoelectric effect experiment is used to calculate the kinetic energy of those electrons with the greatest kinetic energy.

If the intensity of the incident light is increased, with the frequency remaining the same, the number of electrons ejected with maximum kinetic energy _____ but the value of the maximum kinetic energy _____ . If the frequency of the incident light is increased, whether or not the intensity changes, the value of the maximum kinetic energy _____ . Clearly, each of the electrons with maximum kinetic energy receives the same energy from the light, regardless of the intensity of the light, and the energy received is greater for higher frequency light than for lower frequency light.

The classical wave hypothesis cannot explain this phenomenon. Averaged over a cycle, the energy carried by a classical plane wave is spread uniformly over the region in which the wave exists. When the intensity is increased without changing the frequency each electron receives more energy. If this hypothesis were valid you would expect the kinetic energy of the most energetic electron to _____ when the intensity increases. In addition, when the light intensity is low you would expect significant time to pass before electrons are ejected because electrons have such small _____ . Experiments show that even at low intensity electrons are ejected immediately after the light is turned on.

The photon hypothesis explains experimental results as follows. The light is assumed to consist of a large number of photons, each with energy hf. Thus the intensity is propor-tional to Rhf/A, where R is the number of photons per unit time incident on sample area A. Some fraction of the photons interact with electrons and each of those that do transfer energy _____ to an electron and disappear. If the intensity is increased without increasing the frequency then _____ increases and the number of electrons ejected _____ . Since each electron that absorbs a photon gets the same energy, the energy of the most energetic electrons does not change. Now suppose the frequency increases. Then the energy of each photon is greater and each ejected electron gets _____ energy than before. The energy of the most energetic electron is _____ than before.

Carefully note that the fundamental event here consists of a single quantum of light in-teracting with a single particle of matter. Suppose a single photon, associated with frequency f, interacts with one of the most energetic electrons in the sample. If ϕ is the energy needed to remove one of these electrons from the sample then conservation of energy leads to

$$hf =$$

where K_m is the kinetic energy of the electron after leaving the sample. ϕ is called the _____ of the sample. It is different for different materials. It has the SI unit _____ .

You should know something about the experimental arrangement for measuring the max-imum kinetic energy K_m. The sample is enclosed in a metal collector cup, which is held at a negative electric potential relative to the sample. Thus the cup tends to repel electrons. If the potential difference is not too great some electrons reach the cup because _____

Those electrons cause an ammeter in the circuit to deflect. The potential on the cup is adjusted until _____.
If this potential, called the _____ potential, is denoted by V_0 then

$$K_m =$$

gives the maximum kinetic energy.

The Compton effect. In this experiment short-wavelength electromagnetic radiation is scattered by electrons and the wavelength of the scattered radiation is measured. It is found to be longer than the wavelength of the incident radiation and the difference is dependent on the scattering angle. Experiments also show that energy and momentum are transferred from the radiation to the electron.

If the classical wave hypothesis were valid the wavelength of the scattered radiation would be the same as that of the incident radiation. Explain how this would come about: _____

_____

_____

According to the photon hypothesis the fundamental interaction, between a single photon and a single electron, can be treated as a collision in which energy and momentum are conserved. An electron essentially at rest is knocked away from its initial position by a photon. Since the energy of the electron increases in the interaction the energy of the photon must _____. Since the photon energy is related to the frequency by $E = hf$ the frequency of the electromagnetic wave must _____ and since $\lambda = c/f$ the wavelength must _____.

The dependence on scattering angle of the energy transferred and the change in wavelength are such that momentum is conserved in the photon-electron interaction. The diagrams to the right show the situation before and after the collision. Before the collision the electron has a momentum of _____ and a kinetic energy of _____. If the electromagnetic radiation has wavelength λ then the momentum of the photon is given by _____ and its energy is given by _____.

Suppose that after the collision the scattered radiation has wavelength λ' and travels at the angle ϕ, as shown. Then the x component of the photon momentum is _____, the y component is _____, and the energy of the photon is _____. If the electron has speed v and moves off at the angle θ then the x component of its momentum is given by _____, the y component is given by _____, and its kinetic energy is given by _____. Note that relativistic expressions must be used.

Write the conservation laws below.

x component of momentum:

y component of momentum:

energy:

These equations can be solved for the change in wavelength ($\Delta\lambda = \lambda' - \lambda$) in terms of the mass m of the electron and the photon scattering angle ϕ. The result is

$$\Delta\lambda =$$

In the usual experiment many photons (with the same frequency) are incident on many electrons. Photons leave at all angles. A detector is moved around a circle centered at the sample and the wavelength of the radiation is measured for each angular position of the detector.

The maximum change in wavelength occurs for a scattering angle of $\phi =$ _____. For this scattering angle, the change in wavelength is given algebraically by $\Delta\lambda =$ _____ and has the numerical value $\Delta\lambda =$ _____ m (_____ pm).

Notice that the change in wavelength is independent of the wavelength itself. It is the same for visible light, gamma rays, and all other electromagnetic radiation. The reason why the Compton effect is important for gamma rays and short-wavelength x rays but not for visible light or microwaves is _____

_____ .

You can find the prediction of classical physics by setting $h = 0$ in the equation for $\Delta\lambda$. The result is $\Delta\lambda =$ _____ . Classically an electromagnetic wave does not change frequency on scattering.

For scattering from bound electrons and nuclei the change in wavelength is much smaller than for scattering from electrons because _____

_____ .

Cavity radiation. The walls of a cavity (a hollow tube, for example) are heated to a high temperature and the radiation intensity for each of many narrow wavelength ranges is measured. More precisely the spectral radiancy $S(\lambda)$ is measured. This quantity is defined so that $S(\lambda)\,d\lambda$ gives _____ for wavelengths between _____ and _____ . $S(\lambda)$ is independent of the material used to make the cavity, the size of the cavity, and the shape of the cavity. It does, however, depend on the temperature.

On the axes to the right sketch a graph of typical experimental values of $S(\lambda)$ as a function of wavelength. Your graph should show that $S(\lambda) \to 0$ as $\lambda \to 0$ and also as $\lambda \to \infty$. It should also show a decided peak. At moderate temperatures the peak is in the red or infrared portion of the electromagnetic spectrum. As the temperature is raised it moves toward _____ .

$S(\lambda)$

λ

When the classical wave hypothesis is used to calculate $S(\lambda)$ the result is:

$$S(\lambda) =$$

where k is the Boltzmann constant (_____ J/K) and T is the temperature on the Kelvin scale. See Chapter 19. For what wavelengths is the prediction of the classical wave hypothesis close to experimental results? _____
For what wavelengths is the prediction of the classical wave hypothesis in serious error?

_____

When the photon hypothesis is used to calculate $S(\lambda)$ the result is

$$S(\lambda) =$$

This function accurately predicts the spectral radiancy for all wavelengths.

The peak in the spectral radiancy function occurs at a wavelength given by $\lambda_{max}T = 2.898 \times 10^{-3}$ m·K. You can show this by differentiating $S(\lambda)$ with respect to λ and setting the result equal to 0. Do it in the space below:

You should obtain $(-5\lambda kT + hc)\exp(hc/\lambda kT) + 5\lambda kT = 0$. Let $\alpha = hc/\lambda kT$. Then $(-5 + \alpha)e^\alpha + 5 = 0$. The solution to this equation is $\alpha = 4.9651$. Complete the proof in the space below by solving $hc/\lambda kT = 4.9651$ for λT:

In the long-wavelength limit the Planck law becomes the classical expression. This is an example of the what is known as the correspondence principle. Write a general statement of the principle here: _____

_____

_____

Quantization of energy. The electromagnetic energy in a cavity is produced by the oscillating atoms of the walls. Planck assumed that the frequencies present in the radiation matched the frequencies of the atomic oscillations and that the energies of the atomic oscillators were quantized. Explain what the phrase energy quantization means:_____

_____

When an oscillator with frequency f absorbs or emits radiation the energy absorbed or emitted has a value that is given by $E =$ _____, where n is the number of photons absorbed or emitted.

The energies of all oscillators, including pendulums and masses on springs, are quantized and, in each case, the quantum of energy is hf, where f is the frequency of oscillation. Notice that the quantum of energy depends on the Planck constant. When the energy of any oscillator changes it must change by an integer number of quanta. The quantization of energy, however, does not have a measurable effect for macroscopic oscillators because _____

_____

_____.

Line spectra. Atoms emit electromagnetic radiation, the radiation from a given type atom having only certain wavelengths characteristic of that type atom. The set of wavelengths is different for hydrogen atoms and sodium atoms, for example.

Classical electricity and magnetism cannot explain why radiation with a set of discrete wavelengths is emitted. In fact, classical theory predicts that the spectrum of radiation emitted by any atom is _____. Furthermore, classical theory predicts that as an electron in an atom loses energy via radiation it must spiral in toward the nucleus. If this theory were correct there would be no atoms.

Quantum theory predicts that an electron in an atom can exist only in any one of a set of discrete states, each with a definite energy. That is, the energy can have any one of a set of separated, discrete values and it can have no other values. The atom does not radiate as long as it remains in one of these states but it emits radiation when it jumps to a state with lower energy. The energy it loses is carried away by a photon. If E_i denotes the initial energy of the electron and E_f denotes its final energy then, since energy is conserved, the frequency of the radiation is given by

$$hf =$$

Radiation is also absorbed by atoms if the photon energy matches the energy difference of two allowed states. Both emission and absorption spectra are therefore discrete. Look at Fig. 43–9 of the text to see some spectra.

For the electron in a hydrogen atom, the allowed values of the energy are given by

$$E_n =$$

where n is an integer. The ground state ($n = 1$) energy is _____ eV.

Section 43–10 contains Bohr's derivation of this result. Study it carefully. Most of the relationships used in the derivation are classical. The electron is assumed to go around the proton in a circular orbit and is held in its orbit by the electrical attraction of the proton. However, one non-classical assumption is made and it leads to the quantization of energy. Explain what that assumption is: _____

_____

_____

Because the possible values of the energy are discrete, the possible wavelengths of the radiation emitted by a hydrogen atom have only certain discrete values and do not form a continuous spectrum. Look at Fig. 43–10 to see a portion of the spectrum and at Fig. 43–13 to see some allowed transitions. All possible wavelengths are given by the expression

$$\frac{1}{\lambda} =$$

where R is the Rydberg constant. In terms of more fundamental constants $R =$ _____ and its numerical value is $R =$ _____ m^{-1}.

The quantities ℓ and u are integers; ℓ is the quantum number of the _____ state and u is the quantum number of the _____ state. Values of these integers are substituted for n in the expression for E_n to find the energies of the initial and final states of a transition.

The spectral lines of hydrogen are grouped into series. For the Lyman series of transitions, the final state is always the $n =$ _____ state; for the Balmer series, the final state is always the $n =$ _____ state; and for the Paschen series, the final state is always the $n =$ _____ state. The _____ series contains lines that are visible.

II. PROBLEM SOLVING

You should know the relationship between the energy E of a photon and the frequency f of the wave associated with it: $E = hf$. You should also know the relationship between the magnitude of the photon momentum and the wavelength of the wave: $p = h/\lambda$. Both classically and quantum mechanically the momentum and energy are related by $E = pc$. This follows immediately from $f\lambda = c$ and the relationships given above.

You should understand that a collection of N monochromatic photons (in a light beam, for example) has a total energy of Nhf. If a uniform beam of monochromatic photons has cross-sectional area A then the rate at which energy is transmitted through the area is $P = Rhf$, where R is the number of photons that pass through per unit time. The intensity is $I = Rhf/A$.

PROBLEM 1 (7P).   What are (a) the frequency, (b) the wavelength, and (c) the momentum of a photon whose energy equals the rest energy of the electron?

STRATEGY: The rest energy of an electron is given by $E = mc^2$, where m is its mass (9.11×10^{-31} kg) and c is the speed of light (3.00×10^8 m/s). The frequency of the wave associated with the photon is given by $f = E/h$, where h is the Planck constant (6.63×10^{-34} J $\cdot$ s). The wavelength is given by $\lambda = c/f$ and the momentum of the photon is given by $p = h/\lambda$.

SOLUTION:

[ans: (a) 1.24×10^{20} Hz; (b) 2.43×10^{-12} m; (c) 2.73×10^{-22} kg $\cdot$ m/s]

PROBLEM 2 (10P).   A satellite in Earth orbit maintains a panel of solar cells of 2.60-m^2 area at right angles to the direction of the sun's rays. Solar energy arrives at the rate of 1.39 kW/m^2. (a) At what rate does solar energy strike the panel?

STRATEGY: The rate at which energy strikes is the product of the intensity and the area: $P = IA$. Here $I = 1.39$ kW/m^2.

SOLUTION:

[ans: 3.61 kW]

(b) At what rate do solar photons strike the panel? Assume that the solar radiation is monochromatic with a wavelength of 550 nm.

STRATEGY: Since each photon carries an energy of $E = hf$ $(= hc/\lambda)$, the rate R at which photons strike the panel is given by $P = Rhf$. Solve for R.

SOLUTION:

$$[\text{ans: } 1.00 \times 10^{22} \text{ photons/s}]$$

(c) How long would it take for a "mole of photons" to strike the panel?

STRATEGY: Let N_A $(= 6.022 \times 10^{23})$ be the number of photons in a mole. Then $N_A = Rt$, where t is the time. Solve for t.

SOLUTION:

$$[\text{ans: } 60.3 \text{ s}]$$

PROBLEM 3 (12P).   The emerging beam from a 1.5-W argon laser ($\lambda = 515$ nm) has a diameter d of 3.0 mm. (a) At what rate per square meter do photons pass through any cross section of the beam?

STRATEGY: The intensity I in the beam is the power divided by the cross-sectional area of the beam: $I = P/\pi r^2$, where r is the beam radius. It is related to the rate per unit area R_p with which photons pass through a cross section by $I = R_p hf = R_p hc/\lambda$, where f is the frequency and λ is the wavelength. Thus $R_p = \lambda P/\pi r^2 hc$.

SOLUTION:

$$[\text{ans: } 5.5 \times 10^{23} \text{ photons/m}^2 \cdot \text{s}]$$

(b) The beam is focused by a lens system whose effective focal length f_L is 2.5 mm. The focused beam forms a circular diffraction pattern whose central disk has a radius R given by $1.22 f_L \lambda/d$. It can be shown that 84% of the incident power lies within this central disk, the rest falling in the fainter, concentric rings that surround the central disk. At what rate per square meter do photons pass through the central disk of the diffraction pattern?

STRATEGY: Since 84% of the photons go through the central diffraction disk, the number of photons per unit time passing through this disk is given by $0.84P/hf = 0.84P\lambda/hc$ and number passing through per unit time per unit area is given by $0.84P\lambda/\pi R^2 hc$. Use $R = 1.22 f_L \lambda/d$ to calculate the radius of the diffraction disk.

SOLUTION:

$$[\text{ans: } 3.8 \times 10^{30} \text{ photons/m}^2 \cdot \text{s}]$$

The photoelectric effect is described by $hf = K_{max} - \phi$, where f is the frequency of the incident radiation, K_{max} is the kinetic energy of the most energetic photoelectron ejected, and ϕ is the work function of the target material. You should also know that the value of K_{max} is found by measuring the stopping potential V_0 and that $K_{max} = eV_0$.

PROBLEM 4 (16E). (a) The energy needed to remove an electron from metallic sodium is 2.28 eV. Does sodium show a photoelectric effect for red light, with $\lambda = 680$ nm?

STRATEGY: Use $E = hc/\lambda$ to calculate the energy of photon. It must be greater than 2.28 eV for the photoelectric effect to occur.

SOLUTION:

[ans: no, the photon energy is only 1.82 eV]

(b) What is the cutoff wavelength for photoelectric emission from sodium, and to what color does this wavelength correspond?

STRATEGY: When the incident radiation has the greatest wavelength for which photoelectric emission occurs, photoelectrons with the greatest energy just get the surface of the target and have no kinetic energy there. Thus $hf = \phi$, where f is the frequency and ϕ is the work function for sodium. This means $hc/\lambda = \phi$. Solve for λ. Look at Fig. 38–1 to decide on the color.

SOLUTION:

[ans: 544 nm, green]

PROBLEM 5 (17E). Find the maximum kinetic energy of photoelectrons from a certain material if the work function is 2.3 eV and the frequency of the radiation is 3.0×10^{15} Hz.

STRATEGY: Evaluate $K_m = hf - \phi$, where f is the frequency and ϕ is the work function. Either convert the given value for ϕ to joules or convert the value for hf to electron volts.

SOLUTION:

[ans: 1.6×10^{-18} J (10 eV)]

PROBLEM 6 (21E). (a) If the work function for a metal is 1.8 eV, what is its stopping potential for light of wavelength 400 nm?

STRATEGY: Substitute $K_{max} = eV_0$ into $hf = K_{max} - \phi$ and solve for the stopping potential V_0. Convert the value given for ϕ to joules.

SOLUTION:

$$\left[\text{ans: } 1.3\,\text{V}\right]$$

(b) What is the maximum speed of the emitted photoelectrons at the metal's surface?

STRATEGY: If v is the maximum speed then $\frac{1}{2}mv^2 = eV_0$, where m is the mass of an electron. Solve for v.

SOLUTION:

$$\left[\text{ans: } 6.8 \times 10^5\,\text{m/s}\right]$$

PROBLEM 7 (26P).   Photosensitive surfaces are not necessarily very efficient. Suppose the fractional efficiency of a cesium surface (with work function 1.80 eV) is 1.0×10^{-16}; that is, one photoelectron is produced for every 10^{16} photons striking the surface. What would be the photocurrent from such a cesium surface if it were illuminated with 600-nm light from a 2.00-mW laser and all of the photoelectrons produced took part in charge flow?

STRATEGY: The number of photons striking the surface per unit time is given by $R = P/hf = P\lambda/hc$, where P is the power of the laser, f is the frequency, and λ is the wavelength. The number of photoelectrons emitted per unit time is given by $1.0 \times 10^{-16} R$. Multiply by the magnitude of the charge on an electron to find the current.

SOLUTION:

$$\left[\text{ans: } 9.68 \times 10^{-20}\,\text{A}\right]$$

Upon scattering from an electron initially at rest the wavelength associated with a photon changes from λ to $\lambda' = \lambda + \Delta\lambda$, where $\Delta\lambda = (h/mc)(1 - \cos\phi)$. Here m is the mass of an electron, h is the Plank constant, c is the speed of light, and ϕ is the scattering angle (measured from the direction of incidence). The energy of the photon is reduced from $E = hf = hc/\lambda$ to $E' = hf' = hc/\lambda'$. The energy lost by the photon appears as an increase in the kinetic energy of the electron.

PROBLEM 8 (31E).   An x-ray photon of wavelength 0.01 nm strikes an electron head on ($\phi = 180°$). Determine (*a*) the change in wavelength of the photon, (*b*) the change in energy of the photon, and (*c*) the kinetic energy imparted to the electron.

STRATEGY: The change in wavelength is given by $\Delta\lambda = (h/mc)(1 - \cos\phi)$, where h is the Planck constant, m is the mass of an electron, and c is the speed of light. Since $\phi = 180°$, $\cos\phi = -1$. The change in photon energy is given by $\Delta E = hf' - hf = (hc/\lambda') - (hc/\lambda)$, where the primes indicate the values after scattering. Use $\lambda' = \lambda + \Delta\lambda$ to find the wavelength after scattering. Since energy is conserved the kinetic energy imparted to the electron is $-\Delta E$.

SOLUTION:

$$\left[\text{ans: } (a)\ 4.9 \times 10^{-12}\,\text{m}; (b)\ -6.5 \times 10^{-14}\,\text{J}\ (-41\,\text{keV}); (c)\ 6.5 \times 10^{-14}\,\text{J}\ (41\,\text{keV})\right]$$

PROBLEM 9 (37P).   Show that $\Delta E/E$, the fractional loss of energy of a photon during a Compton collision, is given by

$$(hf'/mc^2)(1 - \cos\phi).$$

STRATEGY: The fractional energy loss is $\Delta E/E = (hf - hf')/hf = (hc/\lambda - hc/\lambda')/(hc/\lambda)$, where $f = c/\lambda$ was used. Here the primes indicate values after scattering. A little algebra gives $\Delta E/E = (\lambda' - \lambda)/\lambda'$. Substitute $\lambda' - \lambda = (h/mc)(1 - \cos\phi)$ for the numerator and $\lambda' = c/f'$ for the denominator to obtain the desired result.

SOLUTION:

PROBLEM 10 (38P).   Through what angle must a 200-keV photon be scattered by a free electron so that it loses 10% of its energy?

STRATEGY: Solve $\Delta E/E = (hf'/mc^2)(1 - \cos\phi)$ for ϕ. Use $\Delta E/E = 0.10$ and $hf' = 0.90hf = 0.90 \times 200\,\text{keV} = 180\,\text{keV}$.

SOLUTION:

$$\left[\text{ans: } 44°\right]$$

PROBLEM 11 (39P). Show that when a photon of energy E scatters from a free electron, the maximum kinetic energy of the electron is given by

$$K_{max} = \frac{E^2}{E + mc^2/2}.$$

STRATEGY: The kinetic energy of the electron is given by $K = hf - hf' = (hc/\lambda) - (hc/\lambda')$, where $f = c/\lambda$ was used. The primes refer to values after the scattering. Substitute $\lambda' = \lambda + (h/mc)(1 - \cos\phi)$ and take $\phi = 180°$ for maximum kinetic energy. Then substitute $\lambda = hc/E$ and carry out a little algebra.

SOLUTION:

Many problems that deal with cavity radiation ask you to make use of the relationship between the temperature and the wavelength associated with the peak of the spectral radiancy curve: $\lambda_{max}T = 2898\,\mu\text{m} \cdot \text{K}$. See Exercise 42 of the text and your derivation in the Basic Concepts section above. Some ask you to make use of the relationship between the temperature and the rate of energy radiation per unit area of cavity surface: $P = \sigma T^4$, where $\sigma = 5.67 \times 10^{-8}\,\text{W/m}^2 \cdot \text{K}^4$. See Problem 49 of the text. Don't forget that T represents the temperature on the Kelvin scale.

PROBLEM 12 (45E). In 1983 the Infrared Astronomical Satellite (IRAS) detected a cloud of solid particles surrounding the star Vega, radiating maximally at a wavelength of $32\,\mu\text{m}$. What is the temperature of this cloud of particles? (See Exercise 42.)

STRATEGY: Solve $\lambda_{max}T = 2898\,\mu\text{m} \cdot \text{K}$ for the absolute temperature T.

SOLUTION:

$$[\text{ans: } 91\,\text{K}]$$

PROBLEM 13 (50P). Calculate the rate at which thermal energy is radiated from a fireplace, assuming an effective radiating surface of $0.50\,\text{m}^2$ and an effective temperature of $500°\,\text{C}$. (See Problem 49.)

STRATEGY: According to Problem 49 the rate at which energy is radiated is given by $P = \sigma T^4$, where $\sigma = 5.67 \times 10^{-8}\,\text{W/m}^2 \cdot \text{K}^4$ and T is the temperature in kelvins. Evaluate this expression.

SOLUTION:

$$[\text{ans: } 10\,\text{kW}]$$

PROBLEM 14 (52P).   A cavity at absolute temperature T_1 radiates energy at a power of 12.0 mW. At what power does the same cavity radiate at temperature $2T_1$? (See Problem 49.)

STRATEGY: Since $P_1 = \sigma T_1^4$ and $P_2 = \sigma T_2^4$, the ratio of the powers is $P_2/P_1 = T_2^4/T_1^4$. Substitute $T_2/T_1 = 2$ and solve for P_2.

SOLUTION:

[ans: 192 mW]

Some problems deal with the allowed values of the energy of a hydrogen atom, with the possible wavelengths and frequencies of the radiation emitted when a hydrogen atom makes a transition from one state to a lower energy state, or with the frequency and wavelength of radiation that can be absorbed by a hydrogen atom. The allowed values of the energy are given by $E = -me^4/8\epsilon_0^2 h^2 n^2$, where n is a positive integer. Remember $E = 0$ means the electron is just free of the proton and has no kinetic energy. The energy of the photon is the difference in energy of the two states involved in the transition: $hf = |E_f - E_i|$. $E_f - E_i$ is positive for an absorption event and negative for an emission event.

PROBLEM 15 (63E).   A hydrogen atom is excited from the state with $n = 1$ to that with $n = 4$. (a) Calculate the energy that must be absorbed by the atom.

STRATEGY: The energy that must be absorbed is the difference in the energies of the final and initial states. These can be calculated using $E = hf = hc/\lambda = hcR/n^2$. To obtain this expression $f = c/\lambda$ was substituted for the frequency f, then $1/\lambda = R/n^2$ was used. Here R is the Rydberg constant (0.01097 nm^{-1}).

SOLUTION:

[ans: 12.7 eV]

(b) Calculate and display on a energy-level diagram the different photon energies that may be emitted if the atom returns to the $n = 1$ state.

STRATEGY: The following transitions might take place: $n = 4$ to $n = 3$, $n = 3$ to $n = 2$, $n = 2$ to $n = 1$, $n = 3$ to $n = 1$, $n = 4$ to $n = 2$, and $n = 4$ to $n = 1$. Calculate the energies of these states, then use $\Delta E = E_f - E_i$ to calculate the energies of the photons. Here E_i is the energy of the initial state and E_f is the energy of the final state. Draw the transitions on the diagram below.

SOLUTION:

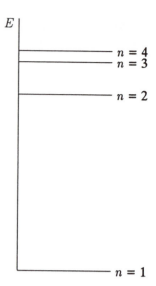

E

_____ $n = 4$
_____ $n = 3$

_____ $n = 2$

_____ $n = 1$

[ans: $4 \rightarrow 3$: 0.661 eV, $3 \rightarrow 2$: 1.89 eV, $2 \rightarrow 1$: 10.2 eV, $3 \rightarrow 1$: 12.1 eV, $4 \rightarrow 2$: 2.55 eV, $4 \rightarrow 1$: 12.7 eV]

PROBLEM 16 (66P). Light of wavelength 486.1 nm is emitted by a hydrogen atom. (*a*) What transition of the atom is responsible for this radiation?

STRATEGY: Use $\Delta E = hc/\lambda$ to calculate the difference in energy of the two hydrogen states involved in the transition. You calculated some hydrogen energies in connection with Exercise 63. Search those to find two with the correct difference.

SOLUTION:

[ans: $\Delta E = 2.55$ eV, this is the $n = 4$ to $n = 2$ transition]

(*b*) To what series does this radiation belong?

STRATEGY: Locate the transition on Fig. 43–13.

SOLUTION:

[ans: Balmer series]

PROBLEM 17 (68P). A hydrogen atom in a state having a *binding energy* (the energy required to remove an electron) of 0.85 eV makes a transition to a state with an *excitation energy* (the difference in energy between a state and the ground state) of 10.2 eV. (*a*) Find the energy of the emitted photon.

STRATEGY: Take the zero of energy to be the energy when the electron is removed without any kinetic energy. Then the initial state has energy −0.85 eV and the ground state has energy −13.6 eV. The final state for the transition considered has energy −13.6 + 10.2 = −3.4 eV. Calculate the difference in energy.

SOLUTION:

[ans: 2.6 eV]

(b) Show this transition on an energy-level diagram for hydrogen, labeling the levels with the appropriate quantum numbers.

STRATEGY: When you solved Exercise 62 you calculated the energies of some low-lying states. Use these results to identify the quantum numbers associated with the initial and final states. Draw the diagram below.

SOLUTION:

[ans: $n = 4$ to $n = 2$]

The derivation given in Section 43–10 of the text shows that once the quantum number n is known you can calculate a great many properties of the electron's motion. As you will see in the next chapter many of these quantities cannot really be known and their semi-classical computation here produces approximate values. Nevertheless the calculations are useful exercises. Here is an example.

PROBLEM 18 (76P).   In the ground state of the hydrogen atom, according to Bohr's theory, what are (a) the quantum number, (b) the electron's orbit radius, (c) its angular momentum, (d) its linear momentum, (e) its angular velocity, (f) its linear speed, (g) the force on the electron, (h) the acceleration of the electron, (i) the electron's kinetic energy, (j) the potential energy, and (k) the total energy?

STRATEGY: The quantum number is $n = 1$ since the atom is in its ground state. The orbit radius is given by $r = n^2 h^2 \epsilon_0 / \pi m c^2$. The electron's angular momentum is given by $L = \sqrt{me^2 r / 4\pi \epsilon_0}$ and the magnitude of its linear momentum is given by $p = L/r$. The angular velocity is related to the angular momentum by $L = mr^2 \omega$ and the linear speed is related to the angular velocity by $v = r\omega$. The force on the electron is the electrical force due to the proton and its magnitude is given by $F = e^2 / 4\pi \epsilon_0 r^2$. According to Newton's second law the magnitude of the electron's acceleration is $a = F/m$. $K = \frac{1}{2} mv^2$ can be used to compute the kinetic energy. The potential energy is due to the electrical interaction between the electron and proton and is given by $U = -e^2 / 4\pi \epsilon_0 r$. Finally, the total energy is $E = K + U$.

SOLUTION:

[ans: (a) $n = 1$; (b) 5.29×10^{-11} m; (c) 1.05×10^{-34} kg $\cdot$ m^2/s; (d) 1.99×10^{-24} kg $\cdot$ m/s; (e) 4.13×10^{16} rad/s; (f) 2.19×10^{6} m/s; (g) 8.24×10^{-8} N; (h) 9.07×10^{22} m/s^2; (i) 2.18×10^{-18} J; (j) -4.36×10^{-18} J; (k) -2.18×10^{-18} J]

III. NOTES

Chapter 44
QUANTUM PHYSICS — II

I. BASIC CONCEPTS

In this chapter you learn that waves are also associated with particles and that a matter wave tells us the probability that a particle is in any given region of space. Pay attention to the mathematical relationship between the wave and the probability. Matter waves give rise to interference and diffraction effects and to tunneling phenomena. Concentrate on what these wave phenomena imply as far as particles and their locations are concerned. The wave associated with a free particle has a wavelength that is related to the momentum of the particle and a frequency that is related to the energy of the particle. Learn these relationships well. Also understand the uncertainty principles that arise when these relationships are applied to non-sinusoidal matter waves.

Matter waves. Particles exhibit interference phenomena. Fig. 44–1 of the text shows a uniform beam of electrons incident on a crystal. Describe in your own words the characteristics of the reflected beam that convince you that wave interference has taken place: _____

_____

_____

Clearly a wave is associated with a particle.

Suppose the wave associated with a particle is sinusoidal, with a definite wavelength λ and a definite frequency f. The wavelength of the wave is related to the _____ of the particle and the frequency of the wave is related to the _____ of the particle. The relationships are

$$\lambda =$$

and

$$f =$$

where h is the Planck constant. These relationships for particles are exactly the same as the relationships between the energy and momentum of a photon on the one hand and the frequency and wavelength of the associated electromagnetic wave on the other.

You should be aware of the magnitudes involved. For macroscopic objects the momentum is usually so large and the wavelength so small that interference and diffraction effects cannot be detected. For atomic particles, on the other hand, the momentum is sufficiently small and the wavelength correspondingly large that interference and diffraction effects are evident. The wavelength of the matter wave associated with a 150-g ball moving at 30 m/s is _____ m, while the wavelength of the matter wave associated with an electron (with a mass of about 9×10^{-31} kg) moving at 10^4 m/s is about _____ m. If the ball is to be diffracted by a grating, the slits should be about _____ m apart. This is clearly impossible. If the electron is to be diffracted, the slits should be about _____ m apart.

Wave functions. The wave associated with a particle has something to do with the position of the particle. Section 44–5 gives some of the details. Quantum mechanics deals with probabilities — the probability a particle is in a certain region at a certain time and the probability the momentum of the particle is in a certain range, for example. If the wave function of a particle moving on the x axis is represented by ψ, a function of position, then _____ gives the probability that the particle is in the small region of width dx. Here the value of the wave function at the position of the region must be used.

Here's how to measure $|\psi|^2\,dx$, in principle. The same experiment is performed many times, always under the same conditions, and the location of the particle is detected. Now pick a small region of the x axis and calculate the fraction of experiments for which the particle is found in that region. In the limit of a large number of experiments this fraction gives $|\psi|^2\,dx$. The quantity $|\psi|^2$ is often called a probability density. This means it is a probability per unit _____ .

Since the particle is guaranteed to be somewhere, the integral of $|\psi|^2$ over the entire x axis has the value _____. This is called the normalization condition on the wave function. A differential equation for the wave function, called the Schrodinger equation, has been developed and allows us to solve for ψ if the forces acting on the particle are known.

Trapped particles. Section 44–5 gives an excellent example of matter waves and their use. Fig. 44–9 of the text shows the potential energy of a particle and is reproduced to the right. We first assume that U_0 is infinite so the particle is trapped in a box that extends from $x = 0$ to $x = L$. No force acts on it when it is in the interior but when it reaches either side of the box an infinite force acts on it, pushing it toward the interior.

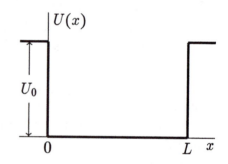

There are many possible wave functions for the particle. They are quite similar to the functions that describe the displacement of a vibrating string fixed at both ends. Inside the well (from $x = 0$ to $x = L$):

$$\psi_n = \sqrt{2/L}\,\sin(n\pi x/L)\,,$$

where n is a positive integer. Outside the well $\psi = 0$. The integer n distinguishes one function from another and is used to label the functions. The functions have been normalized; that is, the integral of any one of them over the x axis from 0 to L has the value 1. Notice that $\psi_n = 0$ at $x = 0$ and at $x = L$.

Strictly speaking, this expression for ψ_n gives only the coordinate-dependent part of the matter wave. ψ_n should be multiplied by a function of time. However, this function has magnitude 1 and $|\psi|^2$ correctly gives the probability density. A lower case psi (ψ) designates a wave function with the time dependent function omitted.

When the wave function associated with the particle is the one labeled n then the probability density is given by

$$|\psi_n|^2 =$$

On the axes below make sketches of the probability density for $n = 1$, 2, and 3:

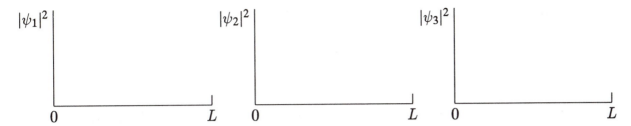

Each of these functions corresponds to a different value for the magnitude of the particle's momentum. The wavelength associated with either of the traveling waves that combine to produce the standing wave $\psi = \sin(n\pi x/L)$ is $\lambda_n = $ _____ and the magnitude of the particle momentum is

$$p_n = \frac{h}{\lambda_n} = $$

Each function also corresponds to a different value of the particle energy. The energy is entirely kinetic and so is given by

$$E_n = \frac{1}{2}mv^2 = \frac{p_n^2}{2m} = $$

where m is the mass of the particle. Notice that the energy increases in proportion to n^2. In terms of L and m the lowest energy is $E_1 = $ _____, the next lowest is _____ times as great, and the next is _____ times as great.

Notice that the quantization of energy is predicted. If the particle wave function is ψ_n then the particle energy is E_n. Since $\psi_n = \sqrt{2/L}\sin(n\pi x/L)$ are the only possible wave functions, $E_n = n^2h^2/8mL^2$ are the only possible values for the energy. The particle cannot have an energy between E_1 and E_2, for example.

Give an argument to show that the particle cannot be at rest in the box: _____
_____
_____

The hydrogen atom. The set of functions appropriate to an electron in a hydrogen atom differs from the set appropriate to an electron in another situation (trapped in a box, for example) because the _____ acting on the electron is different for the two cases. The important quantity for a quantum mechanical calculation of a particle wave function is the _____ function and this is different if different forces act. For hydrogen, which consists of an electron and proton, this function is

$$U = $$

where r is the _____ of the two particles.

The electron is trapped by the proton and quantum mechanics predicts a discrete set of energy levels. They are given by

$$E_n = $$

where n is a positive integer. Notice that even in the lowest energy state ($n = 1$) the electron has energy and is not at rest.

The ground state ($n = 1$) probability density for the electron in a hydrogen atom is

$$\psi^2(r) =$$

where r_B is _____ and has the value _____ m. This is the orbit radius predicted by the Bohr model.

The electron now moves in three-dimensional space, rather than along the x axis and the square of the wave function gives the probability per unit _____ that it is in a small volume. If dV is an infinitesimal volume a distance r from the proton, then the probability the electron is in that volume is given by $\psi^2\,dV$. The <u>radial</u> <u>probability</u> <u>density</u> $P(r)$ is defined so that $P(r)\,dr$ gives the probability of finding the electron in a spherical shell of width dr a distance r from the proton. Since the volume of the shell is $dV = 4\pi r^2\,dr$ the radial probability density is given in terms of the wave function by

$$P(r) = 4\pi r^2 |\psi|^2 .$$

The radial probability density for the ground state wave function of hydrogen is

$$P(r) =$$

Use the axes to the right to sketch a graph of P as a function of the distance r from the proton. Label the r axis to indicate the distance scale. In terms of the Bohr radius the maximum of the function occurs at $r = $ _____ r_B. The fraction of the time the electron is inside a sphere of radius r_B is about _____ and the fraction of the time it is outside the sphere is about _____ .

Barrier tunneling. A matter wave extends into the region beyond any potential energy barriers if the barriers have finite height and width. This means a particle may escape from such a region. Consider the one-dimensional case. The upper diagram to the right shows a potential energy barrier of height U and width L. On the axes below it sketch the probability density for a particle approaching the barrier from the left with energy less than the barrier height. Classically this particle cannot travel to the other side of the barrier. Quantum mechanically it has a non-zero probability of being on either side. If the barrier is made higher or wider the wave function outside the barrier becomes _____ in magnitude.

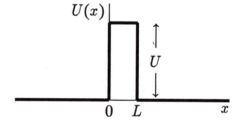

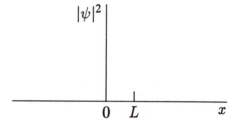

Tunneling through a barrier is described quantitatively by a transmission coefficient T and a reflection coefficient R. The transmission coefficient is defined so that it gives the probability the particle will _____ the barrier, while the reflection coefficient gives the probability the particle does not. The sum of the two is _____ .

If the transmission coefficient is small it is given by

$$T =$$

where

$$k =$$

and E is the energy of the particle. This expression tells us that if the barrier is made wider (L is increased) then T _____ and if the barrier is made higher (U is increased) then k _____ and T _____ . Notice also that the parameter k depends on $U - E$. If the energy of the particle is increased (but is still not greater than U) then k _____ and T _____ .

The text gives some examples in which barrier tunneling plays an important role. List 2 of them here:

1. _____

2. _____

Uncertainty.   If the wave function of a particle is not sinusoidal then the particle does not have a definite momentum. If we measure the momentum a large number of times, with the particle always in the same state, we obtain a distribution of values. Quantum mechanics can be used to predict the probability that the particle momentum will be in any given range but it cannot predict the result of any of the momentum measurements. The situation is quite similar to the measurement of position: quantum mechanics can predict the probability that a particle is in a given region of space but it cannot predict its position at any given time.

The position of a particle can be localized in space. If it is, the wave function is large only over a small region and is zero or nearly zero everywhere else. Consider an electron trapped in a one-dimensional box with infinite potential energy walls. We know the electron is in the box but we cannot know where in the box. The width of the box can be taken as a measure of the uncertainty in its position. The electron can be localized and its position made more certain by reducing the width of the box but if this is done its momentum becomes less certain.

This is the heart of the uncertainty principle. If Δx is the uncertainty in the position of a particle and Δp is the uncertainty in its momentum, then

$$\Delta x \cdot \Delta p \approx$$

The symbol for an approximate equality is used because we have not precisely defined the uncertainties. Similar relationships hold for each coordinate and the corresponding momentum component when the motion is in three dimensions. In words, the principle tells us _____

Carefully note the appearance of the Planck constant h in the uncertainty relations. If the classical limit is obtained, by setting $h = 0$, then the relationship for the x component becomes $\Delta p \cdot \Delta x = 0$. If the same experiment, under the same conditions, is run many times, then classically for every trial it is possible for the particle to have the same position and the same momentum at the same time after the start of the experiment. Quantum mechanically it is not possible.

For an electron trapped in a one-dimensional box of width L and in its ground state ($n = 1$), the magnitude of its momentum is $p = h/\lambda = h/2L$. This is precisely defined but the momentum itself is not because the electron might be traveling in either of two directions. We may take the uncertainty in momentum to be $\Delta p = $ _____ . Clearly the product of $\Delta x = L$ and $\Delta p = h/L$ is h.

You should also understand single-slit diffraction from the point of view of the uncertainty principle. A beam of electrons is incident on a single slit in a barrier and those that pass through are detected at a screen on the other side. The electron waves are diffracted and some electrons reach parts of the screen that are not directly behind the slit. Those that do have a momentum component that is parallel to the slit and since the arrival points are distributed over the screen we know there is a distribution of momentum values. We take the slit width to be a measure of the uncertainty in the position of the electrons as they pass through. We can decrease the uncertainty in position by making the slit width _____ . If we do this the diffraction pattern _____ , indicating that we have _____ the uncertainty in momentum. As the analysis of Section 44–9 shows, the Heisenberg uncertainty principle is obeyed.

In some cases the probability that a particle is in a certain region of space depends on the time. It is possible to construct a wave function that is large in a given region for a only a short time interval. For times before and after this interval the probability that the particle is in the chosen region is small or zero. When the energy of such a particle is measured repeatedly a range of values is obtained; the particle energy is uncertain. The time interval Δt and the uncertainty in energy ΔE are related by an uncertainty principle:

$$\Delta E \cdot \Delta t \approx$$

II. PROBLEM SOLVING

Here are some problems that require you to know the relationship between the energy of a particle and the frequency of its wave ($E = hf$), the relationship between the momentum of a particle and the wavelength of its wave ($p = h/\lambda$), and the relationship between the energy and momentum of a particle ($E = p^2/2m$). The latter expression is valid for a *nonrelativistic* particle of mass m. The analogous expressions for a photon are $E = hf$, $p = h/\lambda$, and $E = pc$. Only the last is different.

PROBLEM 1 (5E).   An electron and a photon each have a wavelength of 0.20 nm. Calculate their (a) momenta and (b) energies.

STRATEGY: The magnitude of the momentum of either particle is given by $p = h/\lambda$. The electron is nonrelativistic and its kinetic energy is given by $E = p^2/2m$, where m is its mass. The energy of the photon is given by $E = pc$.

SOLUTION:

$\big[$ans: (a) 3.3×10^{-24} kg · m/s for each; (b) electron: 6.0×10^{-18} J (38 eV), photon: 9.9×10^{-16} J (6.2 keV)$\big]$

PROBLEM 2 (8E).   If the de Broglie wavelength of a proton is 0.100 pm, (a) what is the speed of the proton and (b) through what electric potential would the proton have to be accelerated to acquire this speed?

STRATEGY: The magnitude of the momentum of the proton is given by $p = h/\lambda$ and its speed is given by $v = p/m$, where m is its mass. Its kinetic energy is given by $K = p^2/2m$ and the electric potential V needed to accelerate it to that speed can be calculated using $K = eV$.

SOLUTION:

$\big[$ans: (a) 3.96×10^6 m/s; (b) 81.9 kV$\big]$

PROBLEM 3 (13P).   The 20-GeV electron accelerator at Stanford provides an electron beam of small wavelength, suitable for probing the fine details of nuclear structure via scattering. What is the wavelength of the electrons and how does it compare with the radius of an average nucleus (about 5.0 fm)? (At this energy it is sufficient to use the extreme relativistic relationship between momentum and energy, namely, $p = E/c$. This is the same relationship used for light and is justified when the kinetic energy of a particle is much greater than its rest energy, as in this case.)

STRATEGY: Use $p = E/c$ and $\lambda = h/p$ to obtain $\lambda = hc/E$. Covert the given value of the energy to joules.

SOLUTION:

$\big[$ans: 0.062 fm, smaller than a typical nuclear radius by a factor of 1.2×10^{-2} $\big]$

PROBLEM 4 (16P).   The highest achievable resolving power of a microscope is limited only by the wavelength used; that is, the smallest detail that can be separated has dimensions about equal to the wavelength. Suppose one wishes to "see" inside an atom. Assuming the atom to have a diameter of 100 pm, this means that we wish to resolve detail of separation of, say, 10 pm. (a) If an electron microscope is used, what minimum energy of the electrons is needed?

STRATEGY: Since the kinetic energy is given by $K = p^2/2m$ and the magnitude of the momentum is given by $p = h/\lambda$, $K = h^2/2m\lambda^2$.

SOLUTION:

$\big[$ans: 2.4×10^{-15} J (15 keV)$\big]$

(b) If a light microscope is used, what minimum energy of photons is needed?

STRATEGY: Use $E = pc = hc/\lambda$.

SOLUTION:

[ans: 2.0×10^{-14} J (120 keV)]

(c) Which microscope seems more practical for this purpose? Why?

STRATEGY: Which requires less energy?

SOLUTION:

Matter waves can exhibit interference and diffraction effects. If the wavelength and direction of incidence are correct they are diffracted by crystals, for example. Eq. 44-1 describes the condition for a strongly diffracted beam of particles.

PROBLEM 5 (20P). In the experiment of Davisson and Germer, (a) at what angles would the second- and third-order diffracted beams corresponding to a strong maximum in Fig. 44–2 occur, provided they are present?

STRATEGY: For a scattering maximum $d \sin \phi = m\lambda$. For first order diffraction $d \sin \phi_1 = \lambda$ and for second order $d \sin \phi_2 = 2\lambda$, so $\sin \phi_2 = 2 \sin \phi_1$. Similarly for third order, $\sin \phi_3 = 3 \sin \phi_1$. In first order the diffracted beam makes the angle $\phi_1 = 50°$ with the normal to the surface.

SOLUTION:

[ans: second and third order diffraction do not occur]

(b) At what angle would the first-order diffracted beam occur if the accelerating potential were changed from 54 to 60 V?

STRATEGY: Now the wavelength has been changed. For the old potential $d \sin \phi_1 = \lambda$ and for the new $d \sin \phi_1' = \lambda'$, so $\sin \phi_1' = (\lambda'/\lambda) \sin \phi_1$. Use $eV = p^2/2m$ to calculate the magnitude of the momentum and $\lambda = h/p$ to calculate the wavelength for each potential. You should find that $\sin \phi_1' = \sqrt{V/V'} \sin \phi_1$. Also use $\phi_1 = 50°$.

SOLUTION:

[ans: 46°]

Some problems deal with the allowed values of the energy and momentum of a particle trapped in a well with infinite barriers at the ends: $E = n^2h^2/8mL^2$ and $p = nh/2L$, where L is the width of the well and n is an integer that designates the quantum mechanical state of the particle. Other problems use a particle trapped in a well to demonstrate some general properties of wave functions. If the particle is in the state n then its wave function is $\psi = \sqrt{2/L}\sin(\pi x/L)$. Remember that $\psi^2(x)\,dx$ gives the probability that the particle is between x and $x + dx$ and that $\int_0^L \psi^2\,dx = 1$ is a statement that the particle is guaranteed to be somewhere in the well.

Similar problems involve the wave functions for the electron in a hydrogen atom. The electron moves in three-dimensional space and $|\psi|^2\,dV$ gives the probability it is in the infinitesimal volume dV. The radial probability density is given by $P(r) = 4\pi r^2|\psi|^2$ and $P(r)\,dr$ gives the probability that the particle is in the spherical shell with inner radius r and outer radius $r + dr$.

PROBLEM 6 (22E).   What must be the width of an infinite well such that the energy of an electron trapped therein in the $n = 3$ state has an energy of $4.7\,\text{eV}$?

STRATEGY: Solve $E = n^2h^2/8mL^2$ for L. When you substitute numbers don't forget to convert the value given for the energy to joules.

SOLUTION:

$$[\text{ans: } 8.5 \times 10^{-10}\,\text{m}]$$

PROBLEM 7 (25E).   An electron, trapped in an infinite well of width $0.250\,\text{nm}$, is in the ground ($n = 1$) state. How much energy must it absorb to jump to the third excited ($n = 4$) state?

STRATEGY: Use the expression $E = n^2h^2/8mL^2$ for the allowed energy values. Calculate the energies of the initial and final states, then find their difference.

SOLUTION:

$$[\text{ans: } 1.45 \times 10^{-17}\,\text{J} \ (90.2\,\text{eV})]$$

PROBLEM 8 (29P).   A particle is confined between rigid walls separated by a distance L. The particle is in the lowest energy state; the wave function for this state is given in Problem 28. Use this wave function to calculate the probability that the particle will be found between the points (a) $x = 0$ and $x = L/3$, (b) $x = L/3$ and $x = 2L/3$, and (c) $x = 2L/3$ and $x = L$.

STRATEGY: The probability that the particle will be found between x_1 and x_2 is given by

$$P(x_1, x_2) = \int_{x_1}^{x_2} |\psi|^2\,dx = (2/L)\int_{x_1}^{x_2} \sin^2\frac{\pi x}{L}\,dx\ .$$

Use the trigonometric identity $\sin^2(\pi x/L) = \frac{1}{2}[1 - \cos(2\pi x/L)]$ to help evaluate the integral. You should get

$$P(x_1, x_2) = \frac{x_2 - x_1}{L} - \frac{1}{2\pi}\left[\sin\frac{2\pi x_2}{L} - \sin\frac{2\pi x_1}{L}\right]\ .$$

Evaluate this expression for the three sets of limits.

SOLUTION:

$$\left[\text{ans:}\;\;(a)\;0.196;\;(b)\;0.608;\;(c)\;0.196\right]$$

PROBLEM 9 (32E).   For the ground state of the hydrogen atom, evaluate the probability density $\psi^2(r)$ and the radial probability density $P(r)$ for the positions $(a)\; r = 0$ and $(b)\; r = r_B$.

STRATEGY: The probability density is given by $\psi^2 = (1/\pi r_B^3)\,e^{-2r/r_B}$ and the radial probability density is given by $P = (4r^2/r_B^3)\,e^{-2r/r_B}$. Put $r = 0$ and use $e^0 = 1$. Then put $r = r_B$.

SOLUTION:

$$\left[\text{ans:}\;\;(a)\;\psi^2 = 1/\pi r_B^3\;(2.15 \times 10^{30}\,\text{m}^{-3}),\;P = 0;\;(b)\;\psi^2 = (1/\pi r_B^3)\,e^{-2}\;(2.91 \times 10^{29}\,\text{m}^{-3}),\;P = (4/r_B)\,e^{-2}\right.$$
$$\left.(1.02 \times 10^{10}\,\text{m}^{-1})\right]$$

Explain what these quantities mean.

SOLUTION:

PROBLEM 10 (34P).   For an electron in the ground state of the hydrogen atom, (a) verify Eq. 44–19 and (b) calculate the radius of a sphere for which the probability that the electron will be found inside the sphere equals the probability that the electron will be found outside the sphere. (*Hint*: See Sample Problem 44-6.)

STRATEGY: (a) You want to evaluate the integral

$$\int_0^\infty P(r)\,dr = (4/r_B^3)\int_0^\infty r^2\,e^{-2r/r_B}\,dr\,.$$

Let $\alpha = 2/r_B$. Then the integral becomes

$$\int_0^\infty P(r)\,dr = (4/r_B^3)\frac{d^2}{d\alpha^2}\int_0^\infty e^{-\alpha r}\,dr\,.$$

The integral can easily be evaluated and the second derivative taken. When you are finished, replace α with $2/r_B$. You should get 1 for the answer.

(b) Use a trial and error method to solve $0.5 = 1 - e^{-2x}(1 + 2x + 2x^2)$ for x, then solve $x = r/r_B$ for r.

SOLUTION:

[ans: (b) $1.34 r_B$ (7.09×10^{-11} m)]

PROBLEM 11 (35P).   For the ground state of the hydrogen atom show that the probability $p(r)$ that the electron lies within a sphere of radius r is given by

$$p(r) = 1 - e^{-2x} \left(1 + 2x + 2x^2\right),$$

in which $x = r/r_B$, a dimensionless ratio.

STRATEGY: You want to evaluate the integral

$$p(r) = \int_0^r P(r)\,\mathrm{d}r = (4/r_B^3) \int_0^r r^2 e^{-2r/r_B}\,\mathrm{d}r\,.$$

As in the last problem, set $\alpha = 2/r_B$ and show that

$$p(r) = (4/r_B^3)\frac{\mathrm{d}^2}{\mathrm{d}\alpha^2} \int_0^r e^{-\alpha r}\,\mathrm{d}r\,.$$

Carry out the integration and the differentiations, then replace α with $2/r_B$.

SOLUTION:

Barrier tunneling problems usually involve a computation of the transmission coefficient $T = e^{-2kL}$, where $k = \sqrt{8\pi^2 m(U - E)/h^2}$. Here L is the width of the barrier (a length), U is the height of the barrier (an energy), and E is the kinetic energy of the particle incident on the barrier. T gives the probability that the particle will tunnel through the barrier. If N particles, all with the same mass and energy, are incident on a barrier, then TN tunnel through. The others are reflected.

PROBLEM 12 (38P).   Consider an energy barrier such as that of Fig. 44–13a, but whose height U is 6.0 eV and whose thickness L is 0.70 nm. Calculate the energy of an incident electron such that its transmission probability is 0.001.

STRATEGY: Use $T = e^{-2kL}$ to calculate k and $k = \sqrt{8\pi^2 m(U - E)/h^2}$ to calculate E. Don't forget to substitute the value for U in joules.

SOLUTION:

$$[\text{ans: } 8.1 \times 10^{-19}\,\text{J } (5.1\,\text{eV})]$$

PROBLEM 13 (39P).   Suppose that a beam of 5.0-eV protons is incident on an energy barrier of height 6.0 eV and thickness 0.70 nm, at a rate equivalent to a current of 1.0 kA. How long would you have to wait — on average — for one proton to be transmitted?

STRATEGY: Use $T = e^{-2kL}$, with $k = \sqrt{8\pi^2 m(U - E)/h^2}$, to find the value of the transmission coefficient. The current through the barrier is given by $i = Ti_0$, where i_0 is the incident current (1.0 kA). If t is the average time interval between protons tunneling through the barrier, then $i = e/t$. Calculate t.

You may have computational difficulty: $2kL$ is about 307 and your calculator may not be able to evaluate e^{-2kL} directly. Evaluate e^{-100} and e^{-7} and write each of them as the product of a number between 1 and 10 and a power of 10. You wish to cube the first result and multiply by the second. Do this by multiplying the small factors and adding the exponents of 10.

SOLUTION:

$$[\text{ans: } 4.8 \times 10^{111}\,\text{s } (1.3 \times 10^{104}\,\text{y})]$$

Here are some examples of problems that deal with the uncertainty principles $\Delta x \cdot \Delta p \approx h$ and $\Delta E \cdot \Delta t \approx h$. Remember that for a particle trapped in a well the uncertainty in its position may be taken to be the width of the well and the uncertainty in its momentum may be taken to be twice the magnitude of its momentum.

PROBLEM 14 (41E).   A microscope using photons is employed to locate an electron in an atom to within a distance of 10 pm. What is the minimum uncertainty in a measurement of the momentum of the electron located in this way?

STRATEGY: Solve $\Delta x \cdot \Delta p = h$ for Δp.

SOLUTION:

$$[\text{ans: } 6.6 \times 10^{-23}\,\text{kg} \cdot \text{m/s}]$$

PROBLEM 15 (44E). Consider an electron trapped in an infinite well whose width is 100 pm. If it is in a state with $n = 15$, what are (a) its energy, (b) the uncertainty in its momentum, and (c) the uncertainty in its position?

STRATEGY: Use $E = n^2 h^2 / 8mL^2$ to calculate its energy. Here h is the Planck constant, m is the mass of a electron, and L is the width of the well. Use $E = p^2/2m$ to calculate the magnitude of its momentum. The uncertainty in the momentum is $\Delta p = 2p$. The uncertainty in its position is L.

SOLUTION:

$$\left[\text{ans: } (a)\ 1.36 \times 10^{-15}\,\text{J} \ (8.46\,\text{keV}); (b)\ 9.94 \times 10^{-23}\,\text{kg} \cdot \text{m/s}; (c)\ 100\,\text{pm} \right]$$

Note that these values do not violate the Heisenberg uncertainty principle. $\Delta x \cdot \Delta p$ is larger than h by a factor of about 15.

PROBLEM 16 (45E). The lifetime of an electron in the state with $n = 2$ in hydrogen is about 10^{-8} s. What is the uncertainty in the energy of the $n = 2$ state? Compare this with the energy of the state.

STRATEGY: Solve $\Delta E \cdot \Delta t = h$ for ΔE. The energy of the state is $-13.6/n^2$, in electron volts.

SOLUTION:

$$\left[\text{ans: } 6.6 \times 10^{-26}\,\text{J} \ (4.1 \times 10^{-7}\,\text{eV}), \text{ a factor of about } 1.2 \times 10^{-7} \text{ times the energy} \right]$$

To wrap up this chapter on matter waves here's a problem to help convince you that you cannot prove an electron has a definite orbit by observing its motion.

PROBLEM 17 (47P). Suppose that we wish to test the possibility that electrons in atoms move in orbits by "viewing" them with photons with sufficiently short wavelengths, say 10.0 pm or less. (a) What would be the energy of such photons?

STRATEGY: Substitute $p = h/\lambda$ into $E = pc$ to obtain $E = hc/\lambda$. Evaluate this expression.

SOLUTION:

$$\left[\text{ans: } 1.99 \times 10^{-14}\,\text{J} \ (1.24 \times 10^5\,\text{eV}) \right]$$

(b) How much energy would such a photon transfer to a free electron in a head-on Compton collision?

STRATEGY: The change in wavelength is given by $\Delta\lambda = (h/mc)(1 - \cos\theta)$, where θ is the scattering angle (180°). The wavelength after scattering is given by $\lambda' = \lambda + \Delta\lambda$ and the energy after scattering is given by $E' = hc/\lambda'$. Finally, the energy transferred to the electron is given by $E - E'$.

SOLUTION:

$$\left[\text{ans: } 6.49 \times 10^{-15}\,\text{J } (40.5\,\text{keV})\right]$$

(c) What does this tell you about the possibility of confirming orbital motion by "viewing" an atomic electron at two or more points along its path?

STRATEGY: Compare, for example, the energy transferred to the ionization energy for hydrogen in its ground state (13.6 eV) to see how dramatically the act of "viewing" changes the electron's state.

SOLUTION:

III. NOTES

Chapter 45
ALL ABOUT ATOMS

I. BASIC CONCEPTS

Here you learn about the structure of atoms and, in particular, about the allowed values of energy, angular momentum, and magnetic dipole moment. Atomic structure is reflected in the periodic table of chemistry and is fundamental to our understanding of the properties of materials. An important new idea is the Pauli exclusion principle. Learn what it is and how it influences the periodic table. The chapter closes with a explanation of how a laser works.

The Schrodinger equation and the hydrogen atom. This is a differential equation for the _____ of a particle. Classically, the motion of a particle is influenced by the forces other particles exert on it. Quantum mechanically, the influence of other particles enters through the _____, which of course is derived from the forces. For an electron in a hydrogen atom this function is

$$U =$$

Of the many solutions to the Schrodinger equation with a particular potential energy function, the physically correct one is chosen by applying the condition _____ _____.

If the particle is bound this condition leads to the quantization of energy. Explain what the term <u>energy quantization</u> means: _____ _____

For an electron in a hydrogen atom the allowed values of the energy are

$$E_n =$$

where n is an integer $(1, 2, 3, \ldots)$ called the _____ quantum number. Since the energy of a photon emitted by an atom equals the difference in energy of two quantum mechanical states, the expression for the allowed energy values leads naturally to a prediction of the energies (and therefore the frequencies) of photons emitted by hydrogen, as discussed in the last chapter.

Angular momentum. The magnitude of the orbital angular momentum of the electron in a hydrogen atom is quantized. The allowed values are given by

$$L =$$

where ℓ is zero or a positive integer, called the _____ quantum number. The constant $\hbar$ is the Planck constant divided by _____. Its value is _____ J·s.

When the atom is in a state with principal quantum number n, the quantum number ℓ can take on the values _____. Thus for a given energy there is a maximum to the angular momentum magnitude. It is _____ .

Each component of the angular momentum is also quantized. The z component, for example, may have only the values

$$L_z =$$

where m_ℓ is an integer (positive, negative, or zero) and is called the _____ quantum number. If the orbital quantum number is ℓ, the possible values of m_ℓ run from _____ to _____ and include all integer values between. Identical expressions are valid for the x and y components. Note carefully that if the z component of the angular momentum has a definite value, the other components do not. The quantum number m_ℓ refers to a single component; you cannot specify all three.

Because the z component of the angular momentum is quantized the angle between the angular momentum vector and the z axis can have only certain values, given by

$$\theta =$$

For a given value of ℓ the smallest angle occurs when $m_\ell =$ _____ .

A magnetic dipole moment is associated with the orbital motion of the electron. At the atomic level, magnetic dipole moments are conveniently measured in units of the Bohr magneton μ_B. In terms of fundamental constants of nature

$$\mu_B =$$

Its value is _____ J/T or _____ eV/T. If an electron is in a state with magnetic quantum number m_ℓ then in terms of m_ℓ and μ_B, the z component of its orbital magnetic dipole moment is

$$\mu_{\ell,z} =$$

Notice that the direction of the dipole moment is opposite to the direction of the angular momentum. This is because the electron is _____ charged.

Spin angular momentum. An electron also has an <u>intrinsic</u> angular momentum, called its spin angular momentum. It is quantized in the same manner as orbital angular momentum. The z component has a value given by $S_z = m_s\hbar$, where the spin quantum number m_s is either _____ or _____ .

A magnetic dipole moment is associated with spin angular momentum. If an electron has spin quantum number m_s then the z component of its spin dipole moment is

$$\mu_{s,z} =$$

Except for a factor of 2 this is the same as the relationship between the z component of the orbital dipole moment and the z component of the orbital angular momentum.

Spin angular momentum and its quantum number are not predicted by the Schrodinger equation but they are predicted by _____ quantum theory.

The Stern-Gerlach experiment. This experiment provides an important experimental verification of space quantization. A beam of atoms passes through a non-uniform magnetic field to a screen. If the field is in the z direction and varies with z then it exerts a force on the atoms in either the positive or negative z direction, depending on whether m_ℓ is positive or negative. If μ_z is the z component of the magnetic moment and dB_z/dz is the rate at which the field varies with z then the force is given by

$$F_z =$$

Suppose the magnetic force is vertically upward for atoms with a positive z component of their magnetic moment and vertically downward for particles with a negative z component. Consider a beam of atoms such that the orbital and spin magnetic moments of all the electrons in each atom cancel except for the moment of a single electron in an $\ell = 0$ state and suppose that half the atoms in the beam are in $m_s = +\frac{1}{2}$ states and half are in $m_s = -\frac{1}{2}$ states. Describe the pattern on the screen: _____

_____

For contrast, describe the pattern you would expect to see if angular momentum were not quantized: _____

_____

Enumeration of hydrogen atom states. The quantum mechanical state of a hydrogen atom is specified by giving the values of 4 quantum numbers. They are

    1. the principal quantum number n, which specifies _____ .

    2. the orbital quantum number ℓ, which specifies _____

    _____ .

    3. the magnetic quantum number m_ℓ, which specifies _____

    _____ .

    4. the spin quantum number m_s, which specifies _____

    _____ .

Wave functions. The text gives several wave functions for the electron in hydrogen, associated with different states. For any of these the quantity $|\psi|^2\, dV$ gives the probability that the electron is _____

_____ .

In terms of the wave function ψ the radial probability density is given by

$$P(r) =$$

The quantity $P(r)\, dr$ gives the probability that the electron is _____

_____ .

The radial probability density associated with the ground state ($n = 1$) of the electron in a hydrogen atom is

$$P(r) = $$

where r_B is _____. Use
the axes to the right to sketch a graph of $P(r)$ as a
function of the distance r from the proton. Label
the r axis to indicate the distance scale. In terms
of the Bohr radius the maximum of the function
occurs at $r =$ _____ r_B. Compare the probability
of finding the electron in a thin spherical shell with
a radius of $2r_B$ to the probability of finding it in a
shell of the same thickness but with a radius of r_B:

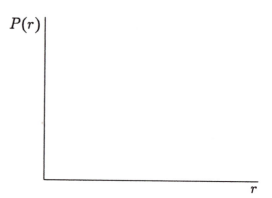

_____

_____

The radial probability density associated with the $n = 2$, $\ell = 0$ state is

$$P(r) = $$

Use the axes to the right to sketch a graph of $P(r)$
as a function of the distance r from the proton.
Label the r axis to indicate the distance scale. In
terms of the Bohr radius the maximum of the function occurs at $r \approx$ _____ r_B. If we take the distance from the proton to the peak in the radial
probability function as an indication of the size of
the atom, it is larger when the electron is in the
$n = 2$, $\ell = 0$ state than when the electron is in the
$n = 1$ state by a factor of about _____.

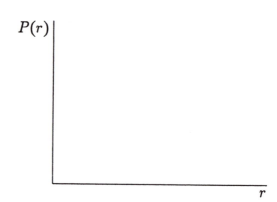

The wave functions for the $n = 2$, $\ell = 1$ states are not spherically symmetric. Look at
Fig. 45–8 to see what the probability densities look like.

Multi-electron atoms and the periodic table.   The same quantum numbers n, ℓ, m_ℓ, and m_s
can be used to label electron states in all atoms, even though the corresponding wave functions
are different for electrons in different atoms.

Electrons obey the Pauli exclusion principle, which states that _____

_____

_____.

Because electrons obey this principle no two electrons have all 4 quantum numbers with the
same values. For example, electrons with the same values of n and ℓ have different values of
m_ℓ or m_s (or both).

Furthermore, when an atom is in its ground state the electrons fill the various states in
such a way that the total _____ is the least possible. You may think of starting with an

atom with atomic number $Z - 1$ and constructing an atom with atomic number Z by adding a single proton to the nucleus and a single electron outside. If the atom is in its ground state the electron goes into the state that gives the atom the lowest possible energy without violating the Pauli exclusion principle. It goes into a previously unoccupied state.

The energies of electrons with the same value of n depend on the orbital quantum number ℓ. In fact, the energy increases with ℓ. This is because electrons with smaller values of ℓ have a higher probability of being near the nucleus than do electrons with greater values of ℓ. When it is near the nucleus the effective charge that interacts with the electron is greater than when it is far away because the nucleus is not screened by as many _____. The electron then has a more negative potential energy.

In discussing many-electron atoms the word <u>orbital</u> means: _____

_____

For given values of the principal quantum number n and orbital quantum number ℓ the number of states in an orbital is _____. The orbital and spin angular momentum vectors point in all possible directions for electrons in a filled orbital. Thus the value of the total orbital angular momentum is _____ and the value of the total spin angular momentum is _____. The contribution of a filled orbital to the magnetic dipole moment of the atom is _____.

The text considers a neon atom, with 10 electrons. _____ of these occupy the $n = 1$, $\ell = 0$ orbital, _____ occupy the $n = 2$, $\ell = 0$ orbital, and _____ occupy the $n = 2$, $\ell = 1$ orbital. We expect a neon atom in its ground state to have a total angular momentum of _____ and a magnetic moment of _____.

A sodium atom has 11 electrons. In the ground state of the atom 10 of them are in the same orbitals as the 10 electrons of a neon atom. The other is in the $n =$ _____, $\ell =$ _____ orbital. In terms of $\hbar$ the magnitude of the z component of the total angular momentum of this atom is _____ and in terms of μ_B the magnitude of the z component of the magnetic dipole moment is _____.

Neon is an inert gas. A neon atom interacts extremely weakly with other atoms and does not enter into chemical reactions. Sodium, on the other hand, is very chemically active. Explain why the addition of a single electron changes the chemical properties so dramatically:

_____

_____

_____

The periodic table of the chemical elements groups atoms with similar chemical properties in the same column, with atoms in neighboring columns of the same row differing by _____ in the number of electrons. Application of quantum mechanics and the Pauli exclusion principle explains the similarities of the properties of atoms in the same column and the variation of properties from column to column.

The number of columns is different for different rows (periods) of the table. Explain why:

_____

_____

_____

For each of the periods list the filled states in the form (n, ℓ) and tell how many chemical elements are in the period:

Filled states for period 1: _____

Number of chemical elements in period 1: _____

Filled states for period 2: _____

Number of chemical elements in period 2: _____

Filled states for period 3: _____

Number of chemical elements in period 3: _____

Filled states for period 4: _____

Number of chemical elements in period 4: _____

Filled states for period 5: _____

Number of chemical elements in period 5: _____

Filled states for period 6: _____

Number of chemical elements in period 6: _____

X-ray spectra. Look at Fig. 15, which shows the spectrum of x rays that are emitted from a molybdenum target when 35-keV electrons strike it. There are two main parts: the relatively broad, low-level, <u>continuous</u> spectrum and the two peaks, called the <u>characteristic</u> spectrum. In addition, an important feature of the continuous spectrum is the <u>cut-off</u> wavelength; no x rays with wavelengths less than λ_{min} are ever emitted when electrons with energy of 35 keV are incident. You should be able to explain the origin of both the continuous and characteristic spectra and explain the reason for a cut-off wavelength.

The basic fact you must know to understand the existence of a continuous spectrum is that an accelerating charge emits electromagnetic radiation. Describe what happens to an incident electron that causes it to decelerate after it enters the target: _____

_____

This radiation is called <u>bremsstrahlung</u>, meaning _____ radiation, appropriately enough.

After an electron is accelerated through a potential difference V its kinetic energy is $K =$ _____ and after it enters the target it may lose any amount from 0 to K in a decelerating event. Radiation with the greatest possible frequency (and so the shortest possible wavelength) is emitted by those electrons that lose _____ their kinetic energy in a single event. The maximum frequency is therefore given by $hf_{max} = eV$, so the shortest possible wavelength is given by $\lambda_{min} =$ _____ . Note that the value of λ_{min} is independent of the target material.

Radiation of the characteristic spectrum is emitted when an incoming electron knocks another electron from a low-energy state, an $n = 1$ state for example. An electron from a state with higher energy makes the transition to the empty state and emits a photon. For most atoms the difference in energy of the two states is sufficient to produce a photon in the x-ray region of the electromagnetic spectrum. The K_α x-ray line is produced when electrons fall from an $n =$ _____ state (labeled an L state) to an $n =$ _____ state (labeled the K shell). The K_β line is produced when electrons fall from an $n =$ _____ state (labeled an M state) to a K state. Radiation with other frequencies is produced when electrons with still higher energies fall into the vacated states.

X-ray energy level diagrams are usually drawn as in Fig. 17 of the text. Carefully note that $E = 0$ represents an atom with all electrons in the states with the lowest possible energies consistent with the Pauli exclusion principle. The energy labeled K represents an atom with an empty K state, the energy labeled L represents an atom with an empty L state, etc. A downward arrow represents a transition in which the energy of the atom decreases and a photon is emitted.

X-ray emissions can be used to determine the position of a chemical element in the periodic table. In his pioneering work, Moseley showed that the frequency of the K_α line, for example, changes in a regular way from element to element in the table. If f is the x-ray frequency associated with this line then a plot of _____ as a function of position in the table produces a nearly straight line. Moseley concluded that the order of atoms in the table depended on the number of _____ in the nucleus.

You should be able to use the Bohr theory to understand this result. If there are Z protons in the nucleus then the effective charge that acts on an electron in a K state is somewhat less than Ze. Explain why: _____

_____

Write $(Z-1)e$ for the effective charge. According to the Bohr theory the frequency associated with K_α-photons is given by

$$f =$$

This shows that $\sqrt{f}$ is proportional to $Z-1$. A Moseley plot is essentially a graph of $\sqrt{f}$ vs. Z and is nearly a straight line. Z is called the atomic number of the chemical element. Bohr theory gives a very poor approximation to the energies of outer electrons in many-electron atoms but it is quite good for the innermost electrons responsible for x-ray emission.

Explain why Moseley plots are of great historical interest: _____

_____

_____

Lasers. List the 4 properties of laser light that distinguish it from light emitted by an ordinary tungsten filament:

1. _____
2. _____
3. _____
4. _____

Laser light is produced in the following process. An electron in a state with high energy is stimulated by an incoming photon to drop to a state with lower energy and emit a photon. The energy of the incoming photon must match the difference in energy of the two states ($hf = E_2 - E_1$) and the energy of the stimulated photon is exactly the same. After the emission there are two photons, identical in every way. If there are sufficient electrons in the higher energy state each of these photons can stimulate the emission of another photon and both the original photon and the new photon can in turn stimulate the emission of more. The number of identical photons quickly multiplies.

Laser beams are highly monochromatic because all the stimulated photons have the same _____ and all the waves associated with them have the same _____. Laser beams are highly coherent because all the waves associated with the stimulated photons have the same _____. Laser beams do not spread significantly because all the stimulated photons travel in the same _____. Actually some spreading does occur because the waves are diffracted as they pass through the window of the laser.

To produce laser light the following conditions must hold:

1. The higher in energy of the two states must be <u>metastable</u>. Explain what this term means.
   _____
   _____

   Explain why it is important that the higher energy state be metastable. _____
   _____

2. The populations of the two states must be inverted. Explain what population inversion is.
   _____
   _____

   Note that an incoming photon with the correct energy to stimulate emission also has the correct energy to be absorbed. Use this to explain why population inversion is important.
   _____
   _____

3. Lasers usually have a mirror at each end. Explain why. _____
   _____

4. One of the mirrors is an excellent reflector while the other is a partial reflector and lets some of the laser light out. This mirror is tilted at the Brewster angle. Explain why.
   _____
   _____

Optical pumping is often used to achieve population inversion. The diagram to the right is for a three-level laser. Light from a non-laser source is used to excite electrons from the state with energy _____ to the state with energy _____. Draw an arrow on the diagram to represent this transition and label it "pumping". Electrons in this state quickly and spontaneously make the transition to the state with energy _____. This state is metastable. Draw an arrow to represent this transition and label it "decay". Laser light is produced when electrons are stimulated to make the transition from the state with energy _____ to the state with energy _____. Draw an arrow to represent this transition and label it "laser light".

_____ E_3

_____ E_2

_____ E_1

Explain how pumping is accomplished in a helium-neon laser: _____
_____
_____

II. PROBLEM SOLVING

For a given value of the principal quantum number n you should know what values of the orbital quantum number ℓ are allowed and for a given value of ℓ you should know what values of m_ℓ are allowed. You should also know the allowed values of the z component of the spin angular momentum. You should be able to calculate the magnitudes of both the spin and orbital angular momenta, their z components, and the angles they make with the z axis. You should also be able to calculate both the orbital and spin contributions to the z component of the magnetic dipole moment.

PROBLEM 1 (5E). If an electron in a hydrogen atom is in a state with $\ell = 5$, what is the minimum possible angle between $\mathbf{L}$ and $\mathbf{L}_z$?

STRATEGY: The cosine of the angle θ between $\mathbf{L}$ and $\mathbf{L}_z$ is given by $L_z = L\cos\theta$, so $\cos\theta = L_z/L$. Now $L = \sqrt{\ell(\ell+1)}\hbar$ and $L_z = m_\ell\hbar$, so $\cos\theta = m_\ell/\sqrt{\ell(\ell+1)}$. The minimum vale of θ corresponds to the maximum value of $\cos\theta$ and thus to the maximum possible value of m_ℓ. This is $m_\ell = \ell$. Solve $\cos\theta = \ell/\sqrt{\ell(\ell+1)}$ for θ when $\ell = 5$.

SOLUTION:

[ans: 24.1°]

PROBLEM 2 (7E). A hydrogen-atom state is known to have the quantum number $\ell = 3$. What are the possible n, m_ℓ, and m_s quantum numbers?

STRATEGY: The value of ℓ must be less than n, m_ℓ can have any integer value from $-\ell$ to $+\ell$, and m_s must be either $-\frac{1}{2}$ or $+\frac{1}{2}$.

SOLUTION:

[ans: $n > 3$; $m_\ell = -3, -2, -1, 0, +1, +2,$ or $+3$; $m_s = -\frac{1}{2}$ or $+\frac{1}{2}$]

PROBLEM 3 (11P). Calculate and tabulate, for a hydrogen atom in a state with $\ell = 3$, the allowed values of L_z, μ_z, and θ.

STRATEGY: The allowed values of L_z are given by $m_\ell\hbar$, where m_ℓ can be $-3, -2, -1, 0, +1, +2,$ or $+3$. The allowed values of μ_z are given by $-m_\ell\mu_B$, where μ_B is the Bohr magneton (9.274×10^{-24} J/T). The allowed values of θ are given by $\cos\theta = L_z/L = m_\ell/\sqrt{\ell(\ell+1)}$.

SOLUTION:

[ans: $m_\ell = -3$: $L_z = -3.16 \times 10^{-34}$ kg $\cdot$ m²/s, $\mu_z = +2.78 \times 10^{-23}$ J/T, $\theta = 150°$; $m_\ell = -2$: $L_z = -2.11 \times 10^{-34}$ kg $\cdot$ m²/s, $\mu_z = +1.85 \times 10^{-23}$ J/T, $\theta = 125°$; $m_\ell = -1$: $L_z = -1.06 \times 10^{-34}$ kg $\cdot$ m²/s,

$\mu_z = +9.27 \times 10^{-24}$ J/T, $\theta = 107°$; $m_\ell = 0$: $L_z = 0$, $\mu_z = 0$, $\theta = 90°$; $m_\ell = +1$: $L_z = +1.06 \times 10^{-34}$ kg $\cdot$ m^2/s,
$\mu_z = -9.27 \times 10^{-24}$ J/T, $\theta = 73°$; $m_\ell = +2$: $L_z = +2.11 \times 10^{-34}$ kg $\cdot$ m^2/s, $\mu_z = -1.85 \times 10^{-23}$ J/T, $\theta = 55°$;
$m_\ell = +3$: $L_z = +3.16 \times 10^{-34}$ kg $\cdot$ m^2/s, $\mu_z = -2.78 \times 10^{-23}$ J/T, $\theta = 30°$]

Find also the magnitudes of **L** and μ.

STRATEGY: The magnitude of **L** is $L = \sqrt{\ell(\ell + 1)}\hbar$ and the magnitude of μ is $\mu = L\mu_B = \sqrt{\ell(\ell + 1)}\mu_B$.

SOLUTION:

[ans: $L = 3.65 \times 10^{-34}$ kg $\cdot$ m^2/s, $\mu = 3.21 \times 10^{-23}$ J/T]

PROBLEM 4 (16P*).   An unmagnetized iron cylinder, whose radius is 5.0 mm, hangs from a frictionless bearing inside a solenoid, so that the cylinder can rotate freely about its axis; see Fig. 45–3. By passing a current through the solenoid windings, a magnetic field is suddenly applied parallel to the axis, causing the magnetic dipole moments of the atoms to align themselves parallel to the field. The atomic angular momentum vectors, which are coupled back to back with the magnetic dipole moments vectors, also become aligned and the cylinder will start to rotate. This is the Einstein-de Haas effect (see Section 45–2). Find the period of rotation of the cylinder. Assume that each iron atom has an angular momentum of $\hbar$ and that the alignment is perfect. (*Hint*: Apply conservation of angular momentum; the answer is independent of the length of the cylinder.)

STRATEGY: Since the total angular momentum is initially zero and is conserved, the angular momentum of the rotating cylinder has the same magnitude as the total atomic angular momentum. The angular momentum of the cylinder as a whole is given by $I\omega$, where I is its rotational inertia and ω is its angular velocity. The total atomic angular momentum is $N\hbar$, where N is the number of iron atoms in the cylinder. Thus $I\omega = N\hbar$. The period T of rotation is given by $T = 2\pi/\omega$. Thus $T = 2\pi I/N\hbar$.

If M is the mass of the cylinder and R is its radius then $I = \frac{1}{2}MR^2$. The number of iron atoms in the cylinder is $N = M/m$, where m is the mass of an iron atom. If M_m is the molar mass, then $m = M_m/N_A$, where N_A is the Avogadro constant.

SOLUTION:

[ans: 6.9×10^4 s (19 h)]

You should be able to use wave functions and radial probability densities to compute the probability of finding the electron in a given region of space, given its quantum numbers.

PROBLEM 5 (18E).   A small sphere of radius $0.10r_B$ is located a distance r_B from the nucleus of a hydrogen atom in its ground state. What is the probability that the electron will be found inside this sphere? (Assume that ψ is constant inside the sphere.

STRATEGY: The probability is given exactly by the integral

$$\text{prob} = \int |\psi|^2 \, dV,$$

over the volume of the sphere. Since the sphere is so small $|\psi|^2$ may be taken to be a constant and

$$\text{prob} = |\psi|^2 V = |\psi|^2 (4\pi R^3/3),$$

where R is the radius of the sphere. The radial probability function for the ground state is given as

$$P = \left(\frac{4r^2}{r_B^3}\right) e^{-2r/r_B} .$$

It is related to the wave function by $P = 4\pi r^2 |\psi|^2$. First show that $|\psi|^2$, evaluated at $r = r_B$, is $(1/\pi r_B^3)e^{-2}$ and then that the probability of being in the sphere is prob $= (4R^3/3r_B^3)e^{-2}$. Put $R = 0.10r_B$, and evaluate this expression.

SOLUTION:

$$\left[\text{ans: } 1.80 \times 10^{-4}\right]$$

PROBLEM 6 (23E).   Using Eq. 45–14, show that , for the hydrogen-atom state with $n = 2$ and $\ell = 0$,

$$\int_0^\infty P(r)\, dr = 1 .$$

What is the physical interpretation of this result?

STRATEGY: For $n = 2$ and $\ell = 0$,

$$P(r) = \left(\frac{r^2}{8r_B^3}\right) \left(2 - \frac{r}{r_B}\right)^2 e^{-r/r_B} ,$$

where r_B is the Bohr radius. You want to evaluate the integral

$$\left(\frac{1}{8r_B^3}\right) \int_0^\infty r^2 \left(2 - \frac{r}{r_B}\right)^2 e^{-r/r_B}\, dr .$$

Let $x = r/r_B$ and use

$$\int_0^\infty x^n e^{-x}\, dx = n! ,$$

an integral that can be found in most tables.

SOLUTION:

PROBLEM 7 (27P). For a hydrogen atom in a state with $n = 2$ and $\ell = 0$, what is the probability of finding the electron between two spheres of radii $r = 5.00 r_B$ and $r = 5.01 r_B$?

STRATEGY: If $P(r)$ is the radial probability density for the $n = 2$, $\ell = 0$ state then the probability of finding the electron between two spheres of radii r_1 and r_2 is given by the integral

$$\text{prob} = \int_{r_1}^{r_2} P(r)\, dr .$$

The difference in the radii of the spheres is so small you may approximate $P(r)$ by a constant. You might take the constant to be the value of the function at the average radius $r = (r_1 + r_2)/2$. Then the integral for the probability becomes

$$\text{prob} = P(r)(r_2 - r_1) = \left(\frac{r^2}{8 r_B^3}\right) \left(2 - \frac{r}{r_B}\right)^2 (r_2 - r_1) e^{-r/r_B} .$$

Evaluate this expression.

SOLUTION:

[ans: 0.0019]

Here's some examples that deal with the Stern-Gerlach experiment. You should know how to calculate the force of a nonuniform magnetic field on a magnetic dipole and be able to interpret a given pattern on the observing screen in terms of the allowed values of the z component of the magnetic dipole moment of the atoms passing through the apparatus. You should also know that states with different values of m_ℓ have different energies in a magnetic field and know how to compute the energy difference.

PROBLEM 8 (31E). Assume that in the Stern-Gerlach experiment described for neutral silver atoms the magnetic field **B** has a magnitude of 0.50 T. (*a*) What is the energy difference between the orientations of the silver atoms in the two subbeams?

STRATEGY: If the magnetic field **B** is in the positive z direction the energy of a magnetic dipole μ in the field is given by $U = -\mu \cdot \mathbf{B} = -\mu_z B$. Suppose the dipole is initially parallel to the field (so μ_z is positive) and it flips (so μ_z is negative). Then change in energy is $\Delta U = 2|\mu_z|B$. For a silver atom $\mu_z = 1.00 \mu_B$, where μ_B is the Bohr magneton (9.274×10^{-24} J/T).

SOLUTION:

[ans: 9.34×10^{-24} J (5.8×10^{-5} eV)]

(*b*) What is the frequency of the radiation that would induce a transition between these two states?

STRATEGY: Since energy is conserved the energy of the absorbed photon equals the difference in energy of the two states, so the frequency f is given by $hf = \Delta U$.

SOLUTION:

[ans: 1.4×10^{10} Hz]

(c) What is its wavelength, and to what part of the electromagnetic spectrum does it belong?

STRATEGY: The wavelength λ and frequency f are related by $\lambda f = c$, where c is speed of light.

SOLUTION:

[ans: 0.021 m, short radio wave]

PROBLEM 9 (32P). Suppose a hydrogen atom (in its ground state) moves 80 cm in a direction perpendicular to a magnetic field that has a gradient, in the vertical direction, of 1.6×10^2 T/m. (a) What is the force on the atom due to the magnetic moment of the electron, which we take to be 1 Bohr magneton?

STRATEGY: If the field is in the z direction and varies with z, the force is also in the z direction and its component is $F_z = \mu_z \, dB_z/dz$. Evaluate this expression with $\mu_z = 1.00\mu_B$.

SOLUTION:

[ans: 1.5×10^{-21} N]

(b) What is its vertical displacement in the 80 cm of travel if its speed is 1.2×10^5 m/s?

STRATEGY: The time taken for the electron to cross the region of the field is $t = d/v$, where d (= 80 cm) is the width of the region. Its displacement along the field is $\Delta z = \frac{1}{2}(F_z/m)t^2$, where m is the mass of the atom (1.67×10^{-27} kg).

SOLUTION:

[ans: 2.0×10^{-5} m]

PROBLEM 10 (35E). An external magnetic field of frequency 34 MHz is applied to molecules of a certain material that contains hydrogen atoms. Resonance is observed when the strength of this applied field equals 0.78 T. Calculate the strength of the local magnetic field at the site of the protons undergoing spin flips.

STRATEGY: Take the magnetic field to be in the z direction. The change in energy that occurs when a spin flips is $\Delta U = 2\mu_s B$, where μ_s is the magnetic moment of a proton and B is the magnitude of the total magnetic field. For a proton $\mu_s = 1.41 \times 10^{-26}$ J/T. The frequency of the field is related to the change in energy by $hf = \Delta U$. Solve for B. Assume the applied and local fields are in the same direction. Then $B = B_{local} + B_{applied}$. Solve for B_{local}.

SOLUTION:

[ans: 0.019 T]

Some problems deal with consequences of the Pauli exclusion principle. Here is an example.

PROBLEM 11 (40P).   Suppose there are two electrons in the same atom, both of which have $n = 2$ and $\ell = 1$. (*a*) If the exclusion principle did not apply, how many combinations of states would conceivably be possible?

STRATEGY: Each electron can be in a state with the combination (m_ℓ, m_s) equal to $(-1, -\frac{1}{2})$, $(-1, +\frac{1}{2})$, $(0, -\frac{1}{2})$, $(0, +\frac{1}{2})$, $(+1, -\frac{1}{2})$, or $(+1, +\frac{1}{2})$. This amounts to 6 different states for each electron. Because the exclusion principle is assumed not to apply, when one electron is in any one of these states the other electron can also be in any one of them. Count the possible combinations, remembering that all electrons are identical. You cannot tell which electron is in which single particle state.

SOLUTION:

[ans: 18]

(*b*) How many states does the exclusion principle forbid? Which are they?

STRATEGY: The exclusion principle forbids states for which all 4 quantum numbers for one electron are the same as the corresponding quantum numbers for the other.

SOLUTION:

[ans: 6 states, the ones for which both electrons have the same value of m_ℓ (-1, 0, or $+1$) and the same value of m_s ($-\frac{1}{2}$ or $+\frac{1}{2}$).]

You should be able to distinguish between continuous and characteristic x-ray spectra, be able to compute the wavelength cutoff for a given energy of the incident electrons, and be able to compute the characteristic wavelengths for a given set of energy levels. You should also know how to use the Bohr theory to predict a characteristic spectrum for a given target material and be able to interpret a Moseley plot.

PROBLEM 12 (44E).   Determine Planck's constant from the fact that the minimum x-ray wavelength produced by 40.0-keV electrons is 31.1 pm.

STRATEGY: The minimum x-ray wavelength is produced when all of an electron's kinetic energy is lost to a single photon. Thus $K = hf$, where K is the original kinetic energy of the electron and f is the frequency associated with the photon. Since the photon wavelength λ is related to the frequency by $f\lambda = c$, $K = hc/\lambda$ or $h = K\lambda/c$. Convert the value of the kinetic energy from electron volts to joules.

SOLUTION:

[ans: 6.65×10^{-34} J $\cdot$ s]

PROBLEM 13 (46P).   A 20-keV electron is brought to rest by undergoing two successive bremsstrahlung events, thus transferring its kinetic energy into the energy of two photons. The wavelength of the second photon to be emitted is 130 pm greater than the wavelength of the first photon to be emitted. (*a*) Find the energy of the electron after its first deceleration.

STRATEGY: If K is the initial kinetic energy of the electron, E_1 is the energy of the first photon, and E_2 is the energy of the second photon, then conservation of energy leads to $K = E_1 + E_2$. The energy of the electron after the first emission is E_2 since it is at rest after it loses this energy.

The energy of a photon is related to its wavelength by $E = hc/\lambda$. Thus $E_1 = hc/\lambda_1$ and $E_2 = hc/\lambda_2$. Let $\lambda_1 = \lambda_2 - \Delta\lambda$, where $\Delta\lambda = 130$ pm. Then $1/E_1 = (\lambda_2 - \Delta\lambda)/hc = (1/E_2) - \Delta\lambda/hc = (hc - E_2\,\Delta\lambda)/E_2 hc$ and $E_1 = E_2 hc/(hc - E_2\,\Delta\lambda)$. The conservation of energy equation becomes

$$ K = \frac{E_2 hc}{hc - E_2\,\Delta\lambda} . $$

Solve for E_2. You should get

$$ E_2 = \frac{(K\,\Delta\lambda + 2hc) \pm \sqrt{(K\,\Delta\lambda + 2hc)^2 - 4Khc\,\Delta\lambda}}{2\,\Delta\lambda} . $$

Evaluate this expression. You must use the negative root to obtain an energy that is less than K.

SOLUTION:

$$ \left[\,\text{ans: } 9.2 \times 10^{-16}\,\text{J } (5.7\,\text{keV})\,\right] $$

(*b*) Calculate the wavelengths and energies of the two photons.

STRATEGY: The energy of the first photon is given by $E_1 = K - E_2$, the wavelength of the first photon is given by $\lambda_1 = hc/E_1$, and the wavelength of the second is given by a similar expression.

SOLUTION:

$$ \left[\,\text{ans: } E_1 = 2.3 \times 10^{-15}\,\text{J } (14\,\text{keV}), \; E_2 = 9.2 \times 10^{-16}\,\text{J } (5.7\,\text{keV}), \; \lambda_1 = 87\,\text{pm}, \; \lambda_2 = 220\,\text{pm}\,\right] $$

PROBLEM 14 (49E).   When electrons bombard a molybdenum target, they produce both continuous and characteristic x rays as shown in Fig. 45–15. In that figure the energy of the incident electrons is 35.0 keV. If the accelerating potential applied to the x-ray tube is increased to 50.0 kV, what new values of (*a*) λ_{min}, (*b*) λ for the K_α line, and (*c*) λ for the K_β line result?

STRATEGY: (*a*) A photon with the minimum wavelength is created when an electron loses all its kinetic energy in a single bremsstrahlung event. Conservation of energy then gives $K = hf = hc/\lambda_{min}$, where K is the original kinetic energy of the electron and f is the frequency of the wave associated with the emitted photon. Thus $\lambda_{min} = hc/K$. Convert electron volts to joules.

(*b*) and (*c*) Recall that the wavelengths of the characteristic photons depend on the target material and not on the energy of the bombarding electrons, provided they are energetic enough to knock electrons in low energy states from atoms of the target.

SOLUTION:

[ans: (a) 24.8 pm; (b) 71 pm (no change); (c) 63 pm (no change)]

PROBLEM 15 (57P). The binding energies of K-shell and L-shell electrons in copper are 8.979 and 0.951 keV, respectively. If a K_α x ray from copper is incident on a sodium chloride crystal and gives a first-order Bragg reflection at 74.1° when reflected from the alternating planes of sodium atoms, what is the spacing between these planes?

STRATEGY: If d is the spacing between the planes, then for first-order Bragg reflection $2d \sin \theta = \lambda$, where λ is the wavelength of the x rays and θ is the angle of incidence, measured from the planes. The angle of reflection, measured from the normal to the planes, is $90° - \theta$. The energy E of a K_α photon is equal to the difference in energy of the K and L shells and is related to the wavelength by $E = hc/\lambda$. Calculate the wavelength, then the plane spacing.

SOLUTION:

[ans: 2.82×10^{-10} m]

PROBLEM 16 (58P). (a) Using Bohr's theory, estimate the ratio of energies of photons due to K_α transitions in two atoms whose atomic numbers are Z and Z'.

STRATEGY: For an atom with atomic number Z the frequency of the K_α line is given by

$$f = \frac{3me^4}{32\epsilon_0^2 h^3}(Z - 1)^2 ,$$

where m is the mass of an electron. The photon energy is

$$E = hf = \frac{3me^4}{32\epsilon_0^2 h^2}(Z - 1)^2 .$$

Find an expression for ratio E'/E in terms of Z' and Z.

SOLUTION:

[ans: $E'/E = (Z' - 1)^2/(Z - 1)^2$]

(b) How much more energetic is a K_α x ray from uranium expected to be than one from aluminum? (c) Than one from lithium?

STRATEGY: The atomic number of uranium is 92, the atomic number of aluminum is 13, and the atomic number of lithium is 3. Substitute into the expression you just derived.

SOLUTION:

[ans: (b) 57.5; (c) 2.07×10^3]

Some problems require you to understand the operation of a three-level laser. You may be asked to calculate the wavelength or frequency of the laser light, given the energy levels or you might be asked to calculate the populations of the levels, in thermodynamic equilibrium or after pumping has occurred.

PROBLEM 17 (65E). Lasers have become very small as well as very large. The active volume of a laser constructed of the semiconductor GaAlAs has a volume of only $200\,\mu m^3$ (smaller than a grain of sand) and yet it can continuously deliver $5.0\,mW$ of power at 0.80-μm wavelength. Calculate the rate at which it produces photons.

STRATEGY: The power is the energy of a single photon multiplied by the rate of photon production. If P is the power and R is the photon production rate, then $P = Rhf = Rhc/\lambda$, where f is the frequency and λ is the wavelength. Thus $R = P\lambda/hc$. Evaluate this expression.

SOLUTION:

$$\left[\text{ans: } 2.0 \times 10^{16}\,\text{photons/s}\right]$$

PROBLEM 18 (70P). An atom has two energy levels with a transition wavelength of $580\,nm$. At $300\,K$, 4.0×10^{20} atoms are in the lower state. (a) How many occupy the upper state, under conditions of thermal equilibrium?

STRATEGY: The probability that an atom occupies a state with energy E is proportional to $e^{-E/kT}$, where T is the temperature and k is the Boltzmann constant. If there are n_ℓ atoms in the lower state and n_u in the upper state then $n_u/n_\ell = e^{-E_u/kT}/e^{-E_\ell/kT} = e^{-(E_u - E_\ell)/kT}$. Solve for n_u. The difference in the energies of the states is the energy of the photon emitted when the atom makes the transition from the upper state to the lower: $E_u - E_\ell = hf = hc/\lambda$, where λ is the transition wavelength.

SOLUTION:

$$\left[\text{ans: } 4.9 \times 10^{-16}\,\text{atoms (essentially 0)}\right]$$

(b) Suppose, instead, that 7.0×10^{20} atoms are pumped into the upper state, with 4.0×10^{20} in the lower state. How much energy could be released in a single laser pulse?

STRATEGY: Population inversion continues until the number of atoms in the upper state is the same as the number of atoms in the lower state. Since there are only 2 states this number is half number of atoms, or $(7.0 \times 10^{20} + 4.0 \times 10^{20})/2 = 5.5 \times 10^{20}$. The largest number of atoms that can make the transition during a single burst is $7.0 \times 10^{20} - 5.5 \times 10^{20} = 1.5 \times 10^{20}$. Multiply this by the energy of single photon (hc/λ).

SOLUTION:

$$\left[\text{ans: } 51\,\text{J}\right]$$

PROBLEM 19 (71P).   The beam from an argon laser ($\lambda = 515\,\text{nm}$) has a diameter d of $3.00\,\text{mm}$ and a continuous-wave power output of $5.00\,\text{W}$. The beam is focused onto a diffuse surface by a lens whose focal length f is $3.50\,\text{cm}$. A diffraction pattern such as that of Fig. 41–9 is formed, the radius of the central disk being given by

$$R = \frac{1.22 f \lambda}{d}$$

(see Eq. 41–11). The central disk can be shown to contain 84% of the incident power. Calculate (*a*) the radius R of the central disk, (*b*) the average power flux density in the incident beam, and (*c*) the average flux density in the central disk.

STRATEGY: (*a*) Evaluate the given expression.

(*b*) The average power flux density in the incident beam is given by $P/A = P/\pi r_b^2$, where A is the area of the beam and r_b is its radius.

(*c*) The power entering the central disk is $P_{\text{disk}} = 0.84P$, where P is the power output of the laser. The average power flux density in the central disk is given by $P_{\text{disk}}/A_{\text{disk}} = 0.84P/\pi r_{\text{disk}}^2$, where A_{disk} is the area of the central disk and r_{disk} is its radius.

SOLUTION:

[ans: (*a*) $7.33 \times 10^{-6}\,\text{m}$; (*b*) $7.07 \times 10^{5}\,\text{W/m}^2$; (*c*) $2.49 \times 10^{10}\,\text{W/m}^2$]

III. NOTES

Chapter 46
CONDUCTION OF ELECTRICITY IN SOLIDS

I. BASIC CONCEPTS

Here the ideas of modern physics are used to understand one of the important properties of solids, their electrical conductivity. As you study the chapter be sure to pay attention to distinctions between metals, insulators, and semiconductors. These materials differ greatly in the fraction of their electrons that participate in electrical conduction and in the changes that occur in their conductivities when the temperature changes. You should understand how the differences come about. Later sections of the chapter are devoted to solid-state devices, so pervasive in modern technology.

Resistivity. Much of this chapter is based on Eq. 46–1 for the resistivity. It is

$$\rho =$$

where m is the _____ of an _____ and e is its _____ . n is _____ of _____ electrons per unit volume and τ is the average _____ between _____ .

The quantity n enters the expression for ρ because the more conduction electrons there are the more will pass through a unit area per unit time when an electric field exists in the sample, all else being the same. That is, if the same electric field exists in two samples for which the mean free time is the same, the one with the greater number of conduction electrons per unit volume will have the larger current. τ enters the expression because it gives the average time that the field accelerates an electron before the electron is stopped by a collision. The greater the mean free time, the greater the drift speed. A greater drift speed means more electrons pass through a unit area per unit time, all else being equal.

In the remainder of the chapter pay close attention to how quantum mechanics is used to determine n and τ.

Energy bands and gaps. The allowed values of the electron energy for a crystalline solid form bands of energy levels, separated by gaps. A <u>crystalline</u> <u>solid</u> is one in which the equilibrium positions of the atoms form a pattern that is repeated with uniform spacing and the same orientation throughout the solid. Look at Fig. 46–1 to see two such patterns. The entire crystal can be generated by placing cubes side-by-side and on top of each other with their edges aligned. A <u>band</u> of levels is _____ _____ _____ .

In one view energy bands arise as atoms are brought close together because the wave functions for electrons originally associated with different atoms _____ . In another view an electron originally associated with one atom experiences electrical forces that arise from charges on neighboring atoms.

The diagram on the right shows several groups of closely spaced energy levels, separated by gaps in which there are no levels. Label the bands B and gaps G. In reality, the number of states in a band is huge, about the same as the number of atoms in the crystal. The true separation between levels cannot be shown on the diagram.

When you have finished reading about the distinguishing properties of metals, insulators and semiconductors, mark on the diagram the highest occupied state for electrons in a metal at the absolute zero of temperature, assuming it to be in the third band from the bottom. Do the same for an insulator or semiconductor. Label these levels so they are meaningful to you when you review.

Metals. For a metal the states in one band, called the conduction band, are partially occupied. States in lower energy bands are completely occupied and states in higher energy bands are all unoccupied. Electrons in the conduction band are responsible for the current when an electric field is turned on.

The energies of these electrons are chiefly kinetic energies, the variations in the potential energy function being small over most of the volume of the crystal. To a first approximation, we may assume that the conduction electrons of a metal are simply trapped in a box the size of the sample and take the potential energy to be zero on the inside and to be infinitely high at the surface. The energy levels are given by an expression similar to Eq. 44–10, but modified to take into account the three-dimensional nature of the box. The levels are extremely close together because the box has macroscopic dimensions (several centimeters, for example). You should realize that this model is applicable only to electrons that are not bound to atoms and that are therefore free to move throughout the solid. For most metals this amounts to roughly 1 electron per atom.

When a system such as a solid has a great many closely spaced energy levels a quantity called the <u>density</u> <u>of</u> <u>states</u> and denoted by $n(E)$ is used to describe them. The quantity $n(E)\,dE$ tells us how many quantum mechanical _____ per unit _____ of sample have energies between _____ and _____. For a free electron metal it is given by

$$n(E) =$$

You should know how to use this function. The number of states with energies between E_1 and E_2, for example, is given by the integral

$$N =$$

where V is the volume of the sample. If the energy interval ΔE (= $E_2 - E_1$) is extremely small the integral can be approximated by $N = n(E)\,\Delta E\,V$ as it is in Sample Problem 46–3 of the text.

Not all states are filled. The probability that at absolute temperature T a state with energy E contains an electron is given by the Fermi-Dirac probability function:

$$p(E) =$$

where E_F is a parameter called the <u>Fermi energy</u>. This function takes the Pauli exclusion principle into account: no state may be occupied by more than 1 electron. The density of *occupied* states n_0 (which is the same as the number of electrons per unit volume of sample per unit energy interval) is given by the product of $n(E)$ and $p(E)$:

$$n_0(E) =$$

At the absolute zero of temperature the probability that a state with energy less than _____ is occupied is 1; these states are guaranteed to be filled. The probability that a state with energy greater than _____ is occupied is 0; these states are guaranteed to be empty. The Fermi energy at $T = 0\,\mathrm{K}$ is determined by the condition that the number of occupied states per unit volume between $E = 0$ and $E = E_F$ is the same as the number of conduction electrons per unit volume in the metal. The determining condition is written mathematically as Eq. 46–5 of the text. Copy it here:

$$n =$$

It leads to the relationship between the Fermi energy and the electron concentration n (see Eq. 46–6 of the text):

$$E_F =$$

When the temperature is raised, an extremely small fraction of the electrons, with energies quite near _____, receive energy. This means that some states with energies slightly above the Fermi energy become occupied while some states with energies slightly below the Fermi energy become unoccupied. The energies of the vast majority of electrons do not change.

When there is no electric field in the metal the average electron velocity is 0 because for every electron traveling in any direction another is traveling with the same speed in the opposite direction. When a field exists, slightly more electrons occupy states with momentum in the direction opposite that of the field than occupy states with momentum in the direction of the field. There is then a current. If we examine the electron distribution when a field exists we see that some states, near the Fermi energy and with momentum opposite the field, are occupied while other states, also near the Fermi energy but with momentum in the direction of the field are unoccupied.

If the electrons did not suffer collisions with atoms of the solid the momenta of the electrons would continue to increase in the direction opposite the field. Electrons with energies near the Fermi energy, however, do suffer collisions. Explain why electrons with less energy do not: _____
_____
_____

The speed of an electron with energy near the Fermi energy is very large, as Sample Problem 46–2 of the text shows. It is this speed that determines the _____ and because the additional speed an electron obtains as a result of its interaction with the field is small, τ is essentially independent of the field. This means _____ law is valid.

On the other hand, the average electron velocity (the drift velocity) is small because most electrons travel with the same speed but in the opposite direction to another electron.

You should know that interactions with ions in a perfectly periodic arrangement, as in a crystal, cannot cause an electron to change its state. Electrons have collisions only when the periodicity is destroyed. This might be because there are impurities or other defects present or simply because the atoms are vibrating. Collisions with vibrating atoms are responsible for the temperature dependence of the resistivity of a metal. As the temperature is raised the amplitude of these vibrations _____ and the mean time between collisions _____. A smaller value for τ means a larger value for ρ. The other quantities in the expression for ρ, including the concentration of free electrons n, are comparatively insensitive to the temperature of a metal. Thus the resistivity _____ as the temperature increases.

Insulators and semiconductors. For both insulators, such as carbon in the form of diamond, and semiconductors, such as silicon and germanium, there are precisely enough electrons in the solid to completely fill all the states in an integer number of bands, with none left over. At $T = 0\,\mathrm{K}$ the most energetic electron is in the highest energy state of one of the bands and is separated in energy from the empty state above it by a gap. The highest filled band is called the _____ band and the lowest empty band is called the _____ band.

At higher temperatures electrons are thermally promoted across the gap from the valence to the conduction band. You use the Fermi-Dirac probability function to find out the probability that a state in the conduction band is occupied and the probability that a state in the valence band is unoccupied. The difference between an insulator and a semiconductor is that the gap for _____ is much larger than the gap for _____, so at any given temperature there are far fewer electrons in the conduction band of _____ than in the conduction band of _____.

You should understand that electrons in a completely filled band do not contribute to an electrical current. This is because _____

_____

Electrons in semiconductors and insulators must be promoted across the gap to the conduction band before an electric field will generate a current. For an insulator exceedingly few electrons cross the gap, even at high temperatures. For semiconductors many more electrons cross the gap, although the number is still far less than the number of free electrons in a typical metal. At room temperature the resistivity of a semiconductor is much greater than the resistivity of a metal because _____

_____

_____.

When electrons have been excited to the conduction band of a semiconductor both the conduction and valence bands are partially filled and both contribute to the current when an electric field is turned on. There are two contributions to the conductivity, one from the conduction band and one from the valence band.

Rather than deal with the contributions of the vast number of electrons in the valence band of a semiconductor, the band may be thought to consist of a much smaller number of fictitious particles, called _____, one for each empty state. These particles have positive charge and are accelerated in the direction of an applied electric field.

As the temperature increases the mean free time decreases for both electrons in the conduction band and holes in the valence band but, unlike for a metal, this does not mean the resistivity increases. Both the number of electrons in the conduction band and the number of holes in the valence band increases dramatically with temperature. Because of this the resistivity of a semiconductor _____ as the temperature increases.

The number of electrons in the conduction band or the number of holes in the valence band can be greatly increased by doping a semiconductor. Explain what a doped semiconductor is: _____

_____

_____

To increase the number of electrons in its conduction band a semiconductor is doped with donor replacement atoms. Each such atom normally has _____ electrons in its outer shell. When a donor atom is substituted for a host atom (silicon or germanium, for example) _____ of these electrons form bonds with neighboring atoms. The other electron is in a hydrogen atom-like state around the impurity. Its energy is in the gap between the valence and conduction bands and it is easily promoted to the conduction band. Semiconductors that are doped with donors are said to be _____-type semiconductors.

To increase the number of holes in the valence band a semiconductor is doped with acceptor atoms. Each of these normally has _____ electrons in its outer shell and can easily accept an electron from the valence band, thereby creating a hole in that band. Semiconductors that are doped with acceptors are said to be _____-type semiconductors.

Semiconductor devices. The basic building block of nearly all semiconductor devices is the p-n junction, consisting of p-type and n-type semiconducting materials in contact. Electrons from the _____-type material diffuse to the _____-type material, where they fall into holes. Holes from the _____-type material diffuse to the _____-type material, where electrons combine with them. The electron and hole currents, called _____ currents, are in the same direction, from the _____ side toward the _____.

Electron and hole diffusion leaves a small region near the boundary on the n side with _____ charged atoms, the donors that have lost their electrons, and a small region near the boundary on the p side with _____ charged atoms, the acceptors that have gained electrons (or lost holes). As a result, an electric field exists near the boundary. It points from the _____ side toward the _____ side. The electric potential on the _____ side is higher than the electric potential on the _____ side. This electric field pushes electrons toward the _____ side and pushes holes toward the _____ side. This is the _____ current. If the circuit is not complete the drift and diffusion currents cancel and the net current is _____. The difference in the electric potential of the two sides is called the _____ potential difference. The region in which the electric field exists is called the _____ region because it has few _____ carriers.

p-n junctions are often used as diodes. Explain what a diode is: _____

_____

An *ideal* diode has _____ resistance for current in one direction and _____ resistance for current in the other direction. Diodes are often used to rectify a current. Explain what this means: _____

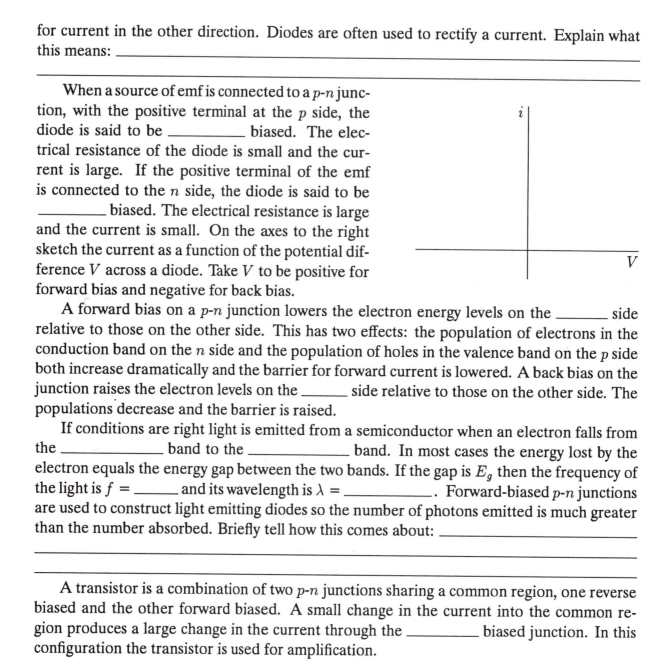

When a source of emf is connected to a p-n junction, with the positive terminal at the p side, the diode is said to be _____ biased. The electrical resistance of the diode is small and the current is large. If the positive terminal of the emf is connected to the n side, the diode is said to be _____ biased. The electrical resistance is large and the current is small. On the axes to the right sketch the current as a function of the potential difference V across a diode. Take V to be positive for forward bias and negative for back bias.

A forward bias on a p-n junction lowers the electron energy levels on the _____ side relative to those on the other side. This has two effects: the population of electrons in the conduction band on the n side and the population of holes in the valence band on the p side both increase dramatically and the barrier for forward current is lowered. A back bias on the junction raises the electron levels on the _____ side relative to those on the other side. The populations decrease and the barrier is raised.

If conditions are right light is emitted from a semiconductor when an electron falls from the _____ band to the _____ band. In most cases the energy lost by the electron equals the energy gap between the two bands. If the gap is E_g then the frequency of the light is $f = $ _____ and its wavelength is $\lambda = $ _____. Forward-biased p-n junctions are used to construct light emitting diodes so the number of photons emitted is much greater than the number absorbed. Briefly tell how this comes about: _____

_____

A transistor is a combination of two p-n junctions sharing a common region, one reverse biased and the other forward biased. A small change in the current into the common region produces a large change in the current through the _____ biased junction. In this configuration the transistor is used for amplification.

II. PROBLEM SOLVING

You should know how to compute the density of states, the occupation probability, and the density of occupied states for the conduction electrons of a metal. You should also know how to compute the Fermi energy. Use

$$n(E) = \frac{8\sqrt{2}\pi m^{2/3}}{h^3} E^{1/2}$$

for the density of states,

$$p(E) = \frac{1}{e^{(E-E_F)/kT} + 1}$$

for the occupation probability,

$$n_0(E) = n(E)p(E)$$

for the density of occupied states, and

$$E_F = \left(\frac{3}{16\sqrt{2}\pi}\right)^{2/3} \frac{h^2}{m} n^{2/3}$$

for the Fermi energy. Some problems may ask you for the Fermi momentum p_F or Fermi velocity v_F, related to the Fermi energy by $E_F = \frac{1}{2}p_F^2/m = \frac{1}{2}mv_F^2$. Here are some examples.

PROBLEM 1 (3P). Calculate the number of particles per cubic meter for (a) the molecules of oxygen gas at $0°$ C and 1.0-atm pressure and (b) the conduction electrons in copper. (c) What is the ratio of these numbers?

STRATEGY: (a) Use the ideal gas law, which is approximately valid for oxygen. The concentration n of oxygen atoms is $n = p/kT$, where p is the pressure and T is the temperature on the Kelvin scale. Covert the pressure to pascals (1 atm = 1.013×10^5 Pa) and temperature to kelvins ($0°$ C = 273 K). (b) The concentration of electrons in copper is given by $n = 16\sqrt{2}\pi m^{3/2}E_F^{3/2}/3h^3$, where m is the mass of an electron and E_F is the Fermi energy of copper (7.00 eV). Convert electron volts to joules (1 eV = 1.60×10^{-19} J).

SOLUTION:

$$\left[\text{ans: } (a)\ 2.69 \times 10^{25} \text{ atoms/m}^3;\ (b)\ 8.38 \times 10^{28} \text{ electrons/m}^3;\ (c)\ 3.1 \times 10^3 \right]$$

(d) What is the average distance between particles in each case? Assume that this distance is the edge of a cube whose volume is equal to the volume per particle (See Sample Problem 28–3.)

STRATEGY: The volume of a cube with edge length d is d^3. Assume there is one particle per cube, so the number of particles per unit volume is $1/d^3$. In each case solve $n = 1/d^3$ for d.

SOLUTION:

$$\left[\text{ans: oxygen atoms: } 3.34 \times 10^{-9} \text{ m, electrons in copper: } 2.3 \times 10^{-10} \text{ m} \right]$$

PROBLEM 2 (4P). The density and molar mass of sodium are 971 kg/m^3 and 23 g/mol, respectively; the radius of the ion Na$^+$ is 98 pm. (a) What fraction of the volume of metallic sodium is available to its conduction electrons?

STRATEGY: A sodium crystal can be constructed of cubic unit cells, with atoms at the corners and centers of the cubes. Each cube has 8 corners and each atom at a corner is shared by the 8 cubes that meet there. Thus there are 2 sodium atoms per cube, one at the corners and one at the center, and the density of sodium is $\rho = 2m/V$, where m is the mass of a sodium atom and V is the volume of a cube. The mass, of course, can be found using $m = M/N_A$, where M is the molar mass (in kg/mol) and N_A is the Avogadro constant. Calculate the cube volume: $V = 2m/\rho$. The volume occupied by 2 sodium ions is $V_{occ} = 8\pi r^3/3$, where r is the radius of an ion. The fraction of the volume that is available for free electrons is $(V - V_{occ})/V$.

SOLUTION:

[ans: 0.900]

(b) Carry out the same calculation for copper. Its density, molar mass, and ionic radius are, respectively, $8960 \, \text{kg/m}^3$, $63.5 \, \text{g/mol}$, and $135 \, \text{pm}$.

STRATEGY: Copper is face-centered cubic, with an atom at each cube corner and an atom at each face center. A cube has 8 corners and each corner atom is shared by eight cubes. A cube has 6 faces and each face-center atom is shared by 2 cubes. Thus there are 4 atoms per cube and the density is given by $\rho = 4m/V$, where m is the mass of a copper atom. Since there are 4 copper atoms per cube the occupied volume is $V_{occ} = 16\pi r^3/3$. The calculation proceeds as before.

SOLUTION:

[ans: 0.125]

(c) For which of these two metals do you think the conduction electrons behave more like a free electron gas?

STRATEGY:

SOLUTION:

[ans: sodium]

PROBLEM 3 (7E). Calculate the density $n(E)$ of conduction electron states in a metal for $E = 8.0 \, \text{eV}$ and show that your result is consistent with the curve of Fig. 46–6a.

STRATEGY: Evaluate

$$n(E) = \frac{8\sqrt{2}\pi m^{3/2}}{h^3} E^{1/2},$$

where m is the mass of an electron. Convert the given value of the energy to joules.

SOLUTION:

[ans: $1.20 \times 10^{47} \, \text{states/J} \cdot \text{m}^3$]

PROBLEM 4 (9E). The Fermi energy of copper is $7.0 \, \text{eV}$. For copper at $1000 \, \text{K}$, (a) find the energy at which the occupancy probability is 0.90.

STRATEGY: The probability of occupancy is given by

$$p(E) = \frac{1}{e^{(E-E_F)/kT} + 1}.$$

Solve for E. You should get

$$E = E_F + kT \ln \frac{1 - p(E)}{p(E)}.$$

Evaluate this expression.

SOLUTION:

$$[\text{ans: } 6.81\,\text{eV} \; (1.09 \times 10^{-18}\,\text{J})]$$

For this energy, evaluate (b) the density of states and (c) the density of occupied states.

STRATEGY: Evaluate

$$n(E) = \frac{8\sqrt{2}\pi m^{3/2}}{h^3} E^{1/2}$$

to obtain the density of states and $n_0(E) = n(E)p(E)$ to obtain the density of occupied states.

SOLUTION:

$$[\text{ans: } (b) \; 1.11 \times 10^{47}\,\text{states/J}\cdot\text{m}^3; \; (c) \; 9.96 \times 10^{46}\,\text{states/J}\cdot\text{m}^3]$$

PROBLEM 5 (12E). Figure 46–7c shows the density of occupied states $n_0(E)$ of the conduction electrons in a metal at 1000 K. Calculate $n_0(E)$ for copper for the energies $E = 4.00, 6.75, 7.00, 7.25,$ and $9.00\,\text{eV}$. The Fermi energy of copper is $7.00\,\text{eV}$.

STRATEGY: Evaluate

$$n_0(E) = n(E)p(E) = \frac{8\sqrt{2}\pi m^{3/2}}{h^3} \frac{E^{1/2}}{e^{(E-E_F)/kT} + 1},$$

where m is the mass of an electron, T is the temperature on the Kelvin scale, and E_F is the Fermi energy.

SOLUTION:

$$[\text{ans: } 8.50 \times 10^{46}\,\text{states/J}\cdot\text{m}^3, \; 1.05 \times 10^{47}\,\text{states/J}\cdot\text{m}^3, \; 5.62 \times 10^{46}\,\text{states/J}\cdot\text{m}^3, \; 5.96 \times 10^{45}\,\text{states/J}\cdot\text{m}^3,$$
$$1.06 \times 10^{37}\,\text{states/J}\cdot\text{m}^3]$$

PROBLEM 6 (16P). Show that the occupancy probabilities of two states whose energies are equally spaced above and below the Fermi energy add up to unity.

STRATEGY: Take the energy of the lower state to be $E_1 = E_F - \Delta E$ and the energy of the upper state to be $E_2 = E_F + \Delta E$. You want to show that

$$f = p(E_1) + p(E_2) = \frac{1}{e^{(E_1 - E_F)/kT} + 1} + \frac{1}{e^{(E_2 - E_F)/kT} + 1}$$

has the value 1. Make the substitutions for the energies to obtain

$$f = \frac{1}{e^{-\Delta E/kT} + 1} + \frac{1}{e^{\Delta E/kT} + 1}.$$

Combine the two terms by using the common denominator $(e^{-\Delta E/kT} + 1)(e^{\Delta E/kT} + 1)$. Multiply the two factors in the denominator, remembering that $e^{-\Delta E/kT}e^{\Delta E/kT} = e^0 = 1$. The numerator and denominator should be the same, so $f = 1$.

SOLUTION:

PROBLEM 7 (18P). Zinc is a bivalent metal. Calculate (a) the number of conduction electrons per cubic meter, (b) the Fermi energy E_F, (c) the Fermi speed v_F, and (d) the de Broglie wavelength corresponding to this speed. See Appendix D for needed data on zinc.

STRATEGY: (a) The number of atoms per unit volume is given by ρ/m, where ρ ($= 7.133\,\text{g/cm}^3$) is the density of zinc and m is the mass of a zinc atom. Use $m = M/N_A$, where M is the molar mass of zinc and N_A is the Avogadro constant, to compute the mass of an atom. Since zinc is bivalent the electron concentration is twice the atomic concentration, so $n = 2\rho/m$. You will need to make some unit conversions. (b) Use $E_F = (3n/16\sqrt{2}\pi)^{2/3}(h^2/m)$ to compute the Fermi energy, relative to the bottom of the conduction band. (c) The Fermi speed v_F is related to the Fermi energy by $E_F = \frac{1}{2}mv_F^2$. Solve for v_F. (d) The de Broglie wavelength is given by $\lambda = h/p_F = h/mv_F$, where p_F is the magnitude of the momentum of an electron with the Fermi energy.

SOLUTION:

$\left[\text{ans: } (a)\ 1.31 \times 10^{29}\ \text{electrons/m}^3;\ (b)\ 1.51 \times 10^{-18}\,\text{J}\ (9.43\,\text{eV});\ (c)\ 1.82 \times 10^6\,\text{m/s};\ (d)\ 4.00 \times 10^{-10}\,\text{m}\right]$

PROBLEM 8 (21P). A neutron star can be analyzed by techniques similar to those used for ordinary metals. In this case the neutrons (rather than electrons) obey the probability function, Eq. 46–7. Consider a neutron star of 2.0 solar masses with a radius of 10 km. Calculate the Fermi energy of the neutrons.

STRATEGY: The Fermi energy is given by

$$E_F = \left(\frac{3}{16\sqrt{2}\pi}\right)^{2/3}\frac{h^2}{m}n^{2/3},$$

where n is the number of neutrons per unit volume. If M is the mass of the star, R is its radius, and m is the mass of a neutron then $n = 3M/4\pi R^3 m$. A solar mass is $1.99 \times 10^{30}\,\text{kg}$ and the mass of a neutron is $1.68 \times 10^{-27}\,\text{kg}$.

SOLUTION:

$\left[\text{ans: } 2.17 \times 10^{-11}\,\text{J}\ (1.35 \times 10^8\,\text{eV})\right]$

Some problems deal with semiconductors. The same probability function is valid and it is important for describing the thermal promotion of electrons across the gap between the valence and conduction bands. The Fermi energy for a pure semiconductor is determined by the condition that the density of occupied states in the conduction band equals the density of unoccupied states in the valence band. Doping with donor or acceptor replacement atoms changes the density of occupied states for the two bands and also changes the Fermi energy. Here are some examples of calculations you should be able to carry out.

PROBLEM 9 (31E).   Pure silicon at room temperature has an electron density in the conduction band of approximately 1×10^{16} m^{-3} and an equal density of holes in the valence band. Suppose that one of every 10^7 silicon atoms is replaced by a phosphorus atom. (*a*) Which type will this doped semiconductor be, *n* or *p*?

STRATEGY: Find phosphorus in the periodic table of the chemical elements and note its column. If the column is greater than 4 a phosphorus atom has more than 4 electrons in its outer shell and is a donor. If the column is less than 4 a phosphorus atom has less than 4 electrons in its outer shell and is an acceptor. In the former case the semiconductor is *n* type while in the latter it is *p* type.

SOLUTION:

[ans: *n* type]

(*b*) What charge carrier density will the phosphorus add? (See Appendix D for needed data on silicon.)

STRATEGY: Each phosphorus atom has 5 electrons in its outer shell so it will add 1 carrier for each phosphorus atom in the solid. First find the number of silicon atoms per unit volume. If ρ (= 2.33 g/cm^3 is the density of silicon and M (= 28.086 g/mol) is the molar mass, then the number of atoms per unit volume is $\rho N_A / M$. Be careful about units here. To find the number of phosphorus atoms per unit volume and the number of additional electrons per unit volume, multiply by 1×10^{-7}.

SOLUTION:

[ans: 5.0×10^{21} electrons/m^3]

(*c*) What is the ratio of the charge carrier density in the doped silicon to that in the pure silicon?

STRATEGY: Since the number of carriers added by the phosphorus is much greater than the number in the pure material, the carrier density for the doped silicon is 5.0×10^{21} carriers/m^3. The carrier density for the pure silicon is $2 \times 1 \times 10^{16}$ carriers/m^3 (counting both electrons and holes).

SOLUTION:

[ans: 2.50×10^5]

PROBLEM 10 (34P). A silicon sample is doped with atoms having a donor state 0.11 eV below the bottom of the conduction band. (*a*) If each of these states is occupied with probability 5.00×10^{-5} at temperature 300 K, where is the Fermi level relative to the top of the valence band? The energy gap in silicon is 1.11 eV.

STRATEGY: Take the zero of energy to be at the top of the valence band. Then the energy of the donor state is $E_d = 1.11 - 0.11 = 1.00$ eV. The probability of occupation is given by

$$p(E) = \frac{1}{e^{(E_d - E_F)/kT} + 1},$$

where E_F is the Fermi energy. Solve for E_F. You should get

$$E_F = E_d - kT \ln\left(\frac{1-p}{p}\right).$$

Use consistent units: either convert the value of E_d to joules or the value of kT to electron volts.

SOLUTION:

$$\left[\text{ans: } 0.744\,\text{eV} \ (1.19 \times 10^{-18}\,\text{J})\right]$$

(*b*) What then is the probability that a state at the bottom of the conduction band is occupied?

STRATEGY: Evaluate

$$p = \frac{1}{e^{(E_c - E_F)/kT} + 1},$$

where E_c (= 1.11 eV) is the energy at the bottom of the conduction band.

SOLUTION:

$$\left[\text{ans: } 7.29 \times 10^{-7}\right]$$

PROBLEM 11 (35P). In a simplified model of an intrinsic semiconductor (no doping), the actual distribution in energy of states is replaced by one in which there are N_v states in the valence band, all of these states having the same energy E_v, and N_c states in the conduction band, all of these states having the same energy E_c. The number of electrons in the conduction band equals the number of holes in the valence band. (*a*) Show that this last condition implies that

$$\frac{N_c}{e^{(E_c - E_F)/kT} + 1} = \frac{N_v}{e^{-(E_v - E_F)/kT} + 1}.$$

(*Hint*: See Problem 17.)

STRATEGY: The number of electrons in the conduction band is the product of the number of states N_c and the probability of occupation $p(E_c)$. The number of holes in the valence band is the product of the number of states and the probability that a state with energy E_v is unoccupied. It is $N_v[1 - p(E_v)]$. A little algebra (see Problem 17) shows that

$$1 - p(E_v) = 1 - \frac{1}{e^{(E_v - E_F)/kT} + 1} = \frac{1}{e^{-(E_v - E_F)/kT} + 1}.$$

Equate the expression for the number of electrons in the conduction band to the expression for the number of holes in the valence band to obtain the desired result.

SOLUTION:

(b) If the Fermi level is in the gap between the two bands and is far from both bands compared to kT, then the exponentials dominate in the denominators. Under these conditions, show that

$$E_F = \frac{1}{2}(E_c + E_v) + \frac{1}{2}kT \ln(N_v/N_c),$$

and therefore that, if $N_v \approx N_c$, the Fermi level is close to the center of the gap.

STRATEGY: If $(E_c - E_F)/kT \gg 1$ and $(E_F - E_v)/kT \gg 1$ then the expression you derived in part a becomes

$$N_c e^{-(E_c - E_F)/kT} = N_v e^{(E_v - E_F)/kT}$$

Solve for E_F. First show that

$$e^{2E_F/kT} = \frac{N_v}{N_c} e^{(E_c + E_v)/kT},$$

then take the natural logarithm of both sides.

SOLUTION:

If $N_v = N_c$ then $\ln(N_v/N_c) = 0$ and $E_F = (E_c + E_v)/2$. This energy is at the center of the gap.

Some problems deal with semiconductor devices. Here is one to help you understand the current characteristics of a p-n junction and another to help you understand photoconduction.

PROBLEM 12 (37P). For an ideal p-n junction diode, with a sharp boundary between the two semiconducting materials, the current i is related to the potential difference V across the diode by

$$i = i_0(e^{eV/kT} - 1),$$

where i_0, which depends on the materials but not on the current or potential difference, is called the *reverse saturation current*. V is positive if the junction is forward-biased and negative if it is back-biased. (a) Verify that this expression predicts the behavior expected of a diode by sketching i as a function of V over the range $-0.12\,\text{V} < V < +0.12\,\text{V}$. Take $T = 300\,\text{K}$ and $i_0 = 5.0\,\text{nA}$.

STRATEGY: Use the axes below.

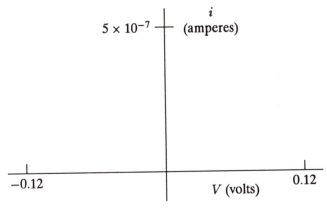

(*b*) For the same temperature, calculate the ratio of the current for a 0.50-V forward-bias to the current for a 0.50-V back-bias.

STRATEGY: For the forward bias substitute $V = 0.50$ V and for the back bias substitute $V = -0.50$ V. The result for the back bias will be negative. Use its magnitude.

SOLUTION:

$$\left[\text{ans: } 2.42 \times 10^8\right]$$

PROBLEM 13 (38E). (*a*) Calculate the maximum wavelength that will produce photoconduction in diamond, which has a band gap of 7.0 eV.

STRATEGY: Assume that the energy needed to promote an electron from the valence band to the conduction band is the difference in energy of the lowest state in the conduction band and the highest state in the valence band. That is, it is $E_g = 7.0$ eV. If λ is the wavelength of the electromagnetic radiation then $E_g = hc/\lambda$, or $\lambda = hc/E_g$.

SOLUTION:

$$\left[\text{ans: } 1.77 \times 10^{-7}\,\text{m}\right]$$

(*b*) In what part of the electromagnetic spectrum does this wavelength lie?

STRATEGY: Locate the wavelength on the chart of the spectrum, Fig. 38–1.

SOLUTION:

$$\left[\text{ans: ultraviolet}\right]$$

III. NOTES

Chapter 47
NUCLEAR PHYSICS

I. BASIC CONCEPTS

This chapter is about the nuclei of atoms. A nucleus takes up only an extremely small fraction of the atomic volume but accounts for most of the mass of an atom. You will learn about the constituents. You will also learn that a nucleus might change into another nucleus by the emission of alpha or beta particles. Pay attention to the energy considerations that allow you to predict whether a nucleus is stable or not and if it is not to find the energy of the decay products. Because nuclear decay is a random process, the rate of decay for a large collection of nuclei can be described in statistical terms. Learn the mathematics of random decay.

The atomic nucleus. Rutherford's experiment determined that the nucleus of an atom contains positively charged particles compressed into an extremely small fraction of the atom's volume. The experiment consisted of firing a beam of _____ particles at a gold foil. The distinguishing feature of the results that led him to his conclusion was _____

_____

You should know some facts about atomic nuclei. They are made up of nucleons, a term that encompasses both _____ and _____. The number of protons is called the _____ number and is denoted by _____, the number of neutrons is called the _____ number and is denoted by _____, and the total number of nucleons is called the _____ number and is denoted by _____. The word <u>nuclide</u> is used to designate a nuclear species. Two nuclides are different if they differ in either their atomic numbers or their neutron numbers. If two nuclides have the same atomic number but differ in their neutron numbers they are said to be _____.

All nucleons attract each other via the _____ force. This force is extremely strong but it also has an extremely short range. Two nucleons must be separated by less than about _____ m to experience the force. As a result, a nucleon interacts with only close neighbors. In addition, protons repel each other via long range electrical forces. As a result, stable nuclei with small mass numbers have the same number of neutrons as protons but stable nuclei with large mass numbers have more _____ than _____. The extra neutrons participate in the _____ interactions, which hold the nucleus together, but not in the _____ interactions, which tend to tear it apart.

The nucleus of an atom contains most of the mass of the atom but takes up only a small fraction of the volume. The surface of a nucleus is somewhat ill defined but an average radius R can be measured. In terms of the mass number A it is given by

$$R = $$

where R_0 is about _____ fm. The radius of a mid-size nucleus is less than the radius of an atom by a factor of about _____. A femtometer (or fermi) is _____ m.

Nuclear masses are commonly measured in <u>atomic mass units</u>, abbreviated _____. 1 u is about _____ kg. The mass of a nucleus in u, rounded to the nearest integer is its _____ number. Because the volume of a nucleus and its mass are both proportional to A the densities of all nuclei are nearly the same, about _____ kg/m^3.

The mass of a stable nucleus is less than the sum of the masses of the constituent nucleons. The difference accounts for the binding together of the nucleons to form the nucleus. If Δm is the mass difference then the binding energy is given by $E_b = $ _____, where c is the speed of light. This is the energy that must be given a nucleus to separate it into its constituent particles, well-separated and at rest. In detail, if m is the mass of a nucleus with Z protons and N neutrons then

$$E_b =$$

where m_p is the mass of a proton and m_n is the mass of a neutron. You should be aware that atomic, not nuclear, masses are usually tabulated. The mass of an appropriate number of electrons must be subtracted from the atomic mass to obtain the nuclear mass or else the atomic mass of a nucleus with Z protons and N neutrons must be compared to the sum of the masses of Z hydrogen atoms and N neutrons.

Fig. 47–6 of the text shows the binding energy per nucleon as a function of mass number. The shape of this curve is important. Nuclides in the vicinity of iron have the greatest binding energy per nucleon and the curve drops as A becomes either larger and smaller. Energy is released when two nuclei with low mass numbers combine to form a single nucleus. This process is called nuclear _____. Energy is also released if a nucleus with high mass number breaks into smaller fragments. This process is called _____.

An internal energy is associated with a nucleus and this energy has a discrete set of allowed values, a different set for each type nucleus. The difference in energy of adjacent low-lying states is on the order of _____ eV and when a nucleus changes state from a higher to a lower energy the photon emitted is in the _____ portion of the electromagnetic spectrum.

Nuclei also have intrinsic angular momenta and magnetic dipole moments. Observed values of the angular momentum are roughly the same as for electrons in atoms but the magnetic dipole moments are smaller by a factor of about _____ because _____

Radioactive decay.   Most conceivable nuclei are not stable. An unstable nucleus may spontaneously turn into another nucleus with the emission of one or more particles. Alpha decay, for example, involves the emission of a _____ nucleus and beta decay involves the emission of either an _____ or a _____. Spontaneous decay occurs only if the mass of the nucleus is _____ than the sum of the masses of the decay products.

Although any nucleus in a collection of identical unstable nuclei might decay in any given time interval, it is impossible to predict which will actually decay. This means that the number that decay in any small time interval Δt is proportional to the number N of undecayed nuclei present and to the interval itself: $\Delta N = -\lambda N \Delta t$, where λ is a constant of proportionality, called the _____ constant. The negative sign appears because ΔN is negative

(N decreases). In the limit as $\Delta t \rightarrow 0$, this equation becomes

$$\frac{dN}{dt} =$$

It has the solution

$$N(t) =$$

where N_0 is the number of undecayed nuclei at time $t = 0$.
   The decay rate R is defined as $R =$ _____ . It also follows an exponential law:

$$R =$$

where R_0 is the decay rate at $t = 0$. Notice that the exponent is the same as the exponent in the law for N.
   A radioactive decay is often characterized by its <u>half-life</u>. Define this term: _____
_____
_____

The half-life τ is related to the disintegration constant λ by

$$\tau =$$

The decay rate R and the number N of undecayed nuclei both decrease by a factor of _____ in every half-life interval. Study Sample Problems 4 and 5 of the text to see how the half-life is calculated from experimental data when it is long and when it is short compared to the observation time.
   The reduction in mass that occurs with a decay, multiplied by c^2, gives the _____ energy and is denoted by _____ . This energy appears as the kinetic energy of the decay products and as the excitation energy of the daughter nucleus, if it is left in an excited state. A nucleus in an excited state decays to its ground state with the emission of one or more photons. The excitation energies involved usually lead to photons in the _____ portion of the electromagnetic spectrum.
   When a nucleus decays by emission of an alpha particle the daughter nucleus has an atomic number that is _____ less than that of parent nucleus, a neutron number that is _____ less and a mass number that is _____ less.
   The half-lives of alpha emitters range from less than a second to times that are longer than the age of universe. The combination of the strong nuclear force of attraction and the electrical repulsion of the daughter nucleus for the alpha particle leads to a high potential energy barrier that tends to hold the alpha inside. Classically it can never escape because its energy is less than the barrier height, but alphas are emitted because quantum mechanical _____ is possible. The half-life depends sensitively on the _____ and _____ of the barrier. The disintegration energy is shared by the alpha and the recoiling daughter nucleus.
   In β^- decay a _____ is converted to a _____ with the emission of an _____ and a _____ . The daughter nucleus has an atomic number that is _____ greater than that

of parent nucleus, a neutron number that is _____ less, and a mass number that is the same. In β^+ decay a _____ is converted to a _____ with the emission of a _____ and a _____. The daughter nucleus has an atomic number that is _____ less than that of parent nucleus, a neutron number that is _____ greater, and a mass number that is the same. Neutron-rich nuclides tend to decay by _____ emission while proton-rich nuclides tend to decay by _____ emission.

The disintegration energy is shared by the β particle, neutrino, and recoiling daughter nucleus, with nearly all of it going the first two. For a collection of identical parent nuclei the energies of the emitted β particles span a wide range but the sum of the β and neutrino energies is always the same. The β particle and neutrino are created in the decay process. They do not exist inside the nucleon before decay.

Neutrinos are rather ephemeral particles. The mass of a neutrino is too small to be measured and may be 0. It interacts only extremely weakly with matter. Enormous numbers, most from the sun but many from other stars, pass through the earth and our bodies every second. They are extremely difficult to detect.

Four different units are used to measure various aspects of radiation from radioactive materials. For each of them tell what quantity it is used to measure and give the value of the unit in more common terms:

curie (abbreviated _____) 1 curie = _____ disintegrations/s
Used to measure: _____

_____.

roentgen (abbreviated _____) 1 roentgen = _____ ion pairs per gram of air
Used to measure: _____

_____.

rad 1 rad = _____ mJ/kg
Used to measure: _____

_____.

rem
Used to measure: _____

_____.

Nuclear models.   Three models are described in the text. Briefly explain how the nucleus is viewed in each.

The liquid drop model: _____

_____

_____

The independent particle model: _____

_____

_____

The collective model: _____

_____

_____

II. PROBLEM SOLVING

Some problems examine details of the Rutherford experiment. Here is one that asks you to find the kinetic energies of the target nucleus and the α particle.

PROBLEM 1 (3P).   When an α particle collides elastically with a nucleus, the nucleus recoils. Suppose a 5.00-MeV α particle has a head-on elastic collision with a gold nucleus, initially at rest. What is the kinetic energy (a) of the recoiling nucleus and (b) of the rebounding α particle?

STRATEGY: (a) Let $v_{\alpha,i}$ be the initial velocity of the α, $v_{\alpha,f}$ be the final velocity of the α, $v_{n,f}$ be the final velocity of the recoiling nucleus, M_α be the mass of the α, and M_n be the mass of the nucleus. Momentum is conserved, so $M_\alpha v_{\alpha,i} = M_\alpha v_{\alpha,f} + M_n v_{n,f}$. Energy is conserved, so $\frac{1}{2}M_\alpha v_{\alpha,i}^2 = \frac{1}{2}M_\alpha v_{\alpha,f}^2 + \frac{1}{2}M_n v_{n,f}^2$. Solve these two equations for $v_{n,f}$ in terms of $v_{\alpha,i}$. First use the momentum equation to find an expression for $v_{\alpha,f}$ and substitute it into the energy equation. Your result for $v_{n,f}$ should be

$$v_{n,f} = \frac{2v_{\alpha,i}}{1 + \frac{M_n}{M_\alpha}}.$$

The kinetic energy of the recoiling nucleus is $K_{n,f} = \frac{1}{2}M_n v_{n,f}^2$. Substitute for $v_{n,f}$ and show that

$$K_{n,f} = \frac{4M_n}{m(1 + \frac{M_n}{M_\alpha})^2}K_{\alpha,i},$$

where $K_{\alpha,i}$ is the initial kinetic energy of the α.

Evaluate this expression. The mass of a gold nucleus is 197 u and the mass of an α particle is 4.00 u. Notice that only the ratio of the masses appears, so any units may be used as long as they are the same. Also notice that the units of $K_{n,f}$ will be the same as those of $K_{\alpha,i}$.

(b) The final kinetic energy of the α is given by $K_{\alpha,f} = K_{\alpha,i} - K_{n,f}$.

SOLUTION:

[ans: (a) 0.39 MeV; (b) 4.61 MeV]

Here are some problems that test your understanding of nuclear properties. You should know the meaning of the terms mass number, atomic number, and neutron number and the relationship between them. You should also know the relationship between the radius of a nucleus and its mass number and how to calculate the binding energy of a nucleus, given its mass and the masses of its constituent nucleons.

PROBLEM 2 (5E).   The nuclide ^{14}C contains how many (*a*) protons and (*b*) neutrons?

STRATEGY: The atomic number Z of carbon can be found in Appendix D. This is the number of protons. The mass number is $A = 14$ and the number of neutrons is $N = A - Z$.

SOLUTION:

[ans: (*a*) 6; (*b*) 8]

PROBLEM 3 (6E).   The radius of a nucleus is measured, by electron-scattering methods, to be 3.6 fm. What is the likely mass number of the nucleus?

STRATEGY: The radius is approximately $R = R_0 A^{1/3}$, where $R_0 = 1.2$ fm and A is the mass number. Solve for A. First cube both sides of the equation.

SOLUTION:

[ans: 27]

PROBLEM 4 (12E).   Calculate and compare (*a*) the nuclear mass density ρ_m and (*b*) the nuclear charge density ρ_q for the fairly light nuclide ^{55}Mn and for the fairly heavy nuclide ^{209}Bi.

STRATEGY: The nuclear mass density is given by $\rho_m = 3M/4\pi R^3 = 3M/4\pi R_0^3 A$, where M is the mass, R is the radius, and A is the mass number. The expression $R_0 A^{1/3}$ was substituted for R. Take M in grams to be A/N_A, where N_A is the Avogadro constant. The charge density is given by $\rho_q = 3Q/4\pi R^3 = 3Ze/4\pi R_0^3 A$, where Ze was substituted for the total charge Q and $R_0 A^{1/3}$ was substituted for R.

SOLUTION:

[ans: (*a*) ^{55}Mn: 2.3×10^{17} kg/m^3, ^{209}Bi: 2.3×10^{17} kg/m^3; (*b*) ^{55}Mn: 1.0×10^{25} C/m^3, ^{209}Bi: 8.8×10^{24} C/m^3]
(*c*) Are the differences what you would expect?

STRATEGY: Since the both the mass and the volume are proportional to the mass number A the mass density should be nearly the same for all nuclides. What about the charge density? The atomic number Z is somewhat less than the value predicted by a linear relationship between Z and A.

SOLUTION:

PROBLEM 5 (21P).   You are asked to pick apart an α particle (^{4}He) by removing, in sequence, a proton, a neutron, and a proton. Calculate (*a*) the work required for each step, (*b*) the total binding energy of the α particle, and (*c*) the binding energy per nucleon. Some needed atomic masses are

^{4}He	4.00260 u	^{2}H	2.01410 u
^{3}H	3.01605 u	^{1}H	1.00783 u
	n	1.00876 u	

STRATEGY: (a) Find the change in total mass when a proton is removed. This the same as the sum of the masses of an ^{3}H atom and a ^{1}H atom minus the mass of an ^{4}He atom. Notice that the masses of two electrons cancel in the subtraction process. The work you must do is the mass difference multiplied by the square of the speed of light. Don't forget to convert mass units to SI units, or easier still, use the energy equivalent of an atomic mass unit: 1 u is equivalent to 932 MeV. Now calculate the change in mass and work required when a neutron is removed from a ^{3}H nucleus. The resulting particles are a ^{2}H nucleus and a neutron. Finally calculate the change in mass and work required when the neutron and proton in the ^{2}H nucleus are separated.

(b) The total binding energy of the α equals the total work required to separate the α into its constituent nucleons and so is the sum of the three works computed in part (a).

(c) The binding energy per nucleon is the result of part (b) divided by 4, the number of nucleons.

SOLUTION:

$$[\,\text{ans: } (a)\ 19.8\,\text{MeV},\ 6.26\,\text{MeV},\ 2.22\,\text{MeV};\ (b)\ 28.3.\,\text{MeV};\ (c)\ 7.07\,\text{MeV}\,]$$

PROBLEM 6 (22P). Because the neutron has no charge, its mass must be found in some way other than by using a mass spectrometer. When a neutron and a proton meet (assume both are almost stationary), they combine and form a deuteron, emitting a gamma ray whose energy is 2.2233 MeV. The masses of the proton and the deuteron are 1.007825035 u and 2.0141019 u, respectively. Find the mass of the neutron from these data, to as many significant figures as the data warrant. (A more precise value of the mass-energy conversion factor than the one presented in the text is 931.502 MeV/u.

STRATEGY: Conservation of energy yields $M_n c^2 + M_p c^2 = M_d c^2 + E_\gamma$, where M_n is the mass of a neutron, M_p is the mass of a proton, M_d is the mass of a deuteron, and E_γ is the energy of the gamma ray. Solve for M_n. Values of the particle rest energies in MeV can be found by multiplying the particle mass in u by the mass-energy conversion factor in MeV/u. Once you have computed $M_n c^2$ in MeV, divide by the conversion factor to find the mass in u.

SOLUTION:

$$[\,\text{ans: } 1.0087\,\text{u}\,]$$

Here are some problems that deal with the number and rate of decays. You should know the relationship between the disintegration constant λ and the half-life τ. The half-life is often given but you need the disintegration constant to carry out the calculation. In an original collection of N_0 nuclei the number that remain undecayed at the end of time t is given by $N = N_0 e^{-\lambda t}$. The rate of decay is given by $R = \lambda N$ and by $R = R_0 e^{-\lambda t}$, where R_0 is the decay rate at $t = 0$.

PROBLEM 7 (32E).   The plutonium isotope ^{239}Pu is produced as a by-product in nuclear reactors and hence is accumulating in our environment. It is radioactive, decaying by alpha decay with a half-life of 2.41×10^4 y. But plutonium is also one of the most toxic chemicals known; as little as 2 mg is lethal to a human. (*a*) How many nuclei constitute a chemically lethal dose?

STRATEGY: The number of nuclei is the mass of a lethal dose divided by the mass of a plutonium nucleus. The latter may be taken to be 239 u. Convert to grams: $1\,u = 1.661 \times 10^{-24}$ g.

SOLUTION:

$$\left[\text{ans: } 5.04 \times 10^{18} \right]$$

(*b*) What is the decay rate of this amount?

STRATEGY: Use $R = \lambda N$, where λ is the disintegration constant and N is the number of plutonium nuclei. Use the relationship $\tau = \ln 2/\lambda$ between the half-life τ and the disintegration constant to compute λ.

SOLUTION:

$$\left[\text{ans: } 4.59 \times 10^6 \, \text{s}^{-1} \right]$$

If you were handling that quantity would you fear being poisoned or suffering radiation sickness?

STRATEGY: The activity is $4.59 \times 10^6 / 3.7 \times 10^{10} = 1.2 \times 10^{-4}$ Ci.

SOLUTION:

PROBLEM 8 (36P).   The radionuclide ^{32}P ($\tau = 14.28$ d) is often used as a tracer to follow the course of biochemical reactions involving phosphorus. (*a*) If the counting rate in a particular experimental setup is initially 3050 counts/s, after what time will it fall to 170 counts/s?

STRATEGY: The decay rate is given by $R = R_0 e^{-\lambda t}$, where R_0 is the rate at $t = 0$ and λ is the disintegration constant, related to the half-life τ by $\tau = \ln 2/\lambda$. Solve for t. You should get $t = -(1/\lambda) \ln(R/R_0)$. If τ is in days, then λ will be in reciprocal days and t will be in days. You do not need to convert units.

SOLUTION:

$$\left[\text{ans: } 59.5 \, \text{d} \right]$$

(*b*) A solution containing ^{32}P is fed to the root system of an experimental tomato plant and the ^{32}P activity in a leaf is measured 3.48 days later. By what factor must this reading be multiplied to correct for the decay that has occurred since the experiment began?

STRATEGY: The experimenters want to determine the amount of ^{32}P that gets to the leaf. Since some decays over the time interval from the beginning of the experiment to the counting, they must multiply by $f = R_0/R$, where R is the final counting rate and R_0 is the initial counting rate. Since $R = R_0 e^{-\lambda t}$, this is given by $f = e^{\lambda t}$, where t is the time interval. Use $\tau = \ln 2/\lambda$ to calculate the disintegration constant.

SOLUTION:

$$[\text{ans: } 1.18]$$

PROBLEM 9 (37P). A source contains two phosphorus radionuclides, ^{32}P ($\tau = 14.3$ d) and ^{33}P ($\tau = 25.3$ d). Initially 10.0% of the decays come from ^{33}P. How long must one wait until 90.0% do so?

STRATEGY: Let R_{32} be the activity of ^{32}P and $R_{32,0}$ be its initial value. Let R_{33} be the activity of ^{33}P and $R_{33,0}$ be its initial value. Initially $R_{33,0} = 0.100(R_{32,0} + R_{33,0})$, so $R_{33,0}/R_{32,0} = 0.111$; at the end of time t, $R_{33} = 0.900(R_{32} + R_{33})$, so $R_{33}/R_{32} = 9.00$. You want to find t. Since $R_{32} = R_{32,0} e^{-\lambda_{32} t}$ and $R_{33} = R_{33,0} e^{-\lambda_{33} t}$, where λ_{32} and λ_{33} are the disintegration constants,

$$\frac{R_{33}}{R_{32}} = \frac{R_{33,0}}{R_{32,0}} e^{(\lambda_{32} - \lambda_{33})t} \, .$$

The solution is

$$t = \frac{1}{\lambda_{32} - \lambda_{33}} \ln \left(\frac{R_{32,0}}{R_{33,0}} \frac{R_{33}}{R_{32}} \right) \, .$$

Each disintegration constant can be found using $\tau = \ln 2/\lambda$, where τ is the half-life.

SOLUTION:

$$[\text{ans: } 209\,\text{d}]$$

Some problems deal with the energetics of α and β decays. You should know that a helium nucleus (2 protons and 2 neutrons) is emitted in an α decay and that either an electron or a positron is emitted in a β decay. In each case you should be able to find the resulting heavy nucleus if the original nucleus is given. You should also be able to calculate the energy Q released in a decay if the various masses are given. Recall that Q is the energy equivalent of the reduction in mass that occurs in a decay. If masses are given in atomic mass units the conversion factor 931.5 MeV/u can be used. Remember that the masses given are often the masses of atoms, not nuclei, and you must, in principle, subtract an appropriate number of electron masses. These may cancel in the equation you are evaluating but you should check in each case. You should also know how to compute the height of the Coulomb barrier to α decay. See the example below.

PROBLEM 10 (50P).   Under certain circumstances, a nucleus can decay by emitting a particle heavier than an α particle. Such decays are very rare and have only recently been observed. Consider the decays

$$^{223}\text{Ra} \rightarrow {}^{209}\text{Pb} + {}^{14}\text{C}$$

and

$$^{223}\text{Ra} \rightarrow {}^{219}\text{Rn} + {}^{4}\text{He}.$$

(a) Calculate the Q values for these decays and determine that both are energetically possible. (b) The Coulomb barrier height for α particles in this decay is 30.0 MeV. What is the barrier height for ^{14}C decay? The needed atomic masses are:

^{223}Ra	223.01850 u	^{14}C	14.00324 u
^{209}Pb	208.98107 u	^{4}He	4.00260 u
^{219}Rn	219.01008 u		

STRATEGY: (a) Let M_{223}, M_{209}, and M_{14} be the masses of the nuclei involved in the first decay. The Q value is given by $Q = (M_{209} + M_{14} - M_{223})c^2$. Calculate the mass difference $M_{209} + M_{14} - M_{223}$ in atomic mass units, then multiply by the energy conversion factor 931 MeV/u to obtain Q in MeV. For the second decay $Q = (M_{219} + M_4 - M_{223})c^2$. Carry out the calculation in the same manner.

(b) The highest point of the Coulomb barrier occurs when the two product nuclei are just touching. The distance between the centers of the nuclei is then the sum of their radii. The electrostatic potential energy is proportional to the product of the charges on the nuclei and inversely proportional to the separation of their centers. For the first decay the barrier height is $U_1 = (1/4\pi\epsilon_0)Z_{209}Z_{14}e^2/(R_{209} + R_{14})$ and for the second it is $U_2 = (1/4\pi\epsilon_0)Z_{219}Z_4e^2/(R_{219} + R_4)$, where Z_{209}, Z_{14}, Z_{219}, and Z_4 are atomic numbers and R_{209}, R_{14}, R_{219}, and R_4 are nuclear radii. The ratio of the barrier heights is $U_1/U_2 = Z_{209}Z_{14}(R_{219} + R_4)/Z_{219}Z_4(R_{209} + R_{14})$. Find the values of the atomic numbers in the appendix and use $R = R_0A^{1/3}$ to find the nuclear radii.

SOLUTION:

$$\left[\text{ans: } (a)\ 31.8\,\text{MeV},\ 5.42\,\text{MeV};\ (b)\ 79\,\text{MeV}\right]$$

PROBLEM 11 (57P).   Some radionuclides decay by capturing one of their own atomic electrons, a K-shell electron, say. An example is

$$^{49}\text{V} + e^- \rightarrow {}^{49}\text{Ti} + \nu, \qquad \tau = 331\,\text{d}.$$

Show that the disintegration energy Q for this process is given by

$$Q = (m_\text{V} - m_\text{Ti})c^2 - E_K,$$

where m_V and m_Ti are the atomic masses of ^{49}V and ^{49}Ti, respectively, and E_K is the binding energy of the vanadium K-electron.

STRATEGY: If m_e is the mass of an electron, then the total mass initially is $m_\text{V} - 23m_e + m_e$. The atomic number of vanadium is 23, so 23 electron masses are subtracted from the mass of a vanadium atom to obtain the mass of a vanadium nucleus. In addition, the K-shell electron that is captured initially has an energy of $-E_K$, where E_K is its binding energy. Similarly the final mass is $M_\text{Ti} - 22m_e$. Calculate Q as the initial energy minus the final energy.

SOLUTION:

PROBLEM 12 (59P).   The radionuclide ^{11}C decays according to

$$^{11}\text{C} \rightarrow {}^{11}\text{B} + e^+ + \nu, \qquad \tau = 20.3\,\text{min}.$$

The maximum energy of the positron spectrum is 0.960 MeV. (*a*) Show that the disintegration energy Q for this process is given by

$$Q = (m_C - m_B - 2m_e)c^2,$$

where m_C and m_B are the atomic masses of ^{11}C and ^{11}B, respectively, and m_e is the mass of a positron and an electron.

STRATEGY: Since the atomic number of carbon is 6 and the atomic number of boron is 5, the initial total mass is $m_C - 6m_2$ and the final total mass is $m_B - 5m_e + m_e$. Multiply the mass difference by c^2.

SOLUTION:

(*b*) Given that $m_C = 11.011434\,\text{u}$, $m_B = 11.009305\,\text{u}$, and $m_e = 0.0005486\,\text{u}$, calculate Q and compare it with the maximum energy of the positron spectrum, given above.

STRATEGY: Multiply the mass difference in u by the energy conversion factor 931.5 MeV/u.

SOLUTION:

$$[\,\text{ans: } 0.961\,\text{MeV, very nearly the same}\,]$$

Here are some problems that deal with the use of radioactivity to date ancient artifacts or geological samples.

PROBLEM 13 (63E).   A 5.00-g charcoal sample from an ancient fire pit has a ^{14}C activity of 63.0 disintegrations/min. A living tree has a ^{14}C activity of 15.3 disintegrations/min per 1.00 g. The half-life of ^{14}C is 5730 y. How old is the charcoal sample?

STRATEGY: Activity is another word for decay rate. As a function of time it is given by $R = R_0 e^{-\lambda t}$, where R_0 is the activity at $t = 0$ and λ is the disintegration constant. Solve for t. If the carbon content of the atmosphere did not change significantly since the charcoal was produced the initial activity (at the time the wood was burned) was $R_0 = 15.3 \times 5.00 = 76.5$ disintegrations/min. Use $\tau = \ln 2/\lambda$ to find the value of the disintegration constant.

SOLUTION:

$$[\,\text{ans: } 1600\,\text{y}\,]$$

PROBLEM 14 (65P).   A rock, recovered from far underground, is found to contain 0.86 mg of ^{238}U, 0.15 mg of ^{206}Pb, and 1.6 mg of ^{40}Ar. How much ^{40}K will it likely contain? Needed half-lives are listed in Problem 60.

STRATEGY: ^{238}U decays to ^{206}Pb. If N_U is the number of ^{238}U nuclei present now and $N_{U,0}$ is the number originally present, then $N_U = N_{U,0} e^{-\lambda t}$, where λ is the disintegration constant for the decay. Solve for t.

To calculate a numerical value for t you must obtain numerical values for N_U and $N_{U,0}$. If M_U is the mass of ^{238}U present now and m_U is the mass of a ^{238}U nucleus, then $N_U = M_U/m_U$. You may use $m_U = 238\,\text{u}$ and convert to kilograms. Since each ^{238}U nucleus that decays produces a ^{206}Pb nucleus the number of ^{238}U nuclei originally present equals the number now present plus the number of ^{206}Pb nuclei now present. Thus

$N_{U,0} = N_U + N_{Pb}$, where N_{Pb} is the number of ^{206}Pb nuclei now present. This quantity can be found using $N_{Pb} = M_{Pb}/m_{Pb}$, where M_{Pb} is the mass of ^{206}Pb present now and m_{Pb} is the mass of a ^{206}Pb nucleus (206 u).

You must also obtain a numerical value for λ. Use $\tau = \ln 2/\lambda$, where τ is the half-life for the decay (4.47×10^9 y).

^{40}K decays to ^{40}Ar. Let N_K be the number of ^{40}K nuclei present now and $N_{K,0}$ be the number present originally. Then $N_K = N_{K,0}e^{-\lambda t}$, where λ is the disintegration constant for this decay. To find its value use $\tau = \ln 2/\lambda$, where τ is the half-life for this decay (1.25×10^9 y).

Now $N_{K,0} = N_K + N_{Ar}$, where N_{Ar} is the number of ^{40}Ar nuclei present now. Thus $N_K = (N_K + N_{Ar})e^{-\lambda t}$. Solve for N_K. Use $N_{Ar} = M_{Ar}/m_{Ar}$, where M_{Ar} is the mass of argon now present and m_{Ar} is the mass of an ^{40}Ar nucleus (40 u).

Once N_K is known you can find the mass of ^{40}K present by multiplying by the mass of a ^{40}K nucleus (40 u).

SOLUTION:

[ans: 2.6×10^{19} nuclei (1.7 mg)]

Some problems deal with radiation dosages. You should be able to distinguish between a curie, a rem, and a rad. Here's a problem designed to help you.

PROBLEM 15 (71P). An 85-kg worker at a breeder reactor plant accidentally ingests 2.5 mg of plutonium ^{239}Pu dust. ^{239}Pu has a half-life of 24,100 y, decaying by alpha decay. The energy of the emitted α particles is 5.2 MeV, with an RBE factor of 13. Assume that the plutonium resides in the worker's body for 12 h, and that 95% of the emitted α particles are stopped within the body. Calculate (a) the number of plutonium atoms ingested, (b) the number that decay during the 12 h, (c) the energy absorbed by the body, (d) the resulting physical dose in rad, and (e) the equivalent biological dose in rem.

STRATEGY: (a) The number ingested is given by $N_{in} = M/m$, where M is the mass of the ingested plutonium and m is the mass of a plutonium atom. Take m to be 239 u and convert to kilograms.

(b) The half-life is very long compared to 12 h so you may approximate the number that decay by $N_{decay} = \lambda N_{in} \Delta t$, where Δt is the time interval. This follows directly from $dN/dt = -\lambda N$, with dN/dt assumed essentially constant over the 12-h interval. Use $\tau = \ln 2/\lambda$, where τ is the half-life, to find the value of λ.

(c) The energy absorbed by the body is $E = 0.95 N_{decay} E_\alpha$, where E_α is the energy of a single α particle.

(d) A rad is 10 mJ/kg. Convert E to units of 10 mJ and divide by the mass of the worker.

(e) Multiply by the RBE factor.

SOLUTION:

[ans: (a) 6.3×10^{18}; (b) 2.5×10^{11}; (c) 0.20 J; (d) 0.23 rad; (e) 3.0 rem]

Some nuclear reactions occur, not spontaneously like decays, but only after an incoming nucleus brings energy to a target nucleus. The total mass of the products is greater than the total original mass, the initial kinetic energy of the incoming nucleus or the excitation energy of the compound nucleus providing the additional energy. Another set of problems deals with the nuclear models discussed in the text.

PROBLEM 16 (76P).   Consider the three decay processes shown for the compound nucleus $^{20}\text{Ne}^*$ in Fig. 47–14. If the compound nucleus is initially at rest and has an excitation energy of 25.0 MeV, what kinetic energy, measured in the laboratory, will (a) the deuteron, (b) the neutron, and (c) the ^3He nuclide have when the nucleus decays? Some needed atomic and particle masses are:

^{20}Ne	19.99244 u	d	2.01410 u
^{19}Ne	19.00188 u	n	1.00867 u
^{18}F	18.00094 u	^3He	3.01603 u
^{17}O	16.99913 u		

STRATEGY: (a) The first decay is $^{20}\text{Ne}^* \rightarrow {}^{18}\text{F} + d$. If M_{Ne} is the mass of the neon nucleus, M_F is the mass of the fluorine nucleus, M_d is the mass of the deuteron, v_F is the velocity of the fluorine nucleus, v_d is the velocity of the deuteron, and E is the excitation energy of the neon nucleus, then conservation of energy leads to $E + M_{\text{Ne}}c^2 = (M_F + M_d)c^2 + \frac{1}{2}M_F v_F^2 + \frac{1}{2}M_d v_d^2$ and conservation of momentum leads to $M_F v_F + M_d v_d$.

Solve the momentum equation for v_F in terms of v_d: $v_F = (M_d/M_F)v_d$. Substitute this result into the energy equation to obtain

$$E + M_{\text{Ne}} = (M_F + M_d)c^2 = \frac{1}{2}\frac{M_d}{M_F}(M_d + M_F)v_F^2 .$$

Notice that the right side contains the factor $\frac{1}{2}M_d v_d^2$, which is the kinetic energy of the deuteron. Replace it with K_d and solve for K_d. The result is

$$K_d = \left[(M_{\text{Ne}} - M_F - M_d)c^2 + E\right]\frac{M_F}{M_F + M_d} .$$

Evaluate this expression. Use the conversion factor 931.5 MeV/u to find the energy equivalents of the masses in the brackets. The masses that appear in the ratio at the right side of the equation may be left in atomic mass units.

(b) Carry out a similar analysis for the decay $^{20}\text{Ne}^* \rightarrow {}^{19}\text{Ne} + n$.

(c) Carry out a similar analysis for the decay $^{20}\text{Ne}^* \rightarrow {}^{17}\text{O} + {}^3\text{He}$.

SOLUTION:

[ans: (a) 3.55 MeV; (b) 7.72 MeV; (c) 3.26 MeV]

PROBLEM 17 (78P). The nucleus ^{91}Zr ($Z = 40$, $N = 51$) has a single neutron outside a filled 50-neutron core. Because 50 is a magic number, this neutron should perhaps be especially loosely bound. (*a*) What is its binding energy? (*b*) What is the binding energy of the next neutron, which would have to be extracted from the filled core? (*c*) What is the binding energy per nucleon for the entire nucleus? Compare these three numbers and discuss. Some needed atomic masses are:

^{91}Zr	90.90564 u		n	1.00867 u
^{90}Zr	89.90471 u		p	1.00783 u
		^{89}Zr	88.90890 u	

STRATEGY: (*a*) The binding energy E_{b1} of the most loosely bound neutron is the difference in energy of the original ^{91}Zr nucleus and the system consisting of a ^{90}Zr nucleus and a neutron. In terms of the masses $E_{b1} = (M_{Zr90} + M_n - M_{Zr91})c^2$. Multiply the mass difference in atomic mass units by the conversion factor 931.5 MeV/u. (*b*) The binding energy of the next neutron is given by $E_{b2} = (M_{Zr89} + M_n - M_{Zr90})c^2$. (*c*) The binding energy of the entire nucleus is given by $(40M_p + 51M_n - M_{Zr91})c^2$. Divide by the mass number (91) to find the binding energy per nucleon.

SOLUTION:

[ans: (*a*) 7.21 MeV; (*b*) 12.0 MeV; (*c*) 8.69 MeV]

The binding energy of the first neutron is obviously much less than the binding energy of the second and even less than the average binding energy of all the nucleons.

III. NOTES

Chapter 48
ENERGY FROM THE NUCLEUS

I. BASIC CONCEPTS

Two important nuclear phenomena, the fission of a heavy nucleus into two lighter nuclei and the fusion of two light nuclei into a heavier nucleus, both convert internal energy associated with nucleon-nucleon interactions into kinetic energy of the products. In each case your chief goal should be to understand the basic process. Learn what the products are, what energy is released, and what inhibits the process. In addition, you will learn about occurrences of the phenomena in nature and about human attempts to produce and control nuclear energy in a sustained fashion.

Nuclear fission.   In the basic fission process a thermal neutron is absorbed by a heavy nucleus to form a compound nucleus and the compound nucleus breaks into two medium-mass nuclei (fission fragments) and one or more neutrons are emitted. Look at Fig. 47–6 and notice that the binding energy per nucleon is greater for nuclei with mass numbers closer to $A = 56$ than for heavier nuclei. The total internal energy of the fragments is less than the internal energy of the original compound nucleus.

Describe what is meant by a thermal neutron: _____

_____

The kinetic energy of a thermal neutron is about _____ eV.

Starting with the same nucleus the mass numbers of the fragments may be different for different fission events and experiments give the relative probabilities for the various possible outcomes. Look at Fig. 48–1 of the text to see the distribution of fragments for the fission of ^{235}U. Notice that the fragments are not usually identical. In fact the probability for identical fragments is near the minimum of the curve. The most likely fragments have mass numbers of _____ and _____ .

Fission takes place in several steps. First the compound nucleus breaks into what are called the primary fragments and these fragments immediately emit one or more neutrons, called the <u>prompt</u> neutrons. The fragments are usually still neutron-rich and they decay by _____ emission. This decay often leaves a fragment nucleus in an excited state and if the excitation energy is sufficient it may decay via the emission of neutrons, although most times the decay is via gamma emission. Neutrons, if they are emitted, are called <u>delayed</u> neutrons because _____

_____ .

The disintegration energy can be computed from the change in total mass. Consider the fission of the heavy compound nucleus F into the fragments X and Y, with the production of b neutrons:

$$F \rightarrow X + Y + b\text{n}.$$

Let m_F be the mass of F, m_X be the mass of X, m_Y be the mass of Y, and m_n be the mass of a neutron. Then the disintegration energy Q is given by

$$Q =$$

Q is positive, indicating the release of energy. Most of this energy appears as the kinetic energy of _____ but some appears as the kinetic energy of the neutrons and if X and Y are the final stable nuclei some appears as the kinetic energies of electrons and neutrinos produced in the β decays.

Not all neutron-rich heavy nuclides fission. As the two fragments pull apart they must overcome the strong mutual attraction of nucleons for each other. The potential energy as a function of the separation of the fragments is shown in Fig. 48–3. Note the barrier height E_b. The neutron absorbed by the target nucleus must supply about this much energy for fission to take place. Once the fragment separation is beyond the peak in the potential energy curve the potential energy decreases with separation because the fragments are both _____ charged and repel each other.

The most energy that can be supplied by a thermal neutron is equal to its _____ energy E_n. This energy can be calculated from the difference in mass of the target nucleus and the so-called _____ nucleus formed by adding a neutron to the target. If this energy is not about the same as the barrier height or greater fission does not occur and the compound nucleus loses the excitation energy by means of _____ emission. On the other hand, fission is favored if $E_n \geq E_b$. A more energetic neutron can sometimes be used to produce a fissionable nucleus for which the barrier is high and the excitation energy is low. Its kinetic energy is transferred to the compound nucleus.

Fission reactors. To produce a large number of fission events in a reactor the fissionable material is made to undergo a <u>chain</u> <u>reaction</u>. Explain what this is: _____
_____
_____

If fission takes place in a solid most of the released energy appears as internal energy and raises the temperature of the solid. In a reactor power generator the energy is used to raise the temperature of _____ and the resulting steam is used to drive a turbine.

The text discusses 3 problems that arise in the design of a fission reactor and their solutions. Carefully read Section 48–4, then describe each of the problems in your own words and briefly tell how it is solved.

1. <u>Neutron leakage</u>: _____
_____
_____
_____

   <u>Solution</u>: _____
_____
_____
_____
_____

2. Neutron energy: _____

_____

_____

_____

Solution: _____

_____

_____

_____

3. Neutron capture: _____

_____

_____

_____

Solution: _____

_____

_____

_____

If the number of neutrons present in a reactor remains constant with time the reactor is said to be _____; if the number decreases it is said to be _____; and if the number increases it is said to be _____. The value of the <u>multiplication</u> factor k indicates the tendency. If N neutrons that will participate in fission events are present at some time, then after the fission events occur there will be _____ neutrons present to participate in the next round of fission events. The value of k is _____ if the reactor is critical, _____ if it is subcritical, and _____ if it is supercritical. For steady power generation $k = $ _____.

The multiplication factor is varied by inserting and withdrawing <u>control rods,</u> which readily absorb _____. To increase the number of fission events per unit time some rods are _____ for a short time, during which k is _____ than 1. The rods are then _____ so k becomes 1 again. Because some neutron emissions are delayed the rods are effective in controlling power generation even though there is a response time for inserting and withdrawing them.

Thermonuclear fusion. Nuclear fusion occurs when two light nuclei fuse together to form a single heavier nucleus. Recall from your study of Chapter 47 that for low mass numbers the binding energy per nucleon is much greater for heavier nuclei than for lighter. The heavy nucleus has much less internal energy and the excess appears as kinetic energy of the products. Products include the final nucleus and perhaps neutrons, β particles, and photons.

The nuclei, of course, are positively charged and repel each other electrically. They must overcome a potential energy barrier before they get close enough for the strong force of attraction to be effective. The barrier height is about _____ keV for two deuterons.

In thermonuclear fusion thermal energy is the chief means by which the light nuclei obtain enough energy to overcome the barrier. Notice that the mean thermal energy need not be nearly as great as the barrier height. There are two reasons: the nuclei can tunnel through the barrier rather than go over the top and in a gas at temperature T there are many nuclei with energies greater than the mean ($\frac{3}{2}kT$). Look at Fig. 21–8 to see the distribution of speeds for molecules in a gas.

Now look at Fig. 48–9 of the text. The curve labelled $n(K)$ gives the number of nuclei per unit energy range as a function of their kinetic energy. The curve labelled $p(K)$ gives the _____ of barrier penetration for protons with kinetic energy K. The kinetic energy that maximizes the product $n(K)p(K)$ and so maximizes the number of fusion events is about _____ keV. At higher energies the probability of barrier penetration is greater but there are fewer protons with sufficient energy. There are more protons with lower energies but the probability of barrier penetration is less.

Fusion processes are responsible for the internal energy of the sun. The proton-proton cycle is diagramed in Fig. 48–10 of the text. Complete the symbolic statements of the process below and tell what happens at each stage.

1. $^1H + {}^1H \rightarrow$
   Two protons fuse to form a _____ nucleus. A _____ (e^+) and a _____ (ν) are also formed. The e^+ annihilates with a free _____ with the result that 2 _____ are produced.

2. $^2H + {}^1H \rightarrow$
   Here a deuterium nucleus and a proton fuse to form a _____ nucleus with the emission of another _____.

3. $^3He + {}^3He \rightarrow$
   Two light helium nuclei fuse to form an _____ particle and 2 _____.

The overall process can be written

$$4{}^1H + 2e^- \rightarrow$$

The disintegration energy for the entire process is about _____ MeV. Some of this energy, about _____ MeV, is carried away by the neutrinos but most is available to maintain the internal energy and keep the sun shining. It is estimated that the proton-proton cycle can continue for about _____ y before the sun's hydrogen is gone.

When all the hydrogen has been fused into helium the sun's core will collapse and as gravitational potential energy is converted to kinetic energy it will get hotter. In the hot dense core helium can fuse into heavier elements. Explain why a higher temperature is required for these fusion events than for the proton-proton cycle: _____

_____

_____

Chemical elements with mass numbers beyond $A =$ _____ cannot be formed in fusion events. Explain why: _____

_____

_____

Controlled fusion. Currently attempts are being made to produce sustained controlled fusion for purposes of generating electric power. The three fusion processes that hold the most promise are

$$^2\text{H} + {}^2\text{H} \rightarrow$$

$$^2\text{H} + {}^2\text{H} \rightarrow$$

$$^2\text{H} + {}^3\text{H} \rightarrow$$

The goal of fusion research is to maintain a high-temperature, high-density gas (a plasma) of particles for sufficient time that a significant number of fusion events take place. If n is the particle concentration and τ is the confinement time then Lawson's criterion for a successful fusion reactor is

$$n\tau \geq$$

In addition, the temperature must be high enough to allow particles to overcome the Coulomb barrier to their fusion. Look at Fig. 48–13 of the text to see the temperatures needed for a given value of the Lawson number $n\tau$. According to this diagram the minimum value of the Lawson number for a sustaining reactor (the ignition curve) is about _____ s·m^{-3}. For this value of $n\tau$ the value of kT must be about _____ keV or T must be about _____ K.

Two types of confinement technologies are presently being developed. Magnetic confinement makes use of a _____ field to maintain high particle concentrations and to keep the hot plasma away from walls. In the inertial confinement method a high-power laser beam is used to ionize and compress small pellets of deuterium and tritium. Fusion takes place in the core of the pellet. Magnetic confinement methods seek to obtain a greater _____ at a smaller _____ than inertial confinement methods but both attempt to reach the critical value of the Lawson number.

II. PROBLEM SOLVING

In a primary fusion event, before any beta-decay of the primary fragments, the number of protons and the number of neutrons are separately conserved. You can use their conservation to predict one of the fission products if the others are known. Here are some examples.

PROBLEM 1 (4E).   Fill in the following table, which refers to the generalized fission reaction

$$^{235}\text{U} + n \rightarrow \text{X} + \text{Y} + bn$$

X	Y	b
^{140}Xe	_____	1
^{139}I	_____	2
_____	^{100}Zr	2
^{141}Cs	^{92}Rb	_____

STRATEGY: ^{235}U has 92 protons and $235 - 92 = 143$ neutrons. The reaction products (on the right side of the equation) must have 92 protons and 144 neutrons. Since ^{140}Xe has 86 neutrons and 54 protons (see Appendix D), Y has $144 - 86 - 1 = 57$ neutrons and $92 - 54 = 38$ protons. According to the appendix it is ^{95}Sr. The other columns in the table are filled in a similar manner.

SOLUTION:

$$[\text{ans: } {}^{95}\text{Sr}, {}^{95}\text{Y}, {}^{134}\text{Te}, 3]$$

The energy released in a fission event can be calculated from the change in the total mass of the system. Don't forget that atomic masses are usually given and you must convert to nuclear masses if the electron masses do not cancel.

PROBLEM 2 (8E). Calculate the energy released in the fission reaction

$$^{235}\text{U} + \text{n} \rightarrow {}^{141}\text{Cs} + {}^{93}\text{Rb} + 2\text{n}.$$

Needed atomic and particle masses are

^{235}U	235.04392 u	^{93}Rb	92.92157 u
^{141}Cs	140.91963 u	n	1.00867 u

STRATEGY: If M_U is the mass of the uranium nucleus, M_n is the mass of a neutron, M_{Rb} is the mass of a rubidium nucleus, and M_{Cs} is the mass of a cesium nucleus, then the energy released is given by $(M_U + M_n - M_{Cs} - M_{Rb} + 2M_n)c^2$. Multiply the mass difference in atomic mass units by the energy conversion factor $931.5\,\text{MeV/u}$. Since the given masses are atomic, not nuclear masses, you should, in principle, also subtract the mass of 92 electrons from both the left and right sides. These masses cancel each other.

SOLUTION:

$$[\text{ans: } 181\,\text{MeV}]$$

The primary fission fragments may decay via the emission of beta particles and neutrons to the final stable products. Here is a problem that asks you to find the number of β-decays required to reach stability.

PROBLEM 3 (11P). Consider the fission of ^{238}U by fast neutrons. In one fission event no neutrons were emitted and the final stable end products, after the beta-decay of the primary fission fragments, were ^{140}Ce and ^{99}Ru. (a) How many beta-decay events were there in the two beta-decay chains, considered together?

STRATEGY: Separately count the original number of protons and neutrons. Don't forget the incoming neutron. Separately count the total number of protons and neutrons in the decay products. You should find that a certain number of neutrons were turned into protons. This is the number of beta-decay events.

SOLUTION:

$$[\text{ans: } 10]$$

(*b*) Calculate Q. The relevant atomic masses are:

^{238}U	238.05079 u	^{140}Ce	139.90543 u
n	1.00867 u	^{99}Ru	98.90594 u

STRATEGY: Calculate the mass difference in atomic mass units, then multiply by 931.5 MeV/u. The initial particles are a ^{238}U nucleus and a neutron. The final products are a ^{140}Ce nucleus, a ^{99}Ru nucleus and 10 electrons. The mass of an electron is 5.4858×10^{-4} u.

SOLUTION:

$$[\text{ans: } 226\,\text{MeV}]$$

The fissioning fragments attract each other via the strong nuclear force and repel each other via the electrostatic force. The maximum in the barrier occurs approximately when their surfaces are touching. You can use the expression for the electrostatic potential energy of two spherical charge distributions to estimate the height of the barrier.

PROBLEM 4 (13P). Assume that just after the fission of ^{236}U* according to Eq. 48–1, the resulting ^{140}Xe and ^{94}Sr nuclei are just touching at their surfaces. (*a*) Assuming the nuclei to be spherical, calculate the Coulomb potential energy (in MeV) of repulsion between the two fragments. (*Hint*: Use Eq. 47–3 to calculate the radii of the fragments.)

STRATEGY: The Coulomb potential energy is given by

$$U = \frac{1}{4\pi\epsilon_0} \frac{(q_{Xe})(q_{Sr})}{R}.$$

where q_{Xe} is the charge on an Xe nucleus ($54e$), q_{Sr} is the charge on an Sr nucleus ($38e$), and R is the distance between the nuclear centers when the nuclei are just touching. It is the sum of their radii: $R = R_{Xe} + R_{Sr}$. Use $R = R_0 A^{1/3}$ to find values for R_{Xe} and R_{Sr}.

SOLUTION:

$$[\text{ans: } 4.04 \times 10^{-11}\,\text{J} \,(252\,\text{MeV})]$$

(*b*) Compare this energy with the energy released in a typical fission event. In what form will the Coulomb potential energy ultimately appear in the laboratory?

STRATEGY: The energy released in a typical fission event is about 200 MeV. The Coulomb potential energy is converted to kinetic energy of the fragments.

SOLUTION:

Here is another problem that asks you to apply your knowledge of proton and neutron conservation in the primary fission event, along with your knowledge of beta-decay.

PROBLEM 5 (20P).   Many fear that helping additional nations develop nuclear power reactor technology will increase the likelihood of nuclear war because reactors can be used not only to produce energy but, as a by-product through neutron capture with inexpensive ^{238}U, to make ^{239}Pu, which is a "fuel" for nuclear bombs. What simple series of reactions involving neutron capture and beta-decay would yield this plutonium isotope?

STRATEGY: A ^{238}U nucleus has 92 protons and $238 - 92 = 146$ neutrons. After an additional neutron is captured there are 147. ^{239}Pu has 94 protons and $239 - 94 = 145$ neutrons. Two neutrons must be turned into protons.

SOLUTION:

$$\left[\text{ans: } {}^{238}U + n \rightarrow {}^{239}U \rightarrow {}^{239}Np + e, {}^{239}Np \rightarrow {}^{239}Pu + e\right]$$

   Some problems deal with the operation of fission reactors. You should know that the multiplication factor k allows you to calculate the change in the number of neutrons from one generation to the next. You should also know how to calculate the power output from the number of neutrons present and the generation time. One of the problems below shows you why light nuclei are used in the moderator.

PROBLEM 6 (23P).   The neutron generation time t_{gen} in a reactor is the average time needed for a fast neutron emitted in one fission to be slowed down to thermal energies by the moderator and to initiate another fission. Suppose that the power output of a reactor at time $t = 0$ is P_0. Show that the power output a time t later is $P(t)$, where

$$P(t) = P_0 k^{t/t_{gen}},$$

where k is the multiplication factor. For constant power output $k = 1$.

STRATEGY: Power output is proportional to the number of neutrons that participate in fusion. This number changes by the factor k in every time interval of duration t_{gen}. In time t there are t/t_{gen} such intervals.

SOLUTION:

PROBLEM 7 (25P).   The neutron generation time t_{gen} (see Problem 23) in a particular reactor is $1.0\,ms$. If the reactor is operating at a power level of $500\,MW$, about how many free neutrons are present in the reactor at any moment?

STRATEGY: If there are N neutrons that participate in fission and each fission event produces energy E then a total energy NE is produced in every time interval of duration t_{gen}. Thus the power output is given by $P = NE/t_{gen}$. Solve for N. Take E to be $200\,MeV$, the average energy released in a fission event.

SOLUTION:

$$\left[\text{ans: } 1.7 \times 10^{16}\right]$$

PROBLEM 8 (27P). (*a*) A neutron, of mass m_n and kinetic energy K makes a head-on elastic collision with a stationary atom of mass m. Show that the fractional energy loss of the neutron is given by

$$\frac{\Delta K}{K} = \frac{4m_n m}{(m + m_n)^2},$$

in which m_n is the neutron mass.

STRATEGY: If v_{n0} is the initial velocity of the neutron, v_n is its final velocity, and v is the final velocity of the atom, then conservation of momentum leads to $m_n v_{n0} = m_n v_n + mv$ and conservation of energy leads to $\frac{1}{2}m_n v_{n0}^2 = \frac{1}{2}m_n v_n^2 + \frac{1}{2}mv^2$. Solve for v_n in terms of v_{n0}. You should get $v_n = (m_n - m)v_{n0}/(m_n + m)$. The energy loss of the neutron is given by $\frac{1}{2}m_n v_{n0}^2 - \frac{1}{2}m_n v_n^2$ and the fractional energy loss is given by

$$\frac{\Delta K}{K} = \frac{m_n v_{n0}^2 - m_n v_n^2}{m_n v_{n0}^2} = \frac{v_{n0}^2 - v_n^2}{v_{n0}^2}.$$

Substitute the expression you found for v_n. You should obtain $\Delta K/K = 4m_n m/(m_n + m)^2$.

SOLUTION:

(*b*) Find $\Delta K/K$ for each of the following as the stationary atom: hydrogen, deuterium, carbon, and lead.

STRATEGY: Substitute values for the masses (see Appendix D). The mass of a neutron is 1.00867 u, the mass of a hydrogen atom is 1.00797 u, the mass of a deuterium atom is 2.01410 u, the mass of a carbon atom is 12.01115 u, and the mass of a lead atom is 207.14 u.

SOLUTION:

[ans: hydrogen: 1.0, deuterium: 0.89, carbon: 0.29, lead: 0.019]

(*c*) If $K = 1.00\,\text{MeV}$ initially, how many such collisions would it take to reduce the neutron energy to thermal values (0.025 eV) if the material (consisting of the stationary atoms) is deuterium, a commonly used moderator? (*Note:* In actual moderators, most collisions are not "head-on".)

STRATEGY: Let $f = \Delta K/K$. If the initial kinetic energy is K_i, then after n collisions the final energy is $K_f = (1 - f)^n K_i$. Solve for n. You should get

$$n = \frac{\ln(K_f/K_i)}{\ln(1 - f)}.$$

Evaluate this expression and round up to the nearest greater integer.

SOLUTION:

[ans: 8]

PROBLEM 9 (29E).   The natural fission reactor discussed in Section 48–5 is estimated to have generated 15 gigawatt-years of energy during its lifetime. (*a*) If the reactor lasted for 200,000 y, at what average power level did it operate?

STRATEGY: If E_{total} is the energy generated in time t then the power level is given by $P = E_{\text{total}}/t$. Substitute $E_{\text{total}} = 15 \times 10^9\,\text{W}\cdot\text{y}$ and $t = 200,000\,\text{y}$ to obtain P in watts.

SOLUTION:

[ans: 75 kW]

(*b*) How much ^{235}U did it consume during its lifetime?

STRATEGY: Assume each fission event generated an energy of $E = 200\,\text{MeV}$. Solve $E_{\text{total}} = NE$ for N. You will need to convert 15 gigawatt-years and 200 MeV to the same energy unit. To find the mass consumed multiply N by the mass of a single ^{235}U nucleus, 235 u. Convert to kilograms.

SOLUTION:

[ans: 1.5×10^{28} nuclei (5.8×10^3 kg)]

You should know how the total energy released by a collection of fissionable nuclei or by a collection of fusing protons depends on the value of Q for a single fission or fusion event.

PROBLEM 10 (34E).   Verify that the fusion of 1.0 kg of deuterium by the reaction

$$^2\text{H} + {}^2\text{H} \rightarrow {}^3\text{He} + \text{n}, \qquad Q = +3.27\,\text{MeV},$$

could keep a 100-W lamp burning for 2.5×10^4 y.

STRATEGY: Take the mass of a deuterium atom to be $m = 2.0\,\text{u}$. The number of atoms in $M = 1.0\,\text{kg}$ is M/m and the number of reactions is $N = M/2m$ since each reaction requires 2 deuterium nuclei. If $E\,(= 3.27\,\text{MeV})$ is the energy released in each reaction the total energy released is $E_{\text{total}} = NE$. The time t that the light remains lit is $t = E_{\text{total}}/P$, where P is the power rating of the light.

SOLUTION:

In order to fuse, two nuclei must have sufficient kinetic energy to overcome their mutual electrostatic repulsion and get close enough to each other for the strong force of attraction to take over. It is not necessary, however, to achieve temperatures that are so high that most protons in a gas fuse because some in the tail of the energy distribution do have enough energy, even at much lower temperatures. Here's a calculation that indicates how the numbers depend on the temperature.

PROBLEM 11 (36P).   The equation of the curve $n(K)$ in Fig. 48–9 is

$$n(K) = 1.13n \, \frac{K^{1/2}}{(kT)^{3/2}} \, e^{-K/kT} \, ,$$

where n is the total density of particles. At the center of the sun the temperature is 1.50×10^7 K and the mean proton energy $\overline{K}$ is 1.94 keV. Find the ratio of the density of protons at 5.00 keV to that at the mean proton energy.

STRATEGY: The ratio of the values of n for two kinetic energies (K_1 and K_2) is

$$\frac{n(K_2)}{n(K_1)} = \left(\frac{K_2}{K_1} \right)^{1/2} e^{(K_1 - K_2)/kT} \, .$$

Evaluate this expression for $K_1 = 1.94$ keV and $K_2 = 5.00$ keV. Convert the energies to joules.

SOLUTION:

[ans: 0.151]

Here is a problem that shows you how to compare the energies obtained by fission and fusion, starting with the same mass of "fuel".

PROBLEM 12 (43P).   Calculate and compare the energy released by (*a*) the fusion of 1.0 kg of hydrogen deep within the sun and (*b*) the fission of 1.0 kg of ^{235}U in a fission reactor.

STRATEGY: (*a*) If M is the mass of material (1.0 kg) and m is the mass of a hydrogen atom (1.0 u = 1.66 $\times$ 10^{-27} kg), then the number of hydrogen atoms is M/m and, since 4 nuclei are required for the proton-proton cycle, the number of fusion events is given by $N = M/4m$. The energy released in each event is $E = 26.7$ MeV and the total energy released is NE.

(*b*) If m is the mass of a ^{235}U nucleus then the number of these nuclei is $N = M/m$. This is the same as the number of fission events since each event requires only one uranium nucleus. The energy released in each event is, on average, about $E = 200$ MeV and the total energy released is NE.

SOLUTION:

[ans: (*a*) 4.0×10^{27} MeV; (*b*) 5.1×10^{26} MeV]

Fusion in the sun and other stars results in a series of nuclear reactions. The text describes in detail the proton-proton cycle, but that is not the only possibility. Here is another. This problem gives you the chance to construct the equivalent reaction, which displays the overall result of a cycle.

PROBLEM 13 (47P).   In certain stars the *carbon cycle* is more likely than the proton-proton cycle to be effective in generating energy. This cycle is

$$^{12}C + {}^{1}H \rightarrow {}^{13}N + \gamma, \qquad Q_1 = 1.95\,\text{MeV},$$

$$^{13}N \rightarrow {}^{13}C + e^+ + \nu, \qquad Q_2 = 1.19,$$

$$^{13}C + {}^{1}H \rightarrow {}^{14}N + \gamma, \qquad Q_3 = 7.55,$$

$$^{14}N + {}^{1}H \rightarrow {}^{15}O + \gamma, \qquad Q_4 = 7.30,$$

$$^{15}O \rightarrow {}^{15}N + e^+ + \nu, \qquad Q_5 = 1.73,$$

$$^{15}N + {}^{1}H \rightarrow {}^{12}C + {}^{4}He, \qquad Q_6 = 4.97.$$

(*a*) Show that this cycle of reactions is exactly equivalent in its overall effects to the proton-proton cycle of Fig. 48–10.

STRATEGY: Sum the left sides of the reaction equations, then the right sides. Cancel any particles or nuclei that appear on both sides of the result. You should obtain $4{}^{1}H \rightarrow {}^{4}He + 3\gamma + 2e^+ + 2\nu$. The two positrons annihilate with two free electrons and each annihilation event produces 2 gamma particles. Add two electrons to the left side and 4 gammas to the right.

SOLUTION:

(*b*) Verify that the two cycles, as expected, have the same Q.

STRATEGY: Add the six Q values given in the problem statement, then add 1.02 MeV for each of the electron-positron annihilation events.

SOLUTION:

$$[\text{ans:}\ Q = 27.6\,\text{MeV}]$$

Here is a problem that helps you relate fusion events in the sun to the total energy production and to the loss in mass.

PROBLEM 14 (49P).   The effective Q for the proton-proton cycle of Fig. 48–10 is 26.2 MeV. (*a*) Express this as energy per kilogram of hydrogen consumed.

STRATEGY: Each fusion cycle requires 4 protons, with a total mass of $4\,\text{u} = 4 \times 1.661 \times 10^{-27}\,\text{kg} = 6.64 \times 10^{-27}\,\text{kg}$. It releases an energy of $26.2\,\text{MeV} = 26.2 \times 10^6 \times 1.60 \times 10^{-19}\,\text{J} = 4.19 \times 10^{-12}\,\text{J}$. To find the energy per kilogram of hydrogen consumed, divide the second result by the first.

SOLUTION:

$$[\text{ans:}\ 6.3 \times 10^{14}\,\text{J/kg}]$$

(b) The power of the sun is 3.9×10^{26} W. If its energy derives from the proton-proton cycle, at what rate is it losing hydrogen?

STRATEGY: If R_H is rate of hydrogen loss (in kg/s) and f is the energy released per kilogram of hydrogen consumed, then the power output of the sun is $P = fR_H$. Solve for R_H.

SOLUTION:

$$[\text{ans: } 6.2 \times 10^{11} \text{ kg/s}]$$

(c) At what rate is it losing mass?

STRATEGY: Since the energy loss ΔE is related to the mass loss Δm by $\Delta E = \Delta m c^2$, the power output P is related to the rate R_M of mass loss by $P = R_M c^2$. Solve for R_M.

SOLUTION:

$$[\text{ans: } 4.3 \times 10^{9} \text{ kg/s}]$$

Account for the difference in the results for (b) and (c).

STRATEGY: Not all the mass is converted to gamma particles and neutrinos. What else is there?

SOLUTION:

(d) The sun's mass is 2.0×10^{30} kg. If it loses mass at the constant rate calculated in (c) how long will it take before it loses 0.10% of its mass?

STRATEGY: The time it takes to lose mass ΔM is given by $t = R_M/\Delta M$, where R_M is the rate of mass loss.

SOLUTION:

$$[\text{ans: } 4.6 \times 10^{17} \text{ s } (1.5 \times 10^{10} \text{ y})]$$

Momentum as well as energy is conserved in fission and fusion events. If there are only two product particles the conservation of momentum determines how the energy is shared between them. Here's a demonstration.

PROBLEM 15 (53P). In the deuteron-triton fusion reaction of Eq. 48–11, how is the reaction energy Q shared between the α particle and the neutron? Neglect the relatively small kinetic energies of the two combining particles.

STRATEGY: If m_n is the mass of the neutron, v_n is its velocity, m_α is the mass of the alpha particle, and v_α is its velocity, then conservation of momentum yields $m_n v_n + m_\alpha v_\alpha = 0$. Conservation of energy yields $Q = \frac{1}{2}m_n v_n^2 + \frac{1}{2}m_\alpha v_\alpha^2$. Solve the first equation to obtain an expression for v_n and substitute the expression into the energy equation. That equation becomes $Q = \frac{1}{2}m_\alpha v_\alpha^2(m_n + m_\alpha)/m_n$. Notice that the first factor is the kinetic energy K_α of the α particle. Solve for it. You should obtain

$$K_\alpha = \frac{m_n}{m_n + m_\alpha} Q.$$

The kinetic energy of the neutron is given by $K_n = Q - K_\alpha$. Evaluate these expressions using $Q = 17.59 \, \text{MeV}$.

SOLUTION:

$$\left[\text{ans: } K_\alpha = 3.5 \, \text{MeV}, K_n = 14.1 \, \text{MeV} \right]$$

III. NOTES

Chapter 49
QUARKS, LEPTONS, AND THE BIG BANG

I. BASIC CONCEPTS

This chapter deals with the ultimate constituents of matter, as far as known, and with the evolution of the universe from its start at the Big Bang. As you will learn, these topics are closely related. Pay attention to the types of particles that exist and to the mechanisms by which they interact. This information can be used to unravel the history of the early universe. Also learn of the evidence for the Big Bang and for dark matter, on which the future of the universe may depend.

Classification of particles. Particles may be classified according to their intrinsic angular momentum (or spin). The maximum value of any cartesian component of the intrinsic angular momentum is given by

$$L =$$

where s is either an integer or half integer and $\hbar$ is the Planck constant divided by 2π.

Particles that have s equal to 0 or an integer are called _____ while particles that have s equal to half an odd integer are called _____ . _____ obey the Pauli exclusion principle; _____ do not. Particles that obey the exclusion principle must be in different _____. Suppose a system has 4 possible states, with energies E_1, E_2, E_3, and E_4 (in increasing order). If there are 3 bosons in the system the least total energy is obtained if _____ particles are in the state with energy E_1, _____ are in the state with energy E_2, _____ are in the state with energy E_3, and _____ are in the state with energy E_4. If the 3 particles are fermions the least total energy is obtained if _____ particles are in the state with energy E_1, _____ are in the state with energy E_2, _____ are in the state with energy E_3, and _____ are in the state with energy E_4.

Give some examples of fermions: _____
Give some examples of bosons: _____

Particles can be classified according to the types of interactions they have with other particles. There are _____ different basic interactions (or forces).

1. Every particle attracts every other particle with a <u>gravitational</u> force, the weakest of all the basic forces. It is important for large macroscopic bodies but it is insignificant for interactions on the subatomic scale. Give some phenomena for which gravity is important:

_____

_____

2. The <u>weak</u> interaction is the next strongest interaction. Give at least one phenomenon for which it is important: _____

_____

3. The next strongest interaction is the <u>electromagnetic</u> interaction. Give some phenomena for which it is important: _____

_____

Particles that carry _____ interact electromagnetically.

4. The <u>strong</u> interaction is the strongest force of all. Give at least one phenomenon for which this interaction is important: _____

_____

Not all particles interact via the strong force. Those that do are called _____ . These particles are further categorized according to their spin. Particles that interact strongly *and* have integer (or zero) spin are called _____ . Give some examples: _____

_____

Particles that interact strongly *and* have half-integer spin are called _____ . Give some examples: _____

_____

Particles that do not interact via the strong force but do interact via the weak are called _____ . Give some examples: _____

_____

An <u>antiparticle</u> is associated with each particle. Tell what is meant by this term: _____

_____

Except for the positron (e^+) an antiparticle is denoted by placing a _____ over the symbol for the particle

Leptons. Counting the neutrinos but not counting the antiparticles there are 6 members of the lepton family. They are: _____ (symbol: _____), _____ (symbol: _____), _____ (symbol: _____), _____ (symbol: _____), _____ (symbol: _____), and _____ (symbol: _____). The _____, _____, and _____ are charged _____ ; the neutrinos have _____ charge. All have intrinsic angular momentum _____ $\hbar$.

None of the leptons participate in the _____ interaction; all participate in the _____ interaction. Charged leptons participate in the _____ interaction, while uncharged leptons do not. No evidence of any internal structure has been observed for any lepton. They are thought to be truly fundamental.

Notice that there is a neutrino associated with each of the other types of leptons (an electron neutrino, a muon neutrino, and a tauon neutrino). Although they have the same charge (0) and spin ($\frac{1}{2}$) and perhaps have the same mass we know they are different particles because _____

_____

_____ .

Mesons.

List some charged mesons: _____

List some neutral mesons: _____

All have antiparticles associated with them, although for some the particle and antiparticle are identical. Not all mesons have an intrinsic angular momentum but for those that do the angular momentum is an integer multiple of _____. The distinguishing characteristic of a meson is that it is an integer spin particle that participates in the strong interaction.

Baryons.

List some charged baryons: _____

List some neutral baryons: _____

Baryons have spin angular momentum that is a odd multiple of _____. The distinguishing characteristic of a baryon is that it is a half-integer spin particle that participates in the strong interaction.

Conservation laws.   There are several conservation laws that hold in every particle inter-action or decay. You have already learned about the conservation of energy, momentum, angular momentum, and charge. Carefully read Section 49–3 to see how these conservation laws are used to determine the properties of particles. In this section two new conservation laws are described.

A <u>baryon number</u> $B = +1$ is assigned to each _____, a baryon number $B = -1$ is assigned to each _____, and a baryon number $B = 0$ is assigned to every other particle. The sum of the baryon numbers before an event is always equal to the sum of the baryon numbers after the event. This means that if a baryon decays, a baryon must be among the products. If an antibaryon decays, an antibaryon must be among the products. If two baryons interact, two baryons, perhaps different from the original baryons, must result.

Carefully note that there is no similar conservation law for mesons. Mesons can be created or destroyed in any number without violating any conservation law.

<u>Strangeness</u> is a particle property that is conserved in strong and electromagnetic inter-actions but not in weak interactions. Particles are assigned strangeness S by observing the results of decays and interactions. K^+ and K^0, for example, have $S =$ _____, K^- and $\overline{K}^0$ have $S =$ _____.

When a decay or reaction takes place via either the strong or electromagnetic interaction the total strangeness of the products must be the same as the total strangeness of the original particles. If the decay or reaction takes place via the weak interaction the total strangeness may change.

The quark model.   A model has been developed to explain the properties of baryons and mesons, the particles that interact via the strong interaction. These particles are thought to be composite—made of smaller, more fundamental particles called _____. Every meson is a combination of a _____ and an _____ whereas every baryon is a combination of 3 _____ and every antibaryon is a combination of 3 _____. Quarks have baryon number _____ and antiquarks have baryon number _____ so the model predicts that the baryon number of a meson is _____, the baryon number of a baryon is _____, and the baryon number of an antibaryon is _____, in agreement with observation.

So far 5 quarks (and the associated antiquarks) have been discovered, in the sense that they are required for the construction of observed mesons or baryons. Many physicists believe a sixth exists. The lowest mass mesons and baryons are combinations of the u (up), d (down), and s (strange) quarks, along with the associated antiquarks. The quark content of the π^- meson, for example, is _____; the quark content of the π^+ meson is _____. Kaons, which are strange mesons, have an s or $\bar{s}$ quark. The quark content of a proton is _____ and the quark content of a neutron is _____. Σ baryons are strange and have s or $\bar{s}$ quarks. Carefully compare Figs. 49–3 and 49–4 to see the quark content of other mesons and baryons.

Quarks are charged and so participate in electromagnetic interactions, but the charge on a quark is a fraction of the fundamental charge e, not a multiple of it. A u quark has a charge of _____ e, a d quark has a charge of _____ e, and an s quark has a charge of _____ e. The charge on an antiquark is the negative of the charge on the associated quark. The charges on quarks that make up a proton sum to _____ e, the charges on quarks that make up a neutron sum to _____ e, the charges on quarks that make up a π^+ meson sum to _____ e, and the charges on quarks that make up a π^- meson sum to _____ e.

Quark number is conserved in strong interactions. That is, the strong interaction can create and destroy quark-antiquark pairs (such as d$\bar{\text{d}}$ or u$\bar{\text{u}}$) but it cannot change one type quark into another. The weak interaction, however, can change one type quark into anther. In a beta decay of a proton ($p \rightarrow n + e^+ + \nu$) a _____ quark is changed into a _____ quark. In a beta decay of a neutron ($n \rightarrow p + e^- + \bar{\nu}$) a _____ quark is changed into a _____ quark.

The J/ψ meson contains a fourth type quark; J/ψ is a combination of a _____ quark and a _____ antiquark.

Messenger particles.   In modern quantum theory the basic interactions are pictured as exchanges of <u>messenger</u> particles. The messenger particles associated with the weak interaction are called _____, the messenger particle associated with the electromagnetic interaction is called _____, and the messenger particles associated with the strong interaction are called _____.

When a particle emits a messenger particle its rest energy is reduced. Nevertheless it retains its character as long as it absorbs a messenger particle a short time later. Energy is not conserved during the interval between emission and absorption. For a given loss in energy ΔE the time interval between emission and absorption must be less than the minimum predicted by the Heisenberg uncertainty principle. Thus

$$\Delta t <$$

This consideration limits the time for events, such as decays, and also limits the range of the interaction.

Quarks interact with each other via the strong interaction, which involves the exchange of gluons. Inside a baryon or meson the interaction energy of two quarks is very weak but the interaction becomes extremely strong if the separation is larger than the particle size. Free quarks have never been observed. When a high-energy bombarding particle strikes a baryon or meson, other quarks are created (in quark-antiquark pairs) and these form combinations with existing quarks and are emitted as other mesons or baryons.

The property of quarks that causes them to emit and absorb gluons (participate in the strong interaction) is called _____. It is somewhat similar to charge, the property that causes particles to absorb and emit photons and thus participate in the electromagnetic interaction. The chief difference is that a gluon carries color away from the quark that emits it and thereby changes the color of the quark. On the other hand, photons do not carry charge.

The three types of color are called _____, _____, and _____. The condition that the net color of quarks in a hadron be white (or colorless) explains the possible combinations of quarks observed (three quarks of different color, three antiquarks of different color, or a quark and antiquark of the same color).

The basic interactions may, in fact, all be different manifestations of a single interaction. The _____ and _____ interactions have already been successfully unified into what is known as the _____ interaction. When particles interact at high energy these two interactions are quite similar. A theory that unifies the strong and electroweak interactions is called a _____ theory (GUT) and a theory that unifies all four interactions is called a theory of _____ (TOE).

Cosmology. You should recognize that when you look at distant cosmological objects (stars, quasars, etc.) you are seeing them as they were when the light left them, perhaps more than 10^9 y ago.

Most physicists believe that the universe started in a state of extremely high density and temperature, perhaps a highly concentrated mixture of quarks, leptons, and messenger particles. A "big bang" initiated an expansion (and cooling) that will continue for some time into the future, perhaps forever. By understanding fundamental particles and their interactions physicists hope to understand the early universe.

Two independent experimental observations lend credence to the idea of the Big Bang. They are the expansion of the universe and the microwave background radiation.

Hubble's law relates the _____ with which a galaxy is receding from us to the _____ between us and the galaxy. Mathematically it is

$$v =$$

where H is the Hubble parameter. Its numerical value is somewhat uncertain but it is about _____ m/(s·ly). An observer in *any* part of the universe sees galaxies at the same distance receding with the same speed.

If the rate of expansion of the universe has been constant the reciprocal of the Hubble constant gives the _____ of the universe. It is about _____ y.

The cosmic microwave background radiation fills the entire universe with an intensity that is nearly the same in every direction. There is no distinguishable source. It is believed to have originated shortly after the Big Bang. Although the background radiation started as a hot gas of highly energetic photons, in the gamma ray region of the electromagnetic spectrum, today it has a thermal spectrum corresponding to a temperature of about _____ K and is chiefly in the _____ region of the spectrum. The decrease in frequency accompanied the _____ of the universe.

Whether the universe will continue to expand indefinitely or will collapse (perhaps in preparation for another big bang) depends on the amount of mass it contains. The universe may contain much more mass than is visible and the postulated invisible matter is called <u>dark matter</u>. The chief evidence for dark matter comes from a study of _____ _____.

The text describes 7 eras in the history of the universe. Tell what you can about each of them.

Big Bang to 10^{-43} s: _____

_____

10^{-43} s to 10^{-35} s: _____

_____

10^{-35} s to 10^{-10} s: _____

_____

10^{-10} s to 10^{-5} s: _____

_____

10^{-5} s to 3 min: _____

_____

3 min to 10^5 y: _____

_____

10^5 y to the present: _____

_____

II. PROBLEM SOLVING

To solve some of the problems you will need to know the relativistic definitions of kinetic energy and momentum. If a particle has velocity **v** then its momentum is given by

$$\mathbf{p} = \frac{m\mathbf{v}}{\sqrt{1 - (v/c)^2}}$$

and its kinetic energy is given by

$$K = \frac{mc^2}{\sqrt{1 - (v/c)^2}} - mc^2 .$$

Its rest energy is mc^2 and its total energy is

$$E = K + mc^2 = \frac{mc^2}{\sqrt{1 - (v/c)^2}} .$$

In addition, its energy and momentum are related by

$$E^2 = (mc^2)^2 + (pc)^2 .$$

Here are some problems that make use of these definitions and relationships.

PROBLEM 1 (6P).   A neutral pion has a rest energy of 135 MeV and a mean life of 8.3×10^{-17} s. If it is produced with an initial kinetic energy of 80 MeV and it decays after one mean lifetime, what is the longest possible track that this particle could leave in a bubble chamber? Take relativisitic time dilation into account. (*Hint*: See Sample Problem 49–1.)

STRATEGY: First solve $E^2 = (mc^2)^2 + (pc)^2$ for the magnitude of the momentum p. The total energy is $E = mc^2 + K = 135 + 80 = 215$ MeV. Next solve $p = mv/\sqrt{1 + (v/c)^2}$ for the speed v. The length of the track is given by $d = vt$, where t is the time as measured in the laboratory. According to the relativistic time dilation equation t is given by $t = \gamma\tau$ where $\gamma = 1/\sqrt{1 - (v/c)^2}$ and τ is the lifetime.

SOLUTION:

$$\left[\text{ans: } 3.1 \times 10^{-8}\,\text{m}\right]$$

PROBLEM 2 (11P).   (*a*) A stationary particle m_0 decays into two particles m_1 and m_2, which move off with equal but oppositely directed momenta. Show that the kinetic energy K_1 of m_1 is given by

$$K_1 = \frac{1}{2E_0}\left[(E_0 - E_1)^2 - E_2^2\right],$$

where m_0, m_1, and m_2 are masses and E_0, E_1, and E_2 are corresponding rest energies. (*Hint*: Follow the arguments of Sample Problem 49–2 except that, in this case, neither of the created particles has zero mass.)

STRATEGY: Conservation of energy yields $E_0 = E_1 + E_2 + K_1 + K_2$ and conservation of momentum yields $p_1 + p_2 = 0$. Since $(E_1 + K_1)^2 = E_1^2 + (p_1c)^2$, $(p_1c)^2 = (E_1 + K_1)^2 - E_1^2$. Similarly, $(p_2c)^2 = (E_2 + K_2)^2 - E_2^2$. The momentum equation, written in the form $(p_1c)^2 = (p_2c)^2$, becomes $(E_1 + K_1)^2 - E_1^2 = (E_2 + K_2)^2 - E_2^2$. Solve for K_2. You should get $K_2 = -E_2 \pm \sqrt{E_2^2 + K_1^2 + 2E_1K_1}$. Use the plus sign to obtain a positive value for K_2.

Substitute for K_2 in the energy equation to obtain $E_0 = E_1 + E_2 + K_1 - E_2 + \sqrt{E_2^2 + K_1^2 + 2E_1K_1}$. Solve for K_1.

SOLUTION:

(*b*) Show that the result in (*a*) yields the kinetic energy of the muon as calculated in Sample Problem 49–2.

STRATEGY: Take m_0 to be the pion, m_1 to be the muon, and m_2 to be the neutrino. Thus $E_0 = 139.6\,\text{MeV}$, $E_1 = 105.7\,\text{MeV}$, and $E_2 = 0$.

SOLUTION:

[ans: $4.12\,\text{MeV}$]

You should be able to apply the principles of the conservation of energy, momentum, angular momentum, charge, baryon number, and strangeness to given decays and reactions. Remember that the conservation of strangeness is not absolute: strangeness need not be conserved in weak interactions.

In most cases insufficient data is given to prove that energy, momentum, and angular momentum are actually conserved in a given situation. You can however show that they *could* be conserved. If the sum of the masses of the initial particles, assumed to be nearly at rest, is greater than the sum of the masses of the final particles, then kinetic energy can account for the difference in rest energy and energy can be conserved. To check on conservation of momentum see if the speeds and directions of the original particles and of the products can be arranged so that momentum can be conserved. The decay of a single particle at rest, for example, to a single moving particle is prohibited by the conservation law. Conservation of angular momentum requires that the sum of the spins of the products must be half an odd integer if the sum of the spins of the original particles is half an odd integer and must be an integer if the sum of the spins of the original particles is an integer.

Related problems ask you to compute the disintegration energy for a decay. It is, of course, the initial total rest energy minus the final total rest energy.

PROBLEM 3 (12E). Verify that the hypothetical proton decay scheme given in Eq. 49–10 does not violate the conservation laws of (*a*) charge, (*b*) energy, (*c*) linear momentum, and (*d*) angular momentum.

STRATEGY: The decay is $p \rightarrow e^+ + \gamma$. (*a*) The total charge of the system before the decay is $+e$ (on the proton) and the total charge after is $+e$ (on the positron). (*b*) The mass of the proton is greater than the mass of the positron. (*c*) The decay yields two particles. Their momenta can be equal in magnitude and opposite in direction, thereby summing to 0. (*d*) The spin of the proton is $\frac{1}{2}$, the spin of the positron is $\frac{1}{2}$, and the spin of the gamma is 1. Angular momentum is conserved if the spin of the gamma is in the same direction as the spin of the proton and the spin of the positron is in the opposite direction.

SOLUTION:

PROBLEM 4 (14P).   The A_2^+ particle and its products decay according to the following schemes:

$$A_2^+ \rightarrow \rho^0 + \pi^+, \qquad \mu^+ \rightarrow e^+ + \nu + \bar{\nu},$$
$$\rho^0 \rightarrow \pi^+ + \pi^-, \qquad \pi^- \rightarrow \mu^- + \bar{\nu},$$
$$\pi^+ \rightarrow \mu^+ + \nu, \qquad \mu^- \rightarrow e^- + \nu + \bar{\nu}.$$

(*a*) What are the final stable decay products?

STRATEGY: Start with the first decay equation on the left and substitute the decay products of ρ^0 and π^+ for those particles. Continue making substitutions until stable particles (e^+, e^-, neutrinos) are reached. You may want to distinguish between electron and muon neutrinos.

SOLUTION:

$$\left[\text{ans: } 2e^+ + e^- + 2\nu_e + \bar{\nu}_e + 3\nu_\mu + 3\bar{\nu}_\mu\right]$$

(*b*) From the evidence, is the A_2^+ particle a fermion or boson? Is it a meson or baryon? What is its baryon number? (*Hint*: See Sample Problem 49–5.)

STRATEGY: All the final decay products are spin $\frac{1}{2}$ particles and there are an even number of them. The spin of the A_2^+ particle must be an even multiple of $\frac{1}{2}$. All of the final decay products are leptons. The total baryon number of the products is 0 and baryon number is conserved in the decay.

SOLUTION:

$$\left[\text{ans: boson, meson, } B = 0\right]$$

PROBLEM 5 (16E).   By examining strangeness, determine which of the following decays or reactions proceed via the strong interaction: (*a*) $K^0 \rightarrow \pi^+ + \pi^-$; (*b*) $\Lambda^0 + p \rightarrow \Sigma^+ + n$; (*c*) $\Lambda^0 \rightarrow p + \pi^-$; (*d*) $K^- + p \rightarrow \Lambda^0 + \pi^0$.

STRATEGY: Calculate the total strangeness before and after the decay or reaction. If it does not change then the decay or reaction could proceed via the strong interaction. If it does change the decay or reaction cannot proceed via the strong interaction. You can find the strangeness of the individual particles in Tables 4 and 5.

SOLUTION:

$$\left[\text{ans: } (a) \text{ cannot be strong; } (b) \text{ can be strong; } (c) \text{ cannot be strong; } (d) \text{ can be strong}\right]$$

PROBLEM 6 (22P). Consider the decay $\Lambda^0 \rightarrow p + \pi^-$ with the Λ^0 at rest. (*a*) Calculate the disintegration energy.

STRATEGY: The disintegration energy is given by $Q = (M_\Lambda - M_p - M_\pi)c^2$. The rest energies are given in Tables 4 and 5.

[ans: 37.7 MeV]

(*b*) Find the kinetic energy of the proton.

STRATEGY: Use the expression you derived in problem 11. Take the proton to be particle 1 and the pion to be particle 2.

SOLUTION:

[ans: 5.35 MeV]

(*c*) What is the kinetic energy of the pion?

STRATEGY: Repeat the calculation with the pion as particle 1 and the proton as particle 2.

SOLUTION:

[ans: 32.4 MeV]

You should be able to use Table 49–6, which gives the properties of quarks, to construct baryons and mesons with given properties. Conversely, given the properties of a particle you should be able to predict its quark content.

PROBLEM 7 (24E). From Tables 49–4 and 49–6, determine the identities of the baryons formed from the following combinations of quarks. Check your answers with the baryon octet shown in Fig. 49–3a: (*a*) *ddu*; (*b*) *uus*; (*c*) *ssd*.

STRATEGY: Find the baryon number, charge, and strangeness of the given quark combinations by summing the properties of the individual quarks in each combination. These properties can be found in Table 49–6. For example *ddu* has baryon number 1, charge 0, and strangeness 0. Now locate in Table 49–4 a particle with appropriate properties.

SOLUTION:

[ans: (*a*) neutron; (*b*) Σ^+; (*c*) Ξ^-]

PROBLEM 8 (28P).   There is no known meson with $Q = +1$ and $S = -1$ or with $Q = -1$ and $S = +1$. Explain why, in terms of the quark model.

STRATEGY: $S = -1$ implies the meson contains a single s quark, with a charge of $-1/3$. The antiquark must have a charge of $4/3$ and a strangeness of 0. A glance at Table 49–6 reveals that no such quark exists. Examine the second particle in the same manner.

SOLUTION:

Here is another problem that allows you to make use of your knowledge of relativity.

PROBLEM 9 (29P).   The spin-$\frac{1}{2}$ Σ^{*0} baryon (see Exercise 27) has a rest energy of 1385 MeV (with an intrinsic uncertainty ignored here); the spin-$\frac{1}{2}$ Σ^0 baryon has a rest energy of 1192.5 MeV. Suppose that each of these particles has a kinetic energy of 1000 MeV. Which, if either, is moving faster and by how much?

STRATEGY: The kinetic energy K and speed v of a particle are related by the relativistically correct formula

$$K = \frac{mc^2}{\sqrt{1 - (v/c)^2}} - mc^2,$$

where m is the mass of the particle. Solve for v. You should get

$$v = \sqrt{1 - \left(\frac{mc^2}{K + mc^2}\right)^2}\, c.$$

The rest energy is, of course, mc^2.

SOLUTION:

[ans: the Σ^0 ($v = 0.386c$) is faster than the Σ^{*0} ($v = 0.361c$) by $0.025c$ (7.5×10^6 m/s)]

You should know how to use Hubble's law to find the distance to an astronomical object, given its speed of recession. The first example below gives Doppler shift data so you can find the speed. In the second example you use Hubble's law to predict the density of mass in the universe that will keep it from expanding forever.

PROBLEM 10 (33P).   The recessional speeds of galaxies and quasars at great distances are close to the speed of light, so that the relativistic Doppler shift formula (see Eq. 42–26) must be used. The red shift is reported as a fractional red shift $z = \Delta\lambda/\lambda_0$. (*a*) Show that, in terms of z, the recessional speed parameter $\beta = v/c$ is given by

$$\beta = \frac{z^2 + 2z}{z^2 + 2z + 2}.$$

STRATEGY: For an object that is receding from the observer the frequency received is given by

$$\nu = \nu_0 \sqrt{\frac{1 - \beta}{1 + \beta}}$$

so the wavelength is given by

$$\lambda = \lambda_0 \sqrt{\frac{1 + \beta}{1 - \beta}}.$$

Substitute this expression into $z = (\lambda - \lambda_0)/\lambda_0$ and solve for β.

SOLUTION:

(*b*) A distant quasar detected in 1987 has $z = 4.43$. Calculate its speed parameter.

STRATEGY: Substitute numerical values into $\beta = (z^2 + 2z)/(z^2 + 2z + 2)$.

SOLUTION:

[ans: 0.934]

(*c*) Find the distance to the quasar, assuming that Hubble's law to valid to these distances.

STRATEGY: Solve $v = Hr$ for r.

SOLUTION:

[ans: 1.65×10^{10} ly]

PROBLEM 11 (34P).   Will the universe continue to expand forever? To attack this question, make the (reasonable?) assumption that the recessional speed v of a galaxy a distance r from us is determined only by the matter that lies inside a sphere of radius r centered on us. If the total mass inside this sphere is M, the escape speed v_e from the sphere is given by $v_e = \sqrt{2GM/r}$ (see Sample Problem 15–8). (a) Show that the average density ρ inside the sphere must be at least equal to the value

$$\rho = 3H^2/8\pi G$$

to prevent unlimited expansion.

STRATEGY: Substitute $M = (4\pi/3)r^3\rho$ and $v = Hr$ into $v = \sqrt{2GM/r}$, then solve for ρ.

SOLUTION:

(b) Evaluate this "critical density" numerically; express your answer in terms of H-atoms/cm^3. Measurements of the actual density are difficult and complicated by the presence of dark matter.

STRATEGY: First find the value of ρ in kg/m^3, then convert to H-atoms/cm^3. The mass of a hydrogen atom is $1.0078\,\text{u} = 1.67 \times 10^{-27}$ kg.

SOLUTION:

$$\left[\text{ans: } 5.77 \times 10^{-27}\,\text{kg/m}^3\ (3.46\,\text{H-atoms/m}^3)\right]$$

In the example below you will explore an important consequence of the cosmic background radiation.

PROBLEM 12 (35P).   Due to the presence everywhere of the microwave background radiation, the minimum possible temperature of a gas in interstellar or intergalactic space is not 0 K but 2.7 K. This implies that a significant fraction of the molecules in space that can occupy excited states of low excitation energy may, in fact, be in those excited states. Subsequent de-excitation would lead to the emission of radiation that could be detected. Consider a (hypothetical) molecule with just one excited state. (a) What would the excitation energy have to be in order that 25% of the molecules be in the excited state? (*Hint:* See Eq. 45–24.)

STRATEGY: Take the energy of the ground state to be 0 and the energy of the excited state to be E. The number of molecules in the excited state is given by $N_e = Ae^{-E/kT}$, where T is the temperature on the Kelvin scale. The number of molecules in the ground state is $N_g = A$. You want $f = N_e/(N_g + N_e)$ to be 0.25. Solve

$$f = \frac{e^{-E/kT}}{1 + e^{-E/kT}}$$

for E. You should get

$$E = -kT \ln \frac{f}{1-f} \, .$$

Evaluate this expression.

SOLUTION:

$$\left[\text{ans: } 4.10 \times 10^{-23} \text{ J } (2.56 \times 10^{-4} \text{ eV})\right]$$

(b) What would be the wavelength of the photon emitted in a transition back to the ground state?

STRATEGY: The energy of the photon is E. Its frequency is given by $E = h\nu$ and its wavelength by $\lambda = c/\nu$.

SOLUTION:

$$\left[\text{ans: } 4.85 \times 10^{-3} \text{ m}\right]$$

The dependence of the rotational period of a star in a galaxy on the distance of the star from the galactic center provides evidence for the existence of dark matter. The following problem shows how the period depends on the distance if dark matter is present and if it is not.

PROBLEM 13 (38P). Suppose that the matter (stars, gas, dust) of a particular galaxy of total mass M, is distributed uniformly throughout a sphere of radius R. A star of mass m is revolving about the center of the galaxy in a circular orbit of radius $r < R$. (a) Show that the orbital speed v of the star is given by

$$v = r\sqrt{GM/R^3} \, ,$$

and therefore that the period T of revolution is

$$T = 2\pi\sqrt{R^3/GM} \, ,$$

independent of r. Ignore any resistive forces.

STRATEGY: The magnitude of the force on the star is given by $GM'm/r^2$, where M' is the mass that lies within its orbit. Since the star moves uniformly in a circular orbit the magnitude of its acceleration is v^2/r and Newton's second law yields $GM'm/r^2 = mv^2/r$. Since the mass of the galaxy is distributed uniformly $M' = (r^3/R^3)M$. Substitute this expression into the second law equation and solve for v. The period is given by $T = 2\pi r/v$. Substitute the expression you found for v.

SOLUTION:

(b) What is the corresponding formula for the orbital period assuming that the mass of the galaxy is strongly concentrated toward the center of the galaxy, so that essentially all the mass is at distances from the center less than r?

STRATEGY: Now the magnitude of the force on the star is given by GMm/r^2. Carry out the same analysis.

SOLUTION:

$$[\text{ans: } T = 2\pi r^{3/2}/\sqrt{GM}]$$

The wavelength and photon energy of the cosmic background radiation depends on the temperature in intergalactic space. This problem explores that relationship.

PROBLEM 14 (39E). From Planck's radiation law it is possible to derive the following relation between the temperature T of a cavity radiator and the wavelength λ_{max} at which it radiates most strongly:

$$\lambda_{max}T = 2898\,\mu\text{m} \cdot \text{K}.$$

(This is Wien's law; see Problem 48 in Chapter 43.) (a) The microwave background radiation peaks in intensity at a wavelength of 1.1 mm. To what temperature does this correspond?

STRATEGY: Solve $\lambda_{max}T = 2898\,\mu\text{m} \cdot \text{K}$ for T.

SOLUTION:

$$[\text{ans: } 2.6\,\text{K}]$$

(*b*) About 10^5 years after the Big Bang, the universe became transparent to electromagnetic radiation. Its temperature then was about 10^5 K. What was the wavelength at which the background radiation was most intense at that time?

STRATEGY: Solve $\lambda_{max}T = 2898 \, \mu\text{m} \cdot \text{K}$ for λ_{max}.

SOLUTION:

$$[\text{ans: } 29 \, \text{nm}]$$

III. NOTES

EXAM SUMMARY

Exam number: _____   **Date:** _____   **Chapters:** _____

Definitions:

QUANTITY   DEFINITION

_____   _____
_____   _____
_____   _____
_____   _____
_____   _____
_____   _____

Physical Laws:

_____
_____
_____

Other Important Relationships:

_____
_____
_____
_____
_____

Important Applications:

_____
_____
_____
_____
_____

Notes:

EXAM SUMMARY

Exam number: _____   **Date:** _____   **Chapters:** _____

Definitions:

QUANTITY   DEFINITION

_____   _____

_____   _____

_____   _____

_____   _____

_____   _____

_____   _____

Physical Laws:

_____

_____

_____

Other Important Relationships:

_____

_____

_____

_____

_____

Important Applications:

_____

_____

_____

_____

_____

Notes:

EXAM SUMMARY

Exam number: _____   **Date:** _____   **Chapters:** _____

<u>Definitions:</u>

QUANTITY   DEFINITION

_____   _____

_____   _____

_____   _____

_____   _____

_____   _____

<u>Physical Laws:</u>

_____

_____

_____

<u>Other Important Relationships:</u>

_____

_____

_____

_____

_____

<u>Important Applications:</u>

_____

_____

_____

_____

_____

<u>Notes:</u>

EXAM SUMMARY

Exam number: _____   **Date:** _____   **Chapters:** _____

Definitions:

QUANTITY   DEFINITION

_____   _____
_____   _____
_____   _____
_____   _____
_____   _____
_____   _____

Physical Laws:

_____
_____
_____

Other Important Relationships:

_____
_____
_____
_____
_____
_____

Important Applications:

_____
_____
_____
_____
_____
_____

Notes:

EXAM SUMMARY

Exam number: _____   **Date:** _____   **Chapters:** _____

Definitions:

QUANTITY   DEFINITION

_____   _____
_____   _____
_____   _____
_____   _____
_____   _____
_____   _____

Physical Laws:

_____
_____
_____

Other Important Relationships:

_____
_____
_____
_____
_____

Important Applications:

_____
_____
_____
_____
_____

Notes:

EXAM SUMMARY

Exam number: _____   **Date:** _____   **Chapters:** _____

Definitions:

QUANTITY   DEFINITION

_____   _____
_____   _____
_____   _____
_____   _____
_____   _____
_____   _____

Physical Laws:

_____
_____
_____

Other Important Relationships:

_____
_____
_____
_____
_____

Important Applications:

_____
_____
_____
_____
_____

Notes: